Applied geography: principles and practice

Essential reading for geography, planning and environmental science students and researchers, and all those concerned with the nature of the relationship between people and the environment, this important volume presents a comprehensive introduction to applied geography. Demonstrating the usefulness of geographic research across the various sub-areas of the discipline, forty-nine leading experts in the field introduce and explore research which crosses the traditional boundary between physical and human geography. A wide range of key issues and contemporary debates are examined within the book's main sections, which cover:

- natural and environmental hazards
- environmental change and management
- challenges of the human environment
- techniques of spatial analysis.

Applied geography is the application of geographic knowledge and skills to identify the nature and causes of social, economic and environmental problems and inform policies which lead to their resolution. The relevance and value of this applied approach, which cross-cuts inter- and intra-disciplinary boundaries, has never been more apparent given the plethora of problem situations which confront modern societies, ranging from extreme natural events (floods, droughts, earthquakes), through environmental concerns (deforestation, disease, desertification), to human issues (crime, poverty, unemployment).

Applied geography: principles and practice offers an invaluable introduction to useful research in physical environmental and human geography, and provides a new focus and reference point for investigating and understanding problem-orientated research from a diverse range of perspectives and disciplines.

The editor of this volume, **Michael Pacione**, is Professor of Geography at the University of Strathclyde, Glasgow, UK.

Applied geography: principles and practice

An introduction to useful research in physical, environmental and human geography

Edited by MICHAEL PACIONE

London and New York

First published 1999
by Routledge
2 Park Square, Milton Park,
Abingdon, Oxon, OX14 4RN

Simultaneously published in the USA and Canada
by Routledge
270 Madison Ave,
New York NY 10016

Routledge is an imprint of the Taylor & Francis Group

Transferred to Digital Printing 2009

Typeset in Bembo by Solidus (Bristol) Limited

Publisher's Note
The publisher has gone to great lengths to ensure the quality of this reprint but points out that some imperfections in the original may be apparent.

British Library Cataloguing in Publication Data

A catalogue record for this book is available from the British Library.

Library of Congress Cataloging in Publication Data

Applied geography: principles and practice/edited by Michael Pacione.
Includes bibliographical references and index.
1. Geography. 2. Natural disasters.
3. Environmental management.
4. Nature – Effect of human beings on.
5. Spatial analysis (Statistics).
I. Pacione, Michael.
G133.A66 1999
910–dc21 98–53104

ISBN 0–415–18268–9 (hbk)
ISBN 0–415–21419–X (pbk)

To CHRISTINE, MICHAEL JOHN and EMMA VICTORIA

Contents

Plates

Figures

Boxes

Tables

Contributors

Professor David Alexander, Department of Geosciences, University of Massachusetts, Amherst, MA 01003-5820, USA.

Dr Michael Barke, Division of Geography and Environmental Management, University of Northumbria, Lipman Building, Newcastle upon Tyne, NE1 8ST, UK.

Professor Tony Barnett, School of Development Studies, University of East Anglia, Norwich, NR4 7TJ, UK.

Professor Peter Beaumont, Department of Geography, University of Wales, Lampeter, Ceredigion, SA48 7ED, UK.

Professor Gerald Blake, Department of Geography, University of Durham, South Road, Durham, DH1 3LE, UK.

Professor John Blunden, Faculty of Social Sciences, The Open University, Walton Hall, Milton Keynes, MK7 6AA, UK.

Dr Keith Boucher, Department of Geography, Loughborough University, Loughborough, LE11 3TU, UK.

Dr Nick Brown, Department of Plant Sciences, University of Oxford, South Parks Road, Oxford, OX1 3RB, UK.

Ms Rosemary Burton, School of Geography and Environmental Management, University of the West of England, Coldharbour Lane, Bristol, UK.

Dr Bruce Carlisle, Geographical Information Systems Unit, School of Agriculture and Horticulture, De Montford University, Lincoln, LN2 2LG, UK.

Professor Tony Champion, Department of Geography, University of Newcastle, Daysh Building, Newcastle upon Tyne, NE1 7RU, UK.

Dr Sylvia Chant, Department of Geography and Environment, London School of Economics, Houghton Street, London, WC2A 2AE, UK.

Dr Gordon Clark, Department of Geography, University of Lancaster, Bailrigg, Lancaster, LA1 4YB, UK.

Dr Graham Clarke, School of Geography, University of Leeds, Leeds, LS2 9JT, UK.

Professor Michael Crozier, Institute of Geography, Victoria University of Wellington, Box 600, Wellington, New Zealand.

Dr Norman Davidson, Department of Geography, University of Hull, Cottingham Road, Hull, HU6 7RX, UK.

Professor Ian Douglas, School of Geography, Manchester University, Manchester, M13 9PL, UK.

Miss Lesley France, 5 Rossway, Whitley Bay, NE26 3EJ, UK.

Dr David Green, Department of Geography, University of Aberdeen, Elphinstone Road, Aberdeen, AB24 3UF, UK.

Dr Cliff Guy, Department of City and Regional Planning, University of Wales, Colum Drive, Cardiff, CF1 3YN, UK.

Professor Martin Haigh, Department of Geography, Oxford Brookes University, Headington, Oxford, OX3 0BP, UK.

Dr Roy Haines-Young, Department of Geography, University of Nottingham, University Park, Nottingham, NG7 2RD, UK.

Professor David Herbert, Department of Geography, University of Wales, Singleton Park, Swansea, SA2 8PP, UK.

Dr Ian Heywood, Centre for Open and Distance Learning, Robert Gordon University, Aberdeen, AB10 1FR, UK.

Professor Brian Ilbery, Department of Geography, University of Coventry, Priory Street, Coventry, CV1 5FB, UK.

Dr Keith Jacobs, School of the Built Environment, University of Westminster, Marylebone Road, London, NW1 5LS, UK.

Professor Ron Johnston, School of Geographical Sciences, University of Bristol, University Road, Bristol, BS8 1SS, UK.

Dr Gavin Jordan, Department of Agriculture and Forestry, Newton Rigg College, University of Central Lancashire, Penrith, CA11 0AH, UK.

Mr Stephen King, Department of Geography, University of Aberdeen, Elphinstone Road, Aberdeen, AB24 3UF, UK.

Dr Philip Kivell, Department of Geography, University of Keele, Keele, Staffordshire, ST5 5BG, UK.

Dr Peter Larkham, School of Planning, University of Central England, Perry Barr, Birmingham, B42 2SU, UK.

Professor Paul Longley, School of Geographical Sciences, University of Bristol, University Road, Bristol, BS8 1SS, UK.

Dr Elena Lopez-Gunn, Department of Environmental Sciences, University of Hertfordshire, College Lane, Hatfield, AL10 9AB, UK.

Dr A.M. Mannion, Department of Geography, University of Reading, Whiteknights, Reading, RG6 6AB, UK.

Professor Adrian McDonald, School of Geography, University of Leeds, Leeds, LS2 9JT, UK.

Professor Andrew Millington, Department of Geography, University of Leicester, Leicester, LE1 7RH, UK.

Dr Stephen Nutley, Department of Environmental Studies, University of Ulster, Coleraine, BT52 1SA, UK.

Professor Michael Pacione, Department of Geography, University of Strathclyde, 50 Richmond Street, Glasgow, G1 1XH, UK.

Professor Edmund Penning-Rowsell, Flood Hazard Research Centre, Middlesex University, Queensway, Enfield, EN3 4SF, UK.

Professor David Phillips, Department of Geography, Nottingham University, University Park, Nottingham, NG7 2RD, UK.

Professor Rob Potter, Department of Geography, Royal Holloway, University of London, Egham, Surrey, TW20 0EX, UK.

Professor Guy Robinson, School of Geography, University of Kingston, Penrhyn Road, Kingston upon Thames, KT1 2EE, UK.

Dr Matthew Smallman-Raynor, Department of Geography, Nottingham University, University Park, Nottingham, NG7 2RD, UK.

Dr Graham Smith, Department of Environmental and Geographical Sciences, Manchester Metropolitan University, Chester Street, Manchester, M1 5GD, UK.

Dr Tom Spencer, Cambridge Coastal Research Unit, Department of Geography, Cambridge University, Downing Place, Cambridge, CB2 3EN, UK.

Dr Brian Turton, Department of Geography, University of Keele, Keele, Staffordshire, ST5 5BG, UK.

Professor Max Wade, Department of Environmental Sciences, University of Hertfordshire, College Lane, Hatfield, AL10 9AB, UK.

Dr Rory Walsh, Department of Geography, University of Wales, Singleton Park, Swansea, SA2 8PP, UK.

Professor Bruce Webb, School of Geography and Archaeology, University of Exeter, Rennes Drive, Exeter, EX4 4RJ, UK.

Preface

Questions on the usefulness of geographical research and the relationship between theory and practice are central to debate over the place and value of geography as an academic discipline for the third millennium. Such issues constitute the core of applied geography – which may be defined as *the application of geographic knowledge and skills to the resolution of social, economic and environmental problems.*

It is important at the outset to identify the place of applied geography within the discipline as a whole. Rather than being considered as a *sub-area* of geography (akin to economic, social or historical geography), applied geography refers to an *approach* that cross-cuts artificial disciplinary boundaries to involve problem-oriented research in both human and physical geography.

The relevance and value of applied geographical research has never been more apparent, given the plethora of problem situations that confront modern societies, ranging from extreme natural events (such as floods, drought and earthquakes) through environmental concerns (such as deforestation, disease and desertification) to human issues (such as crime, poverty and unemployment). An applied geographical approach has the potential to illuminate the nature and causes of such problems and inform the formulation of appropriate responses.

This book offers a comprehensive introduction to the principles and practice of applied geography. It includes coverage of applied geographical research across the traditional boundary between human and physical geography, as well as work in the important fields of environmental geography and computer-based spatial analysis. The book is organised into four main parts. As in all classificatory systems, this structure is employed to impose a degree of order on diversity in the interests of elucidation. This does not, however, imply a demarcation of the themes and issues discussed into discrete 'sub-fields' of applied geography. To do so would be to ignore the complexity of real-world problems, evident in the role of human agency in landscape modification (as in deforestation, desertification and flooding), or conversely, the impact of earthquakes on cities or of coastal erosion on transport routes.

Regrettably, it may be true that a minority of human geographers read papers on physical geography (and *vice versa*), and Stoddart (1987: p. 320) was probably correct to suggest that many geographers 'have abandoned the possibility of communicating with colleagues working not only in the same titular discipline but also in the same department. The human geographers think their physical colleagues philosophically naïve; the physical geographers think the human geographers lacking in rigour.' While it would be absurd to represent applied geography as a Rosetta stone for a divided discipline, one of the strengths of the applied geographical approach is that it rejects artificial academic boundaries and highlights linkages between different geographical phenomena. The fact that some of the chapters in this book might have been accommodated comfortably within more than a single section merely underlines the interrelationship between many of the research foci currently under investigation by applied geographers.

The opening chapter of the book provides an introduction to the principles and practice of applied geography and discusses the definition and development of the approach. Consideration is given to the relationship between 'pure' and 'applied' research, and the particular concept of 'useful knowledge' is introduced. Different approaches to the conduct of applied geography are examined and a general protocol proposed. The question of the value of applied geography for contemporary societies is also addressed. The main part of the book comprises four sections, dealing with *Natural and environmental hazards* (eight chapters), *Environmental change and management* (fifteen chapters), *Challenges of the human environment* (fifteen chapters) and *Techniques of spatial analysis* (five chapters). Each chapter provides a concise authoritative introduction to the field of study, identifies the major causes and consequences of the problem under investigation, presents case study evidence from an international range of settings to illustrate the impacts of the problem, and offers a prospective view for applied geographical research in the context of the particular problem area. Each essay is illustrated with relevant maps, diagrams and photographs and is complemented by a list of references and guide to further reading. The collection of essays in the book illustrates the wealth of research undertaken by applied geographers and provides a comprehensive introduction to the principles and practice of a dynamic and increasingly relevant approach to the study of geography.

REFERENCE

Stoddart, D. (1987) To claim the high ground: geography for the end of the century, *Transactions of the Institute of British Geographers* 12, 327–36.

Acknowledgements

Preparation of this collection of essays dealing with a range of issues that span the discipline of geography was both a challenging and a rewarding experience. The challenge of bringing together and coordinating the efforts of over forty leading authorities working in a wide variety of fields was facilitated by a shared belief in the value of an applied or problem-oriented approach to geographical investigation. My personal academic reward came with the intellectual stimulation provided by exposure to the wealth of knowledge generated by this particular collective of applied geographers.

The end result is a book that illuminates the power of an applied geographical approach to address many of the social, economic and environmental problems that confront modern societies as they enter the third millennium. The book represents a benchmark of the state of applied geography as well as a signpost and catalyst for future work.

Each chapter was written by an acknowledged authority in the field, and I should like to take this opportunity to express my appreciation of the timely and efficient manner in which they each responded to my various editorial requests.

As always, the greatest debt is owed to my family for their forbearance during the preparation of this book – to my wife Christine for her support throughout the course of the project, to my son Michael for his computing skills, and to my daughter Emma for helping with the index.

Introduction

1

In pursuit of useful knowledge: the principles and practice of applied geography

Michael Pacione

THE DEFINITION OF APPLIED GEOGRAPHY

An indication of the nature and content of applied geography may be gained by examining a selection of available definitions of the approach. One of the earliest statements on applied geography was offered by A.J. Herbertson in 1899 in a lecture to the Council of the Manchester Geographical Society. In this he defined applied geography as 'a special way of looking at geography, a limitation and a specialisation of the study of it from one point of view. For the business man this point of view is an economic one, for the medical man a climatic and demographic one, for the missionary an ethic and ethical one' (p. 1). While the second part of this definition presents a somewhat restricted view of the context of applied geography even at the end of the nineteenth century, the opening sentence has proved to be a prescient statement that, as we shall see, remains relevant today.

More recent attempts to define applied geography are also instructive as far as they reflect a particular view of the subject. In reviewing several definitions of applied geography, Hornbeck (1989: p. 15) identified two common factors in that applied geography 'takes place outside the university, and it deals with real world problems'. While the latter observation is apposite, the exclusion of academic research in applied geography reveals an excessively narrow perspective that, in part, reflects the situation in North America, where many applied geographers employ their skills beyond the walls of academia. The extramural focus in applied geographical work is also central to Hart's (1989: p. 15) definition, which saw applied geography as 'the synthesis of existing geographic knowledge and principles to serve the specific needs of a particular client, usually a business or a government agency'. The suggestion of uncritical 'service to a specific client, whether business or public agency' (p. 17) implicit in this definition ignores the volume of critical analysis undertaken by academic applied geographers.

In a more broadly based statement, Sant (1982: p. 1) viewed applied geography as the use of geographic knowledge as an aid to reaching decisions over use of the world's resources. More specifically, Frazier (1982: p. 17) considered that applied geography 'deals with the normative question, the way things should be, a bold but necessary position in dealing with real world problem resolution. In the process, the geographer combines the world of opinion with the world of decision.' This latter perspective is closer to the definition of applied geography favoured here.

In this book, we employ a definition of applied geography that reflects the central importance of normative goals and that acknowledges the involvement of both academic and non-academic applied geographers in pursuit of these goals. Accordingly, applied geography may be defined as *the application of geographic knowledge and skills to the resolution of social, economic and environmental problems.*

The question of how best to attain this goal will be addressed later in the discussion. Here it is appropriate to conclude these introductory comments by examining the academic niche for applied geography, and in particular the question of whether applied geography constitutes a sub-field of geography or an approach to the subject. These issues represent more than a simple question of semantics. In essence, a sub-field of a discipline is expected to generate its own body of theory and methodology, whereas an approach has its rationale founded on a particular philosophy (such as relevance or social usefulness) and can employ appropriate theory, concepts and methodology from across the discipline and elsewhere. Designating the area a sub-field of geography invites criticism of applied geography as lacking a coherent structure and characterised by a pragmatic approach. Johnston (1994: p. 21), for example, concluded that 'there is no central theoretical core or corpus of techniques; rather the sub-field has been characterised by ad hoc approaches to the problems posed, drawing on the perceived relevant skills and information'. This critique, which could be levelled at many sub-fields of geography, is based on a misunderstanding of the appropriate academic niche for applied geography. Identification of a theoretical core or a unified concept (such as the hydrological cycle in hydrology or the energy budget in climatology) is necessary only for a subject area that seeks to establish itself as a distinct sub-field or branch of a discipline. Applied geography does not harbour such parochial ambition and is best viewed as an approach that can bring together researchers from across the range of sub-fields in geography, either in the prosecution of a particular piece of research or in terms of an enduring commitment to the ethos of the approach. For applied geography, the unifying concept is not a specific model or theory but the fundamental philosophy of relevance or usefulness to society. This 'core', which extends beyond the confines of any single sub-field, represents a powerful and clearly articulated rationale. Furthermore, applied geographers would contend that the identification and application of relevant theory, concepts and techniques both from within geography and across disciplinary boundaries is a positive strength, not a weakness, of the applied geography approach. Definitions and critiques that seek to establish applied geography as a branch or sub-field of geography are misplaced. As Herbertson indicated a century ago, applied geography is best seen not as a sub-field but as an approach that can be applied across all branches of geography.

THE CONCEPT OF USEFUL KNOWLEDGE

The concept of useful knowledge will no doubt upset a number of practising geographers. Those who do not see themselves as applied geographers may interpret the subtitle of this book – 'an introduction to useful research in physical, environmental and human geography' – as indicating a corollary in the shape of geographical research that is less useful or even useless. This would be a misinterpretation. The subtitle for the book was selected to express the fundamental ethos of applied geography rather than to annoy 'non-applied' geographers. The choice of subtitle represents a deliberate decision to get off the fence and make explicit the view that some kinds of research are more useful than others. This is not the same as saying that some geographical research is better than other work – all knowledge is useful – but some kinds of research and knowledge are more useful than other kinds in terms of their ability to interpret and offer solutions to problems in contemporary physical and human environments.

We can illustrate this point by comparing the contents of the present volume on *applied geography* with two other geographical agendas, separated by a timespan of fifty years. The first of these is the 'mission statement' delivered by the eminent historical geographer H.C. Darby in his inaugural lecture in the University of Liverpool. In Darby's (1946) view, his goal as a teacher of geography was to help students to learn to read their morning newspapers with greater intelligence and understanding, and to take their evening walks or their Sunday drives with greater interest, appreciation and pleasure. While few modern applied geographers would regard this as an adequate definition of their work, there is a degree of overlap between Darby's agenda and the goals of applied geography in that, from a realist standpoint, Darby's activities could be regarded as emancipatory and an example of critical science.

The second example is taken from a more recent 'call for papers' issued in May 1997 on behalf of the Social and Cultural and the Population Geography research groups of the Royal Geographical Society/Institute of British Geographers. In preparation for a session at the annual conference, offers of papers were requested on the theme of 'the body'. Additional guidelines for prospective contributors were as follows: 'Is the body dead? Has it been 'done'? This joint session seeks to explore current and future critical, geographical perspectives on 'The Body', as a discourse, as a centre for conflict, consensus, rebellion or domination. Participants are encouraged to consider ways in which their own bodies can be used to em-body their presentations. All/any form(s) of (re)'presentation' are welcomed. Clearly, the research topics of interest to participants in this conference session would hold little appeal for many applied geographers. Indeed, some may even be stimulated to recall Stoddard's (1987) impatience with 'so-called geographers . . . who promote as topics worthy of research subjects like geographic influence in the Canadian cinema, or the distribution of fast food outlets in Tel Aviv' (p. 334).

The distinction between the contents of this book on *Applied geography: principles and practice* and the proposed agenda for the 1998 IBG conference session serves as a useful primer for our subsequent discussions. Those who study the kind of topics identified in the call for conference papers might legitimise their agenda by pointing to the eclectic nature of geography and the value of 'pure' research; for these and other geographers, the idea of applied geography or useful research is a chaotic concept that does not fit with the cultural turn in social geography or the postmodern theorising of recent years.

We shall return to this question later but, in the meantime, it is useful to make explicit the views that underlie the kind of applied geography represented in this book. We can do this most clearly by comparing the applied geographical approach with an alternative postmodern perspective. One of the major achievements of postmodern discourse has been the illumination of the importance of difference in society as part of the theoretical shift from an emphasis on economically rooted structures of dominance to cultural 'otherness' focused on the social construction of group identities. However, there is a danger that the reification of difference may preclude communal efforts in pursuit of goals such as social justice. A failure to address the unavoidable real-life question of 'whose is the more important difference among differences' when strategic choices have to be made represents a serious threat to constructing a *practical politics* of difference. Furthermore, if all viewpoints and expressions of identity are equally valid, how do we evaluate social policy or, for that matter, right from wrong? How do we avoid the segregation, discrimination and marginalisation that the postmodern appeal for recognition of difference seeks to counteract. The failure to address real issues would seem to suggest that the advent of postmodernism in radical scholarship has done little to advance the cause of social justice. Discussion of relevant issues is abstracted into consideration of how particular discourses of power are constructed and reproduced. Responsibility for bringing theory to bear on real-world circumstances is largely abdicated in favour of the intellectually sound but morally bankrupt premise that there is no such thing as reality. As Merrifield and Swyngedouw (1996: p. 11) express it, 'intriguing though this stuff may be for critical scholars, it is also intrinsically dangerous in its prospective definition of political action. Decoupling social critique from its political–economic basis is not helpful for dealing with the shifting realities of (urban) life at the threshold of the new millennium.' In terms of real-world problems, postmodern thought would condemn us to inaction while we reflect on the nature of the issue. (As we shall see below, a similar critique may be levelled at the Marxist critique of applied geography that was prevalent during the 1970s and 1980s.)

The views expressed in the above discussion do not represent an attempt to be prescriptive of all geographical research but are intended to indicate clearly the principles and areas of concern for applied geography. It is a matter of individual conscience whether geographers study topics such as the iconography of landscapes or the optimum location for health centres, but the principle underlying the kind of useful geography espoused by most applied geographers is a commitment to improving existing social, economic and environmental conditions. There can be no compromise – no academic fudge – some geographical research *is* more useful than other work; this is the focus of applied geography.

Of course, there will continue to be divergent views on the content and value of geographical research. This healthy debate raises a number of important questions for the discipline and for applied geography in particular. The concept of 'useful research' poses the basic questions of useful for whom? who decides what is useful? and based on what criteria? All of these issues formed a central part of the 'relevance debate' of the early 1970s, which we examine later. The related questions of values in research, the goals of different types of science, and the nature of the relationship between pure and applied research are also issues of central importance for applied geography. These are addressed in the following sections.

THE RELATIONSHIP BETWEEN PURE AND APPLIED RESEARCH

According to Palm and Brazel (1992: p. 342), 'applied research in any discipline is best understood in contrast with basic, or pure, research. In geography, basic research aims to develop new theory and methods that help explain the processes through which the spatial organisation of physical or human environments evolves. In contrast, applied research uses existing geographic theory or techniques to understand and solve specific empirical problems.'

While this distinction is useful at a general level, it overplays the notion of a dichotomy between pure and applied geography, which are more correctly seen as two sides of the same coin. There is, in fact, a dialectic relationship between the two. As Frazier (1982: p. 17) points out, 'applied geography uses the principles and methods of pure geography but is different in that it analyses and evaluates real-world action and planning and seeks to implement and manipulate environmental and spatial realities. In the process, it contributes to, as well as utilises, general geography through the revelation of new relationships.' The conjuncture between pure and applied research is illustrated clearly in geomorphology, where, for example, attempts to address problems of shoreline management have contributed to theories of beach transport; the difficulties of road construction in the Arctic have informed theories of permafrost behaviour; and problems encountered in tunnelling have aided the development of subsidence theory (Brunsden 1985). Applied research provides the opportunity to use theories and methods in the ultimate proving ground of the real world, as well as enabling researchers to contribute to the resolution of real-world problems. More generally, Sant (1982) envisaged theory as essential in applied geography at two levels. First, it provides the framework for asking questions about the substantive relationships embodied in a problem (as, for example, where a model of a hydrological catchment illuminates the potential effects of a proposed flood prevention scheme). Second, social theory provides a normative standard against which current and future social conditions can be judged in terms of defined moral goals (which may address issues such as whether a minimum wage and basic standard of living should be a legal entitlement in advanced capitalist societies).

There is little merit in pursuing a false dichotomy between pure and applied research. A more useful distinction is that which recognises the different levels of involvement of researchers at each stage of the research and specifically the greater engagement of applied geographers in the 'downstream' or post-analysis stages. The applied researcher has a greater interest than the pure researcher in taking the investigation beyond analysis into the realms of application of results and monitoring the effects of proposed strategies. Researcher participation in the implementation stage may range from recommendations in scholarly publications or contracted reports (a route favoured by most academic applied geographers, although not exclusively) to active involvement in implementation (more usually by applied geographers employed outside academia). Between these positions lie a variety of degrees of engagement, including acting as expert witnesses at public inquiries, dissemination of research findings via the media, field involvement in, for example, landscape conservation projects, and monitoring the effects of policies and strategies enacted by governmental and private sector agencies.

The balance between pure and applied research within a discipline varies over time in relation to the prevailing socio-political environment. When external pressures are at their greatest, disciplines will tend to emphasise their problem-solving capacity, while during periods of national economic expansion 'more academic' activity may be pursued in comfort. Taylor (1985) equated these cycles with the long waves of the world economy, and identified three periods in which applied geography was in the ascendancy (in the late nineteenth century, inter-war era and mid-1980s) separated by two periods of pure geography (in the early twentieth century and during the post-1945 economic boom) (Table 1.1).

Our exploration of the link between pure and applied research is not to imply the superiority of one form of knowledge over the other. Rather, it focuses attention on the fundamental question of the use to which the results of geographical research may be put. More specifically, the applied geographer's interest in the application of their research findings is of particular importance given the role of values in the formulation of political decisions. As Harvey (1984: p. 7) observed 'geography is far too important to be left to generals, politicians, and corporate chiefs. Notions of "applied" and "relevant" geography pose questions of objectives and interests served. . . . There is more to geography than the *production* of knowledge' (emphasis added). This conclusion underlines the need for explicit consideration of both the role of values in applied geography and the value of the applied geographical approach. These questions are considered below.

THE VALUE OF APPLIED GEOGRAPHY

A fundamental question for those working within the framework of applied geography concerns the value of a problem-oriented approach. We have examined this issue already in our discussions of useful knowledge and the relationship between pure and applied research, but we return to it here to address the specific critique of applied geography that has emanated from Marxist theorists. While the power of the Marxist critique has been much reduced by its own success in exposing the value bases of research, it still offers a useful perspective on the value of applied research.

The essence of the Marxist critique of applied social research is that it produces ameliorative policies that merely serve to patch up the present system, aid the legitimation of the state and bolster the forces of capitalism, with their inherent tendencies to create inequality. For these radical geographers, participation in policy evaluation and formulation is ineffective, since it hinders the achievement of the greater goal of revolutionary social change.

In terms of praxis, the outcome of this perspective is to do nothing short of a radical reconstruction of the dominant political economy (a position which, as we have seen, may also be

Table 1.1 Cycles of pure and applied geography.

Period	*Characteristics*
First applied period (late nineteenth century)	Geography created as an applied discipline to serve the political, military and commercial interests of the Prussian state.
First pure period (early twentieth century)	Based around the holistic philosophy encompassing both physical and human phenomena and focused on the core concept of the region and regional synthesis.
Second applied period (inter-war)	A period of war, followed by the Depression and war again, demanded that geography demonstrate its usefulness in fields such as land-use planning.
Second pure period (post-1945 boom)	Rejection of ideographic, regionalism replaced by spatial science and the quantitative revolution; demise of holistic approach and emergence of sub-fields within the discipline.
Third applied period (mid-1980s)	Extension of the concept of useful research into new areas of concern relating to social, economic and environmental problems; applied geographers working both in academic and in public and private sectors. Applied geography as an approach rather than a sub-field cross-cuts the artificial boundary between physical and human geography and emphasises the dialectic relationship between pure and applied research. Acknowledgement of the role of human agency and values in research and environmental change, and the need for a pluralist view of science.
Third pure period (?)	Characteristics unknown, but speculatively, a return to a more holistic philosophy reflecting the growing importance of environmental issues and the combinatory perspective of applied geography.

reached from a different direction by postmodernist theorists). Although the analytical value of the Marxist critique of capitalism is widely acknowledged, its political agenda, and in particular opposition to any action not directed at revolutionary social change, finds little favour among applied geographers. To ignore the opportunity to improve the quality of life of some people in the short term in the hope of achieving possibly greater benefit in the longer term is not commensurate with the ethical position implicit in the problem-oriented approach of applied geography.

Neither does the argument that knowledge is power and a public commodity that can be used for good or evil undermine the strength of applied geography. Any knowledge could be employed in an oppressive and discriminating manner to accentuate inequalities of wealth and power, but this is no argument for eschewing research. On the contrary, it signals a need for greater engagement by applied geographers in the policy-making and implementation process provided, of course, that those involved are aware of and avoid the danger of co-optation by, for example, funding agencies.

Furthermore, access to the expertise and knowledge produced by applied geographical research is not the sole prerogative of the advantaged in society but can be equally available to pressure groups or local communities seeking a more equitable share of society's resources. As Frazier (1982: p. 16) commented, applied research 'involves the formulation of goals and strategies and the testing of existing institutional policies within the context of ethical standards as criteria. This should not imply a simple system maintenance approach to problem solving. Indeed, it is often necessary to take an unpopular anti-establishment position, which can result in a major confrontation.' For practical examples of this, we need only refer to the pragmatic radicalism practised by the Cleveland City Planning Commission (Kraushaar 1979), the recommendations of the British Community Development Projects, which advocated fundamental changes in the distribution of wealth and power and which led to conflict with both central and local government, and more recent policy-oriented analyses of poverty and deprivation in which the identification of socio-spatial patterns is used to advance a critique of government policy (Pacione 1990).

VALUES IN APPLIED GEOGRAPHY

At each stage of the research process, the applied geographer is faced with a number of methodological and ethical questions. Decisions are required on defining the nature of the problem, its magnitude, who is affected and in what ways, and on the best means of addressing the problem. All of these require value judgements on, for example, the acceptability of existing conditions (what is an acceptable level of air pollution? or of infant malnutrition?). Values are also central to the evaluation and selection of possible remedial strategies, including comparative analysis of the benefits and disbenefits of different approaches for different people and places. In some cases, the applied geographer may seek to minimise such value judgements by enhancing the objectivity of the research methodology (for example, by employing a classification of agricultural land capability to inform a set-aside policy). In most instances, however, it is impossible to remove the need for value judgement. As Briggs (1981: p. 4) concluded, 'whether objectivity is ever achieved is a moot point. In most cases the subjectivity is merely transferred from the client (for example the politician or the planner) to the research designer.' The impossibility of objective value-free research is now axiomatic.

One issue of particular concern refers to the values that condition the selection, conduct and implementation of research, a dilemma highlighted by the aphorism 'he who pays the piper calls the tune'. J.T. Coppock (1974: p. 9), an advocate of public policy research by applied geographers, expressed this in terms of 'doubts over whether government departments will commission necessary research into the effectiveness and consequences of their own policies and there is a real danger that constraints will be imposed over publication, especially if this contains criticisms of the sponsors or explores politically sensitive areas'. Applied geographers must beware of any restrictions imposed by research sponsors and aware of the ways in which their research results may be used. Applied geographers must seek to ensure that their work contributes to human welfare. In practice, this goal may be approached by careful selection of clients and research projects, by ensuring freedom to disseminate results and, where possible, through engagement in the implementation and monitoring of relevant policy or strategies.

TYPES OF APPLIED RESEARCH

In deciding how to engage in applied geography, practitioners have recourse to three principal kinds of science (Habermas 1974). These are:

1. the empirical–analytical, in which the goal is to predict the empirical world using the scientific methods of positivism;
2. the historical–hermeneutic, with the goal of interpretation of the meaning of the world by examining the thoughts behind the actions that produce the world of experience;
3. the realist–emancipatory, where the goal is to uncover the real explanations governing society and encourage people to seek a superior social formation.

A key feature of Habermas' approach to knowledge is the recognition that different types of science have different goals. Each of these is of relevance for the practice of applied geography. The empirical–analytical approach using positivist scientific explanation remains the principal route to knowledge in applied physical geography, where a primary goal is the understanding, prediction and eventual control of environmental events. Despite the availability of powerful computer algorithms, however, the complexity of many physical environmental processes can confound this prime objective (we need think only of the accuracy of long-range weather forecasts or our primitive attempts at earthquake prediction). In addition, despite a continuing attachment to positivist science, applied physical geographers, in particular those working on environmental problems and management issues, recognise the importance of human agency in environmental change and the role of values in decision making and policy formulation. Slaymaker (1997), for example, argues for a pluralist problem-oriented geomorphology in which the predominant science of positivism is augmented by a realist philosophy that acknowledges the effect of social structures and human geography.

The goal of prediction and control within human geography – often referred to as 'social engineering' or the manipulation of society towards certain ends – is even more problematic (despite the availability of sophisticated macroeconomic models, few governments can claim to control their own economic destiny). Generally, social engineering, such as that attempted in the neighbourhood planning of the early post-war British new towns, has been discredited as both ineffective and ethically unacceptable. Positivist science, although of continuing value in applied physical geography, has limited relevance for applied research in human geography, which draws its methodology from a larger pool.

In applied terms, the goal of historical–hermeneutic science is to increase both self-awareness (by assisting people to reflect on their situation) and mutual awareness (by promoting appreciation of the situations of others). The importance (or usefulness) of inculcating mutual understanding through applied research is seen most clearly in situations where it is lacking – for example within cities, where the stereotyping of areas and social groups can lead to social tension, isolation and conflict. The third route to knowledge, via realist science, builds on the foundations of mutual understanding promoted by historical–hermeneutic or humanistic science and seeks to promote real understanding for people of their position within the socio-political structure and of the factors that condition their lifestyles and living environments. For example, by explaining the factors underlying the closure of a local factory, realist science can provide redundant workers with knowledge of the causal forces behind the event and thereby empower their response in the political arena.

Habermas' three-fold typology of science can be used to characterise applied geographers as technicians, *agents provocateur* or catalysts for social change (Johnston 1986), but this would be an over-simplification. No matter which route to knowledge the applied geographer adopts and irrespective of the methodologies employed, all are moving towards the goal of enhancing human well-being, guided by the shared philosophy of the pursuit of useful knowledge for the resolution of contemporary social, economic and environmental problems.

A PROTOCOL FOR APPLIED GEOGRAPHY

Applied geography is an approach that can be pursued via any of the three main types of science. Accordingly, there is no single method of doing applied geographical research. Nevertheless, it is useful to examine one possible protocol, which, with appropriate methodological modifications to suit the task in hand, can provide a framework for many investigations in applied geography.

The procedure may be summarised as *description*, *explanation*, *evaluation* and *prescription* (DEEP) followed by *implementation* and *monitoring* (Figure 1.1). The 'DEEP' procedure

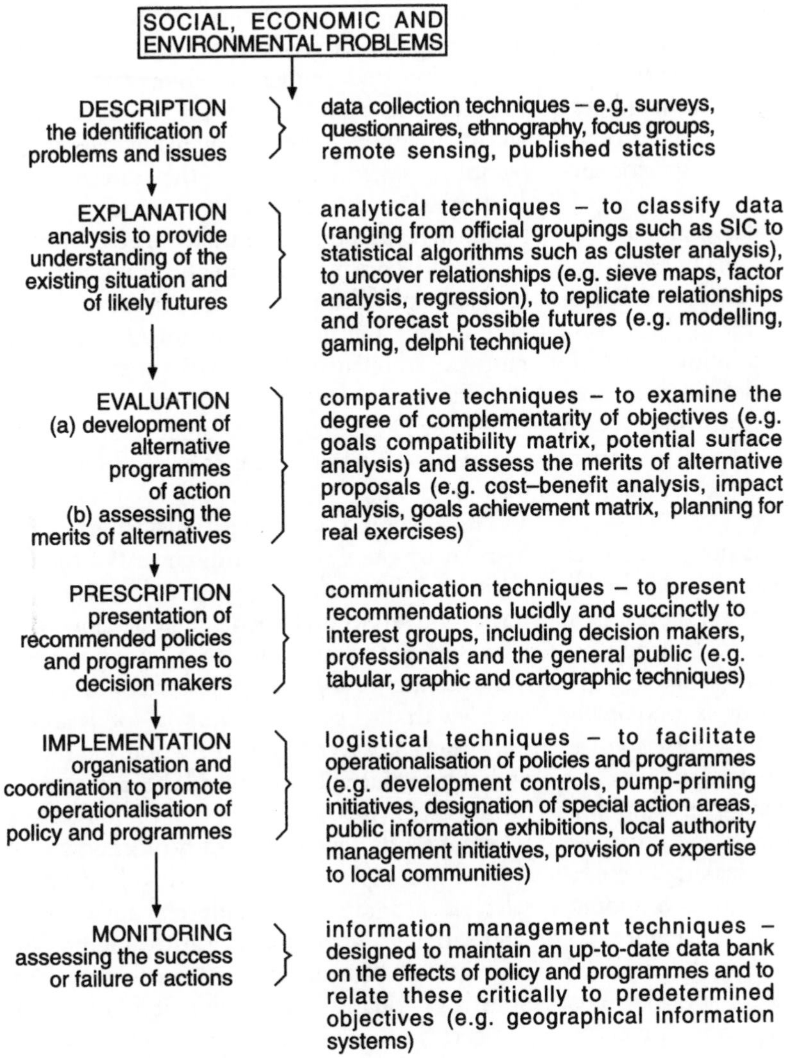

SOCIAL, ECONOMIC AND ENVIRONMENTAL PROBLEMS

Stage	Techniques
DESCRIPTION the identification of problems and issues	data collection techniques – e.g. surveys, questionnaires, ethnography, focus groups, remote sensing, published statistics
EXPLANATION analysis to provide understanding of the existing situation and of likely futures	analytical techniques – to classify data (ranging from official groupings such as SIC to statistical algorithms such as cluster analysis), to uncover relationships (e.g. sieve maps, factor analysis, regression), to replicate relationships and forecast possible futures (e.g. modelling, gaming, delphi technique)
EVALUATION (a) development of alternative programmes of action (b) assessing the merits of alternatives	comparative techniques – to examine the degree of complementarity of objectives (e.g. goals compatibility matrix, potential surface analysis) and assess the merits of alternative proposals (e.g. cost–benefit analysis, impact analysis, goals achievement matrix, planning for real exercises)
PRESCRIPTION presentation of recommended policies and programmes to decision makers	communication techniques – to present recommendations lucidly and succinctly to interest groups, including decision makers, professionals and the general public (e.g. tabular, graphic and cartographic techniques)
IMPLEMENTATION organisation and coordination to promote operationalisation of policy and programmes	logistical techniques – to facilitate operationalisation of policies and programmes (e.g. development controls, pump-priming initiatives, designation of special action areas, public information exhibitions, local authority management initiatives, provision of expertise to local communities)
MONITORING assessing the success or failure of actions	information management techniques – designed to maintain an up-to-date data bank on the effects of policy and programmes and to relate these critically to predetermined objectives (e.g. geographical information systems)

Figure 1.1 A protocol for applied geographical analysis.

represents a useful analytical algorithm. However, the clarity and organisation of the scheme does not imply that simple answers are expected to contemporary social, economic or environmental problems. Normally, in order to understand the nature and causes of real-world problems it is necessary to untangle a Gordian knot of causal linkages that underlie the observed difficulty. In some cases, such as the link between ground slippage and building collapse, cause and effect are relatively straightforward, but in most instances the cause of a problem may be more apparent than real. Thus while the immediate cause of the problems faced by a poor family on a deprived council estate in Liverpool may be a lack of employment opportunities following the closure of a local factory, the root cause of the social and financial difficulties confronting the family may lie in the decisions of investment managers based in London, New York or Tokyo.

As Figure 1.1 indicates, as well as describing the nature and explaining the causes of problems, the applied geographer also has a role to play in evaluating possible responses and in prescribing appropriate policies and programmes that may be implemented by planners and managers in both the public and private sectors, or by the residents of affected communities. In performing these tasks, the applied geographer will be confronted with a variety of potential responses for any problem. The selection of appropriate strategy is rarely straightforward. The decision must be based on not only technical criteria but also on a wide range of conditioning factors, including the views and preferences of those affected by the problem and proposed solution, available finance, and externality considerations or how the strategy to resolve a particular problem (such as construction of flood control levees) may affect other problems (such as increased flooding of downstream communities).

As indicated earlier, applied geographers, in contrast to 'pure' geographers, may also be involved in the implementation stage of the research, normally in a supervisory or consultancy capacity to ensure effective application of a strategy. The nature of any engagement is potentially wide-ranging, for example from overseeing the setting-up of a computer-based route-planning system for a private transport company or public ambulance service to making one's expertise available to community groups seeking to establish a housing cooperative or local economic development initiative. Finally, as Figure 1.1 reveals, applied geographers may be involved in monitoring the impacts of policies and programmes implemented to tackle a problem, and in relating these critically to predetermined normative goals.

THE HISTORICAL DEVELOPMENT OF APPLIED GEOGRAPHY

Applied geography has a long history. As Martin and James (1993) indicated, 'there has never been a time when the search for knowledge about the earth as the home of man has not been undertaken for practical purposes as well as for the satisfaction of intellectual curiosity'.

The earliest geographical research was, of necessity, useful, concerned as it was with describing the nature of the Earth as an aid to exploration and human survival. The applied tradition was central to the Earth measurement and cartographic research of mathematical geographers working under the direction of Eratosthenes at Alexandria (Bunbury 1879). Strabo in his account of the utility of geography in Greek and Roman society identified a central role for geographical knowledge in politics and warfare. During the first millennium AD, the same motives stimulated the development of geographical knowledge and map making in China. In Islamic lands, this was complemented by the production of travel guides as an aid to pilgrims making the *haj*.

These stimuli to applied geography were boosted by the voyages of discovery of the fifteenth and sixteenth centuries at a time when the possession of accurate geographical knowledge bestowed enormous advantage. During the sixteenth century, geographical research was undertaken with the principal purpose of enabling European ships to navigate the world and return with the produce of distant lands for commercial profit (Taylor 1930).

Significantly, few of these early practitioners of applied geography would have described themselves as geographers – explorers, adventurers, sailors, traders, astronomers, cartographers, cosmographers, natural scientists, mathematicians, historians, philosophers, surveyors or topographers, but few outright geographers. (What has in more recent times been referred to as the cocktail party syndrome – 'Oh, you're a geographer! What do you do?' – also has a long history.)

A second point of note is that the early acquisition of geographical knowledge was designed to facilitate domination by merchants or rulers, and the ways in which the knowledge was applied often had negative consequences for those peoples brought within the ambit of the emerging capitalist economic system. While modern applied geographers have been sensitised to the socially regressive consequences of the misuse of geographical research (by the work of anarchist geographers such as Kropotkin and by the Marxist critique of positivist science) at the beginning of the seventeenth century, such concerns were far from the thoughts of those practising geography.

Varenius, one of the founders of geography as a formal academic discipline, justified the subject on three grounds:

1 its value being well suited to man as the dominant species on Earth;
2 its being a pleasant and worthy recreation to study the regions of the Earth and their properties; and
3 'its remarkable utility and necessity, since neither theologians, nor medical men, nor lawyers, nor historians, nor other educated persons can do without knowledge of geography if they wish to advance in their studies without hindrance' (Bowen 1981: p. 282).

The commercial and political nature of applied research continued as an important feature of the discipline throughout the eighteenth and nineteenth centuries. In one of the earliest published references to applied geography, Keltie (1890) sought to demonstrate the importance of geographical knowledge for history and especially industry, commerce and colonisation. Similarly, in North America the early efforts of the American Geographical Society at the turn of the century supported exploration and expeditions in the hope of producing 'not only new scientific data but facts of practical use to the merchant or missionary' (Wright 1952: p. 69).

In the early decades of the twentieth century, the development of applied geography was advanced by A.J. Herbertson (1910), who envisaged the role of geographical prospector mapping the economic value and potential of regions, and by P. Geddes (1915), who was both the founding father of planning and an advocate and exponent of applied geography based on his dictum of 'survey before action'. The scope of applied geography was broadened by 'action-oriented' research undertaken in the 1930s in the UK by G.H. Daysh on the problems of distressed areas; by A.E. Smailes on the conurbations, local administrative boundaries, the concept of a city-region, and, most perceptively, on the possible role of regional parliaments; and by L.D. Stamp, who employed the methods of survey and analysis in his land-use studies of Britain (Stamp 1946).

A similar concern with land-use issues characterised applied geography in North America between the wars. The tradition of resource inventory encapsulated in the numerous explorations and surveys of the American west was continued in the work of C. Sauer on land-use classification. Sauer's land-use survey of Michigan was of the same genre as the First Land Utilisation Survey of Britain, organised by Stamp. In the field of water resource management, between 1935 and 1938, H. Barrows drew up plans for the distribution of user rights to the waters of the upper Rio Grande between the states of Colorado, New Mexico and Texas. The importance of applied geographical research was also demonstrated in economic development planning by the Tennessee Valley Authority, as well as overseas, as in the preparation of a rural land classification and development plan for Puerto Rico.

Geographical research was also applied in the private sector during the inter-war years, notable examples being C. Thornthwaite's use of knowledge of climatology for the benefit of the dairy industry in New Jersey, and the work of W. Applebaum on the location of new retail outlets for the Kroger company. This latter work pioneered the development of marketing geography as an applied field covering issues such as competitive impact analysis and the application of academic models of travel patterns to the business sector (Applebaum 1961).

Applied geographers have also been called into service in times of war and its aftermath. The skills of terrain analysis, air-photograph and satellite imagery interpretation, intelligence gathering, weather forecasting, mapping, route planning and logistics are all of vital importance for military planning. Geographical knowledge and skills are of equal value during the ensuing peace in, for example, adjudicating boundary disputes. The

American geographer I. Bowman played a major role as chief territorial specialist in the Versailles Peace Conference following the First World War, and was involved in the resolution of territorial disputes both within the USA (notably between Oklahoma and Texas) and between a number of Latin American states during the inter-war period, including those between Chile and Peru (1925), Bolivia and Paraguay (1929) and Colombia and Venezuela (1933).

The growing academic importance of applied geography was recognised by the creation in 1964 of an International Geographical Union Commission on Applied Geography. An indication of the concerns of contemporary applied geography is provided by the programme for the 1972 meeting of the commission. This included study of:

- problems relating to the management of resources in developing countries;
- planning for urbanisation;
- forecasting the impact of technology and development programmes in different countries;
- problems of water supply and environmental pollution; and
- exploration of new methods of research using computers in all branches of applied geography.

The continued emphasis on land-use issues is apparent in the IGU agenda, as well as in the main themes of applied geography in British universities identified by Freeman (1972). These documents provide a snapshot of key contemporary issues (such as regional planning), emerging specialisms (e.g. mathematical modelling) and issues of continuing concern (including environmental pollution and conflict over urban sprawl), as well as the notable absence of themes that have come to the fore subsequently (such as poverty and deprivation, the geography of AIDS, and applications of global positioning systems). Probably unwittingly, as he was talking in the particular context of land-use issues, Freeman also gives a hint of the relevance debate that was shortly to impact on geography in his observation that 'no geographical study can have validity unless the wishes of people are taken into full consideration' (p. 41).

The last quarter of the twentieth century saw the greatest change in the practice of applied geography. Foremost among these developments was the emergence of a welfare-oriented socially responsible applied geography. This was stimulated by theoretical and methodological changes within the discipline, and more generally within wider society. After two decades of relative prosperity, the economies of Britain and America began to experience difficulties during the late 1960s as the post-war boom faltered. At the same time, major societal events such as the US involvement in the Vietnam War, racial unrest in American cities, the civil rights movement, feminism and consumer rights and environmental groups contributed to a concern over the general issue of 'quality of life'. In contrast to the optimistic growth-centred outlook of earlier years, poverty and inequality were rediscovered in the American city. 'By the end of the 1960s urban policy in the United States was in disarray, and by any measure the American central city was in severe distress'. (Ley 1983: p. 1). For some radical geographers, these trends provided clear signs that 'the late twentieth century will be a period of continuing and escalating societal crises, the likes of which we have not yet known' (Peet 1977: p. 1). Similar tensions were being experienced to varying degrees in Britain and Europe.

These societal influences were reflected in the changing substantive concerns of applied geography. The direction of change is indicated clearly by the content of papers delivered to annual meetings of the American Association of Geographers in the early 1970s. At the 67th annual meeting, held in Boston in 1971, themes included geographical perspectives on poverty and social well-being, ethnic and religious groups, and urban policy, with a general session devoted to discussion of the problems and strategies facing 'socially responsible geographers'. This trend was continued at the 68th annual meeting in Kansas, where topics included a session on metropolitan spatial injustice, and other action-oriented papers on 'place utility, social obsolescence and qualitative housing change', 'crime rates as territorial social indicators', and 'environmental stress and maladaptive behaviour'. The 'relevance debate' was also taken up by geographers in the UK (Chisholm 1971; Prince 1971; Smith 1971; Dickenson and Clarke 1972; Berry 1972).

One result was that while the pre-war consideration of land-use issues, including urban sprawl, countryside conservation, land use and resource management, continued to attract attention, these were displaced from centre stage by questions relating to the geography of poverty and hunger (Morrill and Wohlenberg 1971), crime (Harries 1974), health care (Shannon and Dever 1974), ethnic segregation (Rose 1971), education (Kirby 1979) and the allocation of public goods (Cox 1973).

While most applied (human) geographers were agreed on the important issues, the social relevance 'movement' was far from united over the question of the best route towards a solution. For some, action within the existing structure of society was preferred, whereas others advocated a more radical approach aimed at a fundamental restructuring of the social order. The liberal approach essentially represented a continuation of the philosophy that underlay much of the applied geography and land-use planning of the inter-war and immediate post-war periods. Work on social issues in the liberal tradition included the mapping of spatial variations in quality of life (Knox 1975) as an input to planning and as a means of monitoring the distributional effects of social policies. Other researchers were more willing to embrace the radical alternatives to liberal formulations. The argument in favour of a Marxist approach was presented by Folke (1972), who considered that geography and the other social sciences are 'highly sophisticated, technique-oriented, but largely descriptive disciplines with little relevance for the solution of acute and seemingly chronic social problems . . . theory has reflected the values and interests of the ruling class' (p. 13). The Marxist critique of capitalism was also a critique of empirical positivist science and, understandably, applied physical geographers found the relevance debate largely irrelevant to the conduct of their research. While we acknowledge the profound and largely beneficial influence of the relevance debate on applied human geography, this does not amount to castigation of applied physical geographers or spatial scientists for a failure to adopt the same precepts. As we have seen, there is more than a single type of science, more than a single route towards knowledge and enlightenment, and all modes of analysis have the capacity to contribute to the applied geographer's goal of addressing real-world problems.

The development of applied geography has been accompanied by debate over the relative merits of pure and applied research. Critics such as Cooper (1966) and more recently Kenzer (1989) warned against the application of geographical methods as a threat to the intellectual development of the discipline. Conversely, Applebaum (1966) took the view that 'geography as a discipline has something useful to contribute to man's struggle for a better and more abundant life. Geographers should stand up and be counted among the advocates and doers in this struggle' (p. 198). In similar vein, Abler (1993) considered that 'too many geographers still preoccupy themselves with what geography is; too few concern themselves with what they can do for the societies that pay their keep' (p. 225). There is no reason why an individual researcher cannot maintain a presence in both pure and applied research. The eminent American geographer C. Sauer was both a 'scholar' who conducted research on agricultural origins and dispersals and an 'applied geographer' who developed a land classification system for the state of Michigan. The terms 'pure' and 'applied' are best seen as the ends of a continuum rather than unrelated polar opposites.

APPLIED GEOGRAPHY: PROSPECTIVE

The practical value of the applied geographical approach has been demonstrated in the foregoing discussion of the principles and practice of applied geography and is illustrated by the wide range of research work presented in this book. Applied geographers are actively engaged in investigating the causes and ameliorating the effects of 'natural' phenomena such as acid precipitation, landslides and flooding. Key issues of environmental change and management also represent a focus for applied geographical research, with significant contributions being made in relation to a host of problems, ranging from the quality and supply of water, deforestation and desertification to a series of land-use issues, including agricultural de-intensification, derelict and vacant land, and wetland and townscape conservation. Applied geographers with a particular interest in the built environment have, in recent decades, directed considerable research attention to the gamut of social, economic and environmental problems that confront the populations of urban and rural areas in both developed and developing countries. Problems of housing, poverty, crime, transport, ill health, socio-spatial segregation and discrimination have been the subject of intense investigation, while other topics under examination include problems ranging from boundary disputes and political representation to city marketing. The application of techniques in applied geographical analysis is of particular relevance in relation to spatial analyses, where the suite of problems addressed by applied geographers ranges from computer mapping of disease incidence to simulation and modelling of the processes of change in human and physical environments.

The list of research undertaken by applied geographers is impressive, but there are no grounds for complacency. While applied geographers have made a major contribution to the resolution of real-world problems, particularly in the context of the physical environment, in terms of social policy formulation in the post-war era the influence of applied geography has been mixed and arguably less than hoped for by those socially concerned geographers who engaged in the relevance debate a quarter of century ago.

Several reasons may be proposed to account for this. The first refers to the eclectic and poorly focused nature of geography and the fact that 'geographical' work is being undertaken by 'non-geographers' in other disciplines. This undermines the identity of geography as a subject with something particular to offer in public policy debate. The very breadth of the discipline, which for many represents a pedagogic advantage, may blur its image as a point of reference for decision makers seeking an informed input. Geographers wishing to influence public policy must compete with other, more clearly identified, 'experts' working on similar themes.

A second reason for the relatively limited influence on public policy may be the apparent reluctance of (human) geographers to 'get their hands dirty' – an attitude redolent of the eighteenth-century distinction between gentlemen, who derived a livelihood from the proceeds of land ownership, and those who earned a living through trade. This applies less to research in physical geography, where a basis in empirical science and positivist methodology has ensured that applied research has attracted support and acclaim more readily both from within the discipline and from external agencies. Significantly, the growth of environmentalism and the accompanying convergence of the philosophy and methodology of physical and human geography has gone some way towards bridging the gap between the two major sub-areas of the discipline and may represent a route for applied geographers to increase their policy influence.

The changing content and shifting emphases of human geography during the last quarter of the twentieth century represent a third factor underlying the limited social impact of applied geography. Over the period, the replacement of the earlier land-use focus in applied human geography by questions relating to the geography of poverty, crime, health care, ethnic segregation, education and the allocation of public goods brought applied geographers into direct confrontation with those responsible for the production and reproduction of these social problems. Unsurprisingly, since policy makers are resistant to research that might undermine the legitimacy of the dominant ideology, social policy remained largely impervious to geographical critique, particularly that which emanated from the Marxist analysis of capitalism.

The failure of applied geography to exert a major influence on social policy, however, does not signal the failure of applied geography to promote any significant improvement in human well-being, which, as we have seen, can be achieved by means other than via public policy. Any assessment of the contribution of applied geography to the resolution of real-world problems must balance the limited success in the specific area of social policy against the major achievements of applied geographers in the large number of other problem areas outlined above. Rather than dwelling on the limited impact to date of applied geographical research in the field of social policy, applied geographers can draw encouragement from their unwillingness to compromise a critical stance in return for public research funds or public acceptability of research findings. Furthermore, much of the applied social research undertaken achieves the goal of addressing real-world problems via its emancipatory power to expose the structural underpinnings of contemporary socio-spatial problems and by encouraging exploration of alternative social arrangements.

Applied geography is an approach whose rationale is based on the particular philosophy of relevance or social usefulness and which focuses on the application of geographical knowledge and skills to advance the resolution of real-world social, economic and environmental problems. As the contents of this book demonstrate, applied geographers are active across the human–physical geography divide and in most sub-areas of the discipline. The range of applied research presented in the book illustrates not only the contribution that applied geography is currently making towards the resolution of social, economic and environmental problems at a variety of geographic scales but also the potential of the approach to address the continuing difficulties that confront humankind.

Applied geography is a socially relevant approach to the study of the relationship between people and their environments. The principles, practice and potential of applied geography to engage a wide range of real-world problems commends the approach to all those concerned about the quality of present and future living conditions and environments on planet Earth.

GUIDE TO FURTHER READING

The changing nature and content of applied geography can be gauged from inspection of the texts by L.D. Stamp (1960) *Applied Geography*, Harmondsworth: Penguin; and J.W. Frazier (1982) *Applied Geography: Selected Perspectives*, Englewood Cliffs, NJ: Prentice-Hall, and by comparing these with the structure and content of the present volume. Insight into the 'relevance debate' of the early 1970s may be gained by examining successive issues of the journal *Area* between 1971 and 1973. For a contemporary view of ongoing research in applied geography, the journals *Applied Geography*, *Progress in Human Geography* and *Progress in Physical Geography* provide regular reports on new research, including work from an applied perspective.

REFERENCES

Abler, R. (1993) Desiderata for geography: an institutional view from the U.S., in R.J. Johnston (ed.) *The Challenge For Geography*. Oxford: Blackwell, 215–38.

Applebaum, W. (1961) Teaching marketing geography by the case method. *Economic Geography* 37, 48–60.

Applebaum, W. (1966) Communications from readers. *Professional Geographer* 18, 198–9.

Berry, B. (1972) More on relevance and policy analysis. *Area* 4, 77–80.

Bowen, M. (1981) *Empiricism and Geographical Thought from Francis Bacon to Alexander von Humboldt*. Cambridge: Cambridge University Press.

Briggs, D. (1981) Editorial: the principles and practice of applied geography. *Applied Geography* 1, 1–8.

Brunsden, D. (1985) Geomorphology in the service of society, in R.J. Johnston (ed.) *The Future of Geography*. London: Macmillan, 225–57.

Bunbury, E. (1879) *A History of Ancient Geography Among the Greeks and Romans from the Earliest Ages Till the Fall of the Roman Empire*. London: John Murray.

Chisholm, M. (1971) Geography and the question of relevance. *Area* 3, 65–68.
Cooper, S. (1966) Theoretical geography, applied geography and planning. *Professional Geographer* 18, 1–2.
Coppock, J.T. (1974) Geography and public policy: challenges, opportunities and implications. *Transactions of the Institute of British Geographers* 63, 1–16.
Cox, K. (1973) *Conflict, Power and Politics in the City*. New York: McGraw-Hill.
Darby, H.C. (1946) *The Theory and Practice of Geography*. London: University of Liverpool Press.
Dickenson, J. and Clarke, C. (1972) Relevance and the newest geography. *Area* 4, 25–27.
Folke, S. (1972) Why a radical geography must be Marxist. *Antipode* 4, 13–18.
Frazier, J.W. (1982) *Applied Geography: Selected Perspectives*. Englewood Cliffs, NJ: Prentice-Hall.
Freeman, T.W. (1972) Applied geography in British universities, in R. Preston (ed.) *Applied Geography and the Human Environment*. Proceedings of the Fifth International Meeting of the IGU Commission on Applied Geography, University of Waterloo, 369–73.
Geddes, P. (1915) *Cities in Evolution*. London: Williams and Northgate.
Habermas, J. (1974) *Theory and Practice*. London: Heinemann.
Harries, K. (1974) *The Geography of Crime and Justice in the United States*. New York: McGraw-Hill.
Hart, J.F. (1989) Why applied geography?, in M. Kenzer (ed.) *Applied Geography: Issues, Questions and Concerns*. Dordrecht: Kluwer, 15–22.
Harvey, D. (1984) On the historical and present condition of geography: an historical materialist manifesto. *Professional Geographer* 36, 1–11.
Herbertson, A.J. (1899) Report on the Teaching of Applied Geography. Unpublished Report to the Council of the Manchester Geographical Society.
Herbertson, A.J. (1910) Geography and some of its present needs. *Journal of the Manchester Geographical Society* 16, 21–38.
Hornbeck, D. (1989) Working both sides of the street: academic and business, in M. Kenzer (ed.) *Applied Geography: Issues, Questions and Concerns*. Dordrecht: Kluwer, 165–72.
Johnston, R.J. (1986) *On Human Geography*. Oxford: Blackwell.
Johnston, R.J. (1994) Applied geography, in R.J. Johnston, D. Gregory and D.M. Smith (eds.) *The Dictionary of Human Geography*. Oxford: Blackwell, 20–5.
Keltie, J. (1890) *Applied Geography: A Preliminary Sketch*. London: G. Philip and Son.
Kenzer, M. (ed.) (1989) *Applied Geography: Issues, Questions and Concerns*. Dordrecht: Kluwer.
Kirby, A. (1979) *Education, Health and Housing*. Farnborough: Saxon House.
Knox, P. (1975) *Social Well-Being: A Spatial Perspective*. Oxford: Oxford University Press.
Kraushaar, R. (1979) Pragmatic radicalism. *International Journal of Urban and Regional Research* 3, 61–80.
Ley, D. (1983) *A Social Geography of the City*. New York: Harper & Row.
Martin, G. and James, P. (1993) *All Possible Worlds: A History of Geographical Ideas*. Chichester: Wiley.
Merrifield, A. and Swyngedouw, E. (1996) *The Urbanisation of Injustice*. London: Lawrence & Wishart.
Morrill, R. and Wohlenberg, E. (1971) *The Geography of Poverty in the United States*. New York: McGraw-Hill.
Pacione, M. (1990) What about people? A critical analysis of urban policy in the United Kingdom. *Geography* 75, 193–202.
Palm, R. and Brazel, A. (1992) Applications of geographic concepts and methods, in R. Abler, M. Marcus and J. Olsson (eds) *Geography's Inner Worlds*. New Brunswick NJ: Rutgers University Press, 342–62.
Peet, R. (1977) *Radical Geography*. London: Methuen.
Prince, H. (1971) Questions of social relevance. *Area* 3, 150–3.
Rose, H. (1971) *The Black Ghetto*. New York: McGraw-Hill.
Sant, M. (1982) *Applied Geography: Practice, Problems and Prospects*. London: Longman.
Shannon, G. and Dever, G. (1974) *Health Care Delivery: Spatial Perspectives*. New York: McGraw-Hill.
Slaymaker, O. (1997) A pluralist, problem-focused geomorphology, in D. Stoddart (ed.) *Process and Form in Geomorphology*. London: Routledge, 328–39.
Smith, D.M. (1971) Radical geography – the next revolution? *Area* 3, 153–7.
Stamp, L.D. (1946) *The Land of Britain and How it is Used*. London: Longman.
Stoddart, D. (1987) To claim the high ground: geography for the end of the century. *Transactions of the Institute of British Geographers* 12, 327–36.
Taylor, E. (1930) *Tudor Geography 1485–1583*. London: Methuen.
Taylor, P. (1985) The value of a geographical perspective, in R.J. Johnston *The Future of Geography*. London: Methuen, 92–110.
Wright, J. (1952) *Geography in the Making*. New York: American Geographical Society.

Part I
Natural and environmental hazards

2

Global warming

Keith Boucher

FIELD OF STUDY AND RELEVANT LITERATURE

Global warming is a term that entered the domain of both popular and scientific literature during the 1980s. It is closely linked with the idea of an increasing greenhouse effect, which was first calculated by a Swedish chemist, Svente August Arrhenius, in 1896 (Arrhenius 1997). It is believed that our atmosphere acts rather like a greenhouse, in which the glass allows solar radiation to pass through, where it is converted into heat. This heat is absorbed by the soil before being radiated out as long-wave radiation and intercepted this time by the glass, which re-radiates some of the energy back into the greenhouse. The atmosphere has properties rather similar to the glass of the greenhouse, hence the 'greenhouse effect', originally postulated by the French mathematician Jean-Baptiste Fourier (1824).

Global warming would seem to imply that the whole atmospheric system is warming up as a result of the greenhouse effect, but this is far from certain. Once the nature of the problem has been outlined, three main areas of investigation will be addressed. First, there is the scientific evidence for global warming; second, the study of the likely impacts and third, the formulation and implementation of strategies to cope with such impacts. The contribution of geographers has chiefly been in the applied field of impact studies.

THE NATURE OF THE PROBLEM

The greenhouse effect

The 'natural' greenhouse effect occurs because some of the gases present in the atmosphere are largely transparent to incoming solar radiation but not to outgoing radiation, which is partially absorbed by water vapour, and the three main greenhouse gases. Their current percentage contributions are 70 per cent for carbon dioxide, 23 per cent for methane and 7 per cent for nitrous oxide. Water vapour is very variable in time and space. CFCs (chlorofluorocarbons) and their interaction with the variable greenhouse gas, ozone, also need to be noted as potential contributors (IPCC 1990). These gases are collectively known as 'the greenhouse gases'. The additional amount of such gases that are present in the atmosphere as a direct or indirect result of human activity, such as power generation and vehicle emissions, leads to an 'enhanced' greenhouse effect.

Although carbon dioxide is not the strongest absorber of outgoing long-wave terrestrial radiation, it is believed that it has the greatest long-term potential for raising global temperatures. Most attention has therefore been directed to the increase in the concentration of carbon dioxide from 280 ppmv (parts per million by volume) at the beginning of the Industrial Revolution in the eighteenth century to current levels of 360 ppmv (Figure 2.1). The annual increase of about 1.8 ppmv adds 3.8 Gt (gigatonnes) to the atmospheric carbon reservoir of

Figure 2.1 Trend line of carbon dioxide concentrations over the past 300 years based on (a) measurements from Antarctic ice core records prior to 1958 (dashed line), (b) direct measurements from the Mauna Loa observatory in Hawaii (solid line).

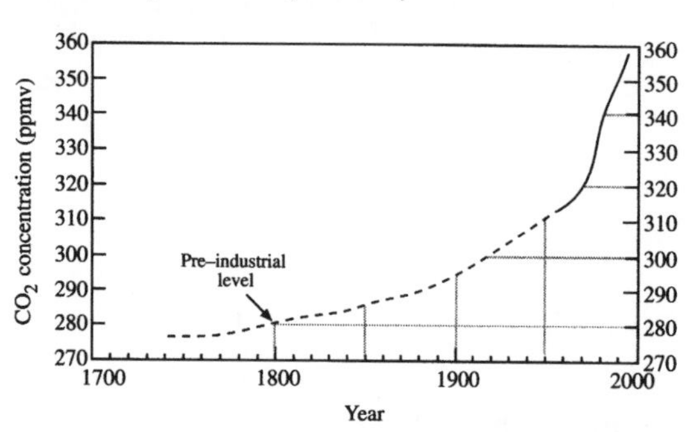

Source: After Houghton 1997 and IPCC WGI 1996.

750 Gt (Houghton 1997). This is about half of the calculated amount of 7.5 Gt being emitted into the atmosphere from forest fires and fossil fuel burning. It is generally accepted that the remainder – about 3.7 Gt. – is absorbed by the oceans.

The atmospheric carbon reservoir forms part of the global carbon cycle in which the natural atmosphere–ocean component has an annual throughput of about 90 Gt, while the biosphere–atmosphere cycle has an annual throughput of about 100 Gt. Both show a small but significant total net loss to the atmosphere of 3.7 Gt. The total annual throughput of carbon, chiefly as carbon dioxide, in the atmosphere is therefore about 25 per cent of the atmospheric reservoir of 750 Gt. The sensitivity to change in the throughput of carbon in the atmosphere needs to be considered in any discussion of global warming, since the climatic effect of relatively small changes in atmospheric carbon could be amplified within the system.

Radiative forcing and equilibrium

Radiative forcing may be defined as a change in average net radiation at the tropopause which is the upper boundary of the troposphere, due to changes in incoming solar or outgoing infrared radiation. Such a change 'perturbs' the Earth's radiation balance, as shown in Figure 2.2, altering the balance between incoming and outgoing radiation. It also alters the nature of the greenhouse effect, which is usually expressed in degrees C (IPCC 1995). Under present climatic conditions and carbon emissions, the natural greenhouse effect amounts to about 33°C if clouds, which act as radiation 'blankets', are included in the calculations (Houghton 1997). The theoretical surface temperature has been calculated to be –

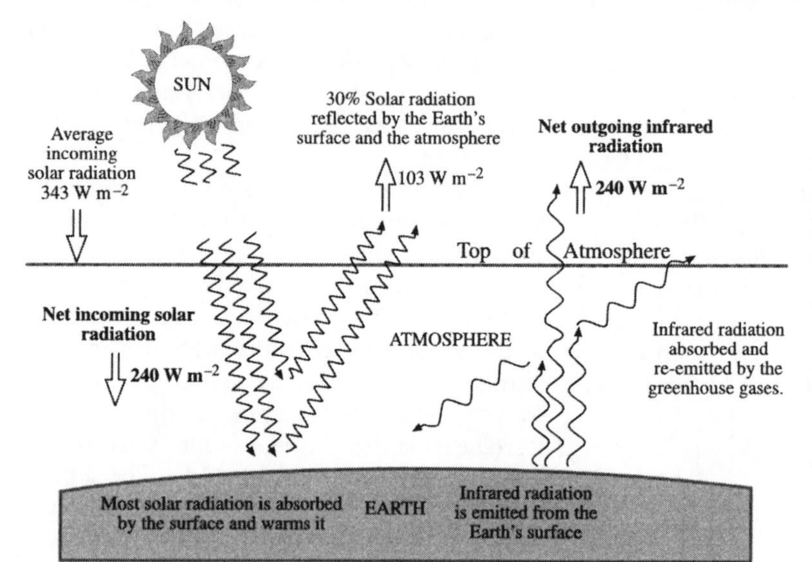

Figure 2.2 Simplified diagram showing the mean annual radiative balance of the atmosphere. Radiative equilibrium is maintained since net incoming solar radiation is balanced by net outgoing infrared radiation (240 W m^{-2}). Total earth–atmosphere albedo is about 30 per cent (103/343 × 100). The natural greenhouse effect (33°C), shown qualitatively as infrared radiation, is absorbed and partially re-radiated back to the surface.

18°C based on a 30 per cent value for the average reflectivity of the Earth and atmosphere. The actual global surface temperature is about 15°C – the difference represents the natural greenhouse effect. Clouds play an important but complex role in this radiation balance (IPCC 1990).

It is customary in studying possible scenarios of a future global climate to model a doubling of the pre-industrial (1750) carbon dioxide concentration of 280 ppmv, or its radiative equivalent. At the current rate of increase in greenhouse gas emissions under a business-as-usual scenario, doubling might occur between 2070 and 2100 (IPCC 1995). Conservative radiation models of the climate system indicate an initial rise of 1.3°C for a doubling of carbon dioxide, which is equivalent to 4 W m^{-2} (Watts per square metre). This would mean an initial fall in outgoing radiation from 240 to 236 W m^{-2} which would need to be restored to 240 W m^{-2} to maintain radiation equilibrium. This assumes that there is no change in the average amount of solar energy reaching the outer edge of the atmosphere, which is about 1,370 W m^{-2} (IPCC 1990). This value is referred to as 'the solar constant'.

One way of indicating the effect on the radiative balance of greenhouse gases and other components is to express this in terms of radiative forcing (W m^{-2}) from pre-industrial times to the present. Figure 2.3 shows estimates of the annually averaged radiative forcing due to human activity over this period. It also shows the natural changes in solar output from 1850 to the present (IPCC WGI 1996). The first column portrays the summed effect of carbon dioxide, methane, nitrous oxide and the halocarbons (CFCs). The radiative forcing is about 2.6 W m^{-2} or 0.8°C, with an error bar of 0.8 W m^{-2}. The error bar indicates the range of current estimates by models, while the confidence level is shown as a subjective assessment by Houghton (1997). It can be seen that the level of confidence is low for both tropospheric aerosols and solar variability. Expressing the radiative forcing effect in this way has led to the concept of global warming potential (GWP), which is defined as the ratio of the enhanced greenhouse effect of any gas compared with that of carbon dioxide. GWP is then used as a basis for 'trade-offs' in which the increase of the GWP of one gas can be offset against a similar reduction in the GWP of another gas (see Box 2.3). This is very similar to the 'bubble' principle

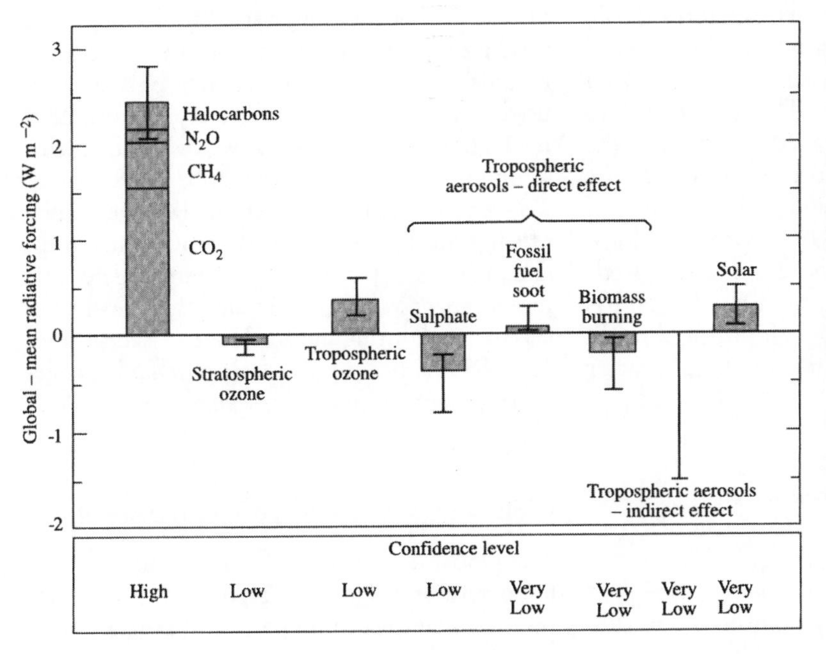

Figure 2.3 Estimates of the globally and annually averaged anthropogenic radiative forcing in W m^{-2} due to changes in concentrations of greenhouse gases and aerosols from a pre-industrial base (CO_2 concentration of 280 ppmv) to 1992. Estimated solar output changes are from 1850 to 1990 (0.02% of the mean solar constant). The shaded rectangular bars represent mid-range estimates of radiative forcing – both positive and negative.

Source: Houghton 1997.

of pollution credits in the USA (Elsom 1992) and an integral part of the 1997 Kyoto Agreement on emission regulation.

Internal feedbacks in the climate system – examples

The debate about internal changes is largely concerned with feedback processes, which may be initiated or enhanced through radiative forcing. Positive feedback tends to increase the rate of a process such as global warming. Long-term positive feedback mechanisms could prove destructive. The IPCC Report (1990) provides an example of a Sahelian drought-type positive feedback, where a drier surface resulting from rising temperatures leads to reduced evaporation, which in turn reduces humidity and cloud cover, promoting greater warming and yet drier surface conditions. If there were no restraining or counter-processes taking place, the Earth might be faced with a 'runaway' greenhouse effect as has occurred on Venus, where surface temperatures of about 525°C have been recorded by the Russian space probes (Houghton 1997). Negative feedback occurs when there are constant checks to the rate of a process. For example, rising sea-surface temperatures (SSTs) would increase evaporation, which in turn would increase the water vapour content in the lowest layers of the atmosphere. This could lead to greater spatial cloud coverage, decreasing the amount of incoming radiation reaching the surface, which in turn would lead to cooling of the Earth's surface. The climate system is dominated by numerous positive and negative feedbacks, some not operating until a threshold is passed such as an SST of 27°C for tropical storm development, others operating in a non-linear or quasi-stochastic manner.

EVIDENCE FOR GLOBAL WARMING

Global data sets: land

The collection and quality control of climatic records, especially the long-term near surface temperature data, have been basic requirements of the scientific community in its task of accumulating evidence for climate change. Since few instrumental records of land surface temperatures existed until about 1850, it is current practice to emphasise the trends in global temperatures from 1861 onwards, when surface observing networks were becoming established in many parts of the world. In the late 1980s, three research groups produced similar analyses of hemispheric land surface air temperature variations – Jones *et al.* (1986a and b) at UEA, UK; Hansen and Lebedeff (1987) at GISS, USA; and Vinnikov *et al.* (1987) of SHI in the former USSR, of which the updated and re-analysed data set of Jones *et al.* has been used in recent studies (IPCC WGI 1996).

Global data sets: oceans

Since the oceans comprise over 60 per cent of the Northern Hemisphere and over 80 per cent of the Southern Hemisphere, it is important to assemble an accurate data set. The collection and correction of global SST data have been undertaken most recently by Folland and Parker (1995) of the UK Meteorological Office. They have produced improved adjustments to temperature records where canvas and wooden buckets were used to sample sea water temperatures from ships from about 1860 to 1941. It is now thought that sampling of SSTs at night might be more reliable. This would avoid any bias due to daytime heating of ships' decks. There are also spatial gaps in the data, particularly over the southern oceans, and temporal gaps, especially in the nineteenth century and during the world wars. Except since 1975, SSTs appear to have lagged behind changes in land temperatures by several years (IPCC 1990).

Global data sets: combined land and ocean

The global data record of surface land and sea temperatures is shown in Figure 2.4. It indicates the two main warming periods that have taken

Figure 2.4 Changes in global average surface air temperature over land and sea (1851–1996) relative to the averaging period 1961–90. The solid curve represents the ten-year RM plotted mid-period. Inset: Comparison of globally averaged temperatures shown as departures from the 1979–94 average for (a) the MSU (Microwave Sounding Unit) channel of the NOAA meteorological satellites sensing temperature in the lower to mid-troposphere – 1–10 km altitude (bold line); (b) radiosonde measurements in the troposphere (thin line); (c) surface air temperature data (dashed line).

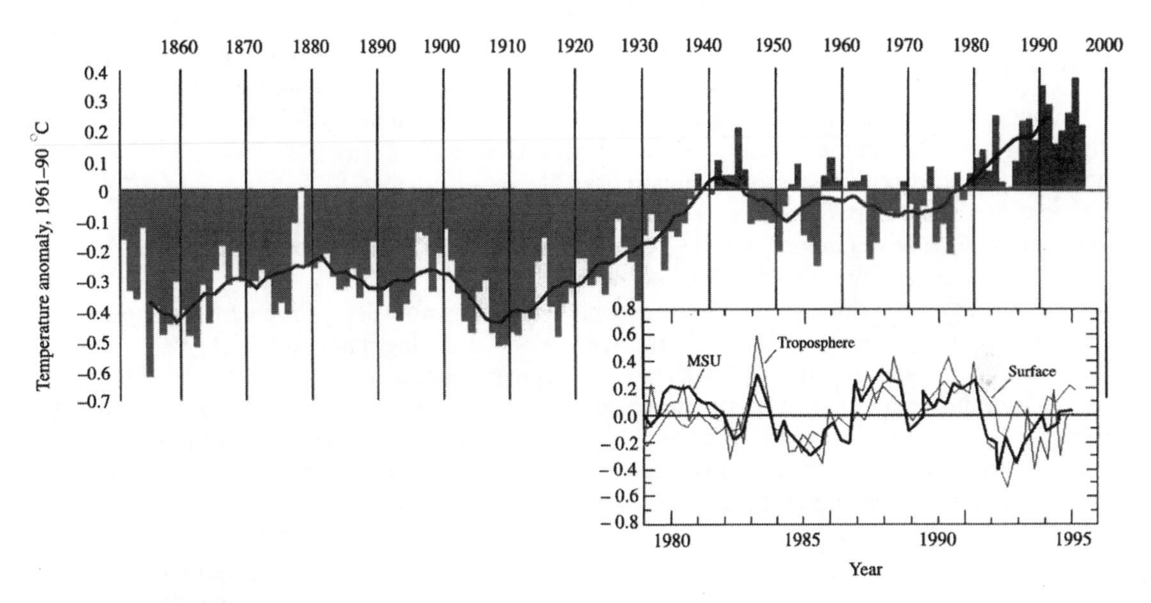

Source: Houghton 1997.

place since 1860, the first from 1910 to 1941 and the second from 1978 to the present. Although not shown on the graph, 1997 was the warmest year to date. The correlation of this record with that of the steady rise in CO_2 (see Figure 2.1) is only approximate. It implies that the climate system is responding in a complex way to direct radiative forcing through increased CO_2.

Over a shorter time scale, Houghton (1997) has provided an interesting comparison of global seasonal temperature anomalies from three different sources – satellites (MSU), radiosondes (from Parker and Cox 1995) and surface air temperatures (see Figure 2.4, inset). While complete agreement could not be expected, the sign of the anomaly (+/–) is similar. For example, each data set shows global cooling, chiefly in 1992, which was associated with the Mount Pinatubo volcanic eruption in the Philippines in 1991.

Regional data sets

Regional collections of temperature data are still important, particularly those that have been compiled with some quality control such as that assembled for central England by Gordon Manley and subsequently extended by the UK Meteorological Office (Jones and Hulme 1997). Ensembles of non-gridded data for other parts of the world have also been produced (Hulme *et al.* 1994), usually with some adjustment for urban heat island effects. Regional time series may also be gridded into boxes, typically 20° latitude × 60° longitude. These show considerable regional variability but also reveal coherent trends between adjacent areas within the same hemisphere (IPCC 1990: pp. 214–5). The published results indicate that moderate cooling (–0.4°C) took place in the Northern Hemisphere in the two decades after 1950, especially in the western sector (0–180° W). During the same period, the Southern Hemisphere showed no coherent trend but the 1970s were marked by warming, and this tendency continued through to 1990 in the eastern sector (0–180° E). Renewed warming is in evidence throughout the Northern Hemisphere since the early 1970s.

Attribution

The question then arises as to whether the general rise in global temperatures may be directly attributed to the greenhouse effect. Opinion has changed over the past two decades, as is evident from the following quotations. Gribbin (1978) reflected the uncertainty in the 1970s of not knowing whether the climate was cooling down or warming up, while the British government Cabinet Office statement was, in part, a reaction to the unexpectedly long Western European drought of 1975–6 and the wish to allay public fears of the impact of even greater climatic anomalies. Pronouncements by geographers have tended to reiterate the broad scientific opinion of the time (see Goudie 1990). However, by the end of 1995, the eight warmest years of the global record (1860–1995) had all occurred within the thirteen-year period 1983–95 (see Figure 2.4). This finally led to the first clear statement of attribution by the IPCC Scientific Working Group at its meeting late in 1995 in Madrid. Even in late 1997 – the warmest year yet in the global instrumental record – there was still some caution in British government publications about admitting attribution (see May 1997; Box 2.1).

> *Box 2.1* Changing attitudes towards global warming
>
> Since the early 1960s, we have seen only too clearly, the shift towards ... a slight cooling of the Northern Hemisphere that we now know signals a return towards the expanded circumpolar vortex conditions of the Little Ice Age.
>
> (Gribbin 1978: p. 54)
>
> Meteorological Office scientists take the view that the variations in weather in recent years are compatible with established climatic patterns. They see no reason to conclude from the historical record that especially large changes are likely in the next few decades.
>
> (Cabinet Office 1980)
>
> The balance of evidence suggests a discernible human influence on global climate.
>
> (IPCC WGI 1996)
>
> Although we do not have data reaching back many hundreds of years, by comparing observations of global mean temperatures with natural variability estimated from climate models, we find the warming has, over the past couple of decades, extended beyond the bounds of our estimates of natural variability.
>
> (May 1997)

MODELLING CLIMATE CHANGE

Modelling now lies at the centre of enquiries into global warming. The demand of governments for climatic forecasts has led to models being used *inter alia* to simulate the effects of doubling CO_2 on future climate. Each set of assumptions within a model run produces what has come to be known as a 'scenario' exemplified by the IS 92a business-as-usual scenario (see Box 2.3 on p. 32). Since about 1994, the more advanced equilibrium models have become 'full' in that they comprise not just the physical atmosphere but the ocean, cryosphere, land vegetation surface and chemical composition of the atmosphere. Despite increasing ability to model complex components, some fundamental problems remain. For example, in order to achieve linkage between atmosphere and ocean models, it is customary to 'spin up' the climate and the ocean component separately before coupling them. An example of a latitudinal performance envelope of nine atmosphere–ocean coupled models is shown in Figure 2.5 for the months December to February. It reveals that some models have rather large 'climatic drifts' (poor agreement) in the higher latitudes, as shown by the wider shaded area. This is mostly due to problems of sensitivity to radiative factors in the models.

Unlike the equilibrium models outlined above, transient models allow for annual adjustments in greenhouse gas concentrations and could be regarded as being closer to reality. Model

Figure 2.5 An envelope (shaded area) enclosing the results of nine ocean–atmosphere coupled models simulating present mean latitudinal surface temperature conditions expressed as departures from observed values for December to February.

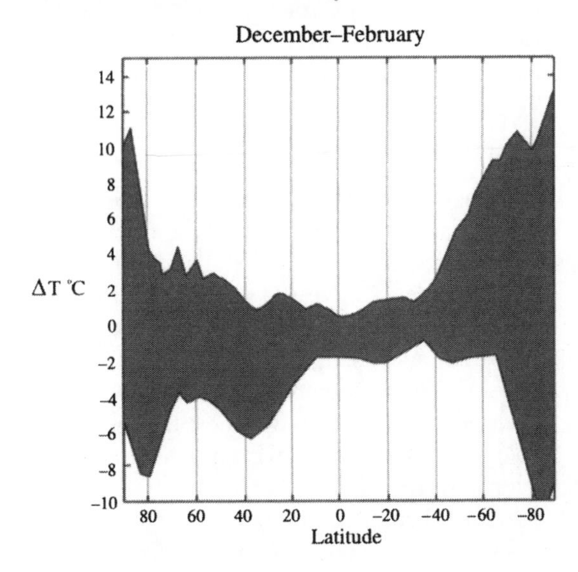

Source: Adapted from Houghton 1997.

results obtained by Hansen and Lebedeff are evaluated by Henderson-Sellers (1994) and the IPCC (1990 to 1996).

THE EFFECTS OF GLOBAL WARMING

Increasing attention is being given to the environmental consequences of climate change as the rise in global temperatures, from whatever cause, becomes more evident. Potentially, the most serious consequences relate to effects on vegetation and changes in sea level. It is these that will be addressed here.

Changing sea levels

Despite problems in measuring the sea level relative to the land over time, there seems little doubt that global sea levels have risen by about 1.2 mm yr^{-1} over the past century (IPCC 1990). Two of the major studies undertaken towards the end of the 1980s agree in the overall increase since 1890, but the study by Gornitz and Lebedeff (1987) employing geological data shows a nearly linear increase over the period, while that of Barnett (1988) suggests a steeper rise of about 1.7 mm yr^{-1} since 1910. Gornitz (1995) has indicated a value closer to 2 mm yr^{-1}, using a post-glacial rebound model (Tushingham and Peltier 1991) to account for the isostatic component in sea level changes. This is very close to the observed long-term record (1851–present) for stations such as San Francisco. The isostatic component, the lowering of relative sea level as calculated from one of the Peltier models, is about 0.5 mm yr^{-1} for a number of locations worldwide, including San Francisco. This may imply that global eustatic rise (non-isostatic component) could be approaching 2.5 mm yr^{-1}. Relatively simple models indicate that the rise in sea level, directly attributable to global warming, has been of the order of 0.28 to 0.52 mm yr^{-1} averaged over the period 1880–1990 (Wigley and Raper 1993). Houghton (1997) postulates that under a business-as-usual scenario, thermal expansion of the oceans is likely to account for about 60 per cent of future sea level rise.

The other main contributor to a rise in sea level is expected to be the continued melting of many glaciers, as indicated by calculations of negative cumulative mass balances. Glacial melt waters have probably provided about 20 per cent of the observed sea level rise this century increasing from about 0.35 mm yr^{-1} around 1900 to 0.5 mm yr^{-1} in the period 1985–93 (Dyurgerov and Meier 1995). In the unlikely event of all the mountain glaciers melting (excepting Antarctica and Greenland), the rise in sea level can be expected to be about 50 cm. Results from a recent model of ice depletion of mountain glaciers produced a rise of 0.5 mm yr^{-1} in sea level over the last 100 years (IPCC WGI 1996). The contribution of sea ice melting remains difficult to estimate. Neither hemisphere has seen significant changes to sea ice extent since 1973, although if data on the total mass of sea ice were available this might be a more meaningful indicator of any change. Manabe and Stouffer (1980) showed that the large horizontal extent and small thickness of sea ice makes it particularly sensitive to climatic change and albedo feedback. First-year ice thickness in the Weddell–Enderby Basin is about 0.5 m while multi-year ice thickness of about 1.4 m has been measured in the Antarctic (IPCC WGII 1996). Scientific opinion has been divided on the degree of global warming needed to semi-permanently melt the Arctic sea ice (Untersteiner 1984).

The total contribution of any melting of the Greenland and Antarctic ice sheets to a rise in sea level is thought to be close to zero at present. Results of models simulating a 1°C rise in global temperatures show a potential sea level rise of 0.3 mm yr^{-1}, caused by the melting of ice over Greenland and a corresponding fall for Antarctica due to ice accumulation (IPCC WGI 1996). The difference in the model estimates between the two ice sheets is a function of the way in which atmospheric feedback mechanisms may operate.

A component that is not so easily modelled is that of calving of the sea ice margins and the production of icebergs, which eventually melt. It is thought that in Greenland, ice loss from surface melting and runoff is of the same order of magnitude as loss from iceberg calving (IPCC WGI 1996). This process is more closely related to the frequency and tracks of storms than to global warming directly. However, there is concern over the possible instability of the West Antarctic Ice Sheet, which some scientists believe could become dislodged from its grounding 2,500 m below sea level. If this massive volume of ice were to melt, it would raise global sea level by 5–6 m, compared with 8 m for the melting of the Greenland ice sheet and 55 m for the east Antarctic ice sheet (Untersteiner 1984). A more likely scenario of West Antarctica ice shelf thinning for a 1°C warming would be a sea level rise of 0.1 mm yr^{-1} through to around the year 2050 (Budd *et al.* 1987).

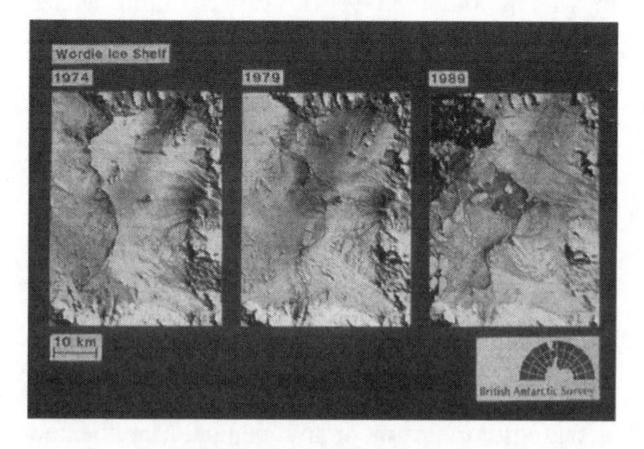

Plate 2.1 Changes in the Wardie Ice Shelf, Antarctica (*photograph: British Antarctic Survey*).

Effects of sea level rise

The biogeophysical effects are quite diverse and not necessarily uniform around the world. They include the inundation of wetlands and lands close to sea level, increased salinity of estuaries, a higher risk of storm flooding and erosion of shorelines, and changes to tidal ranges and the deposition of sediments. Regional responses to sea level rise through geomorphological and ecological systems are proving difficult to identify or forecast (IPCC WGII 1996).

In spite of uncertainties about the degree of rise in sea level, any rise would pose a direct threat to low-lying coastal zones and islands – particularly coral atolls, reef islands and tropical coastal wetlands, where the mangrove ecosystems are under threat (IPCC WGII 1996). Where rising sea level is combined with tectonic subsidence and/or human actions that may exacerbate the problem, the situation is potentially very serious. Table 2.1 indicates the synthesised results from country case studies based on a 1 m rise in sea level by the year 2100 based on the high estimate of global warming under the 1900 business-as-usual scenario. This may be regarded as extreme, but down-scaling still implies substantial problems for countries such as Bangladesh and China, particularly where economic growth in terms of GNP remains at a low level.

Discussion about indirect effects through feedback mechanisms in the climate system is fraught with uncertainties. At present, models show rises in the mean surface air temperature, especially over land, and the majority of models indicate some increase in Asian monsoon rainfall. If the occurrence of tropical cyclones and storm surges increases, or the direction of storm tracks changes, then either could have devastating effects on coastal populations and habitats. Despite the

Table 2.1 Estimates of impact of 1-metre rise in sea level (based on data in table 9.3, IPCC WGII 1996).

Country/source	*People affected*		*Land affected*	
	People in millions	*% of total*	*km²*	*% of total*
China	72	7	35,000	–
Bangladesh	71	60	25,000	17.5
Japan	15	15	2,300	0.6
The Netherlands	10	67	2,165	5.9
India	7.1	1	5,800	0.4
Egypt	4.7	9	5,800	1.0
Nigeria	3.2	4	18,600	2.0
Benin	1.3	25	230	0.2
United States	not known	–	31,600	0.3

impact of individual storms, the evidence from observations and from numerical and theoretical models is inconclusive at present. The storm pattern is made more difficult to interpret due to ENSO (El Niño–Southern Oscillation) events, which tend to swamp a simple global warming interpretation. Almost certainly, an understanding of the incidence of tropical storms will have to be gained through modelling the potential effects of global warming on ENSO.

Vegetation and the nitrogen cycle

Current research suggests that the terrestrial biosphere is currently a carbon sink attaining 2.6 gigatonnes of carbon per year (1992–3), but with a high inter-annual variation. If the tropics are a net carbon source, as seems possible, then the mid/high-latitude sink could even exceed this value. There is also a complex link between nitrogen, principally from soil organic matter, and the storage of carbon in the ecosystem. If the C:N ratio, currently between 10 and 25, were to alter as limiting conditions are reached, this would affect carbon storage. It is also known that high CO_2 conditions stimulate photosynthesis, increasing the ability of plants to fix carbon. The C3 group of plants, including wheat and rice, are well adapted to this fertilisation effect. The C3 'normal' biochemical pathway is characterised by a discrimination of 18 parts per thousand, while C4 photosynthesis is far more complex and is essentially non-discriminatory (IPCC WGII 1996). C4 biomes include tropical grasslands and savannas. Points to note from this brief overview are, first, that small changes in the discrimination value input into net primary production models produce large variations in the carbon sink value; and, second, that there may be an upper concentration limit of CO_2 in the atmosphere at which the fertilisation effect ceases. Both have implications for future levels of CO_2 in the atmosphere.

Agricultural productivity

The view expressed in IPCC (1990) that 'global agricultural production can be maintained relative to baseline production in the face of climate changes likely to occur over the next century' was maintained in IPCC WGII (1996: p. 429) with a medium confidence level. On the other hand, crop yields and productivity would probably be marked by much inter-regional change. Overall, the change is likely to be beneficial, due to the dominance of C3 crops such as barley, wheat, rice and soybeans. The C3 annual crops show yield increases of up to 30 per cent at doubled (700 ppm) CO_2 concentrations under controlled experimental conditions (Table 2.2). This productivity could be further enhanced, since fourteen out of eighteen of the world's worst weeds are C4 plants and would not directly

Table 2.2 Trends in world crop production along with modelled results of the impact of climate change on productivity.

	Crop	*World percentage change in yield 1967–97 (FAO 1998 www)*	*Current world crop yield in 1997 (FAO, 1998 www) tonnes/ha*	*Crop yield model (GISS) % change in yield with 2 × CO_2 climate change only *not GISS, limited regional studies (IPCC (WGII ch13)*	*Crop yield model (GISS) % change in yield with 2 × CO_2 climate change plus direct effects of CO_2 *not GISS limited regional studies (IPCC WGII ch 13)*
C3	Barley	+35.8	2.3	*(–40)	*(–30) Uruguay
	Wheat	+96.1	2.63	–16	+11
	Rice (paddy)	+74.5	3.79	–24	–2
	Soybeans	+56.2	2.11	–19	+16
	Potatoes	+20.3	16.6		*(+20) W Europe
C4	Maize	+70.3	4.13	–20	–15
	Sorghum	+28.2	1.42		*(0) USA
	Millet	+45.0	0.96		*(–66) Senegal
	Sugar cane	+19.6	62.38		*(+9) Australia/Japan

benefit from the CO_2 fertilisation effect. Potentially limiting factors include changes in insect life cycles and an increase in the survival, growth and spread of pathogens. For example, the frequency of outbreaks of powdery mildew and rust on crops is associated with milder mid-latitude winters. Changes in climatic conditions may also affect animal agriculture through feed grain availability, water availability in pasture lands and the incidence of livestock diseases.

Changing vegetation patterns

While there have been numerous studies of plant metabolism under enhanced CO_2, geographers are rather more interested in the redistribution of species and biome adjustment to a changing climatic environment. Webb (1986) has posed the question as to whether vegetation is in equilibrium with climate. One method of assessing the potential for change in the biosphere has been to study vegetation and lake levels since the last glaciation. Some idea may then be obtained of the speed of response of certain biomes. It is assumed that the dynamics of ecosystems will change as

Plate 2.2 Forest fires in the Amazon (*photograph: Panos Pictures*).

Box 2.2 Carbon dioxide released from forest fires

The year 1997 witnessed some of the worst forest and scrub fires ever recorded. There are no accurate estimates yet of how much biomass was converted into CO_2, but an attempt is made to assess the impact of the fires of 1997–8. The process of forest burning in the tropics is an integral part of traditional farming, known universally as 'slash and burn'. Regional terms include *ladang* in Indonesia and *roça* in Brazil. Some governments such as that in Malaysia have attempted to ban or control the burning, but in neighbouring Indonesia the practice has gone unchecked. While research suggests that agroforestry can help carbon sequestration by converting Imperata grasslands into more productive tree-based systems, wholesale burning fails to achieve such an end. There are other dimensions to the problem apart from the use for shifting cultivation, such as the actions of logging companies in allowing or promoting burning, and the need for more land for rice and food crops, as in the permitted burning of 40,000 ha southeast of Palangkaraya (Kalimantan). To these must be added in 1997–8 the active El Niño phase of ENSO, which caused almost unprecedented drought across many parts of equatorial Southeast Asia.

Estimates of C released on burning are based on the following approximations:

- 1 ha contains 200–500 t (tonnes) of biomass, about 50% of which is carbon.
- Mature tropical forests in Indonesia average about 365 t ha^{-1}.
- Burning probably releases two-thirds of carbon into the atmosphere as CO_2.
- The intensity of the fires has caused the peat of the forest floor to ignite, adding to the release of carbon dioxide into the atmosphere by an unknown amount. The value of 365 t ha^{-1} is taken to include this effect in the following estimates. The same figure is used for Brazil.

Comparative data on recent burning and CO2 release:

- The burning of 40,000 ha would therefore release about 0.5 Mt as CO_2.
- Around 2 million ha were lost to fire in 1997 in Indonesia alone, releasing about 0.25 Gt of CO_2 (1 Gt = 1,000 Mt = 10^9 t) (Dudley 1997).
- Estimates by the IPCC of CO_2 emissions from tropical land-use changes – mostly deforestation – in the late 1980s are about 1.6 ± 0.4 Gt per annum. As may be seen from Figure 2.6, this may be compared with 1.75 Gt of CO_2 emissions from fuel combustion in North America in 1996.

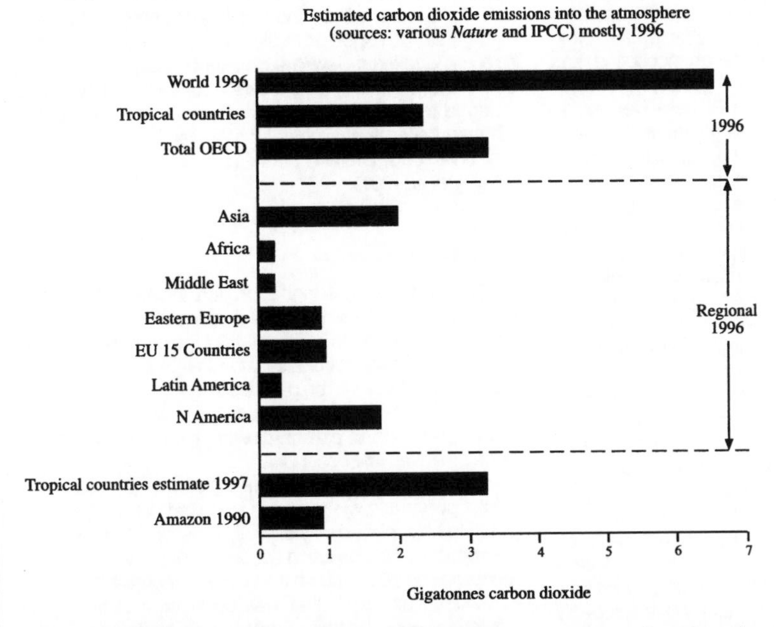

Figure 2.6 Estimated carbon dioxide emissions into the atmosphere.

Sources: Various, mostly from *Nature* and IPCC.

While it is very difficult to obtain estimates of emissions for a specific year, the fires of 1997–8, the worst of which were those in Brazil, must cause concern, since the addition of CO_2 is unlikely to be balanced by sequestration through afforestation in the near future. Serious fires were also reported from Papua New Guinea, Colombia, Peru, Mexico, Australia, parts of Africa and southern Europe.

climate changes during the coming century and that past performance of plant adaptation will apply in the future. There is some anxiety that selected plant species may die back before better-adapted species are established. Such a transient hysteresis effect in vegetation may produce a temporary peak or 'spike' in atmospheric CO_2 concentrations (Smith and Shugart 1993).

Forest fires have periodically posed a threat through the unregulated additions of CO_2 to the atmosphere (see Plate 2.2 on p. 30). Box 2.2 highlights the problems of forest fires during 1997–8.

CONCLUSION

There has been no attempt here to deal with all the complex systems relating to global warming, but this is not to imply that the impacts will be restricted to changing sea levels and vegetation patterns. Changes are also likely in hydrological systems, agricultural practices, associated economic activities, energy options and impact assessments. Policies need to be put in place to mitigate the worst effects of global warming. Some of these are highlighted in Box 2.3. There is,

Box 2.3 Policies for reducing carbon dioxide

The Climate Convention is an attempt by the governments of the world to reach agreement on curbing emissions of greenhouse gases. The UN Framework Convention on climate change was signed by over 160 countries at the Rio Conference on Environment and Development in 1992. Details are provided by Houghton (1997). This was subsequently revised at the 1997 Kyoto Conference after considerable lobbying. It is unclear how effective the Kyoto agreement will be due to problems of monitoring and implementation agreements on greenhouse gas emissions. However, the aim of the Climate Convention was to slow down and then stabilise global emissions to 1990 levels. This was to be achieved by switching to more energy-efficient fuels. Natural gas, for example, generates 40 per cent less CO_2 than coal for the same energy output, but such a switch of fuel is not an option for many countries. There is also a vested interest by a number of transnational companies in increasing the consumption levels of oil in the drive towards greater economic growth. By 1998, changes in CO_2 emissions had less to do with energy saving than with political changes, despite the attempt by some industrial countries to stabilise emissions. Trends during the 1990s indicate that the industrial countries (OECD) showed increases of around 4 per cent, while the increase in developing countries was about 25 per cent but from a lower 1990 emission level. Only a decline of 25 per cent in Eastern Europe, associated with economic collapse, allowed world emissions to remain almost constant.

This 'accidental' stabilisation is unlikely to continue, even if the 'tiger' economies of East Asia suffer economic decline under a 'boom–bust' scenario. Unfortunately, additional uncontrolled emissions (see Box 2.2) have probably renewed the upward global trend seen in the 1980s.

THE TIME COMPONENT

The basic options that are presented in the models of global warming are (1) act now and (2) delay until later – usually termed 'business-as-usual' or BaU. There are possibilities between these two extremes, but problems of economic inertia and political 'short-termism' mean that these are relatively unattractive. In any case, climate models such as the Dutch IMAGE 2 model point to good scientific reasons for acting now. This is because waiting until 2010 to act will be too late in attempting to curb the higher rate of expected emissions. In opposition to this scientific and legal view, the BaU lobby insists that waiting until 2010 is the only viable option for industrial countries (referred to as Annex 1 countries), since new technology will then become available. Non-Annex 1 countries could wait until 2030. There is partial scientific support for BaU (Wigley *et al.* 1997).

OTHER ISSUES

- The USA favours a 'net' approach, in which a country's inventory of carbon emissions will include forest burning as well as sequestration (afforestation) programmes. It also supports a 'global bubble' solution in which the trading of pollution credits would be permitted.
- The EU favours the 'basket' policy, in which the combined effects of carbon dioxide, methane and nitrous oxide are taken into account.
- The G77 developing countries group is opposed to the basket policy and wishes there to be targets for individual gases to be achieved domestically. Most of these countries are opposed to pollution quota trading, want drastic reductions by 2020 and an economic compensation fund.
- Another approach that was considered at the Kyoto Conference was the idea of a per capita emission rate with convergence to a value of 1 tonne CO_2 per capita per year by 2030. This would stabilise CO_2 concentrations at around 450 ppmv. Others favour 560 ppmv, since this represents a doubling of pre-industrial levels and accords with model scenarios.

unfortunately, no sign as yet that most nations of the world will show sufficient resolve or ability to reduce greenhouse gas emissions in the near future. One of the most difficult aspects of global warming is estimating its regional climatic impact (Houghton 1997) and the downstream effect on water and food supplies. The estimated impact may be complicated by other events such as the 1997 ENSO anomaly, which reversed rainfall patterns over wide areas of the tropics. The link between global warming and an enhanced ENSO event is even less easy to establish and is 'not intuitively obvious' (Meehl and Branstator 1992). It is a warning that not only are there many 'actors' on the climate scene but there are also many and possibly unforeseen ways in which the climate system may respond in terms of positive and negative feedbacks. It is hoped that the increasing sophistication of climate models will be a valuable tool in unravelling the global climate system and its response to forcing through global warming. This scientific advance needs to be matched by the better understanding of economic, social and political responsibilities in the stewardship of finite global resources.

GUIDE TO FURTHER READING

The detailed reports of the Intergovernmental Panel on Climate Change (IPCC) provide the findings of a broad consensus of scientists relating to global warming and are listed in the References section. Houghton (1997) has provided a clearly written summary accompanied by notes of the background and results from 1990 to 1997 of the IPCC, of which he has been the chairman. The next IPCC report is due in 2000.

The following books *inter alia* provide a more general survey: Bernard, H.W. (1993) *Global Warming Unchecked*. Indiana University Press; Schneider, S.H. (1990) *Global Warming: Are We Entering the Greenhouse Century?*

There are other specialised texts on related aspects, such as that by Pirazzoli, P.A. (1996) *Sea Level Changes: The Last 20,000 Years*.

For modelling, see McGuffie, K. and A. Henderson-Sellers (1997) *A Climate Modelling Primer*, 2nd edn, with CD-ROM.

Articles may be found in such journals as *Global Environmental Change*, *Climate Change*, *Ambio*, *Earth Surface Processes and Landforms*, with more specialist notes in *Nature* and *Science*. Updates on global warming occur regularly in *Progress in Physical Geography*.

Web sites in the UK include the following (note: these are subject to change):
Climatic Research Unit at the University of East Anglia – Tiempo Climate Cyberlibrary: http://www.cru.uea.ac.uk/tiempo/
The Hadley Centre for Climate Prediction and Research. UK Meteorological Office. http://www.meto.govt.uk/sec5/sec5pg1.html (see 'Global Warming Gallery')

Web sites elsewhere in the world include (links to other sites usually listed):
The Intergovernmental Panel on Climate Change (IPCC): http://www.ipcc.ch/
Official web site of the Climate Change Secretariat: http://www.unep.ch/iucc/
US Environmental Protection Agency: http://www.epa.gov/globalwarming/
Articles, news, etc. (US): http://www.law.pace.edu/env/energy/globalwarming.html
Planetvision: http://www.envirolink.org/orgs/edf/
Virtual Museum on Global Warming (US): http://www.edf.org/pubs/Brochures/GlobalWarming/

REFERENCES

Arrhenius, G. (1997) Carbon dioxide warming of the early Earth, *Ambio* 26, 12–16.

Barnett, T.P. (1988) Global sea level change. In NCPO, *Climate Variations over the Past Century and the Greenhouse Effect*. A report based on the First Climate Trends Workshop, 7–9 September 1988, Washington DC, National Climate Program Office/NOAA, Rockville, Maryland.

Budd, W.F., McInnes, B.J., Jenssen, D. and Smith, I.N. (1987) Modelling the response of the West Antarctic Ice Sheet to a climate warming. In C.J. van der Veen and J. Oerlemans (eds) *Dynamics of the West Antarctic Ice Sheet*, Dordrecht: D. Reidel 321–58.

Cabinet Office (1980) *Climatic Change*, London: HMSO.

Dudley, N. (1997) The year the world caught fire. *World Wildlife Fund International*, Discussion Paper, December, 35 pp.

Dyurgerov, M.B. and Meier, M.F. (1995) Year to year fluctuations in global mass balances of glaciers and their contribution to sea level changes. *IUGG XXI*

Assembly Abstracts, B318, American Geophysical Union.

Elsom, D. (1992) *Atmospheric Pollution, A Global Problem*. Oxford: Blackwell, 422 pp.

Folland, C.K. and Parker, D.E. (1995) Corrections of instrumental biases in historical sea surface temperature data. *Quarterly Journal of the Royal Meteorological Society* 121, 319–67.

Fourier, J.-B, (1824) Mémoire sur les températures du globe terrestre et des espaces planétaires. *Mémorandum Académique Scientifique de l'Institution Française* 7, 569–604.

Gornitz, V. (1995) Sea level rise: a review of recent past and near future trends. *Earth Surface Processes and Landforms* 20, 7–20.

Gornitz, V. and Lebedeff, S. (1987) Global sea level changes during the past century. In D. Nummedal, O.H. Pilkey and J.D. Howard (eds) *Sea Level Change and Coastal Evolution*. Society for Economic Palaeontologists and Mineralogists (SEPM Special Publication No. 4) pp. 3–16.

Goudie, A. (1990) *The Human Impact on the Natural Environment*. Oxford: Blackwell, 388 pp.

Gribbin, J. (1978) *The Climatic Threat*, Glasgow: Fontana/Collins.

Hansen, J. and Lebedeff, S. (1987) Global trends of measured surface air temperature. *Journal of Geophysical Research* 92, 13345–72.

Henderson-Sellers, A. (1994) Numerical modelling of global climates. In C.N. Roberts (ed.) *The Changing Global Environment*, Oxford: Blackwell, pp. 99–124.

Houghton, J. (1997) *Global Warming – The Complete Briefing*. Cambridge: Cambridge University Press, 251 pp.

Hulme, M., Zhao, Z.-C. and Jiang, T. (1994) Recent and future climate change in East Asia. *International Journal of Climatology* 12, 685–90.

IPCC (1990) *Climate Change: The IPCC Scientific Assessment*. Houghton J.T., G.J. Jenkins and J.J. Ephraums (eds). Cambridge: Cambridge University Press.

IPCC (1995) *Climate Change 1994: Radiative Forcing of Climate Change*. Cambridge: Cambridge University Press.

IPCC WGI (1996) *Climate Change 1995: The Science of Climate Change*. Contribution of Working Group 1 to the Second Assessment Report of the Intergovernmental Panel on Climate Change. Cambridge: Cambridge University Press.

IPCC WGII (1996) *Climate Change 1995: Impacts, Adaptations and Mitigation of Climate Change: Scientific–Technical Analyses*. Contribution of Working Group 2 to the Second Assessment Report of the Intergovernmental Panel on Climate Change. Cambridge: Cambridge University Press.

Jones, P.D., Raper, S.C.B., Bradley, R.S., Diaz, H.F., Kelly, P.M. and Wigley, T.M.L. (1986a) Northern Hemisphere surface air temperature variations, 1851–1984. *Journal of Climate and Applied Meteorology* 25, 1213–30.

Jones, P.D., Raper, S.C.B., Bradley, R.S., Diaz, H.F., Kelly, P.M. and Wigley, T.M.L. (1986b) Southern Hemisphere surface air temperature variations, 1851–1984. *Journal of Climate and Applied Meteorology* 25, 161–79.

Jones, P. and Hulme, M. (1997) The changing temperature of 'Central England'. In M. Hulme and E. Barrow (eds) *The Climates of the British Isles*, London: Routledge, pp. 173–96.

Leggett, J. (ed.) (1990) *Global Warming, The Greenpeace Report*, Oxford: Oxford University Press, 554 pp.

Manabe, S. and Stouffer, R.J. (1980) Sensitivity of a global climate model to an increase of CO_2 concentration in the atmosphere. *Journal of Geophysical Research* 85, 5529–54.

May, R. (1997) *Climate Change*, A note by the UK Chief Scientific Adviser. London: Office of Science and Technology, September.

Meehl, G.A. and Branstator, G.W. (1992) Coupled climate model simulation of El Niño/Southern Oscillation: implications for paleoclimate. In H.F. Diaz, and V. Markgraf (eds) *El Niño*, Cambridge: Cambridge University Press, pp. 69–91.

Parker, D.E. and Cox, D.I. (1995) Towards a consistent global climatological rawin-sonde data base. *International Journal of Climatology* 15, 473–96.

Smith, T.M. and Shugart, H.H. (1993) The transient response of terrestrial carbon storage to a perturbed climate. *Nature* 361, 523–6.

Tushingham, A.M. and Peltier, W.R. (1991) ICE-3G: a new global model of late Pleistocene deglaciation based upon geophysical predictions of post glacial relative sea level change. *Journal of Geophysical Research* 96, 4497–523.

Untersteiner, N. (1984) The cryosphere. In J. Houghton (ed.) *The Global Climate*. Cambridge: Cambridge University Press, 121–40.

Vinnikov, K.Ya., Groisman, P.Ya., Lugina, K.M. and Golubev, A.A. (1987) Variations in Northern Hemisphere mean surface air temperature over 1881–1985. *Meteorology and Hydrology* 1, 45–53 (in Russian).

Webb, T., III (1986) Is vegetation in equilibrium with climate? How to interpret the late-Quaternary pollen data. *Vegetation* 67, 75–91.

Wigley, T.M.L. and Raper, S.C.B. (1993) Future changes in global mean temperature and sea level. In R.A. Warwick, E.M. Barrow and T.M.L. Wigley (eds) *Climate and Sea Level: Observations, Projections and Implications*. Cambridge: Cambridge University Press, pp. 111–33.

Wigley T.M.L. *et al.* (1997) Implications of recent CO_2 emission limitation proposals for stabilisation of atmospheric concentrations. *Nature* 390, 267–70.

3

Acid precipitation

A.M. Mannion

INTRODUCTION

Precipitation comprises all solid and liquid forms of water that are deposited on the Earth's surface from the atmosphere. It includes rain, snow, hail, dew and sleet. All forms of precipitation are acid in so far as they have a pH of less than 7; in general, precipitation unaffected by human activity has a pH of 5.6. This naturally acidic state of precipitation is caused by the combination of water and carbon dioxide in the atmosphere to produce carbonic acid. However, the term acid precipitation, or acid rain, is usually applied to precipitation characterised by a pH of less than 5.1 (Elsworth 1984) and that contains sulphurous and nitrous acids. The latter are derived from various sources, among which fossil fuels are the most important.

The phenomenon of acid precipitation was first recognised by Robert Angus Smith, a Scottish chemist, in 1852 following a survey of air pollution in Manchester. Smith coined the term 'acid rain', which he associated with sulphur dioxide emissions from fossil fuels burned in local factories. Various observers subsequently noted the impact of acid precipitation on aquatic and terrestrial ecosystems. For example, Gorham (1958) noted that the chemistry of upland lakes in the English Lake District was affected by acid precipitation from air masses that had passed over Britain's industrial heartland. Despite this recognition of its impact, acid precipitation did not emerge as a major environmental issue until the late 1960s. By this time, Scandinavian ecologists were becoming concerned about declining fish stocks; they were also beginning to recognise transboundary transportation of acid precipitation, i.e. the export of acid precipitation from source areas such as the industrial regions of Europe and the UK and its transport to and deposition in far distant areas such as Scandinavia. In this context, acid precipitation became a political as well as an ecological issue. The polluters were unwilling to recognise this, and the polluted demanded mitigation measures. The impact of acid precipitation manifests in many ways. Both terrestrial and aquatic ecosystems may be adversely affected through reductions in pH, which have repercussions for the biota and water quality; human health may be impaired and building materials may be corroded.

Internationally agreed measures to curb acid precipitation are now in operation in Europe and North America, where the problem is most acute. The first of these was established in 1979. This was the Convention on Long Range Transboundary Air Pollution (CLRTAP), a protocol that was adopted in 1985 and that became known as the '30 percent club' because of the agreement between its thirty-five members to reduce sulphur emissions by 30 per cent of 1980 levels by 1993. Britain, Poland, Spain and the USA declined to subscribe to the convention, although eventually all succeeded in reducing sulphurous emissions to a degree. Another protocol was signed in 1994 in Oslo to tailor targets to polluters rather than to reassert overall objectives.

The measures discussed above have been confined to the Northern Hemisphere, where the impact of acid precipitation has been most intense and most extensive. By *c.* 1750, industrialisation and the large-scale burning of fossil fuel were occurring, so there have been nearly 250 years of uncontrolled emissions of sulphurous and nitrous acids. The impact has been particularly severe in areas of acid bedrock (such as granite) that are in receipt of air masses from industrialised regions. While measures to curb acid precipitation have facilitated a degree of ecosystem recovery in parts of the temperate zone of the Northern Hemisphere, the problem is now spreading into the tropics as developing countries industrialise, especially in Southeast Asia and China. Acid precipitation is thus rapidly becoming an environmental issue of global proportions.

THE NATURE OF THE PROBLEM

The chemistry of acid precipitation

Acid precipitation is produced when oxides of sulphur and nitrogen combine with water in the atmosphere to generate sulphurous and sulphuric acids and nitric and nitrous acids, as shown in Figure 3.1. Although small amounts of these

A Sulphurous and sulphuric acids

SO_2 is emitted from natural and anthropogenic sources and dissolves in cloud water to produce sulphurous acid:

$$SO_2 + H_2O \longrightarrow H_2SO_3 \rightleftharpoons H^+ + HSO_3^-$$

Sulphurous acid can be oxidised in the gaseous or aqueous phase by various oxidants

$$SO_2 \xrightarrow{\text{oxidant}} SO_3$$

Aqueous sulphur trioxide forms sulphuric acid:

$$SO_3 + H_2O \longrightarrow H_2SO_4 \rightleftharpoons H^+ + HSO_4^- \rightleftharpoons 2H^+ + SO_4^{2-}$$

B Nitrous and nitric acids

N_2O is emitted by the process of denitrification and although relatively inert it is a greenhouse gas. NO and NO_2 (collectively designated as NO) are produced by combustion processes and lightning. They are involved in many chemical processes, some of which damage the ozone layer in the stratosphere:

$$O_3 + NO \longrightarrow NO_2 + O_2$$

Other chemical processes may generate ozone in the troposphere, causing photochemical smogs:

$$NO_2 \xrightarrow{\text{light}} NO + O$$
$$O + O_2 \longrightarrow O_3$$

In addition, nitric and nitrous acids may be produced:

$$2NO_2 + H_2O \longrightarrow HNO_3 + HNO_2$$

These acids are components of acid rain along with sulphurous and sulphuric acids

Figure 3.1 The formation (simplified) of the major components of acid rain in the troposphere.

Source: Mannion 1997.

gases are produced naturally through volcanic eruptions, which are fluxes within the natural biogeochemical cycles of nitrogen and sulphur (Figure 3.2), new fluxes between the lithosphere and atmosphere have been created by human activity, notably the combustion of fossil fuels. As Figure 3.2 shows, the anthropogenic influence on these biogeochemical cycles has been considerable (see discussion in Mannion 1997; 1998). For example, fossil fuel combustion fluxes 70–80 Tg yr^{-1} of sulphur from the lithosphere to the atmosphere, which is ten times as much as the volcanic flux. Biomass burning also accelerates the flux of both sulphur and nitrogen from the biosphere to the atmosphere.

Once they have formed in the troposphere (the lower atmosphere), the acids become incorporated into clouds and can produce a pH as low as 2.6. This can have a significant impact on high-altitude ecosystems that experience a low cloud cover or mist for relatively long periods. This is occult deposition. In addition, the entrainment of acids within clouds and air masses means that they can be transported hundreds of kilometres beyond their initial site of production by prevailing winds before they are deposited. This and occult deposition are both forms of wet deposition (Figure 3.3). Alternatively, the dry deposition of oxides of sulphur and nitrogen as gases, aerosols or particulates may take place. This is known as dry deposition (see Figure 3.3) and usually occurs close to the source of nitrous and sulphurous oxide production.

Both wet and dry deposition have an environmental impact, the magnitude of which depends on the capacity of the environment to neutralise, or buffer, the acidic precipitation. For example, in areas of alkaline bedrock such as limestone or chalk, soils have a relatively high pH, usually *c.* 6 to 7.5, and a high cation exchange capacity. They have the capacity to neutralise acidic precipitation because of the presence of substances such as calcium carbonate and magnesium carbonate. Similarly, lakes and rivers in areas of alkaline bedrock are less susceptible to acidification because of the presence of bicarbonate anions (HCO_3^-). Areas of acidic bedrock such as granite or those with acidic soils, peats and lakes are particularly susceptible to further acidification

Figure 3.2 The major reservoirs and fluxes in the global biogeochemical cycle of (A) sulphur and (B) nitrogen.

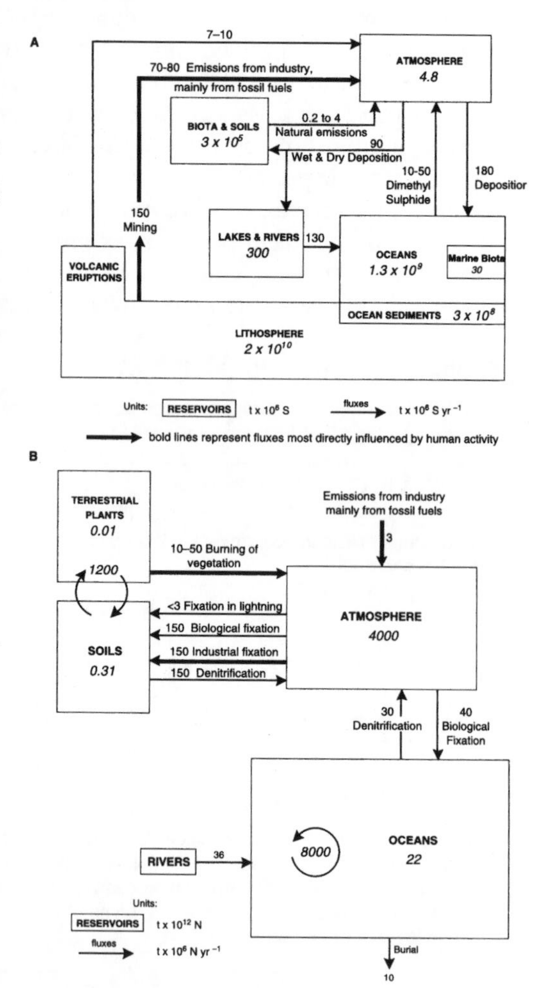

Sources: Based on (A) Charlson *et al.* (1992) and Schlesinger (1997); (B) Jaffe (1992) and Schlesinger (1997).

because there is little or no buffering capacity. As a result, hydrogen ions (H^+) accumulate in the system, while sulphate and nitrate anions (SO_4^{2-} and NO_3^-) freely combine with cations such as sodium and potassium to produce compounds that are readily washed or leached out of the system. With time, soils and peats become progressively impoverished as the cation stock is diminished. The lack of

Figure 3.3 The processes involved in the formation and deposition of acid pollution.

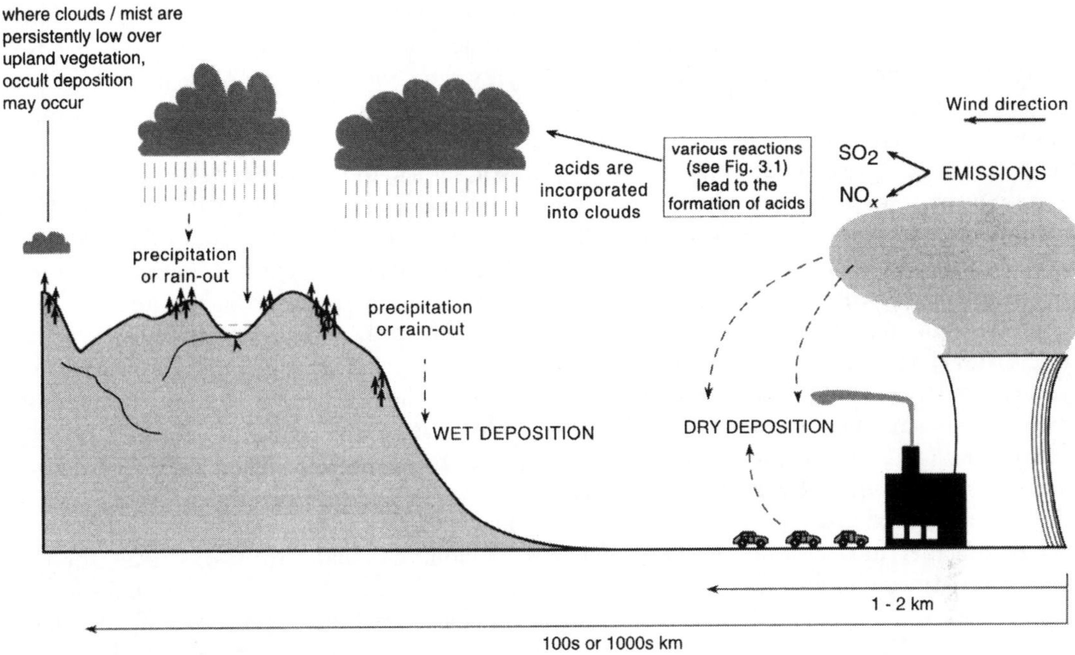

Source: Based on Mannion 1997.

nutrients will limit vegetation growth, leaving soils and peats vulnerable to erosion.

In lakes, the accumulation of hydrogen ions reduces pH. This can have direct and indirect impacts in relation to fish stocks; populations and ranges of species diminish with increasing acidification, which reduces their reproductive success. Moreover, fish kills may be caused when surges of acidified water occasioned by cloud bursts or rapid melts of winter snowfall occur. The indirect effects of acid precipitation are due to its impact on other aspects of environmental chemistry. In particular, aluminium is more soluble at low rather than high pH and is, consequently, removed from soils or peats if they are in receipt of acidic precipitation. Drainage from such soils may eventually enter lake basins, where high concentrations of aluminium will adversely affect fish populations. A combination of pH in the range 5.0–5.5 and high concentrations of aluminium in calcium-deficient water causes physiological changes in fish, particularly in the immature fry, and can cause death. For example, at concentrations of less than 100 $\mu g\ l^{-1}$ the ability of fish to regulate their salt and water content is impaired, and at concentrations above 100 $\mu g\ l^{-1}$ aluminium hydroxide [$Al(OH_3)$] is formed as a gelatinous precipitate on fish gills. Eventually, this causes death through the impairment of respiration (see Wellburn 1994 for details). According to Gleick (1993), fish do not survive at a pH of less than 3.5; salmonids cannot tolerate pH 3.5 to 4.0, although tench, roach, pike and perch can survive, and even at pH 4.0 to 4.5 adult salmonids are likely to be adversely affected, while their fry and eggs will continue to be impaired at pH 4.5 to 5.6. This pH range is suitable for tench, roach and carp, etc., but only at pH above 5 are there no adverse impacts on fish populations.

Aluminium in lake waters may also combine with phosphorus to form chemical complexes, effectively removing phosphorus for uptake by primary producers such as algae. Consequently, primary productivity declines and so, in turn, does secondary productivity. In this context, acid precipitation is influencing the biogeochemical cycle of aluminium. It can also affect the nitrogen cycle; as pH declines to less than 5.7, the activity of nitrifying bacteria is curtailed. This causes ammonia, which is usually oxidised to nitrate by these bacteria, to accumulate in the system. Certain groups of algae can then utilise the ammonia in a complex biochemical process that results in the accumulation of additional hydrogen ions and so compounds acidification.

The overall impact of acid precipitation in terrestrial and aquatic environments is to reduce biodiversity. It does this by altering biogeochemical cycling and generating milieu in which only acidophilous (acid-loving) species can survive. In both terrestrial and aquatic environments, there may also be an accumulation of organic matter as grazing and decomposer food chains/webs become less active as diversity diminishes.

The geography of acid precipitation

Acid precipitation has varied temporally and spatially. Temporally, it began on a large scale in the mid-1700s as industrialisation intensified and spread through Britain and Europe. In the Northern Hemisphere, the impact and timing of acid precipitation has varied according to the buffering capacity of specific environments as well as the location of industry in relation to prevailing wind directions. Its widespread occurrence at the hemispheric scale is reflected in polar ice core records. Ice cores from the Arctic show a substantial increase in acidity from *c.* 1800. In addition, the record of nitrous oxide in pre-industrial times was *c.* 275 ppbv (Raynaud *et al.* 1993) rising to *c.* 312 ppbv in 1994 (Houghton *et al.* 1996), although some of this increase is due to biomass burning and nitrate fertiliser use (see Mannion 1998). Further aspects of the temporal dimensions of acid precipitation have been revealed through studies of lake sediments and peats. Such research provides a record of local and regional pollution histories and is examined below in relation to the UK and the USA.

The spatial distribution of acid precipitation depends on where high emissions of nitrous and sulphurous gases occur; the spatial distribution of the acidification of ecosystems also depends on prevailing wind directions and the buffering capacity of soils and bedrock in areas of deposition (see above). Most of the industrialised world lies in the temperate zone of the Northern Hemisphere, so it is here that the problems of acid precipitation and its deposition are most acute. The regions so far identified (Figure 3.4) as those with the most severe problems are Scandinavia, parts of northern Europe and Russia, parts of China, the northeast USA and eastern Canada. Case studies from these regions are discussed below. In part, the problems arise because these areas are themselves major producers and/or receivers of acid precipitation. The data given in Tables 3.1 and 3.2 show that the major producers of sulphurous and nitrous oxides are the USA, China, Germany and the Russian Federation. As industrialisation proceeds in developing countries, and their consumption of fossil fuel increases, acid precipitation and its deposition will become a problem in tropical regions and in the Southern Hemisphere, as shown in Figure 3.4.

The introduction of measures to curb sulphurous emissions in Europe and North America in 1985 (the '30 percent club') has brought some rewards. As Table 3.1 shows, sulphurous emissions from many countries have declined. This is due partly to the use of less sulphur-rich fuels, energy efficiency programmes, and to the installation of desulphurisation technology in power stations. There are also calls for positive measures of a similar kind to curb nitrous gas emissions, although in general these have fallen as sulphurous emissions have fallen. International agreements to curb carbon dioxide emission through energy-efficiency programmes will also cause acid emission to decline. The reduction of sulphurous emissions has led to ecosystem recovery in some regions, as is discussed below. Amelioration on a short-term basis has also proved successful. This involves the liming of acidified lake ecosystems; however, its impact is

Figure 3.4 Areas currently experiencing problems due to acid precipitation and areas likely to develop problems in the future.

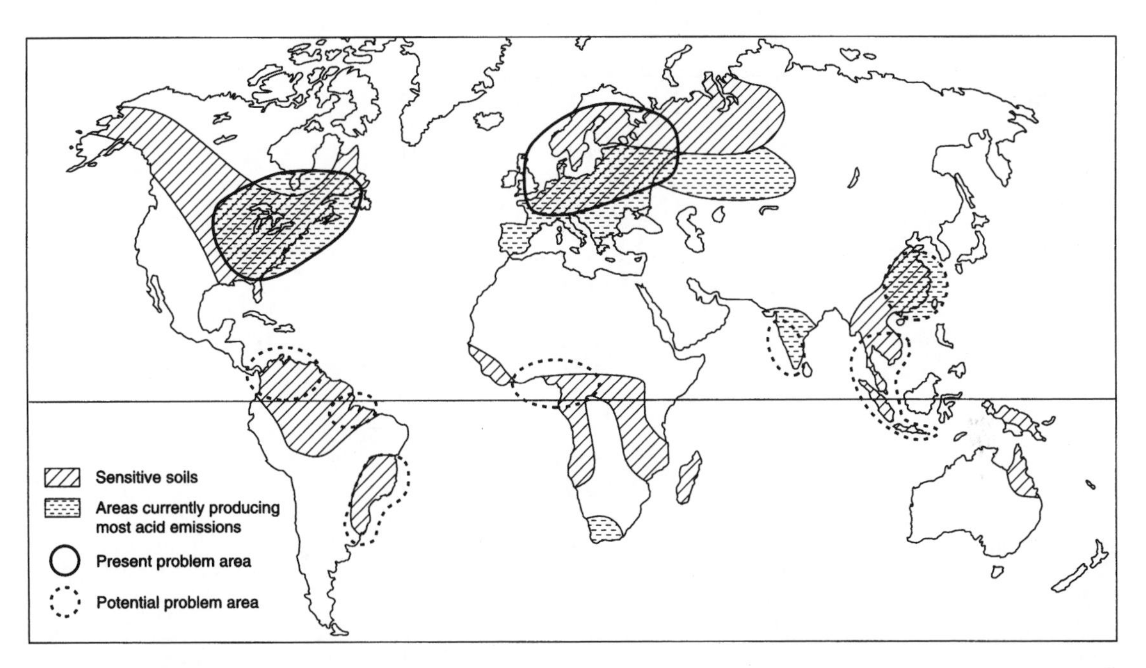

Source: Based on Rodhe and Herrera 1988.

short-lived, a product of treating symptoms rather than the underpinning causes.

Research has also focused on the identification of areas susceptible to acidification; this has led to the formulation of the 'critical load' concept. This is a measure of the amount of acid deposition that can be absorbed by an ecosystem or environment without causing damage. Critical load is the central concept of the Oslo Protocol signed in 1994. Signatories pledged to reduce acidic emissions so as not to exceed critical loads for vulnerable regions. Thus the critical load concept is a management tool and can be used to model future environmental change in response to increased or decreased acid precipitation. The determination of critical load is, however, dependent on information about the volume of acid deposition, the buffering capacity of soils, bedrock, etc., vegetation type, and hydrology. For example, Arp *et al.* (1996) have used a steady-state mass-balance model for calculating critical sulphur and nitrogen loads in upland forests of southern Ontario, Canada. This model used data on wet atmospheric deposition of major cations and anions, the availability for plant use of nitrogen, calcium, magnesium and potassium in the rooting spaces of soils, nutrient uptake and retention in the forest biomass, estimates of soil weathering, mean annual air temperature, precipitation, and evaporation. This reflects the many factors involved in the calculation of critical loads and the importance of environmental monitoring stations to provide the necessary baseline data.

One example of such a monitoring network is that of the UK Acid Waters Monitoring Network (AWMN), which was established in 1988 (Patrick *et al.* 1996). It consists of eleven lake and eleven stream sites in acid-sensitive areas of the UK. The chemical and biological characteristics of these sites are monitored and together with the palaeolimnological record of the lake sediments and data from the UK precipitation monitoring network, they can be

Table 3.1 Emissions of sulphurous gases for selected countries.

A	*Emissions $SO_2 \times 10^3$ tons per year*		
	1980	*1990*	*1995*
Belgium	828	317	253
Bulgaria	2050	2020	1497
Czech Republic	2257	1876	1091
Finland	584	260	96
France	3338	1298	986
Germany (E & W)	7514	5326	2995
Iceland	18	24	24
Italy	3800	1678	1437
Poland	4100	3210	2337
Russia (European)	7161	4460	2983
Spain	3319	2266	2061
UK	4913	3756	2365

B	*Emissions $SO_2 \times 10^3$ tons per year*		
	1975	*1980*	*1987*
Bangladesh	40	57	49
China	10,175	13,372	19,989
India	1652	2010	3074
Indonesia	201	329	485
Japan	2571	1604	1143
Malaysia	193	272	263
Nepal	3.7	4.9	30.9
North Korea	234	271	333
Pakistan	148	198	381
South Korea	1159	1918	1294
Taiwan	609	1036	605
Thailand	224	420	612

Sources: (A) European Monitoring and Evaluation Programme as quoted by Ågren (1997); (B) Quoted in Kato (1996).
Note: Note declines in Europe and increases in Asia.

used to determine acidification trends. Such data can also contribute to the establishment of critical loads. The data so far analysed, for 1988–93, indicate that no regional trends in terms of increasing or decreasing acidification have occurred. It is also recognised that monitoring networks should be established in newly industrialising regions. For example, Yagishita (1995) reports that the Japanese Environment Agency is considering such a network in East Asia; this highlights the need for baseline data before significant acidification occurs and emphasises the international and geographical nature of the problem.

CASE STUDIES

The acidification of lakes

This is a serious problem in the Northern Hemisphere, especially in regions of acid bedrock such as parts of Scandinavia, Scotland, southwest Canada, northern Russia and northeast USA (see Figure 3.4). In addition, numerous studies of lake sediments in these regions provide a means of constructing pollution histories, which have also contributed to the establishment of the causes of acidification. Much of this research has been undertaken as part of two major initiatives: the

Table 3.2 Emissions of nitrous gases for selected countries.

A	*Emissions of nitrogen oxides as $NO_2 \times 10^3$ tons per year*		
	1980	*1990*	*1995*
Belgium	442	352	345
Bulgaria	416	376	266
Czech Republic	937	742	412
Finland	295	300	259
France	1823	1585	1666
Germany (E & W)	3334	2640	2210
Iceland	18	20	23
Italy	1480	2047	2157
Poland	1229	1279	1120
Russia (European)	1734	2675	1995
Spain	950	1178	1223
UK	2416	2897	2295

B	*Emissions $NO_2 \times 10^3$ tons per year*		
	1975	*1980*	*1987*
Bangladesh	46	58	66
China	3727	4907	7371
India	1379	1673	2556
Indonesia	331	465	639
Japan	2329	2132	1935
Malaysia	90	126	177
Nepal	18	21	50
North Korea	325	383	468
Pakistan	101	164	231
South Korea	220	365	555
Taiwan	124	225	325
Thailand	182	255	384

Sources: (A) European Monitoring and Evaluation Programme as quoted by Ågren (1997); (B) Quoted in Kato (1996).
Note: Note increases almost everywhere; even where emission controls are in operation as in Europe they do not apply to nitrous gases.

Palaeoecological Investigation of Recent Lake Acidification (PIRLA) in Canada and the USA; and the Surface Waters Acidification Programme (SWAP) in the UK, Norway and Sweden. The results from these projects have revealed the degree and timing of anthropogenic acidification and have shown it to be a serious and widespread problem.

Much of the PIRLA and SWAP research exploits the sensitivity of species of diatoms (unicellular algae that are the basis of food chains in many freshwater and marine environments) to pH and their widespread occurrence in lake sediments. The pH tolerances of the species recovered as fossils from lake sediments are known from studies of their living equivalents; thus the construction of lake acidification histories is possible based on diatom-inferred pH reconstructions. Acidification has not only influenced primary producers such as the diatoms but has also affected the species composition of other organisms at all trophic levels. The remains of some of these organisms, such as cladocera (microscopic animals occurring in fresh water) and chrysophytes (planktonic algae with an outer covering of scales composed of silica: cysts may also be present in the sediments) have also been used to derive lake pollution histories, along with carbonaceous particles (e.g. soot, charcoal) and the stratigraphic record of heavy metal (e.g. zinc, copper, manganese and iron) deposition.

Table 3.3 gives the results from many of the SWAP and PIRLA sites investigated. It shows that many lakes have become acidified by at least one pH unit. As pH is measured on a logarithmic scale, a decline of one pH unit represents a tenfold increase in hydrogen ion concentration. In the UK, Loch Grannoch in southwest Scotland has been one of the worst affected lakes with a decline in pH of 1.2, although Lake Gårdsjön in Sweden, with a pH decline of 1.5, has been even more severely affected. pH declines of a similar magnitude have been recorded in northeastern USA and Canada. Age estimation of the sediments of these lakes has allowed the timing of acidification to be determined. As Table 3.3 shows, this has varied from site to site. In the Scottish lakes, for example, the varied timing reflects, to a certain extent, the degree of buffering within each catchment. Moreover, most acidification occurred before any afforestation, another factor often associated with freshwater acidification, took place. Numerous similar studies have been undertaken elsewhere (e.g. Austria, Germany and Russia) and the record is similar. An example is given in Box 3.1.

Where emissions of acidic gases have declined, either through the recession of heavy industries or through emission controls, there is some evidence that ecosystem recovery is occurring. For example, Battarbee (1994) reports that the diatom communities in several Scottish lochs are changing in composition as acidic emissions decrease, which, in turn, causes less acidification in lake catchments and waters. Similarly, in eastern Canada, Gunn and Keller (1990) have

Table 3.3 Examples of changes in the pH of lake waters that have occurred since *c.* 1840 in the UK, Europe and North America

Site	*Location*	*pH change*	*Approx. date of initial pH change*
Loch Enoch	SW Scotland	0.9	1840
Loch Grannoch	SW Scotland	1.2	1925
Loch Dee	SW Scotland	0.5	1890
Round Loch Glenhead	SW Scotland	1.0	1950
Loch Laidon	SW Scotland	0.7	1850
Grosser Arbersee	W Germany	0.8	1965
Kleiner Arbersee	W Germany	0.8	1950
Gårdsjön	Sweden	1.5	1960
Hovvatn	Norway	0.75	1918
Holmvatn	Norway	0.5	1927
Malalajärvi	Finland	1.0	1950
Hirvilampi	Finland	0.7	1950
Big Moose Lake	USA	1.0	1950–60
Woods Lake	USA	0.4	1930s
Ledge Pond	USA	0.6	1880s
Beaver Lake	Canada	0.6	1950s
Lake B	Canada	1.6	1955
Lake CS	Canada	1.0	1955

Note: This is only a selection of the many lakes which are known to have become acidified since *c.* 1840.

Box 3.1 The reconstruction of lake pH using diatom analysis

1 Diatoms are unicellular algae that are abundant in most aquatic environments. They form the basis of food chains/webs and many species are particularly sensitive to water pH.
2 The frustules, which are the structural component of diatoms, are composed of silica, which is resistant to decay. Consequently, these frustules become incorporated into lake sediments when the organisms die. As sediments accumulate, a sample of the diatom population is preserved. Changes in the assemblages reflect, among other factors, changes in pH. On the basis of the pH sensitivities of their modern counterparts, the fossil diatoms provide a means of reconstructing the pH history of the lake waters.
3 This has been undertaken for many lakes in acid-sensitive areas, data from some of which are given in Table 3.3. The detailed pH reconstruction of one of those lakes, Round Loch of Glenhead (Battarbee *et al.* 1988) is illustrated below:

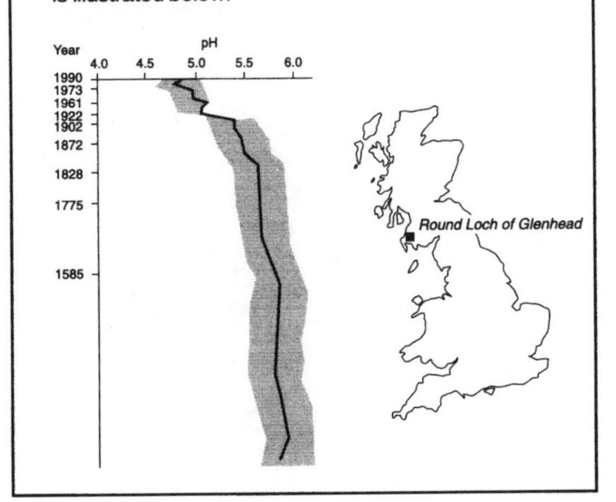

shown that many lakes in the region have lower concentrations of aluminium, heavy metals and sulphate and higher pH than they did in the 1970s; in some cases, trout have even been re-established. Moreover, where liming of lakes or catchment soils has been undertaken, rapid improvement in pH can occur. This is illustrated by the work of Blette and Newton (1996) on Woods Lake in the Adirondack Mountains of the northeast USA and by the work of Dixit *et al.* (1996) on the Aurora trout lakes of Sudbury, Canada. In the latter, water quality in two lakes was improved to such an extent that aurora trout were reintroduced in 1990. However, liming is only a short-term solution because it treats symptoms rather than underlying causes.

What is perhaps of equal concern to the problem of acid precipitation itself is the likelihood of additional affects caused by global warming, as Wright and Schindler (1995) have discussed. They suggest that increases in temperature predicted by general circulation models (GCMs) for the boreal zone could cause increased mineralisation of nitrogen and the oxidation of organic compounds, including those containing sulphur. Thus, despite emission reductions, acidification would continue; the acids produced through mineralisation would continue to mobilise aluminium. Numerous factors may also conspire to allow the penetration of radiation, including the harmful ultraviolet-B (UV-B), to greater than usual depths, where it may have adverse ecological impacts (see discussion in Schindler *et al.* 1996). In addition, it has recently been suggested that declines in dust production, especially that containing bases (e.g. calcium carbonate and magnesium carbonate), may negate the impact of emission controls (Hedin and Likens 1996). The lack of such bases in the atmosphere, due to legislation on air quality, seems likely to cause a decline in neutralisation between acidic and basic components. Thus, despite curbs on acid emissions their environmental impact may be intensified. These issues reflect the complex interplay between numerous biogeochemical cycles and between different forms of pollution.

The impact of acid precipitation on terrestrial ecosystems: forests and peatlands

The impact of acid precipitation on terrestrial ecosystems occurs directly and indirectly. The direct effect involves wet and/or dry and occult deposition on vegetation, which may impair its capacity for photosynthesis and cause a loss of biodiversity as only those least acid-sensitive species survive. Indirect effects are caused by the influence that acid precipitation has on soils, especially on their chemistry and microbiology.

Moreover, and as suggested above, other forms of pollution such as ozone accumulation in the troposphere and global warming may interact with acid precipitation to cause ecosystem damage.

The deposition of nitrous and sulphurous acids increases the hydrogen ion content of the soil and causes declines in essential plant nutrients (see Part II); as microbial activity decreases, fungi dominate the decomposer system. Populations of earthworms and other soil fauna are reduced, which limits soil mixing, drainage and aeration. The loss of bases and other essential nutrients also impairs vegetation growth. This is illustrated by Likens *et al.* (1996), who have monitored soil and water chemistry in the Hubbard Brook Experimental Forest ecosystem, New Hampshire, USA, over many years. There have been substantial declines in total soil calcium and magnesium; this may impair ecosystem recovery as acid emissions decrease. A decline in soil calcium and magnesium is also considered to be a major cause of forest decline in the Kola Peninsula of Russia, where the major emitter of acid precipitation is a nickel smelter (Koptsik and Mukhina 1995). All of these factors and possibly others may be responsible for what is recognised as a substantial decline in forest health in Europe and elsewhere. The case of European forest damage is discussed in Box 3.2 and a detailed study of forest cover change in the Czech Republic is detailed in Ardö *et al.* (1997).

Box 3.2 Forest damage in Europe

1. Acid precipitation is a major contributory cause of forest damage in Europe.
2. Other factors include drought, although drought occurrence may have a greater impact in acid-damaged areas, e.g. southern Poland, than in areas where there is little such damage.
3. The impact of acid precipitation on forest soils is likely to influence the degree of forest damage. Cations, essential nutrients such as magnesium and calcium, may be removed from the ecosystem,while heavy metals, which are also deposited in acid precipitation, may have a toxic effect.
4. Not only is the forest canopy damaged but ground flora may also be altered, with a shift towards acidophilous species such as heather.
5. The distribution of forest damage in Europe is given in the map below. This shows that the greatest damage has occurred in Eastern and Central Europe, with least damage in western maritime regions.

% of trees with defoliation greater than 25%

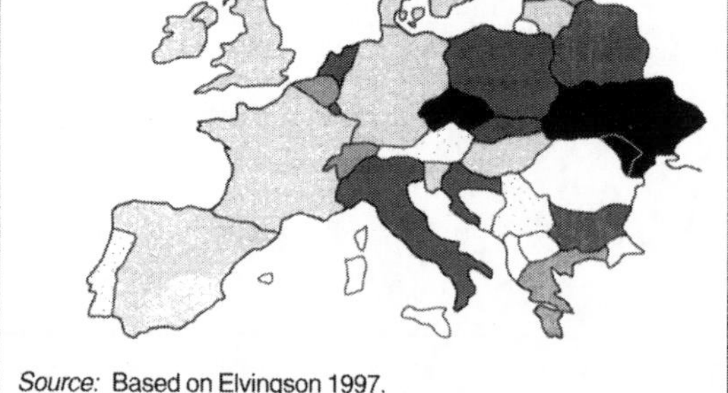

Source: Based on Elvingson 1997.

Most peatlands are naturally acid and thus offer little buffering capacity to acid precipitation. Moreover, where they are in the immediate hinterland of heavily industrialised areas, ecological damage can be considerable. This is exemplified by blanket peats in the Pennines of the UK. According to Lee's (1998) review, the area of peatland between Manchester and Sheffield received considerable quantities of sulphur dioxide from these major industrial centres during the 1800s and early 1900s (prior to control measures on particulate emissions in the 1950s). The impact of this contamination involved a loss of biodiversity, notably of *Sphagnum* species; these bog-forming mosses are particularly susceptible to pollutants because they rely heavily on atmospheric deposition for their nutrient supply. Where species are lost and peatlands become increasingly acidic, erosion may set in as the surfaces lose their protective vegetation cover.

The impact of acid precipitation on urban fabrics, urban air quality and archaeological monuments

Acid emissions do not discriminate between rural and urban environments. Indeed, emissions in cities can be particularly high because of intense motor traffic. The impairment, in effect a form of weathering, of important urban buildings and monuments is apparent in many of the world's major cities. Building materials such as marble, limestone and sandstone may be adversely affected, and so too may metals. All suffer corrosion, which disfigures the detail of architectural design, which is costly to restore. According to McCormick (1997), *c.* 2.5 cm of the Portland stone on St Paul's Cathedral in London has been removed, although the mechanical action of weather has also played a role in this. Similar problems have occurred in Stockholm, Sweden, where damage has defaced the Royal Palace and the Riddarholm church. The calcitic Gotland sandstone, of which both were constructed in the eighteenth century, shows evidence of crumbling and decay. This takes the form of a crumbly, loose surface, gypsum and other salt formation and discolouration.

There are also serious concerns about the condition of many ancient monuments in modern urban centres, as in Athens, Greece, including the Parthenon on the Acropolis. These have been constructed from limestone and are thus susceptible to dissolution as the alkaline limestone neutralises the acidic emissions. Ancient monuments elsewhere have also been defaced by acid rain, as Wilford (1996) reports in relation to the Mayan ruins in Central America and in the Yucatán peninsula of Mexico. The source of the acid precipitation is oil refineries and uncapped oil wells in the region, with an additional contribution from tourist buses. Plate 3.1 shows the damaged detail of the 'Nunnery' at Uxmal, Mexico.

The high concentrations of sulphur dioxide in many former Soviet cities must also be a cause of considerable damage to the urban fabric. For example, Shahgedanova and Burt (1994) report that in 1988 sulphur dioxide emissions exceeded 1000 tonnes per year in twenty-seven towns/cities. In Noril'sk, in Siberia's Taimyr peninsula, emissions were *c.* 2.2×10^6 tonnes, which amounted to 12.4 tonnes per capita; the major source is the copper–nickel smelters of the Noril'sk Mining and Metallurgical Combine. The resulting damage to hinterland forests has increased substantially since 1970, but urban fabrics and human health will not have been spared by such acute levels of pollution.

Plate 3.1 Damage due to acid precipitation on the 'Nunnery', part of the Mayan complex at Uxmal, Mexico (*photograph: Dr M.D. Turnbull*).

As in the case of lake and terrestrial ecosystems, remedial measures must involve the diminution of emissions through increased energy efficiency in buildings and transport systems, and less dependency on the car.

The impact of acid precipitation on human health

Air quality is recognised as an important factor in human health, especially in relation to respiratory diseases. However, acid precipitation may have additional though indirect effects on human health through its impact on soils and water resources. In particular, heavy metals may be released from soils due to acidification; they may then enter aquatic food chains and eventually reach humans.

Respiratory problems caused by poor air quality are the most common diseases to which acid precipitation contributes. McCormick (1997) records the number of deaths caused by serious smogs in the industrialised world pre-1965; among the worst were those in London in 1873 and 1880, which together caused 2500 deaths. These were extreme events linked to industrialisation and little or no air quality control. The old, the young and those suffering respiratory disease were especially vulnerable. Today, such events are characteristic of the newly industrialising countries. Moreover, these were extreme events with an obvious cause and effect. In the developed world, urban air quality may still be poor (see the example of Noril'sk, Siberia, in the previous section) due to gaseous rather than particulate emissions. As McCormick points out, health problems may ensue from regular exposure to air that is characterised by low-level pollution. Although the relationship between air pollution and respiratory problems has eluded precise definition, the huge increase in such problems, especially asthma, reflects the at least partial influence of air quality, although other factors may also contribute. McCormick (p. 32) states 'the World Health Organisation estimates that in Europe alone, excessive levels of sulphur dioxide may be responsible for 6000 to 13000 extra deaths every year among people aged 65 or older, intensified chronic respiratory problems for 89000 to 203000 people and 58000 to 99000 extra cases of diseases in the lower respiratory organs among children.'

The potential for health problems due to the indirect effect of acid precipitation has been discussed by Oskarsson *et al.* (1996). They were particularly concerned with the possible impact of toxic elements such as lead, copper zinc, cadmium, methylmercury and selenium. The latter may become scarce under acidified conditions and cause health problems because of its scarcity; the other metals may cause problems because their mobility is increased under acid conditions, and consequently they enter food chains and webs in increased proportions. In addition, acidic water will cause increased extraction of metals such as lead from domestic plumbing. Oskarsson *et al.* were unable to find unequivocal relationships between health problems due to metal contamination and acidification, although they identified a number of possible links. They suggest that safety margins are small in relation to the exposure of humans to toxic metals and that curbs on acid precipitation are essential before damage to human health becomes apparent.

CONCLUSION

Acid precipitation is a product of the industrial age and represents anthropogenic perturbations to the sulphur and nitrogen cycles. So serious have its consequences been that they have prompted international agreements, with the initial concerted efforts to curb emissions of nitrous and sulphurous gases beginning in 1985. This was one of the first international agreements to tackle a pollution problem, pre-dating the Montreal Protocol on ozone by two years. The effectiveness of the 1985 and later protocols is manifest in many parts of the industrialised world as lake and stream water quality has improved. In some instances, fish populations have been restored. Despite these improvements, acid precipitation remains a significant agent of environmental change in the Northern Hemisphere, and further curbs on emissions, especially nitrous emissions, are necessary. Moreover, the problem of acidification is spreading as many developing countries are industrialising rapidly. Acid precipitation is thus set to become a global problem.

The chemistry of acid precipitation is understood in relation to its production, although once it is deposited its reactions with organic and inorganic substances in soils and water are complex. It influences many other biogeochemical cycles, especially those of the heavy metals, raised concentrations of which may cause further ecosystem change. The most severely affected regions are those on acid bedrock that are either close to emission sources or that receive air masses from industrial centres. Lakes, streams, soils, forests and other vegetation communities may all be damaged by acid precipitation. Society also pays its dues for the wealth generated through the combustion of fossil fuels and industrialisation; the very fabric of urban centres may be disfigured, and there are increased risks to human health through poor air quality.

Perhaps the most disturbing aspect of acid precipitation is its interaction with other environmental problems such as global warming and the as yet unknown reactions and ramifications that will emanate from this relationship. Again, the underpinning mechanisms involve an array of biogeochemical cycles. Who knows where it all may lead? Physical geographers and other applied scientists have a major role to play in finding answers to such questions through their research on environmental processes.

GUIDE TO FURTHER READING

McCormick, J. (1997) *Acid Earth*, (3rd edition). Earthscan: London. A valuable introduction that provides a good overview of the problem from the chemistry to the politics.

Wellburn, A. (1994) *Air Pollution and Acid Rain. The Biological Impact*, (2nd edition). Longman: Harlow. A scientific treatment of air pollution with emphasis on its impact on ecosystems and organisms.

Acid News is a newsletter from the Swedish NGO Secretariat on Acid Rain. It is published by the Swedish Society for Nature Conservation and is available free of charge from the Swedish NGO Secretariat on Acid Rain, Box 7005, S-402 31 Göteborg, Sweden.

REFERENCES

Ågren, C. (1997) Monitoring figures show decline. *Acid News* 4–5, 16–17.

Ardö, J., Lambert, N., Henzlik, V. and Rock, B.N. (1997) Satellite-based estimations of coniferous forest cover changes: Krusné Hory, Czech Republic 1972–1989. *Ambio* 26, 158–166.

Arp, P.A., Oja, T. and Marsh, M. (1996) Calculating critical S and N loads and current exceedance for upland forests in southern Ontario, Canada. *Canadian Journal of Forest Research* 26, 696–709.

Battarbee, R.W. (1994) Diatoms, lake acidification and the Surface Water Acidification Program (SWAP) – a review. *Hydrobiologia* 274, 1–7.

Battarbee, R.W., Flower, R.J., Stevenson, A.C., Jones, V.J., Harriman, R. and Appleby, P.G. (1988) Diatom and chemical evidence for reversibility of acidification of Scottish lochs. *Nature* 332, 530–2.

Blette, V.L. and Newton, R.M. (1996) Application of the integrated lake watershed acidification study model to watershed liming in Woods Lake, New York. *Biogeochemistry* 32, 363–83.

Charles, D.F. and Smol, J.P. (1994) Long-term chemical changes in lakes – quantitative inferences from biotic remains in the sediment record. *Advances in Chemistry Series* 237, 321.

Charlson, R.J., Anderson, T.L. and McDuff, R.E. (1992) The sulfur cycle. In S.S. Butcher, R.J. Charlson, G.H. Orions and G.V. Wolfe (eds) *Global Biogeochemical Cycles*, London: Academic Press, 285–300.

Dixit, A.S., Dixit, S.S. and Smol, J.P. (1996) Long-term trends in limnological characteristics in the Aurora trout lakes, Sudbury, Canada. *Hydrobiologia* 335, 171–81.

Elsworth, S. (1984) *Acid Rain*. Earthscan: London.

Elvingson, P. (1997) Still many trees damaged. *Acid News* 4–5, 14–15.

Gleick, P.H. (ed.) (1993) *Water in Crisis: A Guide to the World's Freshwater Resources*. Oxford University Press: Oxford.

Gorham, E. (1958) The influence and importance of daily weather conditions in the supply of chloride, sulphate and other ions to freshwaters from atmospheric precipitation. *Philosophical Transactions of the Royal Society of London* B241, 147–78.

Gunn, J.M. and Keller, W. (1990) Biological recovery of an acid lake after reductions in industrial emissions of sulphur. *Nature* 345, 431–3.

Hedin, L.O. and Likens, G.E. (1996) Atmospheric dust and acid rain. *Scientific America* 275, 56–60.

Houghton, J.T., Meira Filho, L.G., Callander, B.A., Harris, N., Kattenberg, A. and Maskell, K. (eds) (1996) *Climate Change 1995. The Science of Climate Change*. Cambridge: Cambridge University Press.

Jaffe, D.A. (1992) The nitrogen cycle. In S.S. Butcher, R.J. Charlson, G.H. Orions and G.V. Wolf (eds) *Global Biogeochemical Cycles*, London: Academic Press, 263–84.

Kato, N. (1996) Analysis of structure of energy

consumption and dynamics of emission of atmospheric species related to the global environmental change (SO_x, NO_x and CO_2) in Asia. *Atmospheric Environment* 30, 757–85.

Koptsik, G. and Mukhina, J. (1995) Effects of acid deposition on acidity and exchangeable cations in podzols of the Kola Peninsula. *Water Air and Soil Pollution* 85, 1209–14.

Lee, J.A. (1998) Unintentional experiments with terrestrial ecosystems: ecological effects of sulphur and nitrogen pollutants. *Journal of Ecology* 86, 1–12.

Likens, G.E., Driscoll, C.T. and Buso, D.C. (1996) Long-term effects of acid rain: response and recovery of a forest ecosystem. *Science* 272, 244–6.

Mannion, A.M. (1997) *Global Environmental Change. A Natural and Cultural Environmental History*, (2nd edition). Harlow: Longman.

Mannion, A.M. (1998) Global environmental change: the causes and consequences of disruption to biogeochemical cycles. *Geographical Journal* (164, 162–82).

McCormick, J. (1997) *Acid Earth. The Politics of Acid Pollution*, (3rd edition). London: Earthscan.

Oskarsson, A., Nordberg, G., Block, M., Rasmussen, F., Petterson, R., Skerfring, S., Vahter, M., Glynn, A.G., Öborn, I., Helkensten, M.-L. and Thuvander, A. (1996) Adverse health effects due to soil and water acidification: a Swedish research program. *Ambio* 25, 527–31.

Patrick, S., Battarbee, R.W. and Jenkins, A. (1996) Monitoring acid waters in the UK: an overview of the UK Acid Waters Monitoring Network and summary of the first interpretative exercise. *Freshwater Biology* 36, 131–50.

Raynaud, D., Jouzel, J., Barnola, J.M., Chapellaz, J., Delmas, R.J. and Lorius, C. (1993) The ice core record of greenhouse gases. *Science* 259, 926–34.

Rodhe, H. and Herrera, R. (eds) (1988) *Acidification in Tropical Countries*. Chichester: John Wiley & Sons.

Schindler, D.W., Jefferson Curts, P., Parker, B.R. and Stainton, M.P. (1996) Consequences of climate warming and lake acidification for UV-B penetration in North American boreal lakes. *Nature* 379, 705–8.

Schlesinger, W.H. (1997) *Biogeochemistry: An Analysis of Global Change*. San Diego: Academic Press.

Shahgedanova, M. and Burt, T.P. (1994) New data on air pollution in the former Soviet Union. *Global Environmental Change* 4, 201–27.

Wellburn, A. (1994) *Air Pollution and Climate Change: The Biological Impact*, (2nd edition). Harlow: Longman.

Wilford, J.N. (1996) Saving the pyramids: acid rain accelerates the destruction of the Maya ruins in Mexico's Yucatán Peninsula and Central America: www.stevensonpress.com/acidrain.html accessed 20 April 1998.

Wright, R.F. and Schindler, D.W. (1995) Interaction of acid rain and global changes – effects on terrestrial and aquatic ecosystems. *Water Air and Soil Pollution* 85, 89–99.

Yagishita, M. (1995) Establishing an acid deposition monitoring network in East Asia. *Water Air and Soil Pollution* 85, 273–8.

4

Extreme weather events

Rory Walsh

INTRODUCTION

Extreme weather events and the weather-related events they may induce, such as landslides, floods and storm surges, form an important part of what have been termed 'natural hazards'. They have a major influence not only on the physical landscapes and human societies directly affected by them but also on the wider community through their impact on the insurance industry and the costs of emergency aid or relief. It is increasingly being recognised that their distribution in time and space is dynamic rather than static and significant changes in the frequency of extremes such as heavy daily rainfalls, droughts, extreme heat and cold and tropical cyclones are envisaged in 'global warming' predictions for the next century (IPCC 1996; United Kingdom Climate Change Impacts Review Group 1996; Hulme and Viner 1998). This chapter reviews the roles that geographers have played in examining the climatology of extreme events, their spatial and temporal distribution (including past, current and future changes), their impacts on natural systems and human activities, and the design and effectiveness of strategies aimed at reducing their adverse effects. With particular reference to geographical research on tropical cyclones, it highlights the problems that stem from their inherent rarity, which reduces the sample size upon which to base conclusions and advice, the mismatch between the spatial scale at which one can offer reliable advice and the spatial scale most useful for planning purposes, and the implications and uncertainties associated with recent and predicted future changes in extreme event frequency.

EXTREME WEATHER EVENTS: THE PROBLEMS THEY PRESENT AND THE ROLES PLAYED BY GEOGRAPHERS

Extreme weather events can be divided into *absolute* and *relative* event types (Figure 4.1). Absolute extreme events are considered extreme simply because of their magnitude and nature. Thus tropical cyclones, tornadoes, hailstorms and climate-related events such as avalanches, landslides and floods are considered extreme events, even in areas where they are frequent. Relative extreme events, on the other hand, are considered extreme on the basis of their rarity at the location concerned; they are distinguished from normal events on the basis of their probability of occurrence (Smith 1997) and its inverse, the return period or recurrence interval, which is the average time interval between events of a specified magnitude. If one views events as being statistically distributed around a long-term mean value, then most events lie relatively close to the mean within what Smith terms the socio-economic band of tolerance and can be interpreted as resources, as they constitute the near-normal events on which human activities in that climate are based. Weather extremes such as unusual cold, warmth, drought or wetness are measured relative to (and hence are specific to) the local climate; what constitutes an unusual drought, for example, will therefore vary with location. Extreme events,

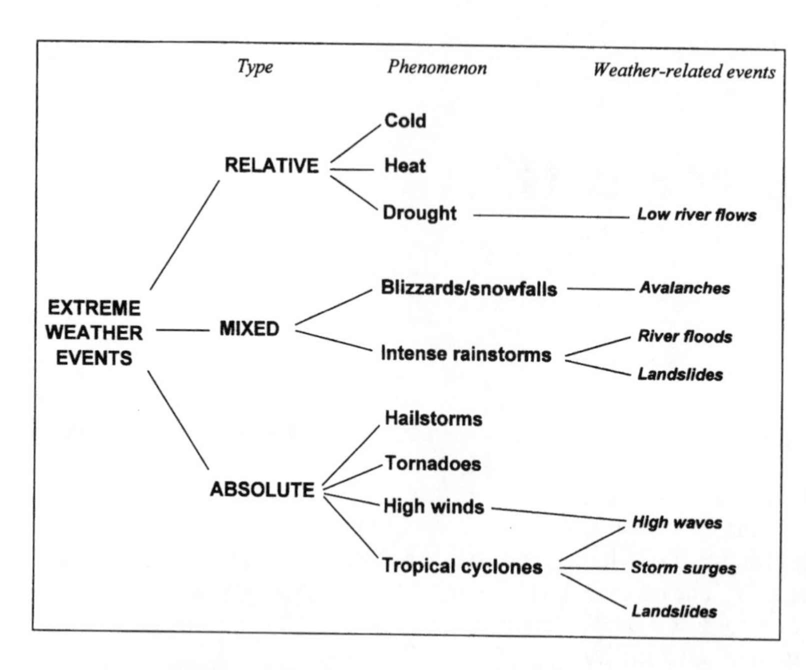

Figure 4.1 A classification of extreme weather and weather-related events.

Source: Based in part on Smith 1997.

whether relative or absolute, are viewed as *climatic hazards* if they impose negative stress on human activities and are considered *disasters* if they cause either considerable loss of life or major economic losses (*ibid.*).

Extreme weather events present several distinct problems in their analysis. First, since by definition many of the events are rare, sample sizes upon which to base calculations and maps of hazard risk are small and become progressively smaller the more extreme the event. This reduces the statistical significance and hence the reliability and utility for planning purposes of hazard analysis and hinders the detection and prediction of changes in the extreme event magnitude-frequency with climatic change. Second, many extreme weather events tend to be (1) spatially localised in nature and impact and (2) variable in size and severity, thus complicating generalisations about their impacts. Third, some present great problems in forecasting either their occurrence and location (as with tornadoes and hailstorms) or their precise track (as with tropical cyclones). Fourth, impacts on landscapes and societies vary not only with event characteristics but also with numerous human variables, such as land use, the level of income and degree of organisation of societies and individuals, the nature of the society, the accuracy and timeliness of forecasting, warning and evacuation systems, and the time since a previous event (and hence the society's preparedness). The design and appropriateness of hazard impact reduction and mitigation strategies will also vary considerably with the nature of the society.

Geographers have provided spatial and temporal perspectives and analytical techniques in quantifying and mapping weather hazards and in assessing changes in risk through time, with increasing use being made of GIS and remote-sensing techniques. For example, in relation to avalanches in the Alps, Gruber and Haefner (1995) used Landsat Thematic Mapper remote-sensing data and a digital elevation model to develop a methodology to produce more reliable avalanche hazard maps for planning purposes and to provide an improved insight into the interactions of forest, snow and avalanche risk in the Alpine landscape.

The findings of Graves and Bresnock (1985) question the wisdom of the traditional return period/probability approach to assessment of extreme events, in which weather hazards are often assumed to be randomly distributed in time with events occurring independently of each other. In the United States, using hail data for a twenty-one-year period at twenty locations and tornado occurrence data for 1916–80, they demonstrated that both hazards were more likely to occur in a particular year if they had been experienced in the previous year. They argued that if the same holds true for a hazard such as early or late frost, then rotating crops to plant a frost-resistant crop in any year following early or late frost damage would save millions of dollars of crop damage.

In drought studies, in addition to the considerable research on drought impacts and human response, particularly in the Sahel (e.g. Trilsbach and Hulme 1984; Hulme 1986; Trilsbach 1987; Walsh *et al.* 1988), one of the most useful contributions of geography has been in developing definitions of droughts and dry periods that are relevant to particular societies or specific physical or human issues. Thus Hulme (1987), in defining wet season length and character, attempted to incorporate farmers' perception and decision making on crop planting dates into his analysis of the impact of extreme drought in semi-arid central Sudan and the response of the rural water supply system. In rain forest environments, where much shorter and less intense periods of low rainfall constitute 'drought', Walsh (1996a) used archival rainfall records and definitions of drought specifically geared to tropical rain forest transpirational demand in identifying changes in drought magnitude frequency in Sabah (Borneo) over the past 120 years and exploring their implications for rain forest dynamics.

Relatively few studies have investigated changes in heavy rainstorm frequency. An early study by Howe *et al.* (1967) linked an increase in flood frequency in the Rivers Wye and Severn since the 1920s to an increase in heavy rainstorm frequency in central Wales. Later work demonstrated an increase in heavy rainstorm frequency since 1925 to be widespread in south Wales and linked to a parallel increase in flood frequency of the Rivers Tawe and Ebbw (Walsh *et al.* 1982). The United Kingdom Climate Change Impacts Review Group (1996), which includes geographers on its panel, used the simulated daily rainfall data outputs of the UKTR model to suggest that return periods of heavy daily rainfalls in the twenty-first century will shorten significantly in summer in the north and in winter throughout the United Kingdom and pointed to the consequences for flooding and soil erosion. Boardman *et al.* (1996) found that soil erosion resulting from a (currently) 1000-year return period rainstorm in which up to 128.7 mm fell over parts of Berkshire and Oxfordshire in central England on 26 May 1993 was particularly severe on fields with spring-planted crops, averaging 66 m^3 ha^{-1} on a maize field. They warned that the current trend towards spring-planted (rather than winter-planted) crops would increase the erosional risk of any increase in large rainstorm frequency.

In the tropics a marked reduction in heavy rainstorm frequency has accompanied the drought epoch since 1965 in the Sudan and has been linked to a decline in *wadi* flows, shallow groundwater recharge and rural water supply (Hulme 1986; Walsh *et al.* 1988), and a decline not only in flood frequency but also in human perception of flood hazards, resulting in large-scale flooding of poorly located squatter settlements of recent migrants in Khartoum and Omdurman in the exceptional rainstorm of August 1988 (Walsh *et al.* 1994). In the eastern Caribbean, significant increases in the return periods of large daily rainstorms have marked the two dry epochs of 1899–1928 and since 1959 compared with the wetter late nineteenth century and 1929–58 periods (Walsh 1998).

Finally, geographers involved in multidisciplinary teams assessing global warming and its impacts have been paying increased attention to formulating scenarios regarding changing frequencies and impacts of weather extremes (IPCC 1996; United Kingdom Impacts Review Group 1996; Hulme and Viner 1998) and Parry and Carter (1997) have incorporated extreme event considerations in their manual on climate impact and adaptation assessment.

CASE EXAMPLES OF GEOGRAPHICAL RESEARCH ON TROPICAL CYCLONES AND THEIR IMPACT

Tropical cyclones, which can be defined as closed-circulation, warm-cored, low-pressure systems with maximum sustained surface wind speeds (1-minute mean) of at least 39 mph, are conventionally divided into two intensity classes: tropical storms (with maximum winds of 39–73 mph) and hurricanes (with maximum winds of at least 74 mph). Hurricanes have been subdivided into five potential damage classes depending on their maximum wind speed, minimum central pressure and storm surge magnitude in what is termed the Saffir–Simpson damage potential scale (Table 4.1) (Simpson and Riehl 1981). Although most work on cyclones has been accomplished by meteorologists, particularly in the United States, geographers have contributed to cyclone research in four principal ways.

The assessment of spatial distribution of the cyclone hazard

Three data problems bedevil objective assessments of the cyclone hazard. First, cyclones vary in intensity and size both between cyclones and during the life-cycle of a single cyclone. The second concerns the varying spatial scales (in terms of size of areal unit) at which cyclone frequency or impacts can be studied and their appropriateness for different purposes. The third problem is that of the increasingly incomplete, imprecise and unreliable data on cyclone occurrences, tracks and characteristics as one goes back into the past; coverage over ocean areas has only become more or less complete during the last three decades with the advent of satellite imagery, thus raising problems about the meaningfulness of many earlier maps of cyclone frequency and assessments of changes in regional cyclone frequency. Geographers have made some significant contributions in designing analytical strategies to accommodate or overcome these problems.

Several geographers have mapped aspects of the spatial distribution of cyclone frequency within macro-regions by adopting a grid-square approach. For example, McGregor (1995) assessed spatial aspects of the cyclone hazard and interannual variations in cyclone activity in the China Sea by constructing and analysing a 2° × 2° grid-square database of six-hourly tropical cyclone position data over the period 1970–89 for the area 105–125° E, 5–27° N. The approach led to maps of the percentage probability that any

Table 4.1 Classification of tropical cyclones with hurricane classes based on the Saffir–Simpson damage potential scale.

Class	*Maximum sustained surface wind (mph)**	*Central pressure (mb)*	*Surge (metres)*	*Damage class***
Tropical Storm	39–73	> 980	< 1.3	n/a
Tropical Hurricane				
Class 1	74–95	> 980	1.3–1.9	Minimal
Class 2	96–110	965–979	2.0–2.5	Moderate
Class 3	111–130	945–964	2.6–3.9	Extensive
Class 4	131–155	920–944	4.0–5.5	Extreme
Class 5	>155	< 920	> 5.5	Catastrophic

Source: After Simpson and Riehl 1981.
Notes: *100 mph = 160.9 kph = 86.88 knots.
** For details of damage associated with each class, see Simpson and Riehl 1981: pp. 366–8.

Figure 4.2 Maps of the cyclone hazard in the South China Sea area: (A) percentage chance of a cyclone entering the China Sea affecting different 2 × 2 degree grid squares; (B) the size of reduction in cyclone frequency (six-hour periods per year) in an ENSO year for 2 × 2 degree grid squares in the China Sea area.

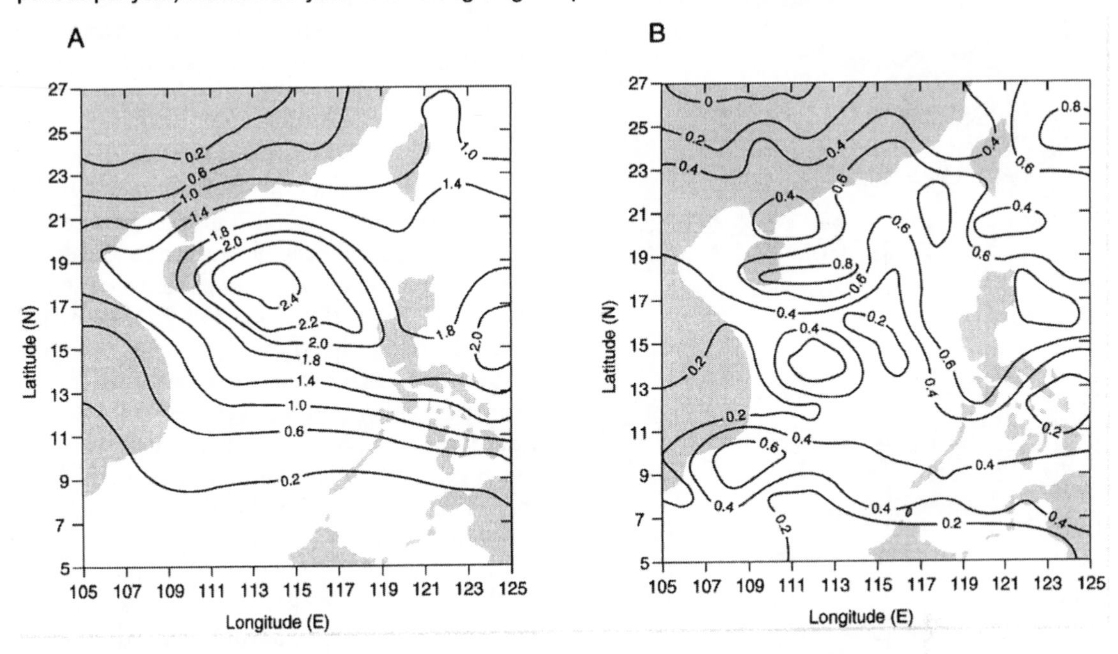

Source: After McGregor 1995.

tropical cyclone entering the South China Sea would pass through any given square (Figure 4.2A). A WNW–ESE oriented zone of peak probability was identified stretching from Hainan, Guangdong and Hong Kong in southern China to central Luzon in the Phillipines. The study also demonstrated how cyclone activity in the region was greatly reduced in all ENSO years during the period (Figure 4.2B), a pattern also found in the Caribbean (Eyre and Gray 1990; Gray and Sheaffer 1991).

Spatiotemporal changes in cyclone magnitude and frequency

Several geographical studies have examined recent or historical changes in the frequency and tracks of cyclones and assessed the question of whether or not global warming, via higher sea-surface temperatures, will result in increased cyclone frequency or severity, as some climatologists and climatic modellers have suggested (e.g. Emanuel 1987). To overcome the problem of increasingly incomplete records back through time, analyses have been restricted to those parts of a macro-region with longer periods of comprehensive data. Figure 4.3 updates to 1995 analyses carried out by several geographers of cyclone frequency for a 5° × 5° grid-square matrix covering the western part of the North Atlantic/Caribbean macroregion using US Weather Bureau charts of cyclone tracks since 1871 (Spencer and Douglas 1985; Walsh and Reading 1991; Reading and Walsh 1995). Regional cyclone frequency was high in 1871–1901 and in 1928–58 but low in 1902–27 and from 1959 to the 1990s, but the spatial distributions of activity in the two peaks and the two troughs differed markedly. Temporal fluctuations varied greatly between different 5° squares (Figure 4.4). Along the Texan coast, cyclone frequency peaked in the late nineteenth century, whereas in Florida and the northern

Figure 4.3 Differences in cyclone frequency in the West Indies grid region for four periods during 1871–1995.

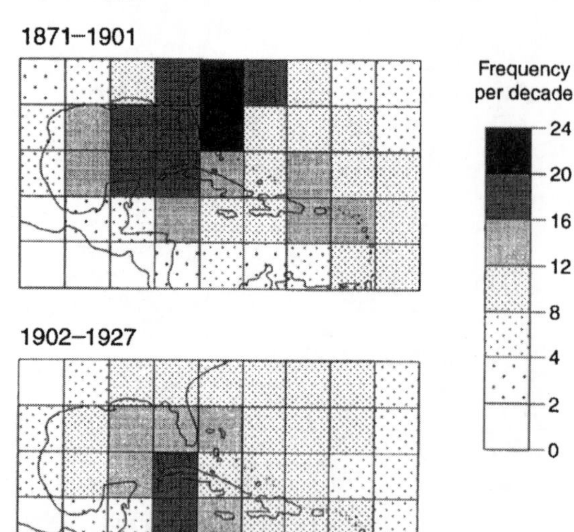

1928–1958

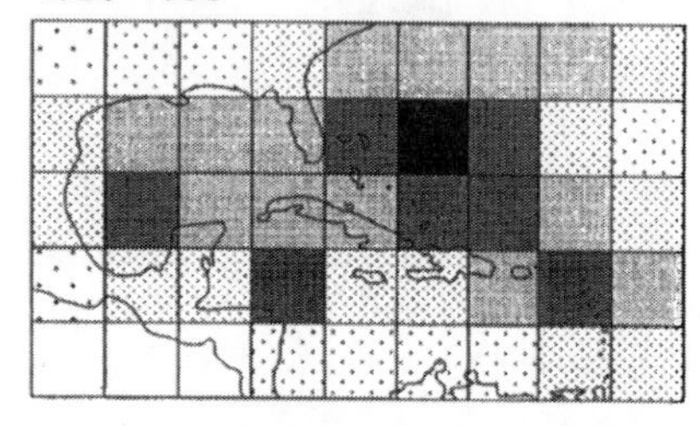

1959–1995

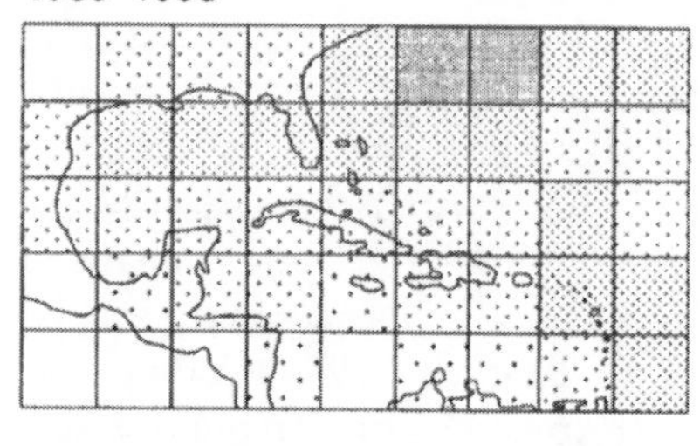

Source: Updated version of Reading and Walsh 1995.

Lesser Antilles frequencies peaked in 1928–58. In Jamaica, peak frequency was actually recorded around 1910, at a time when regional frequency was very low, and cyclone frequency in the southern Lesser Antilles has fallen little in recent years compared with the dramatic declines in Florida, the northern Lesser Antilles and Jamaica.

By focusing on individual land areas in the Caribbean and using a combination of documentary records and several existing chronologies of hurricanes, it also proved possible to examine cyclone frequency changes back to the seventeenth century (Walsh and Reading 1991; Reading and Walsh 1995; Walsh 1998). In the case of the Lesser Antilles, it was demonstrated that cyclone frequency was also high in parts of the late eighteenth and early nineteenth centuries, but much lower than in the twentieth century in 1650–1764, 1794–1805 and 1838–75 (Figure 4.5). What the study also showed, however, was how trends at the island sub-group scale differ both from each other and from the regional pattern. Peak frequency in the Windward Islands occurred in 1876–1901, whereas in the Leeward Islands it occurred in 1765–93 and in the French Islands/Dominica in 1765–93 and 1806–37. The study also demonstrated that changes in frequency cannot be explained solely by changes in sea-surface temperature but appear to be linked more to shifts in key elements of the general circulation and associated changes in the frequency and spatial distribution of low vertical wind shear identified by Gray (1968; 1988) as essential for cyclone development.

Eyre and Gray (1990) examined trends since 1962 in cyclone frequency (as indexed by the number of cyclone-hours) in 5° × 5° degree grids covering three areas: the Caribbean/Atlantic, the eastern Pacific and the southwest Pacific, finding no evidence of any increase in each case (Table 4.2). They also found no evidence of an increase in the intensity or severity of cyclones, a finding since confirmed for the North Atlantic/Caribbean by Landsea (1993). However, other studies by geographers have pointed to recent localised increases in cyclone frequency. Thus Spencer (1994) has reported that cyclones affecting Fiji in the South Pacific have increased from 3.1 per decade in 1941–80 to 11.4 per decade in the 1980s; and Nunn (1990a, b) has reported increased cyclone activity in the Tuvalu, Solomon Islands and Vanuatu areas of the South Pacific. Some of the local increases may reflect an increased

Figure 4.4 Ten-year running means of cyclone frequency for selected 5° × 5° squares in the Caribbean, 1871–1995, and the entire West Indies grid region.

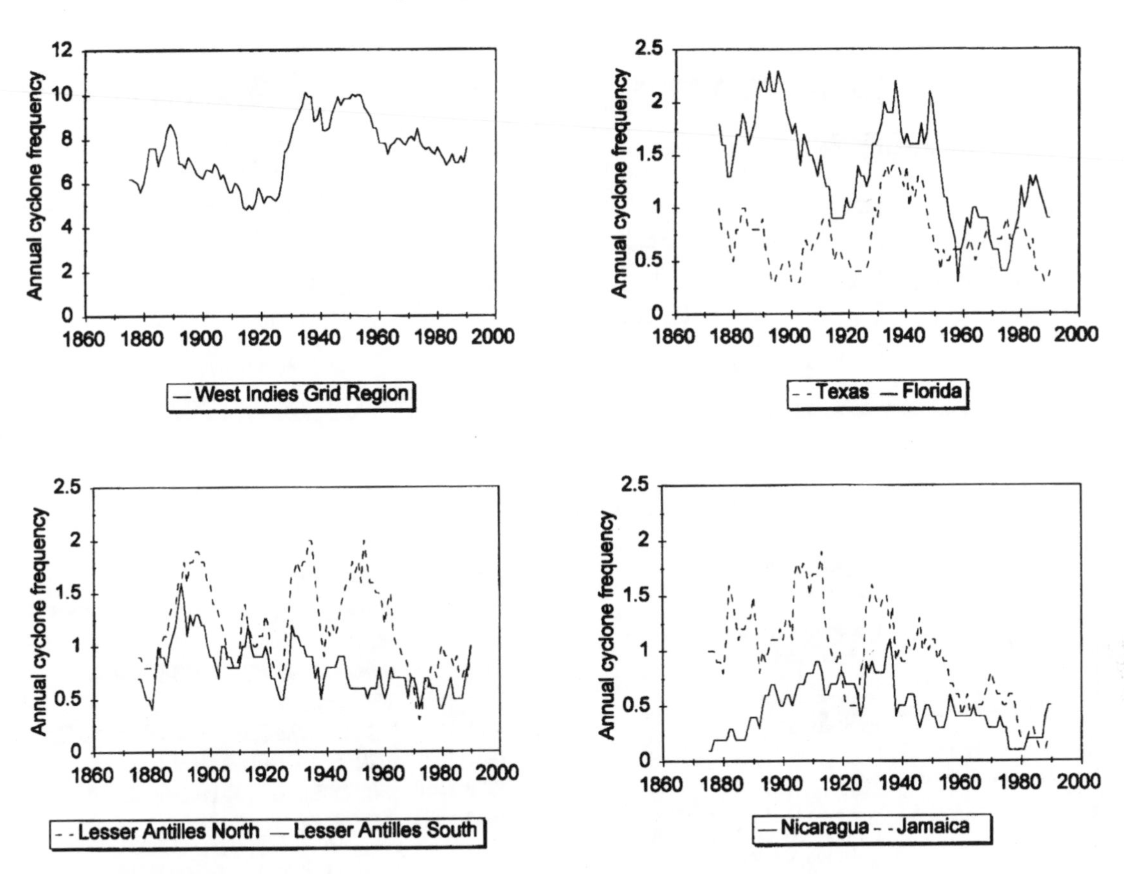

Note: For locations of squares see Figure 4.3.

frequency of very strong ENSO events, as Spencer (1994) reported six cyclones in just five months during the 1982–3 ENSO in a part of French Polynesia that had not been affected by cyclones since 1906. Whether the very high frequency of cyclones in 1995 (Walsh 1998) and in 1998 in the North Atlantic heralds a global warming upturn remains an open question.

Analysing cyclone impacts and ways of mitigating their effects

Cyclone damage is brought about by wind, waves, storm surge (and associated coastal inundation) and heavy rain (which can also lead to landslides and river flooding). The impacts of a cyclone vary with cyclone characteristics (its intensity, size and path in relation to the area affected), the landscape it encounters (topography, vegetation and land use, quality of building stock, etc.), the effectiveness of hurricane warnings, the economic and social character of the population, and not least the previous hurricane experience of the population and landscape. In lowland areas, the main threat to life is the storm surge, but in mountainous landscapes, most danger stems from landslides and river floods, as was the case with Hurricane George in the Dominican Republic and Hurricane Mitch in Nicaragua and Honduras in 1998. A fundamental geographical principle of cyclone damage (both physical and human) is that it tends to be zoned,

Figure 4.5 Changes in cyclone frequency for the Lesser Antilles and its island sub-groups, 1650–1995 (updated to include the 1990s, after Walsh 1998).

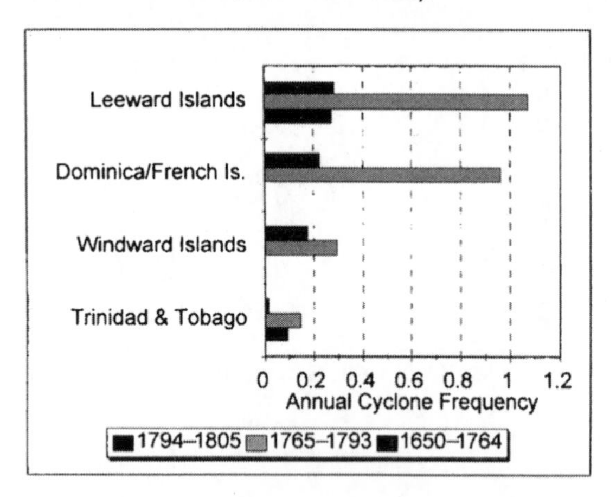

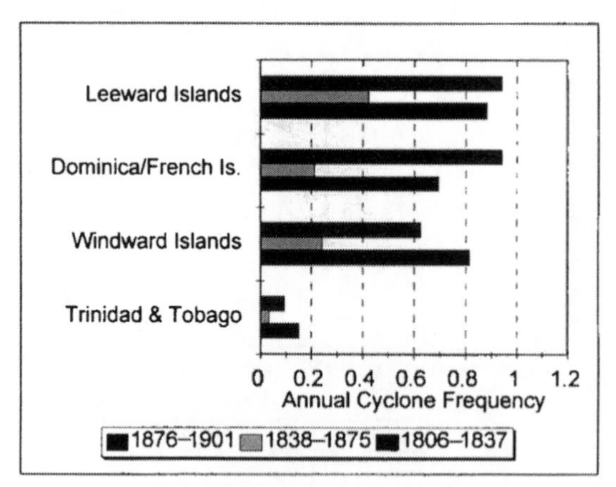

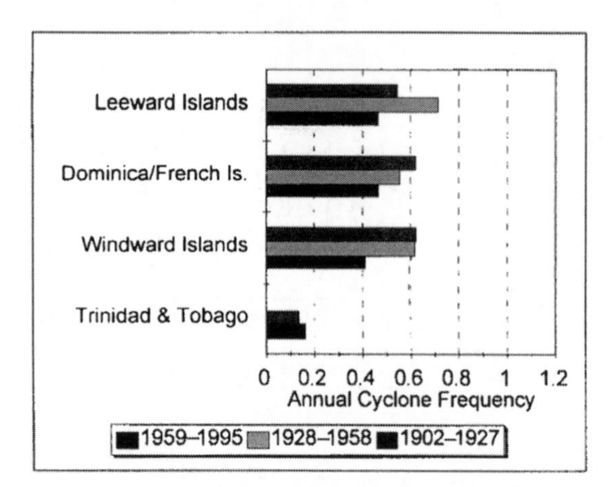

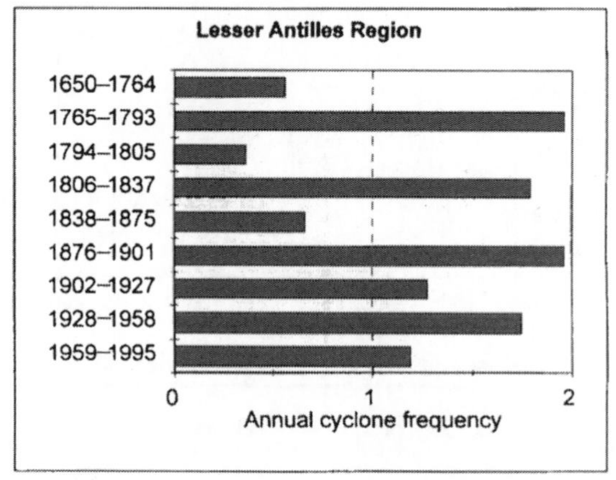

decreasing with increasing distance from the path of the eye of the cyclone (Box 4.1).

The principle of the protective (and land-building) role of mangroves (Box 4.1) formed a cornerstone of plans in the 1980s to help to protect the coastline of the deltas of the Ganges and Brahmaputra in Bangladesh (Stoddart and Pethick 1984; Stoddart 1987), where a 10 m storm surge in a cyclone in 1970 resulted in an estimated 280,000 deaths in the delta region, in part because of the failure of a previous scheme, which had involved the construction of massive polders surrounded by 5 m high earth banks. These proved too expensive and difficult to maintain, and many were breached or overtopped by the storm surge, drowning those who thought they were protected by the scheme. Damage and loss of life were less in the west of the delta, where the traditional mangrove vegetation was more intact. The scheme adopted in the 1980s has been to protect the coastline by restoring the previous natural belt of fringing mangroves, with 65,000 ha planted by 1987 (Stoddart 1987). As well as providing physical protection against waves, wind and storm surge erosion and promoting sedimentation, the mangroves would provide timber and firewood, generate by-products such as honey, prawns and fish and eventually aid the transition of land (with sedimentation) from timber to paddy. Despite these measures, however,

Table 4.2 Percentages of the North Atlantic/Caribbean, east Pacific and Australian (southwest Pacific) cyclone areas showing increases, decreases or no change in cyclone activity (a) over 1962–89 (1962–86 in the Pacific regions) and (b) in 1980–89 (1977–86 in the Pacific regions) compared with the rest of the period.

Cyclone region	*Percentage of cyclone region showing*					
	For the period 1962–89 (or 1962–86)			*1980–89 (or 1977-86) compared with the rest of the period*		
	Persistent increase	*Persistent decrease*	*No consistent change*	*Increase*	*Decrease*	*No change*
North Atlantic/ Caribbean 5–40° N 20–105° W	4	7	89	28	42	30
East Pacific 5–35° N 85–150° W	0	23	77	27	59	14
Southwest Pacific 5–35° S 105–170° E	8	9	83	23	59	18

Source: After Eyre and Gray 1990.

when the area was struck by a cyclone in 1991, much of the region was again inundated by a storm surge, with 200,000 deaths, mostly on the offshore islands of the delta (Smith 1997). The impacts of future cyclones in the region are likely to increase given the high rate of population increase (Stoddart 1987), high sedimentation rates with Himalayan deforestation and predicted sea level rise with global warming (Smith 1997).

Human impacts of a cyclone (as with most extreme climatic events) and human responses and adjustments to cyclones and cyclone risk vary greatly with the socio-economic characteristics of the country (Table 4.4) (Box 4.2). Loss of life, injuries, destruction of homes and livelihood, and susceptibility to longer-term problems of famine, disease and readjustment tend to be greatest in less developed societies with high-density rural populations, such as in Bangladesh; in more developed societies, economic losses in monetary terms are much greater (because families and businesses have so much more to lose) but, because of better warning systems and (in the USA) better evacuation systems and building design, few lives tend to be lost and homes tend to be damaged rather than destroyed (Smith 1997). In assessing rural damage, it is important to consider not only the ability of different land uses to withstand high winds but also the ease with which they can be replanted and come back into production following a hurricane. In the West Indies, one of the advantages of banana production as perceived by the farming community is that, although bananas are the most easily devastated of crops by wind, they are also easily re-established and can produce a regular farmer income less than a year after a hurricane (Walsh 1998). Thus in Dominica the value of exports (mostly bananas) following Hurricane David in 1979 had overtaken 1978 levels by 1981 (Collymore 1995). In more developed countries, 'spread-the-cost' measures tend to be much more effective. Building and crop insurance is more widely adopted, and national subsidies from the rest of the population in terms of emergency aid programmes tend to aid rebuilding and readjustment.

Impacts of extreme events are not always adverse. They can offer a chance for a complete reorientation of an economy. Thus, in Grenada, Hurricane Janet in 1955 hastened the widespread adoption of bananas by destroying the ailing nutmeg and cocoa plantation economy of the island and leading to a sudden substantial

Box 4.1 The zoning of cyclone damage

The zonation concept was described in detail by Stoddart (1971) in his study of the impact of Hurricane Hattie on the vegetation and geomorphology of the offshore reef islands (cays) of Belize in October 1961 (Figure 4.6A). Much of Zone A, especially to the north of the eye and including the capital, Belize City, was affected by a high storm surge. In the high, volcanic island of Dominica following Hurricane David in August 1979, landslide activity and damage to the rain forest vegetation, plantations and property were likewise zoned, but with modifications due to topography (Figure 4.6B) (Walsh 1982; 1996b). Damage was less to the lee (west) of mountain ridges but increased where east–west-oriented valleys funnelled the easterly winds. Both Hattie and David were very severe (Class 5) hurricanes; in less severe

Figure 4.6 Zonation of cyclone damage resulting from Hurricane Hattie in Belize in 1961 (Stoddart 1971) and Hurricanes David and Frederic in Dominica in 1979.

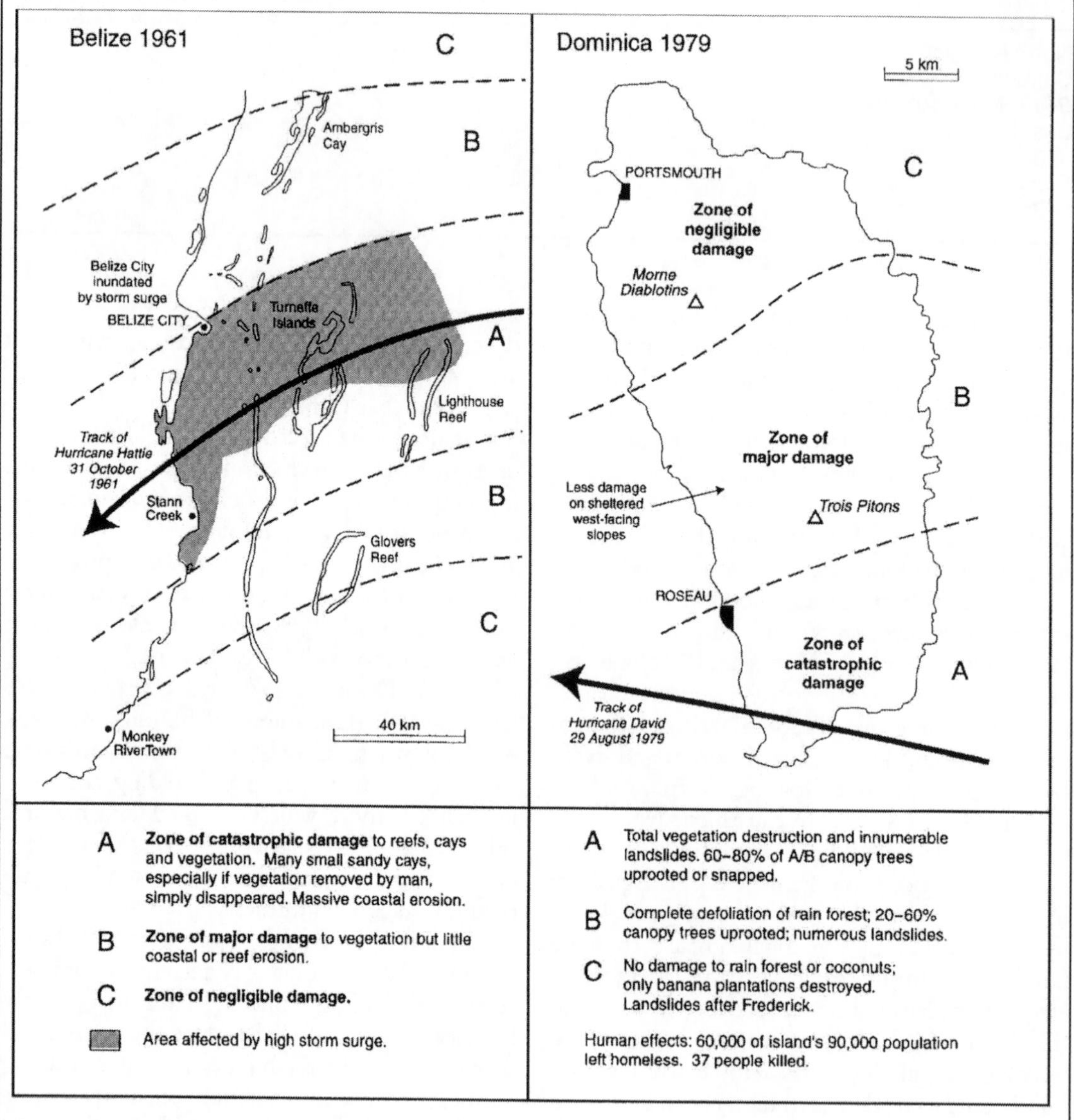

Source: After Stoddart 1971; Walsh 1982; 1996b.

cyclones, the zones of catastrophic and major damage may be absent. Also, different paths of a cyclone (to the north of, south of or directly over a location) will result in different wind (and wave) directions and very different spatial patterns of damage.

Stoddart (1971) demonstrated how land use and island size also affected the scale and nature of reef island damage in Hurricane Hattie (Table 4.3). All fifteen islands that were covered by natural vegetation (a low dense thicket and mangroves) actually aggraded in net terms as a result of the hurricane, demonstrating the essentially 'constructional' role of the extreme event in island building. Most of the islands with coconuts but with a regenerated thicket also grew, whereas islands with no or only low undergrowth suffered major beach retreat or sand-stripping, with two actually disappearing. Small islands were below the threshold for survival, disappearing whether or not they were vegetated. The implied message was clear: if one is to exploit the cays for coconut production, one should maintain a natural thicket understorey and clear as little of the fringing mangrove vegetation as possible.

Table 4.3 Role of vegetation cover in hurricane effects on reef islands in Belize during Hurricane Hattie in 1961.

	Vegetation/land-use type					
Hurricane effect	*Natural vegetation*	*Coconuts with regenerated thicket*	*Coconuts with low undergrowth*	*Coconuts with no undergrowth*	*Small island with small vegetation thicket*	*Unvegetated*
Disappearance	–	–	–	2	10	7
Major surface sand-stripping and channel-cutting	–	2	3	17	–	–
Major beach retreat	–	3	–	10	–	–
Marginal aggradation	15	9	–	–	–	–

Source: After Stoddart 1971.
Note: Figures refer to the number of cays in each category.

Box 4.2 The impacts of Hurricane Gilbert

The impacts that the Class 5 Hurricane Gilbert had on the 'intermediate' society of Jamaica, which it crossed east–west on 12 September 1989, have been analysed by Eyre (1989) and Barker and Miller (1990). Although loss of life was relatively small (forty-five deaths), total damage was estimated at US$800–1000 million, a sum that exceeds the annual value of exports. Landslide activity resulting from the 200–400 mm rainfall was exacerbated by poor agricultural land management, with newly established coffee projects utilising ill-advised monocropping on steep slopes in the Blue Mountains being particularly badly affected. A quarter of the buildings in Jamaica were rendered unusable, at least temporarily, and a further 50 per cent sustained some damage. Over 70 per cent of the private housing and most of the public sector infrastructure were uninsured. Barker and Miller (*ibid.*) found that the hurricane, which was personified as 'a bad aggressive male' by Jamaican society, was also seen as a social leveller, affecting people's property regardless of income and class. This point is interesting, as it demonstrates that in relative economic terms, it is arguably not the very poor who incur the greatest relative loss but intermediate societies such as the wealthier sections of Jamaica, which have more valuable property and goods, but where their buildings are neither hurricane-proof nor insured, as they would be in the more advanced society of the USA.

injection of colonial investment funds (Weaver 1968). In the island interior, the previously economically unproductive rain forest areas, which were largely felled by the hurricane, were replanted with fast-growing commercial Blue Mahot forest plantations. The extra funding also financed modern harbour facilities and a fishing industry based on newly discovered offshore

Table 4.4 Responses to tropical cyclone hazards in relation to stage of development of a territory.

Response class	*Response type*	*Stage of development of territory*		
		Poor (e.g. Bangladesh)	*Intermediate (e.g. Jamaica)*	*Rich (e.g. USA)*
Reduce the loss	Forecasting and warning systems	No	Yes	Yes
	Evacuation systems	No	No	Yes
	Hurricane-proof design of buildings	None	Minority	Many
	River/sea defences	Few	Inadequate	Widespread
	Land-use planning and regulations	None	Poor	Some
Spread the loss	Hurricane insurance	No	Minority	Considerable
	National emergency aid	Poor	Limited	Good
	External relief	Yes*	Yes*	No
Bear the loss	Bear the loss	Dominant	Considerable	Minor

Source: In part based on Collimore 1995 and Smith 1996, 1997.
Note: *In poorer countries, the significance and effectiveness of external relief varies inversely with the size of population of the territory affected.

fishing grounds. Particularly in the case of *small* islands or island states, therefore, hurricanes can focus the world's attention and economic aid on a location that is otherwise an obscure economic backwater.

Finally, some geographers have been directly involved in the development, planning and assessment of mitigation systems. For example Jeremy Collymore, who previously held posts in the Department of Geography in the University of the West Indies, has since worked in the Pan-Caribbean Disaster Preparedness and Prevention Project and then been director of the Caribbean Disaster Emergency Response Agency in Barbados. In an analysis of the advantages and limitations of different approaches to hazard mitigation as practised in the Caribbean (Collymore 1995), he pointed out the limitations of 'rational' approaches such as cost–benefit analysis, when costs such as social dislocation and psychological trauma cannot be easily quantified in monetary terms and when perceptions of risk by people (Caribbean people have for cultural and historical reasons a high tolerance level for risk) do not conform to rational models. Consequently, the 'intermediate' Commonwealth Caribbean territories are characterised by a heavy reliance on information-based mitigation strategies such as hurricane forecasting, warnings and preparedness information, a comparatively poor take-up of insurance, and a lack of comprehensive hazard management planning (see Table 4.4).

Problems and dilemmas in advising planners

A number of problems and dilemmas in giving planning advice arise from recent cyclone research (Walsh 1998). One is the mismatch between the *regional* and *sub-regional* spatial scales, at which geographers provide reliable data (which may be useful for ship insurers), and the *local* ($< 50 \times 50$ km scale), at which data are required for most planning purposes. Not only do temporal patterns in cyclone frequency at the regional, sub-regional and island group spatial scales differ considerably (see Figures 4.3, 4.4 and 4.5), but also the degree of reliability decreases at more local scales as the number of events upon which they are based falls.

A second dilemma concerns the most appropriate cyclone frequency data sets to use for future planning purposes in cyclone regions. In the Caribbean, one is faced with (1) continued predictions of increased cyclone frequency (Emanuel 1987; IPCC 1996; Hulme and Viner 1998) with global warming and higher sea-surface temperatures, but (2) marked changes in cyclone frequency and spatial distribution over the past 125 years, including a recent decline in cyclone frequency. The question therefore arises as to which of four data sets is the most appropriate to use in giving advice as regards future cyclone hazards (Walsh 1998):

1 The entire data set since 1871.
2 The 'worst case' past record (i.e. the 1876-1901 or 1928-58 periods).
3 The most recent period since 1959 – in which cyclones are infrequent.
4 A predicted 'global warming' scenario – with much higher frequencies than at present.

The choice is not made easier given the low level of confidence that one can place in the future predictions of climatic models, particularly as regards extreme events (United Kingdom Climate Change Impacts Review Group 1996) and at local/regional scales. In addition, there is the problem of which rate of sea level rise to adopt for the future, as this will greatly influence coastal impacts of future cyclones regardless of changes in cyclone frequency.

APPLIED GEOGRAPHICAL RESEARCH PRIORITIES FOR THE FUTURE

Extreme weather events remain an acknowledged weak link in climatic-change studies and the modelling of future climate and its impacts (IPCC 1996; Hulme and Viner 1998). Three priorities for future geographical research related to these research needs can be identified.

1 Although there has been much research on the impacts of *individual* extreme events, little is known about their longer-term context and the impacts that different magnitudes and frequencies of events have on natural or human systems. A few studies have been addressing such issues, such as the work of Bayliss-Smith (1988) and Spencer *et al.* (1997) on the impacts of changes in cyclone frequency and El Niño-related bleaching episodes, respectively, on reefs and reef islands; and of ecologists, foresters and geographers on rain forest responses to differences and changes in cyclone magnitude and frequency (e.g. Whitmore 1989; Walsh 1996b) and drought magnitude and frequency (Walsh 1996a; Condit 1998; Whitmore 1998). This research needs to be pursued further, preferably by interdisciplinary groups and incorporating interactions with human systems, if environmental responses to future climatic change are to be predicted.
2 Associated with this is an acute need for more research on the longer-term changes in the magnitude and frequency of some extreme events, particularly cyclones and heavy rainstorms. This is important for three reasons: (1) to provide the long-term data series necessary to establish temporally more robust relationships between atmospheric circulation variables and extreme event frequency and spatial distribution; (2) to aid the formulation of more soundly founded climatic models capable of producing more reliable predictions of future extreme event frequency than at present; and (3) to provide the data to develop and test long-term extreme event impact models.
3 Perhaps the greatest research need is that for the greater involvement of human geographers in extreme event research: (1) in analysing historical documentary records of extreme events and their impacts; (2) in defining climatic extreme event parameters of greatest relevance to human systems; (3) in analysing and modelling impacts of changes in extreme event frequency on contrasting and changing societies; and (4) in designing and assessing mitigation measures within the context of very different cultural and societal settings. At a time when much of research funding is quite rightly giving greater emphasis to 'the socio-economic dimension' of physical environmental problems, it often remains difficult to get academic human geographers involved in geographical or multidisciplinary

research teams. Stoddart's warning of twelve years ago to human geographers: 'Fiddle if you will but at least be aware that Rome is burning all the while' (Stoddart 1987; p. 334), still has much validity.

GUIDE TO FURTHER READING

A recent comprehensive treatment of climatic hazards is given in Smith (1996) *Environmental Hazards: Assessing Risk and Reducing Disaster*, Routledge). A methodology for assessing the impacts of climatic change (including extreme events) and adaptation to them is provided by Parry and Carter (1997). An up-to-date review and discussion of future climatic change in the tropics, including a consideration of tropical storms and hurricanes, is given by Hulme and Viner (1998). The book by Simpson and Riehl (1981) on *The Hurricane and its Impact* remains a classic account of the tropical cyclone hazard. Walsh and Reading (1991) provide a detailed account of historical changes in tropical cyclones in the North Atlantic/ Caribbean, including the methodology used and its limitations.

REFERENCES

Barker, D. and Miller, D. (1990) Hurricane Gilbert: anthropomorphising a national disaster. *Area* 22, 107–16.

Bayliss-Smith, T.P. (1988) The role of hurricanes in the development of reef islands, Ontong Java Atoll, Solomon Islands. *The Geographical Journal* 154, 377–91.

Boardman, J., Burt, T.P., Evans, R., Slattery, M.C. and Shuttleworth, H. (1996) Soil erosion and flooding as a result of a summer thunderstorm in Oxfordshire and Berkshire, May 1993. *Applied Geography* 16, 21–34.

Collymore, J.M. (1995) Disaster mitigation and cost–benefit analysis: conceptual perspectives. In D. Barker and D.F.M. McGregor (eds) *Environment and Development in the Caribbean: Geographical Perspectives,* Kingston, Jamaica: The Press, University of the West Indies, 111–23.

Condit, R. (1998) Ecological implications of changes in drought patterns: shifts in forest composition in Panama. *Climatic Change* 39, 413–27.

Emanuel, K.A. (1987) The dependence of hurricane intensity on climate. *Nature* 326, 483–5.

Eyre, L.A. (1989) Hurricane Gilbert: Caribbean record breaker. *Weather* 44, 160–4.

Eyre, L.A. and Gray, C. (1990) Utilization of satellite imagery in the assessment of the effects of global warming on the frequency and distribution of tropical cyclones in the Caribbean, East Pacific and Australian regions. *Proceedings 23rd International Symposium on Remote Sensing of Environment,* Thailand 365–75.

Graves, P.E. and Bresnock, A.E. (1985) Are natural hazards temporally random? *Applied Geography* 5, 5–12.

Gray, W.M. (1968) Global view of the origin of tropical disturbances and storms. *Monthly Weather Review* 96, 55–73.

Gray, W.M. (1988) Environmental influences on tropical cyclones. *Australian Meteorological Magazine* 36(3), 127–39.

Gray, W.M. and Sheaffer, J.D. (1991) El Niño and QBO influences on tropical activity. In M.H. Glantz, R.H. Katz and N. Nicholls (eds) *Teleconnections Linking Worldwide Climate Anomalies: Scientific Basis and Societal Impact,* Cambridge: Cambridge University Press, 257–84.

Gruber, U. and Haefner, H. (1995) Avalanche hazard mapping with satellite data and a digital elevation model. *Applied Geography* 15, 99–114.

Howe, G.M., Slaymaker, H.O. and Harding, D.M. (1967) Some aspects of the flood hydrology of the upper catchments of the Severn and Wye. *Transactions, Institute of British Geographers* 41, 33–58.

Hulme, M. (1986) The adaptability of a rural water supply system to extreme rainfall anomalies in central Sudan. *Applied Geography* 6, 89–106.

Hulme, M. (1987) Secular changes in wet season structure in central Sudan. *Journal of Arid Environments* 13, 31–46.

Hulme, M. and Viner, D. (1998) A climate change scenario for the tropics. *Climatic Change* 39, 145–76.

Intergovernmental Panel on Climate Change (1996) *Climate Change 1995: The Science of Climate Change: Contribution of Working Group I to the Second Assessment Report of the Intergovernmental Panel on Climate Change,* edited by J.T. Houghton, I.G. Meira Filho, B.A. Callender, N. Harris, A. Klattenberg and K. Maskel. Cambridge: Cambridge University Press.

Landsea, C.W. (1993) A climatology of intense (or major) Atlantic hurricanes. *Monthly Weather Review* 121, 1703–13.

McGregor, G. (1995) The tropical cyclone hazard over the South China Sea 1970–1989. *Applied Geography* 15, 35–52.

Nunn, P.D. (1990a) Warming of the South Pacific region since 1880; evidence, causes and implications. *Journal of Pacific Studies* 15, 35–50.

Nunn, P.D. (1990b) Recent environmental changes on Pacific islands. *The Geographical Journal* 156, 127–40.

Parry, M. and Carter, T. (1997) *Climate impact and adaptation assessment*. London: Earthscan.

Reading, A.J. and Walsh, R.P.D. (1995) Tropical cyclone activity within the Caribbean basin since 1500. In D. Barker and D.F.M. McGregor (eds) *Environment and Development in the Caribbean: Geographical Perspectives*, Kingston, Jamaica: The Press, University of the West Indies, 124–46.

Simpson, R.H. and Riehl, H. (1981) *The Hurricane and its Impact*. Oxford: Blackwell.

Smith, K. (1996) *Environmental Hazards: Assessing Risk and Reducing Disaster*. London: Routledge.

Smith, K. (1997) Climatic extremes as a hazard to humans. In R.D. Thompson and A.H. Perry (eds) *Applied Climatology: Principles and Practice*, London: Routledge, 304–16.

Spencer, T. (1994) Tropical coral islands – an uncertain future? In N. Roberts (ed.) *The Changing Global Environment*, Oxford: Blackwell, 190–209.

Spencer, T., Tudhope, A., French, J., Scoffin, T. and Utanga, A. (1997) Reconstructing sealevel change from coral microatolls, Tongareva (Penrhyn) atoll, northern Cook Islands. *Proceedings of the Eighth International Coral Reef Symposium, Panama*, Vol. 1, 489–94.

Spencer, T. and Douglas, I. (1985) The significance of environmental change: diversity, disturbance and tropical ecosystems. In I. Douglas and T. Spencer (eds) *Environmental Change and Tropical Geomorphology*, London: George Allen & Unwin, 13–33.

Stoddart, D.R. (1971) Coral reefs and islands and catastrophic storms. In J.A. Steers (ed.) *Applied Coastal Geomorphology*, London: Macmillan, 155–97.

Stoddart, D.R. (1987) To claim the high ground: geography for the end of the century. *Transactions, Institute of British Geographers* New Series 12, 327–36.

Stoddart, D.R. and Pethick, J.S. (1984) Environmental hazard and coastal reclamation: problems and prospects in Bangladesh. In T.P. Bayliss-Smith and S. Wanmali (eds) *Understanding Green Revolutions: Ararian Change and Development Planning in South Asia*, Cambridge: Cambridge University Press, 339–61.

Trilsbach, A. (1987) Environmental change and village societies west of the White Nile. In R.I. Lawless (ed.) *The Middle Eastern Village: Changing Economic and Social Relations*, London: Croom Helm, 13–50.

Trilsbach A. and Hulme M. (1984) Recent rainfall changes in central Sudan and their physical and human implications. *Transactions, Institute of British Geographers* New Series 9, 280–98.

United Kingdom Climate Change Impacts Review Group (1996) *Review of the Potential Effects of Climate Change in the United Kingdom*. London: HMSO.

Walsh, R.P.D. (1982) A provisional survey of the effects of Hurricanes David and Frederic on the terrestrial environment of Dominica. *Swansea Geographer* 19, 28–34.

Walsh R.P.D. (1996a) Drought frequency changes in Sabah and adjacent parts of northern Borneo since the late nineteenth century and possible implications for tropical rain forest dynamics. *Journal of Tropical Ecology* 12, 385–407.

Walsh R.P.D. (1996b) Climate. In P.W. Richards (ed.) *The Tropical Rain Forest*, Cambridge: Cambridge University Press, 159–205.

Walsh R.P.D. (1998) Climatic changes in the Eastern Caribbean over the past 150 years and some implications in planning sustainable development. In D.F.M. McGregor, D. Barker and S. Lloyd Evans (eds) *Resource Sustainability and Caribbean Development*, Kingston, Jamaica: The Press, University of the West Indies, 26–48.

Walsh, R.P.D., Hudson, R.N. and Howells, K.A. (1982) Changes in the magnitude–frequency of flooding and heavy rainfalls in the Swansea Valley since 1875. *Cambria* 9(2) 36–60.

Walsh, R.P.D., Hulme, M. and Campbell, M.D. (1988) Recent rainfall changes and their impact on hydrology and water supply in the semi-arid zone of the Sudan. *The Geographical Journal* 154, 181–98.

Walsh, R.P.D. and Reading, A.J.. (1991) Historical changes in tropical cyclone frequency within the Caribbean since 1500. *Würzburger Geographische Arbeiten* 80, 199–240.

Walsh, R.P.D., Davies, H.R.J. and Musa, S.B. (1994) Flood frequency and impacts at Khartoum since the early nineteenth century. *The Geographical Journal* 160, 266–79.

Weaver, D.C. (1968) The hurricane as an economic catalyst. *Journal of Tropical Geography* 27, 66–71.

Whitmore, T.C. (1989) Changes over 21 years in the Kolombangara rain forests. *Journal of Ecology* 77, 469–83.

Whitmore, T.C. (1998) Potential impact of climatic change on tropical rain forest seedlings and forest regeneration. *Climatic Change* 39, 429–38.

5

Earthquakes and vulcanism

David Alexander

INTRODUCTION

On average, twenty-seven earthquakes and four or five volcanic eruptions cause disasters each year. The earthquakes kill about 19,000 people and injure 26,000, while the eruptions kill 1000 and injure fewer than 300. These figures represent less than 10 per cent of natural hazard mortality, although nearly 50 per cent of morbidity; and although more than 2 million people are directly affected each year in some way by earthquakes and volcanic eruptions, they are only 1.5 per cent of all people who suffer the effects of natural catastrophe, as many more are affected by floods and droughts (IFRCRCS 1997).

Earthquakes and vulcanism are a subset of the general interdisciplinary field of natural hazards. The approach to these varies from geophysical to social and psychological. Between these end members there is a broad spectrum that includes studies of hazard, vulnerability, risk, perception, economic conditions, historical aspects, remote sensing, cartography, and the technical aspects of monitoring and warning. Since they were first conceived, natural hazard studies have maintained a strong applied dimension, sustained by the need to make the environment of life safe against extreme natural events and thus to reduce the toll of casualties and damage.

The role of geographers in studying earthquakes and vulcanism has been secondary to that of many other types of scientist and scholar. Many advances have come from seismologists, vulcanologists and other geophysicists. Engineers, architects, geologists and sociologists have also been highly active in this field.

Nevertheless, small numbers of both physical and human geographers have studied the effects of seismicity and eruptions. In this context, the old aphorism is of little use: geography is not merely what geographers do, it is also what the practitioners of other fields accomplish, overtly or unwittingly, with respect to space and place. Thus, geophysicists and engineering geologists have made good use of cartographic methods for both microzonation, the determination of hazard and risk at the local scale, and macrozonation, the plotting of regional damage distributions and hazard levels. Remote sensing has been used by vulcanologists to investigate igneous landscapes and by geologists to elucidate the surface morphology of active faults. Economists have studied earthquake hazards as regional inverse multipliers, and sociologists have looked at the spatial differentiation of perception and organisational behaviour. Geographers have distinguished themselves with studies of volcanic and seismic landforms, earthquake insurance, regional patterns of post-disaster reconstruction, and hazard and risk perception and zonation.

This chapter will focus on applied geographical work on seismic and volcanic hazards regardless of whether it has been carried out by geographers or not, although their work will be highlighted where appropriate. Relevant studies are those that deal significantly with practical problems of landscape, location or the spatial dimension in general, even in the context of other factors such as social relations, economic trends and geological phenomena.

PHYSICAL GEOGRAPHY

The physical geography of earthquakes and volcanoes has been studied in a multiplicity of ways. For example, improvements in the collection of atmospheric data and their analysis as time-series have enabled the climatic impact of volcanic eruptions to be assessed accurately (Handler 1989). At the same time, improvements in remote sensing have led to more comprehensive and accurate assessments of the extent and composition of volcanic aerosols. Erupting and passively degassing volcanoes are estimated to emit the following quantities of gases into the atmosphere each year: 100–200 million tonnes (Mt) of carbon dioxide, 18.7 Mt of sulphur dioxide, 0.4–11 Mt of hydrochloric acid and 0.06–6 Mt of hydrogen fluoride. For the most part, tropospheric gases are easily swept out by precipitation, and CO_2 emissions are dwarfed by those from anthropogenic sources. Hence the main effect of eruptions is to inject large amounts of SO_2 into the stratosphere (e.g. 17 Mt by Mount Pinatubo in 1991), and these may oxidise photochemically to sulphuric acid aerosols, which both add to the acidity of precipitation and reduce global temperatures by up to 0.5°C for a few years by reflecting some insolation back into space. Quantities of hydrogen fluoride (HF) are also significant and may gradually help to change the composition of the Earth's atmosphere.

There has been increasing interest in the hazards to aviation associated with suspended particulates and gas plumes from volcanic eruptions. Acidic aerosols can etch the exterior surfaces of aircraft, especially cockpit windscreens. Thus in the early 1980s the Mexican volcano El Chichón injected 0.5–0.6 km^3 of sulphur-rich products into the stratosphere and one airline found that it had to replace up to forty-two aircraft cockpit windows in a single month at a total cost of US$6.8 million (Bernard and Rose 1990). Moreover, glass pyroclasts can remelt when sucked into jet engines, which increases the operating pressure ratio of turbine compressors and may cause the engines to stall or flame out. Eruptions of Redoubt Volcano in Alaska and Galunggung in Indonesia led to in-flight emergencies and severe damage to aircraft, although fortunately not to casualties. Hence steps have been taken to utilise real-time vulcanological information in flight control (Casadevall 1991). Of particular use are AVHRR images from, for example, the NOAA 10 and 11 satellites, which can help to map the height, distribution, temperature and concentration of plumes and indicate their paths of movement.

About one-tenth of the world's 450–500 active volcanoes may erupt in any single year. Although relatively few are intensively monitored, fifteen are considered to offer such a high risk of disaster that they are the subject of a special monitoring initiative under the auspices of the International Decade for Natural Disaster Reduction (Figure 5.1), and all potentially active sites of vulcanism have been catalogued (Bullard 1984). Vulcanologists were among the first to make intensive use of the Internet for the exchange of data and information, and the Global Volcanism Network is now one of the most active and extensive of such operations. Vulcanologists make considerable use of electronic communications and satellite monitoring, with special emphasis on Total Ozone Mapping Spectrometry and Advanced Very High Resolution Radiometry for the height and composition of atmospheric plumes derived from eruptions, infrared data from the Thematic Mapper for the volume and temperature of emitted radiation, and radar and global positioning system (GPS) data for monitoring the deformation of a volcano's surface.

One aspect of volcanoes that has consistently been studied by a small number of physical geographers is their geomorphology. They have concentrated on the alternation of constructive and destructive land-forming processes, in which magma is extruded at the Earth's surface and its products are remodelled by blast effects and erosion (Ollier 1988). The implications for applied geomorphology are considerable and are developed in works that use the classic geographical method of synthesis across a wide spectrum of disciplines, although vulcanologists

Figure 5.1 Active volcanoes with high disaster potential chosen for intensive monitoring under the auspices of the International Decade for Natural Disaster Reduction.

(Francis 1994) are as likely to be involved as are geographers (Chester 1993). One aspect that distinguishes the latter's contribution is concern for the environmental impact of eruptions (Duncan *et al.* 1981), an interest that is shared with the biologists who have studied the recolonisation of volcanic landscapes by vegetation after the cessation of volcanic activity (Del Moral and Wood 1993).

Less attention has been given by physical geographers to seismic than to volcanic phenomena. Nevertheless, active tectonics have been the subject of some interesting geomorphological studies, including a review of morpho-tectonic process and form in China (Doornkamp and Han 1985) and diffusion models of fault-scarp degradation (Nash 1981). The potential to apply geomorphological methods to practical problems of earthquake mitigation and seismic safety is underexploited. Little has been accomplished in this field by geographers, although some very interesting studies were published in the 1980s by Kenneth Hewitt (e.g. 1983), who sought to connect spatial variations in the hazards of ground shaking with altitudinal patterns of settlement and ecotones. He found that piedmont alluvial fans are especially vulnerable to earthquake damage as they consist of unconsolidated debris that can undergo compaction subsidence and liquefaction failure. In addition, Cooke and Doornkamp (1990: p. 346) argued that tectonic geomorphologists have the potential to make a significant contribution to earthquake prediction studies by helping to identify places where ground stresses are at a maximum. However, there is little evidence that they have yet done so, although they have contributed to seismic risk assessments (Panizza 1989).

MAPPING AND ZONATION

Ever since the formulation of the first intensity scales in the 1700s and 1800s, students of earthquakes have been concerned with the spatial patterns created by seismicity. At the turn of the millennium, efforts are under way to refine the scales used to depict earthquake damage and other effects (MM, MCS, MSK, JMA and GOST scales – see Alexander 1993: pp. 28–31). This will lead to improvements in the depiction of the pattern and extent of earthquake damage and to an enhanced ability to predict the medium- to long-term seismicity of affected regions. A brief exposition of the field methods involved in post-earthquake intensity survey was given by Choudhury and Jones (1996), while Gasparini *et al.* (1992) discussed the statistical methods used to compile isoseismal maps from questionnaire data. Parametric models of the decline in intensity (and therefore damage) from the epicentre of an earthquake have been given in papers by seismologists (e.g. Brazee 1979) and geotechnical engineers (e.g. Seed *et al.* 1976). They tend to show that the attenuation of strong motion with distance from the fault is a function of source characteristics (the most critical factor), transmission path, geometric spreading, energy absorption and local site conditions. However, no geographer has yet risen to the challenge of producing a distributed spatial model of earthquake effects, although some moves in this direction were made by Slosek (1986), who found that deaths and injuries tended to decline remarkably uniformly with distance from epicentre regardless of the shape of the macroseismic field, providing the pattern of building collapse was taken into account. Moreover, using data from several earthquake disasters, I generalised the spatial pattern of earthquake casualties in a series of simple models based on concentric variations, axes and nodes (Figure 5.2; Alexander 1989 – see also Alexander 1993: p. 466). However, more testing and further refinements of such models are needed in order to increase their explanatory power.

In this context, Degg (1989) found that damage in the 1985 Mexico City earthquake varied widely according to type of construction, height of building and type of subsoil. Taller buildings were damaged at greater distances from the epicentre than were lower ones. Yet, Choudhury and Jones (1996) noted that few data exist on the relationship between building failure in earthquakes and the type, severity and distribution of

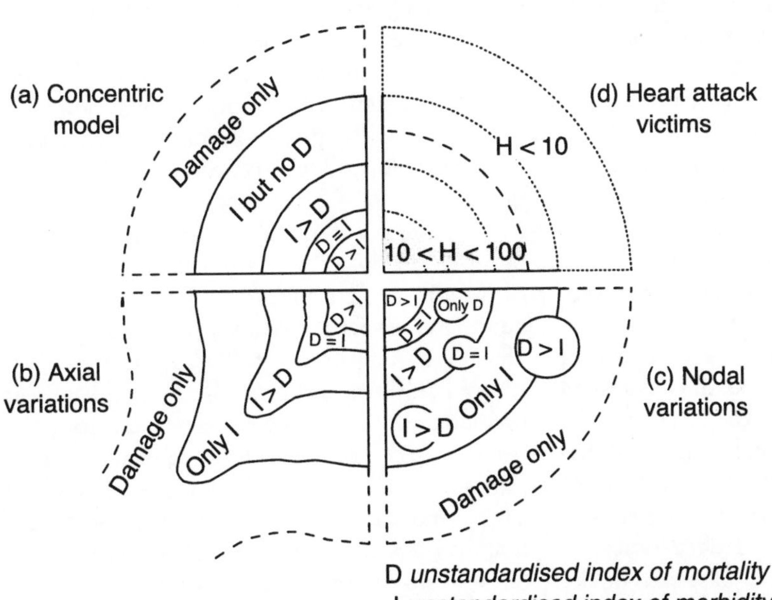

Figure 5.2 Hypotheses of the theoretical spatial distribution of casualties in earthquake disasters.

Source: After Alexander 1989.

injuries. If the relationship were better known, there would be a more adequate basis for locating survivors and knowing how to rescue and stabilise them.

A combination of improved geophysical methods (including the use of magnetometry and radar surveys) and greater integrated data processing has enabled the accuracy of epicentral determinations to be increased (e.g. Qamar and Meagher 1993). This in turn has given rise to improved understanding of the distribution of medium- to long-term seismicity around the world, including better estimates of the return period of earthquakes with given magnitudes or intensities. This in turn has stimulated great increases in the mapping and differentiation of regional seismic hazards, which has been done at various scales. Regional studies have encompassed, for example, East Africa (Mäntyniemi and Kijko 1991), the Mediterranean (Ambraseys 1992), the Middle East (Degg 1992) and the USA (Hays 1984). Studies of individual countries have been published on, among others, Bulgaria (Orozova-Stanishkova and Slejko 1994), France (Levret *et al.* 1994), Greece (Papazachos *et al.* 1993), Iraq (Fahmi and Alabbasi 1989), Jordan (Yücemen 1992), Panama (Muñoz 1989), Slovenia (Lapajne *et al.* 1997), Sweden (Kijko *et al.* 1993), Syria (Malkawi *et al.* 1995) and the United Kingdom (Musson and Winter 1997). Finally, local studies have been made of, for instance, Mexico City (Iglesias 1989), Puget Sound (Ihnen and Hadley 1987), the San Francisco Bay region (Murphy and Wesnousky 1994) and Utah (Gori and Hays 1992). A global seismic hazard assessment programme also promotes mapping initiatives (Giardini 1992).

Hazards can be considered at a variety of scales, and the nature of the risks thus identified may be partly determined by the level of detail inherent in each scale. Regional studies of seismicity are generally considered to be a form of macrozonation. This is invaluable as a means of determining the coefficients needed for local anti-seismic building codes, but adequate earthquake protection also requires there to be more detailed studies of the performance of geological materials, foundations and specific site factors. This is microzonation. Seismic microzonation was first developed for the Los Angeles and San Francisco Bay areas (Kockelman and Brabb 1979) and for Tarcento in the Friuli region of Italy (Brambati *et al.* 1980). It responds to a number of different exigencies. For instance, highly seismic areas that have large deposits of saturated sand or sensitive clay may undergo liquefaction, a spontaneous change from solid to liquid behaviour that usually involves loss of foundation bearing capacity. Liquefaction potential can be mapped in order to determine the most susceptible locations (Kotoda *et al.* 1988). Debris flows, rapid soil flows, rock falls and rock avalanches also constitute a hazard in mountainous areas with large, unstable deposits of clastic material. Indeed, Kobayashi (1981) reported that such events account for more than half of all deaths in earthquakes of magnitude M > 6.8 in Japan, which offers a rather different picture to the common view that building collapse is the principal determinant of mortality in earthquakes (Page *et al.* 1975). Microzonation of susceptible areas must therefore involve hypotheses about the behaviour of unstable rock, soil and debris masses during seismic loading. In the most complex situations, this may necessitate a series of maps that show different hazard levels or effects with different degrees of seismic acceleration of the ground.

The methodology of volcanic hazard zonation is well developed and has been tested for several decades on the Cascades volcanoes of western North America (Crandell and Mullineaux 1975). Smith (1996: pp. 179–81) showed that 1970s maps of Mount St Helens were generally accurate guides to the spatial distribution of impacts of the 18 May 1980 eruption, with allowance for more severe landslides and blast effects than had been predicted. Complex hazard assessments have also been devised for other volcanoes, including Etna (Chester *et al.* 1985) and Vesuvius (see Box 5.1). The former involves a recurrent lava flow hazard that has necessitated a spatial analysis of the density of volcanic vents (at least one per km^2), the length of lava flows (up to 15 km), the location of urban settlements (of which there are

Box 5.1 Volcanic hazards at Mount Vesuvius

Located east of Naples in Campania region, southern Italy, Mount Vesuvius is a 1281 m high strato-volcano composed of layers of basaltic tephrite that are up to 300,000 years old. This complex volcanic edifice sits astride a subduction zone and has a history of occasionally violent eruptive and seismic activity. Major eruptions occurred in 5960 BC, 3580 BC, AD 79 and 1631. The AD 79 event involved the emission of 4 km^3 of magmatic products in 19 hours and the creation of a plinian column of tephra and gas that was 32 km high. About 18,000 people may have died in this event, which deposited 3 m of ash on the city of Pompeii and, by pyroclastic flow, 23 m on Herculaneum.

The last eruption occurred in March 1944, and since then, the volcano has been dormant. The time length of its repose is likely to be correlated positively with the strength of the next eruption. The probability of this has been calculated per ten-year period as 0.099 for a VEI-3 event (a volcanic explosivity index of 3, classified as a 'violent strombolian eruption'), 0.017 for VEI-4 ('subplinian') and 0.003 for VEI-5 ('plinian'). According to vulcanological simulations, the last of these would generate a blast column 11–16 km high, which would consist of 5–10 per cent gas and 20 per cent tephra by weight. It would have an initial temperature of 1000°C, a duration of 3–12 hours and a diffusion rate of 3000 m^2/s^{-1}.

About 700,000 people now live within a 15 km radius of the summit of Mount Vesuvius, mostly in the arc of towns on the southern side, which stretches from Torre Annunziata in the east to the Barra district of Naples in the west. Among these settlements, the urban area of Portici (1990 population 67,824) has a density of more than 17,000 people per km^2, five times that of central Milan. Yet in 1631 4000 people died there in a pyroclastic flow that marked the onset of three centuries of intermittent eruptions.

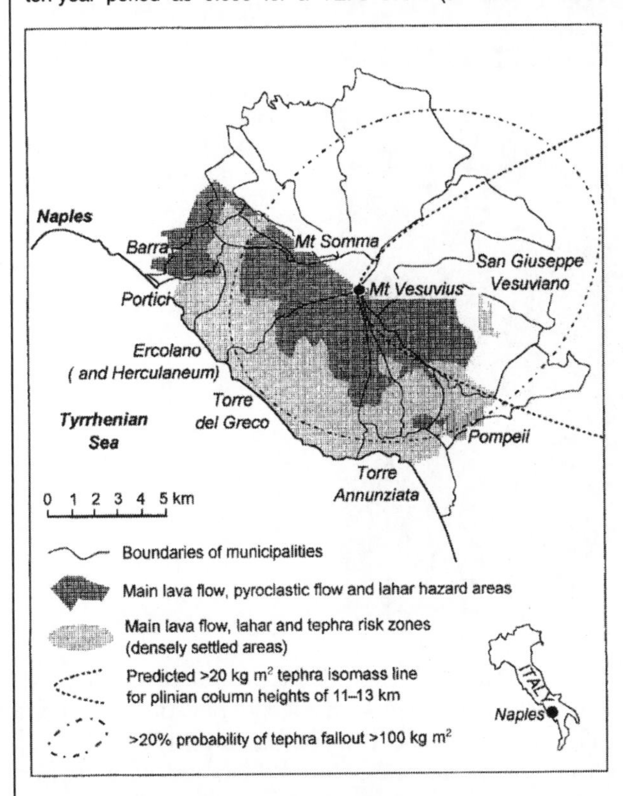

Figure 5.3 Volcanic hazards in the circum-Vesuvian area of southern Italy.

Source: Compiled from various sources – see text and bibliography.

Risk levels depend on the geographical pattern of eruptive effects and human settlement (Figure 5.3). High-velocity winds at altitudes of 8–15 km would blow ash from a vertical column predominantly eastwards. Most buildings would collapse only under weights of deposited ash in excess of 100 kg/m^2, which would be the case in areas that are very limited yet would still encompass several towns with a combined population of at least 51,000. The less populated areas to the north and northwest are shielded by a rampart, which is the remnant of the proto-Vesuvian Somma caldera. On the other hand, the coastal settlements to the south and southeast would bear the full brunt of lava flows, pyroclastic flows, faulting and localised tephra deposition. It has been estimated that an unexpected eruption might take 15,000–20,000 lives, especially as road congestion would probably be immediate and total. However, the volcano is intensively monitored, and detailed plans have been made to evacuate up to 1 million people by every available means, including sea transport.

Plates 5.1 and 5.2 The lower part of the 1992–3 basaltic lava flows on the eastern flanks of Mount Etna. In plate 5.1 (top) the town of Zafferana Etnae can be seen in the distance, approximately 500 m from the distal end of the flow. In the foreground is a collapsed lava tube. Plate 5.2 (bottom) shows where the lava stopped in the garden of a house on the outskirts of the town. Local culture attributes the cessation of the flow to divine intervention by the Madonna of Providence, whose statue was brought in procession to this point, but science attributes it to the lava modification experiments and the end of the eruption that caused the flow.

thirty-seven), and the pattern of valleys down which lava might flow (44 per cent of land below 2000 m is susceptible; Duncan *et al.* 1981). The response to this hazard has included one of the most ambitious and technically demanding lava flow diversion experiments ever mounted. From 14 December 1991 to 30 March 1993, in 473 days of continuous eruption, Etna disgorged 250 million m^3 of lava over an area of 7 km^2. In order to protect the 7000 inhabitants of Zafferana Etnea, artificial channels were dug and lava flows were dammed with 370,000 m^3 of earth. Complicated blocking tactics were used and lava levees were thinned with 7000 kg of explosives in order to retard the flow (Barberi *et al.* 1992).

There are several reasons why volcanic hazards zonation is eminently practicable. One is that many volcanoes offer a clear series of precursory signs of impending eruption, while another is that the characteristic styles, locations and frequencies of eruptions can be deduced from stratigraphic evidence. This means that zonal maps can be constructed on the basis of knowledge of the prevailing volcanic processes that constitute the hazard (Martinelli 1991). Thus hazard maps have been made of ash falls on Etna and Vesuvius in Italy (Gasparini 1993), and of lahars, pyroclastic blasts and ash falls on Nevado del Ruiz, Colombia (Parra and Cepeda 1990). General hazards were mapped at the Soufriere Hills volcano on Montserrat in the Caribbean almost a decade before the repeated eruptions of 1997, which were dominated by pyroclastic flows, vertical blasts and ash falls (Wadge and Isaacs 1988). However, the mere presence of a hazard map does not necessarily mean that there will be an adequate response in terms of land use and civil protection, as the awful case of the November 1985 eruption of Nevado del Ruíz demonstrated so graphically: 23,000 people were killed by lahars, and yet an accurate and comprehensive hazard map had existed for some months previously (Voight 1990).

Remote sensing has proved to be invaluable as a means of generating a mappable overview of seismic and volcanic hazards (Murphy 1994). Further advances have been made using geographic information systems (GIS). For example, Emmi and Horton (1993) used a GIS of Salt Lake County, Utah, to assess the spatial variation of earthquake risk in terms of exposure period, intensity of ground shaking and probability of earthquake occurrence. They related these factors to the vulnerability of the built environment (by constructing an inventory of buildings and a series of earthquake engineering damage functions), the pattern of building occupation and the expected nature of casualties. Simulation of spatial patterns enabled a sensitivity analysis to be conducted in order to assess the model's limitations (Emmi and Horton 1995).

TSUNAMIS

Tsunamis are included here because they are mainly caused by earthquakes and volcanic eruptions. One aspect of tsunami research has been the compilation of catalogues that list known events in the world's marine basins, for example the Pacific (Lockridge 1988) and the Mediterranean (Soloviev 1990). These enable hazard assessments to be refined as new data are added, especially with respect to the recurrence intervals of tsunamis of given magnitudes. For instance, in eastern Honshu (Japan), run-up studies have been carried out since the 1930s, and it is known that 10 m high tsunamis have a return period of only ten years. Even for areas with limited and infrequent tsunamis, such as the Italian peninsula, it has been possible to map the hazard quite successfully (Tinti 1991).

The other pertinent aspect of tsunami research is warning, which is well developed only in the Pacific basin, through a regional mechanism, the Pacific Tsunami Warning System (Pararas-Carayannis 1986), and a rapid reaction local system, Project THRUST (Bernard 1991). Several geographical problems have been experienced with the PTWS. It must cover enormous areas and be able to monitor the progress of waves that travel through open ocean water at the speed of a cruising jet liner but which are not detectable to the naked eye until they make landfall. For adequate monitoring, instruments must be deployed in remote and widely scattered locations, and the network must be dense enough to detect tsunamigenic events rapidly and efficiently in the widest possible variety of seismic, volcanic and tectonic settings. Perhaps it was the apparent remoteness of the problem that led Canada to withdraw briefly from among the twenty-three nations that participate in the PTWS. This proved to be a false economy, and it later rejoined. However, Pacific-wide warning is not particularly effective in the case of near-field tsunamis (i.e. when the interval between tsunamigenesis and the arrival of destructive waves is less than, say, twenty minutes), and for these, Project THRUST was developed on the basis of satellite and microcomputer technology. This is a particularly important development: worldwide, of the 53,000 coastal residents who died in ninety-four tsunamis that occurred over the period 1900–94, 99 per cent were located within 400 km of the point of tsunamigenesis, which was usually an earthquake epicentre. The PTWS gives a minimum alert time of one hour

Box 5.2 Earthquake monitoring at Parkfield, California

In 1976, the prospects for earthquake prediction looked rosy, especially as the Chinese had claimed a major success at Haicheng the previous year. Yet two decades later, little practical progress seemed to have been made, despite vast improvements in both monitoring technologies and the understanding of earthquake source mechanisms.

Statistical studies of Californian earthquakes revealed that the southern part of the state has an 86 per cent chance of experiencing a magnitude 7 tremor by the year 2024. One locality that bears a particularly high risk level is the town of Parkfield (population 34), which sits on a part of the San Andreas fault that periodically accumulates strain and releases it in earthquakes of magnitude 6 or more that have an average recurrence interval of about twenty-two years. The US Geological Survey therefore selected it for intensive monitoring of earthquake precursors. Several hundred instruments were installed in the local area (Figure 5.4), including 130 seismometers and accelerometers, eighty geodolite lines, nineteen alignment arrays, eighteen water well sampling sites, thirteen creep meters, seven dilatometers, six soil hydrogen meters and four tiltmeters. Many of these transmit data continuously to the USGS regional headquarters at Menlo Park near San Francisco.

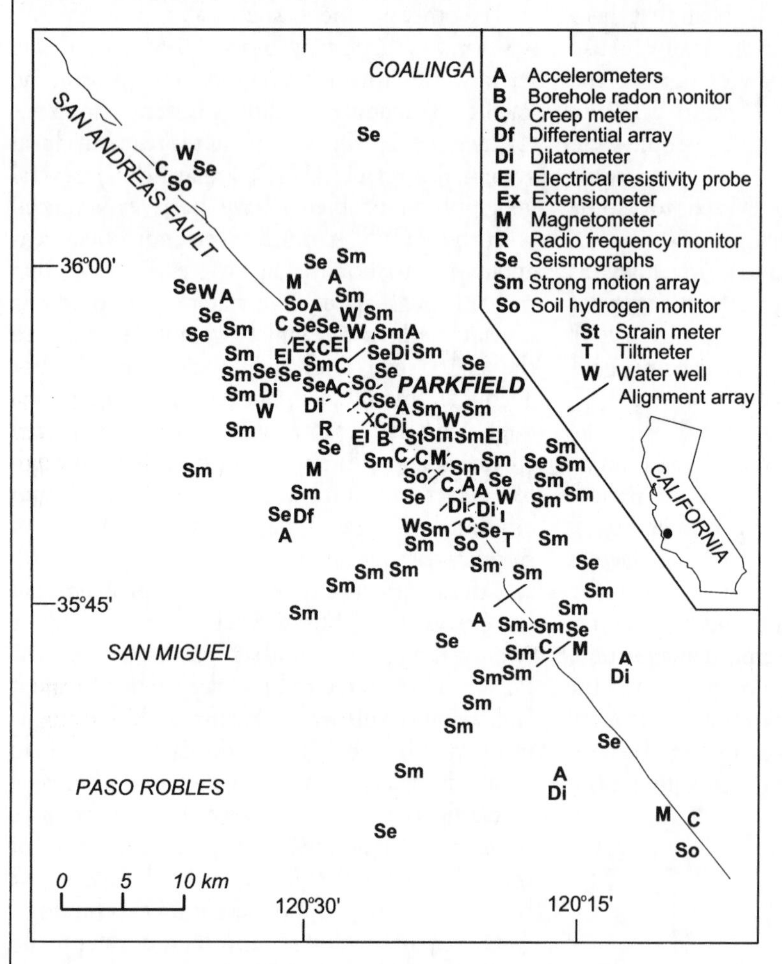

Figure 5.4 Distribution of earthquake-monitoring instruments on the San Andreas Fault in the Parkfield area of California.

Source: After Bakun *et al.* 1988.

The Parkfield Earthquake Prediction Experiment relies on two principles. First, it recognises that seismic precursors are sufficiently complex to require simultaneous interpretation of many geophysical phenomena. Second, earthquake advisories require good communication and understanding between scientists, emergency managers and the public. In 1992 and 1993, considerable mass media interest was stimulated when the USGS issued advisories that warned of an impending earthquake at Parkfield, and substantial preparations were therefore made by the California Office of Emergency Services. However, no earthquake occurred, which is perhaps just as well, as the residents of the

Parkfield area had done little to prepare for the predicted tremors.

Could the Parkfield Prediction Experiment thus be judged a failure? Not exactly: a national working group that was convened to evaluate it argued that Parkfield remains the best place in the country to 'trap' an earthquake and monitor any possible precursors. Although the short-term results had not been encouraging, the long term promises a rich harvest of useful scientific data. However, the lesson of 1992–3 is that communication between scientists and the public needs to be improved, as too much emphasis was given to short-term prediction, and the public had little appreciation of the wider goals of the experiment. Moreover, the impact of false alarms was underestimated.

In conclusion, short-term earthquake prediction remains an elusive goal, as most earthquake faulting mechanisms are unique, complex and, of course, hidden deep in the ground. Yet the January 1995 Great Hanshin earthquake at Kobe in Japan produced anomalies in strain rate, groundwater discharge, and the radon and chlorine content of groundwater, which began three months before the earthquake and were manifest at four locations 20–50 km from the epicentre. It is an open question as to whether they could have been recognised before the main shock and interpreted in such a way as to protect the population.

throughout the Pacific basin and ten minutes at the regional scale. This is insufficient at the local scale, but it is compensated for by THRUST, which has achieved a seventeen-second average response time (Bernard 1991).

MANAGEMENT OF VOLCANIC AND SEISMIC EMERGENCIES

Although earthquakes cannot be predicted accurately in the short term, much effort has been invested in monitoring their precursors (Rikitake 1984), and thus the methodology and technology have improved progressively. If precursors are insufficiently clear for prior warning of major earthquakes, then it is possible that once the shaking begins sensitive equipment, including computer systems and fast trains (Nakamura and Tucker 1988), can be shut down before damage is done.

The prediction of volcanic eruptions is more feasible and relies on a variety of instruments, including infrared sensors for monitoring heat emissions, which are mainly the preserve of satellite remote sensing (Rothery 1992). However, Tilling and Lipman (1993) argued that vulcanologists rely too much on pattern recognition (the empirical approach to predicting eruptions) and not enough on the understanding of source mechanisms. On the other hand, Martinelli (1991) suggested that seismic signals are the best guide to impending eruption, especially if they can be related to other precursors. As monitoring is the key to prediction, the US Geological Survey has developed a portable observation network and associated training scheme that can be used anywhere in the world where volcanic hazards are serious. Be this as it may, when a volcanic emergency occurs, it may last for months and require long-term evacuation (UNDRO/UNESCO 1985), as was the case with the Campi Flegrei bradyseismic activity (i.e. coastal vulcanism without surface eruption, composed mostly of a dome-like uplift) which affected the city of Pozzuoli, west of Naples, over 1983–5 and necessitated the permanent evacuation of 55,000 people (Zelinsky and Kosinski 1991).

SOCIAL RESEARCH

Geographers have played a minor but significant role in the process of planning for volcanic emergencies (Chester 1993), and they have played a more considerable part in planning against earthquakes. This has involved both hazard analysis (Beatley and Berke 1990) and policy studies (Berke and Beatley 1992). The latter have included analyses of the 'window of opportunity' for policy formulation and implementation that opens when a disaster has occurred recently and public opinion demands that something be done (Solecki and Michaels 1994). In the absence of such events, planning can be based on a synthetic form of 'reality' by using scenarios that prefigure the damage and casualties that will arise from an earthquake disaster. Thus Borchardt (1991) used seismic intensity distribution maps as a basis for his Californian scenarios. At the widest scale, earthquake hazards reduction depends on planning at the international level, under the aegis of the International Decade for Natural Disaster Reduction (Bolt 1991). However, seismic planning does not always involve a rational and concerted response to objective risk: Berke *et al.* (1989) surveyed communities in the United States to find out whether they adopted earthquake protection measures, or not. They found that the planning process, rather than the community context, tended to determine whether mitigation would occur. Often the measures were chosen as part of other needs, and hazard mitigation was a secondary consideration.

Social scientists have devoted considerable attention to the political and economic aspects of earthquake hazards. General surveys by Petak and Atkisson (1982) and Alesch and Petak (1986) have clarified the situation in California, especially with respect to the question of how building codes are updated and enforced. It is axiomatic that the major advances in seismic safety follow the principal events that cause damage or highlight risk. Thus, the 1933 Long Beach earthquake led to the first serious attempts to pass and apply anti-seismic building codes in California (it damaged unreinforced masonry buildings very substantially, including schools, where there would have been a considerable death toll if the earthquake had occurred at a different time of day). Also, the near failure of the Lower San Fernando Dam in the earthquake of 1971 (magnitude 6.7), in which liquefaction lowered the crest by 9 metres, led to new rules for dam inspection and certification and a new interest in engineering risk analysis. As a practical result, the Los Angeles Dam was built to withstand forces three times as large as those that its predecessor, the San Fernando Dam, was designed for. In the 1994 Northridge earthquake the older dam, which had been retained as a back-up, was again badly affected, while the new barrage was not seriously damaged (USGS 1996).

Increasing interest in the economics of earthquakes has led to some overall studies that outline the inverse multiplier effects and the taxation burdens associated with a major seismic disaster (e.g. US NRC 1992). Losses are rising considerably, and hence there is also considerable interest in techniques of loss estimation. The Hanshin earthquake of January 1995 at Kobe in Japan is estimated to have cost an order of magnitude more than the 1989 Loma Prieta (California) event (US$131.5 billion against US$12 billion). However, at least part of the increase may be the result of improved accounting procedures that take more heed of the hidden costs than was previously the case. Nevertheless, much soul searching is now going on regarding how to pay for the losses. Only a minor part is likely to be covered by indemnities, but the burden on the insurance and reinsurance industries is becoming difficult to bear. There is thus considerable debate over the best strategy for sustainable earthquake insurance (Mittler 1990), although the question of whether individual householders are motivated to purchase it is a complex one that involves work on both hazard perception and economics. In this context, Palm (1995) found that, although the purchase of insurance by California households has risen substantially and now exceeds 50 per cent in some counties, the pattern of adoption is more related to the perception of risk than it is to the actual severity of the hazard.

According to Ohta and Ohashi (1985), the main factors that govern people's immediate response to earthquakes are seismic intensity, spatial conditions, family or other group composition, age and sex. However, behaviour is also intimately linked to hazard perception. Generally, this is less acute and behaviour less self-protective among poor, disadvantaged and minority groups (Bolin 1990), although it has also been linked to personality factors (Simpson-Housley and Bradshaw 1978). In the heat of the moment, perception can be seriously wrong and hence give rise to maladaptive behaviour, which in extreme cases may lead to injury that otherwise could have been avoided (Alexander 1990).

Perhaps one of the most contentious social

issues is that of public reaction to earthquake predictions, especially if these are made without an adequate scientific basis. Thus, much attention was given to the Iben Browning earthquake prediction in southern Illinois in 1992 (Spence *et al.* 1993). The apparent plausibility of this charlatan prediction, and the willingness of the mass media and a few scientists to take it seriously, undermined the public status of genuine scientific efforts to forecast seismic activity. Official seismological institutions tried to rebut the prediction by ignoring it, which turned out to be a mistake, as the public thought they had something to hide (Stevens 1993). In the same vein, social scientific perspectives on journalistic coverage of earthquakes suggest a lack of faith in the news media's ability to portray events accurately, although some researchers have argued that the media can successfully be induced to play a valuable role as suppliers of emergency management information to the public (Scanlon *et al.* 1985). Although the Western news media's coverage of earthquake disasters in third world countries is not explicitly biased against such places, it seems to be strongly related to the number of reporters on the ground and their contacts in the local area, and to readers' familiarity with the area in question (Gaddy and Tanjong 1986).

Some fine studies of post-earthquake reconstruction have been conducted in the tradition of urban geography. Of particular note are Robert Geipel's longitudinal study of post-earthquake social change in the Friuli region, northern Italy (Geipel 1982; 1990) and William Mitchell's studies of the aftermaths of earthquakes in Turkey (Mitchell 1976; 1977). A theoretical basis to reconstruction was given by Robert Kates' and David Pijawka's comparative study of the earthquakes in San Francisco in 1906, Alaska in 1964 and Nicaragua in 1972 (Kates and Pijawka 1977). Their model charts the progress of recovery through four stages, including one of replacement reconstruction and one of post-reconstruction urban development. It has been taken up again and adapted by development studies specialists (Kirkby *et al.* 1997). But when Hogg (1980) applied this model to the aftermath of the two earthquakes that occurred in Friuli in 1976 she found that the pace, direction and relative success of reconstruction varied geographically in relation to the political and economic ties between communities. It also varied temporally, as in Friuli there were two main shocks separated by six months and hence the reconstruction that had begun at the time of the second earthquake was abruptly set back (see Figure 5.5). Damaged settlements that had effective leaderships and good political ties to the centres of power were the first to be reconstructed. It therefore appeared misleading to use the Kates and Pijawka model to characterise reconstruction in a spatially aggregated way. Moreover, Geipel (1990) found that reconstruction in Friuli led to disillusionment and debt as a result of overambitious planning, which strained the social system and forced the pace of change.

Another factor that characterises the geography of post-seismic reconstruction is geographical inertia. Thus, Mileti and Passerini (1996) argued that there are six reasons why reconstruction tends to occur *in situ*: first, survivors want to return to normal as soon as possible; second, damage is seldom extensive enough to warrant wholesale relocation; third, cultural values bind communities to specific places; fourth, most procedures deal with individual structures and property owners, not with aggregate groups; fifth, funds are seldom sufficient to allow complete relocation; and finally, planning is not usually adequate to the task of complete relocation.

In contrast to earthquakes, relatively little has been done to study the social effects of volcanic eruptions (Peterson 1988). However, thorough studies were made of Mount St Helens after the May 1980 eruption (Perry and Lindell 1990). In a study of ash fall hazards, Warrick *et al.* (1981) found that volcanic risks can be so widely dispersed and so infrequently manifest that they may be poorly perceived by potential victims. This situation is typical of high-consequence, low-frequency hazards in general.

The final aspect of applied social studies of earthquakes and volcanic eruptions concerns

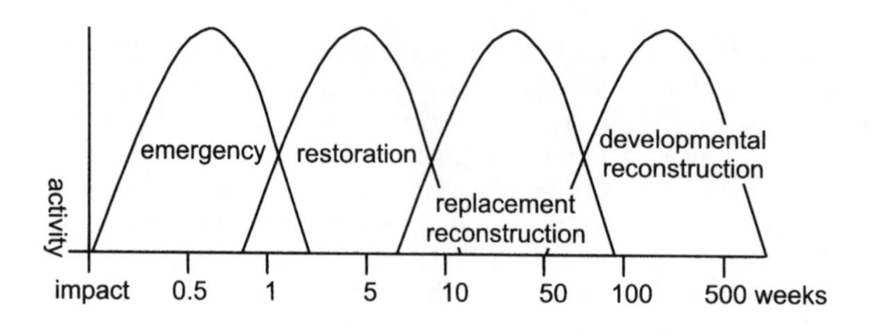

Figure 5.5 Comparison between Kates' and Pijawka's (1977) model of the stages of recovery after disaster (A) and Hogg's (1980) application of this schema in the Friuli region, Italy (B), where in 1976 two damaging earthquakes occurred with a six-month time interval between them.

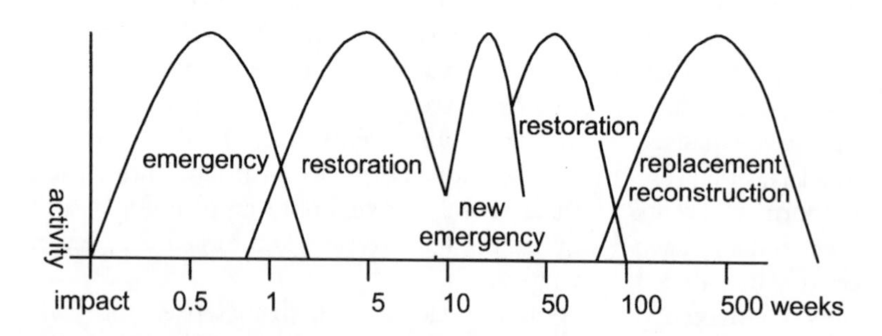

education and training. Both are an important part of the International Decade for Natural Disaster Reduction (IDNDR 1990–2000) and have given rise to a growing methodological debate. One aspect of this is that substantial new prospects have been opened up by the Internet, which has facilitated both distance learning and the acquisition and use of data.

CONCLUSION

So compelling are the phenomena involved that natural hazards studies are almost by definition applied forms of research. Indeed, the field is dominated by practical problems, such as how to provide a safer environment and apportion scarce resources for mitigation and emergency management. In disasters, time is the backbone of events and geographical space is their medium of expression (Alexander 1995). However, even though spatial relationships are fundamental to the interpretation of many processes in natural catastrophe, the role of geographers has been muted. Excellent studies of urban and social change during reconstruction, and of the geomorphology of tectonic and volcanic hazards, have been conducted by geographers, but their work has been somewhat overshadowed by that of geophysicists and sociologists. Yet the human ecological tradition of hazards studies, which stretches back more than fifty years, stems directly from the work of geographers (White 1973).

One suspects that in the field of natural hazards in general, and studies of volcanic and seismic hazards in particular, a great geographical challenge has not been met. There is considerable scope for the formulation of general spatial models that integrate the physical hazard, in terms of the distribution of risks and impacts, with the human response in terms of the patterns of vulnerability, impact and emergency response. Dynamic spatial patterns are thus created by distance decay and the temporal evolution of disaster scenarios. Yet almost no geographers are actively engaged in spatial analysis of data on disaster impacts, at least not with a view to the creation of general models. This is a pity, as spatial regularities are undoubtedly waiting to be discovered, and robust spatial models would greatly aid in forecasting the pattern of impacts, damage and casualties to be expected when the earth shakes or volcanoes erupt.

GUIDE TO FURTHER READING

Alexander, D.E. (1993) *Natural Disasters*. London: UCL Press, and New York: Chapman & Hall. Chapter 2 describes the nature of volcanic and seismic hazards, and the socio-economic reactions to them, while Chapters 5–9 include information on impacts, and their management and mitigation.

Blong, R.J. (1984) *Volcanic Hazards: A Sourcebook on the Effects of Eruptions*. Orlando, Florida: Academic Press. An all-embracing book on the impact of eruptions and what can be done to mitigate them.

Bolton, P.A. *et al.* (eds) (1994–8) The Loma Prieta, California, earthquake of October 17, 1989. *US Geological Survey Professional Papers* 1550–3. A comprehensive modern analysis of a major earthquake disaster, which extends from the seismic to the social, economic and medical aspects of the catastrophe.

Chester, D.K. (1993) *Volcanoes and Society*. London: Edward Arnold. A geographer's eclectic approach to volcanoes, which covers a range of themes that stretches from geophysics to hazard management and economic aspects.

Palm, R. (1995) The Roepke lecture in economic geography: catastrophic earthquake insurance: patterns of adoption. *Economic Geography* 71, 119–131. A summary of a series of geographical field analyses conducted in the western USA in order to study the diffusion process for one non-structural seismic risk mitigation technique.

Tilling, R.I. (1989) Volcanic hazards and their mitigation: progress and problems. *Reviews of Geophysics* 27(2), 237–69. A review of types of volcanic hazard and techniques of monitoring them and forecasting eruptions.

REFERENCES

Alesch, D.J. and Petak, W.J. (1986) *The Politics and Economics of Earthquake Hazard Mitigation*. Monograph no. 43, Boulder, Colorado: Natural Hazards Research and Applications Information Center.

Alexander, D.E. (1989) Spatial aspects of earthquake epidemiology. *Proceedings of the International Workshop on Earthquake Injury Epidemiology for Mitigation and Response*, Baltimore: Johns Hopkins University Press, 82–94.

Alexander, D.E. (1990) Behaviour during earthquakes: a southern Italian example. *International Journal of Mass Emergencies and Disasters* 8(1), 5–29.

Alexander, D.E. (1993) *Natural Disasters*. London: UCL Press, and New York: Chapman & Hall.

Alexander, D.E. (1995) A survey of the field of natural hazards and disaster studies. In A. Carrara and F. Guzzetti (eds) *Geographical Information Systems in Assessing Natural Hazards*, Dordrecht: Kluwer, 1–19.

Ambraseys, N.N. (1992) Long-term seismic hazard in the eastern Mediterranean region. In G. McCall, D. Laming and S. Scott (eds) *Geohazards: Natural and Man-Made*, London: Chapman and Hall, 83–92.

Bakun, W.H. *et al.*, (1988) The Parkfield earthquake prediction experiment in central California. *Earthquakes and Volcanoes* 20(2), 41–91.

Barberi, F., Carapezza, M.L., Valenza, M. and Villari, L. (1992) The control of lava flow during the 1991–1992 eruption of Mount Etna. *Journal of Volcanology and Geothermal Research* 56, 1–34.

Beatley, T. and Berke, P. (1990) Seismic safety through public incentives: the Palo Alto seismic identification program. *Earthquake Spectra* 6(1), 57–79.

Berke, P.R. and Beatley, T. (1992) *Planning for Earthquakes: Risk, Politics and Policy*. Baltimore: Johns Hopkins University Press.

Berke, P., Beatley, T. and Wilhite, S. (1989) Influences on local adoption of planning measures for earthquake hazard mitigation. *International Journal of Mass Emergencies and Disasters* 7(1), 33–56.

Bernard, E.N. (1991) Assessment of Project THRUST: past, present, future. *Natural Hazards* 4(2–3), 285–92.

Bernard, A. and Rose Jr, W.I. (1990) The injection of sulphuric acid aerosols into the stratosphere by the El Chichón volcano and its related hazards to the international air traffic. *Natural Hazards* 3(1), 59–68.

Bolin, R. (ed.) (1990) *The Loma Prieta Earthquake: Studies of Short-Term Impacts*. Monograph no. 50, Boulder, Colorado: Program on Environment and Behavior, Institute of Behavioral Sciences.

Bolt, B.A. (1991) International earthquake hazard reduction program for IDNDR underway. *Stop Disasters* 4, 12.

Borchardt, G. (1991) Preparation and use of earthquake planning scenarios. *California Geology* 44(9), 195–203.

Brambati, A., Faccioli, E., Carulli, G.B., Cucchi, F., Onofri, R., Stefanini, S. and Ulcigrai, F. (1980) *Studio di microzonizzazione sismica dell'area di Tarcento (Friuli)*. Trieste, Italy: Friuli–Venezia–Giulia Autonomous Region and University of Trieste.

Brazee, R.J. (1979) Re-evaluation of modified Mercalli intensity scale using distance as determinant. *Seismological Society of America Bulletin* 69(3), 911–24.

Bullard, F.M. (1984) *Volcanoes of the Earth* (2nd edn). Austin, Texas: University of Texas Press.

Casadevall, T.J. (ed.) (1991) First International Symposium on volcanic ash and aviation safety. *US Geological Survey Circular* 1065, 58 pp.

Chester, D.K. (1993) *Volcanoes and Society.* London: Edward Arnold.

Chester, D.K., Duncan, A.M., Guest, J.E. and Kilburn, C.R.J. (1985) *Mount Etna: The Anatomy of a Volcano.* London: Chapman & Hall.

Choudhury, G.S. and Jones, N.P. (1996) Development and application of data collection forms for post-earthquake surveys of structural damage and human casualties. *Natural Hazards* 13(1), 17–38.

Cooke, R.U. and Doornkamp, J.C. (1990) *Geomorphology in Environmental Management: A New Introduction* (2nd edn). Oxford: Clarendon Press.

Crandell, D.R. and Mullineaux, D.R. (1975) Technique and rationale of volcanic-hazards appraisal in the Cascades Range, northwestern United States. *Environmental Geology* 1, 23–32.

Degg, M.R. (1989) Earthquake hazard assessment after Mexico (1985). *Disasters* 13(3), 237–46.

Degg, M.R. (1992) The ROA Earthquake Hazard Atlas project: recent work from the Middle East. In G. McCall, D. Laming and S. Scott (eds), *Geohazards: Natural and Man-Made.* London: Chapman & Hall, 93–104.

Del Moral, R. and Wood, D.M. (1993) Early primary succession on a barren volcanic plain at Mount St. Helens, Washington. *American Journal of Botany* 80, 981–91.

Doornkamp, J.C. and Han Mukang (1985) Morphotectonic research in China and its application to earthquake prediction. *Progress in Physical Geography* 9, 353–78.

Duncan, A.M., Chester, D.K. and Guest, J.E. (1981) Mount Etna volcano: environmental impact and problems of volcanic prediction. *Geographical Journal* 147(2), 164–78.

Emmi, P.C. and Horton, C.A. (1993) GIS-based assessment of earthquake property damage and casualty risk, Salt Lake County, Utah. *Earthquake Spectra* 9(1), 11–33.

Emmi, P.C. and Horton, C.A. (1995) A Monte Carlo simulation of error propagation in a GIS-based assessment of seismic risk. *International Journal of Geographical Information Systems* 9(4), 447–61.

Fahmi, K.J. and Alabbasi, J.N. (1989) Seismic intensity zoning and earthquake risk mapping in Iraq. *Natural Hazards* 1(4), 331–40.

Francis, P.W. (1994) *Volcanoes: A Planetary Perspective.* Oxford: Clarendon Press.

Gaddy, G.D. and Tanjong, E. (1986) Earthquake coverage by the western press. *Journal of Communication* 36(2), 105–12.

Gasparini, C., De Rubeis, V. and Tertulliani, A. (1992) A method for the analysis of macroseismic questionnaires. *Natural Hazards* 5(2), 169–77.

Gasparini, P. (1993) Research on volcanic hazards in Europe. *Science* 260, 1759–60.

Geipel, R. (1982) *Disaster and Reconstruction: The Friuli (Italy) Earthquakes of 1976* (trans. Wagner, P.). London: Allen & Unwin.

Geipel, R. (1990) *The Long-Term Consequences of Disasters: The Reconstruction of Friuli, Italy, in its International Context, 1976–1988.* Heidelberg: Springer-Verlag.

Giardini, D. (1992) The Global Seismic Hazard Assessment Programme (GSHAP). *Stop Disasters* 8, 11.

Gori, P. and Hays, W.W. (eds) (1992) Assessment of regional earthquake hazards and risk along the Wasatch Front, Utah. *US Geological Survey Professional Paper* 1500A-J.

Handler, P. (1989) The effect of volcanic aerosols on global climate. *Journal of Volcanology and Geothermal Research* 37(3/4), 233–49.

Hays, W.W. (1984) Technical problems in the construction of a map to zone the earthquake ground-shaking hazard in the United States. *Engineering Geology* 20(1/2), 13–24.

Hewitt, K. (1983) Seismic risk and mountain environments: the role of surface conditions in earthquake disaster. *Mountain Research and Development* 3(1), 27–44.

Hogg, S.J. (1980) Reconstruction following seismic disaster in Venzone, Friuli. *Disasters* 4(2), 173–85.

IFRCRCS (1997) *World Disasters Report 1997.* Oxford: International Federation of Red Cross and Red Crescent Societies and Oxford University Press.

Iglesias, J. (1989) The Mexico earthquake of September 19, 1985: seismic zoning of Mexico City after the 1985 earthquake. *Earthquake Spectra* 5(1), 257–71.

Ihnen, S.M. and Hadley, D.M. (1987) Seismic hazard maps for Puget Sound, Washington. *Seismological Society of America Bulletin* 77(4), 1091–109.

Kates, R.W. and Pijawka, D. (1977) From rubble to monument: the pace of reconstruction. In J. Haas, M. Kates and M. Bowden (eds) *Disaster and Reconstruction,* Cambridge, Mass.: MIT Press, 1–23.

Kijko, A., Skordas, E., Wahlström, R. and Mäntyniemi, P. (1993) Maximum likelihood estimation of seismic hazard for Sweden. *Natural Hazards* 7(1), 41–57.

Kirkby, J., O'Keefe, P., Convery, I. and Howell, D. (1997) On the emergence of complex disasters. *Disasters* 21(2), 177–80.

Kobayashi, Y. (1981) Causes of fatalities in recent earthquakes in Japan. *Journal of Disaster Science* 3, 15–22.

Kockelman, W.J. and Brabb, E.E. (1979) Examples of seismic zonation in the San Francisco Bay region. *US Geological Survey Circular* 807, 73–84.

Kotoda, K., Wakamatsu, W. and Midorikawa, S. (1988) Seismic microzonation on soil liquefaction potential based on geomorphological land classification. *Soils and Foundations* 28(2), 127–43.

Lapajne, J.K., Motnikar, B.S. and Zupancvicv, P. (1997) Preliminary seismic hazard maps of Slovenia. *Natural Hazards* 14(2–3), 155–64.

Levret, A., Backe, J.C. and Cushing, M. (1994) Atlas of macroseismic maps for French earthquakes with their principal characteristics. *Natural Hazards* 10(1–2), 19–46.

Lockridge, P.A. (1988) Historical tsunamis in the Pacific basin. In M.I. El-Sabh and T.S. Murty (eds) *Natural and Man-Made Hazards*, Dordrecht: Reidel, 171–81.

Malkawi, A.I.H., Liang, R.Y., Nusairat, J.H. and Al-Homoud, A.S. (1995) Probabilistic seismic hazard zonation of Syria. *Natural Hazards* 12(2), 139–51.

Mäntyniemi, P. and Kijko, A. (1991) Seismic hazard in East Africa: an example of the application of incomplete and uncertain data. *Natural Hazards* 4(4), 421–30.

Martinelli, B. (1991) Understanding triggering mechanisms of volcanoes for hazard evaluation. *Episodes* 14(1), 19–25.

Mileti, D.S. and Passerini, E. (1996) A social explanation of urban relocation after earthquakes. *International Journal of Mass Emergencies and Disasters* 14(1), 97–110.

Mitchell, W.A. (1976) Reconstruction after a disaster: the Gediz earthquake of 1970. *Geographical Review* 66, 296–313.

Mitchell, W.A. (1977) Partial recovery and reconstruction after disaster: the Lice case. *Mass Emergencies* 2, 233–47.

Mittler, E. (1990) Evaluating alternative national earthquake insurance programs. *Earthquake Spectra* 6(4), 757–78.

Muñoz, A.V. (1989) Assessment of earthquake hazard in Panama based on seismotectonic regionalization. *Natural Hazards* 2(2), 115–32.

Murphy, J.M. and Wesnousky, S.G. (1994) A post-earthquake re-evaluation of seismic hazard in the San Francisco Bay area. The Loma Prieta, California, earthquake of October 17, 1989: strong ground motion. *US Geological Survey Professional Paper* 1551A, 255–72.

Murphy, W. (1994) Remote sensing applications for seismic hazard assessment. In Wadge, G. (ed.) *Natural Hazards and Remote Sensing*, London: Royal Society and Royal Academy of Engineering, 34–8.

Musson, R.M.W. and Winter, P.W. (1997) Seismic hazard maps for the U.K. *Natural Hazards* 14(2–3), 141–54.

Nakamura, Y. and Tucker, B.E. (1988) Earthquake warning system for Japan Railways' bullet train: implications for disaster prevention in California. *Earthquakes and Volcanoes* 20(4), 140–55.

Nash, D.B. (1981) FAULT: a Fortran program for modelling the degradation of active normal fault scarps. *Computers and Geosciences* 7, 249–66.

Ohta, Y. and Ohashi, H. (1985) Field survey of occupant behaviour in an earthquake. *International Journal of Mass Emergencies and Disasters* 3(1), 147–60.

Ollier, C. (1988) *Volcanoes* (2nd edn). Oxford: Blackwell.

Orozova-Stanishkova, I. and Slejko, D. (1994) Seismic hazard of Bulgaria. *Natural Hazards* 9(1/2), 247–71.

Page, R.A., Blume, J.A. and Joyner, W.B. (1975) Earthquake shaking and damage to buildings. *Science* 189, 601–8.

Palm, R. (1995) The Roepke lecture in economic geography: catastrophic earthquake insurance: patterns of adoption. *Economic Geography* 71, 119–31.

Panizza, M. (1989) Geomorphological contributions to seismic risk assessment. *Geografia Fisica e Dinamica Quaternaria Supplement* 2, 111–14.

Papazachos, B.C., Papaioannou, Ch.A., Margaris, B.N. and Theodulidis, N.P. (1993) Regionalization of seismic hazard in Greece based on seismic sources. *Natural Hazards* 8(1), 1–18.

Pararas-Carayannis, G. (1986) The Pacific Tsunami Warning System. *Earthquakes and Volcanoes* 18(3), 122–30.

Parra, E. and Cepeda, H. (1990) Volcanic hazard maps of the Nevado del Ruiz volcano, Colombia. *Journal of Volcanology and Geothermal Research* 42(1–2), 117–27.

Perry, R.W. and Lindell, M.K. (1990) *Living with Mount St Helens: Human Adjustment to Volcanic Hazards.* Pullman, Washington: Washington State University Press.

Petak, W.J. and Atkisson, A.A. (1982) *Natural Hazard Risk Assessment and Public Policy: Anticipating the Unexpected.* New York: Springer-Verlag.

Peterson, D.W. (1988) Volcanic hazards and public response. *Journal of Geophysical Research* 93B, 4161–70.

Qamar, A. and Meagher, K.L. (1993) Precisely locating the Klamath Falls earthquakes. *Earthquakes and Volcanoes* 24(3), 129–39.

Rikitake, T. (1984) Earthquake Precursors. In T. Rikitake (ed.) *Earthquake Prediction*, Tokyo: Terra Scientific Publishing, and Paris: UNESCO Press, 3–20.

Rothery, D.A. (1992) Monitoring and warning of volcanic eruptions by remote sensing. In G. McCall, D. Laming and S. Scott (eds) *Geohazards: Natural and Man-Made*, London: Chapman & Hall, 25–32.

Scanlon, J., Alldred, S., Farrell, A. and Prawzick, A. (1985) Coping with the media in disasters: some predictable problems. *Public Administration Review* 45 (special issue), 123–33.

Seed, H.B., Murarka, R., Lysmer, J. and Idriss, I.M. (1976) Relationships of maximum acceleration, maximum velocity, distance from source, and local site conditions for moderately strong earthquakes. *Seismological Society of America Bulletin* 66, 1323–42.

Simpson-Housley, P. and Bradshaw, P. (1978) Personality and the perception of earthquake hazard. *Australian Geographical Studies* 16, 65–72.

Slosek, J. (1986) The spatial distribution of mortality and morbidity caused by Earthquakes. Amherst, Mass.: Department of Geosciences, University of Massachusetts, M.S. thesis.

Smith, K. (1996) *Environmental Hazards: Assessing Risk and Reducing Disaster* (2nd edn). London: Routledge.

Solecki, W.D. and Michaels, S. (1994) Looking through the post-disaster policy window. *Environmental Management* 18(4), 587–95.

Soloviev, S.L. (1990) Tsunamigenic zones in the Mediterranean Sea. *Natural Hazards* 3(2), 183–202.

Spence, W., Herrman, R.B., Johnson, A.C. and Reagor, G. (1993) Responses to Iben Browning's prediction of a 1990 New Madrid, Missouri, earthquake. *US Geological Survey Circular* 1083.

Stevens, J.D. (1993) An association of circumstances: the 1990 Browning earthquake prediction and the Center for Earthquake Research and Information. *International Journal of Mass Emergencies and Disasters* 11(3), 405–20.

Tilling, R.I. and Lipman, P.W. (1993) Lessons in reducing volcano risk. *Nature* 364, 277–80.

Tinti, S. (1991) Assessment of tsunami hazard in the Italian seas. *Natural Hazards* 4(2–3), 267–83.

UNDRO/UNESCO (1985) *Volcanic Emergency Management.* New York: United Nations Press.

USGS (1996) The Los Angeles Dam story. http://www.usgs.gov

US NRC (1992) *The Economic Consequences of a Catastrophic Earthquake.* Washington, DC: Committee on Earthquake Engineering, National Research Council, National Academy Press.

Voight, B. (1990) The 1985 Nevado del Ruiz volcano catastrophe: anatomy and retrospection. *Journal of Volcanology and Geothermal Research* 42(1/2), 151–88.

Wadge, G. and Isaacs, M.C. (1988) Mapping the volcanic hazards from Soufriere Hills Volcano, Montserrat, West Indies, using an image processor. *Geological Society of London Journal* 145(4), 541–52.

Warrick, R.A. *et al.* (1981) *Four Communities Under Ash.* Monograph no. 34, Boulder, Colorado: Natural Hazards Research and Applications Information Center.

White, G.F. (1973) Natural hazards research. In R.J. Chorley (ed.) *Directions in Geography*, London: Methuen, 193–216.

Yücemen, M.S. (1992) Seismic hazard maps for Jordan and vicinity. *Natural Hazards* 6(3), 201–26.

Zelinsky, W. and Kosinski, L.A. (1991) *The Emergency Evacuation of Cities: A Cross-National Historical and Geographical Study.* Savage, Maryland: Rowman & Littlefield.

6

Landslides

Michael Crozier

INTRODUCTION

Landslides are of interest to geographers for three main reasons. First, by eroding, transporting and depositing soil and rock, they represent one of the important geomorphic processes involved in shaping the surface of the Earth. In unstable areas, they may displace up to 2000 m^3/km^2/year (Crozier 1989), severely depleting the soil resource and threatening the sustainability of primary production (Sidle *et al.* 1985). Although they are particularly common in tectonically active mountainous areas, and along river banks and coasts, they may also occur in other areas that have weak material or a susceptible geological structure.

The second reason for geographical interest is that landslides are sensitive indicators of environmental change. As a geomorphic process, a landslide represents a short-term adjustment to disturbance of the natural system. As they take place, they rapidly convert unstable slopes to a more stable condition, allowing other slow-acting processes to assume the role of denudation. In terms of landform evolution, this means that most slopes are stable for most of the time. Thus when landslides occur they are generally responding to some significant change within the natural system. Initiating factors may include tectonic activity, climate change, and natural or human-induced disturbance to the vegetation cover, slope hydrology or slope form. Knowledge of both past and present landslide activity can therefore provide useful information on environmental change. Indeed, there has been a major international research effort aimed at reconstructing past climates and climatic change in Europe, based on landslide evidence preserved in the landscape (Crozier 1997).

The third reason landslides are often studied by geographers is that landslides can present a serious natural hazard (Varnes 1984; Crozier 1996). A full appreciation of hazard requires knowledge not only of the physical process but also of the nature of the threatened society. In a sense, hazards are an aspect of human ecology. They involve interrelationships between physical, social and economic systems; as such, they constitute a field of study in which geographers are able to make a valuable contribution.

This chapter focuses on the principles of landslide hazard and risk assessment. It briefly introduces the physical process and then goes on to discuss different approaches to landslide hazard assessment.

THE PHYSICAL PROCESS

In dealing with landslides, it is important to use a classification that distinguishes between characteristics that are relevant to the intended end-use of the study. The classification should use clearly defined and internationally understood terms. The working party on the World Landslide Inventory (1990) has made an attempt to standardise terminology and defines a landslide simply as: 'the movement of a mass of rock, earth, or debris down a slope.' A more comprehensive definition, which helps to distinguish landslides from the other geomorphological processes, is: 'the downward or outward movement of a mass of slope-forming material under the influence of gravity, occurring on discrete boundaries and taking place initially without the aid of water as a transportational agent.' As this second definition indicates, landslides are more than just a simple downslope movement of material. The three most widely used classifications involving landslides (Sharpe 1938; Varnes 1958; 1978; Hutchinson 1988) separate 'mass movements' (Fairbridge 1968) into two categories: 'subsidence' (which is the vertical sinking of material) and those movements that occur on slopes. These 'slope movements' are then usually divided first into 'landslides', as defined above, and second into the slower, more widespread and ill-defined movements such as 'creep', 'sagging' and 'rebound'. Of all the types of slope movement, it is landslides that have the potential to undergo rapid movement, making them a potentially dangerous form of natural hazard.

Of the many different landslide classifications in existence (Hansen 1984a) the system devised by Varnes (1978) is often preferred because it is simple and easy to apply in the field (Table 6.1). The criteria used to define the landslide types are mechanism of movement, shape of the failure surface, degree of disruption and type of material. These are all characteristics that reflect an aspect of hazard or which need to be known in order to carry out stability analysis. In applying this classification, it is important to remember that the material criterion refers to the original slope material, not to what may subsequently appear in the deposit.

In order to choose the most appropriate method of reducing risk from the landslides (whether prevention, control, avoidance or compensation for loss) it is important to know something about the range of factors that lead to slope failure and how they operate. For example, it may be found that risk can be reduced more cheaply and more effectively by draining groundwater from a slope than by zoning it as unsuitable for use. Comprehensive lists of causative factors are available (Varnes 1958; Cooke and Doornkamp 1990; Crozier 1995), but it is useful to simplify these into categories. One way of doing this is to consider the function that various factors have in changing the conditions of slope stability.

Figure 6.1 introduces some concepts in slope stability. It shows that a slope may pass from a 'stable' condition to a 'marginally stable' condition and finally to the 'actively unstable' condition, where the slope actually fails as a landslide. Whether a slope is likely to move

Table 6.1 Landslide classification

Type of movement			*Type of material*		
			Bedrock	*Engineering soils*	
				Coarse	*Fine*
Falls			Rock fall	Debris fall	Earth fall
Topples			Rock topple	Debris topple	Earth topple
Slides	Rotational	Few units	Rock slump	Debris slump	Earth slump
	Translational	Many units	Rock block-slide	Debris block-slide	Earth block-slide
			Rock slide	Debris slide	Earth slide
Lateral spreads			Rock spread	Debris spread	Earth spread
Flows			Rock flow	Debris flow	Earth flow
Complex	Combination of two or more principal types of movement				

Source: Varnes 1978.

Figure 6.1 Stability factors classified by function.

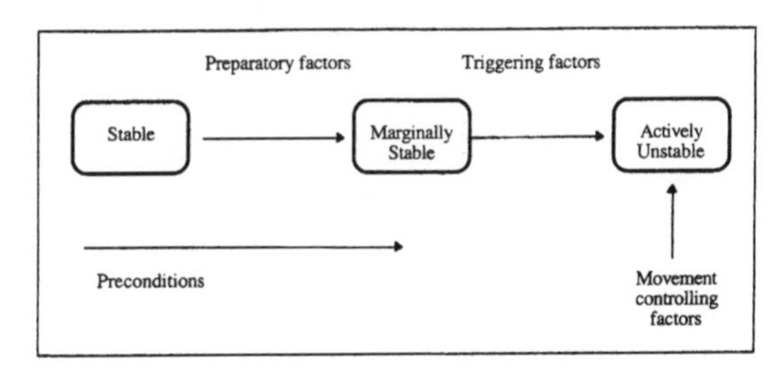

through these stages depends on factors referred to as 'preconditions' (or inherent factors). These may include features of the locality such as susceptible rock structure, weak material and slope form. 'Preparatory factors', on the other hand, are more active and produce changes that make the slope more vulnerable to failure without actually initiating movement. They change the slope from a stable to a marginally stable condition. Some of the most common preparatory factors include deforestation (Plate 6.1), removal of the toe of the slope (either naturally or artificially) and alteration of slope drainage. Some preparatory factors may eventually initiate a landslide, in which case they become 'triggering factors'. The most common triggering factors include rainstorms, earthquake shaking and removal of support from the toe of the slope. Finally, once a landslide begins to occur, 'movement-controlling factors' take over. These may determine, for example, how disrupted the slide becomes, or how far it runs out. All these factors ultimately influence the stress conditions within the slope either by reducing the shear strength or increasing the shear stress (Selby 1993).

IDENTIFYING THE HAZARD

The ultimate goal in landslide hazard assessment is the successful prediction of the place of occurrence of an event, its impact characteristics and its relationship with time (Hansen 1984b). In other words, identifying the threat from landslide hazard means finding out what, where, when and how dangerous? Indeed, this goal is common to the assessment of all hazards. In the conventional

Plate 6.1 Rainfall-triggered soil landslides in a part of New Zealand that has been deforested within the last 100 years. These landslides seriously deplete the soil resource and reduce pasture productivity (*photograph: Hawkes Bay Regional Council*).

definition adopted by United Nations Disaster Relief Organisation, 'time' and 'character' are equated with 'probability' and 'magnitude', respectively, and are taken together to represent hazard for a given place. Thus hazard is defined as the probability of occurrence (frequency) of a given magnitude of event and is incorporated in the hazard/risk equation as:

$$\text{hazard} \times \text{elements at risk} \times \text{vulnerability} = \text{total risk}$$

where, in the case of landslides (Varnes 1984; Crozier 1993): *hazard* is the probability of occurrence (frequency) of a given *magnitude* of failure; *magnitude* refers to the impact characteristics of the process; *elements at risk* are people, property, livelihood and other values; *vulnerability* is the expected degree of loss for a given magnitude; and *total risk* is the expected loss for the time period and place under consideration. The

Box 6.1 The East Abbotsford landslide disaster

IMPACT

At 9.05 pm on the dark and wet winter night of 8 August 1979, a large slice of suburban land in Abbotsford, South Island, New Zealand, suddenly slid downslope, trapping seventeen people, destroying sixty-nine individual homes and displacing 200 people (NZ government 1980). Because early warning signs of instability had been heeded and an efficient emergency management capability was available, nobody was killed, but the costs were high. The total cost from the destruction of houses, urban infrastructure and relief amounted to about NZ$15 million in today's terms (Plate 6.2).

Plate 6.2 Destruction in the suburb of Abbotsford, caused by the block slide of 8 August 1979 (*photograph: Bill Brockie*).

A sophisticated national insurance scheme designed to cope with such disasters, together with government and voluntary relief measures, meant that many of the residents were compensated for much of the direct loss. However, less obvious costs, such as depressed property values in the surrounding area, psychological trauma and the expense of a prolonged commission of inquiry, were not immediately appreciated.

TYPE OF LANDSLIDE

Block slide of sandstone involving bedding plane failure along a weak layer of montmorillonite clay, dipping at 7°. Displacement of 50 m occurred in 30 minutes, leaving a graben of 30 m depth at the head of the slope.

CAUSES

Preconditions

- Unstable geological structure with bedding planes dipping into the valley at angles close to the inclination of the hill slope.
- Permeable material overlying less permeable material, allowing perched water table to develop above the shear plane.
- A very weak montmorillonite-rich layer along the shear plane.

Preparatory factors

- Deforestation within the previous 150 years: lowering evapotranspiration, removing mechanical root reinforcement.
- Urbanisation within the previous forty years: cutting, filling, modification of surface drainage.
- Quarrying of material from the toe of the slope ten years previously, thus removing lateral buttressing support.

Triggering factors

Unknown; possibly a combination of leakage from a city water supply pipeline and rainfall.

HAZARD CHARACTERISTICS

Magnitude: 5.4 million m³.
Rate: initial slow creep, followed by rapid movement of 1.7 m/minute.
Duration: rapid sliding for 30 minutes.
Area affected: 18 ha, with effects over wide area adjacent to slide.
Speed of onset/warning: indications of slow movement, cracking in houses and drains evident at least eleven months prior to slide. Measurements of accelerating creep rate, made six weeks prior to slide, alerted authorities (Figure 6.2).

LESSONS

- Dangerous landslides can occur on very gentle slopes if unfavourable preconditions exist.
- Attention to early warning indicators can enhance preparedness and save lives.
- Human activity can destabilise slopes.
- Low-frequency, high-magnitude events are difficult to predict, but mapping and dating of old landslide features can provide some indication of existing hazard.
- A universal landslide insurance scheme eased the cost burden of victims. However, the event exposed some weakness in the scheme, *viz* that compensation payments would not be made until damage had actually occurred. This meant that money was not available to move houses from threatened areas prior to the main slide occurring. Also, the insurance scheme at the time covered only house and related damage, not land damage.
- A regional landslide hazard assessment should be made where there is evidence of previous landslide activity.

GEOGRAPHERS' ROLE

- Terrain analysis and geomorphic mapping to identify former landslide features.
- Investigation of hazard and event causes by integration of information from multiple sources, including historical records, urban development, climatic records, geology, geomorphology and hydrology.
- Assessment of vulnerability, risks and impacts through analysis of physical, socio-economic factors.

Figure 6.2 Movement rates preceding the Abbotsford landslide of 8 August 1979.

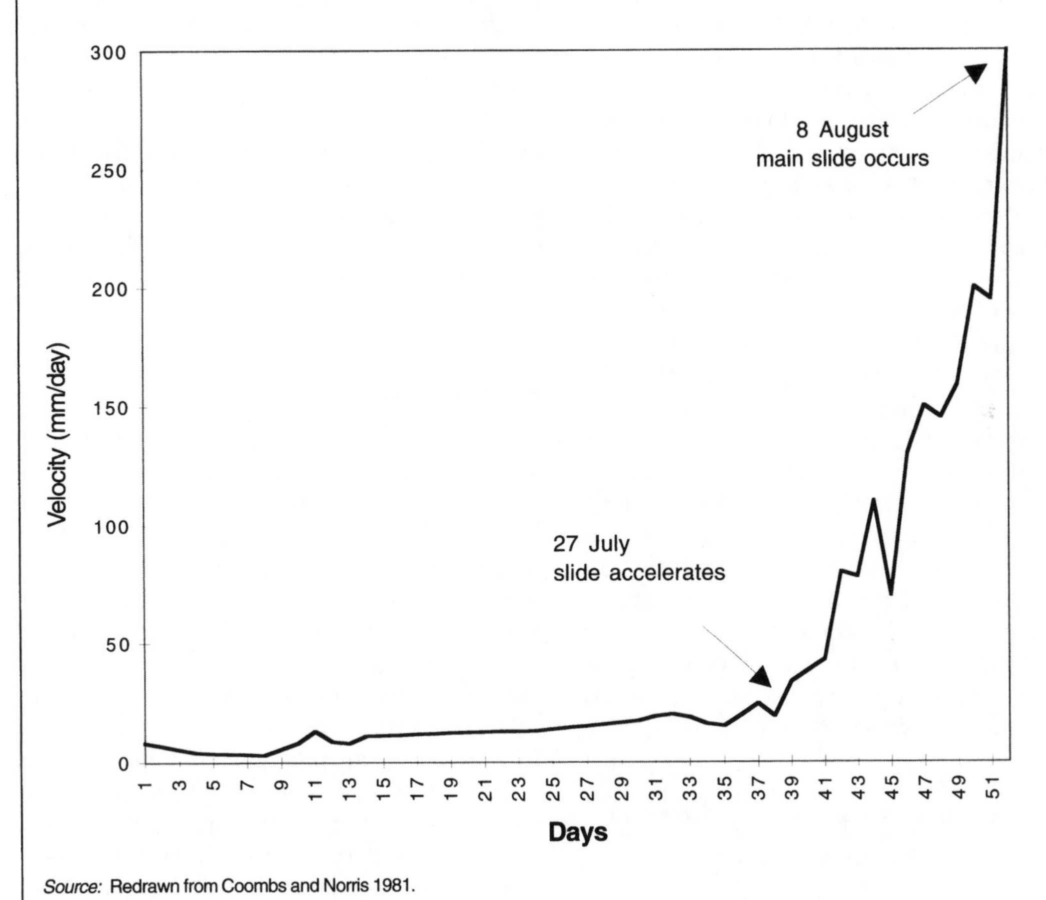

Source: Redrawn from Coombs and Norris 1981.

concepts embodied in this equation acknowledge that risk, in any given place, is a function of the relationship between the physical and human environment.

One of the most important activities in both risk assessment and risk mitigation is the analysis of previous events. The East Abbotsford landslide disaster (see Box 6.1 on p. 86) provides an example of event analysis.

Site assessment

The question of where a landslide may occur may not always be a major factor in landslide hazard analysis. This is because the site of concern may have already been identified by the potential risk or by the fact that a landslide already exists and presents a serious threat. Such situations can occur along highways or reservoirs, or near population centres or valuable assets.

The assessment of hazard at an important site usually takes the form of stability analysis. In the first instance, this may be done by qualitative observations. For example, if landslide features are present, an attempt may be made to determine whether the slide is still active, or when the last movement took place. There are many features that can be used to determine the state of activity (Crozier 1984). If the site has no evidence of previous movement, then stability may be determined by the 'precedence' approach. This involves comparing both stable and unstable slopes in the same terrain in order to identify the threshold conditions (e.g. slope angle and height) that have been associated with landsliding in the past. The site of concern is then compared with these conditions, and if it is found to have similarities to failed slopes, a more detailed quantitative analysis may be required.

Quantitative stability analysis compares the magnitude of resisting forces to the magnitude of shearing forces, expressed as a factor of safety (FoS):

$$\text{FoS} = \frac{\text{resisting stress (i.e. shear strength)}}{\text{shear stress}}$$

At the point of failure, FoS = 1.0. In many engineering studies, stability analysis is the only way in which the probability of occurrence is determined, and hazard is implicitly assumed to be inversely proportional to the factor of safety. This form of analysis requires detailed information on shear strength, slope hydrology, slope geometry, and the shape and position of the potential failure surface. Such information may require expensive subsurface investigation, as well as field and laboratory testing.

Until recently, either assumptions or detailed measurements on pore water pressure were required before stability analysis could be performed. However, computer models are now available that simulate changes in slope hydrology in response to rainfall while continuously analysing the factor of safety on numerous potential shear planes throughout the slope. One of the most successful of these models (the CHASM model) has been developed through the extensive fieldwork carried out largely by geographers (Anderson *et al.* 1988). With such techniques, it is possible to determine the magnitude of rainfall required to produce failure on a given slope. The return period of this threshold value can then be determined from the climatic record to provide a measure of the probability of occurrence of landsliding.

Regional assessment

From a planning and management perspective, territorial authorities in many countries are required to make an assessment of the landslide hazard within their jurisdictions. The scale of this requirement generally precludes the use of detailed geotechnical investigations and stability analysis. Instead, other techniques are employed that require careful analysis of the terrain and use of existing information sources. Geographers have been at the forefront of regional landslide hazard assessment employing techniques such as terrain analysis, geomorphic mapping and geographical information systems (GIS) (Dikau 1989). Many different regional hazard assessment methods are currently in use (Varnes 1984;

Crozier 1995), and they can be classified by their approach into three groups: the parameter method, the stochastic historical method and the triggering threshold method.

Parameter method

The parameter method requires a knowledge of the type, distribution and effectiveness of causative factors for different components of the terrain. The choice of which factors to investigate is determined from prior knowledge or by discriminating between the factors associated with stable and unstable terrain (Gee 1992). Commonly, the initial mapping or investigation units are areas homogeneous for important stability factors such as geology or slope angle. These may be analysed subsequently and perhaps subdivided by other stability factors. Experience can be used to provide a semi-quantitative weighting to stability factors (Sinclair 1992), and summed values can be obtained for ranking each class or areal unit. Usually, factors indicating the presence and activity of any existing landslides are weighted heavily as indicators of the degree of hazard. GIS are of particular value as a tool for analysis, synthesis and computation within the parameter approach to hazard assessment.

Whereas the parameter method is the most common form of regional hazard assessment. it provides only a ranking of susceptibility – not true hazard. By itself, the method does not provide any indication of the probability of occurrence or the magnitude of landslide to be expected. Two examples of the parameter method are presented in Box 6.2.

Stochastic historical method

One of the main difficulties with both stability analysis and the parameter method is the difficulty in obtaining accurate and representative values for the parameters involved. This is because many stability factors exhibit a high degree of spatial variability. The historical method, on the other hand, works on the principle of 'precedence', indicated by the record of previous landslide activity. This is both an advantage, because it does not rely on sample values, and a disadvantage, because to be of use in prediction, it assumes temporal stability in causative factors.

Above all, the method demands a good database. Unfortunately, few countries have standard protocols and procedures for establishing reliable records of landslide activity. Consequently, in employing this approach, much original research of information sources is required, including information from media sources, public organisations, private consultants, etc. The record can be extended beyond the historical period by investigating landslide deposits in the geological record. For example, lake sediments in an unstable area of New Zealand have revealed the occurrence of 395 landslide events in the last 6000 years (Eden and Page 1998). By dividing the number of events by the period of observation, a historical frequency (probability of occurrence) can be determined for different regions.

Triggering threshold method

The triggering threshold approach is more complex than other methods but has greater potential for forecasting landslide activity and determining the mass movement response to climatic change and other triggering factors. This approach couples the forcing process with a process response. For rainfall-triggered landslides, this involves establishing an initiating threshold between rainfall parameters and landslide occurrence (Julian and Anthony 1994; Crozier 1989) (Figure 6.3). For earthquake-triggered landslides, the threshold between non-occurrence and occurrence is usually either a function of shaking intensity or earthquake magnitude (Keefer 1984).

Thresholds established in this way simply measure the susceptibility of the terrain under study to the landslide-triggering process. Clearly, inherent stability conditions and consequently thresholds will vary from place to place. A reliable regional threshold, however, may be used to

Box 6.2 Methods for regional hazard assessment

Two parameter methods for regional hazard assessment are illustrated in summary form. Both methods require the selection of factors that are important in determining the level of hazard in the areas being studied. A number of factors are common to both methods, although expressed in different ways. These include the history of landsliding, the availability of susceptible material, and slope angle. Both methods are relatively easy to apply with limited expertise, and they both provide a simple but clear classification of landslide hazard in a form suitable for use by planners and land managers.

The Tasmania method (Stevenson 1977; Figure 6.3) is hierarchical, with each of the parameters having a different level of importance. It takes the form of a decision tree, with each decision providing a subdivision of a higher-order class. It could, therefore, be used to map an area in anywhere from two to five hazard classes, depending on the resources available and level of detail required. Local knowledge and knowledge of previous landsliding has determined the type of parameters chosen and their relative importance.

The Montrose method (Moon *et al.* 1992; Table 6.2) is designed specifically to assess debris flow hazard in an area where there is a real threat but little recent history of debris flow. Consequently, the selection of parameters is more theoretically based. For example, areas with little outcrop are considered likely to provide a potential supply of debris, while landslides may actively feed into gullies where debris flows can be generated. Steep, high slopes, together with large-volume landslides, ensure a high volume and rapid supply of material to debris flow initiation sites. The hazard assessment classes are decided partly by judgement and partly objectively. Where all hazard factors are present in a catchment, the area is classified as 'high' or 'very high' hazard. The class 'very high' has particularly large volumes of modern landslides. To qualify as 'medium', parameters 1 and 2 must present some degree of hazard, along with hazard recorded for at least one of the other parameters. Other catchments not meeting these criteria are classified as 'low' hazard.

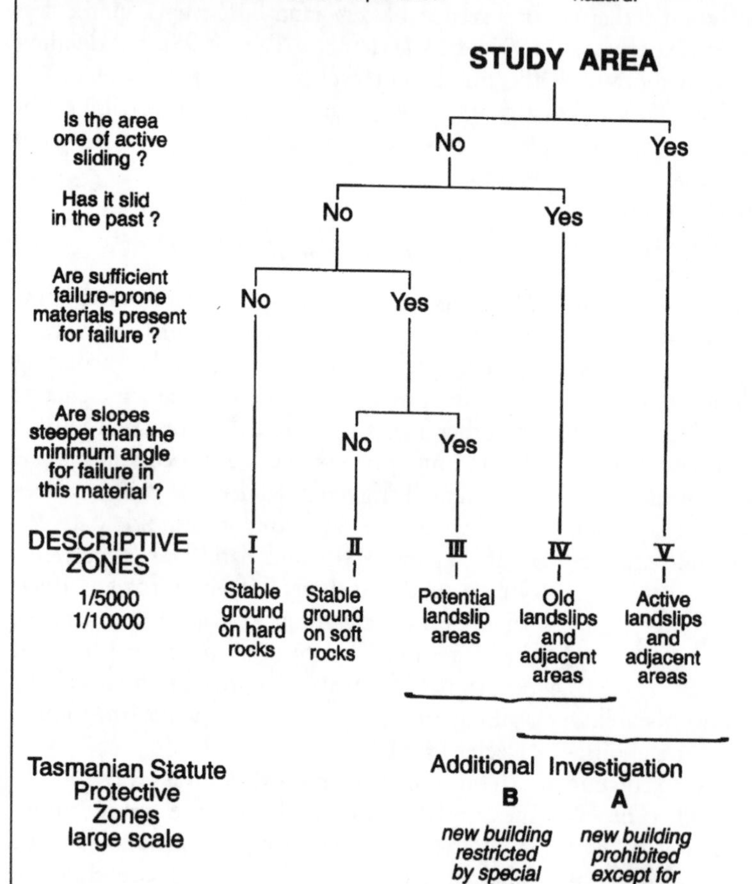

Figure 6.3 Tasmania hazard zonation scheme.

Source: Stevenson 1977.

Table 6.2 The Montrose hazard zonation scheme.

Catchment hazard ranking factor	*Catchment reference*								
	A	*B*	*C*	*D*	*E*	*F*	*G*	*H*	*I*
1 Presence of gullies	●			●	●		●	●	●
2 Outcrop % area	•	●	•	●	•	•	•	•	●
3 Height of slopes > 30°							•	●	●
4 Proportion slopes > 30°							•	●	●
5 Size of colluvial fan			•	•	•	●	●	●	●
6 Number of modern slides		•	•	•				•	●
7 Vol. of modern slides			•		•				●
Catchment hazard ranking assessment									
i Very large debris flow	L	L	L	M	L	L	M	M	X
ii Small or large debris flow					H			H	
iii Debris torrent in gully						H	H		

Source: Moon *et al.* 1992.
Notes: The size of the dot indicates the relative influence of the hazard factor: high, medium or low.
Key to catchment hazard ranking assessment:

Symbol	*Term*	*Probability in 50-year period (%)*
X	Very high	15–39
H	High	5–39
M	Medium	0.5–5
L	Low	< 0.5
VL	Very low	< 0.05

Whereas the Tasmania method provides a susceptibility classification only, the Montrose method has also attempted to identify the probability of occurrence within each class. These probabilities are not just based on recurrence intervals from past landslide activity but also acknowledge that human activity has enhanced the probability of occurrence through landscape modification.

determine the probability of occurrence (statistical frequency) of landslide activity by reference to the frequency–magnitude distribution of events for the triggering agent.

RISK ASSESSMENT

Determining hazard is only one part of identifying the threat from landslides. Together with the hazard assessment in each area, there needs to be information on the elements and values that are at risk, and this in turn needs to be qualified by

Figure 6.4 Landslide-triggered threshold based on daily rainfall and antecedent soil moisture, Wellington City, 1974.

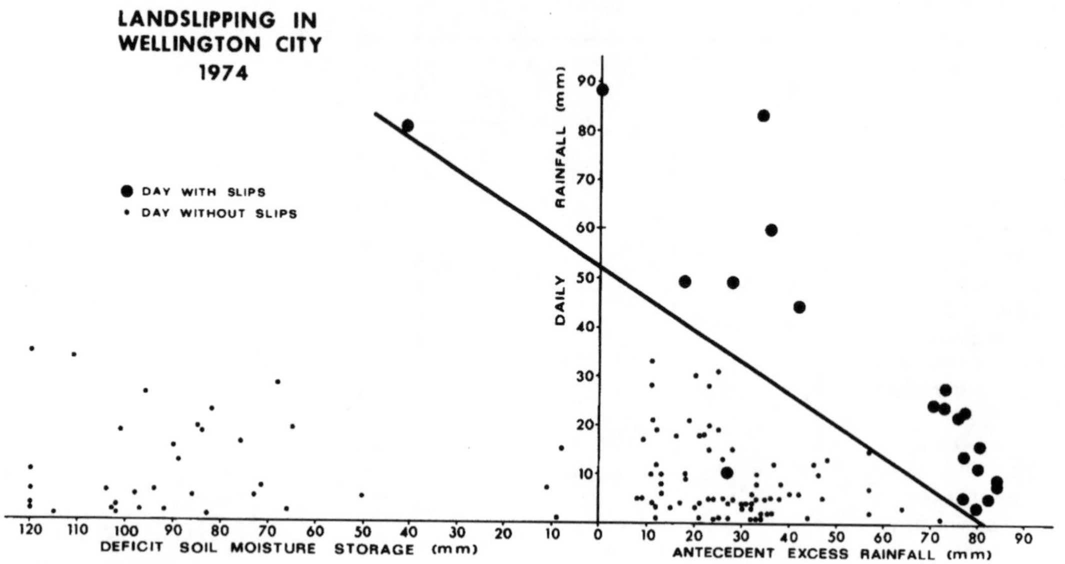

LANDSLIPPING IN WELLINGTON CITY 1974

● DAY WITH SLIPS
• DAY WITHOUT SLIPS

DAILY RAINFALL (mm)

DEFICIT SOIL MOISTURE STORAGE (mm)

ANTECEDENT EXCESS RAINFALL (mm)

Source: Crozier 1989.

the vulnerability of those elements and values.

Elements at risk, in terms of tangible characteristics such as property, buildings and production values, are relatively easy to establish. However, experience has shown that many other short-term and long-term costs are associated with landslide activity (Crozier 1989). These may include expenditure on research and conservation measures, degradation of the soil resource and the consequent cumulative losses from primary productivity. Vulnerability is also difficult to measure and is very dependent on the nature of society affected. This may not just relate to technical factors but also to organisational factors and the distribution of wealth and power.

CONCLUSION

Geographers have a major role to play in landslide hazard and risk assessment. This is because, like all hazards, the risk results from the interrelationships between the human and physical environments – a traditional focus of geographical study. It is not an exclusive role, because there is always a need for the specialists in areas such as soil physics, economics and engineering. However, geographers have also contributed to specialist areas of slope stability research by developing models and providing empirical information from field monitoring.

The landslide process itself is a product of the interrelationships between a number of natural systems, including geological, geomorphological, hydrological, climatic and human land-use systems. Understanding landslides requires an ability to analyse the relationships between these systems. Geographers have been able to make a valuable contribution in this area because they generally examine a wide range of conditions within the landscape from a spatial and temporal perspective at a range of scales.

Above all, geographers treat the process as a component of a human–physical system. This highlights not only the risk and vulnerability of society but also reveals the human factor as a cause of much slope stability.

GUIDE TO FURTHER READING

Brabb, E.E. and Harrod, B.L. (eds) (1989) *Landslides: Extent and Economic Significance.* Balkema, 385 pp.

Brunsden, D. and Prior, B. (eds) (1984) *Slope Instability*. Chichester: John Wiley & Sons, 620 pp.

Dikau, R., Brunsden, D., Schrott, L. and Ibsen, M.-L. (1996) *Landslide Recognition: Identification, Movement and Causes*. Chichester: Wiley, 251 pp.

Turner, A.K. and Schuster, R.L. (1996) *Landslides: Investigation and Mitigation.* Transportation Research Board, Special Report 247, National Research Council, National Academy Press, Washington, DC, 657 pp.

REFERENCES

Anderson, M.G., Kemp, M. and Lloyd, D.M. (1988) Application of soil water finite difference models to slope stability problems. *Proceedings of the Fifth International Landslide Symposium*, Lausanne: 525–30.

Crozier, M.J. (1984) Field assessment of slope instability. In D. Brunsden and D.B. Prior (eds) *Slope Instability*. Chichester: John Wiley & Sons, 103–40.

Crozier, M.J. (1989) *Landslides: Causes, Consequences, and Environment.* London: Routledge.

Crozier, M.J. (1993) Management issues arising from landslides and related activity. *New Zealand Geographer* 49 (1): 35–7.

Crozier, M.J. (1995) Landslide hazard assessment: theme report. In D.H. Bell (ed.) *Landslides: Proceedings of the Sixth International Symposium*, Christchurch, February 1992, 3: 1843–8.

Crozier, M.J. (1996) Magnitude–frequency issues in landslide hazard assessment. In R. Mausbacher and A. Schulte (eds) *Beitrage zur Physiogeographie*. Barsch Festschrift, Heidelberger Arbeiten 104: 221–36.

Crozier, M.J. (1997) The climate landslide couple: a Southern Hemisphere perspective. *Paleoclimate Research* 2: 329–50.

Cooke, R.U. and Doornkamp, J.C. (1990) *Geomorphology in Environmental Management* (second edn). Oxford University Press.

Coombs, D.S. and Norris, R.J. (1981) The East Abbotsford, Dunedin, New Zealand, landslide of August 8, 1979, an interim report. *Bull. Liaison. Labo. P. et Ch. Special* X, January 1981: 27–34.

Dikau, R. (1989) The application of a digital relief model to landform analysis in geomorphology. In J. Raper (ed.) *Three Dimensional Applications in Geographic Information Systems*. London: Taylor & Francis, 51–77.

Eden, D.N. and Page, M.J. (1998) Palaeoclimatic implications of a storm erosion record from late Holocene lake sediments, North Island, New Zealand. *Palaeogeography, Palaeoclimatology, Palaeoecology* 139: 37–58.

Fairbridge, R.W. (ed.) (1968) *The Encyclopedia of Geomorphology*. Reinhold Book Corporation.

Gee, M.D. (1992) Classification of landslide hazard zonation methods and a test of predictive capability. In D.H. Bell (ed.) *Landslides: Proceedings of the Sixth International Symposium*, Christchurch, February 1992, 2: 947–52.

Hansen, M.J. (1984a) Strategies for classification of landslides. In D. Brunsden and D.B. Prior (eds) *Slope Instability*. Chichester: John Wiley & Sons, 1–26.

Hansen, A. (1984b) Landslide hazard analysis. In D. Brunsden and D.B. Prior (eds) *Slope Instability*. Chichester: John Wiley & Sons, 523–602.

Hutchinson, J.N. (1988) General report: morphological and geotechnical parameters of landslides in relation to geology and hydrogeology. *Proceedings of the Fifth International Symposium on Landslides*, 1, Balkema, 3–35.

Julian, M. and Anthony, E.J. (1994) Landslides and climatic variables with specific reference to the Maritime Alps of southeastern France. In R. Casale, R. Fantechi and J.C. Flageollet (eds) *Temporal Occurrence and Forecasting of Landslides in the European Community*. European Community: 697–721.

Keefer, D.K. (1984) Landslides caused by earthquakes. *Bull. Geol. Soc. Am.* 95 (4): 406–21.

Moon, A.T., Olds, R.J., Wilson, R.A. and Burman, B.C. (1992) Debris flow risk zoning at Montrose, Victoria. In D.H. Bell (ed.) *Landslides: Proceedings of the Sixth International Symposium*, Christchurch, February 1992, 2: 1015–22.

New Zealand government (1980) *Report of the Commission of Inquiry into the Abbotsford Landslide Disaster*. Government Printer, Wellington.

Selby, M.J. (1993) *Hillslope Materials and Processes* (second edn). Oxford University Press.

Sharpe. C.F.S. (1938) *Landslides and Related Phenomena*. Pageant.

Sidle, R.C., Pearce, A.J. and O'Loughlin, C.L. (1985) *Hillslope Stability and Land Use*. American Geophysical Union.

Sinclair, T.J.E. (1992) SCARR: a slope condition and risk rating. In D.H. Bell (ed.) *Landslides: Proceedings of the Sixth International Symposium*, Christchurch, February 1992, 2: 1057–64.

Stevenson, P.C. (1977) An empirical method for the

evaluation of relative landslide risk. *Int. Assn. Eng. Geol. Bull.* 16: 69–72.

Varnes, D.J. (1958) Landslides types and processes. In E.B. Eckel (ed.) *Landslides and Engineering Practice*. Highway Research Board Special Report 29, NAS–NRC Publication 544: 20–47.

Varnes, D.J. (1978) Slope movement and types and processes. In R.L. Schuster and R.J. Krizek (eds) *Landslides: Analysis and Control*. Transportation Research Board Special Report 176, National Academy of Sciences, Washington DC, 11–33.

Varnes, D.J. (1984) *Landslide Hazard Zonation: A Review of Principles and Practice.* UNESCO, Paris.

World Landslide Inventory (1990) A suggested method for reporting a landslide. *Bull. Int. Assn. Eng. Geol.* 41, 5–12.

7

Floods

Edmund Penning-Rowsell

INTRODUCTION

Of all the 'natural' hazards to which humans are exposed, floods are probably the most widespread and account for most damage and loss of life (Alexander 1993). Floods also appear to have a special impact on their victims, instilling a fear of the consequences that often exceeds their actual impacts (Green and Penning-Rowsell 1989). They also can have serious secondary impacts on the economy of the regions affected, and they can markedly influence agriculture in disaster-affected areas for some time after the event has passed, by affecting cropping patterns and yields, as dramatically is the case in Bangladesh (Alexander 1993).

Geographers have studied the complexity of such flood hazards for many years and have made significant contributions to their understanding, not least by tackling the interface between physical geography and human geography that is highlighted in the flood situation by the complex relationships between human behaviour and extreme geophysical events.

The foundation of such research was in the 'Chicago' school of hazard geography pioneered by White and others (Burton *et al.* 1978; 1993). This has been followed by the work of Hewitt (1997) and Mitchell (e.g. Mitchell *et al.* 1989) and elsewhere in the world in Australia (Smith 1999), New Zealand (Eriksen 1986), the UK (Penning-Rowsell *et al.* 1986; Arnell *et al.* 1984) and elsewhere (Chan and Parker 1996; Kanti Paul 1997, Pelling 1998).

In addition, geographers have contributed to the hydrology of floods, mainly by evaluating the impact of humans on flood regimes (Hollis 1988), through evaluating spatial flood patterns (Newson 1989) or understanding the geomorphology of floodplain processes (Anderson *et al.* 1996).

From other disciplines has come the sociology of human group interaction in floods and other extreme events (Torry 1979), the psychology of behaviour under risk circumstances and of risk communication (Handmer and Penning-Rowsell 1990), and the configuration of institutions to tackle such hazard phenomena (Hood and Jones 1996). Many of the key debates centre on whether flood and other risk is socially determined rather than physically based, and whether risk is socially divisive (Beck 1992).

THE NATURE OF FLOODING AND FLOOD HAZARDS

Flood types and mechanisms

Floods can be classified into fluvial, coastal and those that result from deficiencies in urban drainage. Fluvial floods occur when river discharge exceeds its bankfull capacity. The return period of out-of-bank flood flow is generally 2.3 years (Newson 1989), and the magnitude of floods and their probability of occurrence are strongly connected, although these relationships are regionally specific and depend on climatic conditions and river catchment character (*ibid.*).

Coastal flooding occurs where tide levels exceed land levels, exacerbated by extreme wave

Box 7.1 Increased coastal flood hazard in Venice, Italy

The historic city of Venice in northern Italy is one of the most famous cases of increasing flood threat. A combination of subsidence in the city and rising sea levels means that the frequency of flooding in St Marks Square in the city centre has risen from seven per year in 1900 to about fifty per year today. A major flood occurred in 1966, causing widespread damage (see below). Projections of sea level rise indicate that this frequency could rise to over 300 per year by the year 2050. Many solutions have been suggested, including a system of gates between the lagoon in which Venice is located and the Adriatic Sea (Bandarin 1994). However, these proposals are controversial in that they appear to tackle only the symptoms of Venice's many problems rather than their causes (Penning-Rowsell *et al.* 1998). They therefore may not solve the flooding problem in the long term – i.e. 50+ years – and do not tackle the associated problems of pollution in the city and the Lagoon, the city's declining population, and the decay of the ancient Venetian buildings. Even the local government *Comune* in Venice has voted against the proposals, which remained mired within the labyrinthine and corrupting Italian political system for years, until vetoed by the Italian Environment Minister in 1998: the problem remains unsolved.

Figure 7.1 Increasing flooding in Venice, 1926–93.

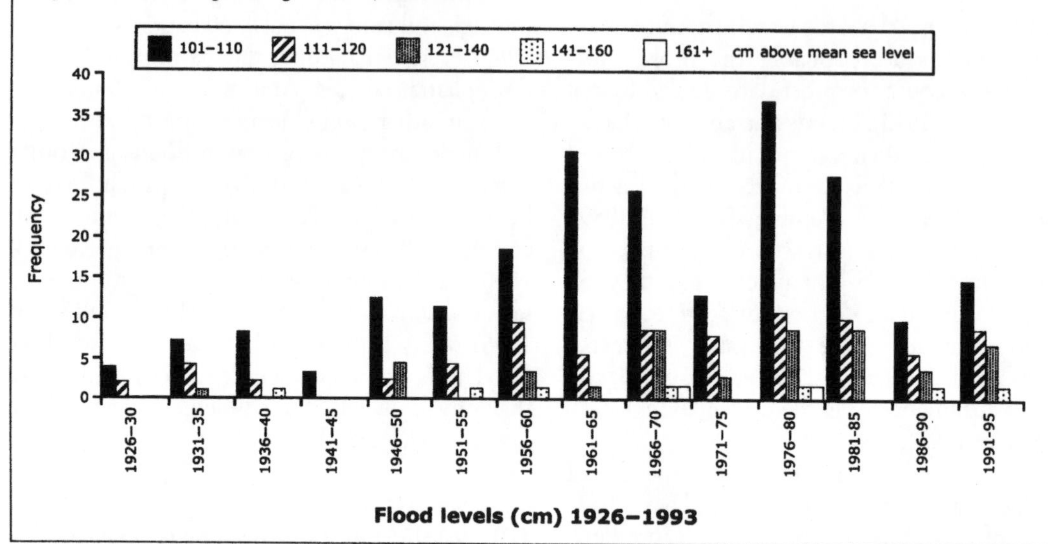

conditions and by sea level surges caused by low-pressure meteorological conditions. Changes in the relative height of land and seas caused by land subsidence or climate change-induced sea level rise also contribute to coastal flooding (Nicholls 1995). Tropical cyclones such as in Southeast Asia also contribute to flooding of coastal regions, as they bring onshore both extreme waves and intense rainfall.

Flooding in urban areas away from the coast or major rivers occurs when summer thunderstorm conditions (or intense cyclonic rain) occur on urban catchments, where infiltration rates are reduced by paved surfaces. The result is rapid and almost complete runoff far exceeding the capacity of drainage and sewer systems. Without deliberately designed storage ponds or other control systems for this runoff, it is liable to cause damage, especially in the basements of buildings or where underground railways or telecoms systems are at risk.

More locally important flood-causing agents are ice-dammed rivers, dam and dike breaks, and tsunamis. In many of these instances, damage is extreme, caused by high water velocities and associated intense storm conditions.

Flood extremes

Table 7.1 gives this data for the twenty-four most

Table 7.1 The twenty-four most severe floods worldwide.

Country	*Gauging station*	*Drainage basin area (km²)*	*Maximum discharge (m³/s)*	*Year of flood*
Mexico	Cithuatlan Paso del Mojo	1370	13,500	1959
Japan	Nyodo Ino	1560	13,510	1963
Japan	Kiso Imujama	1680	11,150	1061
USA	W Nueces, Bracketville	1800	15,600	1959
India	Macchu	1900	14,000	1979
Taiwan	Tam Shui Taipei Bridge	2110	16,700	1963
Japan	Shingu Oga	2350	19,025	1952
USA (Texas)	Pedernales Johnson City	2450	12,500	1952
North Korea	Daeryong Gang	3020	13,500	1975
Japan	Yoshino Iwazu	3750	14,470	1974
Philippines	Cagayan Echague Isabella	4244	17,550	1959
Japan	Tone Yattajima	5,110	16,900	1947
USA (Texas)	Nueces Uvalde	5,504	17,400	1935
USA (California)	Eel Scotia	8060	21,300	1964
USA (Texas)	Pecos Comstock	(9100)	26,800	1954
Madagascar	Betsiboka	11,800	22,000	1927
North Korea	Toedong Gang Mirim	12,175	29,000	1967
South Korea	Han Koan	23,880	37,000	1925
Pakistan	Jhelum Mangla	29,000	31,100	1929
China	Hanjiang Hankang	41,400	40,000	1583
Madagascar	Mangoky Banyan	50,000	38,000	1933
India	Normada Garudeshwar	88,000	69,400	1970
China	Chang Jiang Yitchang	1,010,000	110,000	1870
USSR	Lena Kusur	2,430,000	189,000	1953
Brazil	Amazonas Obidos	4,640,000	370,000	1953

Source: van der Leeden *et al.*

extreme floods as measured by their discharge volumes (van der Leeden *et al.* 1990). This indicates a preponderance of cases in the USA (5), Japan (5), and East Asia and China (8). Figure 7.2 shows the strong correlation between flood magnitude and drainage basin size; the most significant floods in world terms are generated by major rainfall events – or series of events – over large catchments. In each of these cases, the flooding can last many weeks or months, not least when it is the product of seasonal shifts in weather and climate patterns such as the El Niño phenomenon (Penning-Rowsell 1996).

Because emergency relief procedures are now more effective, conditions have changed since 50–100 years ago, when major floods caused massive loss of life (e.g. the Huang He floods in China in 1931 resulted in 3,700,000 deaths). Nevertheless,

Figure 7.2 The relationship between drainage basin area and extreme flood discharge.

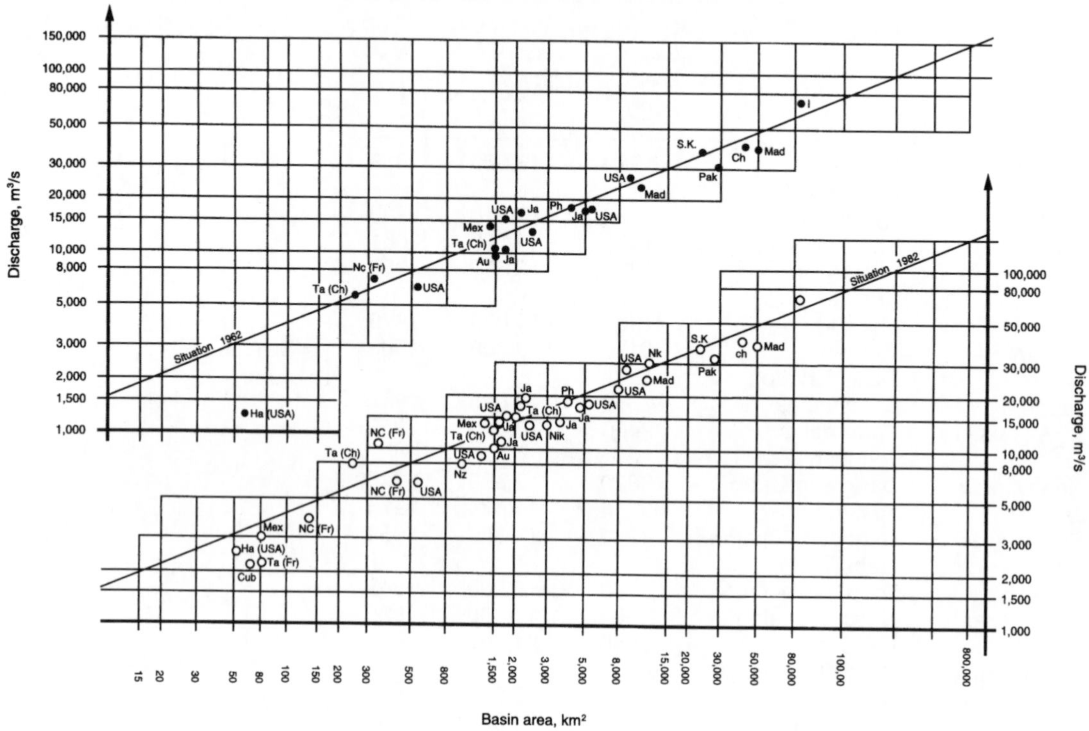

Source: Rodier and Roche 1984.

loss of life is still common, as in Italy in 1998 (135 confirmed deaths) and in Germany, Poland and the Czech Republic in 1997 (a total of 128 casualties in the three countries). And the damage from floods has increased through time as assets in floodplains have risen in value (Parker 1995).

Concepts used to analyse flood hazards

The impacts of floods on humans has been studied from many different perspectives. The *magnitude* of flood events is usually related to their discharge in the fluvial context and the extent of coastal flooding. The *risk* of flooding brings in concepts of probability and *return period* (the average length of time in years between floods of comparable magnitude) (Hood *et al.* 1992). *Hazard* is a combination of the geophysical event (which by itself is not hazardous) and human *vulnerability*. The differential impacts of floods – and other hazards – that come from differential vulnerability are often related to the damage inflicted by the flood event and the ability of flood victims to recover after the floods have subsided. This ability to recover – or *resilience* (Handmer and Dovers 1996) – is often related to the victims' wealth/poverty status, or their experience of similar flood events in the past, or a combination of these factors.

The relationship between flood experience, the impacts of floods and the vulnerability of human populations generated the 'school' of hazard geography studying the 'hazard-response model' of human adjustment to many hazards (Burton *et al.* 1993, Mitchell *et al.* 1989, Hewitt 1997). This model, pioneered by Kates (1962), posits that the response of human populations to reduce the impacts of floods results from their heightened perception of the hazards that they face due to increased knowledge or experience.

Without experience, the population's perceptions of risk and hazard are low, and they do not respond. As a result, they suffer great damage when the floods come. But with experience the population learns to gauge the risk of flooding, and to respond proportionately. In effect, this is a 'dose–response' model dominated by the availability of information to the individual about the risk and hazard that they face and is derived from a paradigm that emphasises the choice of adjustments that individuals can make in reducing their own vulnerability.

Criticisms of the hazard-response model are numerous. They point to the difficulty that people have in responding when they do not have the necessary resources (Torry 1979; Beck 1992), and also to different traditions in countries other than the USA, which emphasise the use of collective (state) action to reduce vulnerability within a welfare state tradition rather than relying on individuals (Parker and Penning-Rowsell 1983). Sociologists have stressed that some potential hazard victims cannot respond because they do not have the power to do so, and others have emphasised the way that risk creates social divisions between those who are vulnerable and those who are not (Beck 1992; Quarantelli 1998).

The trend in conceptualisation is towards a recognition that hazard and risk are socially constructed (i.e. they are a product of society and the way that society is configured) rather than an emphasis on either physical processes or individual decision making and its determinants. In this respect, the study of hazards is maturing and bringing many different disciplines to bear in a complex mixture of the natural and social sciences.

Flood alleviation strategies

A central theme in hazard geography has been the choice of hazard mitigation 'adjustments', because such choices illuminate the way in which humans and their hazardous environments interact.

The conventional view is to differentiate between structural flood alleviation measures, principally concerned with engineering works, and 'non-structural' alternatives (Figure 7.3). Much

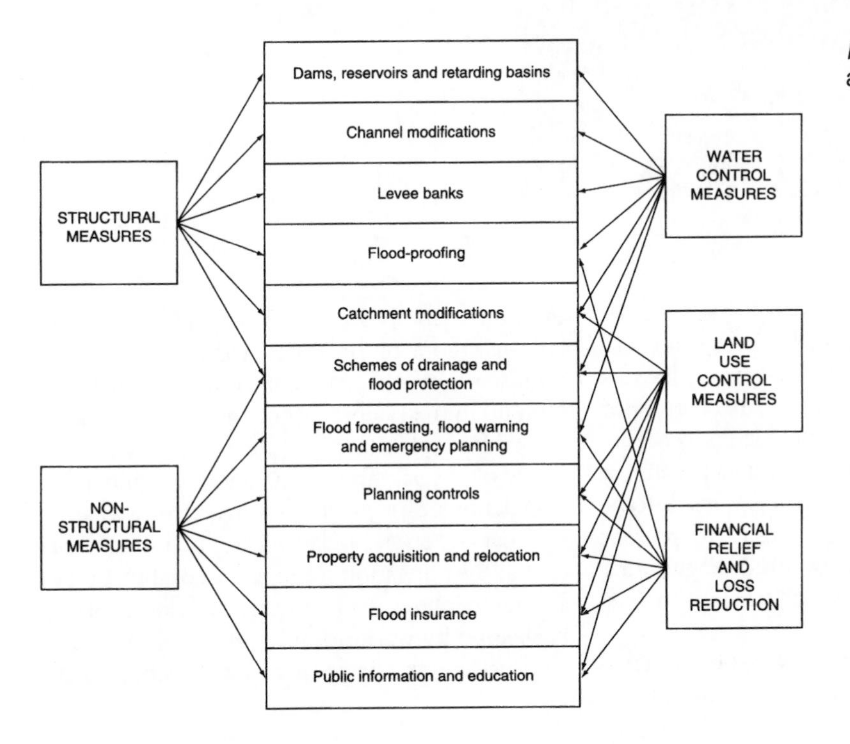

Figure 7.3 Alternative flood alleviation strategies.

Box 7.2 House raising as a flood alleviation strategy: an example from Australia

Most attention is given by engineers and governments to flood alleviation strategies that involve major construction, but in many cases the individual can make adjustments to their property and behaviour to minimise flood damage potential.

The example below shows a case of house raising in Lismore, NSW, Australia. Many hundreds of the town's properties have been raised in this way, thus reducing the direct damage that the frequent flooding brings. The frequency of this flooding has approximately doubled in the last fifty years along this north Australian coastline, and many of the houses were originally constructed at ground level during a period of infrequent flood events.

Research has shown that this house raising is a rational strategy for the individual house owner, in terms of the costs of the raising compared with the benefits of flood damage avoided (Penning-Rowsell and Smith 1987). The main problem that remains is that the households are isolated at times of flood, and this can cause distress and the dislocation of the inhabitants' lives. They will also suffer poor access to medical and other facilities that may be needed in the days or weeks that the flood waters surround the building.

Plate 7.1 House raising against flooding in Lismore, NSW, Australia. Both houses were originally at ground level; the far one was raised 3 m to above flood levels.

criticism of the former approach focuses on the environmental damage that such works can bring, and the way that certain strategies – mainly levees – can create rising flood losses if they are overtopped or breached. The structural approach is generally reactive: a flood hazard has been created by human occupancy of a flood-affected area and needs to be tackled at the location of the hazard to prevent future flood damage, either by major engineering river works or with more minor schemes such as protecting individual properties.

The non-structural alternatives are both reactive – flood relief and flood insurance being classic examples – and pro-active. The latter is shown by land-use control measures either designed to reduce runoff through controlling flood flows high up in the catchments or by controlling land use in floodplain areas to deter 'encroachment' there of damageable property and vulnerable populations. In most circumstances, what is optimal in terms of damage reduction and the cost of the mitigation strategy is a combination of several measures, such as structural flood control backed up by warning and insurance systems, or land-use control supported by emergency relief.

GAUGING THE URBAN FLOOD HAZARD IN MANCHESTER, UK

Flooding mechanisms and impacts

The River Irwell rises in the hills above Bolton in Lancashire, UK, and flows southwestwards into the Mersey and the Irish Sea.

The flood-generating mechanism is frontal rain on these hills from Atlantic winter depressions. The effect is flooding downstream in the Salford area of Greater Manchester (Figure 7.4). This flooding has been exacerbated because, first, the channel of the river was moved fifty years ago to allow for urban and industrial development, and second, because the river is now 'hemmed in' by low embankments to maximise the usable area of valuable land in a congested urban location. Some out-of-bank flooding nevertheless occurs at return periods as low as five years.

The area has experienced some severe flooding in the past, notably in 1866, 1946, 1954 and 1980 (see Figure 7.3). However, there is no recent history of significant flooding, and the residents and industrialists using the floodplain are generally very unaware of the risks that they face. As a consequence, and supporting the hazard-response model, there is very little attempt by the people at risk to protect themselves against future flooding. Many of the industrial properties are occupied by ephemeral 'ethnic' clothing manufacturing businesses, thus exacerbating the general ignorance of flooding problems in the area.

Damage potential

A survey has gauged the likely damage that would occur were the properties on the floodplain to be flooded from flood severities up to and including the 250-year event. This survey was designed to assess the benefits of providing a range of flood protection measures, including

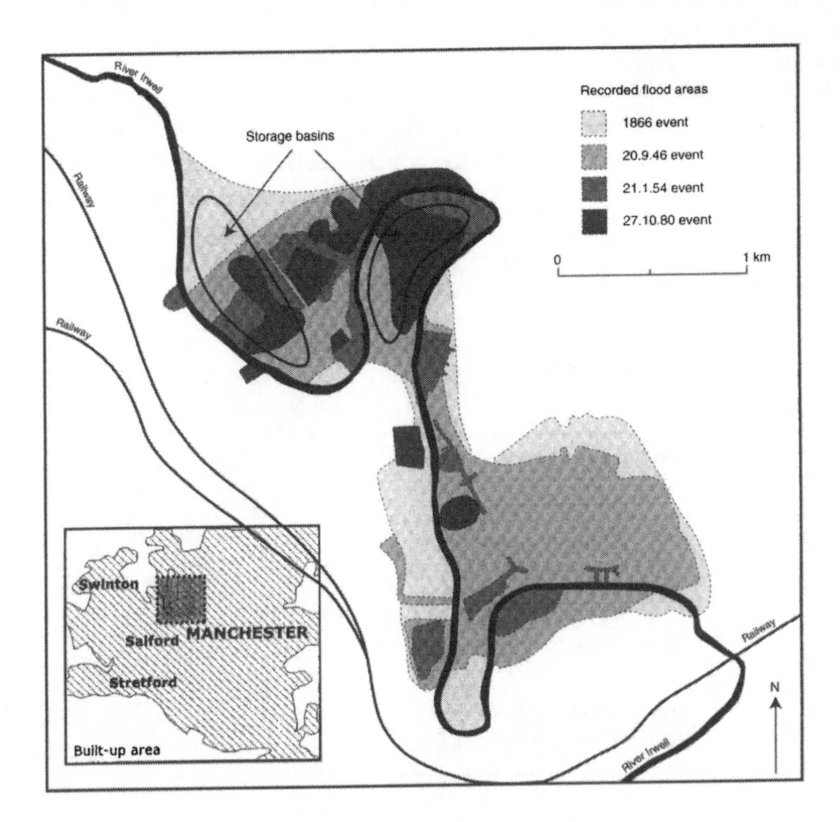

Figure 7.4 Flooding and flood hazard solutions for the River Irwell, Manchester, UK.

major engineering works to the river and its floodplain, within a framework of investment appraisal that compares the costs of such works with the benefits that they generate (Penning-Rowsell *et al.* 1994).

The issue at stake here is whether a major scheme can be justified given the relative poverty of the population in this somewhat deprived area of northern Britain and the fact that such cost–benefit analysis favours protecting richer people and more valuable property since their more valuable assets at risk from flooding generate higher calculated benefits against which to compare costs.

Table 7.2 gives the results from this survey, and these are not untypical of similar calculations in other major urban areas. Up to 4823 properties could be damaged by a flood with a return period of 250 years, and up to 1000 properties could be affected by the fifty-year event. Damage to these properties would be £94.5 million and £13.6 million, respectively. The majority of damage in the 250-year event here would be suffered by householders (48 per cent) rather than the industrial concerns (11 per cent). For the fifty-year flood, 73 per cent of the damage would be residential, since most of the industry is located at the upper margins of the floodplain.

Engineering options and standards

By any standards, this is a serious flood problem, with over 4000 houses and probably 12,000 people at risk. Many would not be insured, since insurance is often not taken up by those with lower than average incomes. A flood warning system is already in place, and the additional solutions proposed are limited to engineering schemes. However, these can be implemented at different standards and costs:

Scheme A: Two flood storage basins in the floodplain, plus river channel works designed to give protection to 3269 properties against the 100-year event, costing £11.3 million.

Scheme B: A single flood storage basin and the same river channel works, protecting approximately 2000 properties against the 74-year event and costing £7.1 million.

Scheme C: Just the works to contain flood waters within a dredged and thereby enlarged channel, giving only a 1 in 39-year standard of protection, to only about 800 properties, but at a cost of only £2.9 million.

Which gives better value for money? As part of government moves in Britain to control public expenditure, which in turn are a part of moves to rein back the role of the state, these schemes are subject to rigorous economic appraisal. Application of the decisions rules developed by the British government in association with Middlesex

Table 7.2 The potential flood damage to different land uses in the River Irwell floodplain.

Land-use sector		*Return period (years)*					
		10	*25*	*50*	*100*	*150*	*250*
1	Residential property	686	709	957	2832	3074	4328
4	Hotels, etc.	0	0	0	4	4	4
5	Retail shops	18	18	25	304	308	341
6	Offices	2	2	3	20	21	27
7	Public buildings	11	11	14	36	36	43
8–9	Industry etc.	0	0	0	73	75	80
Total properties		717	740	999	3269	3518	4823
Event damage (£)		7.4m	8.4m	13.5m	54.9m	63.9m	94.5m

University (MAFF 1998) gives the results in Table 7.3.

These results show that the marginal benefits of Scheme A exceed its marginal costs, such that the incremental increase in benefits is greater than the incremental increase in costs by a factor of 1.083. Scheme C by itself is the best value, but Scheme A is still highly worthwhile and provides the standard of protection appropriate to such a large and important urban area. A variant of Scheme A is being implemented.

THE FLOOD EMERGENCY IN NORTHERN EUROPE IN SUMMER 1997

The floods in context

The floods in Germany, Poland and the Czech Republic in 1997 caused immense dislocation and destruction, and significant loss of life. They must therefore represent one of the most serious flood events ever in Europe, comparable to (but not as serious as) the North Sea floods in 1953 (Penning-Rowsell and Fordham 1994).

The flooding was more severe than any previously recorded for most locations along the River Oder, and was caused by rainfall events of exceptional rarity (Figure 7.5). The flood was unusual in that most of the previous 'floods of record' were winter rather than summer floods, as is usual in the region.

Recent research for Green Cross UK (Penning-Rowsell 1998) showed that the economic impact was severe. This was especially the case in Poland, where 25 per cent of the country's population was affected (Bednarz 1997); 23 per cent of the nation's buildings are in the flood-affected regions, and 26 per cent of Poland's GDP is generated there. Agriculture, transport, trade and tourism were the sectors most seriously affected. The reinsurance company MunichRe estimated that the losses in Poland and the Czech Republic reached 10 billion DM (£3.4 billion).

The environmental impacts were also severe, mainly from chemical pollution spilled from flooded and damaged factories. In addition, many domestic sewage treatment works in the Polish area affected were destroyed, and large agricultural areas may be damaged for years to come. Strategic oil reserves in Poland unwisely located in storage tanks on the floodplain were also breached by the flood waters, and heavy metal contaminants resulting from industrialisation in the past were flushed from sediments within the

Table 7.3 Economic appraisal of alternative flood alleviation standards for the River Irwell (£ million).

	Option (see text)				
	Do nothing	*Do minimum*	*C*	*B*	*A*
Costs [PVs}	–	0.320	2.891	7.068	11.320
Flood damage [PVd]	72.755	56.168	23.794	19.322	14.717
Flood damage avoided [PVda]	–	16.587	48.961	53.433	58.038
Total benefits [PVb]	–	16.587	48.961	53.433	58.038
Net present value [NPV]	–	16.267	46.070	46.365	46.718
Average cost–benefit ratio	–	51.834	16.936	7.560	5.127
Incremental cost–benefit ratio	–	–	12.592	1.071	1.083

Box 7.3 The Czech, Polish and German floods in July, August and September 1997

Beginning of July:

- Record heavy rainfall on the Czech–Polish border

By 9 July:

- Almost a quarter of the Czech Republic was affected, with thousands of families evacuated: 16,000 were evacuated from the Ostrava region when a reservoir had to be discharged to avoid its collapse
- North of the mountains, in Poland, river levels up to 3 m above alarm levels were being recorded on the Oder
- Forty thousand Czechs were without electricity or drinking water

By 11 July:

- The number of evacuees in Poland had reached 40,000, and about 100 towns and villages were cut off

By 14 July:

- Fifty thousand Czechs had been evacuated from the severely affected regions of Bohemia and Moravia
- Twenty-nine people had died in the floods
- Flood waters were starting to recede in the northeast, and recovery was starting in southern Poland

By 16 July:

- It was clear that the flood was a major event, the largest in the region for a century or longer
- River levels were still rising on the Oder, downstream of Wroclaw and on the Vistula
- In all, some 320 towns in Poland were affected by the flooding

By 21 July:

- The death toll had reached 49 in Poland and 46 in the Czech Republic
- Of the 137,000 Poles who had been evacuated, 70,000 were still not able to return to their homes
- Flooded factories were releasing pollutants into the flood water
- Many tens of thousands of farm animals (cattle, pigs and poultry) had been drowned

By *c.* 23 July:

- The Polish government re-allocated US$1 billion of its budget for rehabilitation
- Offers of immediate assistance came from many other countries
- The death toll continued to rise to 128 (in all three countries combined)
- A dyke collapsed on the German (east) bank of the Oder, and 2300 people had to be evacuated

By 26 July:

- The situation was stable in Poland and the Czech Republic as the flood wave moved further downstream, although it was still worsening in Germany.

Oder's channel. The long-term effects of these dangerous pollutants are unknown.

Many Polish towns and cities were flooded. The political dimension to the events here was that some of these had been sited as new settlements on the floodplain as a strategic political move since 1945 to ensure that the border between Poland and Germany was well 'defended' through the area supporting a relatively dense population. The flood damage situation in Poland was also much more serious than in Germany because money was spent more quickly in Germany on repair and rehabilitation.

Emergency procedures were often deficient. Countering the flood mostly focused on protecting the embankments bordering the river from breaching. The techniques were quite simply sandbags and earth-moving equipment. But there were significant inefficiencies and delays and, on the Polish side, local authority representatives waited in vain on occasions for the central government to act, due to long traditions of bureaucratic governance there during the Communist era. The embankments were successfully protected in most instances, although some were breached (e.g. just downstream from Frankfurt an der Oder). Much was done to restrict loss of life, but loss of life did occur. Contingency planning was rudimentary on both sides of the river, but this is perhaps not surprising given the rarity of the event and the lack of experience of this severity of flooding in the last twenty-five years.

Lessons learned and future plans

The floods were poorly forecast. It is clear that many lessons have been learned with regard to the meteorological and hydrological systems involved in the flood forecasting and warning system. These are now being improved, although there will be difficulties here given the inter-

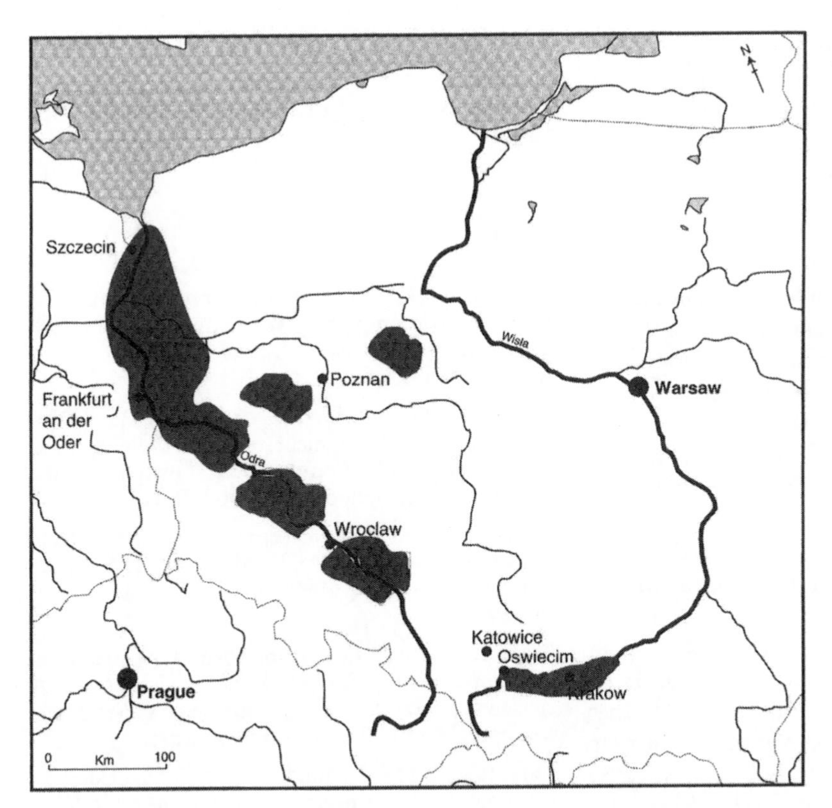

Figure 7.5 Flooding along the River Oder in Eastern Europe, summer 1997.

national character of the Oder (involving coordination across Poland, the Czech Republic and Germany).

Lessons have also been learned by the emergency services, particularly about the need to restore and further protect the dykes. But it would not appear that lessons have been learned about floodplain management or about pollution storage in flood risk areas, since no plans are yet in preparation for new thinking here. What needs to be done is a more systematic approach to restraining encroachment of urban areas into the floodplain, and better contingency planning for the major floods that occur. As far as the environmental damage is concerned, it would also appear that no lessons have been learned and no data have been collected; priority has been given – perhaps understandably in the context – to humanitarian relief and economic recovery.

One engineering 'solution' seeks to integrate the Oder with the west German navigation, and thereby with the whole of the European inland waterway system, through the construction of river training works up the upper Oder to contain the flood waters in an enlarged channel. This could bring economic growth to southern Poland, since the Oder is currently not navigable by the barges using the European waterways. But such a scheme might be ecologically damaging. Alternative scenarios would restore the upper River Oder and its floodplain to a more natural state, perhaps by setting back the existing system of embankments, and thereby decreasing the river's flood conveyance capacity, thus reducing the vulnerability of urban communities downstream.

But much will depend on political will and on resources. Again the situation is worst in Poland, being a poorer country than Germany. A grant of US$300 million was provided to Poland by the World Bank after the 1997 floods, but even this apparently large sum is not likely to result in any significantly reconfigured river or major change in the way that the floodplain is used. Thus there is a real chance that floods like those in 1997 will recur and it is not clear that a coherent plan for flood damage mitigation will be in place soon enough to counter the damage and dislocation that will result.

CONCLUSIONS

Natural hazards are complex interactions of social and physical forces, and cannot be understood fully without a multidisciplinary approach. The broad vision of the geographer can contribute significantly here, through having an understanding of physical processes, human impacts and the potential for sustainable plans and sensible action.

The human dimension is that floods create social exclusion by demonstrating the separation of the 'haves', who receive support and protection, from the 'have-nots', who do not. On an international scale, this is shown by the huge national response to the damaging floods that occurred in the USA in 1993 compared with the relative neglect of the millions of people affected annually by floods in Bangladesh (Alexander 1993). In Poland and Germany, above, the same was true: the impoverished Poles suffered more than the affluent Germans.

The physical dimension is that each flood has a different character and that 'template' solutions do not always apply. Physical space is limited. Solutions to many flood hazards are difficult to apply when people have unwisely occupied floodplain and coastal areas, because this means that there is insufficient room for the natural overflow of rivers on to floodplains without the damage and dislocation that this causes.

But also, as with many natural hazards, floods are intensely political phenomena as well as being complex geophysical events. The political character of floods derives from the damage and loss of life that they cause, and governments and their agencies are called to account for these effects. Floods also create emergencies and the associated emotional circumstances when governments and international organisations spend money – on relief or alleviation measures. These resource flows can be the subject of intense political bargaining (Penning-Rowsell *et al.* 1998) and sometimes corrupt practices (Penning-Rowsell 1996).

GUIDE TO FURTHER READING

Alexander, D. (1993) *Natural disasters*. London: UCL Press. This is a comprehensive text on natural disasters, both from a human and a physical perspective, and on an international scale.

Anderson, M.G., Walling, D.E. and Bates, P.D. (1996) *Floodplain Processes*. Chichester: John Wiley & Sons. An important collection of research material on the geomorphology of floods and floodplains.

Beck, V. (1992) *Risk Society: Towards a New Modernity*, trans. M. Ritter. London: Sage. The 'cutting edge' of provocative 1990s thinking about risk and society.

Hewitt, K. (1997) *Regions of Risk: A Geographical Introduction to Disasters*. London: Longman. A challenging account of the changing geographical conceptualisation of hazards and disasters.

Kanti Paul, B. (1997) Flood research in Bangladesh in retrospect and prospect: a review. *Geoforum* 28(2), 121–31. This is an excellent synthesis of research in what must be the world's most flood-prone nation.

Penning-Rowsell, E.C. and Fordham, M. (1994) *Floods Across Europe: Hazard Assessment, Modelling and Management*. London: Middlesex University Press. This volume surveys research at a European scale into the complex processes of risk assessment and flood hazard adjustment.

Useful flood hazard-related web sites:
http://www.yahoo.com/Science/Earth_Sciences/Meteorology/Weather_Phenomena/Floods
Provides up-to-date information and reports on major events around the globe as well as a selection of links to other sites dealing with floods.
http://www.fema.gov/fema/flood.html
The US Federal Emergency Management Agency (FEMA) web pages on floods provide basic information about flood hazards and flood damage.
http://www.dartmouth.edu/artsci/geog/floods
The Dartmouth College Flood Remote Sensing Page is the home page of the Global Flood Monitoring and Analysis Project.
http://www.floodplain.org
Includes sections containing full-text articles, a calendar of upcoming events, an index of publications,

and a list of US contacts in floodplain management.
http://www.colorado.edu/hazards/sites/sites.html
The Natural Hazards Center at the University of Colorado, Boulder, USA, includes an extensive annotated list of useful sites on floods and other natural hazards.
http://www.mdx.ac.uk/www/gem/fhrc.htm
Middlesex University's Flood Hazard Research Centre has information on flood damage, hazard mitigation strategies, river and water management, and lists of publications.
http://www.environment_agency.gov.uk
The Environment Agency has responsibility for flood defence in England and Wales. See this page for details of its role.
http://www.cres.anu.edu.au
The Centre for Resources and Environmental Studies (CRES) at the Australian National University is the foremost centre in the Southern hemisphere for flood hazard studies.

REFERENCES

Arnell, N.W., Clark, M.J. and Gurnell, A.M. (1984) Flood insurance and extreme events: the role of crisis in prompting changes in British institutional response to flood hazard. *Applied Geography* 4, 167–81.

Bandarin, F. (1994) The Venice project: a challenge for modern engineering. *Proc. Inst. Civ. Eng.* 102, 163–74.

Bednarz, E. (1997) Poland after the flood. *Warszawa* September, 42–3.

Burton, I., Kates, R.W. and White, G.F. (1978) *The Environment as Hazard*. Oxford: Oxford University Press.

Burton, I., Kates, R.W. and White, G.F. (1993) *The Environment as Hazard* (2nd edition). Oxford: Oxford University Press.

Chan, N.W. and Parker, D.J. (1996) Response to dynamic flood hazard factors in peninsular Malaysia. *Geographical Journal* 163(3), 313–25.

Ericksen, N.J. (1986) *Creating Flood Disasters?* Water and Soil Miscellaneous Publication No. 77. Wellington, New Zealand: National Water and Soil Conservation Authority.

Green, C.H. and Penning-Rowsell, E.C. (1989) Flooding and the quantification of intangibles. *Journal of the Institution of Water and Environmental Management* 3(1) 27–30.

Handmer, J. and Dovers, S.R. (1996) A typology of resilience: rethinking institutions for sustainable development. *Industrial and Environmental Crisis* 9(4), 482–511.

Handmer, J. and Penning-Rowsell, E.C. (eds) (1990) *Hazard and the Communication of Risk*. Aldershot, UK: Gower.

Hollis, G.E. (1988) Rain, roads, roofs and runoff: hydrology in the cities. *Geography* 73, 9–18.

Hood, C. and Jones, D.K.C. (eds) (1996) *Accident and Design: Contemporary Debates in Risk Management*. London: UCL Press.

Hood, C., Jones, D.K.C., Pigeon, N.F., Turner, B.A., Gibson, R., Bevan-Davies, C., Funtowicz, S.O., Horlick-Jones, T., McDermid, J.A., Penning-Rowsell, E.C., Ravetz, J.R., Sime, J.D. and Wells, C. (1992) Risk management. In Royal Society Study Group: *Risk: Analysis, Perception and Management*. London: Royal Society.

Kates, R.W. (1962) Hazard and choice perception in floodplain management. Research Paper No. 78, Department of Geography, University of Chicago. Chicago: University of Chicago Press.

MAFF (Ministry of Agriculture, Fisheries and Food) (1998) *Project Appraisal Guidance*. London: MAFF.

Mitchell, J.K., Devine, N. and Jagger, K. (1989) A contextual model of natural hazard. *Geographical Review* 79(4), 391–409.

Newson, M.D. (1989) Flood effectiveness in river basins: progress in Britain in a decade of drought. In K. Beven and P. Carling (eds) *Floods: Hydrological, Sedimentological and Geomorphological Implications*. Chichester: John Wiley & Sons pp. 151–69.

Nicholls, R.J. (1995) Coastal megacities and climate change. *GeoJournal* 37(3), 369–79.

Parker, D.J. (1995) Floodplain development policy in England and Wales. *Applied Geography* 15(4), 341–63.

Parker, D.J. and Penning-Rowsell, E.C. (1983) Flood hazard research in Britain. *Progress in Human Geography* 7(2), 182–202.

Pelling, M. (1998) Participation, social capital and vulnerability to urban flooding in Guyana. *Journal of International Development* 10, 469–86.

Penning-Rowsell, E.C. (1996) Flood hazard in Argentina. *Geographical Review* 86(1), 72–80.

Penning-Rowsell, E.C. (1998) *The Floods in Poland and Germany in 1997*. London: Green Cross UK (University of Kingston upon Thames).

Penning-Rowsell, E.C., Chatterton, J.B. and Winchester, P. (1994) *River Irwell Flood Control Scheme: Benefit–Cost Assessment*. London: Middlesex University Flood Hazard Research Centre.

Penning-Rowsell, E.C., Parker, D.J. and Harding, D.M. (1986) *Floods and Drainage: British Policies for Hazard Reduction, Agricultural Improvement and Wetland Conservation*. London: Allen & Unwin.

Penning-Rowsell, E.C. and Smith, D.I. (1987) Self-help flood hazard mitigation: the economics of house-raising in Lismore, N.S.W., Australia. *Tijdschrift voor Economische en Sociale Geografie* 78(3), 176–89.

Penning-Rowsell, E.C., Winchester, P. and Gardiner, J.L. (1998) New approaches to sustainable hazard management for Venice. *Geographical Journal* 164(1), 1–18.

Quarantelli, E.L. (ed.) (1998) *What is a Disaster? Perspectives on the Question*. London: Routledge.

Rodier, J.A. and Roche, M. (1984) *World Catalogue of Maximum Observed Floods*. Wallingford, UK: International Association of Hydrological Sciences (publication 143).

Smith, D.I. (1999, in press) *Water in Australia: Resources and Management*. Oxford: Oxford University Press.

Torry, W.I. (1979) Hazards, hazes and holes: a critique of 'The Environment as Hazard' and general reflections on disaster research. *Canadian Geographer* 23(4), 368–83.

van der Leeden, F., Troise, F. and Todd, D.K. (1990) *The Water Encyclopedia*. Michigan, USA: Lewis.

8

Coastal erosion

Tom Spencer

INTRODUCTION

In 1985, Bird reported on a project undertaken by the International Geographical Union's Commission on the Coastal Environment: this found 70 per cent of the world's sandy coastline undergoing net erosion. As 60 per cent of the global population (or nearly 3 billion people) live in the planet's coastal zones, and two-thirds of the world's cities with populations of 2.5 million or more are located in open coast or estuarine locations (Viles and Spencer 1995), Bird's (1985) statistic identifies a major environmental issue. It is an issue already strongly imprinted on many local, and national, consciences. Strong conflicts can arise in the tackling of coastal erosion between local residents; local, regional and national regulatory bodies and interest groups; and consultant scientists: the interaction of physical processes and economic, social and political forces makes coastal erosion a strongly geographical problem. Furthermore, any coastal study must take account of the great diversity of coastal settings and of the role of environmental change over the last 10,000 years in determining contemporary shoreline morphology (see Box 8.1). There has been a growing concern in the last decade that coastlines are at risk and under pressure. Two broad sets of processes, both potentially accelerating, have been identified.

The first set concerns the impact of sea level rise over the next 100 years consequent upon human-induced global climate change. The figure of *c.* 100–150 cm of sea level change by AD 2100 is still widely found in the literature, but current 'best guess' estimates for this period are 49 cm, a considerable downward revision on earlier figures (see Warrick *et al.* 1996 for detailed discussion and French *et al.* 1995a for geomorphological and ecological implications). However several caveats should be applied to this apparently comforting reduction. First, very large uncertainties remain in the predictions of global environmental change. Second, it is not clear how the primary effect of sea level rise might influence a range of secondary effects, such as changing tropical cyclone magnitudes and frequencies and mid-latitude wave climates, which might in themselves have greater impact on coastal communities than sea level rise *per se*. Third, although the expected sea level rise for the next 100 years is now much lower than previously envisaged, it still represents a significant increase on the previous 100 years. The magnitude of expected sea level rise converts to an average rate of sea level rise of 4.5 mm a^{-1}. Although it is difficult to provide a single figure for the rate of sea level rise over the last 100 years, Warrick *et al.* (1996) suggest an average rate of rise of 1.8 mm a^{-1}. Thus future rates are currently expected to be 2.5 times those of the last 100 years.

Any future sea level change will be played out against the backdrop of the second set of major processes to affect the world's coastlines. This is the creation of highly modified, 'artificial' shorelines as a result of long-continued, but now larger-scale, human modification and utilisation of the coastal zone. Typically 30–40 per cent of open coasts in developed countries (e.g. USA, England, Japan) have protection against flooding,

Box 8.1 Coastal classification and coastal erosion

Coastal dynamics, including erosion, are controlled at the large scale by two sets of factors, one historical (in the broadest sense) and one contemporary. In the first, coastal type is controlled by plate tectonic setting, and this provides a broad emergence/submergence categorisation. This in turn is overlain by more immediate historical factors, particularly the nature of sea level change over the last 10,000 years of the postglacial transgression. Site-specific variations in sea level in this period have resulted from the varying contribution of regional isostatic (affecting the movement of land surfaces) and global eustatic (affecting the volume of the oceans) factors to sea level change. Thus the different sea level histories of, for example, Australia, where present sea level was reached 6000 years ago, and Arctic Canada, where sea levels have been falling since the start of the deglaciation, are a component in explaining shoreline morphology. This historical backdrop is then worked upon by current global variations in wave energy, tidal regime and, for vegetated coastal ecosystems, biogeographic patterns and processes. In 1952, Valentin brought these factors together in an elegant coastal classification scheme. In Valentin's diagram, the *y* axis represents the historical and the *x* axis the contemporary components. Varying combinations of these two factors give coastal advance or retreat. The solid line running through the centre of the diagram from top left to bottom right separates all retreating coastlines (bottom left) from all advancing ones (top right).

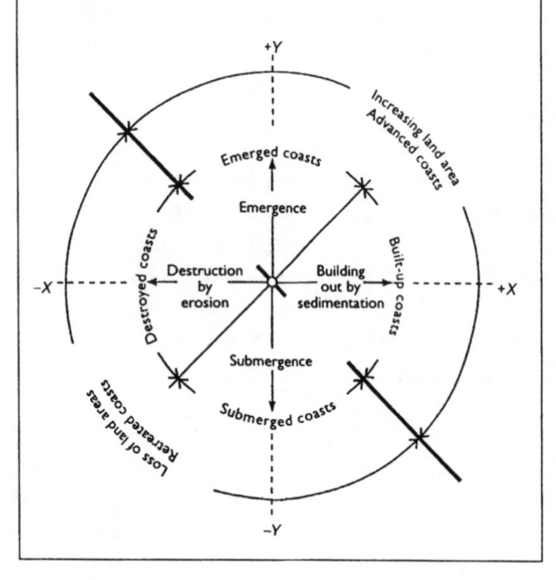

erosion and navigation hazard (Nordstrom 1994), with this figure reaching close to 100 per cent in some localities (e.g. the coastline of Belgium). The profound implications of these engineering structures for coastal processes and erosion are discussed below.

This chapter takes the view that dealing with coastal erosion requires an assessment of its position in the wider scheme of coastal dynamics (erosion is just one component of a sedimentary budget for any coastline) and that only by understanding, and allowing for, changing coastal position will truly sustainable coastal management be achieved. This review, limited by space, concentrates upon problems on sandy and muddy coastlines in the developed world (Plate 8.1); however, the pressures generated by coastal erosion in developing countries are enormous. Thus, for example, in Bangladesh a huge and rapidly growing, poor, rural population must contend, not only with a tectonically active, highly mobile floodplain coast, subject to both major riverine floods and catastrophic storm surges, but also the environmental problems resulting from the construction of hard-engineered flood defence lines imposed by external agencies (e.g. Brammer 1993). The diverse coastal problems faced by developing countries are well covered by Nicholls and Leatherman (1995).

Plate 8.1 A developed coast: Waikiki, Honolulu, Hawaiian Islands. How does one devise a sustainable coastal management strategy for such a coastline? (*photograph: T. Spencer*).

THE NATURE OF THE PROBLEM

Beaches and their adjacent nearshore zones are highly effective buffers to incoming wave energy, with offshore bathymetry reducing incident wave energy by 95–99 per cent (Carter and Woodroffe 1994). As the ratio of force to resistance is high, and changing inputs result in demonstrable morphological change, the application of a process-based approach to coastal studies (largely promulgated through oceanography, coastal engineering and sedimentology rather than geomorphology) has been highly successful, developing from the studies of the Beach Erosion Board in the United States in the immediate post-Second World War period and reaching its fullest expression in the encyclopaedic Shoreline Protection Manual (CERC 1984). Thus the forecasting of deep-water wave conditions from meteorological and other variables, the modification of deep-water waves by nearshore bathymetry, the nature of breaking waves, and water and sediment movements within the breaker zone are now well known (e.g. Komar 1976).

The changing state of intertidal and shallow subtidal nearshore profiles has long been recognised, with the use of such terms as 'winter' and 'summer' beaches (e.g. Shepard's (1950) classic studies in southern California) or 'storm' and 'swell' profile types (Komar 1976). The idea of shifts between equilibrium beach states has proved particularly attractive when extended to the possible effects of sea level rise. In particular, the so-called 'Bruun rule' (Bruun 1962; Dean 1991) has proved to be a flawed but remarkably persistent concept. It begins from the premise that there is an equilibrium depth of water offshore. With sea level rise, water depth increases and sea bed deposition must take place to restore the equilibrium depth; this is achieved by shore erosion and shoreline retreat (Figure 8.1). These parameters can be easily cast into simple mathematical form and thus the degree of shoreline shift calculated from estimates of sea level rise. Despite the difficulty of establishing both upper and, particularly, offshore points of closure for studied profiles; of accepting sudden rather than progressive changes in sea level; and in the absence of allowing for longshore processes in controlling profile characteristics, (and see Pilkey *et al.* 1993, for example, for additional problems) the Bruun rule has been widely quoted, albeit usually with reference to a Great Lakes study, where the local geology, topography, bathymetry and lake climate are peculiar and not easily transferable to open coasts.

More sophisticated analyses have attempted to establish what associations exist between beach states and breaking wave characteristics; what the range of these states might be; and how the rate and direction of change between states might be predicted. Largely through the efforts of the 'Australian School', six characteristic domains have been identified (Figure 8.2). The term 'reflective' describes one end-member: steep, narrow, coarse-grained beaches, often typified by rhythmical longshore topography, where wave

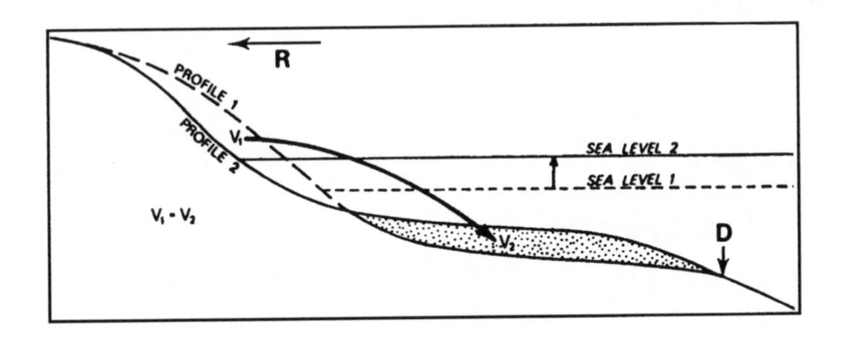

Figure 8.1 The 'Bruun Rule'. On sea level rise (sea level 1 to 2) sediment is transferred from the shoreface sediment store (V_1) to the nearshore zone (V_2) to re-establish an equilibrium water depth. The coastline will retreat (R) until stability is re-established. Point D is the 'closure depth', the outer limit of profile adjustment.
Source: After Bird 1993.

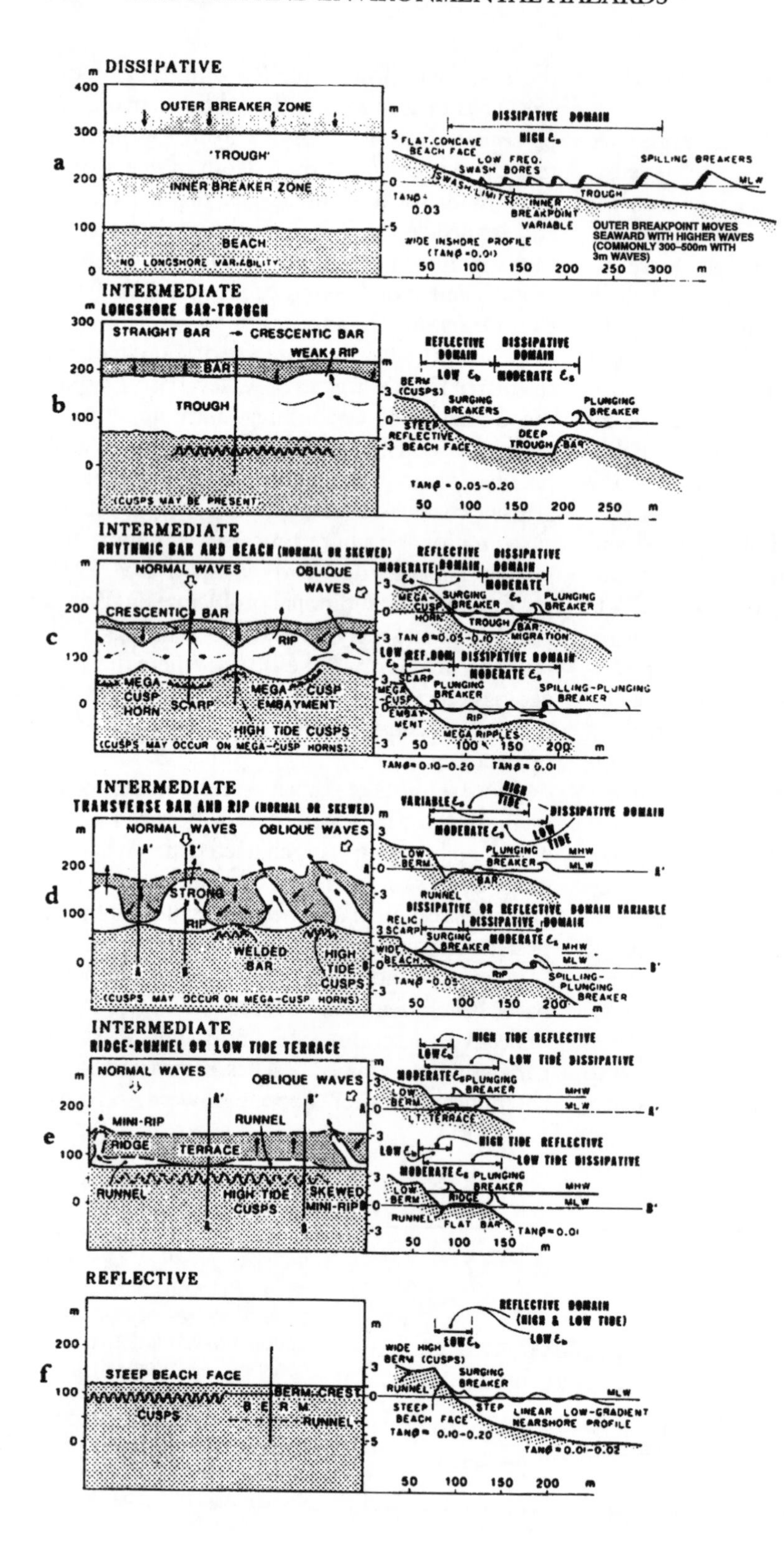

Figure 8.2 Plan and profile configurations of the six major beach states observed on the Australian coast. a = dissipative; b = longshore bar-trough; c = rhythmic bar and beach; d = transverse bar and rip; e = ridge runnel; f = reflective.

Source: After Wright and Short 1984.

energy is reflected back from the shore and trapped inside the breaker zone. At the other extreme, 'dissipative' beaches exhibit low-angle, wide, finer-grained surf zones that strongly attenuate incoming wave energy (Wright and Short 1984; Masselink and Short 1993). Examination of long time-series has shown that these end-member states are quite rare and unstable and that most moderate- to high-energy coasts alternate between the four intermediate domains (Wright *et al.* 1985; and see Sonu and James (1973) for transition matrices and Markov chain modelling of these processes). Useful though this morphodynamic approach has proved, it is best suited to relatively short-term coastal behaviour and has rarely been extended along long stretches of coastline. In this regard, the concept of the coastal or littoral cell, open to energy throughflow but more closed with respect to sediment transfers, becomes of particular value (Carter and Woodroffe 1994).

Cell boundaries may be fixed at topographic limits (such as headlands, estuaries or deltas) or, more contentiously, be mobile with free boundaries delimited by shore-parallel wave flux or littoral drift rates (Carter 1988) on open coasts. As wave directions change on such shorelines, so cell structure changes with migrations and mergings (Figure 8.3; *ibid.*). Over time, cells may reach a static equilibrium, either the coast being in all places parallel to the approaching wave refraction (or 'swash-aligned'), or where longshore currents are nullified ('drift-aligned') or where beach sediments are graded in such a way that the threshold for particle entrainment is never reached (*ibid.*). However, most cells exhibit patterns of sediment exchange over time, between beaches and beach ridge and dune systems, between estuaries and coasts, and between coastal and offshore environments. These exchanges may be formalised through the calculation of sediment budgets, although this is easier in theory than in practice; few published budgets appear to balance.

As one moves into larger temporal scales, then the role of extreme events becomes important: thus at Moruya Beach, southeast Australia, the 'normal' range of beach response was forced into

Figure 8.3 Variations in coastal cell structures and boundary positions with varying wave approach, Magilligan Point, Northern Ireland.

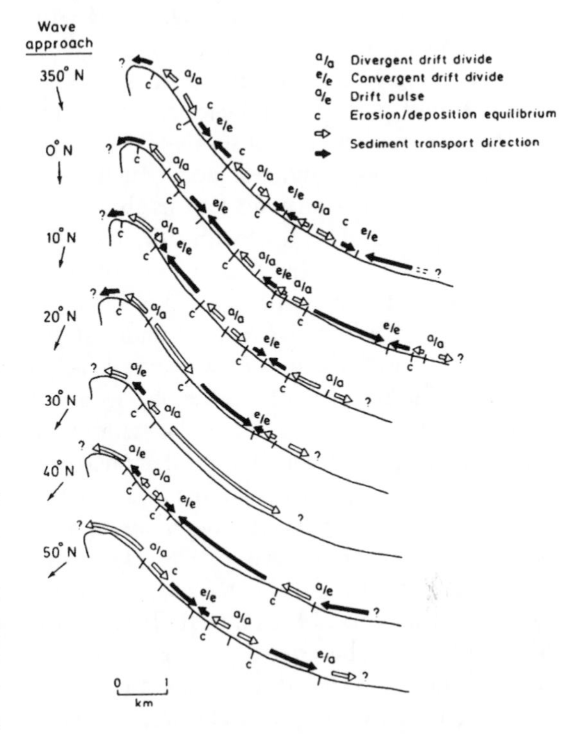

Source: After Carter 1988.

a quite different state, from which recovery took six years (Thom and Hall 1991). However, this work still suggests some form of equilibrium. Recent research, however, has begun to investigate the idea that shoreline behaviour may be non-linear (e.g. Phillips 1992), focused around a set of ideas under the heading of large-scale coastal behaviour (LSCB; de Vriend *et al.* 1993), applied to timescales of decades and distances of tens of kilometres where boundary conditions are set by geological time (and refer back to Box 8.1). Where high-quality data sets are available, such as the Dutch JARKUS bathymetric database, such approaches really begin to provide a rigorous, detailed picture of the space–time complexities in the nearshore zone (Wijnberg and Terwindt 1995) and, by implication, erosion and deposition patterns at the shore.

Cells, and sediment movements within and

between cells, provide a useful framework within which to consider human disruptions to natural processes. Sand mining (Earney 1990) removes sediments directly, for building aggregates, soil improvement and precious-mineral mining (e.g. diamond mining on the Namibian coast). However, the most pervasive influence at the coast is the interference between onshore–offshore and alongshore processes and nearshore structures.

Coastal morphodynamics tells us that beach profiles expand and contract with changing wave energy inputs. However, the emplacement of sea walls prevents the natural expansion of the shore profile to a fully dissipative state under storm conditions. Large volumes of water on beaches lead to a dominance of backwash, sediment down-combing and landslide-like failure of saturated beach sediments. The resultant narrowing and lowering of beaches backed by structures by comparison with profiles on adjacent non-defended shores has been extensively documented (e.g. Kraus and Pilkey 1988; Louisiana, USA: Nakashima and Mossa 1991; Texas, USA: Morton 1988). As beach sediment loss continues, so walls are threatened by undermining, so initiating a cycle of progressively larger defences with bases to lower and more seaward levels on the shore profile (e.g. history of sea walls at Porthcawl, UK: Carter 1988). A further problem is that accelerated erosion is often characteristic of the downdrift end of a defence structure; the temptation, therefore, is to extend the structure progressively downdrift, ultimately until perhaps hundreds of kilometres of coastline are so protected. Elsewhere in alongshore directions, patterns of deposition and erosion result from the interruption of longshore sediment transport; these are well known (see Viles and Spencer 1995 for review). Jetties (shore-normal structures that extend seawards beyond the breaker zone) induce updrift accumulation and downdrift erosion. More shoreward, on the beach itself, shore-normal groynes are used to conserve remaining beach volumes or capture sediments moving alongshore. Groyne fields can similarly lead to problems of down-drift erosion, although once 'full', longshore drift is re-established as sediment bypasses the structure at either its landward or seaward limits. Groyne fields are thus an essential component in beach conservation on many coasts (e.g. UK: Bray *et al.* 1992; Japan: Walker and Mossa 1986). A rather different interruption to longshore drift occurs with the presence of offshore breakwaters: these reduce wave energy, and thus longshore sediment transport rates, leading to sediment accumulation in their lee (and the need for dredging at port and harbour installations).

The frequent consequence of this human-induced erosion is to attempt to alleviate such problems by artificially re-establishing natural sediment movements. Examples include the pumping of sediments across inlets to maintain downdrift sediment transport (e.g. Fort Worth, Texas, USA) and the dredging and dumping of offshore sediments to replenish beaches deprived of sediment supply by updrift barriers (e.g. Santa Barbara, California, USA). In recent decades, beach nourishment schemes have become a particularly favoured solution to beach volume loss (see following section). However, such activities must also be seen in the context of sediment delivery to coastlines. Estimates for contemporary fluvial inputs to the coastal zone are not well known, but perhaps 10–15 per cent of the total input of 10^{16-17} t may contribute to coastal aggradation. Regional variations are great, partly controlled by plate tectonic controls on drainage basin characteristics and partly by climatic regime (Milliman and Meade 1983; Milliman and Syvitski 1992). However, a further set of controls has been, and continues to be, human activity within feeder drainage basins. The timing and style of catchment modification has played an important role in determining sediment supply and coastline response. Thus, for example, in the Mediterranean, it appears that the sediment pulse was initiated in pre-Classical times and accelerated with Greco-Roman landscape modification – thus northern Mediterranean shores have been largely fossil for a thousand years or more (Vita-Finzi 1964) and consequently heavily managed to retain what sediment is present. The sediment flush/sediment starvation pulse in North America has been much more recent (IGBP 1993). Even more recently, in Egypt, the earlier expansion of the Nile delta has been replaced by rapid recession with the removal of sediment supply following the completion and closure of the Aswan High Dam in 1964 (Stanley and Warne 1993). Finally, at the present time, tropical deforestation is creating the latest geographical focus in the developing tropics for sediment pulses and coastal progradation. It seems likely that the familiar pattern of shoreline problems will follow: initial difficulties with navigation and siltation as river sediments disrupt coastal settlements and infrastructure; the construction of sediment control measures; and the erosion that follows the combination of the cessation of accelerated sediment inputs and the interference from structures in the natural processes of sediment redistribution.

CASE STUDIES

Coarse sediments

It is clear from the preceding discussion that a substantial beach is an excellent form of coastal defence and that there are considerable management benefits from maintaining or improving a beach frontage through beach nourishment/replenishment. The advantages of this approach are three-fold. First, such natural defences are morphodynamically active and hence can adjust to changing wind and wave conditions in a way that is not possible for a fixed defence line. Second, reduced wave run-up and flooding behind defence lines, as a result of wave energy reflection or dissipation, is achieved without the need to emplace sediment-retaining structures, which then create down-coast sediment starvation and erosion problems. Third, beach nourishment often provides the additional benefit of recreating a valuable recreational and amenity asset. The attractiveness of this form of shoreline improvement to operating authorities can be seen from the rapid increase in the adoption of nourishment schemes, first introduced in the USA in the 1920s, from the 1970s onwards (Pilkey 1995). However, the performance of individual beach replenishment efforts has been mixed (Davidson *et al.* 1992) and not all these differences can be explained by large-scale regional differences in shoreline type or sea level change (Leonard *et al.* 1990). Thus, of 110 such schemes in the Gulf of Mexico, 23 per cent were found to have persisted for over five years, 54 per cent for one to five years and 23 per cent for less than one year (Dixon and Pilkey 1989).

It is instructive to look at failed beach nourishment projects; many identify a lack of knowledge of coastal geomorphology as a key component in failure. The great majority of beach nourishment schemes are undertaken because of the loss of the existing beach; thus it is essential to know why the beach is being eroded and where the sediment sink is located, whether to landward, to seaward or alongshore. Thus any individual beach must be set in its regional context, in terms of both shoreline morphology (both natural and artificial) and wind, wave and tidal regimes (including the role, if any, of extreme events). Such pre-planning should help to determine the nature, volume and location of sediment emplacement. A second lesson from scheme failure is the need to monitor the performance of beach recharge after emplacement, as the following case study illustrates.

Bournemouth beach, southern England

Bournemouth beach forms the central section of the 30 km long broad embayment of Poole Bay on the coast of southern England (Figure 8.4A). The predominant wave direction is from the southwest, coinciding with the maximum fetch from the Atlantic Ocean. The bay is backed by low, erodible cliffs stabilised to varying degrees; the characteristic beach sediment is fine sand to the west, but it coarsens to gravels towards Hengistbury Head in the east (May 1990). The town of Bournemouth relies upon its sandy beaches to help to sustain an economically important tourist industry and has traditionally used the standard coastal protection measures of seawalls and groyne fields to prevent urban

Figure 8.4 Beach replenishment at Bournemouth: (A) location map with numbered coastal groynes; beach profiles fall within this groyne field; (B) changes in monitored and predicted beach volumes, derived from detailed beach profile data set, and determination of critical beach volumes under different sea-level rise scenarios (see text for further discussion).

A

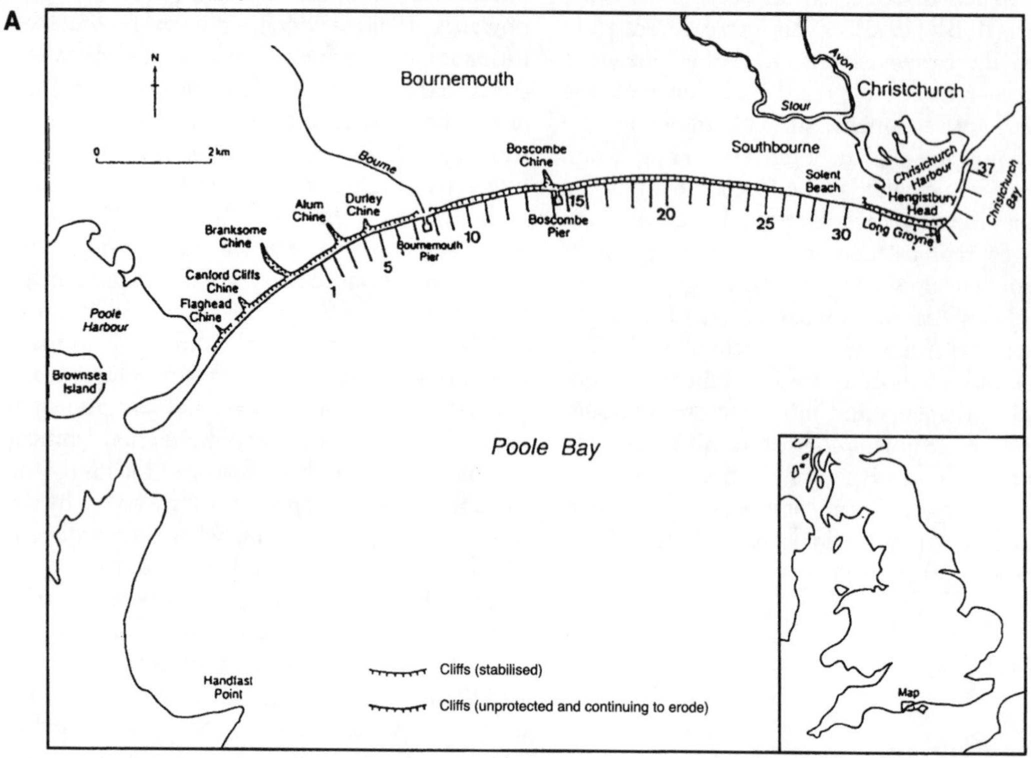

B

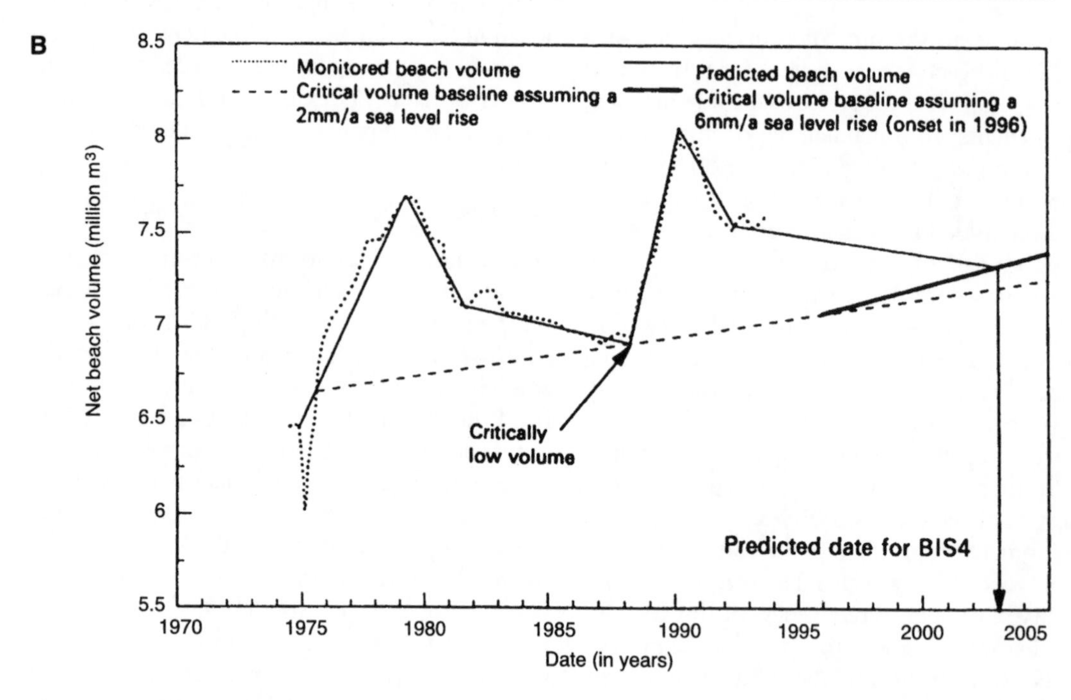

Source: After Harlow and Cooper 1996.

flooding and retain beach sediments, respectively. However, in the 1960s this system began to fail, with falling sea levels and increasing damage to seawall structures. A pilot beach nourishment scheme (Beach Improvement Scheme (BIS) 1) was implemented in 1970, with 84.5×10^3 m^3 of dredged sand emplaced along a 1.8 km frontage. This was followed by a large-scale scheme (BIS 2) in 1974–75, when 654×10^3 m^3 of sand was pumped directly onto beaches over an 8.5 km frontage, with an estimated additional 749×10^3 m^3 of material, as 'leakage' from the pumping process, being deposited in the shallow nearshore zone. A third emplacement (BIS 3) was undertaken between 1988 and 1990, adding 999×10^3 m^3 of fill dredged from the entrance to Poole Harbour (Harlow and Cooper 1996). The reasons for this third phase of beach replenishment are clear from the synthesis of data on beach dynamics gathered from almost forty beach profile lines on some fifty occasions since 1974 (Figure 8.4B). The total beach volume increased from a minimum of 6×10^6 m^3 in 1975 to a peak of 7.7×10^6 m^3 in 1979 as a result of both the direct emplacement of beach material and the onshore migration of 'leaked' material. Thereafter, beach volume decreased to 6.9×10^6 m^3 in 1988, the accompanying narrowing of the beaches and the onshore migration of the mean high water mark necessitating the BIS 3 scheme. This restored beach volume to over 8×10^6 m^3 in 1990. By late 1993, beach volume had decreased to 7.9×10^6 m^3, suggesting the necessity for the next phase of beach replenishment, with a factor included for potential near-future sea level rise, before 2003 (Figure 8.4B). The important conclusion from this study is the need for not only sensible pre-emplacement planning of nourishment but also the value of ongoing monitoring once emplacement has taken place. Monitoring must be regular and sustained to establish trends in beach volume change from the high degree of variability between successive individual surveys, and extend far enough offshore to encompass shallow subtidal as well as intertidal changes. It is only through an ongoing management commitment of this kind that the technique of beach nourishment can be 'fine-tuned' for optimal performance, thus avoiding the traditional, and costly, crude design practice of 'over-filling' (often by 40 per cent; Verhagen 1992) nourished beaches to allow for unmonitored losses.

Fine sediments

It is not only beaches that provide important natural coastal buffer zones. Intertidal mudflats and vegetated mangrove swamps and salt marshes also perform similar energy-dissipating functions (Pethick 1992).

At the coastal scale, salt marshes within estuaries reduce tidal range and flooding potentials through frictional drag on water sufaces and by allowing the high tide stage of water on marsh surfaces (Burd 1995). Recent field measurements have shown that marsh surfaces can dissipate between 47 and 99 per cent of incident wave energy over distances of 200 m or less (Moeller *et al.* 1996), thus reducing wave run-up and overtopping risk on marsh-fronted sea defences. However, on many coasts land reclamation has at worst removed and at best reduced the width of such fronting marshes. Furthermore, there is evidence in many localities that remaining marshes have been subjected to accelerated loss in recent decades, perhaps as a result of sea level rise. On natural coastlines, marshes will migrate landwards as sea level rises to maintain their position in the tidal frame. However, on protected coastlines, this migration is prevented by a landward defence line, and marsh volumes cannot be preserved. The loss of fronting salt marsh both results in increased wave action against the defence and removes mechanical support from its toe; the likelihood of undermining and collapse thus generates a demand for costly re-engineering of the defence line.

However, an alternative approach is that provided by 'managed retreat' (also known as 'coastal setback' or 'shoreline realignment') whereby the defence line is repositioned in a more landward position. Such schemes are attractive for physical reasons, as they create an immediate energy-dissipating zone in front of the new defence line, and for ecological/conservation reasons, as replacement intertidal habitat compensates for tidal wetlands lost elsewhere. At the present time, therefore, managed retreat trials are being implemented in several locations, including the UK and the Californian and Pacific northwest coasts of the USA. In the UK, the county of Essex in eastern England is protected by 430 km of seawalls, a defence system strengthened in the aftermath of the disastrous floods of 1953 but one now reaching the end of its design life. Some 60 per cent of these seawalls are protected by fronting salt marsh, but erosion of these marshes has accelerated since the 1970s. Thus several managed retreat trials, including monitoring studies on water, sediment and nutrient exchanges; soil physical and chemical changes; vegetation re-establishment; and sedimentation/accretion processes, are being undertaken in this region. One such experiment is at Tollesbury Fleet on the Blackwater estuary, a 21 ha site re-flooded in August 1995 after a controlled breach of the old sea wall (Plate 8.2). A former drainage channel to a now defunct sluice on the eastern side of the site divides the site into a lower, northern area previously sown with clover from a higher, sloping southern area formerly under cereal crops (Figure 8.5).

There are considerable unknowns as to the long-term performance of such schemes. One crucial set of questions revolves around the observation that usually new tidal exchange reactivates former salt marsh surfaces converted to agricultural use on enclosure from the sea. Dewatering, compaction and soil chemistry changes on isolation from the marine environment mean that re-flooded surfaces are likely to be much altered from natural marsh surfaces (Portnoy and Giblin 1997). Furthermore, neighbouring marshes outside reclaimed areas will have continued to accrete vertically, and thus there are likely to be height differences – perhaps of the order of 1.0–1.5 m in systems dominated by inorganic sediments (Pethick and Burd 1995) – between higher natural and lower reactivated marsh surfaces. One of the challenges for managed retreat site design, therefore, is to ensure that these height differences are eliminated by rapid sedimentation. This can only be achieved through a proper understanding of salt marsh development processes and how they can be manipulated for management purposes. At Tollesbury, natural rates of elevation change, measured on naturally vegetated salt marsh surfaces outside the managed retreat site, have averaged *c.* 5 mm a^{-1} since monitoring began in 1995 (sites 1 and 2, Figure 8.5; Table 8.1), a figure consistent with the surfaces keeping pace with

Plate 8.2 Managed realignment of an estuarine shoreline: Tollesbury Fleet, Blackwater estuary, Essex, England. The view is bisected by the now breached old seawall defence line. To the left foreground, natural salt marsh is just being flooded by the rising tide. To the right in the distance, small boats have sailed through the breach into the lower, already well-flooded managed retreat site (*photograph: T. Spencer*).

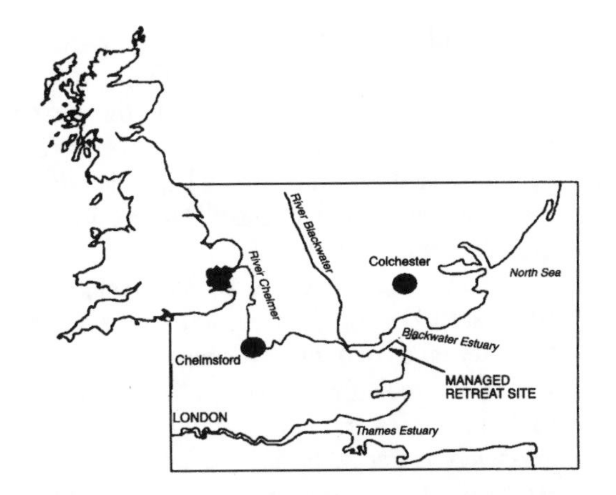

Figure 8.5 Location of Tollesbury Fleet managed retreat site. Station numbers refer to Table 8.1.

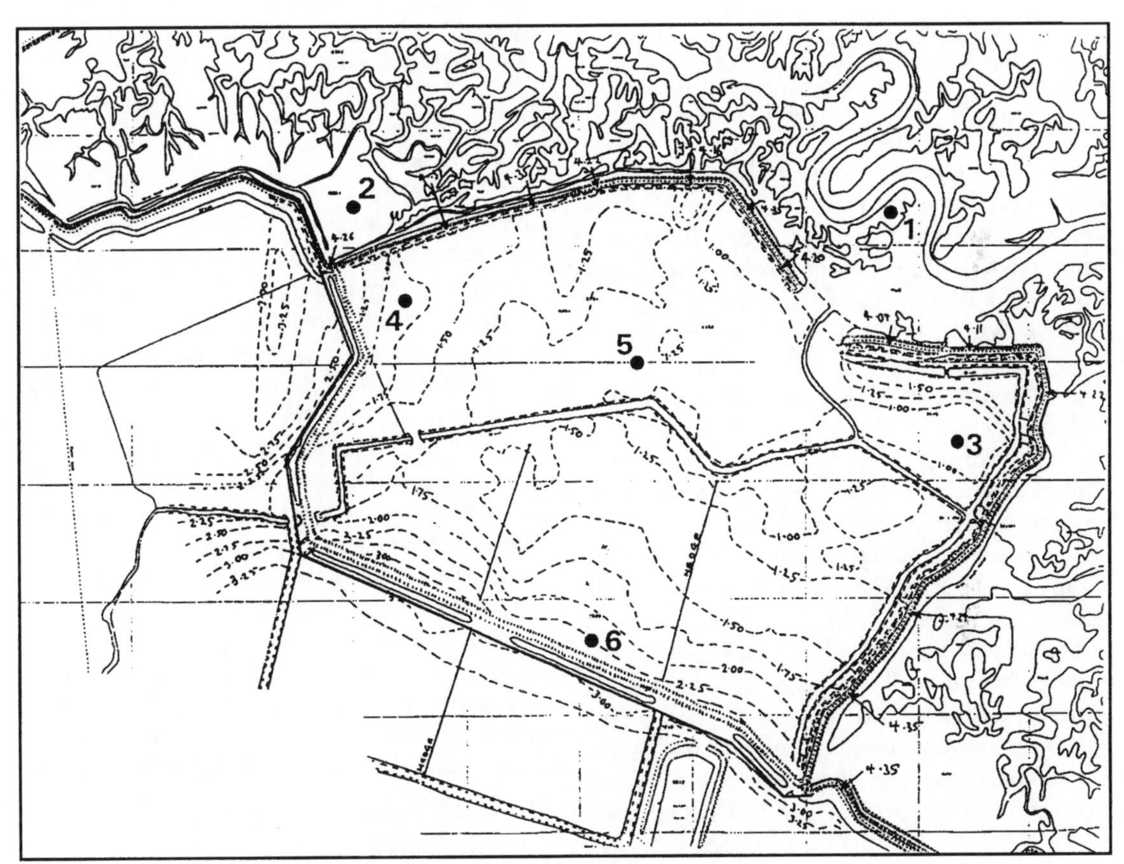

local estimated sea level rise (M. Herman, personal communication 1998). Inside the site, rates of elevation change have been highly variable (Table 8.1); how might these differences be explained?

In marsh systems where vertical accretion is largely driven by externally derived inputs of inorganic sediments (as opposed to peat-dominated tidal marshes, where *in situ* decomposition and accumulation of organic material is

Table 8.1 Average rates of surface elevation change, 1995–7, outside and inside the managed retreat site at Tollesbury Fleet, Blackwater estuary, Essex, UK.

*Station**	*Outside/inside the site**	*Rate of elevation change (mm a^{-1})***
1	Outside	6.6
2	Outside	4.2
3	Inside	35.7
4	Inside	16.1
5	Inside	8.5
6	Inside	–0.4

Notes: *See Figure 8.5 for station locations.
**Negative numbers indicate surface lowering.
Work forms part of an ongoing collaborative project with the US Geological Survey (Dr D. Cahoon), Louisiana Universities' Marine Consortium (Dr D.J. Reed) and University College, London (Dr J.R. French).

the prime input), studies of long-term marsh development show rapid vertical height gains, which slow with time (often after *c.* 100 years) as progressively higher marsh surfaces are flooded by fewer and fewer tidal incursions (French 1993). Within this behaviour, however, recent studies at single-tide to tidal-monthly timescales have been able to resolve the smaller-scale detail of sedimentation dynamics (e.g. French *et al.* 1995b; Leonard 1997), showing that proximity to tidal feeder channels, or 'creeks', is a prime control on patterns of sediment deposition. Thus one way to increase sediment delivery to, and within, a managed retreat site may be to consider either the re-excavation of old tidal creeks or to construct a new network of artificial creeks, taking design rules from fluvial geomorphology. As salt marsh channels also aid plant establishment through substrate drainage and dewatering, then such approaches should be seriously considered (French 1995). However, such work does need to be undertaken with care: poorly designed networks may promote erosion rather than sedimentation (e.g. Haltiner *et al.* 1997), and channel design may also have implications for within-channel hydrodynamics and nutrient exchanges (e.g. Emmerson *et al.* 1997). At Tollesbury, it appears that one of the problems of site design is the difficulty with which the remoter parts of the site receive sediment inputs (see Figure 8.5, Table 8.1). Continued monitoring over the next few years should provide interesting information on changing temporal and spatial patterns in sedimentation and on plant establishment and subsequent vegetation dynamics.

Finally, and while the above discussion has concentrated upon the need to better understand the internal dynamics of managed retreat sites, it should be recognised that the wider uptake of such schemes, and the establishment of progressively larger managed retreat areas, will generate a need to understand how such schemes will impact on broader-scale, whole-estuary hydrodynamics and sedimentation.

CONCLUSIONS

It should be clear from this review that there are fundamental discrepancies between the way in which many coastlines have been modified and subsequently maintained and the way in which natural processes operate; these discrepancies often manifest themselves in severe coastal erosion problems. These difficulties are bad enough without the additional problems introduced by administrative frameworks that tend to fragment coastal management into small units and support local interests above regional and larger-scale concerns. Countries, and indeed regional economic groupings, should therefore in the first instance work towards larger-scale, integrated coastal management plans. Clearly with limited resources, some form of prioritisation will be required. This must, for any extensive stretch of coastline, decide between one of three strategies:

1 'Hold the line' by providing robust and reliable defences;
2 'Accommodate' shoreline change by allowing continued occupancy but with adaptive measures, including adjustments to periodic flooding by modifying buildings and access routes or by the acceptance of periodic inundation; and

3 'Managed realignment' of the coastline by progressively abandoning land and defence structures and recreating ecologically valuable intertidal habitats previously lost through coastal erosion.

Until now, for political reasons there has been a marked reluctance for governments at all levels in developed countries to embrace option 3. However, the huge cost of strengthening existing sea defences in the face of accelerated sea level rise over the next century may force this option to be considered more seriously than hitherto. If so, then the coastal geomorphologist should have an important role to play in formulating the design rules for new, 'semi-natural' coastlines by building into them an understanding of the space–time dynamics of coastal processes.

The wider social and economic evaluation of all these options is, of course, not straightforward. Evaluations are strongly time- and area-specific: thus, for example, a scheme that may be sustainable on a twenty-year planning horizon may not be acceptable on a two-year timescale, and action that may be sensible in the context of regional shoreline change may produce unacceptable local impacts. Implementation will also bring with it a need to prioritise coastal management options on a year-to-year basis; one way to establish a priority list would be to evaluate levels of risk for coastal communities. In addition, there will be a need for the formulation of long-term adaptive policies to lessen the risk to persons and property from flooding and coastal erosion. Such plans will need to offer compensation to those individuals who sacrifice property and land assets in order to improve natural flood defence elsewhere. The building of such a framework will clearly be a major challenge at the start of the twenty-first century for all those concerned with understanding coastal landscapes and their peoples.

GUIDE TO FURTHER READING

Two well-organised and well-argued texts on coastal geomorphological processes are those by Komar (1976) and Pethick (1984), although both are now starting to show their age. Carter (1988) is highly comprehensive, while Viles and Spencer (1995) covers tropical environments; both include a strong engagement with coastal management issues. For a more detailed exposition of the complexities of shoreline evolution, Carter and Woodroffe (1994) contains much excellent material. The *Journal of Coastal Research* is the most accessible and generally useful periodical for keeping up to date with coastal research issues. The Sixth Report of the UK House of Commons Select Committee on Agriculture (1998) provides a fascinating insight into contemporary coastal management issues in the UK: whether it acts as a catalyst for change or lies unopened in unvisited archives remains to be seen.

REFERENCES

Bird, E.C.F. (1985) *Coastline Changes: A Global Review*. Chichester: J. Wiley.

Bird, E.C.F. (1993) *Submerging Coasts*. Chichester: J. Wiley.

Brammer, H. (1993) Geographical complexities of detailed impact assessment for the Ganges–Brahmaputra–Meghna delta of Bangladesh. In R.A. Warwick, E.M. Burrows and T.M.L. Wigley (eds) *Climate and Sea Level Change: Observations, Projections and Implications*. Cambridge: Cambridge University Press, 246–62.

Bray, M.J., Carter, D.J. and Hooke, J.M. (1992) Coastal sediment transport study. *Report to SOPAC*, Portsmouth: University of Portsmouth.

Bruun, P. (1962) Sea level rise as a cause of shore erosion. *Journal of Waterways and Harbor Division, Proceedings of the American Society of Civil Engineers* 88, 117–30.

Burd, F. (1995) *Managed Retreat: A Practical Guide*. Peterborough: English Nature.

Carter, R.W.G. (1988) *Coastal Environments*. London: Academic Press.

Carter, R.W.G. and Woodroffe, C.D. (1994) *Coastal Evolution: Late Quaternary Shoreline Morphodynamics*. Cambridge: Cambridge University Press.

Coastal Engineering Research Center (1984) *Shore Protection Manual* (4th edition). Washington DC: US Government Printing Office.

Davidson, A.T., Nicholls, R.J. and Leatherman, S.P. (1992) Beach nourishment as a coastal management tool: an annotated bibliography on developments associated with the artificial nourishment of beaches. *Journal of Coastal Research* 8, 984–1022.

de Vriend, H.J., Capobianco, M., Chesner, T., de

Swart, H.E., Latteux, B. and Stive, M.J.F. (1993) Approaches to long-term modelling of coastal morphology: a review. *Coastal Engineering* 21, 225–69.

Dean, R.G. (1991) Equilibrium beach profiles: characteristics and applications. *Journal of Coastal Research* 7, 53–84.

Dixon, K. and Pilkey, O.H. (1989) Beach replenishment on the U.S. coast of the Gulf of Mexico. *Proceedings, Coastal Zone '89*, 2007–20.

Earney, F.C.F. (1990) *Marine Mineral Resources* London: Routledge.

Emmerson, R.H.C., Manatunge, J.M.A., Macleod, C.L. and Lester, J.N. (1997) Tidal exchanges between Orplands managed retreat site and the Blackwater estuary, Essex. *Journal of the Chartered Institution of Water and Environmental Management* 11, 363–72.

French, J.R. (1993) Numerical modelling of vertical marsh growth and response to rising sea-level, Norfolk, UK. *Earth Surface Processes and Landforms* 18, 63–81.

French, J.R. (1995) Function and optimal design of saltmarsh channel networks. *National Rivers Authority Project Record 480/1/SW*, Bristol: National Rivers Authority/Sir William Halcrow and Partners, 85–95.

French, J.R., Spencer, T. and Reed, D.J. (1995a) Geomorphic response to sea level rise – existing evidence and future impacts. *Earth Surface Processes and Landforms* 20, 1–6.

French, J.R., Spencer, T., Murray, A.L. and Arnold, N.S. (1995b) Geostatistical analysis of sedimentation in two small tidal wetlands, north Norfolk, U.K. *Journal of Coastal Research* 11, 308–21.

Haltiner, J., Zedler, J.B., Boyer, K.E., Williams, G.D. and Callaway, J.C. (1977) Influence of physical processes on the design, functioning and evolution of restored tidal wetlands in California. *Wetlands Ecology and Management* 4, 73–91.

Harlow, D.A. and Cooper, N.J. (1996) Bournemouth beach monitoring: the first twenty years. In C.A. Fleming (ed.) *Coastal Management: Putting Policy into Practice*, London: T. Telford, 248–59.

International Geosphere Biosphere Programme (IGBP) (1993) *Land–Ocean Interactions in the Coastal Zone – Science Plan*. Stockholm: IGBP.

Komar, P.D. (1976) *Beach Processes and Sedimentation*. Englewood Cliffs, New Jersey: Prentice-Hall.

Kraus, N.C. and Pilkey, O.H. (1988) Effects of seawalls on the beach. *Journal of Coastal Research, Special Issue* 4.

Leonard, L.A. (1997) Controls of sediment transport and deposition in an incised mainland marsh basin, southeastern North Carolina. *Wetlands* 17, 263–74.

Leonard, L.A., Dixon, K.A. and Pilkey, O.H. (1990) A comparison of beach replenishment on the U.S. Atlantic, Pacific and Gulf coasts. *Journal of Coastal Research, Special Issue* 6, 15–36.

Masselink, G. and Short, A.D. (1993) The effect of tidal range on beach morphodynamics and morphology: a conceptual beach model. *Journal of Coastal Research* 9, 785–806.

May, V.J. (1990) Replenishment of the resort beaches at Bournemouth and Christchurch, England. *Journal of Coastal Research, Special Issue*, 6, 11–15.

Milliman, J.D. and Meade, R.H. (1983) Worldwide delivery of river sediments to the oceans. *Journal of Geology* 91, 1–21.

Milliman, J.D. and Syvitski, J.P.M. (1992) Geomorphic/tectonic control of sediment discharge to the ocean: the importance of small mountainous rivers. *Journal of Geology* 100, 525–44.

Moeller, I., Spencer, T. and French, J.R. (1996) Wind wave attenuation over saltmarsh surfaces: Preliminary results from Norfolk, England. *Journal of Coastal Research* 12, 1009–16.

Morton, R.A. (1988) Interactions of storms, seawalls and beaches of the Texas coast. *Journal of Coastal Research, Special Issue* 4, 113–34.

Nakashima, L.D. and Mossa, J. (1991) Responses of natural and seawall-backed beaches to recent hurricanes on the Bayou Lafourche headland, Louisiana. *Zeitschrift für Geomorphologie* 35, 239–56.

Nicholls, R.J. and Leatherman, S.P. (1995) Potential impacts of accelerated sea-level rise on developing countries. *Journal of Coastal Research, Special Issue*, 14.

Nordstrom, K.F. (1994) Developed coasts. In R.W.G. Carter and C.D. Woodroffe (eds) *Coastal Evolution: Late Quaternary Shoreline Morphodynamics*. Cambridge: Cambridge University Press, 477–509.

Pethick, J.S. (1984) *An Introduction to Coastal Geomorphology*. London: E. Arnold.

Pethick, J.S. (1992) Saltmarsh geomorphology. In J.R.L. Allen, and K. Pye (eds) *Saltmarshes: Morphodynamics, Conservation and Engineering Significance*. Cambridge: Cambridge University Press, 41–62.

Pethick, J.S. and Burd, F. (1995) Sedimentary processes under managed retreat. *National Rivers Authority Project Record 480/1/SW*, Bristol: National Rivers Authority/Sir William Halcrow and Partners, 14–26.

Phillips, J.D. (1992) Nonlinear dynamical systems in geomorphology: revolution or evolution. *Geomorphology* 5, 219–29.

Pilkey, O.H. (1995) The fox guarding the hen house. *Journal of Coastal Research* 11, iii–iv.

Pilkey, O.H., Young, R.S., Riggs, S.R., Smith, A.W.S., Wu, H. and Pilkey, W.D. (1993) The concept of shoreface profile of equilibrium: a critical review. *Journal of Coastal Research* 9, 255–78.

Portnoy, J.W. and Giblin, A.E. (1997) Effects of historic tidal restrictions on salt marsh sediment geochemistry. *Biogeochemistry* 36, 275–303.

Shepard, F.P. (1950) Beach cycles in southern California. *Technical Memorandum 20*, Beach Erosion Board, US Army Corps of Engineers.

Sonu, C.J. and James, W.R. (1973) A Markov model for beach profile change. *Journal of Geophysical Research* 78, 1462–71.

Stanley, D.J. and Warne, A.G. (1993) Nile delta: recent geological evolution and human impact. *Science* 208, 628–34.

Thom, B.G. and Hall, W. (1991) Behaviour of beach profiles during accretion and erosion dominated periods. *Earth Surface Processes and Landforms* 16, 113–27.

UK Government (1998) *Sixth Report of the U.K. House of Commons Select Committee on Agriculture*. London: HMSO.

Valentin, H. (1952) Die Kusten der Erde. *Petermanns Geographische Mitteilungen* 246, 118.

Verhagen, H.J. (1992) Method for artificial beach nourishment. *Proceedings, 23rd International Conference on Coastal Engineering* 2474–85.

Viles, H.A. and Spencer, T. (1995) *Coastal Problems: Geomorphology, Ecology and Society at the Coast*. London: E. Arnold.

Vita-Finzi, C. (1964) Synchronous stream deposition throughout the Mediterranean area in historical times. *Nature* 202, 1324.

Walker, H.J. and Mossa, J. (1986) Human modification of the shoreline of Japan. *Physical Geography* 7, 116–39.

Warrick, R.A., Le Provost, C., Meier, M.F., Oerlemans, J. and Woodworth, P.L. (1996) Changes in sea level. In J.T. Houghton, L.G. Meira Filho, B.A. Callander, N. Harris, A. Kattenberg and K. Maskell (eds) *Climate Change 1995: The Science of Climate Change*. Cambridge: Cambridge University Press, 359–405.

Wijnberg, K.M. and Terwindt, J.H.J. (1995) Extracting decadal morphological behaviour from high-resolution, long-term bathymetric surveys along the Holland coast using eigenfunction analysis. *Marine Geology* 126, 301–30.

Wright, L.D. and Short, A.D. (1984) Morphodynamic variability of surf zones and beaches: a synthesis. *Marine Geology* 56, 93–118.

Wright, L.D., Short, A.D. and Green, M.O. (1985) Short-term changes in the morphodynamic states of beaches and surf zones: an empirical model. *Marine Geology* 62, 339–64.

9

Physical problems of the urban environment

Ian Douglas

THE NATURE OF PHYSICAL ENVIRONMENTAL PROBLEMS IN CITIES

Cities are where people have most transformed nature, replacing vegetation with roofed and paved surfaces, burying stream channels, creating indoor climates, and making huge artificial transfers of energy water and materials. Expanding cities transform hydrological relationships, changing the magnitude and frequency of flooding. Rising land prices often mean that homes are built on relatively unstable slopes or on the floodplains of rivers. The poor, especially in dense squatter settlements in third world cities, often have no choice but to occupy hazardous sites on steep slopes, close to rivers, or near polluting factories. All too often, their settlements are vulnerable to road collapse, water pipe breakages and sewer failures and to floods, landslides or subsidence.

Two aspects of this vulnerability are of special significance to geographers: the differing vulnerability of social groups and communities within the city; and the way in which expanding cities increase in vulnerability through time as they spread across more hazardous sites and occupy more unstable terrain. Knowledge of urban hydrology and urban geomorphology is not only a key to good urban planning but should also be available to every house purchaser. The home builder or buyer should 'know the ground being built upon'.

Nature is widespread in cities. Only a small part of an urban area is completely paved and roofed. In European and North American cities, many suburban areas are dominated by green vegetation. Urban areas often have greater biodiversity and wildlife protection than adjacent intensively cropped farming country, 3000 different species having been recorded in a single suburban garden in Leicester, England.

Land prices and policies encouraging house building on existing urban land lead to denser occupation of urban land. Gardens are partially changed to impermeable, paved surfaces modifying natural air, water, materials and energy flows. Relatively little of the urban area is completely covered by roofs or paved surfaces and thus totally impervious. Satellite imagery showed only 0.9 km^2 of the Bolton urban area in England to be entirely impervious, 6.5 km^2 was 82 per cent impervious, while 25 km^2 was 45 per cent impervious (Adi 1990).

URBAN VULNERABILITY TO CLIMATIC EXTREMES

Cities have little direct impact on the global radiation balance, but the internal urban climate produced by the absorption and subsequent reradiation of heat from the surfaces of the built enrivonment, and by the emission of artificial heat through combustion create an urban heat-island effect. Cities are warmer than the surrounding countryside at night and often, especially in high latitudes, during the daytime. In Tokyo, anthropogenic heat raises the urban surface temperature by around 1.5°C in summer and 2.5°C degree in winter, and urban land-use effects raise temperatures by about 1°C in both seasons (Ichinose 1997). In large mid-latitude cities during summer heatwaves, the release of warm air from cooling machinery makes city streets extremely hot and excess deaths occur, as in Greece and the eastern United States in the summer of 1998.

While most cities experience fogs and storm rains, the rarer tornadoes, cyclones, heat waves, droughts, bush fires and haze problems are more difficult to plan for. Part of the growing risk arises from the rapid growth of cities in areas affected by severe tropical revolving storms, known locally as cyclones, hurricanes or typhoons. Before 1987, no single event produced urban storm damage claims of over a billion dollars. Since then, claims have included those from Cyclone Iniki in 1992 ($1.4 billion), Hurricane Hugo in 1989 ($5.8 billion) and Hurricane Andrew in 1992 ($20 billion). Such huge losses and possible changes in world weather patterns may lead to rises in premiums and even insurance company collapses. The building design and urban land use zoning needed to reduce cyclone damage benefits from the use of Geographical Information Systems (GIS). However, geographers also contribute to the behavioural studies that examine how people's preparedness for, and response to cyclones may be improved.

URBAN VULNERABILITY TO AIR POLLUTION

Three general types of air pollution have affected cities in the twentieth century:

- sulphur dioxide (SO_2) and soot from coal burning;
- lead emissions from motor vehicles;
- oxides of nitrogen and fine particulate matter from motor vehicles.

In Britain, as trends in Greater Manchester demonstrate (Figure 9.1), the worst effects of the first two have largely been overcome by technical advances and legislation, but the effects of traffic-related pollution remain a major concern. Elsewhere in the world, all three forms of air pollution can be found.

Lead from motor vehicles has caused much concern for the health of children, especially in school playgrounds close to main roads. Lead concentrations are high close to the kerb but 10 to 20 m away from the road are much lower. Concentrations are also high in the soil close to traffic lights, where vehicles are temporarily stationary. Atmospheric lead began to decline in Europe with the introduction of lead-free petrol, development of more efficient vehicle engines

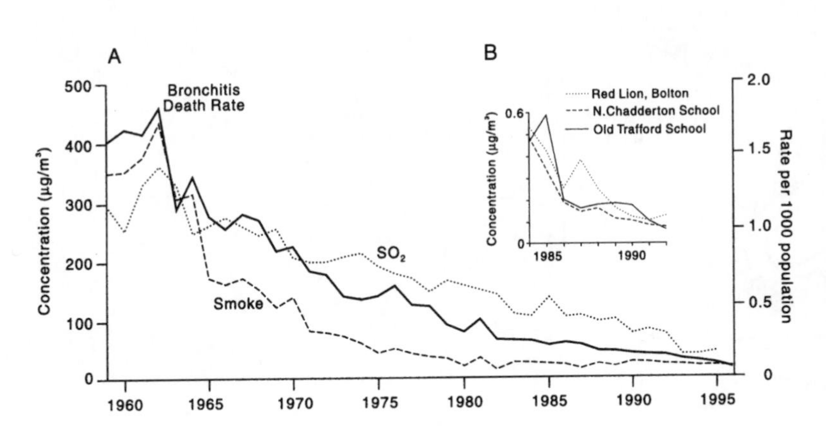

Figure 9.1 The decline in smoke, sulphur dioxide and lead (A) in the air in Greater Manchester and the fall in the death rate due to bronchitis. Smoke and sulphur dioxide data (B) are for central Manchester, bronchitis death rate for the city of Manchester as a whole and lead data for the places indicated in different parts of Greater Manchester. *Source:* Environment and Development Department, Manchester City Council.

and taxation policies favouring lead-free petrol. However, such changes have not yet occurred in many cities of lower latitudes, where children remain at risk.

Air pollution arising from particulates and oxides of nitrogen emissions from motor vehicles is most severe in cities with stable, dry anticyclonic weather conditions for large parts of the year, such as Los Angeles, Athens, Tehran and Mexico City, where it produces photochemical smog and high concentrations of low-level atmospheric ozone. In Britain, such smogs may occur during infrequent stable atmospheric conditions in both summer and winter. In Athens, Paris and Mexico City, their incidence has been so severe that restrictions on the use of motor vehicles are imposed. Under extreme circumstances, especially in Mexico City, factories are asked to reduce or halt production until atmospheric pollutant levels decrease. Geographers in Sydney, Australia, and Los Angeles have traced the diurnal migration of the

Box 9.1 Air pollution by smoke and sulphur dioxide in Sheffield, UK, in the 1950s and 1960s; and Chongqing, China, in the 1980s and 1990s

Alice Garnett's fine air pollution studies (1967) in Sheffield in the 1950s and 1960s showed that the heavy steel industries along the Don valley downstream of the city emitted waste heat equivalent to approximately 20 per cent of the incoming solar radiation and released sulphur at rates exceeding 1000 t km2 y^{-1}. Atmospheric SO_2 concentrations often reached levels of 1500 µg m^{-3} in the industrial area but fluctuated widely with weather conditions.

In China, where coal remains a prime source of energy, the city of Chongqing has severe SO_2 pollution affecting 2.5 million people along the Jialin and Yangtze Rivers (Figure 9.2). Major steelworks and heavy engineering plants emit smoke, which moves down valley towards the city centre and cannot disperse because of the overhead inversion layers and fogs, which occur naturally. Chongqing has the highest SO_2 levels of sixty Chinese cities and rainfall pH averaging 4.1, producing acid rain, which causes severe corrosion of many metal structures in the city. Domestic emissions will fall as natural gas replaces coal briquettes in cooking stoves, but factory emissions will continue, while inland Chongqing fares less well than coastal cities in China's economic expansion.

In both Sheffield and Chongqing, topography and their regional situation play a role in determining the location of the most severe air pollution. In planning the location of industries, or traffic concentrations with high emissions, the relief and regional climate must always be considered if the most severe pollutant concentrations are to be avoided.

Figure 9.2 Sulphur dioxide levels (SO_2 in mg m^{-3}) in Chongqing (average for 1981) and Sheffield (daily average on 14 December 1964) showing the down-valley concentration in both cities, but with the central district of Chongqing affected by higher levels than the centre of Sheffield.

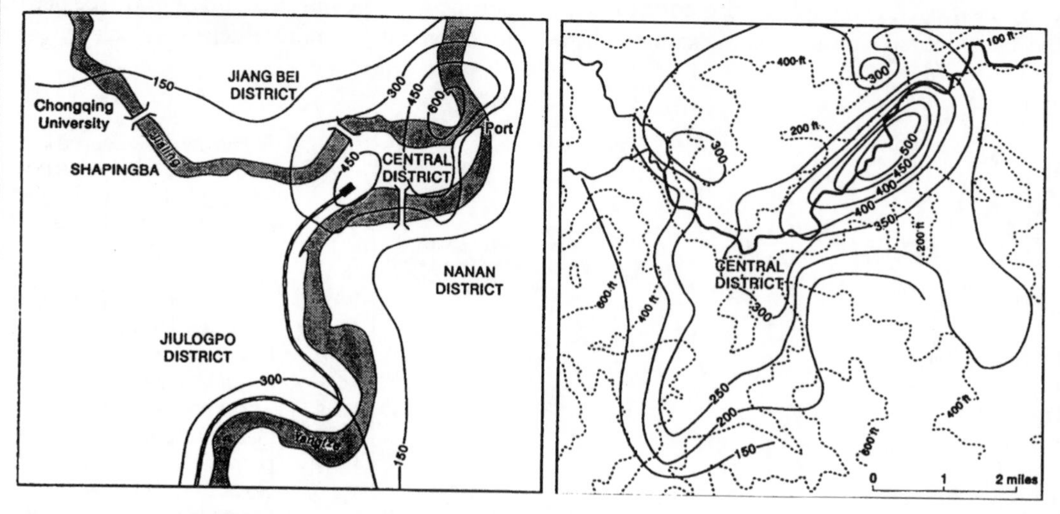

Source: Chongqing University, and Garnett 1967.

smog from the intense morning concentration around the central business districts to the suburbs further inland, carried by the sea breeze, and then a seaward migration again in the evening when the land cools and the sea stays relatively warm.

Similar air pollution problems prevail in many equatorial cities, even though frequent rains usually wash pollutants out of the atmosphere. Vehicular emissions account for more than three-quarters of the air pollutants in and around Kuala Lumpur, Malaysia.

During August 1991, the suspended particulates content of the air increased by 13 per cent per day from 120 μg m^3 on 16 August to over 400 μg m^3 on 27 August, then dropping rapidly when the weather changed (Samah 1992). However, in 1997, haze over Kuala Lumpur was associated with forest fires in Sumatra and Borneo, the dust particles from which had been carried across the Malacca Straits and the South China Sea. Causes of pollution events may thus be local or regional, and probably the worst conditions arise when there is a combination of the two. These haze episodes tend to produce or aggravate respiratory and eye problems. People susceptible to related diseases were warned by the Health Department to stay indoors. Much work remains to be done on the possible relationship between conditions such as asthma and air pollution. In Britain, severe air pollution is believed to aggravate existing respiratory diseases, which are probably caused by allergens in the indoor environment.

WATER PROBLEMS IN CITIES

A city has a dual hydrological system: the people-modified natural hydrological cycle of rainfall, runoff and river discharge; and the artificial water supply and waste water disposal system. The nature of the urban surface is particularly important in the disposal of rain-water and snow falling on a city. Drainage systems have to take account of the percentage of the surface that is impermeable and that will yield water directly to drains and artificial drainage channels. In North American and Australian cities, about 33 per cent of industrial and commercial areas and 25 per cent of residential areas are impervious (Nouh 1986). Urbanisation modifies the hydrological cycle in four main ways:

1 increase in storm runoff;
2 reduced infiltration to groundwater aquifers;
3 changes in water quality;
4 changes in the hydraulic amenities of streams.

The growth of cities affects the flow of small streams in two ways:

- an increasing percentage of the surface becomes impervious to infiltration as it is covered by buildings, driveways, pavements and parking lots;
- the introduction of storm sewers brings storm runoff from paved and roofed areas directly to stream channels for discharge. Runoff travel time to streams is shortened, while the impervious area increases runoff volume. These two changes in combination reduce the time from peak rainfall to peak stream flow (the lag time) and raise the peak storm stream flow. Many rapidly expanding suburban communities are finding that low-lying, formerly flood-free, residential areas now experience periodic flooding as a result of upstream urbanisation. The size of small, frequent floods is increased many times by these processes, but large, rare floods that cause serious damage usually result from conditions that saturate entire catchment areas and are little, if at all, affected by urban land uses (Hollis 1975).

Urban floods can be classified into four broad categories:

Localised flash floods: usually produced by short-duration, extremely intense thunderstorms, which produce rapid runoff of large volumes of water, which exceed the capacity of small stream channels. This causes localised flooding of roads and houses and in urban areas is often related to limitations imposed by culverts or old bridges on suburban streams. The flooding may last from twenty minutes to a few hours.

Catchment-wide flooding from prolonged heavy rain: usually produced in the UK by depressions whose passage is blocked by immobile high pressure to the east; these drop large volumes of water over a wide area, causing rivers to burst their banks and flood large areas. This is the type of flooding affecting the River Severn and the River Mersey about once every two to three years (Plate 9.1). In tropical regions, this would be associated with the passage of typhoons (hurricanes or cyclones) or disturbances within a monsoonal air flow. The flooding lasts from a few hours to two or three days, depending on catchment size.

Regional flooding: usually produced by an unusual combination of prolonged rain over a large, already wet area: typical of the floods that have occurred in the Mississippi and its tributaries in recent years, the most severe floods on the great rivers of Asia, and the floods that take weeks to pass down the Murray–Darling river system in Australia. Towns in the path of such floods can expect widespread inundation lasting for many days to two or more weeks.

Plate 9.1 Flooding of Flixton Road, Carrington, Greater Manchester, in December 1991 as a result of catchment-wide flooding from prolonged heavy rain.

Snow melt flooding: seasonal snow melt river regimes occur on many rivers draining mountain regions, such as the Rhine, Rhône and Danube in Europe. While these annual high water levels are usually coped with well, exceptional combinations of weather conditions can produce high volumes of warm rainfall, which cause extremely rapid snow melt, creating disasters like those that affected the Guil valley in the French Alps in 1957, when whole villages on old alluvial fans were destroyed.

Urban areas often experience localised flooding through poor design and planning of developments, or simply through the thoughtless throwing of debris into small streams, as when in Manchester a mattress and other debris blocked a culvert entrance and caused the flooding of an adjacent housing estate. An undersized drain at Llandudno Junction, Wales, caused thirty new houses in a small flood basin to be flooded to a depth of 1 m by the Afon Wydden in both October 1976 and February 1977 (Parker and Penning-Rowsell 1980).

The classic method of regulating streams and rivers in cities was to either turn them into concrete or stone culverts, or embank them, as along the Rio Guadelmedina at Malaga, Spain (Plate 9.2), or divert them around city centres. Since 1980, much attention has been paid to designing rivers for multiple use, with both adequate flood storage and control and with major environmental benefits. Attempting to 'tame' any natural alluvial river is now seen as undesirable (Thorne 1998). Single-purpose levee and channelisation projects exclude other possible uses for stream and flood plains, such as preservation of riparian woodlands, creation of greenways for urban parks, conservation of stream fishery habitat, and storage of flood waters in floodplains (Riley 1998). In the USA, complex overlapping federal, state and municipal responsibilities have often led to slow decision making about stream channel improvement, but now participatory, multi-agency 'watershed councils' are often proving effective. In Portland, Oregon, the city council adopted a management plan with objectives ranging from

Plate 9.2 The aggraded, embanked channel of the Rio Guadelmedina at Malaga, Spain, in 1980. Sand and gravel brought down from the mountains have raised the bed of the confined channel above the level of the adjacent streets, so putting adjacent buildings at risk of flooding.

fish habitat restoration to flood damage reduction prepared by citizen and agency members of the Johnson Creek Watershed Council working together over five years (*ibid.*).

In the UK, flood alleviation works have been strongly influenced by the Wildlife and Countryside Act of 1981, Section 48 of which requires engineers to undertake river works in an environmentally sensitive manner. An investigation into floodplain management along the River Colne, west of London, involved a comprehensive holistic assessment of environmental resources in order that the ecological and recreation benefits of this valley could be retained and enhanced (Driver and Pepper 1996).

DANGERS OF INADEQUATE URBAN WATER SYSTEMS

Inadequate water systems lead to four main sets of problems:

Health problems from lack of fresh water and inadequate waterborne sanitation

Geography enters into urban water management in a more complex way when considering the implications of water supply for human health. Often the wealthiest people in cities are able to ensure good piped supplies or have the finance to buy bottled water. For the poor things are very different. When cities lack the financial resources to maintain their water systems, human health is threatened. In 1995, five of the 500,000 inhabitants of Nikolayev, Ukraine, died from cholera due to lack of funds to repair the collapsing water and sewerage system. Others had died in 1991 and 1994. Cholera probably comes from people swimming in, or eating fish from, the Southern Bug River, into which enters partly treated sewage carrying the cholera bacterium from the people to the river. All cities affected by war, civil strife or natural disasters face similar risks, but economic difficulties can make them long-lasting.

Competition for water between rich and poor communities in cities

In 1995, water tankers appeared in wealthy suburbs of south Delhi (Goldenberg 1995) as the city administration cut back supplies by 20 per cent and brought in rationing. Although a virtually permanent problem for the poor of Delhi, the impacts on the wealthy suburbs as the temperature rose to 45°C led to political action, with Haryana state upstream on the Jumna River releasing extra water to avert the crisis.

The geographical injustice of urban water access and management is a fruitful field for applied geography.

Competition for groundwater resources, depletion of aquifers and subsidence

Delhi illustrates how most low-latitude cities have outgrown the municipal water distribution systems. Many rich people sink wells, saving water charges, but lowering the water table and forcing people to dig deeper. This again tends to deprive the poor of access to cheap, good-quality water and may also induce the type of subsidence that has affected Bangkok and Mexico City.

Competition between rural and urban areas for water supplies

Another controversy arises in trying to balance urban and rural water needs. With a *per capita* consumption of around 220 litres per day, Delhi uses twice as much water per head as Bombay. Rural people protest that water is used for swimming pools in Delhi, while they do not have enough for their crops. Within the city, water is used to keep politicians' lawns green, while the middle class do not get enough pressure for a real shower and the slums are not connected to the water mains at all.

THE ROLE AND SIGNIFICANCE OF URBAN GEOMORPHOLOGY

Every climatic, topographical and geological situation creates problems for the construction and maintenance of cities. Some of these are due to processes occurring at present, and some are legacies from the past, such as the remobilisation of ancient landslides, or foundation problems due to complex subsurface conditions in sediment laid down in past climates, greatly different from those of the present. Many cities face earthquake, volcanic, tsunami or avalanche risks. Possible worldwide sea level rise threatens many millions of urban people, especially in poor countries that cannot afford coastal protection works. Subsidence affects many parts of urban areas.

In 1997, a garden in Ripon in Yorkshire, UK, disappeared into a 40 m deep hole. Within hours, a double garage disappeared, along with the children's sandpit. The house involved and two neighbouring dwellings had to be evacuated. They were built above soluble Triassic gypsum, which extends along the eastern edge of the Pennines and also occurs in Cheshire. Such karstic solubility phenomena have to be considered when planning new housing estates.

Over much of southern Britain, and in many other countries, the problem of cracking clay soils provides a major constraint on urban development. Many clay soils expand when wet and shrink and crack during long dry periods. Such 'shrink–swell' phenomena cause differential shifts of parts of the structure, such that floors tilt slightly and windows and doors no longer close properly. Such natural subsidence may be covered by household insurance, premiums for which were previously based on the pattern of past claims. Now geographers are assisting companies in mapping the hazard due to cracking clays (Doornkamp 1995).

Any form of land development alters the form of the slopes and the passage of water over the ground and into the weathering profile. Landslide and soil erosion potentials are altered. Developments in steep granitic terrain in Hong Kong, the Rio de Janeiro area of Brazil and peninsular Malaysia have led to large-scale landslides, severe gullying and soil erosion. Impacts can be divided into two phases of the development process: those during project construction and those in the period after completion of building work. Geographers have helped to develop guidelines for the recognition of landslide potential and for erosion control.

Six key contributions can be made by geomorphologists to urban environmental management (Hart 1986):

1 making initial reconnaissance surveys to select suitable sites for urban development within a region or a country;

2 mapping potentially hazardous zones within the selected areas;
3 giving advice on the beneficial use of topography and surface materials in planning specific traffic routes and subdivision street layouts;
4 providing advice on site slope stability and erosion control works;
5 analysing weathering problems likely to affect foundations and building materials;
6 dealing with post-construction problems such as urban stream channel maintenance, causes of damage by hazards and the monitoring of geomorphological change.

GEOTECHNICAL MEASURE TO MINIMISE IMPACTS

Geomorphological mapping, developed by geographers, helps to designate areas of potential slope instability and land suitability for different types of construction. Applications range from building control in Hong Kong, to the planning of developments at Suez City, Egypt, and earthquake hazard zonation in San Francisco. Mapping also assists in assessing multiple hazards, such as landsliding, flooding and debris flows.

Erosion control guidelines suggest that construction should be carried out in phases to avoid disturbing too much of the land at any one time. No unnecessary clearing should be undertaken. Immediately below any cleared area, detention ponds should be constructed to retain any sediment washed off the site and to hold back stormwater runoff so that peak discharges in streams below are not increased.

Particular attention should be paid to the design of construction roads, and later of permanent roads. Road design should be governed by four basic principles:

1 Minimise the amount of disturbance caused by road construction by (a) controlling the total mileage of roads; and (b) reducing the area of disturbance on the roads that are built.
2 Avoid construction in areas of high erosion hazard.
3 Minimise erosion on areas that are disturbed by road construction by a variety of practices designed to reduce erosion.
4 Minimise the off-site impacts of erosion.

ENGINEERING PROBLEMS IN KARST TERRAIN

Soluble rocks have subsurface conditions that pose problems that are particularly acute in the limestone areas of Southeast Asia (Box 9.2) and the Caribbean, including Florida, where cave systems below ground developed during low Pleistocene sea levels. In Florida, many cases of subsidence due to sinkhole collapse have involved law suits in which geomorphologists have acted as expert witnesses. Knowledge of the subsurface, of past and present geomorphic processes, is essential if fair judgements are to be obtained and future problems avoided. The key role here for applied geography is to assist people to 'know the ground they build upon'.

CONCLUSIONS: TASKS FOR APPLIED PHYSICAL GEOGRAPHY IN THE URBAN ENVIRONMENT

Urban areas continue to grow, and within them land use intensifies. The crowding, consumption and congestion leads to increasing emissions and waste, greater vulnerability of the city's people and more and more significance for the quality of urban management. The risks associated with waste disposal and the legacies of past industrial activity, the opportunity to apply biogeographical principles in planning urban nature conservation, and the issues of wildlife and pests within the city are too complex to be dealt with here, but they are all part of the urban ecosystem, part of the daily lives of most of the world's people.

To avoid risks to health, lifestyles and commerce, good information on the state of urban areas is needed. On the one hand, work towards

Box 9.2 Subsidence beneath Kuala Lumpur, Malaysia

Beneath the northeastern part of central Kuala Lumpur is a highly cavernous buried limestone karst, with many voids (Figure 9.3). The buried karst plain is overlain by sands and clays deposited by rivers as the sea level rose after the last ice age. The sands contain alluvial tin deposits, the mining of which created pits, which were later filled. The mining produced two types of subsidence: that due to the general lowering of the groundwater table, especially by the pumping of water out of mine pits;and that due to the sudden collapse of sediments over sinkholes. Some of the filled pits have been reclaimed for low-cost housing. In a few instances, fill has subsided into sinkholes with, in one place, a row of low-cost houses collapsing into a reopened hole.

The buried karst now poses serious problems for civil engineering works (Bergado and Sebanayagan 1987) New high-rise buildings require deeper foundations than the low-rise buildings that sufficed until the 1970s. In Kuala Lumpur, the low-rise structures had their foundations on the stiff clay layer within the alluvium. Taller multistorey structures require piling into the underlying limestone. However, the irregularity of both the karst surface and the cavities within the buried karst means that foundation investigations have to be particularly circumspect. Drill holes may strike limestone, unaware as to whether it is buried rockfall material or a pinnacle, while a neighbouring hole might pass through several more metres of alluvium before hitting limestone. Special precautions had to be taken when constructing the Petronas Twin Towers, the world's tallest building at the time.

Figure 9.3 Schematic cross-section through a buried karst plain such as that underlying part of Kuala Lumpur, showing the hidden cave systems and the problems faced when using deep foundations.

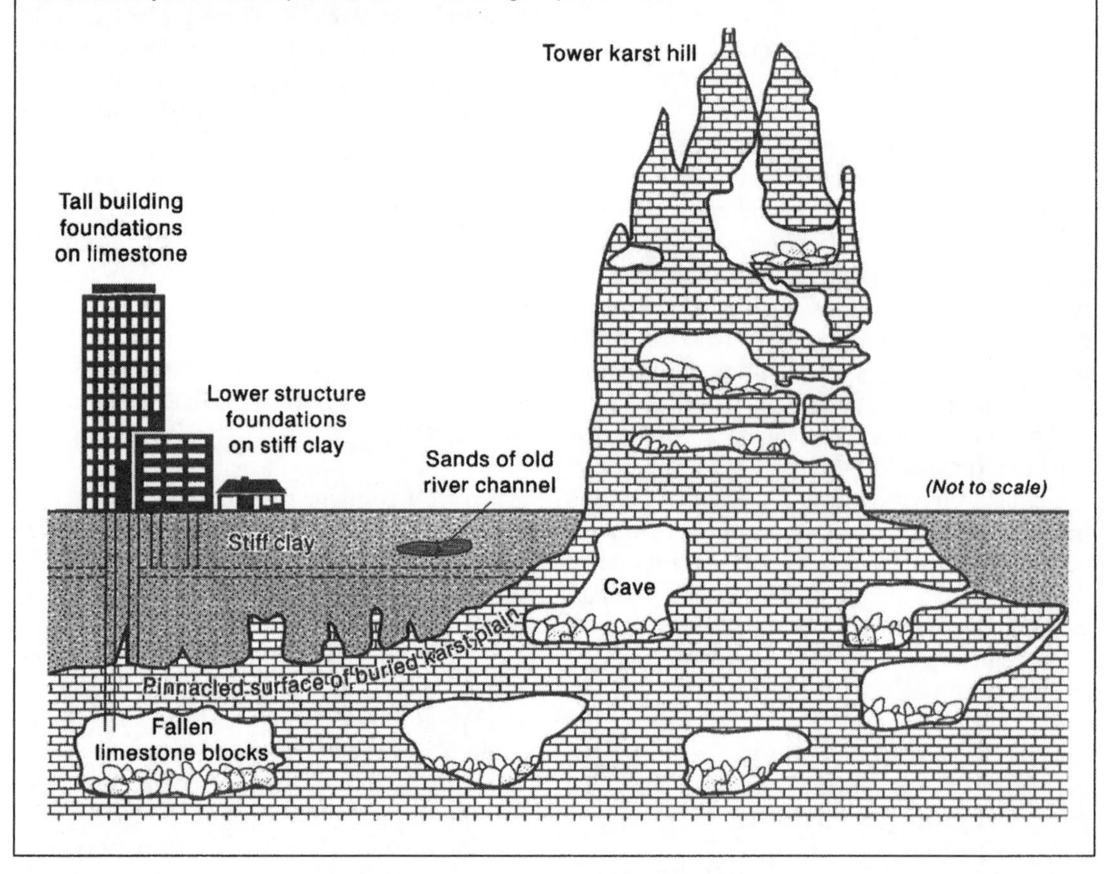

making cities more sustainable and more environmentally friendly, particularly through Local Agenda 21, is using indicators of the state of the city. On the other, the different databases on gas and water pipes, telephone and electricity cables, postcodes, housing stock, census data, health status, locations of services and recreational facilities for individual cities are becoming more integrated, so that everything from air quality to ground conditions and waste disposal patterns can be accessed through a single database or GIS. Such 'smart' cities can also become 'safe' cities by the effective management of the information system to both give warnings of, and plan to manage, impending disasters. The need to expand such urban information systems has been identified as a priority in the European Commission fifth Framework Research and Development Programme.

Applied geography will be a major contributor to the development of the 'information' city. Current geographical research in this direction includes projects to:

- use GIS to map areas of energy loss from buildings using infrared photography so that better strategies for investment in insulation can be developed;
- use GIS and real-time weather data to predict areas and populations at risk in the event of major chemical releases to the atmosphere in urban areas;
- use GIS to record areas of contaminated soil at different depths below the surface to look at sources of substances that might contaminate groundwater or that may develop into future 'chemical time bombs';
- use GIS to integrate geomorphological information in the planning of new urban areas and identify those areas on which new building should not be allowed as a contribution to the type of work already done in the Hong Kong Geotechnical Control Office or in landslide hazard zoning by the local authorities around San Francisco Bay.

However, the work goes beyond sheer considerations of wise planning and safety. The conservation of the built and natural heritage of urban areas for enjoyment and education benefits from good geographical analysis, including considerations of access, relationship to other facilities and areas and links with the community. Examples of this approach are to be found in:

- discussions about the integration of urban nature reserves and other open spaces of natural vegetation, perhaps with cultural heritage buildings or remains, into urban biosphere reserves;
- collaborative work on urban river valley improvements, where researchers work with local government, voluntary organisations and community groups to enhance open space use for recreation and nature conservation.

Many such tasks aim to improve both the quality of people's surroundings and the quality of life for individuals. Resolving or alleviating physical problems as part of a new urban human ecology – a new science of human settlements – is at the heart of the United Nations Human Settlements Programme (HABITAT) and of the work of NGOs like the Commonwealth Human Ecology Council. Applied geography is geography at the service of the world's people. Since most of those people will be living in urban areas in the twenty-first century, there is no more urgent need than to put geography at the service of those who have to survive in and manage cities.

GUIDE TO FURTHER READING

General syntheses of urban environmental problems and their management include:

Detwyler, T.R. and Marcus, M.G. (1975) *Urbanization and Environment*. Belmont: Duxbury.

Douglas, I., (1983) *The Urban Environment*. London: Arnold.

Hardoy, J., Satterthwaite, D. and Mitlin, D. (1992) *The Environmental Problems of Cities in Developing Countries*. London: Earthscan.

Girardet, H. (1993) *The Gaia Atlas of Cities – New Directions for Sustainable Urban Living*. Stroud: Gaia Books.

White, R.R. (1994) *Urban Environmental Management*. Chichester: Wiley.

Good urban environmental histories, which examine how and why past problems were or were not successfully remedied, paying particular attention to the roles of individual entrepreneurs, lobbyists, city managers and municipal politicians, are provided by:

Tarr, J.A. (1997) Searching for a 'sink' for an industrial

waste. In C. Miller, and H. Rothman (eds) *Out of the Woods: Essays in Environmental History*. Pittsburgh: University of Pittsburgh Press 163–80.

Hays, S.P. (1998) *Explorations in Environmental History*. Pittsburgh: University of Pittsburgh Press.

The major journals specifically tackling urban environmental problems are *Environment and Urbanisation*; *Urban Ecosystems*; *Urban Nature Magazine* and *Atmospheric Environment*. A highly valuable introductory bibliographical source is provided by:

Dawe, G. (ed.) (1990) *The Urban Environment: A Sourcebook for the 1990s*. Birmingham: Centre for Urban Ecology.

REFERENCES

Adi, S. (1990) The influence of urbanization on flood magnitude and storm runoff in the Bolton area, Greater Manchester. Unpublished MSc thesis, University of Manchester, U.K.

Bergado D.T. and Sebanayagan, A.N. (1987) Pile foundation problems in Kuala Lumpur limestone, Malaysia. *Quarterly Journal of Engineering Geology* 20, 159–75.

Doornkamp, J.C. (1995) Perception and reality in the provision of insurance against natural perils in the UK. *Transactions Institute of British Geographers* NS 20, 68–80.

Driver, A. and Pepper, A.T. (1996) The Lower Colne improvement scheme: environmental costs and benefits. *Journal Chartered Institution of Water and Environmental Management* 10, 79–86.

Garnett, A. (1967) Some climatological problems in urban geography with special reference to air pollution. *Transactions Institute of British Geographers*, 42, 21–43.

Goldenberg, S. (1995) Delhi's water crisis now a VIP problem. *Guardian*, June 12 1995.

Hart, M.G. (1986) *Geomorphology: Pure and Applied*. London: Allen & Unwin.

Hollis, G.E. (1975) The effect of urbanization on floods of different recurrence interval. *Water Resources Research* 11, 431–35.

Ichinose, T. (1997) website: http://www.urban.rcast.u-tokyo.ac.jp/ues/hanakilab/ichinose/abs.html

Nouh, M. (1986) Effect of model calibration in the least-cost design of stormwater drainage systems. In C. Maksimovic and M. Radjokovic (eds) *Urban Drainage Modelling*, Oxford: Pergamon, 61–71.

Parker, D.J. and Penning-Rowsell, E.C. (1980) *Water Planning in Britain*. London: George Allen & Unwin.

Riley, A.L. (1998) *Restoring Streams in Cities: A Guide for Planners, Policymakers, and Citizens*. Washington DC: Island Press.

Samah, A.A. (1992) Investigation into the haze episodes in the Kelang Valley, Malaysia. In A.J. Hedley, I.J. Hodgkiss, N.W.M. Ko, T.L. Mottershead, J. Peter and W.W.-S. Yim (eds) *Proceedings Seminar on the Role of the ASAIHL in Combating Health Hazards of Environmental Pollution*. Hong Kong: The University of Hong Kong, 221–27.

Thorne, C.R. (1998) *Stream Reconnaissance Handbook*. Chichester: Wiley.

Part II

Environmental change and management

10

Water supply and management

Adrian McDonald

INTRODUCTION

Water is fundamental to life and is overall the largest component of resources that have to be actively sought. Air we need continuously but, for the time being at least, is available at all locations. Water supply and management mean different things to different communities, clients, industries and suppliers. For the purposes of this analysis, water supply will be taken as the supply of water for consumption delivered through a system. In forming this definition, we have excluded waters abstracted from rivers and groundwater without treatment for use by people and industry.

The management of the supply of water involves three requirements:

1 The delivery of an adequate volume of water at the time and locations required.
2 The provision of water of an adequate quality.
3 The provision of the water at an acceptable price to consumer and supplier.

Figure 10.1 shows the stylised structure of the

Figure 10.1 Stylised operation of the water industry, the place of regulators and the main actions and therefore research areas in the system.

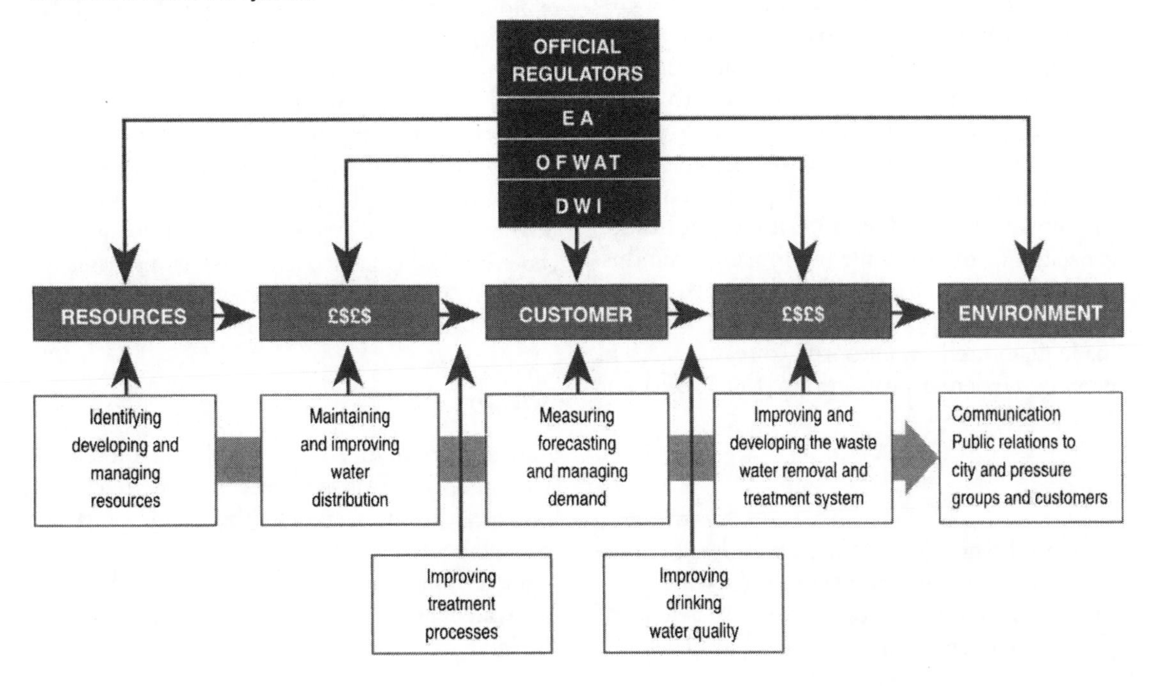

water supply system and incorporates the three requirements above. Water supply cannot be separated from water quality issues. There are inherent and demanding standards of quality for potable waters. After use, potable waters are discharged into the sewer system and after treatment to the natural environment. At each stage from resource development to return of effluent to the natural environment the supply system is formally monitored by several agencies, which are also shown in Figure 10.1. In this section, environmental water quality is not considered but it is treated elsewhere; see Chapter 11.

The structure of the water industry in the United Kingdom has evolved over the years. Broadly, that evolution can be considered as falling into four phases, namely:

- pre-1850
- 1850–1974
- 1974–89
- post-1989.

Pre *c.*1850

Before 1850, the water supply in the UK was small-scale, with local communities being supplied from the local river or well. Such a situation had existed for 1500 years, and any lessons that might have been learned from the relatively sophisticated Roman water supply systems had been lost to the indigenous community. The small, dispersed population had, for the most part, relied on the natural purification capacities of the natural environment. As industrialisation and urbanisation increased, cities' local water supplies became more and more inadequate and waterborne disease reached, literally, epidemic proportions (McDonald and Kay 1988).

1850–1974

The real driver to change then, was disease in the cities, the government response to which was focused through a Victorian sense of civic pride and independence. City corporations developed their own water undertakings, each city being supplied by its own reservoir system independently from the resource developments of adjacent cities. Reservoir developments required an act of parliament and involved the construction of both supply reservoirs and reservoirs to compensate downstream mill owners for the changed water conditions to which they, as riparian owners, were being subjected. For over 100 years, the situation remained unchanged, although the scale of the reservoirs increased and the size of catchwaters and aqueducts increased dramatically. In essence, however, the country had water provided by a set of independent unlinked source–sink resource exploitation systems. Each supply relied on four lines of defence, the so-called multiple barrier concept:

- a pristine catchment;
- long storage in a large reservoir;
- filtration; and
- disinfection.

Certainly, the first two lines of defence have eroded over the years with the pressure of demand (1) for land, (2) for access, (3) for increased water sport and (4) for preferential exploitation of cheap upland waters. The final line of defence, while not eroded, has come under growing criticism because of problematic side effects. The use of chlorination is criticised because it generates trihalomethanes when applied to the organic-rich peat-derived waters from British uplands. Chlorination is widely accepted in the UK but is not popular in Europe. The generation of organic-rich water is discussed in Mitchell (1990) and management implications in Mitchell and McDonald (1995).

1974–89

Concern about the fragmented and piecemeal structure of the British water industry, which had hundreds of suppliers and thousands of waste management agencies, led to the creation of the regional water authorities in England and Wales. Ten water authorities covered the country, nine in England and a further Welsh Water Authority. A form of the previous system remained in Scotland and Northern Ireland, such that during this period the UK was characterised by different administrative regions. The water authorities planned and implemented water management and pollution control, flood management and the water environment. The water authorities, while applauded throughout the world as an example of integrated catchment-based water management, suffered from two significant problems: (1) they had a very constrained capability to raise funds for investment; and (2) they were seen as self-regulating, both the gamekeeper and poacher in the same organisation.

Post-1989

To solve both of the problems outlined above, and in keeping with its political ideology, the Conservative government of the time disbanded the water authorities, privatising the supply and waste treatment functions as water plcs, and separating out the regulatory function through the creation of the National Rivers Authority. Both the authority and the plcs started with the same spatial and management basis but, particularly since 1994, a number of significant changes have taken place. The NRA has become the Environment Agency and has taken on board responsibility for waste management and integrated industrial pollution control (historically air pollution). NRA regions have been amalgamated to reduce the ten NRA regions to eight EA regions. The water plcs have seen amalgamations (takeovers) with other water undertakings (Northumbrian Water owned by Lyonaise des Eaux), combinations with other non-water utilities within a region (United Utilities for electricity and water in the northwest) or outside the region (Scottish Power ownership of Southern Water).

The water industry faces research challenges associated with every element in Figure 10.1, namely

- measuring, managing and forecasting demand;
- identifying, developing and managing resources;
- improving treatment process, control and management;
- maintaining and improving water distribution;
- maintaining and improving drinking water quality;
- improving and developing the waste water removal and treatment system.

But perhaps more importantly, the industry has to overcome all the challenges while maintaining the confidence of customers and shareholders and delivering on all these fronts, not individually but as an integrated response to an interrelated system. Herrington (1996) has examined the effect of climate change on water demand, and Clarke *et al.* (1997) and Likeman *et al.* (1995) consider new demand prediction and resource management technologies, respectively. The problems of continued climate variability (Wigley and Jones 1987) bring new planning challenges to the water industry, since it is difficult to plan to a changing and apparently less predictable base, and this is recognised by Ofwat. Economic controls, perhaps more responsive to shortage, are elaborated in Ofwat (1996) but are confounded by the economically unresponsive and politically sensitive issue of leakage (Lambert 1994).

CONTROLLING PROCESS

Process is the domain of the chemical engineer, although no longer solely so. Raw waters, whether of high or low quality, are transformed through a sequence of chemical, physical and biological processes, into potable waters. The particular selection of processes will depend on raw (environmental) water quality. Herein is a major research opportunity. The industry requires methods for the rapid assessment of likely typical and extreme characteristics of the waters draining from catchments that it might wish to utilise for supply after treatment. There are examples in many regions of the country of ill-designed treatment strategies and processes that have failed to deliver the expected potable water quality because the raw water quality did not have the expected characteristics.

Within the process, all critical parameters need to be monitored using reliable, well-designed and easily maintained instruments. The data from such monitoring systems, when properly analysed, will allow process modification and management and decision support such that the expected levels of service can be delivered at a reasonable cost, with failure risk minimised to customers, shareholders and regulators.

This area of the industry, reliant as it is on communication and remote-monitoring technology, will benefit most from shared understandings of common problems. To gather data in a manner that will allow its effective integration will require a setting of protocols and standards for data collection, aggregation and transmission. It will also require a changed approach by university researchers away from the groupings of like-minded researchers (say several hydrologists from different institutions) to innovative and 'unnatural' groupings (water engineer, electronics engineer, psychologist, groundwater hydrologist, archaeologist and global positioning system analyst) if, for example, ground penetrating radar (GPR) is to be developed for tracing leakage in the water industry effectively.

DISTRIBUTION SYSTEMS

The distribution system connects the customer to the treatment works through a structure consisting of a dendritic pipe network incorporating service reservoirs. The system has to deliver, efficiently, a reliable and secure supply of water of an appropriate quantity, quality, flow and pressure. In the UK, many of these systems are well over 100 years old. They have been added to as a response to urban growth rather than as planned engineering extensions. Both the age and the nature of the development present the water manager with serious challenges.

Water leaves the treatment works at near pharmaceutical quality. It is contaminated when the system integrity is impaired through leakage, repairs, standpipes or back siphoning. Even when the pipes remain whole, dead-ends lead to chlorine decay, stagnation and sedimentation. Lining material leaches from mains and from lead supply pipes. Research is needed both on technical and procedural improvements to pipe renovation and usage to minimise contaminant ingress and on the determination of environmental factors (age, soils and topography) that influence pipe failure. Geophysical techniques such as GPR offer powerful techniques to locate pipes and leakage, particularly for deep non-metallic pipes, where correlators are less effective.

Pressures, quantity and flow (PQF) rates are subject to standards set by the Ofwat director general but have been the subject of intense political interest because it is claimed that pressure reduction is being used as a tool to achieve the leakage reductions required by the government. PQF has to be sufficient not only to cope with peak demand but also to cope in the face of planned and unplanned (mains failures) interruptions and exceptional fire-fighting requirements. This presents network management and network information modelling and simulation opportunities as well as contingency planning improvement to deal with emergency supplies in the event of failure.

Leakage *is* the most topical characteristic of distribution systems. Leakage occurs as two types: (1) a steady background leakage due mainly to poor jointing and corrosion; and (2) intermittent failures due to ground movement often induced by weather and traffic and third party damage from other companies having or installing underground assets such as cable TV. In practice, there is a gradation between these two types. Bursts, particularly if unseen by the public and of a magnitude that does not interfere with supply to an area, will be viewed as part of the background. Leakage has become a political issue, and leakage targets have been set by the government as part of John Prescott's 'Water Task Force' (see Table 10.1).

All systems leak (e.g. schools have truants and electricity is lost as heat in transmission), and in most situations this is accepted. The setting of

Plate 10.1 A mains failure in the Headingley district of Leeds in 1991 resulted in the loss of potable water supply as well as serious flooding to a number of properties. Emergency water was provided from bowsers since standpipes were unavailable.

Table 10.1 Water management priorities. Largely in response to the drought of 1995–6 but exacerbated by a series of public relations failures by the water companies, John Prescott, the deputy prime minister, chaired a task force, which established these priorities, 1997.

- Mandatory leakage targets
- Free leak detection and repairs by water companies
- Statutory duty to conserve water
- Companies to promote better usage in the home, and offer free efficiency audits
- Review of the charging systems and implications of introducing more domestic metering
- Compensation payments to customers affected by the drought
- Companies reveal performance and financial information
- Encouraging 'best practice' programmes for industry and agriculture
- Review of environmentally damaging licences permitting water companies to abstract water from rivers and boreholes
- Environment task force could be called in where companies persistently fail

absolute leakage reduction targets as a political response to criticism of the water industry is strangely at odds with the Ofwat-derived policy of setting an economic level of leakage in which absolute levels will differ according to water supply and leakage repair 'costs'. Leakage presents an opportunity for research in questions of both policy and practicality:

- replace or repair
- renovation technologies
- area prioritisation
- find and fix technologies
- spatial characterisation of leakage
- forecasting of leakage
- cost–benefit analysis of leakage.

For some time, companies have been anxious to differentiate between the leaks from pipes for which they have responsibility and those that belong to the customer. Today, the approach to leakage reduction has been integrated with water quality improvement and customer satisfaction. Many companies now offer customers free repairs to their installations and are also replacing service pipes, which may leak or contaminate potable water through, for example, dissolved lead.

Service reservoirs provide a 'local' supply of treated water and serve to balance the supply and demand and to provide an emergency reserve of water. Service reservoirs also have negative characteristics. They allow chlorine to decay; they are a focus for microbial growth; they act as chemical reactors for disinfection by-products; and they leak. Management will seek to minimise residence time by maximising turnover; to ensure internal circulation and to monitor chlorine levels and chlorine decay products such as trihalomethanes. Researchers will provide technology and systems to optimise this management. Social and computational scientists will have the opportunity to contribute to the most worrying problem: that of risk of deliberate contamination as a result of vandalism, extortion or terrorism. Risk analysis and vulnerability assessments to address these issues are becoming a part of management planning.

Distribution systems are complex spatially organised systems. To optimise the performance of these assets and to achieve the tightening performance standards in the face of stringent cost constraint requires improved monitoring, data transfer, modelling and communication to identify trends in performance and to indicate potential problems and network sensitivities. It also allows operational staff to be trained in 'what if' scenario planning. Such network models have been credited with the success of Yorkshire Water in delivering uninterrupted water to all customers in the face of a 1 in 600-year drought. It could do nothing to address the PR failures in that situation.

MAINTAINING AND IMPROVING DRINKING WATER QUALITY

From the privatisation of the water industry in 1989 until the drought of 1995, the quality of drinking and environmental waters was at the top of the political, public and company agendas. It remains near the top of the agenda today. Drinking water must be 'wholesome' and fit for human consumption. In practical terms, these very positive objectives are interpreted through the negatives: (1) that the water does not contain anything detrimental to public health; and (2) it does not exceed defined standards. Examples of the key prescribed standards and values (PSVs) are given in Table 10.2.

The standards set have a mixed pedigree. They are growing in number and are becoming more stringent. They often lack a scientific basis, and the significance of non-compliance (an understanding of the dose–response relationship) is infrequently addressed. In most cases, a wide safety margin is incorporated into the individual standards implicitly to compensate for increases in toxicity associated with aggregate element effects. Drinking water quality is assessed through the analysis of random water samples from the households served. While the individual companies review the results against the prevailing standards and take action if appropriate, the results also go to the Drinking Water Inspectorate to assess compliance and trends in compliance.

Research challenges are of two types: those that relate to the composition of the PSVs and those that relate to the sampling and interpretation of the PSVs. In both cases, researchers must be aware of the sensitivity of the research topic and results. Table 10.3 suggests key research areas.

FORECASTING AND PRICING

Forecasting and the research needs associated with that practice have increased in importance in the last ten years. At a time of population,

Box 10.1 Managing risk at Cod Beck Reservoir

This case study is chosen because it is unexceptional because it is typical of the risk recognition, evaluation and management problems that occur throughout the industry.

THE CONTEXT

Cod Beck is a small reservoir on the eastern edge of the North York Moors National Park. It supplies small communities unconnected to the water grid. Water quality problems from this catchment would be significant, because there was no plan for quality control, no means of diverting waters, no alternative raw water sources to the single treatment works, limited treatment capacity, and no alternative treated water supplies. Complex land ownership and the amenity requirements of a site on the edge of a national park exacerbate management.

THE PROBLEMS

- Extensive vandalism e.g. telemetry and fencing
- Informal barbecue fires between water and forest
- Forest edge used as informal toilets
- Litter problems
- Parking problems
- Forest windblow
- High forest fire potential
- Sediment flows
- Forest operations spillages

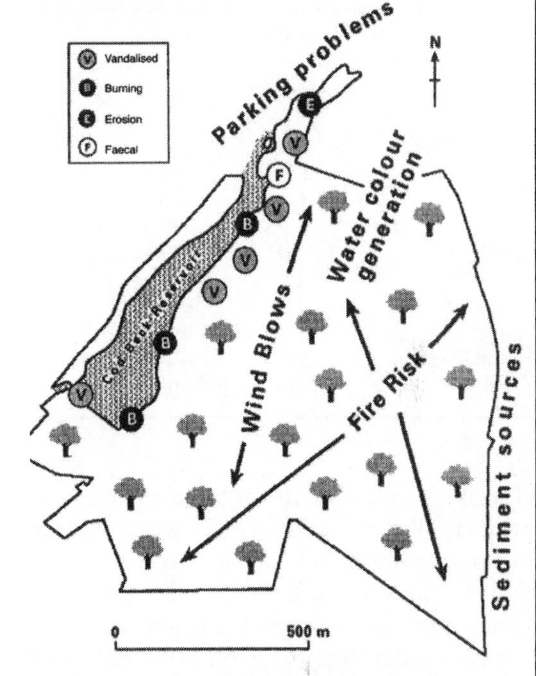

THE SOLUTIONS

Clearfell forest back from the water to wind-firm, amenity-planted forest edge. This will reduce forest usage for fuel and toileting. Clearance is taken up the side stream valleys to provide an extensive buffer zone and to open the waters edge to surveillance.

Forest-induced drought conditions and future clearfelling of even-aged monoculture forests exacerbate colour problems from organic-rich soils on low slopes. A programme of thinning, patch clearance and replanting with diverse species over the next twenty years will reduce the impact of harvesting on colour release and satisfy amenity considerations.

Fire risk is reduced in an irregular, mixed-species forest and through the reduction in the number and intensity of recreational fires (both promoted by the initiatives above). Brashing and clearance of forest residues in the remaining sites should be introduced.

Block parking on erosion-prone sites at the head of the reservoir. Substitute (but do not add to) parking space below the dam and provide toilets, interpretative boards and guided walks. In effect, accept this is a popular recreational site and seek to manage it.

demand and climate stability, it could be assumed that the water needs of a region would not change dramatically from year to year. A slow evolutionary need for increase would prevail, and that increase would be easily satisfied in a system that promoted a large quantity safety margin. Today, although there is still some debate about causes, few would disagree that climate change is underway. The last thirty years has also seen a substantial change in personal habits and national economy, both influencing water demand. So we live in and forecast a time of change.

Changes in the funding of the water industry have tied accurate forecasting to the plans and

Table 10.2 Examples of the required standards of potable waters. The achievement of these standards is tested by sampling of tap water and the formal reporting of results to the Drinking Water Inspectorate. The standards do not apply in all circumstances. Derogations can be sought to exempt companies if local circumstances always cause failures naturally or if there is a time-limited specific agreement to resolve a problem.

Parameter	*Value*	*Units*
Colour	20	mg/l
Taste	3	Dilution number
Temperature	25	°C
Aluminium	200	mg/l
Turbidity	4	FTU
Faecal coliforms	0	*N*/100 ml
Lead	50	µg/l
Arsenic	50	µg/l
Cyanide	50	µg/l
PAHs	0.2	µg/l
Pesticides	0.1	µg/l
THMs	100	µg/l

charging regimes of the water companies. The price companies can charge for water is set in a periodic review, which is conducted from time to time by the water industry regulator at Ofwat. Companies have to present a business plan that contains statements (forecasts) of future water resource demands and availability. To achieve the business plans under the conditions of demand and resources forecast, the companies also present to the regulator an asset management plan (AMP). The AMP will identify the required expansion and renewal in the underground and overground assets of the individual water company over the time period of the business plan. This expansion and renewal will require capital and a capital expenditure plan (CAPEX), as well as operating expenditure plans (OPEX), are presented to Ofwat.

This collection of forecasts, if believed by the regulator, will be a key consideration in the assignment of the K factor, an amount, modified by the retail price index, that the company can charge its customers for the service provided. Clearly, the plans should be considered as the opening proposals in a bargaining game, but the role of the forecast to both sides is crucial. Errors in either direction will, on the one hand, damage profitability to the detriment of shareholders and customer levels of service and, on the other hand, damage the authority of the regulator and promote overcharging of the customer.

Prices are set for a ten-year period but are reviewed after five. The next review, is in 1999. This will be a more complex and sophisticated review as the regulator learns from past experience. The components (intimated in open letters to the industry as actions that the director general is 'minded' to do) are shown in Table 10.4.

The K factor refers to a 'basket of goods and services', for example effluent treatment, sewerage management, and domestic and commercial water supply. The water companies have been permitted to charge what they wish for each item in the basket as long as they adhere to the overall K value. Ofwat is encouraging companies towards a balanced application of K. The current single requirement is that the unmetered customer who uses the 'average' amount of water be charged, through the unmetered tariff, the same as the metered customer using that average volume of water. Further parity constraints across the basket may be expected. Water companies are expected to recover costs. There can therefore be no 'loss leaders' in this industry. Therefore the Urban Wastewater Treatment Directive has resulted in higher K values for sewage treatment than for water supply. This has an impact on the large number of small, water-only, companies.

Certainly, the future will be more financially fraught for the water industry. The additional costs of the windfall levy, announced in the 1997 Budget, will not be allowed to be passed on to customers. Further, Ofwat also believes that the water industry underestimated 'efficiency gains' achievable under AMP2 in 1994 and that the corresponding price limits set under AMP2 were too generous. Consequently, Ofwat proposes a one-off cut in the K factor for 2000. While Ofwat is promoting a downward pressure on K, the Environment Agency is lobbying Ofwat for an increase in K (or a least a smaller drop) in order

Table 10.3 The prescribed standards and values which drinking water must achieve pose many general research themes and specific questions, some examples of which are suggested here.

Research area	*Research questions*	*Research examples*
Standards	New issues	Endocrine disrupters Plasticisers Algal toxins Viruses
	New parameters	Faecal streptococci Solvents Treatment polymers
	Standard changes	Standard validity Treatment developments for, say, lead and THMs Catchment based 'pre-treatment'
Sampling	Representative samples Timely and effective analysis	New techniques for rapid assay of bacterial and biological material
Interpretation	Statistical investigations	Permitted non compliance Validity of derogations Mean, median, percentiles basis
	Communication	Closed or open access to information
	Significance	Cost of compliance Cost–benefit balances Health gains GIS-based presentation and analysis

Table 10.4 The probable components of 'K' to be employed in periodic review 3 in 1999.

Component	*Effect*
X	Expected efficiency in the future
Q	Investment to improve 'quality of service'
Po	Delivery of efficiency gains
S	Supplementary investment in customer interest
V	Increased security of supply

that funds are available to address environmental issues such as the Shellfish Directive, the Bathing Water Quality Directive, the Urban Wastewater Treatment Directive and the Statutory Water Quality Objectives.

One of the key forecasts needed is the likely demand for water. It is complicated by the differing interpretations of demand made by different sections of the water industry. Expressed simply, perhaps crudely, the water industry has only recently and incompletely started to differentiate supply and demand. True water demand, that is water use by people and industry in a manner that is more or less responsive to price, is the definition of demand that is used by those sections of the industry that deal with pricing structures, tariff policy and similar functions. Resource engineers, on the other hand, refer to 'demand on sources' or 'demand on works', in other words, supply. In addition to the price-sensitive demands for individuals and companies is a series of other water uses, some of which may be highly intermittent. These are identified in Table 10.5.

In making demand forecasts for the major component of water use, domestic demand, the regulator has required companies to express demand as a demand for water per person, usually referred to as pcc (*per capita* consumption) by the water industry, and herein lies an unnecessary complexity of marked geographical significance. Water is supplied to households, bills are for households, the census deals with households, and any remotely sensed data is attributed to households and not to individuals. We have never effectively monitored a representative individual and probably never will. Therefore there is a considerable body of research devoted to linking geographical expertise in census information manipulation concerned with the individual make-up of households to the water use of these households. Such work draws on microsimulation and has been reported in Clarke *et al.* 1997.

Unmetered demand is measured in two fundamental ways: by area meters; and by domestic consumption monitors. Both bring the water manager to confront the diversity and complexity of populations, industry and urban structure. An area meter is simply that: a meter on the potable water distribution system, post treatment works, which measures water delivered to an area. Interpretation of the raw data from an area meter presents many challenges:

Table 10.5 Components and possible amplitudes of true and resource demand.

Component	*Nature of resource*	*Amplitude*
Metered domestic demand	True demand	Daily/Weekly
Unmetered domestic demand	True demand	Daily/Weekly
Industrial demand	True demand	Daily/Annual
Leakage	Resource demand	Seasonal/Intermittent
Fire-fighting	Resource demand	Highly Intermittent
Illegal use	Resource demand	Daily/Annual
Meter misreading	Resource demand	Systematic

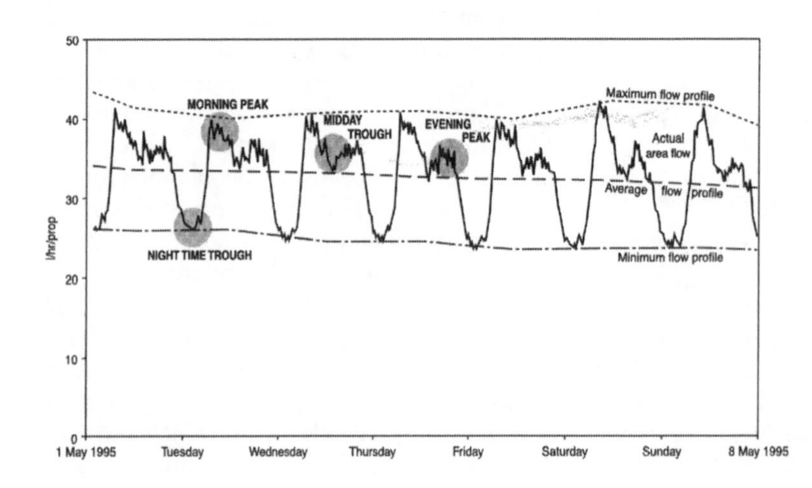

Figure 10.2 Graphical representation of output from an area meter that measures the flow of water to a small area having a mix of industry and domestic users. The minimum flow, less an allowance for legal use at night, is the nightline. By subtracting metered industrial use of water the area leakage can be identified. Readers will appreciate that this is a simplified outline of the leakage estimation procedure.

Box 10.2 To meter or not to meter, that is the question
(with apologies to W. Shakespeare)

In the UK, water has traditionally *not* been metered for the domestic market, a market that dominates the potable water sector. Such an approach has several outcomes.

1 The real demand for water is unclear.
2 Water management is made more difficult. 'If you can't measure it you can't manage it!' has real meaning.
3 Water balances cannot be established, since several elements are unknown.
4 Water managers will continue to confuse demand with supply – often referring to demand on works rather than customer demand.
5 The public will continue to regard water as unlimited (and unpriced). They will therefore continue to regard unlimited use as a right and will be resistant to change.

How much water do we use? 150 litres per person per day perhaps! might be the most honest answer. We don't really know. People who have selected metering are unrepresentative. Area meters have many unknowns in the aggregate figures (leakage being the most obvious) and where 'area meter' areas are simple (new development, cul de sacs, isolated communities) they are not representative. Domestic consumption monitors (based on representative households that are metered but that pay on an unmetered tariff) age rapidly (through births, deaths, job status and migration) and if not maintained will be erroneous before they accumulate enough data to provide the answer they were developed to provide.

Industry is now at a crossroads: both routes have advantages and pitfalls for suppliers and consumers.

THE CASE AGAINST METERS

- Water is not a trading good that can be done without.
- Water is required for life (and this applies to existence and lifetime!).
- Clean water is required for heath (for both hygiene and consumption).
- Retrospective payment meters (credit meters) run the risk that users will run up big bills.
- Prepayment meters typical of electricity and gas supplies have been declared illegal. Does this constitute court recognition that water is a special-case commodity? That water has no substitute, unlike gas, substitute electricity, substitute coal or wood.
- Expensive national retrofitting.
- The least able (big families, older facilities) will bear the financial burden.

THE CASE FOR METERS

- Reveals demand and thus quantifies leakage.
- Allows price control of water.
- Encourages water efficiency.
- Reduces resource exploitation and environmental damage.
- Aids demand prediction.
- Focuses demand on households rather than the current artificial *per capita*.

- The area changes as water operators in the front line open and close valves to correct such things as low pressure.
- The meter measures the total water entering the area and thus an estimate for leakage, industrial use and other constants has to be made and, on occasion, a measure for intermittent use in fire-fighting is also required.
- The metered area does not match with population information areas. The population of the area is not precisely known and in any event is dynamic.

An added complication in this system of multiple estimated components is that the same monitoring system is used to determine leakage. Figure 10.2 shows area water flows to a part of a major UK city. The line joining the lowest trough each day is called the nightline (the night-time water use). From this is subtracted a nominal, and convenient, 1 l/hh/hr to represent legitimate night use, and the remainder is deemed to be leakage if it is accepted that the district has no industrial or commercial use and that there is no illegal use.

In an effort to counteract these deficiencies, water companies have selected and developed particular areas for area meters. 'Golden' districts (where data quality is high – complete and reliable), *cul de sacs* and detached communities all have attributes that are attractive to the analysts, for they present simple, uncomplicated and 'isolated' areas. Sadly, these very characteristics compromise their validity as a basis for estimation, because they are unrepresentative.

Domestic consumption monitors (DCMs), on the other hand, are composed of a large number (~2000) of households that are metered but that pay on a non-metered tariff. For the geographer, DCMs must have housing and household characteristics that are typical of the area they represent.

DCM managers must appreciate the dynamic characteristics of urban populations and take positive steps to maintain the DCM. A properly maintained and analysed DCM will yield detailed information about water use. Such information can then be used with small-area population forecasts to forecast water demand.

WASTE WATER MANAGEMENT

Waste water management concerns the management of water after use by the customer prior to return to the environment. It is considered within this section because waste waters are the financial responsibility of the water companies. A fuller treatment of water quality issues is given in Chapter 11.

Sewer networks remove organic wastes from houses and wastes from factories. Sewerage networks remove street waters after rainfall and as such are a replacement for the hydrological network disrupted in urban development. In the United Kingdom sewage and sewerage networks are mainly combined. They have developed piecemeal and for the most part are long-established under a variety of regulatory regimes. Such regimes were initially local and have relatively recently become national. The networks usually lead to a treatment works, the effluent from which discharges into the natural environment. Treatment works are designed primarily to remove organic wastes, and these works cannot remove exotic chemicals. Therefore the operators of the works have a right to choose which discharges, other than domestic discharges, they will accept into the network. This choice is exercised through a Trade Effluent Permit. The control of effluents from the treatment works to the natural environment is controlled by a Consent to Discharge, which is set by the Environment Agency in the UK. Both consents and permits set conditions on the nature and concentration of the effluents, and both set financial requirements on the holder.

Because the UK system is combined, a rainstorm of moderate size in an urban area, concentrated through the system, could easily flood the water treatment works, so sewer overflows exist to divert the high flows away from the works and directly into the river. This occurs, in theory, at about six times the dry weather flow, but changes in rain characteristics, in the density and extent of the network, in the condition of the network, etc. has resulted in combined sewer overflows (CSOs) occurring too frequently and over an excessively prolonged period. The sewer system, like the potable water supply system, is out of sight and was built by predecessor organisations from which records of type, location and characteristics have not always been available. The resulting ignorance of the basic system raises a number of important and apparently simple research questions, which are listed below. They are questions of great importance since replacement of the underground sewer assets is (gu)estimated to cost in excess of £100billion.

1 Where is the pipe network?
2 What are its characteristics and condition?
3 Which section will fail next?
4 How can repair, maintenance and replacement be carried out without above-ground disruption?

All of these questions will prompt the development of remote, preferably non-invasive, monitoring technologies, probably deriving as much from the geophysical as the engineering domain. Assessments of condition and such things as sewage exfiltration will be required while operational use continues and will preferably be made through external surrogate data. The water manager will increasingly use intelligent technologies to monitor and repair the network, particularly to ensure that problems such as blockages are removed and not simply displaced. A greater use of modelling will be required to predict and prioritise maintenance and replacement. This modelling will need to be highly integrated with the forecasts for other elements of the water system, such that the effects of water conservation, grey water reuse and roof rain retention on the sewer flows can be determined.

Wastewater treatment works (like the water

treatment works discussed earlier) are primarily the domain of the chemical engineer. The wastewater treatment manager today is under four key pressures from customer expectation, sustainability, cost efficiency and new standards (Figure 10.3). Toxicity-related consents will require the development of rapid testing technologies, both to control the process and to trace 'polluters' who may not have, or do not adhere to conditions in, trade effluent permits. Endocrine disrupters and oestrogen mimics give rise to great concern, particularly because considerable elements of the budget for these chemicals derives from domestic properties, the effluent from which the companies are required to accept. Progress on detection, analysis, treatment and source control is urgently needed. Tighter standards, for example in relation to nitrogen and phosphate, and new standards relating to ubiquitous substances like washing powder brighteners, will bring further new challenges.

CONCLUSIONS

The water industry is operating at a time of change. The industry has no agreement with its raw material supplier and worse, the natural resource, which is its raw material, appears to be likely to be more erratic in delivery. Following the droughts in the UK in 1976, 1984 and 1995, the industry must re-evaluate resource reliability and the likely impact of present policies on the rivers and groundwater. All environmental impacts on water bodies, whether from abstraction or from discharges, must be reviewed in the light of the heightened public and political concern for the environment. The water industry has failed in its relations with the public. It has focused on 'compliance', forgetting that this might not accord with customer views on water quality. In relation to droughts and to leakage, the customer has been dismissed as emotional and ignorant; both more than likely largely true, but now the companies will be forced to levels of leakage control and standards of service that are not perhaps strictly economic. The industry appears to have a short memory. Customer-oriented management and recognition of public perception through, for example, the introduction of more water efficiency programmes has been a short-term response. Already some companies are

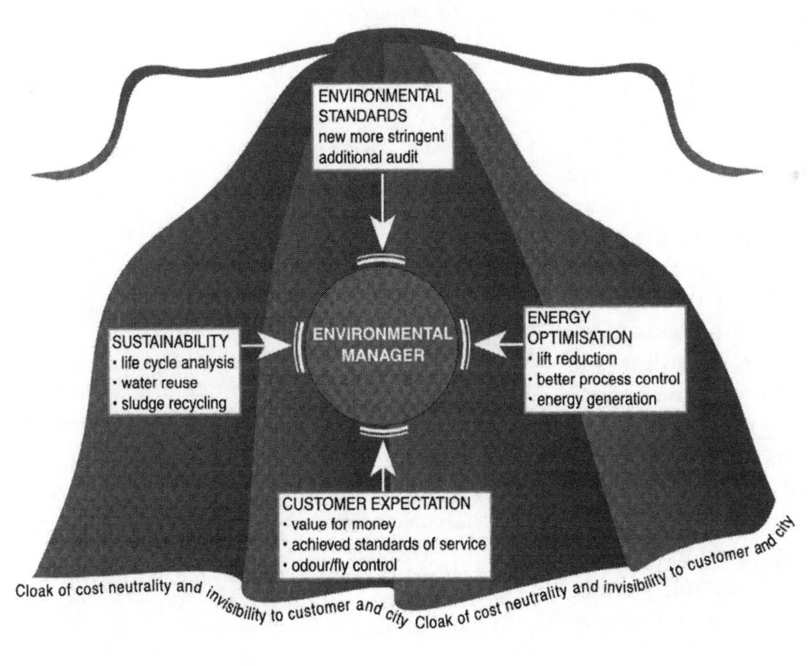

Figure 10.3 Challenges to the water manager. The water manager is pressured by several, sometimes contradictory, challenges. Most actions taken to respond to these pressures have to be invisible to the customers and have minimum effect on price.

introducing internal competition that will bring them into conflict with customers and regulators again as, for example, will occur when the competitive separation of potable and waste water sides of the business results in sub-optimal environmental management.

The companies will be forced towards a changed social and environmental attitude. Already the courts have ruled against prepayment meters, an implicit recognition of the fundamental importance of water. The companies would be wise to evaluate carefully winners and losers before embarking on any development and expansion of the metering programme. Whether or not the companies are provided with funding from Ofwat via 'K', the Environment Agency can require environmental improvements. During the present period, the agency was finding its feet as an organisation with enlarged responsibilities. Over the next few years, it will concentrate more effectively on discharging its considerable responsibilities. In the face of company intransigence in the past regarding such issues as low flows, the companies should now anticipate a revisiting of Abstraction Licences and Consents to Discharge.

Although still secondary to engineering, the applied geographer has a significant and growing role in both research and application for improved water supply and management. That role peaks where there is a direct industry contact with the environment or clients or where there is strong spatial element of analysis. Geographers have contributed to the understanding of processes that modify raw water quality and to the interventions, such as land-use change, that promote improved water quality. Applied geography has helped to analyse in a spatial sense the demand for potable waters in the industrial and domestic sectors and the contribution of leakage and conservation to the overall water balance. In turn, such work has laid the foundations for evaluations of the sustainability of water management strategies and for the determination of ecological footprints for water resource activities. Progress to a more sustainable water system will without doubt draw applied geography to contribute further to resolution of the challenges that face the water industry in the next millennium.

GUIDE TO FURTHER READING

As we appproach the millennium, we must accept that more information is appearing daily on the web. For relatively recent information on the views of the key government agencies controlling water supply and management in the UK, review the web pages listed below. The director general of Ofwat is already using the web as a vehicle for dissemination of his possible views on financial regulation.

Department of the Environment, Transport and the Regions – http://www.environment.detr.gov.uk
Environment Agency – http://www.environment-agency.gov.uk
Open Government – http://www.open.gov.uk
Ofwat – http://www.open.gov.uk/ofwat

REFERENCES

McDonald, A. and Kay, D. (1988) *Water Resources Issues and Strategies*. Longman: Harlow, UK.

Clarke, G.P., Kashti, A., McDonald, A.T. and Williamson, P. (1997) Estimating small area demand for water: a new methodology. *Journal of the Chartered Institute for Water and Environmental Management* 11, 186–92.

Herrington, P. (1996) *Climate Change and the Demand for Water.* Department of the Environment. HMSO.

Lambert, A. (1994) Accounting for the losses: the burst and background concept. *Journal of the Institution of Water and Environmental Management* 8, 205–14.

Likeman, M.J., Field, S.R., Stevens, I.M. and Fleming, S.E. (1995) Applications of resource technology in Yorkshire. *British Hydrological Society Proceedings of the 5th National Hydrology Symposium.*

Mitchell, G. (1990) Natural discolouration of freshwater: chemical composition and environmental genesis. *Progress in Physical Geography* 14, 317–34.

Mitchell, G. and McDonald, A.T. (1995) Catchment characterisation as a tool for upland water quality management. *Journal of Environmental Management* 44, 83–95.

Ofwat (1996) *1996–97 Report on the Tariff Structure and Charges.* Office of Water Services, Birmingham, 72 pp.

Russac, D.A.V., Rushton, K.R. and Simpson, R.J. (1991) Insight into domestic demand from metering trial. *Journal of the Institution of Water and Environmental Management* 5 (3), 342–51.

Wigley, T.M.L. and Jones, P.D. (1987) England and Wales precipitation: a discussion and an update of recent changes in variability and an update to 1985. *Journal of Climatology* 7, 231–44.

11

Water quality and pollution

Bruce Webb

INTRODUCTION

Water in every phase of the hydrological cycle, from precipitation through terrestrial surface and groundwater systems to the marine environment, has a quality dimension that can be described by reference to numerous physical, chemical and biological properties, and is controlled by a myriad of natural factors and human influences. Water quality is of fundamental importance in the provision of potable supplies to sustain human life and in the health of aquatic ecosystems. It also significantly affects a wide range of human uses of water in industry, agriculture, transport and recreation. At the same time, these uses and other human activities, directly or indirectly, provide manifold sources of water contamination. Where the consequences or side-effects of human scientific, industrial and social habits result in conditions within the water environment that are harmful or unpleasant to life, the term 'water pollution' is used (Sweeting 1994). Acute water quality problems, however, may also arise from natural climatic or geological conditions.

Problems of freshwater pollution, on which the present chapter focuses, have a long history and have changed in character as world population has grown and human technological capability has increased and become more complex. Local contamination of the aquatic environment has been recognised for at least two millennia, and in some countries legal means were taken to prevent water pollution as early as medieval times. In the UK, for example, laws were passed in the thirteenth century to prohibit washing the products of charcoal burning in the River Thames (*ibid.*). Water pollution, and its deleterious consequences for human and ecosystem health, has accelerated since the nineteenth century with the increasing urbanisation and industrialisation of human society and intensification of agriculture to support an ever-growing population (UNEP/WHO 1988). One of the first documented examples of the inimical effects of bad water quality concerned the outbreaks of cholera in London, which were traced by John Snow in 1854 to the gross pollution of the River Thames by raw sewage. Problems of faecal contamination of rivers used for public water supply in developed economies were subsequently largely solved by the invention of sand filtration and the use of chlorination. The sequence of problem occurrence and perception followed by the application of control measures is one that has been repeated, especially during the last fifty years (Figure 11.1). Over this period, increases in public awareness of pollution, in the ability to develop remedial measures and in the political will to implement strategies to control water contamination have to a certain extent paralleled the rapid emergence of a succession of water quality problems.

A conceptual model of water pollution occurrence and control has been proposed by Meybeck *et al.* (1989) using the example of the history of domestic sewage contamination in Western Europe over the last two centuries (Box 11.1). This model can also be applied to other types of pollution and to countries that have different patterns of economic development. In the latter

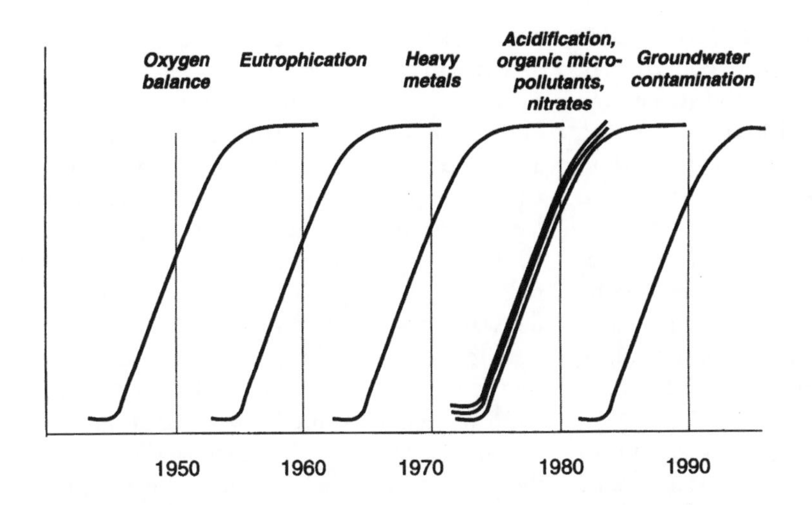

Figure 11.1 Sequence of occurrence and perception of severe water pollution problems in Europe.

Source: After Meybeck *et al.* 1989.

Box 11.1 A conceptual model of water pollution occurrence and control

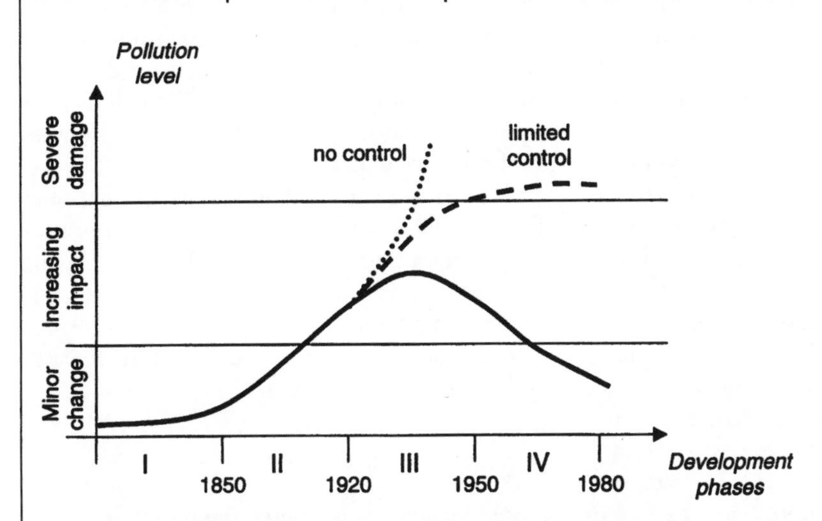

Four phases were recognised in the conceptual model control devised by Meybeck *et al.* (1989) that relates water pollution occurrence and control to economic development. The first phase is typical of an agricultural society when levels of pollution are low and tend to increase linearly with population growth. In the second phase, which characterises newly industrialised countries, pollution increases exponentially with industrial production, energy consumption and agricultural intensification. The third phase, which occurs in highly industrialised countries, sees containment of pollution problems as a result of implementing effective control strategies (mechanical and biological wastewater treatment), while the fourth phase is the desired ultimate situation where contamination is reduced to a level that is ecologically tolerable and does not interfere with water use. Where no pollution control is enacted, levels continue to rise rapidly and severe damage to the environment occurs. If limited controls (mechanical sewage treatment) are employed pollution rises more slowly but may eventually cause severe damage.

case, however, the time scale of the phases is different to that experienced in Western Europe. In rapidly developing nations such as Brazil, China and India, for example, population growth and transition from an agricultural to an industrialised society is taking place at a much faster rate than occurred in Western Europe and North America. In consequence, the sequence of pollution problems is emerging over decades rather than centuries. In many countries of Eastern Europe, industrialisation and agricultural intensification have proceeded at a similar pace to that of Western Europe, but pollution has risen to the level of severe damage because environmental regulation has been slow or lacking. Water pollution and control is affected not only by conditions within a particular country but also by factors operating at an international scale, such as the effects of long-range transfer of pollutants in the atmosphere, climate change and nuclear accidents (Peters *et al.* 1998).

It is increasingly being recognised that issues of water quality and pollution are as critical as issues of water scarcity, although freshwater vulnerability may be most severe when problems of water contamination are combined with water shortage. Reference has been made recently to an emerging global crisis of water quality. The dimensions of this crisis include the death of five million people annually from waterborne diseases, the loss of biodiversity and the occurrence of ecosystem dysfunction, the contamination of freshwater and marine ecosystems from land-based activities, the pollution of groundwater resources, and global contamination by persistent organic pollutants (Ongley 1996). Many experts now believe that freshwater quality will become the principal limiting factor for sustainable development in many countries early in the twenty-first century (Ongley 1999), and already in China the aggregate cost of water pollution to the national economy has been estimated as 0.5 per cent of GDP (Smil 1996).

THE NATURE OF WATER QUALITY PROBLEMS

Table 11.1 summarises major issues of freshwater quality and highlights the great diversity of pollution problems that currently affect the surface and groundwaters of the Earth. Water quality varies naturally in space and time in response to climatic, geological, pedogenic, biotic and hydrological factors (e.g. Walling 1980; Walling and Webb 1986; Meybeck 1996), and natural ecological and climatic conditions may give rise to contamination by parasites, salts and metals, especially in groundwater reservoirs. A greater number of water quality issues, however, are related to human activities, which may cause freshwater pollution at varying spatial scales from local (less than 10^4 km^2) and regional (10^5–10^6 km^2) to global (10^7–10^8 km^2). The severity of pollution is usually inversely related to the size of the water body impacted. Temporal scale is also significant in water quality issues, because the time it takes for freshwater to become polluted and the period required for remediation of the contamination varies according to the source of the problem and the hydrological environment affected (see Table 11.1). Thus, accidental chemical spillage into a river will have an almost instantaneous effect, but the transit time of the pollutant from headwaters to the mouth of the system, even in major rivers, will be of the order of weeks or months, and the system will recover rapidly. In contrast, it may take several decades for fertiliser pollutants to migrate from the soil profile to an underlying aquifer (Burt and Trudgill 1993), while retention or absorption of the pollutant by soils and bedrock may make the clean-up period more protracted than the contamination phase.

A number of major sources of freshwater pollution can be identified.

Organic pollution

Domestic sewage is one of the most significant and widespread sources of organic matter added to freshwaters by human activity, and it causes pollution at local, regional and continental scales (UNEP/WHO 1988). It has been estimated that for major European rivers, such as the Rhine, growth in human population since pre-industrial times has been associated with a three-fold increase in the organic carbon burden (Zobrist and Stumm 1981). Today, domestic sewage remains a major cause of river pollution in developed countries (Sweeting 1994), while lack of sanitation and inadequate waste management amplify this problem for a large percentage of the

Table 11.1 Major water quality issues excluding ecological quality.

Major causes/issues	Major related issues[1]	Space scale	Time scale (years)		Major controlling factors	
			Contamination[2]	Clean-up[3]	Biophysical	Human
Population	Pathogens	Local	<1	<1		Density and treatment
	Eutrophication*	Regional	<1	1–100		Treatment
	Micropollutants	Regional	<1	1–100		Miscellaneous
Water management[4]	Eutrophication*	Regional	<1	10 >100		Flow
	Salinisation	Regional	10–100	10 > 100		Water balance
	Parasites	Regional	1–10	>100		Hydrology
Land use	Pesticides	Local–regional	<1	1–100		Agrochemicals
	Nutrients (NO_3^-)	Local–regional	10–100	>10		Fertiliser
	Suspended solids*	Local–regional	1–10	10–100		Construction/ clearing
	Physical changes	Local	<1-10	>100		Cultivation
Long-range atmospheric transport	Acidification*	Regional	>10	10		Cities
	Micropollutants	Regional	>10	1–100		Smelting
						Fossil fuel emissions
Concentrated pollutant sources:						
Megacities	Pathogens	Local	<1			Population and treatment
	Micropollutants	Local–regional	<1			
Mines	Salinisation	Local–regional	10–100			Types of mines
	Metals	Local–regional	<1			
Nuclear industry	Radionuclides	Local–global	<1			Waste management
Global climate change	Salinisation	Global	>10	>100	Temperature and precipitation	Fossil fuel emissions
Natural ecological conditions	Parasites*	Regional	Permanent	Permanent	Climate and hydrology	
Natural geochemical conditions	Salts	Regional	Permanent	Permanent	Climate and lithology	
	Fluoride**	Local–regional				
	Arsenic**, metals**	Local–regional			Lithology	

Source: Peters *et al.* 1998.

Notes: 1 * is relevant primarily to surface water and ** is relevant primarily to groundwater.

2 Space scales: local — <10,000 km^2; regional — 10^5 to 10^6 km^2; and global — 10^7 to 10^8 km^2.

3 Lag between cause and effect.

4 Longest time scale is for groundwater, followed by lakes, and shortest for rivers and streams.

world's population (Peters *et al.* 1998). Effluents from industrial processes such as canning, milling, tanning, and manufacture of textiles, pharmaceuticals and coffee, may also introduce large quantities of organic wastes into watercourses (Meybeck *et al.* 1989). Point source domestic and industrial organic waste inputs into watercourses are biodegraded, with well-known consequences for the downstream dissolved oxygen balance, water chemistry and aquatic organisms (Figure 11.2).

Figure 11.2 Effects of an organic effluent on the quality and ecology of a river.

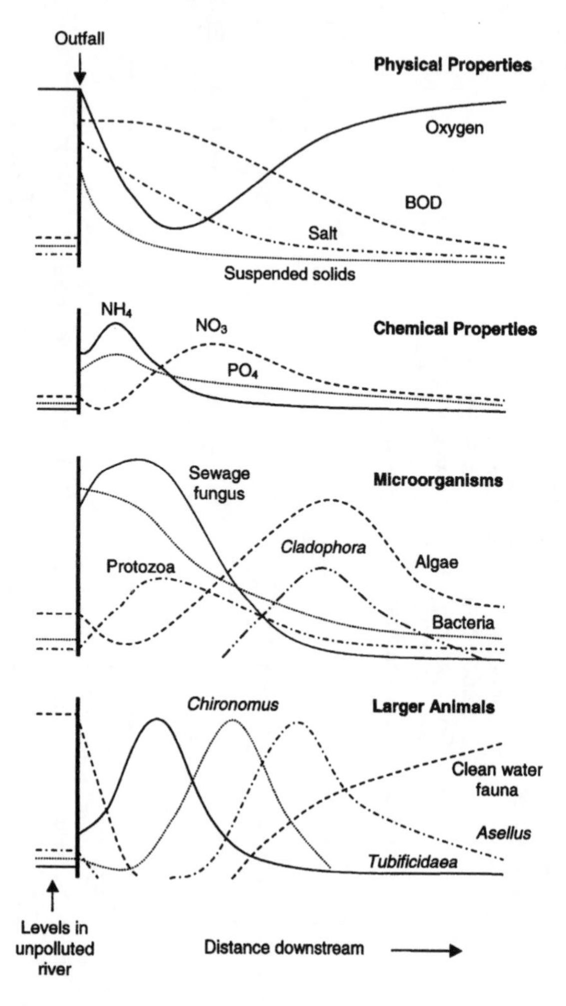

Source: After Mellanby 1980.

Agricultural practices may also provide large amounts of organic material in the form of animal slurry, silage liquor, sewage sludge spread on to agricultural land and effluents from dairies, abattoirs and vegetable-processing plants (National Rivers Authority 1992). Major pollution incidents involving organic farm wastes are often associated with poor containment of slurry or silage liquors and spillage following failure of storage areas. These wastes can be very damaging if they enter watercourses because of their very high biochemical oxygen demand (BOD), which for silage liquor is typically 60,000 mg l^{-1}, compared with 350 mg l^{-1} for untreated human sewage and <5 mg l^{-1} for clean river water (*ibid.*).

Pathogens

It has recently been highlighted that the microbiological safety of water is in global decline (Ford and Colwell 1996), and water-related disease is one of the most important worldwide human health concerns in terms of morbidity, mortality and cost. For example, more than half the world population has suffered diseases resulting from drinking polluted waters (Barabas 1986) and more than 10 million deaths occur each year from water-related diseases. Furthermore, the social cost of gastro-intenstinal illness not requiring physician consultation or a period in hospital was estimated at more than US$19.5 billion in 1985 for the United States (Garthright *et al.* 1988).

Agents of waterborne disease in freshwater bodies include bacteria such as *Shigella* and *Salmonella*, viruses such as hepatitis A and enterovirus, protozoans such as *Giardia* and *Entamoeba*, and parasitic worms such as the beef tapeworm (*Taenia saginata*) and blood flukes (*Schistosoma* spp.) (Meybeck *et al.* 1989). New water-transmitted pathogens are continually being discovered, such as *Helicobacter pylori*, which is now believed to be the cause of many types of ulcer (Klein *et al.* 1991). The major source of freshwater pathogens is faecal contamination from infected humans, pets, farm animals and wildlife (Geldreich 1997). Sewage discharges often introduce a variety of pathogens to watercourses, and the occurrence of major floods, which lead to sewage contamination of drinking water systems, increases exposure to disease agents. Population growth, increasingly concentrated husbandry of domestic animals and declining water supplies have exacerbated the occurrence of waterborne disease. In the future, changes in the distribution of freshwater pathogens and the vectors of water habitat diseases, such as snails, mosquitoes and flies, can be expected through increased water temperatures associated with future global warming (Webb 1997) and through other water quality trends such as increased eutrophication.

Pollution of freshwaters by pathogens can be combated by waste water and water supply treatment, and water bodies have the potential for self-purification. However, even in developed countries, traditional water treatment processes may not effectively eradicate newly recognised pathogens such as *Cryptosporidium* (Department of the Environment and Department of Health 1990; 1995), which has caused health problems recently in the UK and in the United States. MacKenzie *et al.* (1994), for example, report that in 1993 more than 400,000 people were infected with cryptosporidiosis in Milwaukee, Wisconsin, as a result of contamination of the drinking water supply from Lake Michigan. The outbreak killed more than 100 people and led to hospitalisation of more than 4000 others.

Nutrients

Nitrogen (N) and phosphorus (P) are regarded as key nutrients in surface and groundwaters (Heathwaite *et al.* 1997). Human activities have increased the concentration and loading of these elements in freshwaters, leading to pollution problems at regional and potentially wider scales. Considerable attention has been focused on nitrates in recent years (Royal Society 1983; Burt *et al.* 1993) because of concerns that high concentrations in drinking water may be responsible for methaemoglobinaemia (blue baby syndrome) and stomach cancer. Monitoring has revealed rising nitrate concentrations in European and North American rivers, groundwaters and lakes since the 1930s, and especially over the last thirty years (Figure 11.3). Although industrial and

Figure 11.3 Examples of rising nitrate concentrations in (A) rivers, (B) groundwaters and (C) lakes.

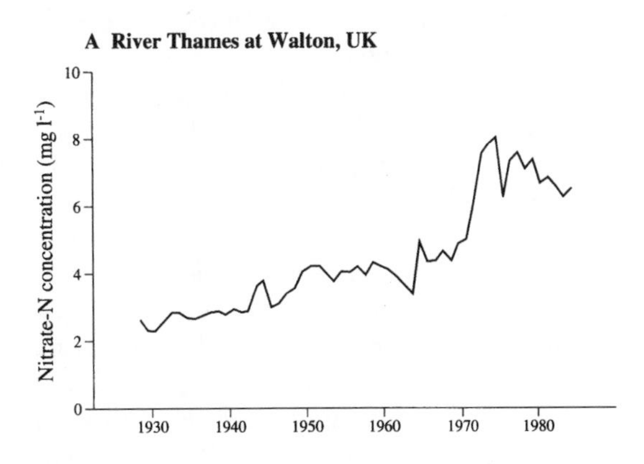

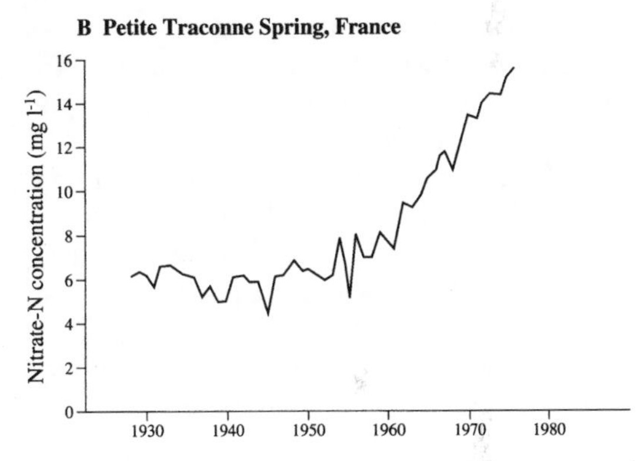

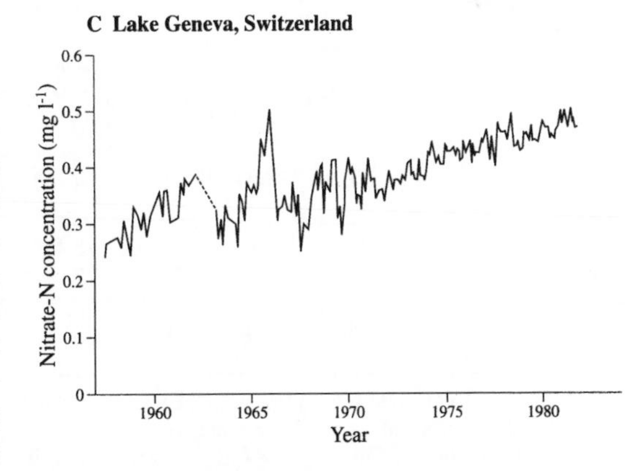

Sources: Roberts and Marsh 1987; CIPEL 1986.

sewage effluents can contain high nitrate concentrations, much of the nitrogen reaching surface waters originates from diffuse rather than point sources. Increasing nitrate concentrations in freshwaters are therefore thought to reflect intensification and expansion of agricultural production, but the linkages between land-use change, increased fertiliser application and rising nitrate levels are not always direct or simple (Heathwaite *et al.* 1993).

Less information is available about total nitrogen and the relative importance of different N species in freshwaters, especially for river systems. A recent study of four contrasting UK rivers (Russell *et al.* 1998) showed that total nitrogen loads were dominated by the dissolved total oxidised fraction (nitrate and nitrite), which accounted for 76 to 82 per cent of the annual flux in 1995 (Table 11.2). Dissolved organic N and particulate N did not contribute more than 16 and 8 per cent of the annual load, respectively, and dissolved inorganic nitrogen in the form of ammonium was a minor component of total N in these predominantly rural and not highly polluted systems. In contrast, information for UK lakes (Heathwaite 1993) shows that the organic fraction often dominates the dissolved component of nitrogen, especially in upland environments, where lentic waters are poor in nutrients and in plant life.

The particulate fraction is of more importance in the transport of phosphorus in freshwater systems, because erosion of sediment from agricultural land provides an important pathway by which P bound to the soil matrix can be introduced to river systems. In the study of the four UK catchments referred to above, total particulate P, comprising both inorganic and organic fractions, accounted for 26–75 per cent of the annual total P flux (see Table 11.2). Increased use of fertilisers containing inorganic P, which has for example tripled in the USA between 1945 and 1993 (Puckett 1995), and greater dissolved P inputs from sewage, livestock wastes and detergents (Withers 1994) have led to increased phosphorus concentrations in rivers and lakes of Europe and North America over the last fifty years (Heathwaite *et al.* 1997). However, improvements to waste water treatment facilities and the banning of phosphate-based detergents have markedly reduced total phosphorus concentrations during the last 15 years in some rivers and lakes (e.g. Edmonson 1985; Peters *et al.* 1997).

Enrichment of water bodies with plant nutrients, especially phosphorus and nitrogen, which leads to changes in biological structure and function, is the process of eutrophication (Vollenweider 1968; Harper 1992; Rast and Thornton 1997). This process takes place naturally over geological time but may be greatly accelerated by human disruption of catchment nutrient cycles. Excessive algal and rooted plant

Table 11.2 The percentage contribution of individual fractions to nutrient fluxes in selected UK drainage basins.

River	*% contribution of fraction to total nitrogen yield*					*% contribution of fraction to total phosphorus yield*					
	TDN	*NH_4-N*	*TON*	*DON*	*TPN*	*TDP*	*DIP*	*DOP*	*TPP*	*PIP*	*POP*
Severn	96.7	1.1	80.4	15.2	3.3	56.8	50.6	6.2	43.2	27.5	15.7
Avon	95.4	0.3	82.0	13.1	4.6	73.7	70.3	3.4	26.3	20.3	6.0
Exe	92.2	1.2	77.1	13.9	7.8	32.2	25.9	6.3	67.8	43.1	24.7
Dart	92.0	0.6	75.7	15.7	8.0	24.6	15.4	9.2	75.3	47.1	28.2

Notes: TDN = total dissolved nitrogen; NH_4-N = dissolved ammoniacal nitrogen; TON = dissolved total oxidisable nitrogen (nitrate and nitrite); DON = dissolved organic nitrogen; TPN = total particulate nitrogen; TDP = total dissolved phosphorus; DIP = dissolved inorganic phosphorus; DOP = dissolved organic phosphorus; TPP = total particulate phosphorus; PIP = particulate inorganic phosphorus; POP = particulate organic phosphorus.

growth occurs, which in turn results in impairment of water quality, recreational potential, fisheries and flow of the affected water body (Meybeck *et al.* 1989). Some blue-green algal blooms, most notably those of *Microcystis*, *Aphanizomenon*, *Anabaena* and *Oscillatoria*, form toxic scums, which may cause skin rashes, eye irritation, vomiting, diarrhoea, fever and muscle and joint pains in humans who swim in affected waters. The toxins can also cause severe illness and death in wild, farm and domestic animals (Environment Agency 1997a).

It should be noted that eutrophication is not always viewed as a pollution problem, especially when nutrient enrichment increases the biological productivity of waters in developing countries. Resulting high yields of fish and crustaceans provide a valuable source of protein and of cash (Rast and Thornton 1997).

Salinisation

One of the most significant and most widespread forms of groundwater pollution is increasing salinity (Meybeck *et al.* 1989), and concentration of salts in soils and waters may make land and water resources unusable. Although high salinity may arise naturally through evaporation, mineral dissolution, airborne sea salt and juvenile and connate waters, a range of human activities may greatly exacerbate the problem (Table 11.3). Irrigation agriculture is a major cause of salinisation, which particularly affects semi-arid and arid climatic zones (e.g. Chaouni Alia *et al.* 1997; Rimawi and Al-Ansari 1997). Evaporation of the water applied to crops causes the accumulation of salts in the soil, which may be leached to groundwater. Additionally, waterlogging through rising water tables in irrigated areas leads to further salinity deterioration, either by encouraging salt dissolution or by bringing already saline groundwater closer to the land surface.

Irrigation may also cause problems in other climatic regions, such as Washington, western USA, where it is that estimated irrigation return flows have increased Na+ concentrations in the Lower Yakima River by fifty-fold (Sylvester and Seabloom 1963). Over-exploitation of fresh groundwater reserves, which promotes intrusion of saline groundwaters, is also an important cause

Table 11.3 Principal sources of salinity caused by human activity.

Source	*Rivers*	*Lakes and reservoirs*	*Groundwater*
Irrigated agriculture			
waterlogging and salinisation			****
irrigation return flows	**	*	*
excessive river water withdrawals	**	***	
overpumping groundwater			**
Saline intrusion			****
Mining activities	**	*	**
Disposal of oilfield brines			**
Upconing of connate water			*
Highway de-icing	*		*
Landfill leachates	*		**
Leaking sewers			*

Source: Meybeck *et al.* 1989.
Notes: Blank space – unimportant as a source of salinity.
*, **, ***, **** – general assessment of relative importance of sources.

of salinisation, especially in coastal and oceanic island environments, where sea water intrusion poses a threat to drinking water supplies (e.g. Contractor and Yong 1997).

Acidification

The deposition of a cocktail of acidifying compounds on terrestrial and aquatic environments, in wet form associated with rain, snow, mist and fog or in dry form as gases and particles, has become a concern of global significance (Park 1987; Meybeck *et al.* 1989). While freshwater acidification may arise through upland afforestation, heathland regeneration and acid mine drainage, most attention has focused on the impact of increasing SO_2 and NO_x emissions as a consequence of fossil fuel combustion in power stations, factories, houses and vehicles. It is well documented that industrialisation of Western Europe and North America during the last 150 years has increased the acidity of precipitation over Scandinavia, many other European countries, northeastern USA and eastern Canada (e.g. Brimblecombe and Stedman 1982; Varheyli 1985). Emissions of ammonia from agricultural sources, particularly from livestock wastes (Buijsman *et al.* 1987), may also contribute significantly to freshwater acidification because nitrification of ammonium compounds generates hydrogen ions (Meybeck *et al.* 1989).

Measurements of pH in the three decades following the Second World War have revealed significant acidification of lakes and rivers in regions where precipitation is known to have become more acidic over the same period (Park 1987). For example, measurements of 187 lakes in southern Norway during the period 1923–49 and in the 1970s reveal a ten-fold increase in acidity (Wright 1977), while the pH of the River Klaralven in Sweden fell from 6.9 in 1965 to about 6.4 in 1974 (Oden 1976). Acid surges, involving sudden and dramatic reductions in pH levels, can occur when snowmelt, or the first rainfall after prolonged drought, flushes out sulphur and nitrogen compounds accumulated through dry deposition. Acidified freshwater systems experience biological changes that often involve a decline in population size and diversity for most species, damage to macrophytes and a decline in microscopic plants and animals (Park 1987; Mason 1991). Fish deaths may also be caused by the mobilisation of aluminium and other heavy metals from soils by infiltrating acidic precipitation.

The impact of acid deposition reflects the buffering capacity of soils and bedrock of affected catchments (Glass *et al.* 1982). Areas underlain by chalk, limestone or other deposits rich in calcium carbonate will not be adversely affected by acid precipitation, while regions with acid igneous or quartz-rich sedimentary rock types will be very sensitive to acidification. It is possible to calculate 'critical load' for acidity in different ecosystem types (Critical Loads Advisory Group 1994) and to determine whether that load is exceeded by current rates of acidic deposition. Maps of critical load exceedence for freshwater have been calculated for the UK and reveal acidification to be most problematic in localised areas of the Pennines, western Scotland and Wales (Figure 11.4).

Concern over surface water acidification in northwest Europe has resulted in international agreements to reduce emissions of S and N oxides (UNECE 1994) through implementation of control strategies, which started for some countries in the early 1970s. Evidence from the UK as to whether freshwaters are responding to reduced emissions of SO_2 over the last decade is conflicting. Studies of the Cairngorm Mountains in Scotland (Soulsby *et al.* 1997) suggest some reversal of acidification in the period 1983–94. However, an investigation of upland catchments in mid-Wales (Robson and Neal 1997) provided no indication of reduced acid deposition rates or falling acidity of runoff over a very similar time period (1983–93). Whatever the immediate effects of controls in the developed countries, it should be remembered that S and N emissions continue to rise in the rapidly industrialising countries of Asia and South America and threaten to extend the regions of the world affected by acidification problems (Meybeck *et al.* 1989).

Figure 11.4 Exceedence of critical loads for acidity by deposition of sulphur and nitrogen to freshwater.

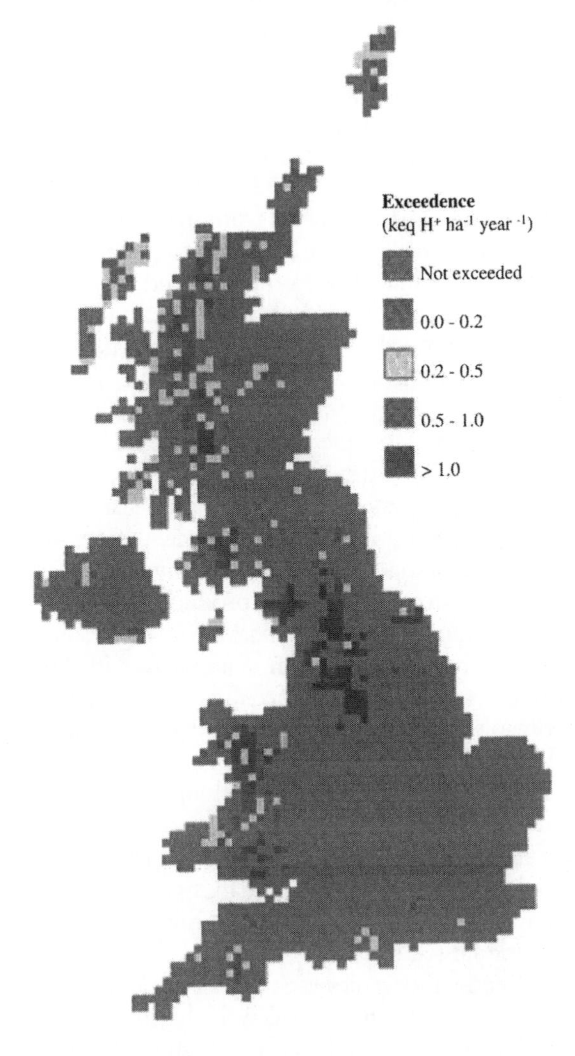

Source: Critical Loads Advisory Group 1994.

Heavy metals

Metal pollution of freshwaters may pose serious problems because dissolved heavy metals, such as arsenic, lead, mercury, cadmium and selenium, are toxic to humans and animals in low concentrations. Natural sources of heavy metals include volcanic and geothermal activity and geological weathering, but human activities have added greatly to these, and it is calculated have caused global enrichment of selenium, silver, cadmium, selenium and mercury by 2.8, 3, 7.6, 11.3 and 24.1 times, respectively (Nriagu and Davidson 1986). Copper smelting at around 7000 BC is one of the earliest instances of human exploitation of heavy metals (Renfrew and Bahn 1991), but heavy metal extraction and processing have shown greatest increase since the Industrial Revolution and particularly during the present century. Anthropogenic release of heavy metals is associated with transport, industry, mining, municipal wastes, agriculture, geothermal development and waste dump leaching (Foster and Charlesworth 1997). For some metals, the dominant source of anthropogenic enrichment has changed during the last 200 years from mineral extraction and processing to releases associated with fossil fuel combustion and product use (Tarr and Ayres 1990).

Understanding the movement and storage of heavy metals in fluvial systems is made more complicated by the fact that a very high proportion of river transport takes place for many metals in association with sedimentary particles, especially the silt and clay fractions. Concentrations of heavy metals in fine-grained bed sediments may be five orders of magnitude greater than concentrations dissolved in the water column (Horowitz 1991). For some metals, such as iron, the particulate-associated fraction has been shown to account for more than 99 per cent of the total transport in major rivers, such as the Amazon and Mississippi (Salomons and Förstner 1984). Over the short term, flushing of sediment and heavy metals in urban drainage during the early part of storm events may lead to high metal concentrations as contaminants accumulated on roofs, in storm drains and especially on road surfaces (Table 11.4) are removed (e.g. Ellis *et al.* 1986; Xanthopolous and Hahn 1993). Over the longer term, fine sediments with high heavy metal concentrations may go into floodplain storage through overbank deposition and channel accretion, or be deposited in lake, reservoir or estuarine environments. However, this material

Table 11.4 Concentrations ($\mu g l^{-1}$) of selected metals in roof, street and stormwater runoff recorded in the Karlsruhe/Waldstadt region of Germany.

Metal	*Runoff*		
	Roof	*Street*	*Stormwater*
Lead	104	311	5
Cadmium	1.0	6.4	1.5
Zinc	24	603	5
Copper	235	108	1.5
Nickel	–	57	5

Source: Xanthopolous and Hahn 1993.

may be remobilised in the fluvial system at a later date by, for example, channel reworking of floodplain deposits (e.g. Bradley 1995).

Data on heavy metals in freshwaters are not routinely collected in many countries (Meybeck *et al.* 1989) and few long-term records of river concentrations exist. Information on sediment-associated metal concentrations at sampling sites on the Rhine and other rivers in the Netherlands (Salomons and Förstner 1984) reveal a steady increase during the twentieth century until 1975, when controls on waste water inputs of metals led to a decline. Reductions in atmospheric emission of heavy metals from coal-burning power stations and other sources over the last fifteen years in the UK and elsewhere (e.g. Quality of Urban Air Group 1993) may also be a significant cause of lower freshwater concentrations in recent years in some systems.

Organic micropollutants

It has been estimated that about 4 million chemical compounds are in existence (Meybeck *et al.* 1989), and more than 100,000 commercial chemicals are known or suspected of causing health problems in humans, animals and plants (Peters *et al.* 1998). Recently, there has been growing concern that some chemicals in river and drinking water may cause hormone disruption in aquatic wildlife and adversely affect the reproductive health of humans (Environment Agency 1997a). Particularly hazardous organic chemical substances have been identified in lists drawn up by the World Health Organization, the US Environmental Protection Agency and the European Community on the basis of toxicity, persistence, biological accumulation and presence in the environment (e.g. EEC 1982; WHO 1984).

Toxic organic substances enter freshwaters through point and diffuse source inputs. The former comprise effluents from major industrial activities, such as petrol refining, coal mining and the manufacture of synthetic products, while the latter include organic pollutants carried in runoff from urban and agricultural areas, often in sediment-associated form. Organic micropollutants originating from household and industrial use include volatile organic substances employed as extraction, degreasing and dry-cleaning solvents and as aerosol propellants, the halogenated derivatives of methane, ethane and ethylene, and polycyclic aromatic hydrocarbons (PAHs) from petroleum products and their combustion. These substances have a high rate of environmental dispersion compared with the amount produced (Meybeck *et al.* 1989). Other organic micropollutants originating from mainly industrial processes have low rates of environmental dispersion and include chlorinated derivatives of benzene, naphthalene, phenol and aniline, which are used in dyestuff manufacture, and polychlorinated biphenyls (PCBs), which are constituents of heat-exchange fluids and dielectric substances.

Synthetic organic pesticides used in agriculture and in horticulture have a high potential to pollute water from direct runoff, spray drift and their storage, handling and disposal (Environment Agency 1997a). A wide range of chemicals are employed in pesticides. For example, about 450 active ingredients are currently approved for use in the UK, and they vary greatly in their physico-chemical characteristics, including water solubility. Pesticides have been grouped into organochlorine insecticides, organophosphate insecticides, herbicides of the plant hormone type, triazines, substituted ureas and others (Meybeck *et al.* 1989). Many countries ban or

Table 11.5 Pesticides on list I (Black List) of the EC Dangerous Substances Directive (76/464/EEC).

List I Pesticide	*Use approved*
Hexachlorocyclohexane (HCH) (lindane)	Yes
DDT	No
Pentachlorophenol (PCP)	Yes
Aldrin	No
Dieldrin	No
Endrin	No
Isodrin	No
Hexachlorobenzene	No

control the use of most organochlorine compounds, such as DDT, because of their toxicity and environmental persistence. Eight pesticides are included in List I (Black List) of the EC Dangerous Substances Directive (76/464/EEC), and although the use of six of these is no longer approved in the UK (Table 11.5), long-term monitoring of river sites in England and Wales demonstrates that older chemicals, such as the insecticide dieldrin, are very persistent in the fluvial environment (Environment Agency 1997a).

Other pollutants

A range of other pollutants may also cause water quality problems. These include heated effluents, especially from power generation facilities (Langford 1990), and oil spills, such as the accident on the Monongahela River, Pennsylvania, USA, which resulted in the release of 3.5 million litres of diesel fuel from a riverside tank in January 1988 and caused major disruption to water supplies, factory closures, cessation of river traffic and high wildlife mortality (Mason 1991). Radioactivity may be added to freshwaters by cooling waters and other effluents from nuclear power stations and reprocessing facilities, and there is concern about future contamination by leakage from land disposal sites and in connection with decommissioning of redundant nuclear plants. A much greater threat of pollution by radioactivity is associated with nuclear accidents, such as the one that occurred at the Chernobyl nuclear power station in April 1986, which released 2×10^{18} Bq of radioactive materials into the atmosphere and subsequently led to contamination of freshwaters and fluvial sediments with radiocaesium over wide areas of Europe and beyond (e.g. Walling *et al.* 1992).

Sediment transport in rivers, especially in suspended form, may also cause serious water quality problems (Walling 1988). Accelerated soil degradation and loss from agricultural land has led to increased sediment yields in many river systems (Walling 1995), which often result in both 'in-stream' and 'off-stream' effects (Clark *et al.* 1985; Tim and Jolly 1994). The former include biological impacts through for example infiltration of fine sediment into spawning gravels, recreational impacts leading to restrictions on swimming, boating and fishing, sedimentation impacts, which cause reduction in reservoir capacity and navigation problems, abrasion impacts relating to damage of HEP turbines, and aesthetic impacts through degradation in the visual quality of the river environment. The latter involve flood damage when flood water cannot pass silted and aggraded river channels, sedimentation problems in irrigation and drainage channels, increased water treatment costs because more time is required to clarify turbid waters, and problems for the industrial use of river water because sediment-laden flows are less efficient for cooling and lead to abrasion of plant. In addition to the physical effects of turbid water, many chemical contaminants are adsorbed by the silt and clay fraction of river sediments and are transported through the fluvial system in particulate-associated form.

PATTERNS, TRENDS, PROTECTION AND REMEDIATION – THE EXAMPLE OF ENGLAND AND WALES

Data collected by the Environment Agency of England and Wales and the statutory bodies that preceded its formation provide information for a

Western European society on the variation in water quality between urban and rural areas, the relative importance of different types and sources of contamination and the changes in pollution status over recent decades. Rivers and canals in England and Wales are currently classified under the General Quality Assessment Chemical scheme, which is based on measurements of dissolved oxygen, biochemical oxygen demand and ammonia over a three-year period, into six categories from very good (Class A) to bad (Class F). Results for the period 1994–96 show that for England and Wales as a whole almost 60 per cent of the rivers had a chemical quality that was good or very good, whereas nearly 10 per cent of the watercourses were classified as poor or bad (Table 11.6). National averages, however, obscure significant regional contrasts in water quality within England and Wales. The highest proportion of very good and good quality rivers are found in the predominantly upland, less intensively farmed and high-runoff areas of Wales and the southwest, whereas the highest percentages of poor and bad quality reaches are encountered in the industrialised regions of the northwest and northeast (Table 11.6). Rivers tend to be dominated by fairly good or fair chemical quality in the intensively farmed and low-runoff Anglian region, and in the heavily populated Midlands and Thames regions (Table 11.6).

Environment Agency records show that the total number of reported pollution incidents in England and Wales has risen from around 13,000 in the early 1980s to over 30,000 in the late 1990s (Figure 11.5A). This rise, to a certain extent, reflects an increased public awareness of water quality issues. Furthermore, many of the incidents are relatively minor and cannot subsequently be substantiated, and the proportion of reports where no pollution could be found on further investigation has risen from around 25 per cent in the early 1990s to over 35 per cent in 1996 and 1997 (Figure 11.5A). Substantiated incidents are classified into three categories depending on their severity (Table 11.7). Analysis of all substantiated incidents in 1997 by type of pollution (Figure 11.5B) showed that the most common contaminants were oil and sewage, whereas organic wastes, mainly in the form of cattle slurry, chemicals from paints and dyes, and silt were each responsible for less than 10 per cent of the incidents. However, for major (Category 1) incidents sewage was less important than contamination by oil or chemicals (Figure 11.5B).

Table 11.6 Percentage length of rivers in different chemical quality classes in England and Wales and regions, 1994–6.

	Chemical Quality Class					
	(A) Very good	*(B) Good*	*(C) Fairly good*	*(D) Fair*	*(E) Poor*	*(F) Bad*
Northeast	28.6	29.2	17.0	10.7	12.4	2.1
Anglian	5.6	31.6	31.1	18.9	12.2	0.6
Thames	13.1	35.8	26.9	15.3	8.6	0.2
Southern	14.2	40.2	26.0	10.0	9.0	0.6
Northwest	20.2	32.3	21.0	10.7	13.1	2.7
Southwest	40.4	36.1	15.5	4.4	3.3	0.3
Welsh	73.0	18.5	5.2	1.6	1.6	0.1
Midlands	12.6	32.9	30.9	12.9	9.9	0.9
England and Wales	**27.1**	**31.5**	**21.2**	**10.3**	**8.8**	**1.0**

Source: Environment Agency 1997a.

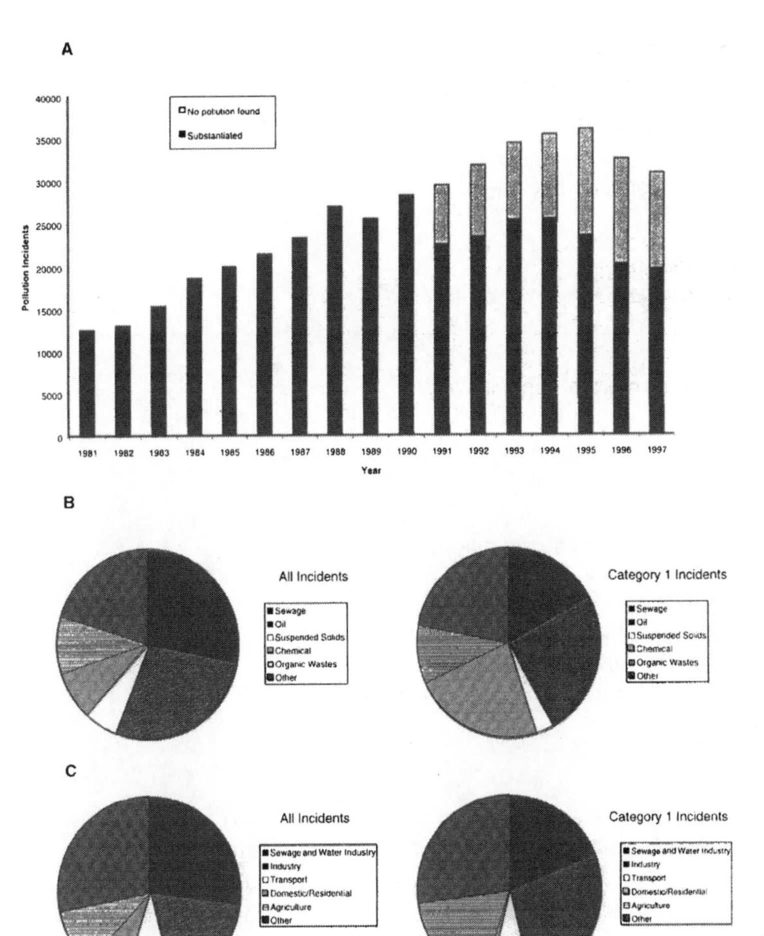

Figure 11.5 Total number of water pollution incidents in England and Wales, 1981–97: (A) and distribution of substantiated pollution incidents in 1997 classified by type (B) and by source (C).

Source: Environment Agency 1997a, b.

Classification of pollution by source (Figure 11.5C) showed that the sewage and water industry, especially as a result of uncontrolled discharges from sewerage systems, accounted for the highest percentage of all substantiated incidents, but industry was responsible for the greatest proportion of Category 1 contamination.

A range of UK and European legislation affects water policy in England and Wales and protects the river environment from pollution (Table 11.8). It is an offence to cause or knowingly permit polluting matter to enter rivers, groundwaters and other water bodies without permission (UK Department of the Environment, Transport and the Regions 1998). The Environment Agency grants consents, which set conditions regarding concentration, volume and other aspects, in order to control the discharge of effluents into the fluvial system and to prevent pollution. If pollution occurs, including the breaking of a discharge consent, the polluter may be prosecuted, fined and made to clean up the contamination. In 1997, for example, the Environment Agency brought prosecutions relating to sixty-seven incidents, and during the year a further 104 cases from 1996 were also heard, and the highest fine imposed was £12,000. Considerable efforts are also made by the Environment Agency and others to prevent pollution occurring in the first place. These included advice

Table 11.7 Definition of pollution incident categories.

Category 1
A major incident involving one or more of the following:
- potential or actual persistent effect on water quality or aquatic life
- closure of potable water, industrial or agricultural abstraction necessary
- extensive fish kill
- excessive breaches of consent conditions
- extensive remedial measures necessary
- major effect on amenity value

Category 2
A significant pollution which involves one or more of the following:
- notification to abstracters necessary
- significant fish kill
- measurable effect on invertebrate life
- water unfit for stock
- bed of watercourse contaminated
- amenity value to the public, owners or users reduced by odour or appearance

Category 3
Minor suspected or probable pollution that, on investigation, proves unlikely to be capable of substantiation or to have no notable effect

Source: National Rivers Authority 1993.

given through codes of practice, guidance notes, pollution prevention campaigns and site visits; establishment of regulations designed to prevent pollution, such as the setting of minimum standards for the construction of stores to house agricultural waste; and serving of notices on consent holders requiring action to prevent breaches of a consent (*ibid.*). Licences issued by the Environment Agency regarding abstraction of surface waters also have regard to the quality of the resulting reduced volume of flow.

Various financial, legislative and other initiatives have been taken in order to improve water quality in England and Wales. For example, the package of measures to reduce nutrient inputs to the aquatic environment (*ibid.*) include:

- Sewage treatment improvement.
- Encouraging farmers to follow codes of good practice.
- Designation of sixty-nine Nitrate Vulnerable Zones, where farmers will be required to reduce leaching of nitrate.
- Declaration of thirty-two Nitrate Sensitive Areas to encourage further reductions in nitrate leaching (Box 11.2).
- Identification of thirty-three river reaches and lakes as Eutrophic Sensitive Areas, where the water industry will be required to install phosphorus removal at qualifying sewage treatment works.

FUTURE DIRECTIONS AND CONCLUSIONS

Water quality and pollution are multifaceted and complex issues, which increasingly are becoming critical both socially and economically at national and international levels. While an enormous amount of effort has been devoted to the monitoring, modelling and management of freshwater contamination, especially in developed countries, many problems of water pollution remain to be solved. It has been suggested by some experts (e.g. Ongley 1994) that especially in developing countries, but even in developed nations, existing

Table 11.8 Recent legislation affecting water policy in England and Wales.

Legislation	*Effect*
The Environmental Protection Act 1990	Established statutory provisions for a range of environmental protection purposes, including integrated pollution control for dangerous processes.
The Water Resources Act 1991	Consolidated previous water legislation in respect of both the quality and quantity of water resources.
The Water Industry Act 1991	Consolidated legislation relating to the supply of water and the provision of sewerage services.
The Environment Act 1995	Established the Environment Agency and introduced measures to enhance protection of the environment, including further powers for the prevention and remediation of water pollution.
The EC Urban Waste Water Treatment Directive	Sets requirements for the provision of collecting systems and the treatment of sewage according to the size of discharge and the nature of the receiving water.
The EC Dangerous Substances Directive	Together with the Water Resources Act 1991, requires control over inputs of dangerous substances into water.
The EC Groundwater Directive	Related to the Dangerous Substances Directive and applies to groundwater protection.
The EC Freshwater Fish Directive	Aimed at protecting the health of freshwater fish by designating waters in need of protection and setting standards for those waters.
The EC Nitrate Directive	Requires member states to reduce the nitrate pollution in waters that arises from agricultural inputs.
The EC Surface Water Abstraction Directives	Set quality objectives for the surface water sources from which drinking water is taken.

Source: UK Department of the Environment, Transport and the Regions 1998.

Box 11.2 Details of the Nitrate-Sensitive Area Scheme in England

The Nitrate Sensitive Area (NSA) Scheme allows farmers in selected areas of England to opt to receive payments for changing their farming practices on a voluntary basis in order to stabilise high and/or rising nitrate levels in key sources of public water supplies. There are thirty-two NSAs, comprising a total of about 35,000 ha of eligible agricultural land, and all are designated as Nitrate Vulnerable Zones under the EC Nitrate Directive (91/676/EEC). Three different types of voluntary measure are offered. The Premium Arable Scheme involves conversion of arable land to extensive grass, but there are different options regarding fertilisation, grazing and the use of woodland. The Premium Grass Scheme entails extensification of existing intensively managed grass, while the Basic Scheme develops low-nitrogen arable cropping on the basis of restricted and standard crop rotations. Farmers enter land into a scheme for five years on a field-by-field basis, and payments range from £590 to £80 per hectare, depending on location of the NSA and which scheme and option are selected. In 1996, 359 farmers were participating in the NSA scheme with the areas in the basic, premium grass, and premium arable schemes being 15,529, 460 and 3622 ha, respectively (MAFF 1998).

monitoring programmes, with an emphasis on the analytical accuracy of chemical determinations, fail to provide appropriate information on water quality for interpreting ecosystem and public health effects, for establishing the presence of many critical toxic chemicals and assessing levels of toxicity and for underpinning either scientific understanding or management decisions. Many developing nations are unable to mount reliable programmes for monitoring water quality. There is, therefore, a critical lack of data not only to guide pollution abatement planning and investment in some countries (Ongley and Kandiah 1998) but also to make international compilations for addressing global problems relating, for example, to biodiversity and to contamination by atmospheric transport over long distances (Ongley 1999).

In order to meet the data crisis and to advance understanding of water pollution, especially at a global level, there is a general need in many countries to enhance the political awareness of water quality issues, to modernise the technical, institutional, legal, training and support aspects of water quality programmes (Ongley 1997), and to build personnel and institutional capacity in applying those programmes to real-world management problems (Ongley 1997; Peters *et al.* 1998).

GUIDE TO FURTHER READING

Park, C.C. (1987) *Acid Rain Rhetoric and Reality*. London and New York: Methuen. A comprehensive overview of the science, social and other issues associated with a major source of water pollution.

Meybeck, M., Chapman, D. and Helmer, R. (1989) *Global Freshwater Quality. A First Assessment*. Oxford: Basil Blackwell. A seminal global synthesis of the state of water quality and river pollution.

Burt, T.P., Heathwaite, A.L. and Trudgill, S.T. (1993) *Nitrate: Processes, Patterns and Management*. Chichester, UK: John Wiley & Sons. Provides an excellent synthesis of the processes that affect the nitrogen cycle in river catchments, the spatial patterns and temporal trends in water pollution that have arisen because of this major nutrient, and the legislative and other means by which nitrate problems can be ameliorated.

Webb, B.W. (1997) *Freshwater Contamination*. Proceedings of Rabay Symposium S4, April–May 1997, IAHS Publication No. 243. Highlights, through a collection of conference papers, current research in a wide range of water pollution fields.

Environment Agency (1997) *State of the Environment*. http://www.environment-agency.gov.uk/gui/ is a rich, attractively presented and easily accessed source of information on the Internet concerning a wide range of water quality and pollution issues in the UK.

Peters, N.E., Bricker, O.P. and Kennedy, M.M. (1997) *Water Quality Trends and Geochemical Mass Balance*. Chichester, UK: John Wiley & Sons. Collects together recent research on past and future water quality trends.

REFERENCES

Barabas, S. (1986) Monitoring natural waters for drinking-water quality. *WHO Stat Q* 39, 32–45.

Bradley, S.B. (1995) Long-term dispersal of metals in mineralised catchments by fluvial processes. In I.D.L. Foster, A.M. Gurnell and B.W. Webb (eds) *Sediment and Water Quality in River Catchments*, Chichester, UK: John Wiley & Sons, 161–77.

Brimblecombe, P. and Stedman, D. (1982) Historical evidence for a dramatic increase in the nitrate component of acid rain. *Nature* 298, 460.

Buijsman, E., Mass, H.F.M. and Asman, W.A.H. (1987) Anthropogenic NH_3 emissions in Europe. *Atmos. Environ.* 21(5), 1009–22.

Burt, T.P., Heathwaite, A.L. and Trudgill, S.T. (1993) *Nitrate: Processes, Patterns and Management*. Chichester, UK: John Wiley & Sons.

Burt, T.P. and Trudgill, S.T. (1993) Nitrate in groundwater. In T.P. Burt, A.L. Heathwaite and S.T. Trudgill (eds) *Nitrate: Processes, Patterns and Management*, Chichester, UK: John Wiley & Sons, 213–38.

Chaouni Alia, A., El Halimi, N., Walraevens, K., Beeuwsaert, E. and de Breuck, E. (1997) Investigation de la salinisation de la plaine de Bou-Areg (Maroc nord-oriental). In B.W. Webb (ed.) *Freshwater Contamination* (Proceedings of Rabay Symposium S4, April–May 1997), IAHS Publication No. 243, 211–20.

CIPEL (1986) Commission Internationale pour la Protection des Eaux du Leman contre la Pollution Annual Reports Avenue de Chailly, Lausanne, Switzerland.

Clark, E.H., Haverkamp, J.A. and Chapman, W. (1985) *Eroding Soils: The Off-farm Impacts*. The Conservation Foundation: Washington DC.

Contractor, D.N. and Yong Qi (1997) Effects of turbulent flow on saltwater intrusion in the island of Guam. In B.W. Webb (ed.) *Freshwater Contamination* (Proceedings of Rabay Symposium S4, April–May 1997), IAHS Publication No. 243, 221–6.

Critical Loads Advisory Group (1994) *Critical Loads of Acidity in the United Kingdom, Summary Report*. Department of the Environment.

Department of the Environment and Department of Health (1990) Cryptosporidium *in Water Supplies: The Report of the Group of Experts*. London: HMSO.

Department of the Environment and Department of Health (1995) Cryptosporidium *in Water Supplies: Second Report of the Group of Experts*. London: HMSO.

EEC (1982) Communication from the Council on dangerous substances that could be placed in List I of Council Directive 76/464/ECC. *Official Journal of the European Communities*, 14 July, C176/7–C176/10.

Edmonson, W.T. (1985) Recovery of Lake Washington from eutrophication. In *Proc. Int. Congress on Lake Pollution and Recovery, European Water Pollution Control Association*, Rome, 15–18 April, 228–34.

Ellis, J.B., Harrop, D.O. and Revitt, D.M. (1986) Hydrological control of pollutant removal from highway surfaces. *Water Research* 20, 589–95.

Environment Agency (1997a) *State of the Environment*. http://www.environment-agency.gov.uk/gui/

Environment Agency (1997b) *Water Pollution Incidents in England & Wales 1997 Report Summary*.

Ford, T.E. and Colwell, R.R. (1996) A global decline in microbiological safety of water: a call for action. *A Report from The American Academy of Microbiology*, Washington, DC.

Foster, I.D.L and Charlesworth, S.M. (1997) Heavy metals in the hydrological cycle: trends and explanation. In N.E. Peters, O.P. Bricker and M.M. Kennedy (eds) *Water Quality Trends and Geochemical Mass Balance*, Chichester, UK: John Wiley & Sons, 103–37.

Garthright, W.E., Archer, D.L. and Kvenberg, J.E. (1988) Estimates of incidence and costs of intestinal infectious diseases. *Public Health Reports* 103, 107–16.

Geldreich, E.E. (1997) Pathogenic agents in freshwater resources. In N.E. Peters, O.P. Bricker and M.M. Kennedy (eds) *Water Quality Trends and Geochemical Mass Balance*, Chichester, UK: John Wiley & Sons, 191–209.

Glass, N.R., Arnold, D.E., Galloway, J.N., Henry, G.R., Lee, J.J., McFee, N.W., Norton, S.A., Powers, C.F., Rambo, D.L. and Schofield, C.L. (1982) Effects of acid precipitation. *Environmental Science Technology* 16(3), 162A–169A.

Harper, D. (1992) *Eutrophication of Freshwaters. Principles, Problems and Restoration*. London: Chapman & Hall.

Heathwaite, A.L. (1993) Nitrogen cycling in surface waters and lakes. In T.P. Burt, A.L. Heathwaite and S.T. Trudgill (eds) *Nitrate: Processes, Patterns and Management*, Chichester, UK: John Wiley & Sons, 99-140.

Heathwaite, A.L., Burt, T.P. and Trudgill, S.T. (1993) Overview – the nitrate issue. In T.P. Burt, A.L. Heathwaite and S.T. Trudgill (eds) *Nitrate: Processes, Patterns and Management*, Chichester, UK: John Wiley & Sons, 3–21.

Heathwaite, A.L., Johnes, P.J. and Peters, N.E. (1997) Trends in nutrients. In N.E. Peters, O.P. Bricker and M.M. Kennedy (eds) *Water Quality Trends and Geochemical Mass Balance*, Chichester, UK: John Wiley & Sons, 139–69.

Horowitz, A.J. (1991) *A Primer on Trace Metal Sediment Chemistry*, second edition. Chelsea: Lewis.

Klein, P.D., Gastrointestinal Physiology Working Group, Graham, D.Y., Gaillout, A., Opekun, A.R. and Smith, E.O. (1991) Water source as a risk factor for *Heliobacter pylori* infection in Peruvian children. *Lancet* 337, 1503–6.

Langford, T.E.L. (1990) *Ecological Effects of Thermal Discharges*. London: Elsevier Applied Science.

MacKenzie, W.R., Hoxie, N.J., Proctor, M.E., Gradus, S., Blair, K.A., Peterson, D.E., Kazmierczak, J.J., Addiss, D.G., Fox, K.R., Rose, J.B. and David, J.P. (1994) A massive outbreak in Milwaukee of *Cryptosporidium* infection transmitted through the public water supply. *N. Engl. J. Med.* 331, 161–7.

MAFF (1998) *Nitrate Sensitive Areas*. http://www.maff.gov.uk/environ/envsch/nsa.htm

Mason, C.F. (1991) *Biology of Freshwater Pollution*, second edition. UK: Longman.

Mellanby, K. (1980) *The Biology of Pollution*, second edition. The Institute of Biology's Studies in Biology No. 38, London: Edward Arnold.

Meybeck, M. (1996) River water quality. Global ranges, time and space variabilities, proposal for some redefinitions. *Verh. Int. Verein. Limnol.* 26, 81–96.

Meybeck, M., Chapman, D. and Helmer, R. (1989) *Global Freshwater Quality. A First Assessment*. Oxford: Basil Blackwell.

National Rivers Authority (1992) *The Influence of Agriculture on the Quality of Natural Waters in England and Wales*. Water Quality Series No. 6, Bristol: National Rivers Authority.

National Rivers Authority (1993) *Water Pollution Incidents in England and Wales – 1992*. Bristol: National Rivers Authority.

Nriagu, J.O. and Davidson, C.I. (1986) *Toxic Metals in the Atmosphere*. Wiley Ser. Adv. Environ. Sci. Technol. 17, New York: J. Wiley & Sons.

Oden, S. (1976) The acidity problem – an outline of concepts. *Water, Air and Soil Pollution* 6, 137–66.

Ongley, E.D. (1994) Global water pollution: challenges and opportunities. In *Proceedings: Integrated Measures to Overcome Barriers to Minimizing Harmful Fluxes from Land to Water*, Publication No. 3, Stockholm Water Symposium, August 10–14, 1993, Stockholm, Sweden, 23–30.

Ongley, E.D. (1996) *Control of Water Pollution from Agriculture*. FAO Irrigation and Drainage Paper No. 55, Rome: FAO.

Ongley, E.D. (1997) Matching water quality programs to management needs in developing countries: the challenge of program modernization. *European*

Water Pollution Control 7(4), 43–8.
Ongley, E.D. (1999) Water quality: an emerging global crisis. In S.T. Trudgill, D.E. Walling and B.W. Webb (eds) *Water Quality: Processes and Policy*, Chichester, UK: John Wiley & Sons (in press).
Ongley, E.D. and Kandiah, A. (1998) Managing water pollution from agriculture in a water scarce future. In *Proceedings of the 1997 Stockholm Water Symposium*, August 11–14, 1997 (in press).
Park, C.C. (1987) *Acid Rain Rhetoric and Reality*. London and New York: Methuen.
Peters, N.E., Buell, G.R. and Frick, E.A. (1997) Spatial and temporal variability in nutrient concentrations in surface waters of the Chattahoochee River basin near Atlanta, Georgia, USA. In B.W. Webb (ed.) *Freshwater Contamination* (Proceedings of Rabay Symposium S4, April–May 1997), IAHS Publication No. 243, 153–65.
Peters, N.E., Bonell, M., Hazen, T., Foster, S., Meybeck, M., Rast, W., Schneider, G., Tsirkunov, V. and Williams, J. (1998) Water quality degradation and freshwater availability – need for a global initiative. In *Water: A Looming Crisis* (Proceedings of International Conference, June 1998), Paris: UNESCO.
Puckett, L.J. (1995) Identifying the major sources of nutrient water pollution. *Environ. Sci. Technol.* 29, 408–14.
Quality of Urban Air Group (1993) *Urban Air Quality in the United Kingdom*. London: HMSO.
Rast, W. and Thornton, J.A. (1997) Trends in eutrophication research and control. In N.E. Peters, O.P. Bricker and M.M. Kennedy (eds) *Water Quality Trends and Geochemical Mass Balance*, Chichester, UK: John Wiley & Sons, 171-89.
Renfrew, C. and Bahn, P.G. (1991) *Archaeology, Theories, Methods and Practice*. London: Thames & Hudson.
Rimawi, O. and Al-Ansari, N.A. (1997) Groundwater degradation in the northeastern part of Mafraq area, Jordan. In B.W. Webb (ed.) *Freshwater Contamination* (Proceedings of Rabay Symposium S4, April–May 1997), IAHS Publication No. 243, 235–43.
Roberts, G. and Marsh, T. (1987) The effects of agricultural practices on the nitrate concentrations in the surface water domestic supply sources of western Europe. In *Water for the Future: Hydrology in Perspective*, IAHS Publication No. 164, 365–80.
Robson, A.J. and Neal, C. (1997) Water quality trends at an upland site in Wales, UK, 1983–1993. In N.E. Peters, O.P. Bricker and M.M. Kennedy (eds) *Water Quality Trends and Geochemical Mass Balance*, Chichester, UK: John Wiley & Sons, 59–79.
Royal Society (1983) *The Nitrogen Cycle of the United Kingdom: A Study Group Report*. London: Royal Society.
Russell, M.A., Walling, D.E., Webb, B.W. and Bearne, R. (1998) The composition of nutrient fluxes from contrasting UK river basins. *Hydrological Processes* 12, 1461–82.
Salomons, W. and Förstner, U. (1984) *Metals in the Hydrocycle*. Berlin: Springer-Verlag.
Smil, V. (1996) *Environmental Problems in China: Estimates of Economic Costs*. East–West Center Special Reports No. 5, Honolulu, Hawaii: East–West Center.
Soulsby, C., Turnbull, D., Langan, S.J., Hirst, D. and Owen, R. (1997) Reversibility of surface water acidification in the Cairngorm Mountains, Scotland. In B.W. Webb (ed.) *Freshwater Contamination* (Proceedings of Rabay Symposium S4, April–May 1997), IAHS Publication No. 243, 15–26.
Sweeting, R.A. (1994) River pollution. In P. Calow and G.E. Petts (eds) *The Rivers Handbook Volume 2*, Oxford: Blackwell Scientific Publications, 23–32.
Sylvester, R.O. and Seabloom, R.W. (1963) Quality and significance of irrigation return. J. *Irrig. Drain. Div. Proc. ASCE* 89, 1–27.
Tarr, J.A. and Ayres, R.V. (1990) The Hudson Raritan River Basin. In B.L. Turner, W.C. Clark, R.W. Kates, J.F. Richards, J.T. Mathews and W.B. Meyer (eds) *The Earth as Transformed by Human Action*, Cambridge: Cambridge University Press with Clark University, 623–41.
Tim, U.S. and Jolly, R. (1994) Evaluating agricultural nonpoint-source pollution using integrated geographical information systems and hydrologic/water quality model. *Journal of Environmental Quality* 23, 25–35.
UK Department of the Environment, Transport and the Regions (1998) *Water Quality: A Guide to Water Protection in England and Wales*. http://www.environment.detr.gov.uk/wqd/guide/water.htm
UNECE (1994) *Protocol to the 1979 Convention on Long Range Transboundary Air Pollution*. Geneva: United Nations.
UNEP/WHO (1988) *Assessment of Freshwater Quality*. London.
Varheyli, G. (1985) Continental and global sulphur budgets 1: anthropogenic SO_2 emissions. *Atmos. Environ.* 19(7), 1029–40.
Vollenweider, R.A. (1968) Scientific fundamentals of the eutrophication of lakes and flowing waters, with particular reference to nitrogen and phosphorus as factors in eutrophication. Tech. Rep. DAS/CSI/68.27, Environmental Directorate, OECD, Paris.

Walling, D.E. (1980) Water in the catchment ecosystem. In A.M. Gower (ed.) *Water Quality in Catchment Ecosystems*, Chichester, UK: John Wiley & Sons, 1–47.

Walling, D.E. (1988) Erosion and sediment yield research – some recent perspectives. *Journal of Hydrology* 100, 113–41.

Walling, D.E. (1995) Suspended sediment yields in a changing environment. In A.M. Gurnell and G.E. Petts (eds) *Changing River Channels*, Chichester, UK: John Wiley & Sons, 149–76.

Walling, D.E., Quine, T.A. and Rowan, J.S. (1992) Fluvial transport and redistribution of Chernobyl fallout radionuclides. *Hydrobiologia* 235/236, 231–46.

Walling, D.E. and Webb, B.W. (1986) Solutes in river systems. In S.T. Trudgill (ed.) *Solute Processes*, Chichester, UK: John Wiley & Sons, 251–329.

Webb, B.W. (1997) Trends in stream and river temperatures. In N.E. Peters, O.P. Bricker and M.M. Kennedy (eds) *Water Quality Trends and Geochemical Mass Balance*, Chichester, UK: John Wiley & Sons, 81–102.

WHO (1984) *Guidelines for Drinking-Water Quality. Vol. 1 Recommendations.* Geneva: World Health Organisation.

Withers, P.J.A. (1994) The significance of agriculture as a source of phosphorus pollution to inland and coastal waters in the UK. unpublished report to MAFF, London.

Wright, R.F. (1977) *Historical Changes in the pH of 128 Lakes in Southern Norway and 130 Lakes in Southern Sweden over the Years 1923–1976.* Oslo: SNSF Project TN34/77.

Xanthopolous, C. and Hahn, H.H. (1993) Anthropogenic wash-off from street surfaces. In J. Marsalek and H.C. Torno (eds) *Proceedings VIth Int. Conf. Urban Storm Drainage, Niagara Falls, Ontario, Canada*, British Columbia: Seapoint Publishing, 417–22.

Zobrist, J. and Stumm, W. (1981) Chemical dynamics of the Rhine catchment area in Switzerland, extrapolation to the 'pristine' Rhine river input to the ocean. In *River Inputs to Ocean Systems*, UNEP/UNESCO/IOC/SCOR New York: United Nations, 52–63.

12

Irrigation

Peter Beaumont

INTRODUCTION

Irrigation is the addition of water to the soil to produce near optimum soil moisture conditions for crop growth in regions of water scarcity (Rydzewski and Ward 1989). Irrigated agriculture appears to have commenced on a small scale soon after the domestication of cereals in the Middle East. This domestication probably began about 12,000 years ago, yet by 10,000 years ago archaeological excavations at Jericho have revealed that groundwater from a nearby spring was being used for irrigation (Kenyon 1969–70). However, irrigated agriculture is perhaps most closely associated with the development of urban civilisations in the great river valleys of the Old World, including the Tigris–Euphrates, Indus and Nile (Adams 1965; Wittfogel 1957). The key factor is that irrigation of these huge floodplains using large-scale diversion structures and canals generated the wealth necessary for the construction of complex urban systems and the founding of empires.

Irrigation, by providing more assured crop production and by increasing yields by up to four times those of their rain-fed equivalents, meant that much larger populations could be supported from relatively small areas of cultivated land. In turn, this creation of wealth from agriculture led to the division of labour within society and the growth of specialist occupations, such as soldiers, enquirers and accountants, which are taken for granted today. Irrigation development can, therefore, be regarded as a vital element in the evolution of human societies.

THE NATURE OF IRRIGATION – METHODS AND PRACTICES

There is a continuum of irrigation practices in terms of the volume of water added, from crops such as wheat, which require relatively little water, even in arid environments, to crops like rice, which can demand very large quantities of water even in humid environments (Burns 1993). The earliest methods of irrigation were probably furrow and flood types. Both employ low technology and require only small amounts of human or animal labour to carry them out. With furrows, a series of v-shaped channels are fashioned across the irrigated area using a form of plough-like implement. The furrows slope in a downslope direction to ensure that the water moves along them under the influence of gravity. With flood irrigation, small banks are constructed to delimit roughly rectangular flat areas of various sizes into which water can be led to flood the ground to a depth of several centimetres. Water is then usually led from one small basin to another in a downslope direction. Basin or flood irrigation is best known from the Nile valley (Hamdan 1961).

Since the end of the Second World War, new methods of irrigation have been developed. They are usually described as pressure systems, as the water is delivered to the plant/crop through a series of pipelines. All these systems require a much higher level of technology than is the case with the traditional furrow and flood systems, and the infrastructure costs associated with them are high (Plusquellec *et al.* 1994). The commonest of these methods is the sprinkler system, which delivers water to the crop from overhead in droplets that are similar in form to rainfall (Plate 12.1). These can vary in size from small individual 'gun-like' structures that have to be moved by hand around a field to self-propelled centre-pivot systems with overhead arms up to 500 m or more in length. Other widely used methods are trickle and drip systems, which are employed for higher-value crops. Flexible or rigid pipes are laid along crop rows and the water trickles or drips from holes in the pipes into the soil adjacent to the crops. In some cases, the pipes can actually be buried into the soil so that the water is delivered directly to the root zone of the growing crop. With trickle and drip systems, the volumes of water used can be strictly controlled, so water wastage rates are very low. Fertilisers and other chemicals can also be added directly to the water before it is delivered to the crops.

Poorly designed or badly operated irrigation projects can cause serious environmental problems (ESCAP 1989; Gelburd 1985). The main cause of problems is usually the application of excessive amounts of water. The water that enters the soil is quickly evaporated from the ground surface, leaving behind salts in the upper layers of the soil, which reduce crop growth or may even completely prevent it (Misak *et al.* 1997). Once soils are affected by salinisation, it is extremely difficult for them to be reclaimed, and costs are always high. Soviet Central Asia is a region that has been particularly badly affected by poor irrigation practices, with the result that thousands of hectares of land have had to be abandoned (O'Hara, 1997). Similar problems have occurred in the Murray–Darling basin of southeast Australia, although here the main causes have been the removal of the natural vegetation and the high salinity levels of local soils and waters (Mackay and Landsberg 1992; Simpson and Herczeg 1991).

For irrigation to be possible, a reliable source of water is essential. This can be obtained from either surface or groundwater sources. The very large volumes of water that are necessary for irrigation, often in excess of 10,000 m^3 per hectare, have meant that in traditional societies groundwater from wells has not been a significant means of obtaining water except for small gardens (Beaumont 1993). Water is extremely heavy, so its extraction from the ground is highly energy-intensive.

In many of the drier parts of the Middle East and Central Asia, the extraction of groundwater for irrigation has been achieved by the use of *qanats*. These are underground tunnels that lead water from beneath the water table to the ground surface under the influence of gravity. They are

Plate 12.1 Modern centre-pivot irrigation system, southern Spain.

commonly found on large alluvial fans, where over 100 *qanats* may be present (Beaumont, 1971; 1973). *Qanats* can take many years to construct, yet they often supply only relatively small volumes of water. This provides an indication of just how valuable irrigation water is in such environments. The great advantage of the *qanat* is that once it is constructed, water is delivered to the ground surface without the need for any other energy source.

Since the beginning of the twentieth century, water for irrigation has been stored and supplied by high-technology methods. With surface waters, the most important infrastructure feature has been the major dam and associated reservoir. These structures have permitted the storage of large volumes of water, which could then be distributed, over hundreds of kilometres if necessary through pipelines and canals to where the irrigation water is needed. The modern development of groundwater reserves has been much slower. Indeed, it really only began after the Second World War, when reliable mechanised pumps became widely available. The problem with the use of pumps is that it is possible to withdraw water from the aquifer at a rate that is in excess of natural recharge. Under such conditions, the groundwater is said to be being 'mined' and the inevitable consequence is that the water table will fall.

An interesting example of the potential conflict between modern and traditional methods of groundwater extraction for irrigation can be seen on the Varamin Plain of Iran (Beaumont 1968) (Plate 12.2). The traditional pattern of water use was to extract groundwater by qanats. The crucial point here is that the volume of water produced by the *qanat* depends on the height of the water table (Figure 12.1). When the water table is high, the water-producing section of the *qanat* is longer and hence discharge increases. With a falling water table the converse is true. Therefore, under fluctuating climatic conditions *qanat* outputs vary. During the 1960s, over 100 pumped wells were sunk on the Varamin Plain to develop new areas of land for cultivation. The result was that the regional water table of the plain began to decline, and as it did so the discharges of many of the *qanats* fell as well. Indeed, a number dried up completely, causing considerable social dislocation in the affected villages.

GLOBAL USE OF IRRIGATION

FAO data suggest that the total irrigated area in the world has grown from 167 million hectares in 1970 to 255 million hectares in 1995, representing a 53 per cent increase. With an average water use of around 10,000 m^3 per hectare, this irrigated area increase implies an extra water usage of 880,000 million m^3, or the equivalent of ten times the flow of the River Nile. Currently, the overall pattern of growth appears to be a steady one, and there is no indication that the observed rate of increase is slowing down (Figure 12.2).

The distribution of irrigated agriculture reveals marked variations from one country to another at the present day. The countries with the largest irrigated areas by a considerable margin are India and China, which both possess around 50 million hectares of irrigated land (Table 12.1). These two countries alone account for around 39 per cent of the world's irrigated lands. Pakistan is next with 17 million hectares, followed a long way behind by Iran, Mexico, the Russian Federation and Thailand. In total, twenty-three countries have irrigated areas of more than 2 million hectares, and these countries together account for approximately 75 per cent of all the irrigated land in the world.

The relative importance of irrigated land within an individual country depends largely upon climatic conditions. The drier the country the more dependent it tends to be upon irrigation. FAO statistics indicate that in 1995 four countries – Bahrain, Egypt, Kuwait and Qatar – were totally dependent on irrigation for any crop growth (Table 12.1). Perhaps surprisingly, only fourteen countries were dependent upon irrigation for more than two-thirds of their cropped lands, and of these only Egypt, Uzbekistan and Pakistan can be classed as countries with large irrigated areas. Even if all countries with more

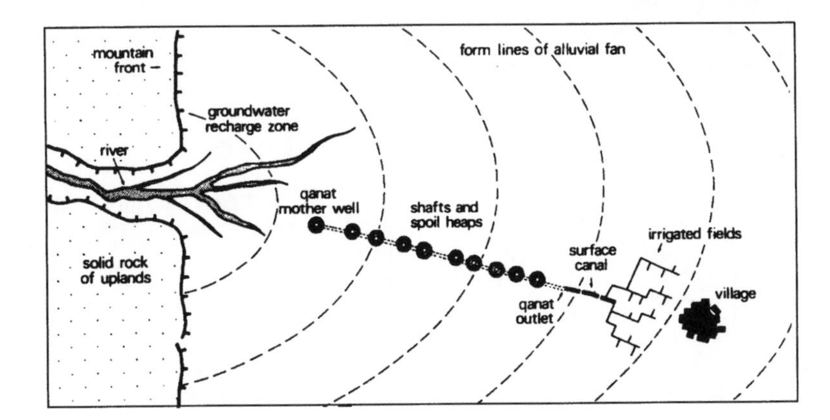

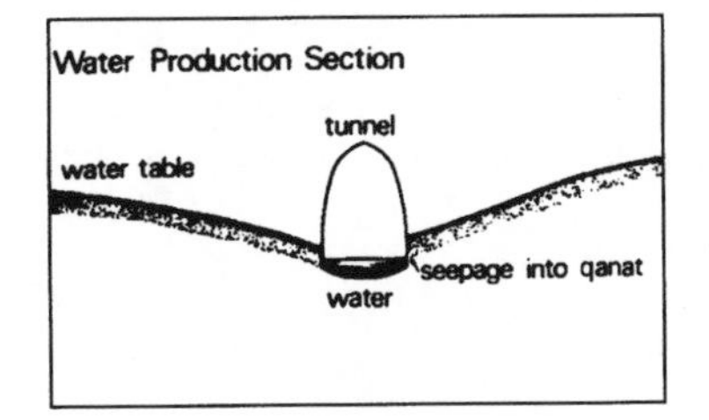

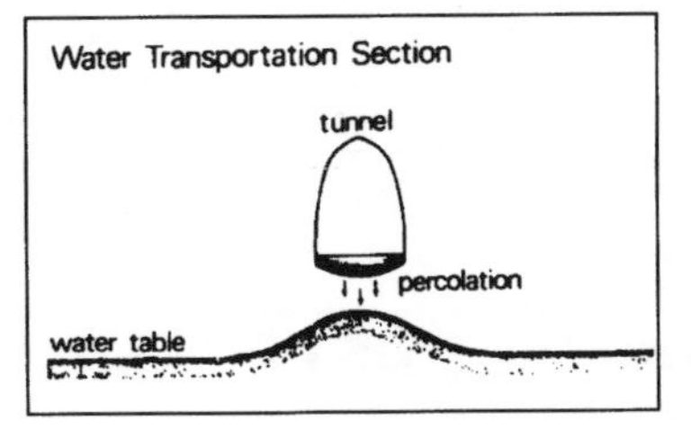

Figure 12.1 Cross-section and plan of a *qanat*.

than one third of their cultivated area under irrigation are counted, the number only increases to thirty-three.

IRRIGATION AS A DEVELOPMENT TOOL

In the modern world, the political aspects of irrigation development must not be forgotten, as national governments have often used irrigation as a tool in their policy options. This has been true of both developed and developing countries. No better example is seen than in the USA. In the early 1900s, the federal government committed itself to a policy of opening up the west through the development of irrigated agriculture by the Bureau of Reclamation.

The work of the Bureau of Reclamation has to be seen in terms of the overall pattern of irrigation growth in the USA (see Box 12.1). Data from the USA register a steady growth in the area of irrigated land from less than 1.62 million hectares in

Plate 12.2 Traditional irrigated agriculture along the Zayandel River, to the west of Isfahan, Iran.

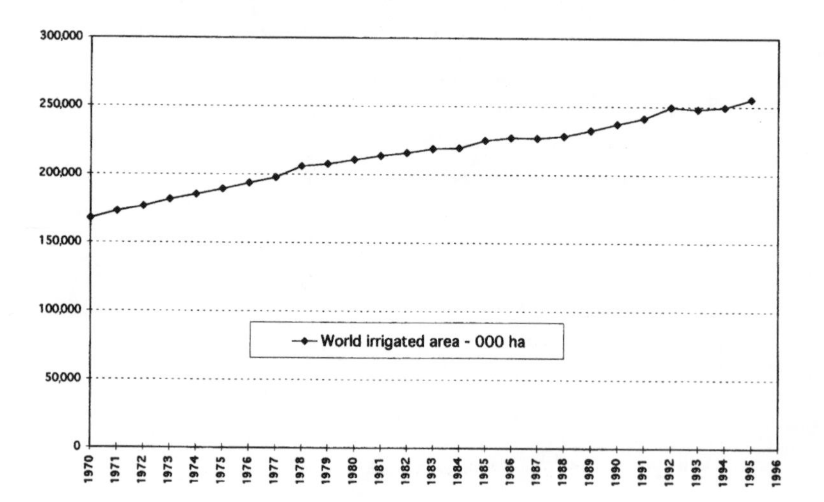

Figure 12.2 World growth of irrigated land.

1890 to a maximum of 20.2 million hectares in 1978. Since then, the irrigated area has fluctuated slightly from census to census, but the high value of 1978 has never been reached again (Figure 12.3).

Irrigation was also used both as a political and strategic tool by the government of Israel following the establishment of the state in 1948 (Shuval 1980; Lonergan and Brooks 1994). The policy of the government was to spread Jewish settlement over as much of the land of Palestine as was possible so that sabotage and infiltration activity by dispossessed Arabs would have a minimal impact on the new state. The way in which this was to be achieved was by the widespread development of irrigated agriculture and the establishment of rural settlements, perhaps the most famous of which are the *kibbutz* and the *moshav*. The key element with this early policy was the pumped well, which permitted the exploitation of the coastal aquifer system. Overpumping over two decades led to falling water tables and the landward penetration of sea water

Table 12.1 World irrigation by country, 1995.

Country	*Irrigated area (1000 ha)*	*Irrigated land as a % of cultivated land*	*Population (millions) 1995*
India	50,100	29.5	929.0
China	49,857	52.0	1220.2
Pakistan	17,200	79.6	136.2
Iran	7264	39.3	68.4
Mexico	6100	22.3	91.1
Russian Fed.	5360	4.0	148.5
Thailand	5004	24.5	58.2
Indonesia	4580	15.2	197.5
Turkey	4186	15.4	60.8
Uzbekistan	4000	88.9	22.8
Spain	3527	17.5	39.6
Iraq	3525	61.3	20.1
Egypt	3283	100.0	62.1
Bangladesh	3200	36.8	118.2
Brazil	3169	4.8	159.0
Romania	3110	31.4	22.7
Afghanistan	2800	34.8	19.7
Italy	2710	25.2	57.2
Japan	2700	61.7	125.1
Ukraine	2586	7.5	51.8
Kazakhstan	2380	7.4	16.8
Australia	2317	4.8	17.9
Vietnam	2000	29.6	73.8

Source: FAO Production Yearbook 1996.

into the aquifer, but at least it allowed the state to create a pattern of Jewish rural settlement that spread to cover a large area of the new state. With the completion of the National Water Carrier in the late 1960s, water from the River Jordan was used to recharge the depleted and overused aquifer systems.

As a result of this policy, the area of irrigated land grew rapidly and with it the demand for water. Water used for agriculture increased from less than 300 million m^3 per annum in 1949 to around 1400 million m^3 by the end of the 1980s (Figure 12.4). Since then, it has begun to decline as water has been diverted away from agriculture to other sectors of the economy. Even in the 1990s, it is interesting to note that Israel continues to produce 'realities on the ground' by its continued

Box 12.1 The Bureau of Reclamation and the American West

In 1902, the Reclamation Service, later to be known as the Bureau of Reclamation, was established through the Reclamation Act. The main contribution of the Bureau of Reclamation to the opening up of the west was that it injected large capital sums, introduced modern engineering techniques and possessed a bureaucracy that was capable of building large-scale irrigation projects and advocating the construction of others. As an organisation, it was sufficiently large and powerful to overcome any objections that might be put in its way. Between the Reclamation Act in 1902 and 1920, 0.89 million hectares were developed into irrigated lands, and by the end of the Second World War this figure had grown to 1.68 million hectares. However, the really rapid increase in the Bureau's activities took place between 1945 and 1965, when 1.62 million hectares of irrigated land were commissioned, producing a total of 3.24 million hectares (National Water Commission 1973). Thereafter, the pace of development slowed down and by 1981 had only reached 4.09 million hectares.

The Reclamation Reform Act of 1982 attempted to redress many of the problems that had arisen over the years as the scale of irrigation operations had increased. However, well before the enactment of the Reclamation Reform Act, times were changing for irrigation in the American west. The last major authorisation of funds for major construction works by the Bureau of Reclamation took place in the late 1960s at a time when a growing environmental movement in the USA was expressing opposition to water development projects. From then on, it effectively ceased to be a major construction organisation and instead became more concerned with the operation and maintenance of existing facilities. The bureau wrote: 'The arid West essentially has been reclaimed. The major rivers have been harnessed and facilities are in place or are being completed to meet the most pressing current water demands and those of the immediate future.' These changes were acknowledged between 1988 and 1994 by a major reorganisation and slimming down of the Bureau of Reclamation at a time when the last major projects, which had been authorised in the 1960s, were finally being completed.

Although the Bureau of Reclamation is now a much smaller organisation than previously, it still operates around 180 projects in the seventeen western states of the USA, which represents a total investment of around US$ 11 billion. Water from Bureau of Reclamation projects continues to irrigate 3.68 million hectares (1992), which amounts to about 20 per cent of the total irrigated area in the region. The overall impact of the Bureau of Reclamation on irrigation development in the American west has been enormous. Its own projects are impressive in their own right, but it is important to remember that the mere existence of the bureau has had an important catalytic effect on the development of irrigation in general throughout the southwest by private individuals as well. In reality, the Bureau of Reclamation has been an 'irrigation flagship' with a reputation that has spread around the world.

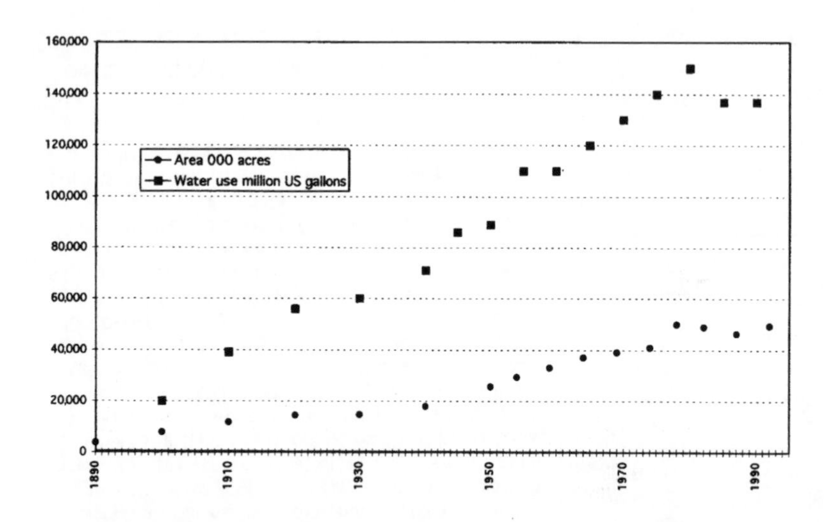

Figure 12.3 Irrigated area and water use in the USA.

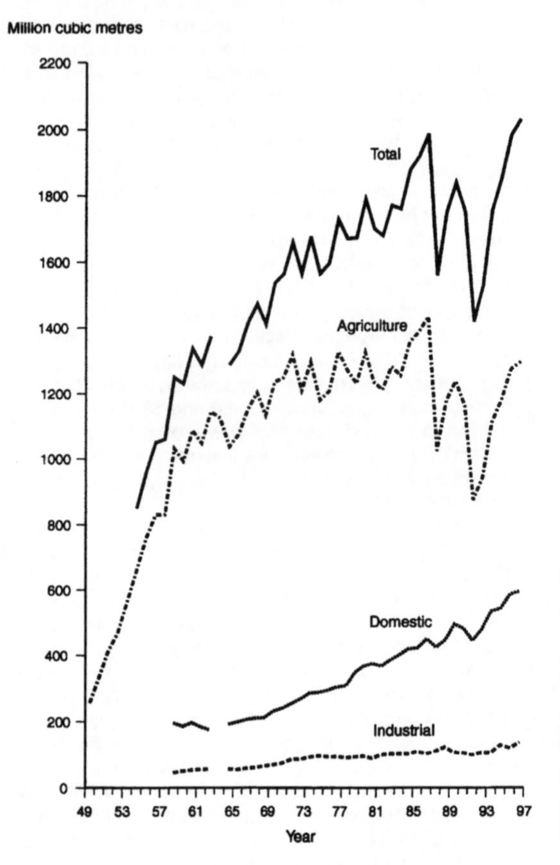

Figure 12.4 Water use in Israel.

policy of encouraging the establishment of settlements, many of which are based on irrigated agriculture, in the occupied parts of the West Bank.

UNSUSTAINABLE IRRIGATION – EXCESSIVE RIVER ABSTRACTION AND THE MINING OF GROUNDWATER

The pace of technological change in the twentieth century has meant that human societies now possess the ability to utilise water resources for irrigation and other purposes at a rate in excess of the natural regeneration of water supplies. Perhaps the most spectacular example of the depletion of surface water resources is found within the watershed of the Aral Sea in Central Asia (Levintanus 1992; Precoda 1991). This is a basin of inland drainage ringed by high mountains to the south and south-west. In these mountains of the Tien Shan and Pamir are found the headwaters of the two main rivers – the Amu Dar'ya and the Syr Dar'ya. Precipitation here is between 800 and 1600 mm each year, but downstream on the plains around the Aral Sea precipitation falls to less than 100 mm per annum. In the period since the Second World War, and especially since the 1960s, there has been massive irrigation development on these plains using waters from the Amu Dar'ya and Syr Dar'ya. So great has been the expansion of irrigated cotton and rice that near total diversion of the waters of the two rivers has occurred. Most of the water that is used for irrigation is lost as evaporation.

Because the Aral Sea is a basin of inland drainage it has been particularly sensitive to the depletion of its water supplies. The lake level has fallen greatly and at present the water area is less than half of what it was in 1960. As the lake level has declined and the water become more saline, large areas of former lake bed have been exposed. These lacustrinal sediments contain high salt levels and when eroded by the wind have been deposited on adjacent irrigated land causing severe salt pollution. The irrigation projects themselves have also created ecological problems. Many schemes were provided with inadequate drainage systems, so salinisation of the soil has become widespread. A wide range of chemicals and fertilisers were utilised in the production of the cotton and rice, and these have now polluted the surface and groundwater resources. The overall result is that the natural ecosystem has changed so significantly that it is not too extreme to describe it as an ecological disaster (Micklin 1988). Various plans have been put forward to obtain extra water supplies for the basin, including the diversion of water from the large rivers flowing into the Arctic Ocean. However, the huge cost of such projects makes it unlikely that they will ever come to fruition. Equally, it would appear unrealistic to reduce irrigation water volumes significantly, as in excess of 40 million people are today dependent on the agricultural production of the basin (Kirmani and Le Moigne 1997).

In recent years, since the advent of the pumped well, many large aquifer systems throughout the world have been overdeveloped or mined to produce prosperous irrigated regions. A striking example is the Ogallala aquifer underlying the High Plains of Texas (Green 1973). The pace of development of irrigation in the region was quite remarkable, and from 1945 to 1970 the irrigated area grew from less than 0.2 million hectares to almost 2 million hectares (Figure 12.5) (Beaumont 1985). After a number of years of high levels of extraction the water table began to fall, in some areas by over a metre a year, and the costs of obtaining the water grew as a result. The situation was made even worse by rapidly increasing energy prices from the late 1960s onwards. Inevitably the profitability of irrigated agriculture declined, and by 1978 the irrigated area had peaked. Thereafter it began to decline.

The response of the farmers to the fall in the water table was to try to get the state to provide extra water by importing it from outside the state. This idea was first put forward in the Texas Water Plan of 1968, but at that time and on later occasions the voters of the state voted down the proposition for providing the funds necessary to build it. As a result, the farmers of the High Plains were forced to resort to new strategies to cope with declining water volumes from their wells and increased pumping costs. One of these was to select crops that were better suited to the dry conditions of the High Plains. Another was to introduce moisture conservation techniques so that the natural precipitation of the region could be maximised. The overall result was that much more use was made of the natural precipitation, and irrigation was used only in a supplemental form.

Despite all the changes that have taken place, the irrigated area on the High Plains has continued to contract quite markedly. The basic problem has been that groundwater mining of the Ogallala aquifer is not a sustainable activity that can continue indefinitely. The farmers have chosen to exploit the water as a short-term resource, in much the same manner as oil might be pumped out of the ground. All of them would now accept that the water is a finite resource, and once it has been pumped out it will not be replaced in the short term. It is interesting to note, though, that the water from the Ogallala has been able to produce a prosperous agricultural system on the High Plains that has already lasted fifty years and with careful stewardship of the remaining resources might well continue for a number of decades.

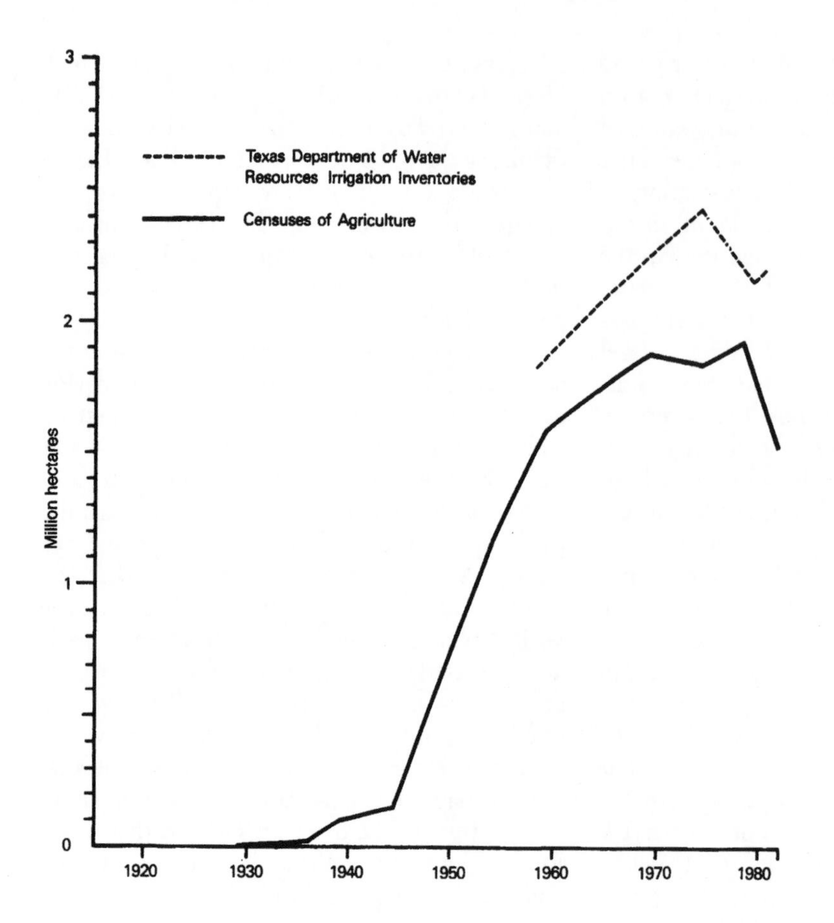

Figure 12.5 Growth of irrigated agriculture on the High Plains of Texas, USA.

TRANSBOUNDARY RIVERS AND IRRIGATION

Disputes with regard to the use of water for irrigation have or seem set to occur on transboundary rivers like the Colorado, Jordan, Nile and Tigris–Euphrates (Kirmani and Rangeley 1994; Kliot 1994). The problem with many of these rivers is that in recent years the irrigation demand for their waters has grown tremendously as population numbers have increased and standards of living, at least in urban regions, have started to rise. On the Colorado, developments since the 1930s have meant that all of the water has effectively been committed to beneficial uses (Fradkin 1984). The result is that today the discharge of the river across the Mexican border is little more than a trickle. However, the division of the waters of the River Colorado between the USA and Mexico has been agreed through negotiation, while within the USA complex legal agreements between the interested states govern water usage from the river.

Elsewhere this is not the case, and disputes over water look likely to increase. Some claim that armed conflict over water may take place in the future, while others dispute this prognosis (Beaumont 1994; Starr 1991). On the River Euphrates, the crisis is immediate. With the Euphrates the watershed is located in three countries – Turkey, Syria and Iraq. Until as late as the 1970s, only Iraq was a significant user of the waters of the river. Iraq's use dates back at least 5500 years to the time when the first great irrigation civilisations were established (Beaumont 1978). The irrigation that was practised was based on making use of the snowmelt flood wave that

Box 12.2 Groundwater mining and food security in Saudi Arabia

Another case of rapid groundwater development for irrigation is found in Saudi Arabia. In the 1980s, as a result of government policy to increase food self-sufficiency, large subsidies were made available to farmers for wheat production. At this time, Saudi Arabia was dependent upon large imports of wheat annually. Given the arid nature of the Saudi Arabian climate and the lack of surface runoff, the growth of wheat was only feasible through irrigation waters supplied from deep wells. Speculative development of the groundwater reserves took place through a well-drilling programme and as a result the area of wheat rose from 60,000–90,000 hectares in the late 1970s to a peak of over 900,000 hectares in 1992 (Figure 12.6). The actual production of wheat increased over the same period from around 90,000–150,000 tonnes to more than 4 million tonnes. So great was this production that in 1992 Saudi Arabia was able to export a staggering total of more than 2 million tonnes of wheat.

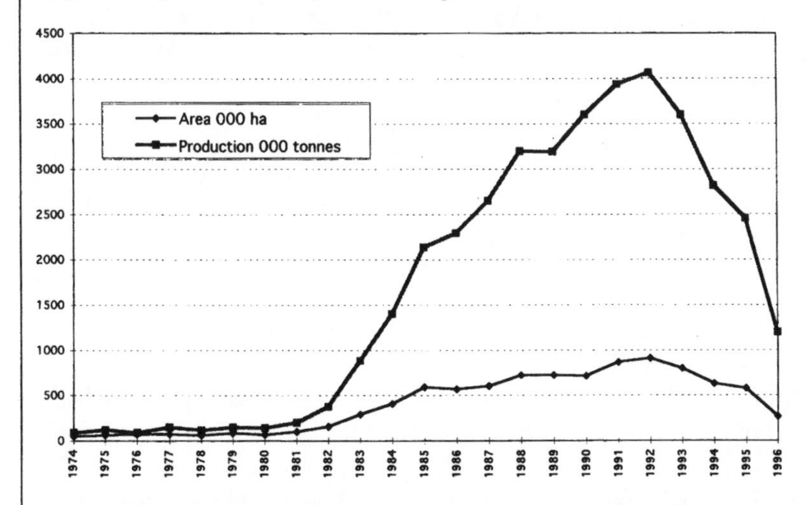

Figure 12.6 Area and production of irrigated wheat in Saudi Arabia.

Eventually, the Saudi Arabian government came to realise that the subsidies, which were often more than four times the price of wheat on the world market, made little economic sense, and in recent years they have been gradually reduced. As a result, by 1996 the cropped area for wheat has been reduced to 265,000 hectares, and wheat production now looks set to fall to around 1.2 million tonnes. However, it has to be stated that this experiment with irrigated wheat production has resulted in the severe depletion of the aquifer beneath Saudi Arabia and a wastage of water that could have been used for more productive purposes.

came down the river in April and May. However, only a small portion of the water was abstracted, and most of it flowed unused into the Persian Gulf.

An interesting aspect about the Euphrates is that at least 88 per cent and possibly over 95 per cent of the flow of the river is generated by precipitation falling over Turkey. The remaining flow comes from within Syria. No flow at all is provided by Iraq. As a result, there is a growing conflict between the upstream and downstream states as to the uses of the water. Turkey's position is that as almost all of the water of the Euphrates is generated within its boundaries it has a strong claim to make use of as much of the waters of the river as it wishes. On the other hand, Iraq's position is that since it has been using the waters of the river for over 5000 years it has a right to go on using them without hindrance.

With the construction and opening of the Ataturk Dam in the early 1990s Turkey has become a major user of Euphrates water for irrigation purposes. Turkey stated that it would guarantee a downstream flow of at least 500 m^3 per second. However, this represented only about half the natural flow of the river on an annual basis, and so both the downstream states of Syria and Iraq reacted strongly against this declaration. Turkey was not willing to change its position and went ahead with its irrigation schemes on the Harran Plain and adjacent areas (Figure 12.7). One of the potential problems with these planned irrigation projects is that the return waters may well prove to have a high saline content and so reduce the quality of the waters of the main stream itself (Beaumont 1996).

There can be no doubt that the development of new irrigation projects in Turkey and Syria will have a very great impact on irrigation in Iraq. Once the Turkish and Syrian schemes are fully commissioned in the early years of the twenty-first century, the volume of water in the River Euphrates available to Iraq may be only around 5000 million m^3 each year, compared with a figure of around 30,000 million m^3 in the early 1960s before the projects along the river began to be constructed. As a result, Iraq will have to abandon large areas of land that it was once able to irrigate. This is bound to cause severe social problems, as many of the villagers along the Euphrates will no longer be able to earn their livelihood by irrigated agriculture.

The situation on the River Nile is not yet as critical as on the Tigris–Euphrates, although potentially it could become even worse. The main user of the water from the river at the present day is Egypt, which uses approximately 55,000 million m^3 each year out of a total of around 84,000 million m^3 (Howell and Allan 1990). This represents almost two-thirds of the flow of the river. However, none of the flow of the river is actually generated within Egypt. In recent years upstream countries, like Ethiopia, in which the majority of the precipitation that feeds the river occurs, are now beginning to plan for major irrigation schemes of their own. If these come into fruition, then the main sufferer will be Egypt and its irrigated agriculture. Given the large population of Egypt and its rapid growth, this will undoubtedly cause many problems.

THE FUTURE OF IRRIGATION

One of the key issues to consider is whether irrigated agriculture is a sustainable activity or

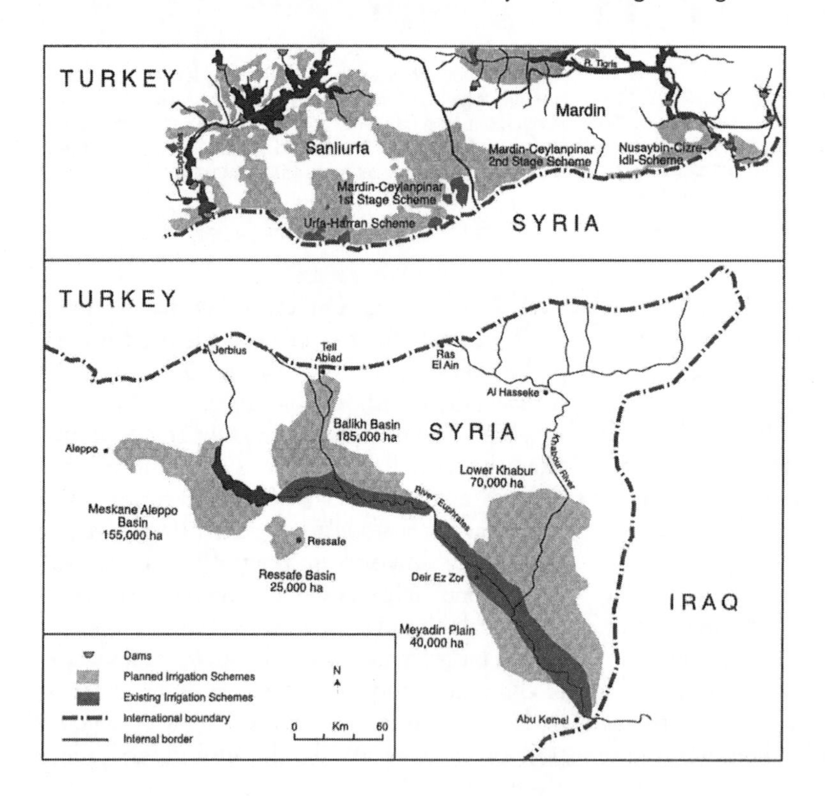

Figure 12.7 New irrigated areas in the Euphrates basin of Turkey.

merely represents a mining of resources (van Schilfgaarde 1990). There is evidence to suggest that under ideal environmental conditions irrigation can be practised for long periods of time with few problems developing. An example that is often quoted is the Nile valley, where continuous cultivation at the same point does seem to have occurred for at least hundreds, if not thousands of years. However, the floodplain of the Nile is blessed with ideal under-drainage conditions in the form of gravel layers, which have minimised the build-up of saline soil conditions. Elsewhere, however, there is evidence that under certain environmental conditions long-term irrigation can lead to a lowering of crop yields as water tables rise to such an extent that economically viable crop production is not possible. This might well have been the case in the Tigris–Euphrates valley (Adams 1978; Jacobsen and Adams 1958).

It is very difficult to determine the exact nature of modern irrigation projects as few have been in existence for more than 100 years. However, many schemes do appear to begin successfully but then experience growing difficulties after a number of years as soil conditions deteriorate (Gardner and Young 1988; Tanji 1990). Such declines in productivity may show themselves after a few years on some projects but may take many decades on others. Overall, it is difficult to state categorically just how irrigation should be regarded. After all, even traditional dry farming is capable of producing considerable environmental degradation when practised over long periods, as the Mediterranean region clearly demonstrates. Perhaps too often in the past it has been assumed that irrigated agriculture does not have significant environmental costs, as it appears to be such a controlled and artificial environment. In the future, it might be expedient to think of irrigation as a mechanism for causing potentially serious environmental stress when it is practised without careful consideration of all the possible consequences (Umali 1993).

As the twenty-first century approaches, another basic question that has to be posed is whether large volumes of water can continue to be used for what is essentially low-value irrigated food and fibre production when demands for water for more economically valuable uses are growing rapidly (Beaumont 1994; 1997). The issue, therefore, is about how water can be utilised for the maximum benefit of society. The likely answer to the question will depend on the level of development of the society being considered. For example, in traditional societies such as the Palestinians on the occupied West Bank, the use of water in irrigated agriculture seems to be one of the few ways in which wealth for the whole community can be generated. In contrast, in the adjacent country of Israel, which now has a much more mature urban/industrial society, the continued use of water for irrigation appears a very wasteful use of available resources. For Israel, which is already suffering from severe water shortages, the water problem in future can only be alleviated by the reallocation of water from irrigated crop production to more valuable urban/industrial uses.

In the United States, the relationships between irrigated agriculture and the urban/industrial uses of water have always been complex. Surprisingly, competition between the two water uses began as early as the latter years of the nineteenth century, when the irrigation boom in the USA was just beginning. One of the best examples is to be found in terms of the water supply of Los Angeles (Hoffman 1981; Walton 1992). What happened here was that the city fathers realised at the end of the nineteenth century that indigenous water resources were very limited and that something had to be done to ensure that future water resources were available for the city. Their solution to the problem was to buy up agricultural water rights in the Owens Valley in a period of little over a decade. The means employed at times seem to have been close to being illegal, but the net result was that the city of Los Angeles did obtain a large volume of water to supply its future needs. As far as the Owens valley was concerned, what had once been a fertile and green valley floor supported by irrigation was turned into a region of brown, semi-arid scrubland.

A more recent manifestation of the same phenomenon has been the development of water markets since the 1960s. Although the potential for the market transfer of water resources from irrigation to urban usage in the USA appears to be considerable, as yet little has been achieved (Brajer *et al.* 1989; Colby 1990; Howe *et al.* 1986; Willey 1992). However, the position is beginning to change and in a recent article it is claimed that 'In most western states [of the USA] market-based water transfers are now expressly protected as a recognised beneficial use of water' (Graf and Yardas 1998: p. 167). Some of the difficulties that have hindered water transfers have been of a legal nature associated with the 'prior appropriation' doctrine for western USA water use. This implies that the first water user in time has a right to continue with a particular water abstraction, provided that it is being utilised for beneficial purposes, irrespective of the water needs of other individuals or organisations. In effect, this has led to the concept of 'use it or lose it'. As a result, many farmers have continued to use excessive amounts of water compared with their actual needs in an attempt to safeguard their legal rights.

An example of water reallocation from irrigation to urban /industrial use is provided by the Metropolitan Water District (MWD) of Southern California, which supplies water to approximately 15 million people. In 1989, an agreement was reached between MWD and IID, the largest irrigation district in California. As a result, IID allocated MWD a long-term, although not in perpetuity, right to utilise 100,000 acre-feet of water each year in exchange for an investment of more than US$100 million in water conservation and related works in the Imperial Valley (*ibid.*). MWD has also negotiated what might be termed a water emergency policy with Palo Verde Irrigation District. With this, when the MWD is facing water shortages in times of drought, it has been agreed that the irrigation district will withdraw land from irrigation temporarily so that the water can be transferred for urban/industrial use.

What then is the future of irrigated agriculture in drylands (Rosegrant and Binswanger 1994; Zwarteveen 1997)? In many regions it would appear to be bleak, as competition for water from urban industrial complexes seems set to grow as increasing populations and rising standards of living demand ever more water. The basic problem is that irrigation can produce relatively small amounts of money for each cubic metre of water consumed. With high-technology irrigation, using drip systems to grow high-value crops like flowers, the returns to the farmer can be considerable. However, the market for such products is limited, and most other farmers are forced to produce relatively low-value crops. In contrast, a cubic metre of water used for industrial production, especially if that industry is of the modern high-technology type, can produce very high returns indeed.

In many dryland regions, all available water resources are already committed and any future development will require new sources to be found. In some countries though, extra water reserves are just not available, apart from the expensive solution of desalination. This means that any future development of the economy can occur only if there is a reallocation of existing water resources. This does not necessarily need to cause as many problems as might be suggested. In most dryland countries, around 70 to 80 per cent of all water used is for irrigation purposes. Usage rates for irrigation are high and often reach values in excess of 10,000 m^3 per hectare. This means that if only a relatively small area of irrigated land is withdrawn from cultivation, a large volume of water can be allocated for other uses. For example, irrigation water for 100 hectares will supply the urban/industrial needs of 10,000 people each year (Beaumont 1997).

It is, however, quite obvious that all decisions about water use are not made from an objective viewpoint. Certainly during the twentieth century it would seem that all countries that practise irrigation have subsidised their irrigation systems. This has been achieved in many different ways, including low interest rates on capital for infrastructure development and subsidised power rates for water-pumping purposes. This has meant that it is extremely difficult to assess the economic viability of existing projects and would suggest that in future governments and other organisations are not necessarily going to make unbiased decisions with regard to irrigation projects.

In the more advanced Western economies and in dryland countries where pressure on the water resource base is already high, it seems likely that irrigated areas may already have reached their peak values and may even be in decline. In Israel, the movement is already downwards, while in the USA it seems to be past the highest point, but no major decline has yet been registered. Elsewhere, the statistics are not reliable enough to be sure exactly what is happening. On the other side of the coin, it does appear likely that in many developing countries the irrigated area will continue to expand for perhaps another decade or so before even here the relentless pressure of population growth and associated water demand will bring this increase to an end.

GUIDE TO FURTHER READING

Beaumont, P. (1997) Water and armed conflict in the Middle East – fantasy or reality? In N.P. Gleditsch (ed.) *Conflict and the Environment*, Dordrecht: Kluwer Academic Publishers, 355–74. Examines the increasing pressures on water resources in the region and the necessity for restructuring water usage, with irrigation water being reallocated to more economically productive uses.

Beaumont, P. (1971) Qanat systems in Iran. *Bulletin of the International Association of Scientific Hydrology*, 16, 39–50. Guide to a traditional method of groundwater irrigation which played a key role in the agricultural development of the drylands of Asia from China to the Mediterranean.

Graf, T.J. and Yardas, D. (1998) Reforming Western water policy: markets and regulation. *Natural Resources and Environment*, 12(3), 165–9. Studies the growing impact of water markets in the USA and the transfer of water from irrigation to urban/industrial use.

Howell, P. and Allan, T. (eds) (1990) *The Nile – Resource Evaluation, Resource Management, Hydropolitics and Legal Issues*. London: School of Oriental and African Studies, University of London. Detailed assessment of irrigation water use in the Nile basin.

Lonergan, S.C. and Brooks, D.B. (1994) *Watershed – The Role of Fresh Water in The Israeli–Palestinian Conflict*. Ottawa: International Development Research Centre. Discusses the conflict over water use for irrigation in both Israel and the occupied Palestinian West Bank.

Rydzewski, J.R. and Ward, C.F. (eds) (1989) *Irrigation: Theory and Practice*. London: Pentech Press. Comprehensive textbook on irrigation.

REFERENCES

Adams, R.M. (1965) *Land Behind Baghdad*. Chicago: Chicago University Press.

Adams, R.M. (1978) Strategies of maximisation, stability and resilience in Mesopotamian society, settlement and agriculture. *Proceedings of the American Philosophical Society* 122, 329–35.

Bauer, C.J. (1997) Bringing water markets down to earth: the political economy of water rights in Chile, 1976–95. *World Development* 25(5), 639–59.

Beaumont, P. (1968) Qanats on the Varamin Plain, Iran. *Transactions of the Institute of British Geographers* 45, 169–79.

Beaumont, P. (1971) Qanat systems in Iran. *Bulletin of the International Association of Scientific Hydrology* 16, 39–50.

Beaumont, P. (1973) A traditional method of ground water extraction in the Middle East. *Ground Water* 11, 23–30.

Beaumont, P. (1978) The Euphrates river – an international problem of water resources development. *Environmental Conservation* 5, 35–43.

Beaumont, P. (1985) Irrigated agriculture and groundwater mining on the High Plains of Texas, USA. *Environmental Conservation* 12(2), 119–30.

Beaumont, P. (1993) *Drylands – Environmental Management and Development*. London: Routledge.

Beaumont, P. (1994) The myth of water wars and the future of irrigated agriculture in the Middle East. *International Journal of Water Resources Development* 10(1), 9–21.

Beaumont, P. (1996) Agricultural and environmental changes in the upper Euphrates catchment of Turkey and Syria and their political and economic implications. *Applied Geography* 16(2), 137–57.

Beaumont, P. (1997) Water and armed conflict in the Middle East – fantasy or reality? In N.P. Gleditsch (ed.) *Conflict and the Environment*, Dordrecht: Kluwer Academic Publishers, 355–74.

Brajer, V. *et al.* (1989) The strengths and weaknesses of water markets as they affect water scarcity and sovereignty interests in the West. *Natural*

Resources Journal 29, 489–509.

Burns, R. (1993) Irrigated rice cultivation in monsoon Asia: the search for an effective water control technology. *World Development* 21, 771–89.

Colby, B. (1990) Transaction costs and efficiency in Western water allocation. *American Journal of Agricultural Economics* 72, 1184–92.

Economic and Social Commission for Asia and the Pacific (ESCAP) (1989) *Desertification in Indus Basin due to Salinity and Waterlogging: A Case Study*. Bangkok, Thailand: ESCAP.

Fradkin, P.L. (1984) *A River no more – The Colorado River and the West*. Tucson: University of Arizona Press.

Gardner, R.L. and Young, R.A. (1988) Assessing strategies for control of irrigation induced salinity in the upper Colorado river basin. *American Journal of Agricultural Economics* 70(1), 37–49.

Gelburd, D.E. (1985) Managing salinity lessons from the past. *Journal of Soil and Water Conservation* 40(4), 329–31.

Graf, T.J. and Yardas, D. (1998) Reforming Western water policy: markets and regulation. *Natural Resources and Environment* 12(3), 165–9.

Green, D.E. (1973) *Land of the Underground Rain*. Austin: University of Texas Press.

Hamdan, G. (1961) Evolution of irrigated agriculture in Egypt. In *History of Land Use in Arid Regions, Arid Zone Research* 17, New York: UNESCO, 119–42.

Hoffman, D. (1981) *Vision or villainy: origins of the Owens Valley – Los Angeles water controversy*. College Station: Texas A & M Press.

Howe, C. *et al.* (1986) Innovative approaches to water allocation: the potential for water markets. *Water Resources Research* 22, 439–45.

Howell, P. and Allan, T. (eds.) (1990) *The Nile – Resource Evaluation, Resource Management, Hydropolitics and Legal Issues*. London: School of Oriental and African Studies, University of London.

Jacobsen, T. and Adams, R.A. (1958) Salt and silt in ancient Mesopotamian agriculture. *Science* 128, 1251–8.

Kenyon, K.M. (1969–70) The origins of the Neolithic. *The Advancement of Science* 26, 1–17.

Kirmani, S. and Le Moigne, G. (1997) *Fostering riparian cooperation in international river basins – the World Bank at its best in development diplomacy*. World Bank Technical Paper No. 335, World Bank, Washington, DC.

Kirmani, S. and Rangeley, R. (1994) *International inland waters – concepts for a more active World Bank role*. World Bank Technical Paper No. 239, World Bank, Washington, DC.

Kliot, N. (1994) *Water Resources and Conflict in the Middle East*. London: Routledge.

Levintanus, A. (1992) Saving the Aral Sea. *Water Resources Development* 8(1), 60–4.

Lonergan, S.C. and Brooks, D.B. (1994) *Watershed – The Role of Fresh Water in the Israeli–Palestinian Conflict*. Ottawa: International Development Research Centre.

Mackay, N. and Landsberg, J. (1992) The health of the Murray–Darling river system. *Search* 23(1), 34–7.

Micklin, P.P. (1988) Desiccation of the Aral Sea: a water management disaster in the Soviet Union. *Science* 241, 1170–6.

Misak, R.F., Abdel Baki, A. and El-Hakin, M.S. (1997) On the causes and control of the waterlogging phenomenon, Siwa Oasis, north western desert, Egypt. *Journal of Arid Environments* 37(1), 23–32.

National Water Commission (1973) *Water policies for the future*. Washington, DC: US Government Printing Office.

O'Hara, S.L. (1997) Irrigation and land degradation: implications for agriculture in Turkmenistan, central Asia. *Journal of Arid Environments* 37(1), 165–79.

Plusquellec, H., Burt, C. and Wolter, H.W. (1994) *Modern water control in irrigation – concepts, issues and applications*. World Bank Technical Paper No. 246, World Bank, Washington, DC.

Precoda, N. (1991) Requiem for the Aral Sea. *Ambio* 20(3–4), 109–14.

Rios, M. and Quiroz, J. (1995) *The market of water rights in Chile: major issues*. World Bank Technical Paper No. 285, World Bank, Washington, DC.

Rosegrant, M. and Binswanger, H. (1994) Markets in tradable water rights: potential for efficiency gains in developing country water resource allocations. *World Development* 22, 1613–25.

Rydzewski, J.R. and Ward, C.F. (eds) (1989) *Irrigation: Theory and Practice*. London: Pentech Press.

Shuval, H.I. (Ed.) (1980) *Water Quality Management under Conditions of Scarcity – Israel as a Case Study*. New York: Academic Press.

Simpson, H.J. and Herczeg, A.L. (1991) Salinity and evaporation in the River Murray basin, Australia. *Journal of Hydrology* 124(1–2), 1-11.

Starr, J.R. (1991) Water wars. *Foreign Policy* 82 (Spring), 17–36.

Tanji, K.K. (ed.) (1990) *Agricultural Salinity Assessment and Management*. New York: ASCE.

Umali, D.L. (1993) *Irrigation induced salinity – a growing problem for development and the environment*. World Bank Technical Paper No. 215, World Bank, Washington, DC.

Van Schilfgaarde, J. (1990) Irrigated agriculture: Is it sustainable? In K.K. Tanji (ed.) *Agricultural Salinity Assessment and Management*. New York: ASCE, 584–94.

Walton, J. (1992) *Western Times and Water Wars: State, Culture and Rebellion in California*. Berkeley: University of California Press.

Wittfogel, K.A. (1957) *Oriental Despotism*. New Haven, Conn.: Yale University Press.

Willey, Z. (1992) Behind schedule and over budget: the case of markets, water and the environment. *Harvard Journal of Law and Public Policy* 15, 391–425.

Zwarteveen, M.Z. (1997) Water: from basic need to commodity: a discussion on gender and water rights in the context of irrigation. *World Development* 25(8), 1335–49.

13

Desertification

Andrew Millington

INTRODUCTION

Desertification is a strongly emotive term with significant negative environmental overtones. Much of this emotion can be defused by considering desertification as the net result of a set of processes that can result in land degradation. Each of these processes occurs naturally, but short-term climatic fluctuations, long-term climatic desiccation, human activities or a combination of these factors can accelerate the rates of these processes. None of these factors is new – climates have always fluctuated (although maybe not always as rapidly as at present) and there is a long history of the impacts of human occupation in drylands (although, again, not at current levels). We may then reason that, because of the high contemporary population levels in drylands and the rapidity of climate change, the processes that constitute desertification are currently operating at higher than normal rates.

The importance of desertification is underlined by the fact that drylands (hyper-arid, arid, semi-arid and dry sub-humid regions) comprise about approximately a third of the world's land area, and are home to over 900 million people (Toulmin 1997). The aims of this chapter are to show:

- desertification is so important that it is recognised globally as one of the world's major environmental issues;
- the processes that comprise desertification cannot be considered simply but only in a holistic, complex manner;
- responses to it, like its causes, are far from straightforward; and
- geographers have an important role to play in future desertification research.

The International Convention to Combat Desertification

While deconstruction of the term desertification has approached some kind of academic epitome, the term is widely used (in academia, among practitioners and in the media) and, most importantly considering its global importance, has been a key environmental and development focus within the United Nations (Toulmin 1997).

Stimulated by the Sahelian droughts of the 1970s, the UN organised a major conference on desertification in 1977 in Nairobi, which produced the Global Plan of Action to Combat Desertification. The United Nations Environmental Programme (UNEP) was given the mandate to execute this plan, which mainly focused on national action plans, surveys of the extent of desertification and the establishment of a donor funding mechanism. Its success was limited, particularly in terms of the national action plans (which were often unfeasible and did not tackle the most important issues on the ground) and donor funding (which continued along bilateral lines). In the negotiating period leading up to the UN Conference on Environment and Development (the Earth Summit) in Rio de Janeiro in 1992, African states called for a further UN initiative on desertification. This call arose because they felt that their main concerns (drought, poverty and food security) were being sacrificed at the altar of the two main concerns of developed nations (biodiversity and climate change). Post-Rio negotiations in 1993 and 1994 led to the International Convention to Combat Desertification (CCD), which came into force when the fiftieth country to ratify it, Chad, signed in December 1996. The first conference of the parties was held in 1997. Particular emphasis in this convention has been placed on national action programmes and urgent action for Africa, the continent most affected by drought and desertification.

Desertification as a global issue has been officially recognised by the UN for over two decades. What then, in this time, has been achieved in terms of recognising its scope, investigating its causes and effects and implementing remedial solutions?

DESERTIFICATION – CAUSES AND EFFECTS

Desertification processes

There is now general agreement that the scope of desertification is broad, as can be seen from the Article 1 of the CCD, which defines it as:

> Land degradation in arid, semi-arid, and dry sub-humid areas resulting from various factors, including climatic variations and human activities.

Desertification includes environmental impacts such as soil erosion, reduced soil fertility, loss of vegetation cover and loss of species, and human impacts such as increased vulnerability of people to drought, reduced levels of national food security and, in the worst cases, malnutrition and starvation. It is possible to divide these processes into three broad areas:

- specific soil degradation processes (water and wind erosion, salinisation, waterlogging, topsoil induration, and reduced soil fertility);
- broader natural resource degradation issues (vegetation destruction (e.g. Plate 13.1), loss of biodiversity, localised changes in microclimate, mesoscale land–climate feedback mechanisms, and impacts on surface and groundwater quality and availability); and
- impacts of economies and societies (increased vulnerability to drought from the household to national levels; population displacement;

Plate 13.1 The first stages in a complex fuelwood supply chain in Pakistan. When I interviewed this wood collector in 1991 in the Baluchistan Desert he had arrived at a wood market on the main road near Turbat, over 500 km from Karachi. He had five similarly laden camels. The journey from his home to one of the few wooded areas in the desert, the time it had taken him to cut the trees, and his journey to market had taken three days. He received approximately US$10 for the five camel loads of wood from the market's owner. The price of this wood would have increased three to four times due to labour and transport costs before it was sold to Karachi.

differential gender, poverty and health impacts; and famine, malnutrition and starvation).

It is evident from studies of desertification that many of the processes listed above are not only effects of a wide range of processes but that they also act as causes. There are no simple cause-and-effect relationships in desertification, only a complex web of interrelated processes that act as both cause and effect. These causes and effects overlap on the ground; nonetheless, it is important to try to discriminate between causes to enable policies and interventions to be developed that address the issues (Warren and Khogali 1992). The main demographic change and policy issues are population growth, population movements, agricultural change and modernisation, warfare, and political change. In terms of climatic forcing there are two main areas: first, unpredictable, short-term fluctuations in climate – droughts and floods – and, second, longer-term climatic desiccation. However, it is not adequate to ascribe the main cause of desertification in an area to either climate forcing or demographic and policy issues. For example, Livermann (1990), working in northern Mexico, argues that drought impacts are not simply a function of drought severity but are also influenced by the political, economic and technical characteristics of the region affected; and Glantz (1994) provides a number of detailed case studies in this area.

Demographic change, policy issues and desertification

The interactions between human activities and natural resources in the context of desertification are a complicated area. In some cases, high rates of population directly increase pressure on the natural resource base. This may lead to high cultivation densities, which reduce fallow periods and lead to soil nutrient depletion (Warren and Khogali 1992). Alternatively, it may lead to rangeland overstocking. In this area though, ideas have changed. In the 1960s and 1970s, the prevalent paradigm in range science was to match stock numbers to range resources using the concept of carrying capacity; a shift in the research paradigm now suggests that range potential is a mainly rainfall-driven parameter. Moreover, issues surrounding stock numbers and land degradation are complicated because of degradation by stock concentrations around watering points and the use of feed supplements, e.g. in eastern Jordan (Campbell and Roe 1998). However, high population density does not always lead to degradation. Boserup's argument that high population densities lead to agricultural innovation can be seen in the development of indigenous soil and water conservation systems in drylands (Reij 1991) and in the management of trees in areas of intense pressure from fuelwood collection.

Large-scale population movements are another aspect of demographic change that affect desertification. These movements occur because of land shortages, land alienation and warfare. Movements can be rural-to-urban or rural-to-rural. The first type of migration is known to increase pressure on rural production systems, particularly biofuel production systems, which have a dominant rural-to-urban flow in drylands (Floor and Gorse 1988; see Plate 13.1). In Kenya, people have moved from the fertile Kenyan highlands to the drier, northern regions because of land alienation, while in Jordan, the limited amount of highly productive land (in the Jordan valley and the western highlands) combined with the large numbers of Palestinian refugees has shifted the frontier of cultivation eastwards into desert regions with mean annual rainfall totals of less than 250 mm. In the West African Sahel, pastoralists have moved southwards due to drought, while cultivaters have moved north due to lack of land, thereby causing an uneasy mix of pastoralism and cultivation in a zone that is economically marginal in terms of cultivation (Warren and Khogali 1992). Population concentrations around refugee camps leads to pressure on natural resources (e.g. biofuels) in their immediate environs (Munslow *et al.* 1989).

Areas of outmigration also suffer from desertification because the labour shortages created lead to a lack of maintenance of conservation and land management infrastructure (Plate 13.2). An interesting example of the effects of high population growth rates, combined with the effects of outmigration, can be found in Yemen (Box 13.1).

Changes in the political control of the resource base is a further factor that affects both land management practices and population dynamics and, as a consequence, desertification. These changes may be brought about by economic polices, such as the adoption of neo-liberal economic models and GATT, or through fundamental political change. Important contemporary examples of the breakdown in the control of land management in dryland areas due to a shift in the political systems can be found throughout Central Asia in the post-Soviet era (Box 13.2).

The consequences of demographic and policy triggers often initially manifest themselves in terms of vegetation destruction – due to overgrazing and overbrowsing, vegetation clearance for cultivation, or biofuel collection. The subsequent breakdown of the complex vegetation–soil–water relationships found in drylands means that vegetation's many environmental benefits are diminished or lost altogether. The resulting soil degradation and reduced soil water availability lead to a reduction in soil fertility (used here as a qualitative term covering a wide range of specific soil characteristics, but which provides a useful guide to land productivity in terms of potential crop or forage production). The reduction in potential land productivity has clear implications for household livelihoods and national food security.

Climatic forcing and desertification

The second scenario occurs when climatic forcing (in terms of droughts, floods and longer-term desiccation) affects the resource base. While climate has direct affects on the soil–water–vegetation environment (e.g. in terms of increased evapotranspiration or vegetation destruction by flooding), the greatest impact of climatic forcing factors is probably indirectly through their impacts upon dryland populations and their economic activities.

There is general agreement that desiccation of most of the world's drylands has been occurring since the middle of the last century (Goudie 1990). This is important for understanding contemporary dryland problems in two ways. First, it provides an important underpinning to current distributions of population, agricultural activities and urban areas. Second, it can provide indications of the likely impacts of future climate change in drylands. Parry (1990), in reviewing the

Plate 13.2 Ridge-top villages, such as the one in this photograph, are common throughout northern Yemen. They are typically surrounded by flights of agricultural terraces. A decline in rural labour has meant that some terraces have not been maintained (see immediately below the village). This causes the terrace walls to disintegrate and water runoff and topsoil erosion to increase, resulting in a decline in land productivity in the uplands and increased flooding in the *Tihama* (the Red Sea lowlands).

Box 13.1 The natural resource crisis in Yemen

The southwestern Arabian peninsula has long been home to farmers who lived in isolated villages and relied on a relatively rich natural resource base of fertile volcanic soils, a normally adequate wet season rainfall and well-wooded slopes to sustain their rural production systems. Linked to these villages by paths and camel trails were large highland cities, e.g. San'a' and Tai'zz (Figure 13.1)

The mainstays of the rural production system were

Figure 13.1 Areas with significant reserves of woody vegetation and woodfuel supply chains to main cities in northern Yemen.

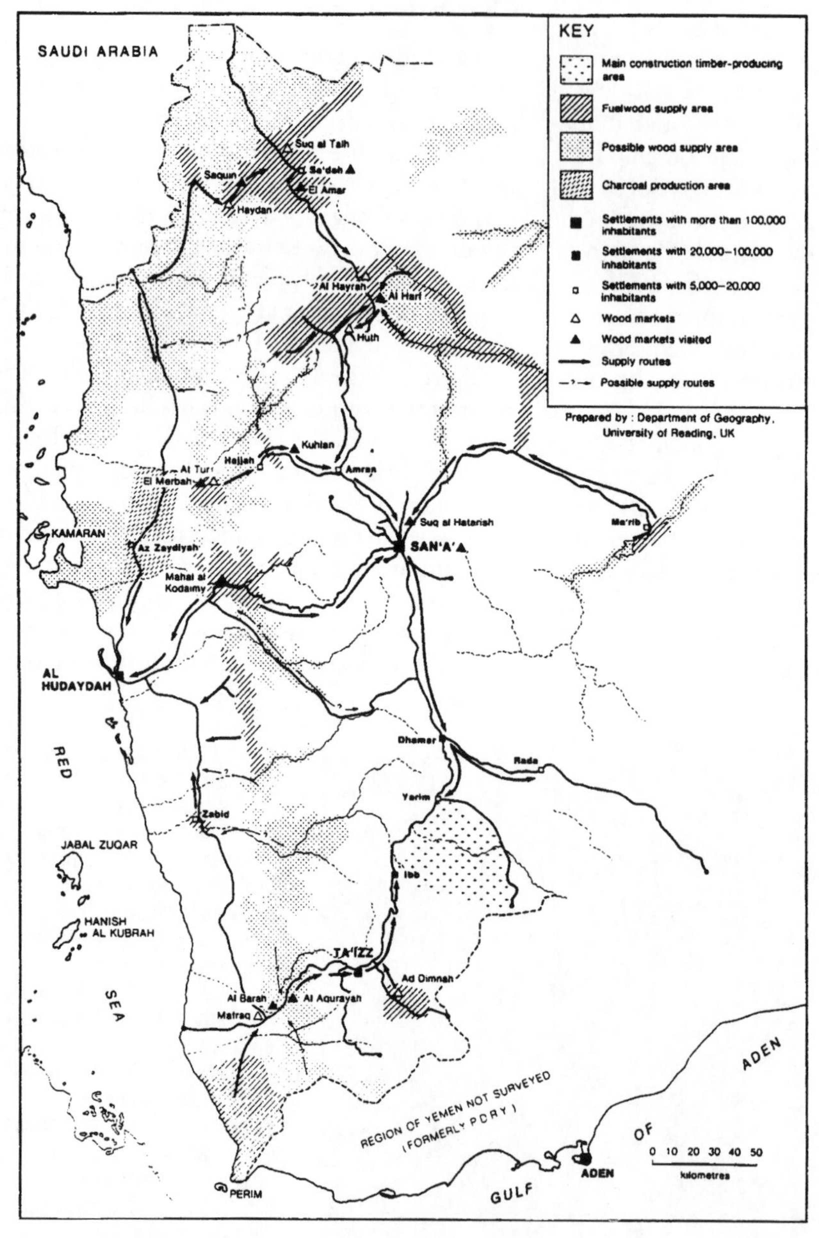

Source: From Millington and Crosettti 1989.

extensively terraced hill slopes. The terraces had the important role of conserving the fertile soils, thereby allowing farmers to grow subsistence cereal crops during the wet season as well as tree crops (e.g. coffee) as cash crops. These terraces also formed part of an extensive water control system that allowed the farmers in the mountains to control the flow of water down on to the Red Sea coastal plains, or Tihama. In the Tihama, the farmers managed the movement of water through their fields using channels and bunds, but always under the control of each village's 'water master'. This water carried vast quantities of fertile silt from the mountains, which naturally fertilised the fields in the Tihama. This is a highly seasonal system, typical of semi-arid environments, which is dependent on the reliability of the annual rains and careful water management. The release of water from the mountains was not only controlled by the action of the terraces but also by the rain-green subtropical woodlands that formed the natural vegetation in the western mountain ranges of Yemen and southwest Saudi Arabia (see Figure 13.1)

The combination of two factors – rural depopulation and domestic energy demand from urban areas – has compromised this centuries-old soil and water management system, which was central to Yemeni life. Rural depopulation is a recent phenomenon that has been stimulated by labour demands created by the oil industry in Saudi Arabia and the Gulf states. The affect of rural depopulation on the rural production systems has been twofold. First, peak labour demands at various times in the farming calendar have not been met. Consequently, some terraces have been abandoned and are not maintained. Once the terraces fall into disrepair, their soils and water-retaining functions are compromised. More violent floods occur throughout the entire water management system, and soil erosion rates on the mountain slopes increase markedly. Second, capital remittances from the migrant workers have promoted the purchase of small pumps. These are now extensively used to irrigate for year-round cultivation using groundwater reserves in small intermontane basins. The immediate impact of pumping on groundwater levels has been to lower the water table. In the Dhamar region, groundwater levels declined at a rate of 30 cm/year during the 1980s. A further, less obvious impact is that as farmers have turned their attention to irrigation they have abandoned yet more terraces.

Most of the forest that the Yemeni mountains supported were lost many centuries ago. The remaining subtropical woodlands have been more or less destroyed over the past few decades by biofuel (fuelwood and charcoal) extraction. Biofuel dependency is very high in Yemeni households (Table 13.1), and fuel wood supplied 54 per cent of final energy consumption in 1988 (ESMAP 1991). The greatest demand exists in the cities, and biofuel supply chains have been developed between wood collection areas and the cities. The key characteristics of these supply chains are markets where main roads cut through or run close to wooded areas, transport along the main roads from supply area markets to cities, and urban markets. Combined, these provide mark-ups of between 33 and 52 per cent of the retail price of fuel wood in urban markets (*ibid.*). The high costs of wood fuels in Yemen, compared with other western Asian and African countries, has stimulated this trade. However, an even greater stimulation has been provided by the construction of the road network in Yemen during the 1970s and 1980s, which brought many wooded areas within an economically viable zone of wood extraction for the cities (Millington and Crosetti 1989).

Table 13.1 Household fuel use for the northern governorates of Yemen, 1998, in thousand tonnes oil equivalent.

	Fuelwood	*Charcoal*	*Cattle dung*	*Crop residues*	*Other fuels*[1]	*Total*
Cooking[2]	1800	2	96	152	101	2151
Lighting					59	59
Space heat	23	2			3	28
Water heat	29			5	13	48
Water pipes		25				25
Other	7	1	7		9	24
Total	1860	29	103	157	184	2333

Source: Adapted from ESMAP, 1991.
Notes: 1 Includes electricty, LPG and kerosene.
2 Includes baking.

The destruction of these woodlands not only causes problems in meeting the domestic energy requirements in Yemeni cities. The loss of woodland on non-terraced slopes increases soil erosion and runoff rates on these slopes, thereby contributing to increased, uncontrolled flooding in the Tihama, and it reduces infiltration rates, thereby further reducing groundwater recharge.

Box 13.2 Political change, desertification and population dynamics in Central Asia

The break-up of the USSR provides important insights into the links between the changing political control over natural resource management, population dynamics and desertification. Drylands form a large proportion of the newly formed Central Asian republics and southern Russia (Table 13.2).

Extensive environmental degradation exists in these areas, particularly salinisation, wind and water erosion, deteriorating pasture quality, and water pollution. Data for four of the areas (Table 13.3) reveals the spatial extent of these processes.

Table 13.2 Dryland areas in the Central Asian countries and southern Russia republics.

Country	*Russia*[1]	*Kazakhstan*	*Turkmenistan*	*Uzbekistan*	*Other countries*[2]
Area (million ha)	175	45	35	20	25
As a proportion of land area of the countries (%)	1.2	64.4	92.2	78.2	58.4

Source: After Babaev 1991.
Notes: 1 Only includes data for Astrakhan and Kalmykia republics.
2 Azerbaijan, Kirgizistan and Tajikistan.

Table 13.3 Areas of the main types of environmental degradation in some parts of Central Asia.

	Wind-eroded area ('000 ha)	*Water-eroded area ('000 ha)*	*Area of salinised soils ('000 ha)*	*Degraded pasture ('000 ha)*
Tajikistan	600	3600	300	No data
Uzbekistan	2260	2990	1970	900
Azerbaijan	200	1370	1200	No data
Kalmykia (Russia)	840	610	No data	4300

Source: Glazovsky and Shestakov 1996.

Degradation of the natural resource base as a consequence of state-controlled planning is exemplified by the situation around the Aral Sea basin. State planners diverted significant amounts of water from the Amu Dar'ya and Syr Dar'ya, the rivers that feed the Aral Sea, to irrigate land for cotton production. The resulting reduced water flows into the Aral Sea have caused severe shrinkage in terms of volume and areal extent. Salinity increased from around 12–14 per cent in the early 1970s to 23 per cent in the late 1980s, and the subsequent decline in phytoplankton has propagated itself through the food chain. Of particular economic importance is the fact that only five of the twenty fish species that existed in this once important inland fishery remain. Relict and endemic plants in the floodplain forests are threatened with extinction. Lake sediments contaminated with untreated waste, fertilisers, and organochlorine and organophosphate pesticides, herbicides and defoliants have, as the lake has dried out, deflated and posed severe health problems in the adjacent areas. For example, child mortality in the Bazataus Rayon in Karakalpakstan exceeds 110 per 1000, and over 60 per cent of children examined have medical abnormalities.

The wider socio-economic and demographic relevance of these data lies in the responses of the 33 million inhabitants of this region to desertification. Glazovsky and Shestakov (1996) argue that the acute nature of degradation in the region has led to a serious decline in food production and the number of domestic animals, land-use changes, modified employment profiles, increased infant mortality, and population migration. The migration of 'environmental refugees' from the region is the end-result of the environmental degradation process. Demographic patterns that existed prior to the break-up of the USSR and the migration patterns that have emerged since then as part of the dynamic political situation in the region confuse this line of argument. This is characterised by the (re)formation of political units outside and within Russia, and growing levels of nationalism, ethnic discrimination, economic instability and armed conflicts. Consequently, although there is a long and well-documented history of environmentally induced population migrations in Central Asia dating back to Neolithic times (Gumilev 1992; Glazovsky and Shestakov 1996), in terms of contemporary migration it is difficult to separate the environmental, economic, social and political forces governing it.

Trying to resolve the contribution of dryland degradation to population displacement requires a comparison of pre- and post-Soviet demographic data. During the late Soviet era, the key demographic features of Central Asia were:

- High natural population growth rates (ranging from 14.0 in Kazakhstan to 27.6 in Uzbekistan).
- High infant mortality (ranging from 10.8 in Kalmikya to 54.7 in Tajikistan).
- Migration to rural areas (which resulted in a 210 per cent increase in the region's rural populations between 1967 and 1986), matched by a low urbanisation rate.

There has also been a significant movement of people from and within Central Asia since the 1970s, i.e. before the break-up of the Soviet Union. These movements fall into three categories:

1. *Inter-regional migration*, from desertification-prone areas in the south (in what are now the Central Asian republics) to more ecologically resilient areas to the north (in what is now Russia).
2. *Migration within republics*. Internal population redistribution within Kazakhstan from the severely degraded environmental zones in the south, as well as out-migration from Kazakhstan (i.e. inter-regional migration) has been identified.
3. *Rural to urban migration*. This has been particularly acute around the Aral Sea, although it is offset to a certain extent by the high natural population growth rates of the ethnic populations in that area.

It seems clear from the Soviet-era data that environmental issues probably had a significant role in leading to migration away from desertification-prone areas. Using this argument, Glazovsky and Shestakov (1996) ascribe a dominant environmental influence over the last ten to fifteen years to contemporary out-migration from the following regions:

- Aral Sea basin (Karakalpakstan in Uzbekistan, and Aktubisk and Kzyl-Orda in Kazkhstan); and
- parts of southern European Russia – Kalmykia (in particular Khalmg-Tangch) and Astrakhan. The deteriorating environmental situation in Kalmykia has become so severe that it caused a state of emergency to be declared in 1993 (Gabunshina 1997).

Nonetheless, in most cases the poor socio-economic conditions associated with uneven development in these regions has also been a contributing factor to out-migration.

While it can be postulated that a prime factor in migration in the 1970s and 1980s was the worsening environmental situation in the southern Soviet Union, the influence of political change in the 1990s has had an additional effect. Between 1986 and 1990, 736,500 people left Central Asia. Approximately 481,500 of these people came from Kazakhstan, mainly migrating to central and eastern Russia (Population of Russia 1993). Since independence, over 1 million people per year have emigrated from Kazakhstan. A simple migration model exists in which Kazakhs move within Kazakhstan, and Russians migrate from Kazakhstan to Russia. The situation will worsen as the increase in rural population density, combined with a decrease in productive land due to desertification, is predicted to lead to decline in the amount of available land per person from 0.23 ha/person in the late 1980s to 0.15 ha/person in 2010.

For countries such as these, which are at the cutting edge of desertification, ratification of the CCD should be almost automatic. However, Sievers and Tsaruk (1997) reported that only Uzbekistan and Turkmenistan had ratified the CCD by May 1997. They ascribe the lack of ratification of the CCD by the other states in the region to:

- an inability to fund internally activities related to international conventions;
- political considerations, such as civil unrest and maintenance of the delicate regional power balances; and
- bureaucratic blockages, which restrict information sharing.

evidence for the impact of climate change on agriculture, shows that productivity over a large proportion of the world's drylands, using the outputs of *most* general circulation models, is likely to decline in the first part of the next century (Table 13.4).

Far greater prominence (at least in the media, and therefore among people at large and policy makers) has been given to the impacts of droughts on drylands and their inhabitants. The first drought to stir the imagination of the world's scientists and policy makers was the Sahelian drought of the 1970s; the first that really nagged at the global conscience was the Ethiopian drought and famine of the 1980s. In reality, droughts are part of the normal (high) variability of dryland climates, and they are a recurrent feature of the instrumental record as well as in the archaeological and palaeoenvironmental records. Acknowledgement of the importance of droughts and their links to desertification has arisen from the increased coverage of droughts by the global media as well as from the impact that they have had on larger numbers of people. A pertinent

Table 13.4 Predicted changes in soil moisture in drylands under a 2 × CO_2 scenario from three general circulation models.

Season	*Region*	*Models predicting a decrease in soil moisture levels*		
		CCC	*GFDL*	*UKMO*
December–February	Canadian Prairies			
	US Prairies	X		X
	California	X	X	
	Texas & N Mexico	X	X	X
	NE Brazil	X	X	X
	Altiplano (Central Andes)	X	X	
	Pampas (Argentina & Uruguay)		X	
	S Europe	X		X
	N Africa	X	X	X
	Sahara	X	X	
	Horn of Africa	X		
	Southern Africa	X	X	X
	W Asia	X		X
	Arabian Peninsula	X	X	X
	Middle East & Pakistan			X
	India		X	
	Central Asia	X		
	N China	X	X	X
	Australia		X	
July–August	Canadian Prairies	X	X	X
	US Prairies	X	X	X
	California			X
	Texas & N Mexico	X		X
	NE Brazil	X	X	X
	Altiplano (Central Andes)	X	X	X
	Pampas (Argentina & Uruguay)	X		X
	S Europe	X	X	X
	N Africa	X	X	X
	Sahara	X		X
	Horn of Africa	X		X
	Southern Africa	X	X	X
	W Asia	X	X	X
	Arabian Peninsula		X	X
	Middle East & Pakistan		X	X
	India			X
	Central Asia	X	X	X
	N China	X	X	X
	Australia	X	X	

Source: Data from Parry 1990.
Notes: CCC = Canadian Climate Centre model.
GFDL = Geophysical Fluid Dynamics Laboratory model.
UKMO = UK Meteorological Office model.

question to ask is whether droughts have greater impacts on people now than in the past. The answer to this question appears to be yes. In particular, traditional societies in drought-prone regions of Africa had developed sustainable drought-coping mechanisms, which have been compromised by post-1960s modernisation and development. The Kenyan Akamba provide a notable example. Their homeland is centred on the Machakos Hills to the east of Nairobi. This is a semi-arid region with about 300–600 mm of rainfall, most of which falls in two wet seasons; the area is drought-prone. Traditional Akamba drought-coping mechanisms were to:

- have a mixed agro-pastoral farming system that minimised the risk of exposure to the severe climate fluctuations by not relying on a single crop but spreading their risks by cultivating a variety of crops and having stock as well;
- reduce cattle numbers in drought years by sending them temporarily to other villages in regions not suffering drought and receiving cattle from others in wet years; and
- diversify the household/village economic base by actively trading; this was possible by virtue of historical trade routes between the East African highlands and Indian Ocean coast, which passed through their territory.

These traditional drought-coping mechanisms have broken down with the economic and social changes that have characterised Kenya in the late colonial and post-independence eras, leaving the region's inhabitants less able to cope with drought.

Our understanding of the causes of drought has advanced because of developments in climate modelling. In particular, research into global teleconnections between sea-surface temperatures and drought is important. The El Niño Southern Oscillation (ENSO) has been linked to droughts, e.g. in India (Kiladis and Sinha, 1991), Australia (van Dijk *et al.* 1983), northeast Brazil (Chu 1991) and the Sahel (Lamb and Peppler 1991). Less prominence has been given to the impacts of severe flooding in drylands, some of which can be linked to ENSO events (e.g. the coastal drylands of Peru in 1998). This is probably because water surplus in not considered to be a serious issue in drylands by the wider public. However, flooding is a significant hazard and does lead to land degradation through increased soil erosion by water and waterlogging.

FUTURE GEOGRAPHICAL RESEARCH ON DESERTIFICATION

Geographers have an important role in desertification studies, first, because its complex nature provides fertile territory for holistic research approaches and, second, because it often focuses on population–environment relationships. However, both the holistic and people–environment approaches rely on an integrated model of geography, which currently appears at odds with a discipline that tends towards increased specialisation in many of its sub-disciplines and the growing engagement with qualitative approaches in human geography.

Far from being a problem, this diversification within geography strengthens our hand in the multidisciplinary research approaches that characterise the largest, well-funded international initiatives in desertification in three ways. First, the holistic theme will remain strong in geography, providing geographers with a competitive advantage over many other disciplines, particularly in our ability to form and lead multidisciplinary teams. Second, the rise of qualitative research methods is potentially important for household and village-level studies on natural resources use in dryland communities and on the impacts of resource degradation, which have often relied on shaky statistical data in the past. Third, many investigations require specialist knowledge within a holistic framework. To list all of the specialists required to investigate a desertification issue is akin to listing the academic departments in a large university! Nonetheless, a number of specialist areas within geography are recognised internationally for their work on desertification-related problems. Those with the greatest (research) impact are:

- wind and water erosion of soils
- monitoring vegetation dynamics using remotely sensed data
- land cover and land degradation monitoring using remotely sensed data
- biofuel production systems
- agricultural production systems, in particular rain-fed cultivation systems
- drought climatology and related (surface) water resource issues
- population–environment relations
- development studies, especially with reference to gender and poverty.

In some of these specialist areas geographers are active globally, e.g. soil erosion. In other areas, rates of research activity among geographers are high in some countries but low in others with equally large numbers of geographers. A case in point is drought climatology, which is an important research area among US geographers but is less so in the UK. This is partly due to the greater relevance of drought climatology over much of North America but is also an outcome of the relationships of geographical sub-disciplines and allied disciplines, in this case climatology and meteorology, in different countries.

The main role of geographers in 'desertification studies' since 1977, and that which will be carried out in response to the CCD, will be in:

- surveying the extent of the problem, by applying our spatial analytical and mapping skills to remotely sensed data and in using GIS; and
- assessing the impact of desertification on the physical environment (in particular soils and vegetation) and people's livelihoods (at a variety of scales but concentrating on the household/village level).

Geographers are not very active in researching the influences of changing environmental and economic policies on desertification, nor the impacts of the declining resource base and the increased vulnerability of people to drought on a wide range of policies and issues at both national and regional scales. Nonetheless, geographers have the skills to make significant theoretical and applied contributions here. In addition, geographers have only a minor voice in the formulation of national action plans. This is particularly so in desertification-prone countries in the developing world, with the exception of some of the Central Asian republics, India and some countries in southern Africa. This situation is partly historical and relates to which (European) model of geographical investigation has been adopted in a particular country. It has also arisen because of inadequate investment in universities and research institutes in developing countries. This has, for instance, reduced the opportunities for geographers to develop skills in key areas such as remote sensing and GIS, which are critical selling points of the discipline in the development work carried out in such countries.

GUIDE TO FURTHER READING

Beaumont, P. (1989) *Drylands. Environmental Management and Development*. London: Routledge. Written from a geographical perspective, this is probably the best general, introductory review of the full range of environmental management issues in drylands.

Glantz, M.H. (1994) *Drought Follows the Plow*. Cambridge: Cambridge University Press. A series of historical and contemporary case studies of desertification from around the world. These studies stress the anthropogenic influences in desertification.

Mortimore, M. (1989) *Adapting to Drought: Farmers, Famines and Desertification in West Africa*. Cambridge: Cambridge University Press. A wealth of research on the behaviour of West African farmers. The droughts (and famines) that characterise the Sudanian and Sahelian zones of West Africa are summarised in this book. Although set in a West African context, much can be gathered about how people and households respond to drought in other parts of the world.

Thomas, D.S.G. and Middleton, N. (1995) *Desertification: Exploding the Myth*. Chichester: John Wiley. This books argues that desertification is less of an objective scientific fact and more of a socio-political construct. It examines the origin of the 'desertification myth' and uses new research findings to show that desertification might be a smaller problem and less locally significant than many authors argue. Global in scope.

Tiffen, M., Mortimore, M. and Gichuki, F. (1994) *More People, Less Erosion: Environmental Survey in Kenya*. Chichester: John Wiley. Though focused on one country – Kenya – the findings in this book are widely applicable to drylands in developing countries. The book's strength lies in its detailed analysis of population–degradation relationships, and in debunking the myth that increased population always leads to increased degradation.

REFERENCES

Babaev, A.G. (1991) *Desertification – Threat to Human Society*. Moscow: Vestnik AS. [in Russian]

Campbell, D. and Roe, A. (1998) Results of a preliminary survey of livestock owners. In R.W. Dutton, J.I. Clarke and A.M. Battikhi (eds) *Arid Land Resources and Their Management*. London: Kegan Paul International, 189–96.

Chu, P.-S. (1991) Brazil's climate anomalies and ENSO. In M.H. Glantz, R.W. Katz and N. Nicholls (eds) *Teleconnections Linking Worldwide Climate Anomalies*, Cambridge: Cambridge University Press, 43–71.

ESMAP (1991) *Republic of Yemen: household energy strategy study, Phase I. A preliminary study of the northern governorates*. Report 126/91. 125 pp. Washington DC: UNDP/World Bank Energy Sector Management Assistance Programme (ESMAP).

Floor, W. and Gorse, J. (1988) Household energy issues in West Africa. In World Bank (ed.) *Desertification control and renewable resource management in the Sahelian and Soudanian zones of West Africa*. World Bank Technical Paper 70, Washington DC: World Bank, 10–27.

Gabunshina, E. (1997) Desert reaches Europe. *Our Planet* (UNEP Newsletter), 8.5. [http://www.ourplanet.com/txtversn/85/gabun.html – accessed 22 June 1998]

Goudie, A.S. (ed.) (1990) *Techniques for Desert Reclamation*. Chichester: John Wiley.

Glazovsky, N.F. and Shestakov, A.S. (1996) *Environmental migration caused by desertification in Central Asia and Russia*. LEAD CIS Publications. [http://ntserver.cis.lead.org/infores/publications.htm – accessed 22 June 1998]

Glantz, M.H. (1994) *Drought Follows the Plow*. Cambridge: Cambridge University Press.

Gumilev, L.N. (1992) *Ancient Russia and the Great Steppe*. Moscow: Moscow. [in Russian]

Hepper, F.N. and Wood, J.R.I. (1979) Were there forests in the Yemen? *Proceedings of the Seminar for Arabian Studies* 9, 65–71.

Kiladis, G.N. and Sinha, S.K. (1991) ENSO, monsoon and drought in India. In M.H. Glantz, R.W. Katz and N. Nicholls (eds) *Teleconnections Linking Worldwide Climate Anomalies*, Cambridge: Cambridge University Press, 431–58 .

Lamb, P.J. and Peppler, R.A. (1991) West Africa. In M.H. Glantz, R.W. Katz and N. Nicholls (eds) *Teleconnections Linking Worldwide Climate Anomalies*, Cambridge: Cambridge University Press, 121–89.

Livermann, D.M. (1990) Drought impacts in Mexico: climate, agriculture, technology, and land tenure in Sonora and Puebla. *Annals Assoc. American Geographers* 80(1), 49–72.

Millington, A.C., Mutiso, S.K., Kirby, J. and O'Keefe, P. (1989) Soil erosion – nature undone and the limitations of technology. *Land Rehabilitation and Reclamation* 1, 45–60.

Millington, A.C. and Crosetti, M. (1992) Rapid appraisal of biomass resources – a case study of northern Yemen. *Biomass and Bioenergy* 3(2), 93–104.

Munslow, B., Katarere, Y., Ferf, A. and O'Keefe, P. (1988) *The Fuelwood Trap: A Study of the SADCC region*. London: Earthscan.

Parry, M. (1990) *Climate Change and World Agriculture*. London: Earthscan.

Reij, C. (1991) *Indigenous Soil and Water Conservation in Africa*. IIED Gatekeeper Series, 27. 32pp. London: International Institute for Environment and Development.

Russian Federation, Statistical Bulletin (1993) *Migrations and natural movements of population in the Russian Federation in the first half of 1993*. Moscow: Goscomstat RF. [in Russian]

Sievers, E. and Tsaruk, O. (1997) The convention to combat desertification (CCD): An NGO perspective from Central Asia. *Arid Lands Newsletter 41*. [http://ag.arizona.edu/OALS/ALN/aln41/aln41toc.html – accessed 1 June 1998]

Toulmin, C. (1997) The Desertification Convention. In F. Dodds (ed.) *The Way Forward: Beyond Agenda 21*, London: Earthscan, 55–64.

van Dijk, M., Mercer, D. and Peterson, J. (1983) Australia's drought and the southern climate. *The New Scientist* 7 April, 30–2.

Warren, A. and Khogali, M. (1992) *Assessment of Desertification and Drought in the Sudano-Sahelian Region, 1985-1991*, Paris: United Nations Sahelian Office.

14

Deforestation

Martin Haigh

INTRODUCTION

Deforestation is an environmental problem that threatens the survival of the entire current biosphere. It is counted among the most important environmental crises facing our planet, not least because of its role in reducing biodiversity, increasing global warming and expanding deserts. However, most humans live in degraded forest landscapes, agricultural and urban landscapes. These landscapes, like those of most long-settled areas, both within and outside the tropics, have been claimed from forest. They demonstrate that when the trauma of forest conversion is past, many – but not all – former forest lands may be managed sustainably and productively. Forest conversion is not, inherently, a bad thing for human society or even for the biosphere. Excessive forest conversion is another matter.

DEFORESTATION AS ENVIRONMENTAL CRISIS

The world's forests, especially the tropical forest belt, have been conceived as part of the planet's cooling system (Lovelock 1990). Although less significant than the oceans, (estuaries and) tropical forests are key regulators. The forests remove atmospheric carbon dioxide and transform it into wood, soil, perhaps eventually peat, coal, etc. Forests also pump water into the atmosphere. The moisture produced creates clouds, which cool both the forests and the planet by reflecting solar radiation. The rain from these clouds helps to sustain more forest in areas that would otherwise be too dry (Dickinson 1987).

Deforestation may deprive the Earth of 2.5 Mg $\times$ 10^9 (Gigatonnes) of above-ground biomass each year (World Resources Institute 1994). In the 1980s, tropical deforestation was thought to add 4.6 Mg $\times$ 10^9 of carbon dioxide to the atmosphere, equivalent to more than 90 per cent of US emissions from energy and cement production and perhaps 20 per cent ($\pm$ 10 per cent) of global emissions from fossil fuel (*ibid.*). Fearnside (1997), argues that although 90 per cent of the planets forest biomass remains, continued losses of forest and *cerrado* savannah could dump 275 Mg $\times$ 10^6 of carbon dioxide into the atmosphere through the next century. Deforestation is also reducing the planet's ability to cool itself by removing carbon dioxide from the atmosphere. Using satellite data, Jang *et al.* (1996) estimate the impact of deforestation on the decline of global net primary productivity (NPP). Between 1986 and 1993, approximately 19 per cent (2,600,000 km^2) of the high-NPP regions (>2000 g m^{-2} yr^{-1}), mainly tropical rain forests, were reduced to intermediate-NPP regions (500–1500 g m^{-2} yr^{-1}) mainly savannah and cultivated land (Jang *et al* 1996).

Lovelock (1991) warns that the loss of the tropical forests could contribute to a sudden and dramatic failure of the planet's current system of climatic regulation, regarded as already being close to the margins of its stability. Using analyses based on the mathematics of dynamic systems theory, and images from the liveliest traditions of environmentalism, Lovelock predicts a future of unprecedented and violent environmental fluctuation (Lovelock 1990). This would be followed by an abrupt jump to an equilibrium state, with a very much higher stable planetary temperature. Lovelock (1990) links the destabilising influences of deforestation and accelerated soil erosion together as a planetary disease, 'exfoliation'. The guru's advice to environmentalists is to plant trees and to minimise the release of greenhouse gases to the atmosphere, even if that means using nuclear energy rather than burning fossil hydrocarbons (Lovelock 1990). Indeed, NASA-based Noever and team (1996) calculate that, to preserve its current temperature regime, the planetary system already requires more forest and less desert.

On the smaller landscape scale, where most geographers work, deforestation affects most of the issues that concern geographers. It impacts on both macro- and micro-scale climatic patterns (Reading *et al.* 1995). It leads to dysfunctions in landscape systems, which are caused by the interrelated degradation of its climatic, hydrological, edaphic and biological components. Deforestation allows increased soil erosion, increased landslide activity, sediment pollution, changes in fluvial geomorphology, and changes in the hydrological, biogeochemical and climatic regime (Haigh 1984). Tropical forests are enormously complex and highly stable systems. After deforestation, they are replaced with systems that are much simpler and have much reduced biodiversity and much lower stability (Reading *et al.* 1995). These replacements are, in general, much less efficient in the tasks of self-preservation, they retain and recycle nutrients less efficiently, but they may recover more rapidly from disturbances such as destruction by fire.

Given its significance, current estimates of deforestation are alarming. Of course, these estimates, in common with those for similar global issues – soil erosion, soil degradation, biodiversity loss and the rate of species going extinct – are often suspect. Their range is huge and shifts according to the definition of deforestation applied, the techniques used for estimation and their efficiency, the method employed to convert remotely sensed data into deforestation data, and the geographical areas selected to 'ground-truth', that is calibrate empirically the remotely sensed data (Grainger 1993; Parisi and Glantz 1992; *cf.* Skole and Tucker 1993). This is quite apart from any shift caused by the bias and ambitions of those who would use such data. However, recent years have seen a convergence of estimates (Downton 1995).

Forest, of some description, covers about 40 per cent of the Earth's land surface. The FAO Forest Resource Assessment suggests that the world's forests cover 3454 million ha (1995), a little more than half of which lies in developing nations. The FAO (Food and Agriculture Organisation of the United Nations) definition includes forests with a greater than 10 per cent crown canopy cover in the developing world and 20 per cent cover in the developed world (World Resources Institute 1996). The FAO (1997) also suggests that 15.4 million ha of tropical forest is lost annually. Myers (1993) suggests that the loss to the entire biome is about 2 per cent per year. Murali and Hegde (1997) prefer 1.8 per cent and that the rate is greatest in the smaller nations, especially in Africa. The World Resources Institute (Washington) suggests that the true rate is 0.8 per cent per year. It adds that, between 1960 and 1990, Asia lost nearly a third of its tropical rain forest and Africa and Latin America a sixth each. However, the rate of increase in the area deforested declined everywhere except Latin America, where agricultural extension continued to accelerate (World Resources Institute 1994).

Around 50 per cent of the surviving tropical rain forests are found in the Amazon basin. Skole and Tucker (1993) used Landsat images to determine that, between 1978 and 1988, the deforested area had expanded from 78,000 to 230,000 ha and that threatened or degraded had increased from 208,000 to 588,000 ha. Elsewhere, others offer more dramatic figures. For Venezuela, Centeno (1996) argues that during the 1980s, deforestation affected 6 million ha and that the average rate of deforestation (1.2 per cent per year) was twice that of Brazil and three times that of Peru. Myers (1993) identifies fourteen main deforestation fronts affecting a quarter of all tropical forests and 43 per cent of deforestation losses (Table 14.1).

However, these global statistics do not quantify forest degradation, fragmentation, biodiversity loss, loss of forest vitality or loss of 'naturalness' due to ecological invasions. As pressure builds on the remaining forestland, mature forests are replaced by immature. Processes include accelerated recycling in stressed shifting agriculture systems. Forest edges are nibbled away by local timber harvesting and by agricultural extension. Occasionally, larger tracts are cleared through the 'mining' of a timber concession, or through some official resettlement or infrastructure development venture. Everywhere, forests are broken into patches and made more accessible by road construction. Air photographs from the Himalaya and Amazonia show how forest clearance spreads to either side of new roadways. Globally, the geo-ecological problems caused by forest fragmentation and degradation may be as serious as those of deforestation itself.

However, the balance sheet is not entirely negative. In the five years to 1995, the world's forests lost 65.1 million hectares (m ha) but gained 8.8 m ha. The forests in much of the developed world are gaining ground (Haigh 1999). Even in the developing world, the area of plantations has doubled since 1980 and now reaches 80 m ha (World Resources Institute 1996). In China, afforestation exceeded deforestation. This was thanks to massive afforestation projects, not least the planting of the Great Green Wall, designed to halt the spread of deserts in the north (Lai 1985).

DEFORESTATION AND DEVELOPMENT

Deforestation remains the hallmark of the onset of human civilisation (*cf.* UNCED 1992). It is now, and always has been, the means by which modern human societies 'develop' a landscape. It is the process by which they transform land into the habitat that most favours human society's agricultural and economic processes. To deny the developing world deforestation is also to deny it the right to development.

Deforestation is a negative, a pejorative, term for a process that others describe as 'development'. The term 'deforestation' is applied to a

Table 14.1 Main deforestation fronts.

Deforestation front	*Forest loss (% – 1991 data)*	*Forest area (million ha – 1991 data)*
Southern Mexico	10.0	5.0
Madagascar	10.0	2.0
Thailand (N and NE)	9.6	5.1
Philippines	6.7	4.5
Vietnam	6.6	5.3
Nigeria (E) and Cameroon (SW)	5.3	5.4
Central America	5.1	8.0
Myanmar (E)	4.8	12.5
Malaysia (E)	4.4	9.8
Amazonia (W)	3.0	20.0
Colombian Choco	3.0	5.0
Bolivia (N)	2.6	6.6
Indonesia (Sumatra and E. Kalimantan)	2.3	37.5
Amazonia (S and E)	1.6	61.2

Source: After Myers 1993.

large number of processes. These include the creation of infrastructure such as roads, reservoirs, pipelines, cables, mines, factories and settlements; the replacement of old-growth forest by new-growth secondary forest; and the conversion of forest to plantations, to agricultural land, or to grazing land of various grades. Only in extreme cases do these grades range down to near desert. However, there are many celebrated examples. Until the first two centuries of the Common Era (i.e. AD), the barren loess plateau of China was well covered by grass and trees. Subsequently, it suffered deforestation, induced in part by human activities and in part by climatic change. This result was intense erosion, which turned a rolling tableland into gullied wasteland (Fang and Xie 1994). Similar tales are told in other ravine lands like those of South Asia (Haigh 1984).

However, land destruction is not the most usual consequence, and Europe provides some excellent illustrations (Darby 1956 and *cf.* Linnard 1982). Here, the primeval 'wild-wood' is gone, although a few hectares survive on the borders of Poland. Old-growth forests are common, but less common as one moves into more densely inhabited areas. Now, most woodland is planted or naturally regenerated secondary growth. The land has been turned over to agriculture, grazing, heath, or human settlement and its infrastructures. In some places, former forest land has degraded so far that it is now almost desert, as in parts of the Mediterranean basin (Johnson and Lewis 1995). However, across most of Europe, this process has been buried by history and only fragments remain. Madeira is one. In 1419, J.G. Zarco brought the first colonists to this Portuguese island. Zarco's colonists used fire to clear the native forest. According to tradition, the island's forests burned uncontrollably and it was shrouded in smoke for five or seven years. Shortly after, Madeira's harvests from newly introduced sugar cane, and later vines, grown in ashes from the big fires, began to bring the pioneers prosperity. Today, the island's 460 km² area supports more than 300,000 people, and its deforested landscapes are much admired by tourists. In sum, the phenomenon of the big burn is not new. Madeira's early pall of smoke differs little from the smog that has shrouded Southeast Asia and Latin America in recent years. The point is that in the tropics, the process is current (Box 14.1).

Box 14.1 Fire! Fire!

In 1997 and 1998, forest burning in Southeast Asia reached epidemic proportions and added a new term to the language. 'The Haze' was the name given to the severe air pollution that hung over Indonesia and Malaysia, affecting 70 million people. In Indonesia's East Kalimantan, 300,000 acres of forest burn had left 5000 people suffering from smog-related diseases. In 1997 alone, 'Haze'-related losses due to health impacts, loss of tourist trade, and disruptions of airlines and industry exceeded US $1.4 billion, with the cost to Indonesia alone reaching US $1 billion, 90 per cent attributable to short-term health costs (World Wildlife Fund 1998a).

On 3 April, 1998, the World Wildlife Fund (1998b) reported on the worst fire ever in the Brazilian Amazon, a 400 km long line of fire that destroyed 15 per cent of Roriama state's forest and savannah. However, *The Economist* (30.5.1998) called the fires in Central America yet more damaging, with many fire fronts extending for 20 up to 50 km. Nicaragua, its forest area halved to 40,000 km² since 1960, was losing 500 km² each week to the burn. Near Tegucigalpa, Honduras, 70 per cent of the highland pine forest had burned.

The article adds that the big problem is people, not the El Niño weather: 'big ranchers clearing land for cattle, timber companies taking more trees than allowed, peasant farmers – often displaced to the hills by expanding agribusiness or by war – clearing forest for agriculture.' The message here, as in other areas, is that it is not the traditional peasant farmer who is the driving force for deforestation. Most of it is accomplished, either directly or indirectly, by government policy.

Resources: Guidelines on Fire Management in Tropical Forests are published by the International Tropical Timber Organisation (*http://www.itto.or.jp*), as are guidelines on sustainable forest and biodiversity management. The International Association of Wildland Fires (*http://ltpwww.gsfc.nasa.gov/geowarm/dbtoc/project0099.htm*) maintains a database of 40,000 sources on all aspects of wildland fire management. For a different viewpoint, the Gaia Forest Archive can be found at (*http://forests.org/forests/susforest.html*), and see Dudley *et al.* (1995).

DEFORESTATION – NOTES FROM A FIELD DIARY

Let me take you on a short journey. This trip runs a couple of dozen kilometres from the north coast of Honduras in Central America and, in the process, it runs through the entire deforestation story. It begins at La Ceiba, a bustling coastal city on the narrow coastal plain. From there, the route runs south via the Cangrejal and Cuero river valleys to the new settlements of Toncontin and El Urraco. South of the city, a dirt track leads into the mountains. It runs alongside a wild river in a narrow valley and through a rain forest reserve. The junction between the reserve and the inhabited coastal plain is abrupt, then the forest is tall and dense, interrupted only by occasional intrusions from the huts and boats of 'Ecotourist' enterprises. Further into the hills, these disappear and the forest is unbroken. Eventually, the road runs out of the reserve. A few kilometres further on, the first signs of deforestation appear. Quickly the scene turns to one of total devastation: fallen timbers, burnt trees, muddied slopes, landslides, rampant erosion, scene's from an environmentalist's hell. Further on, the road passes through this deforestation front. The landscape becomes calm and it transforms into something very familiar. Clumps of old and secondary forest remain on the hilltops and steeper slopes. Between, there are villages with well-kept gardens and carefully tended fields. The familiar maize, beans and banana agricultural landscapes of Honduras have begun.

The message is this: forest conversion represents a very dramatic landscape transformation. This transformation affects most of the fundamental topics addressed by geography: climate, ecology, hydrology, landforms, soils, and the habitats and economy of human communities. It also affects subjects more usually studied by other environmental scientists, such as biogeochemical cycling, water quality, biodiversity and the distribution of pathogens. In addition, it involves those major changes in human communities that exercise cultural and socio-economic geographers, anthropologists, and the other social scientists who watch on as the communities of the original forest are transformed by inroads from the communities of deforesters and their followers. However, deforestation is, beyond all argument, an aspect of the process called 'development'.

A

B

C

Plate 14.1 Deforestation in northern Honduras, 1996; (A) before; (B) during; (C) after.

CAUSES OF DEFORESTATION

Development and scarcity

In detail, many reasons are cited as causes for tropical deforestation. Myers (1993) offers population expansion, poverty, national indebtedness, migration, economic development and the expansion of road networks, and political policies that offer perverse subsidies to destructive practices (*cf.* Repetto and Gillis 1988). Tole (1998), analysing data from ninety developing nations, suggests that the twin processes driving deforestation are development and scarcity. Scarcity factors include population growth/density, economic stress, value of extractable forest products, pressures on the land encouraging conversion to pasture or arable land, and *per capita* energy consumption because wood is a traditional fuel (Bawa and Dayanandan 1997). For development, Kaimowitz (1996) identifies the seven processes that resulted in the replacement of forest by pasture in Central America, 1979–94. They were (1) favourable markets for livestock products, (2) subsidised credit, (3) road construction, (4) land tenure policies, (5) policies that reduced timber values, (6) reduced levels of political violence, and (7) improved cattle husbandry.

Rudel and Roper (1997) recast these two major explanations for tropical deforestation as 'immiserisation theory' and 'frontier theory. 'Immiserisation theory' attributes most deforestation to expanding rural populations who, having few economic resources, must clear additional land for agriculture to meet their needs. Reviewing deforestation in the Yaque del Norte, Dominica, Kustudia (1998: 32) writes 'Given their circumstances, farmers have acted rationally, that is they have done what they had to do to survive in the short run.' These people live on the margins of their society and it is that society that has 'fomented a climate in which agriculture has been practised at the expense of the watershed's forests' (*ibid.* : 32). Internationally, deforestation happens because large landowners clear more land for cattle; timber companies mine forest reserves – logging indiscriminately and often illegally; agribusiness and settlements expand: governments support land tenure systems that favour dispossession; and because war happens (Myers 1995). Such factors displace poor farmers, who are forced to the margins and into the forest to eke out a living (Dudley *et al* 1995).

'Frontier theory' links deforestation to agricultural coloniation and the expansion of a developing nation's economy into its wilderness areas, often a process of trial and error. Chomitz

Plate 14.2 Deforested hillsides surround a new hill station, Khasauli, Himachal Pradesh (notice the exposed rocks in the deforested zone – the original forest sustains up to a metre deep layer of soil).

and Gray (1996) model the relationship between new roads, land use, and deforestation in southern Belize, an area experiencing rapid expansion of both subsistence and commercial agriculture. They suggest that road building in areas with agriculturally poor soils and low population densities may be a 'lose–lose' proposition, causing habitat fragmentation and providing low economic returns. However, as the frontier moves, so deforestation accelerates with the expansion of infrastructure, trade, debt and investment in people, and resource-based economic expansion.

Rudel and Roper (1997) continue to suggest that frontier theory best describes deforestation in large forest zones and immiserisation theory that in areas with small fragmented forests. Curiously, they ignore the third major set of explanations, which blame deforestation on global capitalism acting through the agencies of (multinational) corporations, abetted by ineffective and/or corrupt national administrations.

Government

In fact, it may be facile to blame national governments for failing to preserve tropical forests. Simply preserving these forests may not be in their best local interests. According to a World Bank analysis, the within-country economic benefits of protecting tropical forests are often small enough for most governments to ignore (Chomitz and Kumari 1996). Certainly, any economic benefit to be gained from preserving tropical rain forest is highly sensitive to local circumstance and, when the alternative use is agroforestry or plantation crops, the preservation of natural forests may yield no direct domestic benefit. Second, the hydrological benefits of forest preservation versus deforestation are poorly understood and highly variable (Bruijnzeel 1990). Third, estimates of forest value based on non-wood forest products are often faulty and, where domesticated or synthetic substitutes exist (e.g. rubber and kapok), the benefits from non-wood forest products from natural forests tend to zero. Thus, in Ecuador's highlands, where the conversion of forests into pastures is the dominant long-term land-use change, the conversion benefits the country because it increases the area's commercial integration with urban markets. Consequently, public institutions tend to support deforestation for pasture through credit and tenure incentives and the toleration of poor forest administration (Wunder 1996). In sum, from the viewpoint of a national government, the main economic benefits to be derived from tropical forests are gained by exploiting their reserves of timber and the reservoir of undeveloped agricultural soil that lies beneath them.

Shifting cultivation

Governments and other outsiders benefit from the destruction of tropical forests. However, it is often the forest people who get the blame (Myers 1994: 28). Traditionally, the main agricultural system, and main sustainable economic activity of most forest peoples, is 'shifting cultivation' (Dembner 1996). Today, 'shifting cultivators' are counted responsible for up to 50 per cent of tropical deforestation (Angelsen 1995). In fact, 'shifting cultivation' and 'slash-and-burn' are, like 'deforestation', pejorative terms. In this case, they are used to misrepresent a sophisticated and self-sustainable system of agriculture. The reality is that traditional agriculturists have managed forest lands for millennia through the application of an organic, energy-efficient, low-input/high-output, long-rotation forest fallow, agricultural farming system.

However, it is true that many of these systems face increasing pressures, often due to immigration and commercial forest exploitation. Under these conditions, indigenous systems of shifting cultivation cause forest degradation, though rarely full deforestation. Shifting cultivation does not involve sudden or dramatic clearances. It works a forest in patches and achieves deforestation by slow, progressive forest change (Saxena and Nautiyal 1997). What happens is that as pressure on the land increases, the length of the cultivation cycle decreases and the forests that are cut become younger and younger. In parts of the Golden Triangle of Southeast Asia and in Meghalaya, India, the cycle has declined from fifty to around eight years and, arguably, the fallowing process has become less effective (Haigh 1990; 1985). However, the share of responsibility for deforestation given to the forest's traditional cultivators is easily exaggerated (Angelen 1995). Few in number, not usually politically powerful, often belonging to minority nationalities, often on the socio-economic margins of a society, forest peoples make easy scapegoats. It may be convenient for governments to overestimate the impact of shifting cultivation in order to smokescreen other deforestation agencies (*cf.* Brothers 1997; Kustudia 1998).

In truth, most deforestation is externally driven. This is true even within the communities of shifting cultivators. In Colombia, primary montane rain forests are threatened by the illegal cultivation of narcotics such as poppies for opium, morphine and heroin. The area that has already been deforested for this crop may be 50,000 ha, and the deforestation rate associated with opium is around 0.5 per cent per year. In Sumatra, increased rubber planting and expansion into primary forest occurs in response to increased rubber profitability and expected land scarcity (Bulte and van Soest 1996). The local people seek to establish property rights in the face of government land claims and thus become embroiled in a self-reinforcing land race.

The situation leads Bulte and van Soest to argue that encroachment by shifting cultivators may have beneficial effects for the conservation of primary forests. The threat of encroachment acts as a natural brake on the pace at which governments allow concessionaires to open up primary forest areas. Hence, the ecological damages of encroachment are restricted to secondary forest areas. Once again, the example shows how deforestation is driven by government policies and by external commercial elites (Dove 1993).

Population pressure

Another traditional explanation for tropical deforestation is that the forests are being swept back by the rising tide of population growth (*cf.* Myers 1993). Population pressure, the lack of economic opportunity, dispossession, war, famine, landlessness and fuelwood dependency also foster forest encroachment (Tole 1998; *cf.* McNeil 1982). Murali and Hegde (1997) analyse the relative contribution of demographic and economic factors to deforestation, using 1990 data on the total geographical area, forest area, population density, the export of forest products and population growth rate of 141 countries. They agree that population pressure does not correlate with greater levels of deforestation globally, but it may in some of the nations of Africa and Asia. Here too other factors intervene. In Africa, small countries had higher deforestation rates than larger countries, perhaps owing to a proportionately greater need to export natural resources to generate foreign exchange. Forests produce more than 10 per cent of the GDP of eighteen African nations (Department for International Development 1998). By contrast, Europe has high population densities, but preserves low deforestation rates through its ability to import from tropical countries instead of using its own resources (Murali and Hegde 1997).

Fuelwood scarcity

Forests provide cooking fuel for 2000 million people. Several hundred millions rely on forests for other products, ranging from timber to medicine (Department for International Development 1998; Dudley *et al.* 1995). However, the role of fuel scarcity in deforestation is often overstated. Indeed, the topic has a mythological status equal to that of the 'population pressure' and 'slash-and-burn' agriculture theories of deforestation. Certainly, fuelwood and charcoal remain the major energy sources of much of the developing world. However, wood extraction is mainly achieved through forest degradation. It involves lopping and thinning more than wholesale forest clearing (Haigh 1994).

Nevertheless, in Magu District, Tanzania,

Ishengoma *et al.* (1995) found a 5 per cent annual deforestation rate, giving the area's forests a ten-year half-life caused by cutting for charcoal. Further, nearly 90 per cent of the charcoal consumption was local and domestic. In Rwanda, even before the war, there was a fuelwood deficit of around 3×10^6 m^3 yr^{-1}. Subsequently, more than 50,000 ha of forests and plantations have been destroyed. Similar problems have emerged in the former Socialist countries, where economic decline and hardship have forced people to steal trees to provide winter fuel.

By contrast, in Ecuador, fuelwood is a temporary resource, used only in the earliest phases of agricultural frontier expansion (Wunder 1996). Deforestation in highland Ecuador is dominated by agricultural colonisation and cattle ranching, and only rarely is fuel-wood a motivation.

Agricultural extension

Certainly, deforestation is driven by land scarcity and by resource-poor migrants. In Latin America, many people are being driven from their homes by the expansion of the modern economy and forced to find land on the forest margin. Nicholson *et al.* (1995) argue that the root causes are poor economic opportunities among the rural poor and their inability to manage new lands sustainably. Nygren's (1995) study of Costa Rica's Central Valley argues that the 'frontier' farmers, unlike their traditional forest-farming counterparts, do not conceive the forest as a renewable resource and do not believe that forest management is a worthwhile activity within the context of peasant agriculture (*cf.* Bunch 1984). Changing such perceptions provides a major objective for development in this region.

One cause of deforestation is the replacement of peasant farming systems with agribusiness, a process that has forced displaced peasants to clear forests on steep marginal lands (Bunch 1984). At my colleague Jon Hellin's steep-slope research-site in southern Honduras, we try to evaluate the benefits of soil conservation technologies, working with resource-poor local farmers to grow maize (Hellin and Larrea 1997). The test plot looks down across a prosperous melon farm on the fertile plains. Owned by a foreign company, this highly capitalised agribusiness produces cash crops for export, earns foreign exchange for the government and provides seasonal cash income for local pickers and packers. It is also why local farmers cultivate steep slopes. However, agricultural extension often involves steep marginal lands. In India, forests in Uttarakhand's Pranmati Gadh declined from 75 to 65 per cent, while the cultivated area increased from 12.5 to 18.5 per cent. Most of the new arable land was located on slopes of 20–35° (Haigh *et al.* 1998).

Thus the most important direct reason for deforestation is the permanent conversion of forests to agricultural land (Bulte and van Soest 1996). Of Kerala's 256,000 ha of forest cleared between 1964 and 1984, 60 per cent has been converted to agriculture, while most of the remainder has gone for irrigation and hydro-electric works (Haigh *et al.* 1998). In Venezuela, deforestation is primarily due to the expansion of agriculture; just 20 per cent is related to timber extraction (Centeno 1996).

These figures reinforce the popular image of deforestation created by media images of forest burning and clear felling for the creation of arable or grazing land. So too does the case history of deforestation in the Madagascar highlands. This begins around AD 600, when Indonesian settlers moved into the land and began to degrade the forest through long rotation forest fallow agriculture. After AD 1000, the introduction of zebu cattle from Africa encouraged farmers to expand their pastures by burning. By AD 1600, aided by poor regeneration, most of the forest had gone. The consequences included massive erosion, floods, drought, faunal extinction and economic collapse as the land degenerated to impoverished grassland (Gade 1996). Similar illustrations litter the world. In the West, it has been possible to reverse such catastrophes through concerted investment in reforestation, soil conservation and hydrological regulation, as in the Yazoo Highlands of Mississippi (Duffy and Ursic 1991). In the developing world, and through history, when land conversion goes too far or goes wrong, the human society that caused the problem has no recourse. In Sri Lanka and in Central America, there are ancient cities reclaimed by forest.

Active forest suppression

This, at least, demonstrates that forests are resilient systems. Given time, forests regenerate and reclaim their land, even if they are not actively replanted. In many cases, deforested land has to be kept deforested by the regular suppression of regeneration, often through grazing and burning. An outer zone of fire-resistant species characterises many forest islands in Africa's savannah regions. The last patch of native prairie in northern Illinois is preserved because local environmentalists, in defiance of Chicago city ordinances, ensure that the site is burned regularly. The uplands of Wales and much else in Europe are kept deforested by overgrazing sheep. Even in Amazonia, clearance is not a one-way process: for every 3 hectares cleared, perhaps one is reclaimed by forest regeneration (Skole and Tucker 1993).

Western capitalism

Murali and Hegde (1987) are among those who point out that much tropical deforestation is deforestation exported from the developed world. Unwilling to consume their own forest resources, developed nations are happy to use their corporate muscle and economic wealth to plunder the resources of their poorer neighbours in the developing world. The World Resources Institute suggests that, of the 15.4 million ha of tropical forest lost each year to deforestation, 5.9 million ha (38 per cent) is due to logging, and of this 4.9 million ha (83 per cent) comes from the timber mining of primary forest reserves (World Resources Institute 1994). The situation is a far cry from the ambition of the International Timber Trade Organisation's resolution to have all timber extraction derived from sustainable, that is replanted, sources by the year 2000. Meanwhile, environmental organisations throughout the Developed World conduct a largely futile campaign to restrict the import of tropical timbers (Dudley *et al.* 1995). Ecological certification is one strategy. It assumes that consumers will pay a premium for wood from sustainable forest operations. A survey of timber merchants actively involved in US markets for certified wood products, indicates a growing market (Merry and Carter 1997). However, there is little evidence of a green premium being applied to final products, and the main change is that new entrants to the market are purchasing from domestic, not international sources.

The problem is not restricted to timber. Colleague Norman Myers memorably found himself sued by that most litiginous of international hamburger chains on account of a misplaced comment in his exposure of the 'hamburger connection', which links tropical deforestation to the demands of Western consumers for cheap meat. Recently, Nicholson *et al.* (1995) analysed the argument that extensive cattle production causes deforestation in Central America. They assessed the idea that the problem could be reduced by the use of more intensive cattle production methods. However, intensification was found unlikely to affect deforestation rates, because consumer demand for livestock products was not the driving factor. They argue that there is a need to find systems of intensification that respect the logic of existing extensive cattle systems and that balance trade-offs between objectives of producers and policy makers.

Many different factors

In most cases, deforestation is the result of a combination of factors (*cf.* Brown and Pearce 1994). Brothers (1997) presents a useful case study from the microcosm of a single village in Dominica. Here, rapid deforestation was associated with the construction of access roads, agricultural colonisation and pasture conversion. Logging also played an important role in opening up the forest to agricultural settlement. Pasture conversion was not due to the expansion of large ranches by wealthy landowners but a local response to the economic advantages of cattle raising. As always, government actions also strongly influenced deforestation. In this case, the problem was neither colonisation schemes nor economic subsidies for cattle ranching but the monopolistic practices of the Dictator Rafael Trujillo and the chaos following his assassination. Brothers concludes that the national government bears a major responsibility for deforestation and contradicts that government's suggestion that the problems are caused by farmers (*cf.* Kustudia 1998). The continued opposition between the government and the rural population also militates against any remedy to the problem.

ENVIRONMENTAL IMPACTS OF DEFORESTATION

Japan's 'Basic Plan for Forest Resources, 1996' ascribes five main functions to forests: conservation of water resources, environmental conservation, timber production, health/cultural activities and disaster prevention (Price 1998). Trees play a major role in mitigating extreme events: floods, avalanches and landslides (Charoenphong 1991; Duffy and Ursic 1991). When the trees are gone, negative environmental changes may follow. However, the message offered by the section that follows is that not all of the environmental impacts of deforestation need be disastrous and that many can be mitigated by careful environmental management, by good land and water husbandry (Shaxson 1995; Shaxson *et al.* 1989; Pereira 1989).

Landslides

Deforestation has its greatest geomorphological impact in mountain regions. Price (1998) argues that the best way of minimising risks from natural hazards in mountains is to ensure a stable forest cover. In steep lands, where there is rapid uplift and/or stream channel incision, hill slopes evolve to a condition that lies close to their margins of stability. In geotechnical terms, these slopes have a 'factor of safety' close to unity. This means that the forces that preserve the slope and resist failure, such as the strength of the rock, tree roots or soil, are in balance with those that encourage its failure, such as gravity, depth of soil, weight of trees, etc. Where rock and debris slopes lie on the very threshold of failure, they exhibit behaviour associated with 'self-organised criticality' (Bak 1997; Noever 1993; Haigh *et al.* 1988). Vulnerable to very small disturbances, these slopes fail unpredictably in individual cases, but predictably in statistical terms. Their vulnerability may be predicted as a characteristic log linear association between landslide volume and landslide frequency (Haigh 1988).

Trees are increasingly employed by bioengineers for the stabilisation of steep slopes (Barker 1984). The reason is that tree roots have a tensile strength that can be equivalent to, or greater than, steel. In forest environments, tree roots provide a significant component in the capacity of a hillside to resist failure. Tropical forest data are not to hand, but research in Alaska's declining yellow cedar forests suggests that tree roots can be as important a control on landsliding as pore water pressure on shallow soils (Johnson 1993). Here, the application of simple map overlay techniques demonstrated that steep slopes suffering cedar decline have three times as many landslides as steep slopes in adjacent forests (Johnson and Wilcock 1998). These authors investigated the possibility that increased landslide activity was due to increased soil saturation following the reduced transpiration of the dying cedars but, ultimately, rejected this possibility for the more conventional explanation of loss of root strength. Where trees had been dead for 14–51 years, 70–90 per cent of roots with diameters from 1–30 mm had decayed. This reduced soil cohesion by up to 80 per cent (4.6 to 0.9 kPa). The effect of root decay on the factor of safety of the slope was as great as that of changes in pore water pressure, and greater on slopes with shallow soils.

The bulk of the landslide activity that follows deforestation is the generation of shallow slumps of the deep soils and debris covers that forests preserve on even quite steep slopes. The process is affected by slope angle. Studies at Taranaki, New

Zealand, find that 28° is the threshold for post-deforestation landslides, and most occur on slopes steeper than 32°. Average surface lowering was 0.2 m in 10 years. However, in many situations, the relationship between such environmental controls, forest cover and landslide activity is difficult to prove. The author's studies of the impacts of new hill roads in the Himalaya found that forested slopes were more prone to landslide activity than those where the forest was replaced. The reason was that, in this environment, forest survived only on those slopes that proved too unstable for other development (Haigh *et al.* 1995).

Channel response

Landslide sediments often have an important impact on stream channels in mountain regions. Right across the Himalaya and its Siwalik fringe, deforestation and development are associated with changing the dominant processes in stream channels from incision to sedimentation. Affected streams may be converted to bedload channels. They tend to become wider (sometimes braided), more shallow and less sinuous. Frequently, surface flows decrease and become restricted to flood conditions, and the water flows through the channel bed's fill. In the Himalayan fringes, there have been dramatic increases in stream width, and it is likely, but not yet proven, that this has led to further undercutting and destabilisation of hill slopes (Froelich and Starkel 1993; Haigh 1994). Elsewhere, in situations where the sediment supply to the rivers is not much increased, the increased runoff, especially the increased severity of the mean annual flood, that accompanies deforestation, coupled with reduced vegetative protection and increased compaction of the soil surface, may lead to exactly the reverse effect. In semi-arid environments, deforestation often leads to stream channel trenching, decreases in the width/depth ratio, and increased channel sinuosity. The impact of the affected channels depends upon the balance and the character of the local change in the supply of water and sediment.

Hydrological balance

It is widely accepted that deforestation results in increased flood hazard. However, the case is more easily argued than proved. The hydrological impacts of deforestation are diverse and, to date, do not register on the record from the world's largest forest basin, the Amazon (Marengo 1995). The role of deforestation in causing the increasing flood problems of north India is debated (Haigh 1994).

However, the human impact, including deforestation, is thought to have an important impact on sediment yields from the major rivers of South and East Asia. It also has an impact on the flow of smaller streams (Bruijnzeel 1990). Theory suggests that deforestation leads to increased annual water yields and greater groundwater recharge and greater dry season flow. Certainly, it is usually associated with increased flood discharge, if only because the volume of water lost to evapotranspiration, commonly 40 per cent of that received in precipitation, remains on the ground. Nevertheless, in the seasonally dry tropics, deforestation is often linked to dry springs and streams (Bartarya 1991; Valdiya 1998). Sandstrom (1995) attempts to show that the hydrological consequences of deforestation vary with land management, hydroclimate and landscape characteristics. Others believe that the key is the depth of the soil (Haigh 1994). Forests create deep open-textured soils, with a large water-holding capacity, even on steep slopes. When the trees are cleared, this soil becomes reduced and compacted. More rainwater is converted to runoff and near-surface flow and less enters the soil or groundwaters (*cf.* Shibano 1998). Where soils are deep, through flow is the dominant supplier of storm flow. It is unusual for rainfall to exceed the infiltration capacity of forest soils, except where there has been compaction, but saturated overflow is not unusual.

Forests return a great deal of the rainfall they receive to the atmosphere. Annual evapotranspiration in tropical moist lowland forests ranges up to 1500 mm y^{-1}, with transpiration accounting for a maximum of 1045 mm y^{-1}

(Bruijnzeel 1990). In most environments, these figures are very much lower. Hence, when forests are removed, a great deal more water remains in the environment than before, and the total discharge of affected streams tends to increase. In the tropics, increases of water yield equivalent to between 110 and 825 mm y^{-1} are reported in the year following clearance. After reviewing 145 studies, Sahin and Hall (1996) estimate for southwest Australia that, given a 10 per cent reduction in cover, removing deciduous hardwoods would increased annual yield by 17–19 mm yr^{-1}; removing conifers, 20–25 mm yr^{-1}; but eucalypts, only 6 mm yr^{-1}. Afforestation by scrub would effect a 5 mm yr^{-1} decrease in yield.

In Nigeria, Lal (1997) reports results from a seven-year study of a secondary tropical rain forest watershed (44.3 ha) that yielded 2.2 to 3.1 per cent of its annual rainfall as runoff. Deforestation of 7 per cent of the watershed area increased water yield to 7.0 per cent of annual rainfall. Total water yield following deforestation and conversion to agricultural land use rose from 9.6 to 21.3 per cent. Base flow increased after deforestation by 5–18 per cent of annual rainfall, while peak flow decreased by 2.3–6.2 per cent. Previously, dry season flows decreased as the dry season progressed, but in the years following deforestation it increased. Forest lysimeters showed higher seepage losses than those in cropland.

Back in Western Australia, Bari *et al.* (1996) used paired experimental catchments, with a six-year calibration, to examine the effect of logging. This resulted in an increase in groundwater levels. Deep, permanent groundwater levels rose for four years to a maximum of 5 m and then began to decline. Springs that flowed two–three months each year before logging flowed five–six months afterwards. Stream flows increased dramatically one year after logging due to decreased interception and evapotranspiration, increased recharge, decreased soil moisture deficit and consequently an increase in through flow. The increase in baseflow was twice that of quick flow.

Water chemistry

Deforestation affects many aspects of water quality (Anderson and Spencer 1991). These include its load (suspended, dissolved and bed), its turbidity and organic matter content, its thermal and hydrological regime, and its ecology – often shifting the ecosystem from heterotrophy to autotrophy. Naturally, these impacts affect the water chemistry. Deforestation is traditionally linked to a decline in dissolved organic matter and a flush of released plant nutrients (*ibid.*: 15). An attempt to apply the MAGIC model to a central Amazonian catchment indicated that, in a thirty-year period, deforestation would lead to the release of SO_4^{2-}, inducing acidification and a general increase in ionic concentration (Forti *et al.* 1995). In the longer term, the stream water ionic concentrations might be expected to reach new equilibria at levels higher than before deforestation and leading to eventual acidification.

When Nigerian lowland alfisols are converted to agriculture, their soil chemical quality and pH decreases cultivation time (Lal 1996). At 0–5 cm, the rate of decrease at 0–5 cm was 0.23 pH units per year, 0.05 per cent per year for SOC, 0.012 per cent per year for total nitrogen, 0.49 cmol kg yr^{-1} for Ca^{2+}, 0.03 cmol kg yr^{-1} for Mg^{2+}, 0.018 cmol kg yr^{-1} for K^{+}, and 0.48 cmol kg yr^{-1} for CEC. At 5–10 cm depth, while Mn^{3+} concentration and phosphate increased due to fertilisation; there was also an increase in total acidity.

However, even acid rain does not cause acidification of streamwaters in Japan's mountain forests. They seem to be immune and only recently has some explanation for this begun to emerge (Box 14.2).

Soil quality and erosion

Deforestation exposes vulnerable tropical soils to erosion and fosters their degradation through the loss of soil nutrients and reduced soil vitality (Theng 1991). Logging may have especially dramatic impacts. From Malaysia, Lal *et al.* (1985) report cases of erosion climbing to 13 Mg ha^{-1} $month^{-1}$ from almost nothing, and runoff climbing to 190 m^3 ha^{-1} $month^{-1}$ from 2 m^3 ha^{-1} $month^{-1}$

Box 14.2 No acidification in Asia's mountains

Why is it that Asia's tropical forest catchments suffer acid rain, but not stream water acidification or acid-induced deforestation like that experienced in the West? It is well established that trees are effective scavengers of acidic atmospheric compounds and that this process increases acid deposition in forest catchments to levels several times those experienced in open fields (Neal *et al.* 1992). However, despite receiving inputs similar to basins in industrial Europe and in defiance of years of careful monitoring, Japan's forest basins do not show signs of acidification. Review of the international literature suggests that they are not alone. Similar findings come from other mountain basins in Asia (Ohte *et al.* 1998). Indeed, there seems to be a geographical regionalisation of the way catchments respond to acid deposition. In Asian experimental basins, rainwater of pH 4 to 6 is converted to stream water of pH 6 to 7. In North American and European (except Mediterranean) basins, rainwater of pH 4 to 5 is converted to stream water of pH 4 to 6 (*ibid.*).

Investigations at Kirya watershed, Japan, suggest that its hydrochemical processes are controlled by two factors: a biological factor and a geochemical factor (*cf.* Finley and Drever 1997). Biological reactions in the upper soil increase the pH from 5.5 to 6.0 during the unsaturated vertical infiltration of throughfall and temporary saturated lateral infiltration from the hill slope. Two mechanisms are active: cation (Ca^{2+} and Mg^{2+}) exchange with organic acids near the soil surface, and proton consumption by weathering reactions, using acid supplied by the CO_2 dissolution–dissociation reaction. Thus the acidity, gained from throughfall and from the production of organic acids in the soil, is mitigated by exchanges in the soil and during infiltration to groundwater.

The balance between these processes is affected by forest succession (Asano *et al.* 1998). Study of three steeply sloping (average 20–34°) basins on granite bedrock confirms that the major hydrochemical controls are differences in the interactions between biological and geochemical factors. Biological NO^{3-} is the major source of H^+ under the mature forest, geochemical SO_4^{2-} under the younger forest, while pH changes little during infiltration in the deforested control basin. There was an imbalance between cation supply and demand in the younger forest basin, where the trees are active in the acceleration of geochemical weathering.

However, neither tree growth nor geological factors can explain the differences between these basins and those of Europe. The Japanese researchers suggest that the cause lies in the relative immaturity of the soils in the Japanese and other Asian mountain headwaters. Ohte *et al.* (1998) point out that as forest soils develop, the cation-leached layer becomes thicker and the unleached layer with high buffering capacity becomes thinner. Typically, Europe's podsolic soils generate acid from deep surface organic accumulations, and infiltration proceeds through a soil horizon that is entirely leached. Asian forest catchments on steep hillsides preserve immature soil profiles through soil refreshment by erosion and deposition. In Asia's steep lands, erosion and landsliding often bury or eliminate litter accumulations. In addition, in many young forest soils, the infiltration layer includes a rapidly developing C-horizon of weathered but not yet much leached bedrock. The final pH of the infiltrated water depends on the relative effectiveness of the acid-producing litter layer and the unleached C-horizon. The differences in buffering capacity between the Western and Asian case studies may thus be related to local slope steepness, slope instability and the age of the forest soil.

on newly constructed log skid roads. However, soil and runoff losses from new agricultural lands may also be high. In Sri Lanka, where agricultural extension reduced forest cover from 70 per cent to 20 per cent between 1900 and 1995, there was an increase in the rainfall/runoff ratio estimated as 0.7–1.4 per cent per year. In areas cultivated to tobacco and upland annual crops, soil loss became 25–70 times that in forest (Haigh *et al.* 1998). Deforestation is generally associated with dramatic increases in soil loss and sediment yield (Derose *et al.* 1993).

A review of about eighty studies of surface erosion in natural forest and tree-crop systems finds median soil loss in natural forests to run at about 0.3 (range 0.03–6.2) Mg ha^{-1} yr^{-1} (Wiersum 1984). Plantations and tree crops with ground cover/mulch suffer losses of about 0.6–0.8 (range 0.02–6.2) Mg ha^{-1} yr^{-1}. A small number of studies of the cropping phase under long-rotation forest fallow suggest soil losses of around 2.8 (range 0.4–70) Mg ha^{-1} yr^{-1}. In the western Himalaya, comparison of bedload sediments trapped from parallel streams draining four, steep 1 km^2 micro-catchments found that the sediment loads from undisturbed forest were five–seven times smaller than from deforested areas covered by grass and scrub. The depths of soils on the deforested areas were significantly smaller, and there were larger patches of exposed bedrock (Haigh *et al.* 1998; Harding and Ford 1993).

Conversion to arable cropping often accelerates erosion. On poorly managed agricultural steep lands and sites recently cleared of ground cover, huge soil losses can be reported. Bruijnzeel (1990) reports figures from tens to hundreds

Mg ha^{-1} yr^{-1}, including a maximum of around 500 Mg ha^{-1} yr^{-1} for fields planted to onions, tilled up- and down-slope in Java. Soil conservationists consider recently up- and down-slope filled bare fallow to be the most erodible condition for an arable field. The losses from the same land fallowed under forest may be anything up to three orders of magnitude smaller, depending on the slope angle, soil type, climatic regime and scale of the study.

One of the reasons that soil losses increase after deforestation is that less water soaks into the soil. This is often due to compaction, which leads to higher bulk soil densities and reduced infiltration rates. In the arid Zagros Mountains, Iran, conversion of a coarse silty calcixerollic xerochrept soil from oak (*Quercus brontii*) forest to field crops resulted, after twenty years, in almost a 20 per cent increase in bulk soil density, a 50 per cent decrease in organic matter and total nitrogen content, and a 10–15 per cent decrease in soluble ions compared with soil under undisturbed forest (Hajabbasi *et al.* 1997). In humid Nigeria, studies near Ibadan found that deforestation and cultivation increased bulk soil density and penetration resistance but decreased mean weight diameter of aggregates (Lal 1996). The infiltration rate declined with deforestation and time under cultivation. Soon after deforestation, saturated hydraulic conductivity and equilibrium infiltration rate in cleared and cultivated land declined to only 20–30 per cent of that under forest (*ibid.*).

In Brazil's Rio Grande de Sul, conversion of 90 per cent of the forest allowed soil losses to climb to a regional level of 7 Mg ha^{-1} yr^{-1} and was followed by the emergence of surface flow pathways (Mendiondo *et al.* 1998; Castro *et al.* 1997). Natural fallowing resulted in a ten-fold increase in infiltration rate within five years. Well-managed agricultural lands had soil losses comparable with those under natural forest. On steep slopes, the effect may be more dramatic: cleared sites in the Western Ghats, India, shed 120 Mg ha^{-1} in a single season. However, erosion rates decline rapidly subsequently. In the Ghats, erosion from sites cultivated to pepper were less than 3.5 Mg ha^{-1} yr^{-1}. In Rio Grande de Sul, after conversion to no-till agriculture, soil losses declined to levels too small to record (Mendiondo *et al.* 1998).

A similar message comes from studies of lands converted to pasture. De Moraes *et al.* (1996) examined the consequences of forest conversion to pasture on soils in the southwestern part of the Brazilian Amazon basin. After pasture installation, soil bulk densities were higher in the 0–5 cm soil layer, with small changes detected in deeper layers. Soil pH increased, while total soil carbon (0 to 30 cm depth) was 17 to 20 per cent higher in 20-year-old pastures than in original forest sites. However, another Rondonian study comparing soil inorganic N concentrations, net mineralisation and nitrification rates in forests and pastures (>3 years old) on ultisols/oxisols found that forest soils had higher extractable NO_3–N and total inorganic N concentrations than pasture soils. Rates of net N mineralisation and net nitrification were higher in the forest soils (Neill *et al.* 1997; Anderson and Spencer 1991: 40). Nitrogen and nitrification rates remained the same across the six forest plots tested, suggesting that the controls are similar. The low mineralisation and nitrification rates in pasture soils suggest that annual nitrogen losses from deforested landscapes may be lower than from the original forest (Neill *et al.* 1997). In the Barbudal Reserve, Costa Rica, tests have confirmed that soil C, N and K were lower, while many base cations and micro-nutrients were higher in grassland plots than in forest plots (Johnson and Wedin 1997).

The removal of tree biomass by logging removes on averages 30 and 15 Mg ha^{-1} of carbon from seasonally moist and seasonally dry forests, respectively. The rotting of roots may add 50 Mg ha^{-1} (Anderson and Spencer 1991). The amount of C that is lost to the system varies with many environmental factors as well as with the way the land is cleared and used subsequently. In soils under grazed pastures converted from forest 18–25 years previously were investigated in Costa Rica. The net loss of C was 2180 g m^2 for a hapludand and 150 g m^2 for the humitropept soil (van Dam *et al.* 1997).

Gaston *et al.* (1998) produced a GIS-based

macro-scale study of changes in ecosystem carbon pools caused by land conversion in Africa's tropical forests. They estimate that the above-ground forest biomass accounts for 75 per cent of the total carbon, below-ground forest biomass for 21 per cent, and grass/shrub savannahs for 4 per cent. Mean biomass C densities are reported as 180 Mg ha^{-1} for lowland moist forests, 82 Mg ha^{-1} for all forests, and 6 Mg ha^{-1} for grass savannahs. Forest conversion, 1980–90, caused a 13 per cent decrease in the above-ground forest carbon pool, including 5.6 per cent due to deforestation and 7.4 per cent to biomass reduction by other human activities (*ibid.*). In Brazil, deforestation and burning has increased carbon monoxide and ozone concentrations in the lower atmosphere (Kirchoff 1996).

The microbial coenoses of tropical forest soils remain largely unexplored. A recent Amazonian study described 100 sequences of genes, ninety-eight of which were bacterial and two archaean. No duplicate sequences were found, and none of the sequences had been previously described (Bornemann and Triplett 1997). Eighteen per cent of the bacterial sequences could not be classified in any known bacterial family. There were significant microbial population differences between a mature forest soil and an adjacent pasture soil.

Vesicular arbuscular mycorrhiza are endophytic fungal symbionts that aid plant growth by increasing the uptake of soil nutrients. Johnson and Wedin (1997) examined mycorrhizae during conversion of dry tropical forest to grassland in Costa Rica. They found that while the beta diversity of mycorrhizal spore communities was lower in the grassland plots than in the forest plots, total spore density and alpha diversity of mycorrhizal spore communities were unaffected by conversion to pasture or by subsequent burning. These results suggest that forest regeneration would not be constrained by any lack of mycorrhizal symbionts. Johnson and Wedin suggest that the grasslands are sustainable, alternative stable states for these former forest areas. Positive feedback between the grassland vegetation, fire and nutrient cycling systems reinforce this condition.

BIOGEOGRAPHICAL IMPACTS OF DEFORESTATION

It is well known that tropical rain forests are large and complex ecosystems. They rank among the world's greatest reserves for biodiversity and offer huge pools of potentially useful species and genes (Reading *et al.* 1995: 151–5; Wilson 1992). Environmentalists argue that we may be losing 50–200 species each day, and that current rates of extinction are five orders or magnitude above the geological norm (Myers 1995: 179 *et seq.*). Myers continues to suggest that the biodiversity of the Earth could be halved by the middle of the next century and that, if this happens, tropical deforestation will have been the reason. Naturally, these numbers are contested.

However, loss of biodiversity is not the only biogeographical impact. Biogeographical changes also result from ecological invasions and habitat fragmentation. For example, in the Himalaya, mined areas, roadsides and degraded vegetated areas have provided avenues for the invasions of exotic species such as *Celosia argentea*, *Lantana camara*, and *Eupatorium glandulosum* (Rajwar 1998). Once established, *Lantana* spreads into shrub land and forest, and *Eupatorium* into montane grassland. Their competitive advantage is secured by high primary productivity and by non-palatability to grazing animals (*ibid.*). Once established in disturbed sites, these exotics often spread into less disturbed areas. Many more dramatic tales come from tropical Oceania and Australasia, where ecological invasions by species like the cane toads in Queensland and mynah birds on Rarotonga have had major impacts on the numbers and range of indigenous species.

Deforestation can also influence the epidemiology of disease. For example, when coffee prices doubled in 1986, this prompted large-scale deforestation for coffee plantations in southern Thailand. The increased area of standing water favoured *Anopheles minimus*, a highly efficient vector of malaria. The new plantations also attracted migrants from endemic and non-endemic areas. In 1986–7, an epidemic wreaked havoc among the non-immune migrants (Kondrashin 1991).

Walsh *et al.* (1993) review the relationship between deforestation and several other vector-borne diseases apart from malaria, including the arboviruses; Chagas' disease, leishmaniasis, loiasis, lymphatic filariasis, onchocerciasis, and schistosomiasis.

REFORESTATION

A huge volume of work is published on the problems and processes of forest reconstruction (Lamb *et al.* 1995). The environmental impacts of reforestation are not the reverse of those due to deforestation. Many of the impacts of deforestation occur because of the loss of the forest soil and litter layers. New forests mine the environment for nutrients. They accelerate weathering and fracture rocks for anchorage. New forests may have very different environmental contexts from the forests they replace. Schreier *et al.* (1998) had to resort to GIS modelling to deconstruct Nepal's record in afforestation. Among other things, in Nepal, new plantations tend to be located on relatively gentle slopes, while deforested areas are more often on erodible steep lands (*ibid.*). In fact, the results from research into the impacts of reforestation are full of surprises. Awaiting publication are findings showing that on steep grazed pastures, there is a positive correlation between trees and erosion, because animals congregate beneath trees, and trees planted on steep banks may reduce erosion by three-quarters, even in the absence of forest soil and litter layers. Finally, results from Japan demonstrate that reforestation in mountains can lead to long-term increases in streamflow, despite the increased losses to evapotranspiration – perhaps due to rain harvesting or more likely due to changes in deep seepage to groundwater (Shibano 1998).

CONCLUSION

The causes of deforestation are complex. The research literature is gigantic, complex, multidisciplinary, scattered and often hard to evaluate. However, the main cause of deforestation remains clear. Deforestation is a byproduct of development. Deforestation's foot soldiers may be the rural poor who, lured by promises of economic gain or driven by scarcity, clear more and more forest land. Their lieutenants may be national administrations eager to expand their national economies and win foreign exchange. However, neither the rural poor nor their struggling national governments are capable of preventing deforestation. As the World Bank analysts prove, their economic best interests usually lie elsewhere (Chomitz and Kumari 1996). Deforestation might be halted by concerted international action although, at present, despite the noise of the environmentalists, the political will needed to effect change simply does not exist. However, international action cannot halt the most important cause of deforestation. The developing world must develop. Deforestation is the hallmark of human civilisation. It marks the conversion of a traditional economy to a modern agricultural and urban economy. It has affected everywhere that civilised society has taken root. The main factor that divides the progress of deforestation in the developing world from that in the developed world is that, in the latter, deforestation has already taken place.

Certainly, deforestation is also an agency of environmental degradation, and it can be the nemesis of development. On the loess plateau of China, deforestation converted an economically and culturally advanced society into an impoverished and backward one (Fang and Xie 1994). Similar tales are told for places as diverse as North Africa, Central America and Easter Island. However, deforestation continues, and will continue, because it is in the immediate rational self-interest of those who destroy forests to do so. Preventing deforestation will require a major change in cultural values, social attitudes and, most especially, the economic rules of play. This is possible. In northern Europe and China, the rate of forest increase exceeds extraction, albeit for different reasons. Meanwhile, the developed world conspires towards the deforestation of the underdeveloped. It still deploys more economic muscle in favour of those who deforest than towards those who manage their forests sustainably. Deforestation is driven by international trade, by government policies, by regional priorities and local economics.

Still, the situation is not wholly bad. True, deforestation affects many aspects of the environment adversely. Forests, especially tropical forests, are major and largely unexplored reserves of biodiversity. Deforestation effects a transformation of the ecology, soils and waters of affected lands, which is usually negative. It has major impacts on climatic and biogeochemical processes at various scales from micro- to macro-, which are mainly damaging. Large tropical forests, especially the Amazon rain forest, play a key role in the geophysiological regulation of the Earth's atmosphere and climate. Forests are major sinks of CO_2 and forests play an important role in the moderation of global warming (Lovelock 1991). However, at the local scale, deforestation is not necessarily a disaster. In many cases, it is possible to convert forest land to new, sustainable and more economically productive land uses. Even where this fails, then, if tree growth is not actively prevented then, in most cases, the forests, especially the tropical forests, can and will regenerate.

GUIDE TO FURTHER READING

Bruijnzeel, L.A. (1990)*Hydrology of Moist Tropical Forests and Effects of Conversion: A State of Knowledge Review.* Paris: UNESCO/IHP: 224 pp. An impartial overview of the state of the art in tropical forest hydrological research.

Haigh, M.J. (1994) *Deforestation in the Himalaya.* In Roberts, N. (ed.) *The Changing Global Environment. Oxford: Blackwell*, 440–62. A review of the three main viewpoints on the causes, degree, physical consequences and management implications of deforestation in the Himalaya.

Haigh, M.J. and Krecek, J. (eds) (1991) Special feature on headwater management. *Land Use Policy* 8(3): 171–205. Three key case studies of the impacts of deforestation in headwater regions including field evidence of the drying up of springs in the Himalaya, the conseqences of over-zealous timber extraction in Thailand, and a classic case of environmental reconstruction after deforestation from the Yazoo Highlands of Mississippi.

Lovelock, J.E. (1991) *Gaia: The Practical Science of Planetary Medicine. Stroud: Gaia Books.* This is an introductory overview of the current state of the planetary system, including the role of the tropical forests. Written by one of the greatest and most progressive scientific thinkers of our time, this book also sketches out a new way of conceiving and interacting with Planet Earth.

Murali, K.S. and Hegde, R. (1997). Patterns of tropical deforestation. *Journal of Tropical Forest Science* 9, 465–76. An overview of tropical deforestation written by and for the foresters in the tropical nations.

Myers, N. (1993) Tropical forests – the main deforestation fronts. *Environmental Conservation* 20(1), 9–16. A sample broadside from one of the great environmentalist campaigners, author of 'The Sinking Ark' (Oxford: Pergamon, 1980) who has carved his niche in the world by acting as spokesperson for the world's guilty conscience and by shouting out truths that others might rather not hear.

These days, it is easier to access a web site than find a book. The United Nations system and some of the larger environmental organisations have excellent web resources that offer an up-date of both the international actions and assessments of the deforestation problems in the years after Rio and its Agenda 21. Here follow some key sources:

UNCED (1992) *Report of the United Nations Conference on Environment and Development* (Rio de Janeiro, 3-14 June 1992). Chapter 11: Combating Deforestation. New York: United Nations a/Conf. 151/26 (Vol. II) 14 pp. [gopher.//gopher.un.org:70/00/conf/unced/English/a21_11.txt]

FAO (Food and Agriculture Organisation of the United Nations) (1997) *Agenda 21 Progress Report.* Sustainable Development Dimensions/Environmental Policy, Planning and Management Special, Chapters 10: Land Resources, 11: Deforestation, 14: Sustainable Agriculture. [http://www.fao.org/sd/Epdirect/Epre033htm]

World Resources Institute (1996) Forests and land cover. In *World Resources 1996-1997: Guide to the Global Environment*: 9. Washington: World Resources Institute http://www.igc.org/wri/wr-96-97/lc_txt2.html]

REFERENCES

Anderson, J.M. and Spencer. T. (1991) *Carbon, nutrient and water balances of tropical rainforest ecosystems subject to disturbance – management implications and research proposals.* MAB Digest 7, Paris: UNESCO.

Angelen, A. (1995) Shifting cultivation and 'deforestation': a study from Indonesia. *World Development* 23(10), 1713–29.

Asano, Y., Ohte, N., Katsuyama, M. and Kobashi, S. (1998) Changes of hydrochemical formation processes by forest succession stages. In: Haigh, M.J., Krecek, J., Rajwar. G.S. and M.P Kilmartin (eds) *Headwaters: Hydrology and Soil Conservation.* Rotterdam: A.A. Balkema, 85–97.

Bak, P. (1997) *How Nature Works: The Science of Self-ordered Criticality.* Oxford.

Bari, M.A., Smith, N., Ruprecht, J.K. and Boyd, B.W., (1996) Changes in streamflow components following logging and regeneration in the southern forest of Western Australia. *Hydrological Processes* 10, 447–61.

Barker, D. (ed.) (1984) *Vegetation and Slopes: Stabilisation, Protection and Ecology.* London: Institution of Civil Engineers.

Bartarya, S.K. (1991) Watershed management strategies in Central Himalaya: the Gaula river basin, Kumaun, India. *Land Use Policy* 8(3), 177–84.

Bawa, K.S. and Dayanandan, S. (1997) Socioeconomic factors and tropical deforestation. *Nature* 386: 562–3.

Bornemann, J. and Triplett, E.W. (1997) Molecular microbial diversity in soils from eastern Amazonia: evidence for unusual microorganisms and microbial population shifts associated with deforestation. *Applied and Environmental Microbiology* 63(7), 2647–53.

Brothers, T.S. (1997) Deforestation in the Dominica Republic: a village-level view. *Environmental Conservation* 24(3), 213–23.

Brown, K. and Pearce, D.W. (eds) (1994) *The Causes of Tropical Deforestation.* London: University College of London Press.

Bulte, E., and van Soest, D. (1996) Tropical deforestation, timber concessions, and slash-and-burn agriculture – why encroachment may promote conservation of primary forests. *Journal of Forest Economics* 2(1), 55–66.

Bunch, R. (1984) *Two Ears of Corn: A Guide to People Centred Agricultural Improvement.* Oklahoma City: World Neighbors.

Castro, N.M.R., Auzet, A.V., Bordas, M.P., Chevallier, P., Leprun, J.-C. and Mietton, M. (1997) Ecoulement et transfer des sediments dans les bassins versants de grande culture sure basalte due Rio Grande do Sul (Bresil). *International Association of Hydrological Sciences, Fifth Scientific Assembly (Rabat, Morocco)* S6, 65–73.

Centeno, J.C. (1996) Deforestation – out of control in Venezuela. *Global Biodiversity* 5, 4–8.

Charoenphong, S. (1991) Environmental calamity in southern Thailand's headwaters: causes and remedies. *Land Use Policy* 8(3), 185–88.

Chomitz, K.M and Gray, D. (1996) Roads, land use, and deforestation: a spatial model applied to Belize. *World Bank Economic Review* 10, 487–512.

Chomitz, K.M. and Kumari, K. (1996) *The domestic benefits of tropical forests: a critical review emphasizing hydrological functions.* World Bank Policy Research Working Paper WPS1601, 1–41.

Darby, H.C. (1956) The clearing of the woodland in Europe. In Thomas, W.L. (ed.) *Man's Role in Changing the Face of the Earth I.* Chicago: University of Chicago Press, 183–216.

Dembner, S.A. (ed.) (1996) Forest-dependent people. *Unasylva* 47(3), 2–59.

De Moraes, J.F.L., Volkoff. B., Cerri, C.C., and M. Bernoux, M. (1996) Soil properties under Amazon forest and changes due to pasture installation in Rondonia, Brazil. *Geoderma* 70(1), 63–81.

Derose, R.C., Trustrum, N.A. and Blaschke, P.M. (1993) Post-deforestation soil loss from steepland hillslopes in Taranaki, New Zealand. *Earth Surface Processes and Landforms* 18(2), 131–44.

Department for International Development (1998) *Forests Matter: The DFID Approach to Forests.* London: Department for International Development.

Dickinson, R.E. (ed.) (1987) *The Geophysiology of Amazonia.* New York: Doubleday.

Downton, M.W. (1995) Measurement of tropical deforestation: development of the methods. *Environmental Conservation* 22(3), 229–40.

Dove, M.R. (1993) A revisionist view of tropical deforestation and development. *Environmental Conservation* 20(1), 17–25.

Dudley, N., Jeanrenaud, J.P. and Sullivan, F. (1995) *Bad Harvest? The Timber Trade and the Degradation of the World's Forests.* London: Earthscan.

Duffy, P.J. and Ursic, S. (1991) Land rehabilitation success in the Yazoo Basin, USA. *Land Use Policy* 8(3), 196–205.

Fang, J.Q. and Xie, Z.R. (1994) Deforestation in pre-industrial China: using the loess plateau as an example. *Chemosphere* 29(5), 983–99.

Fearnside, P.M. (1997) Greenhouse gases from deforestation in Brazilian Amazonia: net committed emissions. *Climatic Change* 35(3), 321–60.

Finley, J.B. and Drever, J.I. (1997). Chemical mass

balance and rates of mineral weathering in a high elevation catchment, Wyoming. *Hydrological Processes* 11, 745–64.

Forti, M.C., Neal, C. and Jenkins, A. (1995) Modelling perspective of the deforestation impact in stream-water quality of small preserved forested areas in the Amazonian rain forest. *Water, Air and Soil Pollution* 79, 325–37.

Froelich, W. and Starkel, L. (1993) The effects of deforestation on slope and channel evolution in the tectonically active Darjeeling Himalaya. *Earth Surface Processes and Landforms* 18(3), 285–90.

Gade, D.W. (1996) Deforestation and its effects in highland Madagascar. *Mountain Research and Development*. 16(2), 101–16;

Gaston, G., Brown, S., Lorenzini, M. and Singh, K.D. (1998) State and change in carbon pools in the forests of tropical Africa. *Global Change Biology* 4(1), 97–114.

Grainger, A. (1993) Rates of deforestation in the humid tropics: estimates and measurements. *Geographical Journal* 159(1), 33–44.

Haigh, M.J. (1984) Ravine erosion and reclamation in India. *Geoforum* 15(4), 542–61.

Haigh, M.J. (1985) Agrobotanic research among the shifting cultivators of Xishuangbanna, Southwest China: lessons for the Himalaya. *Himalayan Research and Development* 4 (1), 1–4.

Haigh. M.J. (1990) Shifting agriculture (jhum) and environmental devastation: the search for a solution. In NK. Sah, S.D. Bhatt and RK. Pande (eds) *Himalaya: Environment, Resources and Development.* Almora, UP, India: Shree Almora Book Depot, 37–76.

Haigh. M.J. (1999) Headwater control: despatches from the research front. In Haigh, M.J. and Krecek, J. (eds) *Environmental Regeneration in Headwater Areas.* NATO ASI (2. Environment) Series Dordrecht, Kluwer, 9–37.

Haigh, M.J., Rawat, J.S. and Bartarya, S.K. (1988) Entropy minimising landslide systems. *Current Science* 57(18), 1000–2.

Haigh, M.J., Rawat, J.S., Rawat, M.S., Bartarya, S.K. and Rai, S.P. (1995) Interactions between forest and landslide activity along new highways in the Kumaun Himalaya. *Forest Ecology and Management* 78, 173–89.

Haigh, M.J., Singh, R.B. and Krecek, J. (1998) Headwater control: matters arising. In Haigh, M.J. Krecek, J., Rajwar, G.S. and Kilmartin, M.P. (eds) *Headwaters: Water Resources and Soil Conservation*, Rotterdam: A.A. Balkema, 3–24.

Hajabbasi, M.A., Jalalian, A. and Karimzadeh, H.R. (1997) Deforestation effects on soil physical and chemical properties, Lordegan, Iran. *Plant and Soil* 190, 301–8.

Harding, K.A. and Ford, D.C. (1993) Impacts of primary deforestation upon limestone slopes in northern Vancouver Island, British Columbia. *Environmental Geology* 21(3), 137–43.

Hellin, J. and Larrea, S. (1997) Live barriers on hillside farms: are we really addressing farmer's needs? *Agroforestry Forum* 8(4), 17–21.

Ishengoma, R.C., Gillah, P.R. and Kiwale, A.Y. (1995) Charcoal production, consumption and deforestation in Magu District, Mwanza, Tanzania. *Annals of Forestry* 3(2), 138–46.

Jang, C.J., Nishigami, Y. and Yanagisawa, Y. (1996) Assessment of global forest change between 1986 and 1993 using satellite-derived terrestrial net primary productivity. *Environmental Conservation* 23(4), 315–21.

Johnson, A.C. (1993) The association between landslides and yellow cedar decline in southeast Alaska. *Eos (American Geophysical Union)* 74, 315.

Johnson, A.C. and Wilcock, P. (1998) Effect of root strength and soil saturation on hillslope stability in forests, SE Alaska. In Haigh, M.J., Krecek, J., Rajwar, G.S. and Kilmartin, M.P. (eds) *Headwaters: Water Resources and Soil Conservation.* Rotterdam: A. A. Balkema, 381–90.

Johnson, D.L. and Lewis, L.A. (1995) *Land Degradation: Creation and Destruction.* Oxford: Blackwell: 335 pp.

Johnson, N.C. and Wedin, D.A. (1997) Soil carbon, nutrients, and mycorrhizae during conversion of dry tropical forest to grassland. *Ecological Applications* 7, 171–82.

Kaimowitz, D. (1996) Livestock and deforestation in Central America in the 1980's and 1990's: a policy perspective. *Ciencias Veterinarias Heredia.* 1996 (1–2), 113–61.

Kirchoff, V.W.J.H. (1996) Increasing concentrations of CO and O_3: rising deforestation rates and increasing tropospheric carbon monoxide and ozone in Amazonia. *Environmental Science and Pollution Research International* 3(4), 210–12.

Kondrashin, A.V. (1991) Deforestation for agriculture and its impact on malaria in southern Thailand. Sharma V.P., Suvannadabba S. (eds) *Forest Malaria in Southeast Asia: Proceedings of an Informal Consultative Meeting: February 1991.* WHO-MRC, 221–6.

Kustudia, M. (1998) Cornucos, campesinos and the contested cordillera. *Forests, Trees and People, Newsletter* 36/37, 26–33.

Lai, H.G. (1985) To stop the Yellow Dragon with a Green Ocean. *Commonwealth Forestry Review* 64(3), 251–8.

Lal, R. (1996) Deforestation and land-use effects on soil degradation and rehabilitation in western Nigeria. I. Soil physical properties and soil erosion

II. Soil chemical properties. III. Runoff and nutrient loss. *Land Degradation and Development*. 7, 19–45, 87–98, 99–119.

Lal, R. (1997) Deforestation effects on soil degradation and rehabilitation in western Nigeria. IV. Hydrology and water quality. *Land Degradation and Development* 8, 95–126.

Lal, R., Sanchez, P.A. and Cummings, D.J. (1985) *Land Clearing and Development in the Tropics.* Rotterdam: A.A. Balkema.

Lamb, D., Howell, S., Read, T. and Broekstra, L. (1995) *Rehabilitation of Degraded Tropical Forests: An Annotated Bibliography*. Jakarta: UNESCO.

Linnard, W. (1982) *Welsh Woods and Forests: History and Utilisation*. Cardiff: National Museum.

Lovelock, J.E. (1990) *Visionary Voices.* VV506. Forres, Scotland: Whole World Productions.

Marengo, J.A. (1995) Variations and change in South American stream flow. *Climatic Change* 31(1), 99–117.

McNeil, J. (1982) *Mountains of the Mediterranean World*. Cambridge University Press.

Mendiondo, E., Castro, N., Auzet, A.V. and Chevallier, P. (1998) Surface flow pathways in subtropical agricultural headwaters: a case study from Southern Brazil. In Haigh, M.J., Krecek, J., Rajwar. G.S., and M.P. Kilmartin (eds) *Headwaters: Hydrology and Soil Conservation*. Rotterdam, A.A. Balkema, 285–93.

Merry, F.D. and Carter, D.R. (1997) Certified wood markets in the US: implications for tropical deforestation. *Forest Ecology and Management* 92, 221–8.

Myers, N. (1995) *Ultimate Security: Environmental Basis of Political Security*. New York: W.W. Norton.

Myers, N. (1994) Tropical deforestation: rates and patterns. In Brown, K. and Pearce, D.W. (eds) *The Causes of Tropical Deforestation.* London: University College of London Press, 27–40.

Neal, C., Smith, C.J. and Hill, S. (1992) *Forestry impact on upland water quality*. Institute of Hydrology (Wallingford, UK.) Special Report 119, 1–50.

Neill, C., Piccolo, M.C., Cerri, C.C., Steudler, P.A., Melillo, J.M. and Brito, M. (1997) Net nitrogen mineralization and net nitrification rates in soils following deforestation for pasture across the southwestern Brazilian Amazon Basin landscape. *Oecologia*. 110, 243–52.

Nicholson, C.F., Blake. R.W. and Lee, D.R. (1995) Livestock, deforestation, and policy making: intensification of cattle production systems in Central America revisited. *Journal of Dairy Science* 78(3), 719–34.

Noever, D. (1993) Himalayan sandpiles. *Physical Review* E47(1), 724–5.

Noever, D., Brittain, A., Matsos, H.C., Baskaran, S. and Obenhuber, D. (1996) Effects of variable biome distribution on global climate. *Biosystems* 39(2), 135–41.

Nygren, A. (1995) Deforestation in Costa Rica: an examination of social and historical factors. *Forest and Conservation History* 39(1), 27–35.

Ohte, N., Asano Y. and Tokuchi, N. (1998) Geographical variation in acid buffering processes in forest catchments. in Haigh, M.J., Krecek, J., Rajwar, G.S. and M.P Kilmartin (eds) *Headwaters: Hydrology and Soil Conservation.* Rotterdam: A.A. Balkema, 69–84.

Parisi, P. and Glantz, M.H. (1992) Deforestation and public policy. *The World and I* (November 1992): 270–7.

Pereira, H. C. (1989) *Policy and Practice in the Management of Tropical Watersheds.* Boulder: Westview, London: Belhaven.

Price, M. (1998) Sustainable mountain development: the roles of forests. In Haigh, M.J., Krecek, J., Rajwar. G.S. and M.P Kilmartin (eds) *Headwaters: Hydrology and Soil Conservation,* Rotterdam, A.A. Balkema, 443–50.

Rajwar, G.S. (1998). Changes in plant diversity and related problems for environmental management in the headwaters of the Garhwal Himalaya. In Haigh, M.J., Krecek, J., Rajwar, G.S. and M.P Kilmartin (eds) *Headwaters: Hydrology and Soil Conservation,* Rotterdam, A.A. Balkema, 335–44.

Reading, A.J., Thompson, R.D. and Millington, A.C. (1995) *Humid Tropical Environments.* Oxford: Blackwell.

Repetto, R. and Gillis, M. (eds) (1988) *Public Policies and Misuse of Forest Resources*. Cambridge University Press.

Rudel, T. and Roper, J. (1997) The paths to rain forest destruction: cross national patterns of tropical deforestation, 1975–90. *World Development* 25, 53–65.

Sahin, V. and Hall, M.J. (1996) Effects of afforestation and deforestation on water yields. *Journal of Hydrology* 178(1–4), 293–309.

Sandstrom, K. (1995) Differences in groundwater response to deforestation – a continuum of interactions between hydro-climate, landscape characteristics and time. *Geojournal* 35(4), 539–46.

Saxena, A.K. and Nautiyal, J.C. (1997) Analyzing deforestation: a dynamic systems approach. *Journal of Sustainable Forestry* 5 (3–4), 51–80.

Schreier, H., Brown, S., Carver, M. and Shah, P.B. (1998) Nutrient and sediment transport in a degraded middle mountain watershed in Nepal. In Haigh, M.J., Krecek, J., Rajwar. G.S. and Kilmartin, M.P. (eds) *Headwaters: Hydrology and Soil Conservation*, Rotterdam: A.A. Balkema, 315–28.

Shaxson, T.F. (1995) Principles of good land husbandry. *Enable* 5, 4–13.

Shaxson, T.F. *et al.* (1989) *Land Husbandry: A Framework for Soil and Water Conservation.* Ankeny, Iowa: WASWC/SWCS.

Shibano, H. (1998) Effects of forest regeneration on the water budget of small catchments in the mountains of Central Japan. In Haigh, M.J., Krecek, J., Rajwar. G.S. and M.P Kilmartin (eds) *Headwaters: Hydrology and Soil Conservation.* Rotterdam: A.A. Balkema, 273–83.

Skole, D. and Tucker, C. (1993) Tropical deforestation and habitat fragmentation in the Amazon: satellite data from 1978–1988. *Science* 260, 1905–10.

Theng, B.K.G. (1991) Soil science in the tropics – the next 75 years. *Soil Science* 151(1), 76–90.

Tole, L. (1998) Sources of deforestation in tropical developing countries. *Environmental Management* 22(1), 19–33.

Valdiya, K.S. (1998) *Dynamic Himalaya.* Hyderabad: Universities Press.

Vvan Dam, D., Veldkamp, E. and van Breemen. N. (1997) Soil organic carbon dynamics: variability with depth in forested and deforested soils under pasture in Costa Rica. *Biogeochemistry* 39(3), 343–75.

Walsh, J.F. *et al.* (1993) Deforestation: effects on vector-borne disease. *British Society for Parasitology, Symposium* 30, 55–75.

Wiersum, K. (1984) Surface erosion under various tropical agroforestry system. In O'Loughlin, C. and Pearce, A.J. (eds) *Effect of Forest Land Use on Erosion and Slope Stability, Proceedings.* IUFRO, 131–9.

Wilson, E.O. (1992) *The Diversity of Life.* Cambridge, Mass.: Belknap Press.

Wunder, S. (1996) Deforestation and the uses of wood in the Ecuadorian Andes. *Mountain Research and Development* 16, 367–82.

World Resources Institute (1994) The problem of forest loss. Washington: World Resources Institute: *[http://www.igc.org/wri/biodiv/opp-ii.html].*

World Wildlife Fund (WWF) (1998a). Haze damage from 1997 Indonesian forest fires exceeds $1.3 billion study shows. World Wildlife Fund. *[http://www.worldwildlife.org/new/fires/dam.htm].*

World Wildlife Fund (WWF) (1998b). Burning again: this year's fires in the Amazon are the worst ever. Washington: 3 April 1998: World Wildlife Fund. *[http://www.worldwildlife.org/new/fires/home.htm].*

15

Maintaining biodiversity

Nick Brown

INTRODUCTION

Conservation is a management discipline that has traditionally looked to the biological sciences for its paradigms. However, many of the problems that have been encountered during the last decade have revealed that a strictly biological approach to conservation has proved inadequate. Conservation biology has, until quite recently, failed to take a geographical perspective and has no theoretical framework within which the role of human societies can be incorporated. Conservation needs to take account of both the geographical and social context within which management operations are occurring. This chapter will attempt to illustrate how conservation of biological diversity is a problem that geography is uniquely positioned to address.

TYPES AND SCALES OF BIODIVERSITY

Although it is common to think of biodiversity as synonymous with species richness, this is only one level at which variety is measured in organisms. There is a hierarchy of organismic variation from the genetic through the species and population levels to diversity in ecosystems. Functional, structural and age diversity may also be of great importance to the way that ecosystems operate. For example, structural diversity in British woodlands is one of the most important determinants of animal species richness.

Biological diversity is so widely accepted as being valuable that the reasons why it is valued are rarely carefully analysed. Table 15.1 lists a crude classification of the reasons for valuing biological diversity and examines the scales at which these might best operate. Not all levels of variation are equally valuable for all reasons. For example, the diversity of species in an ecosystem may fluctuate, while the diversity of functional or morphological types may remain relatively constant. Hawthorne (1993) has shown that logging of a Ugandan rain forest may actually increase species diversity. High levels of disturbance allow widespread pioneer species, including common agricultural weeds, to colonise along logging tracks and in clearings. These new arrivals more than compensate for the immediate loss of some forest species. Although the total species diversity increased, there was considerable turnover in composition, and functional and morphological diversity may well have decreased.

Mares (1992) has pointed out that hotspots of species diversity may not necessarily coincide with areas of the highest genetic diversity. Tropical rain forests contain large numbers of very closely related species. For instance, on the island of Borneo there are approximately 267 species in the family Dipterocarpaceae. All are trees with very similar ecology. Closely related species may differ in less than 5 per cent of their nuclear DNA sequences. Conservation of all species of dipterocarp may preserve less genetic information than the conservation of many fewer but more distantly related species. In contrast, it is well established that many marine ecosystems have extraordinarily high levels of phylum diversity. As a consequence, although they may

Table 15.1 The values of biological diversity.

Value	*Description*	*Level of operation*
Commodity value	The utilitarian value of an organism to humans as food, medicine etc. Populations that are not used may possess genes that improve the value of exploited populations.	Population and species diversity
Amenity value	The value to humans of visiting, viewing or learning about natural communities.	Species and ecosystem diversity
Ecological integrity value	The importance of maintaining all functionally critical aspects of an ecosystem.	Functional diversity
Ethical value	The moral obligation on humans not to drive other species to extinction.	Species and ecosystem diversity
Option value	The future evolutionary potential of a species or the potential for future exploitation of a species' commodity value.	Genetic and population diversity

be less species-rich than many terrestrial ecosystems it is probable that they contain considerably greater total genetic diversity.

All measures of diversity seem to be scale-dependent. The total number of species can be measured at a point in space (often referred to as α-diversity). But community composition will vary from point to point across a habitat. For example, Newbery *et al.* (1992) found that in thirty-two adjacent 0.25 ha sub-plots in a Borneo rain forest, only 20 per cent of the species occurred in more than half of the plots. The turnover in species composition from point to point within a habitat is usually referred to as β-diversity and is related to the size of species ranges. An α-diverse community may, nevertheless, be composed of widespread species. There will be a high degree of compositional similarity from place to place, and β-diversity will be low. The Ugandan tropical rain forest described by Hawthorne (1993) had high α-and β-diversity. After logging had driven some forest species to local extinction and replaced them with widespread weeds, α-diversity remained high but β-diversity almost certainly declined.

ECOLOGICAL PROCESSES GENERATING AND MAINTAINING DIVERSITY

There are clear patterns in species diversity at a global scale. Diversity tends to decrease with altitude and latitude. Isolated or geographically restricted areas have low diversity. However, some patterns are unexpected. Diversity often peaks but then declines along productivity and successional gradients. Unfortunately, despite decades of research activity there is still no widely accepted view as to which processes are primarily responsible for creating these patterns of diversity. Without a sound understanding of how biologically diverse communities arise or are maintained any attempts at their conservation are likely to be largely haphazard.

Ideas abound on the causes of high species diversity. To a considerable extent other levels of variation have been ignored. This undoubtedly reflects the comparative ease with which species diversity can be quantified and not that this level of biological diversification is of greater importance. The development of a rigorous classification of the functional attributes of an ecosystem presents a significant challenge for the environmental sciences.

Species richness depends on the number of species that are able to maintain populations and reflects the balance between processes increasing and decreasing the number of individuals per species. Some of these processes have been the focus of a great deal of ecological research. They can be crudely characterised as falling into two groups: theories that describe ecological processes that occur at a local scale, and those that invoke biogeographical processes occurring at a regional scale or larger.

Local-scale processes

Research by the Russian zoologist G.F. Gause during the 1930s demonstrated experimentally that a population would grow exponentially until it exhausted the supply of an important resource. The rate of supply of this resource would then set the equilibrium population size. Furthermore, Gause showed that if the populations of two or more species were limited by the same resource they would inevitably compete for supplies. The species that was better able to capture the scarce resource would increase its population at the expense of the poorer competitor. Gause concluded that this weaker species would be driven to local extinction and therefore that species could not coexist if they had a common limiting resource. The corollary of this idea was that of the 'niche', that all coexisting organisms must have a unique combination of resource requirements in order to avoid competitive interactions. Ecologists hypothesised that the more species that live in a particular habitat the narrower their individual resources bases will be. They argued that there would be a limit to how narrow niches could become. Beyond a certain point, species populations would become so small that they would be highly likely to go locally extinct. This limit to how similar species can be in their resource use means that the total amount of resource available in a particular habitat sets the upper limit to the number of species that can coexist. In summary, local conditions, ecologists argue, determine the maximum number of species that are found there. Phillips *et al.* (1994) concluded, from a study of twenty-five tropical forest sites, that the more dynamic forests are (measured from their mortality and recruitment rates) the greater their species diversity. They hypothesised that high levels of spatial and temporal variability in dynamic forests provided opportunities for a wide range of species. Clinebell *et al.* (1995) have recently demonstrated that annual rainfall and rainfall seasonality alone account for over 60 per cent of the variation in species richness in sixty-nine neotropical forests spread across the continent.

The idea of ecosystems existing in equilibrium with environmental conditions has been elaborated to include the effects of interactions between species. One species population may become a resource for another to exploit, for example. Additionally, predator–prey relationships may prevent competition between prey species leading to extinctions. Janzen (1970) proposed that predators might prevent one species from driving competing species to extinction by feeding predominantly on those species that are most abundant. The more potential prey of a particular species there are around, the more likely it is that they will be eaten. This is termed a density-dependent process, since the rate of loss to predators depends directly on the abundance of prey. Competitively superior species populations may therefore be limited in size by losses to predators. Predation may thus help to maintain species diversity by preventing extinctions due to interspecific competition. Trophic interactions can therefore mediate the equilibrium number of species found in a particular environment.

The strength of interactions between species is known to differ. Some species appear to be disproportionately important in determining how their communities are structured. Terborgh (1986) has shown how only twelve plant species in a flora of approximately 2000 species at Cocha Cashu reserve in Peru sustain nearly the entire fruit-eating animal community for three months of the year. Although the climate is tropical, flowering and fruiting are strongly seasonal and there are periods of the year when there is very little food available for frugivorous animals. Such species have become known as 'keystone' species. A keystone is the wedge-shaped stone at the apex of an arch that locks the whole structure together. If this stone is removed the whole arch will fall. If such species exist, clearly they should be a conservation priority.

One prediction that stems from the competition hypothesis concerns the nature of changes that take place in a vegetation community in response to disturbance. Catastrophes such as storms, floods and fires kill off the dominant plants and create opportunities for pioneer plants to colonise. As succession proceeds, increasingly competitive plants invade and grow to dominate the community while the less competitive organisms are driven to local extinction. Ultimately, the community will consist of a stable mixture of the most competitive species, each with its own unique niche. A hypothesis put forward by J.H. Connell (1978) suggested that this equilibrium condition rarely, if ever, occurs. Connell pointed out that most natural environments are repeatedly disturbed by events of a range of magnitudes and at a variety of time-scales. A high magnitude or frequency of disturbance will result in a community with a large proportion of pioneer and early successional plants. In contrast, the most competitive plants will dominate environments with rare or low-magnitude disturbance. Intermediate levels of disturbance will allow opportunities for both guilds of plant to coexist and are therefore likely to maximise ecological diversity. Natural environmental disturbance may act in a very similar way to predation in preventing interspecific competition from leading to local extinctions.

Regional-scale processes

Many ecologists are uncomfortably aware that the standard ecological models assume that local-scale processes (such as competition and local disturbance regimes) determine the level of species diversity. However, as Ricklefs (1987) has pointed out, if these models were accurate areas that have very similar physical environments should support similar number of species. Community composition should correspond to limits set by local conditions alone, but there is a worrying lack of evidence for local determinism. Ricklefs argues that local diversity is strongly governed by much larger-scale biogeographical processes. Ecology, he claims, has ignored these scales because they are less amenable to experimentation and analysis. Preoccupation with trophic and competitive interactions has misled many ecologists into assuming that closed population processes alone can explain local population structure. A biogeographical perspective is required before population dynamics can be properly understood.

Ricklefs and Schulter (1993) have proposed that species diversity is influenced by a hierarchy of processes, each one acting at a different temporal and spatial scale (Figure 15.1). At the very largest scale, global and regional processes both play a fundamental role in generating the pool of species from which regional communities are drawn. Regional processes determine how species disperse over space. All environments are patchy. As a direct consequence, the abundance of individuals is not constant across a species range. It is likely that there will be peaks of abundance where conditions are particularly favourable, and the population will decline or even disappear in less suitable habitat. Space, therefore is not continuously occupied by individuals; rather, they exist in a series of small, interacting sub-populations. Andrewartha and

Figure 15.1 A hierarchy of processes determining species diversity.

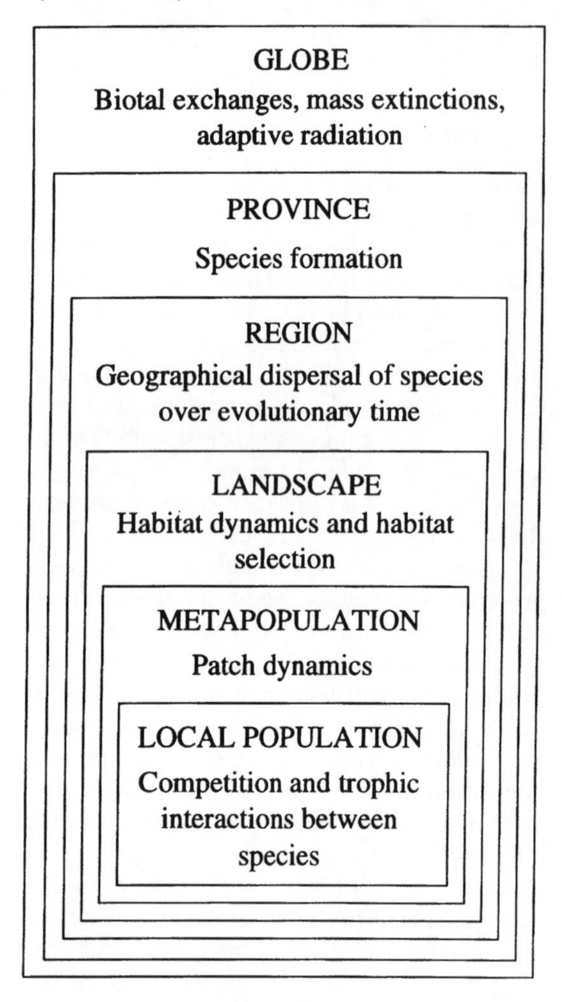

Source: Ricklefs and Schulter 1993.

Figure 15.2 Although the small local populations (A) frequently go extinct they combine to produce a metapopulation (B), which is relatively stable over time.

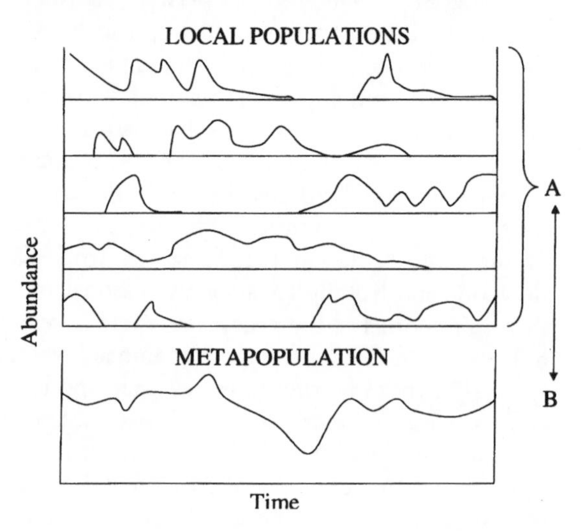

Birch (1954) proposed that since many of these local populations were small they would commonly go extinct, only to be re-established sometime later by the arrival of new individuals from elsewhere. They viewed the 'metapopulation' as rather like the lights on a Christmas tree. The tree is lit all the time, but individual lights flash on and off (Figure 15.2). This gives an important spatial structure to populations that ecology has failed to recognise.

Many of the changes that occur in a local population cannot be understood without reference to regional processes of immigration and extinction. One of the very few ecological models to take such processes into account was the equilibrium theory of island biogeography developed by MacArthur and Wilson (1967). Their model attempted to account for the number of organisms found on offshore islands in terms of the size and degree of isolation of the island. These two variables determined the equilibrium between the rate at which new species arrived and at which established species became extinct (Figure 15.3). These ideas were adopted, rather uncritically, by conservation biology. Rapid rates of habitat fragmentation are a major cause of decline in many species populations, and it was believed that residual habitat fragments might behave in a similar way to islands (Plate 15.1) (Simberloff and Abele 1976). This triggered a prolonged debate over what has become known as the SLOSS (single large or several small) problem; do a few large reserves maximise the chances of long-term survival for more species than a larger number of small reserves? Regrettably, this debate ignored the important contribution of metapopulation dynamics to the understanding of the behaviour of fragmented habitats.

Figure 15.3 Immigration and extinction rates determine the equilibrium number of species that inhabit an island. Immigration rates vary with proximity to the mainland. Extinction rates vary with the size of island.

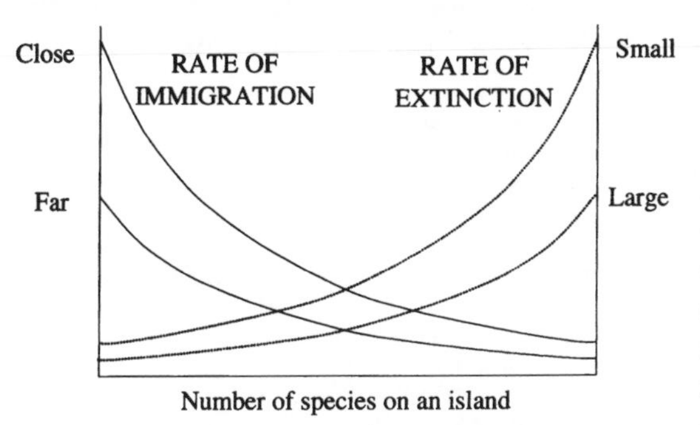

Source: MacArthur and Wilson 1967.

Although specific extinction and dispersal rates are important in maintaining local populations, they may not be critical for the survival of the metapopulation as a whole. The metapopulation is much more dependent on the spatial arrangement of habitat fragments. A small habitat patch, for example, may be too small to support a viable local population but may be a crucial stepping stone in the recolonisation of other patches. A key difference is that not all patches need to be occupied all the time by all species in order to serve a very important conservation function (Hanski 1996). Some high-quality habitat areas may act as sources for emigration to inferior sink areas where mortality exceeds the birth rate. As a consequence, although such sink patches may be occupied they may not necessarily be able to support a viable local population. Any attempts at species conservation would need to recognise that many local populations may not be at equilibrium, and regional processes may be crucial in sustaining the metapopulation.

Plate 15.1 Regional metapopulation processes may be crucial in sustaining species populations in habitat fragments such as small woods in an agricultural landscape.

Landscape ecology has developed some indices of spatial pattern (O'Neill *et al.* 1988) (Figure 15.4). Dominance describes the abundance of a particular habitat type in the landscape. Contagion expresses the degree to which habitat fragments are clustered or dispersed. Fractal dimension describes the complexity of patch shape. Little progress has yet been made in exploring the relationship between these indices and metapopulation dynamics.

Figure15.4 Indices of spatial structure in landscape ecology.

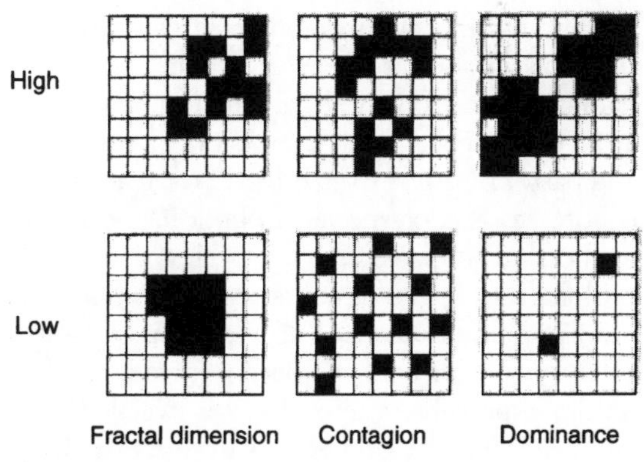

Source: Adapted from O'Neill *et al.* 1988.

Box 15.1 Determining the species status of the green frog

Globally, there has been a recent and alarming decline in populations of many amphibians. Populations may show high natural variation, and therefore there has been considerable debate over the extent to which widespread decline reflects a serious conservation problem or just a stochastic fluctuation. The only way that a satisfactory answer can be found is to mount long-term monitoring of species numbers. Many amphibians live in isolated habitat patches that may interact as part of a large metapopulation. Hecnar and M'Closkey (1997) monitored the distribution and abundance of a common green frog (*Rana clamitans mellanota*) in 160 ponds in southwestern Ontario, Canada, at a variety of scales in order to discover how the population was changing.

The geen frog is capable of inhabiting a wide range of permamanent ponds. Subadults leave the pond where they are born and disperse up to five km to a new site. Turnover in the population is high, with very few adults living longer than five years. Wetlands in southern Ontario covered over 60 per cent of the land area up to 100 years ago. Extensive drainage for agriculture means that wetlands now cover less than 10 per cent.

In this survey, the number of green frogs at each pond in three regions was counted. At a geographic scale, there was very little change in the number of ponds occupied by frogs during the three-year survey period. However, there was considerable variation from region to region, with frogs occupying under half the ponds surveyed in one region and all of the ponds in another. At sub-regional and local scales, it was found that in some areas the number of ponds occupied by frogs was declining, while it was increasing in others. The number of adult green frogs remained stable at a geographic scale but was increasing in one region, declining in another and stable in the third. At the sub-regional and local scale, trends in the abundance of frogs were extremely variable. Abundance of frogs was stable at 20 per cent of ponds, increasing at 18 per cent, declining in 14 per cent and showing no consistent trend in the remainder.

This study illustrates how the status of a species is highly scale-dependent. Local-scale studies may come to highly misleading conclusions about trends in the distribution and abundance of the green frog if extrapolated to a geographic scale. Variance in both occupancy of ponds and adult numbers increased as the spatial scale of the survey decreased. This study confirmed that the smaller a local population the more likely it is to become extinct. Extinctions occurred only where there were fewer than ten adults in a pond. Frog populations were spatially dynamic, with common extinctions and recolonisations. The implications for conservation biology are clear. First, it is not adequate to assess the status of a species at a restricted spatial scale. It also emphasises the importance of facilitating species dispersal in order to ensure that a local population can re-extablish itself after stochastic extinction. Small, high-quality habitat patches may make an important contribution to increasing species dispersal, even though they may not be able to sustain a local population over a long time period. Conservation needs to understand the spatial dynamics of a species in order to prevent decline at a geographic scale.

MANAGEMENT FOR BIODIVERSITY

It is almost a truism to point out that the best method for maintaining biodiversity is to control the most important causes of decline. These are habitat destruction and fragmentation, primarily as a consequence of the expansion of cultivated and pastoral areas. Habitat degradation also contributes to loss of species. Pollution, the introduction of invasive exotic species and over-exploitation of natural ecosystems are some of the most important causes of degradation. Unfortunately, governments have been reluctant to take action, especially where controls would have significant impacts on people's livelihoods. As a consequence, strategies for biodiversity conservation are frequently limited to attempts to exclude such influences from protected areas and attempts to rescue highly endangered species from extinction.

Protection of endangered species

Ex situ conservation strategies attempt to preserve representatives of highly endangered species outside their native ranges, usually in a zoological or botanical garden. In principle, individuals can then be reintroduced to their natural environment. Controversy continues to rage in the conservation world about the ethics of reintroducing a species to sites where they are known to have existed in the past but have recently gone extinct. This may be desirable where a habitat has been restored to the point where it may, once again, adequately support the species, or where some systemic extinction pressure has been removed. It may be necessary if the species in question has a poor colonising ability and is unlikely to arrive unaided. Some species may have a significant influence on ecosystem processes, including the regulation of other species. There may be an argument for reintroductions of these keystone species when it is anticipated that they will have wider conservation value. There is, for example, considerable interest in the reintroduction of the European beaver (*Castor fiber*) to Britain (Macdonald *et al.* 1995). Where beavers dam stream channels they may create habitat for aquatic species that inhabit pools and stagnant water.

Reintroductions may also be important where a population of a species has declined to the point where it is no longer considered viable. Bringing in more individuals from a larger population elsewhere may reinforce the recipient population. Although this may often prevent the deleterious effects of inbreeding depression in an isolated population by bringing in new genotypes it may also result in the dilution of important local varieties. Some imported genotypes may be poorly adapted to the local environmental conditions and do little to enhance the viability of the recipient population. There is also a risk of introducing new diseases when infected individuals are used to supplement a local population. One of the most important considerations, however, in any species reintroduction is the biogeography of the resultant population. Many species, particularly where conditions are marginal for population survival, behave as metapopulations. There may be considerable cost but little conservation value in supplementing a local sink population. A good understanding of the spatial dynamics of a regional population may be crucial to ensuring the success of a reintroduction programme (Hodder and Bullock 1997).

Zoos and botanical gardens may serve a very important education function that far exceeds their somewhat doubtful value for reintroductions. They provide an opportunity for society to marvel at the diversity and beauty of organisms and can alert the public to the threats of biodiversity loss.

In situ conservation measures attempt to halt the decline of species populations within their natural range. Determining the status of a species population is therefore of fundamental importance. Lack of reliable information on population levels can often confound conservation efforts. Changes in the abundance of an organism can only be assessed against knowledge of prior levels. Unfortunately, there are very few communities or even taxa for which accurate data have been recorded over time. A further problem is that where records have been kept they show that the abundance of many organisms fluctuates naturally from year to year. Long-term trends are difficult to detect against this highly variable background. For example, the abundance of larvae of the larch bud moth (*Zeiraphera diniana*) was monitored for over thirty years at a site in Switzerland (Baltensweiler 1984). The population density of moths varied by five orders of magnitude during the period of the study (Figure 15.5). High population variance and strong autocorrelation in population size from year to year makes detection of long-term trends very difficult.

Even when long-term patterns are detectable, it is difficult to extrapolate from data collected at a single site to a more ecologically meaningful scale. Populations of a species fluctuate not only in time but also in space. An alternative and perhaps more sensitive approach to monitoring species populations ignores the size of a population but looks instead at its geographical distribution. Shaffer *et al.* (1998) have proposed that museums, herbaria and other historic archives offer a valuable source of information on geographical distributions of organisms over time. Historical records of the presence or absence of species at a particular site are often more reliable than data on their abundance. The latter are very dependent on the sampling effort and method used. If records or collections are treated as random samples from the entire range of an organism, then changes over geographically relevant scales can be assessed and population fluctuations at individual locations ignored. In this way, geographical research may reveal more about patterns and changes in biodiversity than detailed ecological studies of population dynamics.

Although campaigns to save highly endangered species stimulate considerable popular support, in reality very few have proved

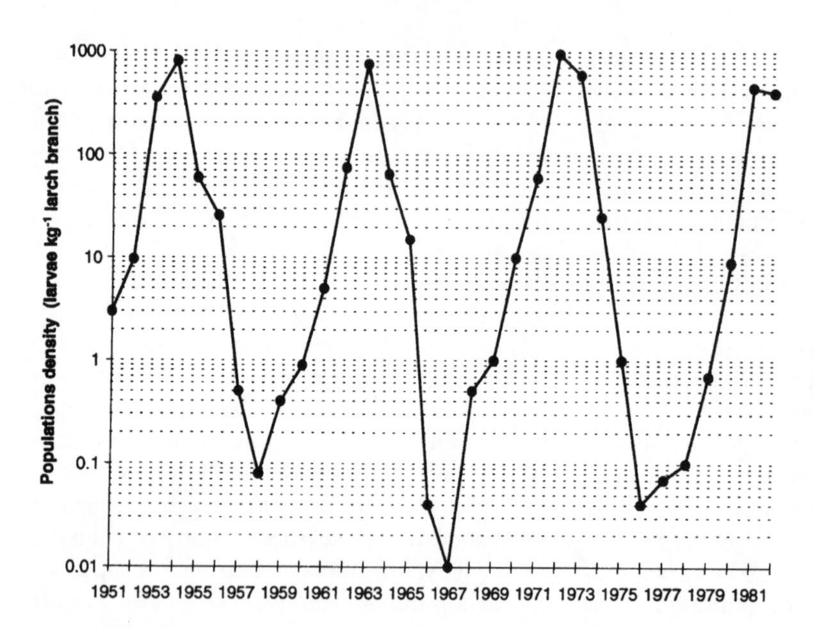

Figure 15.5 Densities of larvae of the larch bud moth (*Zeiraphera diniana*) on larch in Switzerland.

Source: Baltensweiler 1984.

successful. They also risk diverting resources to those species that are least likely to survive in the long term. It would be both less costly and probably more effective if conservation projects sought to protect viable populations of more common species within their native range to prevent them ever reaching endangered status. It should not be assumed that rarity is synonymous with vulnerability to extinction. Many species are naturally rare, especially at the limits of their range. Rarity is also scale-dependent; local rarity does not always imply that an organism is regionally rare.

Selection of protected areas

As attempts to protect endangered species have so frequently proved inadequate, an alternative strategy for biodiversity conservation is one that protects the greatest variety of species and habitats. There are two distinct problems that face conservation in its efforts to select areas for conservation. The first is to ensure that all valuable facets of diversity are protected. This involves some form of biodiversity inventory. The second problem is to make provision for their protection in perpetuity.

The most logical way to identify which species and habitats are inadequately protected would be to draw up species distribution maps and compare these with maps of existing protected areas. This procedure has been termed 'gap analysis' (Scott *et al.* 1993) and is an approach that has been made considerably more tractable by the development of sophisticated geographic information systems. Such an approach makes good sense, because most species are distributed in a strongly non-random pattern. Similar biogeographical histories for many species mean that their distribution patterns coincide. The effectiveness of protected areas in protecting species and habitats can be increased significantly by the selection of diversity hotspots, where the ranges of large numbers of species overlap.

Part of the rationale for protecting biodiversity is the belief that it may contribute to future evolutionary potential. Giving priority for conservation to areas that encompass the greatest possible species diversity may not be the best way of retaining the greatest capacity for further evolutionary development (Brooks *et al.* 1992). Areas that contain a diversity of habitats will often contain large numbers of species. However, close juxtaposition of habitats within a restricted area may mean that they are at marginal ecotones and no one habitat may be adequately represented. Areas that have been the focus of evolutionary development in the past (areas of endemism) may not currently contain the greatest number of species but are likely to be the focus for future species evolution.

Gap analysis protocols can now incorporate information not only on species richness but also on which species are well conserved within non-protected landscapes, on rarity, on restricted range species and on endemism hotspots. A great deal more information is needed on the degree to which the distributions of endangered species, diversity hotspots and centres of endemism may coincide, a job for which geography is ideally suited.

How can priority areas for conservation be identified quickly and cheaply? It is a horrifying truth that most species of organism on Earth remain undescribed. This places a very serious constraint on any attempts to map diversity, particularly in the super-diverse regions of the tropics, where rates of habitat loss may be at their highest. The description of the geographical distribution of biological diversity is one of the most pressing conservation problems. Several approaches have been tried in order to overcome this problem. Given that at large scales there would appear to be good correlation between the diversity of different taxa, one method might be to use the diversity of well-known groups, such as birds, as an indicator of all organism diversity. Unfortunately it now seems that such correlations break down at finer geographic scales. Prendergast *et al.* (1993) identified diversity hotspots for five different taxonomic groups in the UK and found only 34 per cent overlap. An alternative approach is to use the diversity at higher taxonomic levels. Non-experts can easily identify the families of many organisms, and family diversity is very highly correlated with species diversity (Blamford *et al.* 1996). A third approach is to use abiotic data (Belbin 1993). This may be particularly amenable to geographical analysis given existing detailed maps of environmental variables such as climate, topography and soils.

In reality, conservation rarely works in a rational or systematic way. Most nature reserves are declared on the residual land that has been left behind by the pastry cutter of development. Historically protected areas have been established on an *ad hoc* basis, either when an opportunity fortuitously presents itself or when an area perceived to be valuable comes under threat.

Integrating conservation and sustainable development

There is a growing belief among conservation professionals that protected area systems alone cannot adequately conserve the world's biodiversity (WRI/IUCN/UNEP 1992). Attention has refocused on the potential for integrating rather than segregating conservation in land-use planning.

As much as 50 per cent of global biodiversity may be located in tropical regions. However, this is part of the globe that contains some of the world's poorest countries and some of the most rapidly developing. The cost of preserving large areas of wilderness may be prohibitive for countries where a large proportion of the population is dependent on subsistence agricultural production and exploitation of natural resources. The greatest costs may be associated with the forgone development potential of the land.

Local people are often perceived to be a threat from which nature should be protected. Protected areas are often designed to exclude or even remove local people. For those communities that have traditionally made use of such areas as part of their livelihood systems, the consequences of this exclusion may be devastating and vehemently resented (Plate 15.2). Such communities are faced with the costs of substituting for the goods and services that they would have obtained free from natural ecosystems. Experience has shown that without the consent and active participation of local people, many conservation projects may be severely undermined. Indeed, where that popular consent is lacking there are serious ethical questions as to why and for whom conservation of biodiversity is taking place. Participatory conservation projects that engage local communities have many advantages.

Box 15.2 Social forestry in the Western Ghats of India

The Western Ghats, a 1500 km long mountain range running down the southwestern coast of India, is one of the world's most important biodiversity hotspots (over 3500 species of plant have been identified) and has extraordinarily high levels of endemism (25–60 per cent of recorded species). Biogeographically, the Western Ghats has long been isolated from the vast Southeast Asian humid forest tract and thus protects a relict pocket of an evolutionarily distinct biota. Heterogeneous geology, soils and climate also contribute to promoting high biodiversity. Forests of the regions vary from dry deciduous to wet evergreen types. The forests have very high local use value, particularly for firewood, pasture and as sources of non-timber forest products, including leaf compost.

India has a longer tradition of legal protection of nature than any other nation, and legislation has often favoured conservation over other forms of land use. During the 1970s, as India's population expanded rapidly, the effects of environmental degradation began to be felt. In the Western Ghats, a protected area network (wildlife sanctuaries and national parks) was created. Protected areas managed by the government are often large (a few hundred square kilometres), consisting of a mosaic of landscape elements including forests. They are protected by legal authority, but there are frequent conflicts since they impose land uses that compromise the livelihood requirements of the local populaion. The protection programme was imposed with little community participation, and local concerns were rarely addressed. Despite attempts at a 'guards and guns' approach to protection, local people continued to remove firewood and graze cattle within the reserves. It was estimated that nearly 80 per cent of firewood in some areas was taken from reserves. Many protected forests had little or no seedling or sapling regeneration as a consequence of grazing and fire.

It was quickly realised that social conditions in the Western Ghats meant that gathering of firewood and grazing cattle could not be prevented. Conservation strategies that ignored the social and economic context were doomed to failure since they would inevitably end in conflict with local communities. Projects were therefore redesigned to integrate biodiversity conservation into a strategy of landscape rehabilitation and sustainable use. Part of the reason for intense pressure on protected areas was the severe degradation of communal woodland and grazing areas. A revised conservation strategy aimed to improve the productivity of extensive areas of degraded land in the hope that this would relieve pressure on protected areas. Regrettably, although the objectives of the new social forestry project were sound, the project failed to engage local communities in the design and implementation. Participation can be passive or interactive. In the former, external experts identify the problem and define their solutions. Local people are expected to comply with management decisions that arise from these. In interactive participation, local communities participate in joint analysis of the problems and develop their own action plans. The social forestry project failed because many of the outcomes of the project failed to solve the real problems and indeed even exacerbated some of them. For example, community woodlots were established to provide fuelwood. However, trees were planted by the Forestry Department at a close spacing conventional in a timber plantation. As a consequence, grasses failed to thrive between the trees, and local communities found that their grazing land was even further reduced. Harvesting of woodlots produced large quantities of timber at infrequent intervals, but local people dependent on forests for firewood required regular supplies of small quantities.

The Western Ghats Forestry Project began in 1991 and attempted to address some of the earlier limitations. Project planning and implementation was achieved through the establishment of Joint Forestry Management, a project management system developed elsewhere in India. The core of JFM is the village forest committee in each community, which has a legal right to make joint decisions with the Forestry Department on how management is carried out. This method allowed local people, for the very first time, to participate in the management of forests with the Forestry Department and to have a legal right to share in the benefits of this operation. Sharing of the management and usufruct with the people has important conservation and development implications. Part of the income from forestry operations goes into a village forestry development fund and can be used for community projects.

They may be a powerful motor for sustainable development, avoiding many of the pitfalls of conventional development models that lead to environmental degradation. Costs of sustainable use of natural ecosystems may be significantly lower than those associated with their total protection. Species are protected across the whole landscape rather than in restricted reserves.

Although there is a growing trend for devolution of responsibility for conservation to local communities where detailed knowledge of local conditions and appropriate priorities lie, there are many challenges ahead. The benefits of biodiversity conservation accrue mainly at the national or global level. Decentralisation of conservation projects means that the costs are borne locally. Ecosystems vary in the degree to which conservation and local use are compatible. The Gir National Park is home to one of the most important populations of Asiatic lions in India; however, it is likely that the present population of lions is too large for the park. They depend on

Plate 15.2 'Wild man committing forest offence'. Photograph *c.* 1920 of an Indian forest officer arresting a man for illegal cutting of timber (*Oxford Forestry Institute Collection*).

predating the livestock of surrounding villages, and attacks on humans are not infrequent. Over 160,000 people live in and around Gir, and many view it as an important pasture that they have traditional rights to use. It is far from clear how the development aspirations of the local people and conservation of a high profile but endangered species can be reconciled. Although participatory approaches to conservation are always preferable, there are circumstances when it is neither moral nor responsible for the developed world to leave the burden of biodiversity conservation to local communities in developing nations.

CONCLUSIONS

Conservation is a land-use management problem that has been the exclusive purview of biology for too long. It is improbable that any attempts to maintain biodiversity will be effective without both the spatial and social perspectives that geography is so able to integrate. Biology has singularly failed to address either of these, lacking both the analytical tools and the research experience.

Much remains to be done where geographers can take a lead. First and foremost, conservation needs an adequate methodology for the rapid assessment of the functional values of diverse communities of organisms. A biology-based conservation science has been overly preoccupied with species diversity and has ignored the fact that other levels of organismic variation may be of equal or even greater importance. Although there are established methods for monitoring relationships between biodiversity and biogeochemical cycling, hydrology, the atmosphere and vulnerability to natural and man-made hazards, the identification of functional groups of organisms is still under-researched. No adequate classifications exist to assist practical conservation.

Geography will undoubtedly play an important role in the development of methods for estimating biodiversity in non-sampled areas. A great deal of research effort is currently underway investigating the potential for the remote sensing of correlates with biodiversity.

Metapopulation dynamics offers an attractive theoretical framework within which to analyse the spatial dynamics of species populations. However, there remains a gulf between the theory and its practical application. Conservation needs to develop methods for using the geographical tools of spatial analysis to make empirical studies of spatial dynamics and to assess the consequences of different management options. One particularly important current problem is the prediction of species, responses to human-induced environmental change. Large-scale processes will act over an entire species population.

Finally, geographers may bring to conservation an ability to synthesise both ecological and social science perspectives.

Conservation biologists have little or no training in the theory or methods of social science research. The lack of training or understanding of practitioners who have had a science education can explain much of the naivety of many conservation projects and the slow rate at which methods for participatory conservation have been adopted. Geography, with its spatial and interdisciplinary perspectives, is the conservation discipline of the future.

GUIDE TO FURTHER READING

Maurer, B.A. (1994) *Geographical Population Analysis: Tools for the Analysis of Biodiversity*. Blackwell Scientific Publications, Oxford. An advanced-level text that discusses the spatial dynamics of species populations and describes the mathematical techniques for their description and analysis.

Rosenzweig, M.L. (1995) *Species Diversity in Space and Time.* Cambridge University Press, Cambridge. An excellent analysis of the ecology of species diversity with particular emphasis on the spatial and temporal dynamics of populations.

Scott, J., Davis, F., Csuti, B., Noss, R., Butterfield, B., Groves, C., Anderson, H., Caicco, S., D'Erchia, F., Edwards, T., Ulliman, J. and Wright, R. (1993) Gap analysis: a geographic approach to protection of biological diversity. *Wildlife Monographs* 123, 1–41. A detailed analysis of a technique for identifying gaps in conservation provision whereby spatially explicit data such as vegetation types and wildlife habitats are superimposed on maps of protected areas.

Sutherland, W.J. (ed.) (1998) *Conservation science and action*. Blackwell Science, Oxford, UK. This book brings together contributions from authors from a wide range of disciplines to create a state-of-the-art review of conservation biology. Chapters by Kevin Gaston on Biodiversity and William Adams on Conservation and Development are particularly valuable.

REFERENCES

Andrewartha, H.G. and Birch, L.C. (1954) *The Distribution and Abundance of Animals*. Chicago: University of Chicago Press.

Baltensweiler, W. (1984) The role of environment and reproduction in the population dynamics of the larch budmoth *Zieraphera diniana* Gn. (Lep.: Torticidiae). in W. Engels, W.H. Clark, A. Fischer, P.J.W. Olive, and F.F. Went (eds) *Advances in Invertebrate Reproduction*, 3, 291–301.

Belbin, L. (1993) Environmental representativeness: regional partitioning and reserve selection. *Biological Conservation* 66, 223–30.

Blamford, A., Green, M.J.B. and Murray, M.G. (1996) Using higher-taxon richness as a surrogate for species richness: I Regional tests. *Proceedings of the Royal Society of London, Series B* 263, 1267–74.

Brooks, D.R., Mayden, R.L. and McLennan, D.A. (1992) Phylogeny and biodiversity: conserving our evolutionary legacy. *Trends in Ecology and Evolution* 7(2), 55–9.

Clinebell, R.R., Phillips, O.L., Gentry, A. H., Starks, N, and Zuuring, H. (1995) Prediction of neotropical tree and liana species richness from soil and climatic data. *Biodiversity and Conservation* 4, 56–90.

Connell, J.H. (1978) Diversity in tropical rain forests and coral reefs. *Science* 199, 1302–10.

Hanski, I. (1996) Metapopulation ecology. In Rhodes, O.E., Chesser, R.K. and Smith, M.H. (eds) *Population Dynamics in Ecological Space and Time*, Chicago: University of Chicago Press.

Hawthorne, W.D. (1993) *Forest regeneration after logging: findings of a study in the Bia South Game Production Reserve, Ghana. ODA Forestry Series No. 3*, Overseas Development Administration, London.

Hecnar, S.J. and M'Closkey, R.T. (1997) Spatial scale and determination of species status of the green frog. *Conservation Biology* 11, 670–82.

Hodder, K.H. and Bullock, J.M. (1997) Translocations of native species in the UK: implications for biodiversity. *Journal of Applied Ecology* 34, 547–65.

Janzen, D. (1970) Herbivores and the number of tree species in tropical forests. *American Naturalist* 93, 338–9.

MacArthur, R.H. and Wilson, E.O. (1967) *The Theory of Island Biogeography.* Princeton, New Jersey: Princeton University Press.

Macdonald, D.W., Tattersall, F.H., Brown, E.D. and Balharry, D. (1995) Re-introducing the beaver to Britain: nostalgic meddling or restoring biodiversity? *Mammal Review*, 25,161–200.

Mares, M.A. (1992) Neotropical mammals and the myth of Amazonian biodiversity. *Science* 255, 976–9.

Newbery, D.M., Campbell, E.J.F., Lee, Y.F., Ridsdale, C.E. and Still, M.J. (1992) Primary lowland dipterocarp forest at Danum Valley, Sabah, Malaysia: structure, relative abundance and family composition. *Philosophical Transactions of the Royal Society London, Series B* 335, 341–56.

O'Neill, R.V., Krummel, J.R., Gardner, R.H., Sugihara, G., Jackson, B., DeAngelis, D.L., Milne, B.T., Turner, M.G., Zygmnuht, B., Christensen, S.W., Dale, V.H. and Graham, R.L. (1988) Indices of landscape pattern. *Landscape Ecology* 1, 153–62.

Phillips, O.L., Hall, P., Gentry, A.H., Vásquez, R. and Sawyer, S. (1994) Dynamics and species richness of tropical rain forest. *Proceedings of the National Academy of Sciences USA* 91, 2805–9.

Prendergast, J.R. (1993) Rare species, the coincidence of diversity hotspots and conservation strategies. *Nature* 365, 335–7.

Ricklefs, R.E. (1987) Community diversity: relative roles of local and regional processes. *Science* 235, 167–71.

Ricklefs, R.E. and Schulter, D. (1993) Species diversity: regional and historical influences. in R.E. Ricklefs and D. Schulter (eds) *Species Diversity in Ecological Communities*, Chicago: University of Chicago Press.

Scott, J., Davis, F., Csuti, B., Noss, R., Butterfield, B., Groves, C., Anderson, H., Caicco, S., D'Erchia, F., Edwards, T., Ulliman, J. and Wright, R. (1993) Gap analysis: a geographic approach to protection of biological diversity. *Wildlife Monographs* 123, 1–41.

Shaffer, H.B., Fisher, R.N. and Davidson, C. (1998) The role of natural history collections in documenting species declines. *Trends in Ecology and Evolution* 13(1), 27–30.

Simberloff, D.S. and Abele, L.G. (1976) Island biogeographic theory and conservation practice. *Science* 191, 285–6.

Terborgh, J. (1986) Keystone plant resources in the tropical forest. In M.E.Soulé *Conservation biology: the science of scarcity and diversity*, Sinauer Associates, 330–44.

WRI/IUCN/UNEP (1992) *Global Biodiversity Strategy: Guidelines for Action to Save, Study and Use Earth's Biotic Wealth Sustainably and Equitably*. World Resources Institute, Washington DC, World Conservation Union, Gland, and United Nations Environment Programme, Nairobi.

16

Landscape evaluation

Rosemary Burton

INTRODUCTION

The physical landscape consists of two elements, the landform landscape and the land-use landscape. The former is made up of land and drainage systems and is a product of the interaction between geology, climate and tectonics expressed through geomorphological processes. The land-use landscape consists of the land surface, which in most climatic zones is dominated by the flora and fauna, and is the product of ecological processes. In a truly 'natural' landscape, these processes are unmodified by man. However, throughout the world man is an agent of rather more rapid environmental and landscape change, either directly as a consequence of past and present exploitation of natural resources, or indirectly through man-induced climate change.

Therefore it could be argued that there are few if any areas in the world where landscapes are totally free of man's influence. Indeed, landscape ecologists include man as an integral part of the landscape (Neveh and Leiberman 1989). Landscapes that result from the interaction between people and land are termed 'cultural' landscapes.

Early geographical interest in landscapes concerned their analysis rather than their evaluation, and one strand of geographical research has continued to describe, analyse, classify and map landscape character, most recently with the use of geographical information systems (GIS) (Jeurry Blankson and Green 1991; Brabyn 1996).

Such studies aim to be objective and non-evaluative; they provide a database that can inform the implementation of spatial landscape policies but do not contribute directly to the policy debate. However, it is applied research in that it provides tools and techniques of immediate practical use in policy implementation.

Landscape evaluation research, on the other hand, is by implication policy-related because it is concerned with the values that different people attach to landscapes. Landscapes can be valued for different things, such as their ecological characteristics, their visual qualities, and their cultural and historical meanings. Evaluation of the ecological aspects of landscape is generally left to expert ecologists, because such judgements are made on criteria such as biodiversity, rarity and complexity rather than on criteria related to visual characteristics, and ecologically based landscape planning on other characteristics of natural systems (Selman 1993). While the separation of the 'natural' and 'cultural' is to be regretted (Phillips 1998), in practice landscape evaluation research has been concerned more with investigating the visual, aesthetic, cultural and heritage values of landscape.

THE NATURE OF THE PROBLEM

The designation of the world's first national parks in the nineteenth century (e.g. Yellowstone, USA, in 1872, the Royal in Australia, 1879) indicates that the protection of valued landscapes is not a recent addition to the planning agenda. The process of identifying and designating valued landscapes for protection has continued throughout the twentieth century, but the rapid growth of recreation and tourism in the Western world since the 1960s has increasingly focused attention on landscape as a leisure resource, and on the potential conflicts between landscape protection and its leisure use. Also, processes of globalisation have both spread and accelerated landscape change worldwide, and increased the perceived urgency for action to conserve on a world scale.

In response, policy-related landscape research has developed rapidly since 1960. Policy makers and researchers were confronted with three major policy areas that not only presented many practical problems to decision makers but also raised a host of difficult research questions for the academics. These policy areas were:

1 The rationale for the designation and management of protected landscapes.
2 The planning and management of recreational and tourist use of landscapes.
3 The regulation of man-induced landscape change.

In the simplest terms, the policy makers and landscape managers needed to know which landscapes were the 'best', the 'most preferred' or 'most desired', so that landscapes could be designated for protection and shielded from undesirable change. Planners perceived a need to select, improve and create landscapes suitable for leisure use, and at the same time to protect landscapes from recreational impacts. In practice, many landscape designations were made and implemented under the existing systems (Box 16.1) before rigorous research on landscape values was available.

However, this does not invalidate such

Box 16.1 World systems of landscape designation

THE RATIONALE FOR DESIGNATION AND MANAGEMENT OF PROTECTED LANDSCAPES

Under the auspices of the UN, the IUCN has developed a system of classification with six categories of different types of protected area. In only two of these categories is the character or quality of the landscape one of the criteria for designation.

IUCN category 2 covers national parks, selected primarily from ecosystems that are not materially altered by man and are predominantly unspoilt natural landscapes. Their management would be preoccupied with the problems of maintaining the integrity of their ecosystems rather than with landscape quality, because a natural landscape is the product of the unconstrained operation of these natural processes. However, additional criteria for national park designation include their recreational or scientific interest, or their great landscape beauty. Thus, according to the IUCN, national parks must be natural areas but may also be beautiful. Landscape value is not the main criterion for selection and designation.

It is only in IUCN category 5 ('Protected landscapes') that the character and national value of the landscape is the primary criterion for designation. According to the IUCN, the purpose of this designation is 'to maintain nationally significant natural landscapes which are characteristic of the harmonious interaction of man and the land, while providing opportunities for public enjoyment through recreation and tourism within the normal life style and economic activity of these areas.' Although the term 'natural landscape' appears in this definition, these landscapes are primarily cultural landscapes. It should be noted that UK national parks fall into this category.

The managers of such areas are faced with the problem of identifying the 'national significance' of the landscape, and protecting the landscape characteristics that represent or embody that significance. This requires a clear view of the value of such landscapes to the nation concerned. The aesthetic qualities of the landscape will almost certainly contribute to its national significance, but other values, cultural, historic, recreational and social, may also play a part. The protection of landscape features that characterise these values may conflict with the interests of the economic activity in the area. Different groups in the community may value the landscape for different reasons and to varying extents. Thus the managers of these 'protected landscapes' face even more contentious policy issues than the managers of international national parks.

The UN World Heritage Convention has led to another international system of protective designation, which is quite independent of the IUCN system. World heritage areas can be designated for their natural features and must meet at least one of four criteria, only one of which, that of containing 'superlative natural phenomena, or areas of exceptional natural beauty or aesthetic importance', concern their landscape quality. In this case, the emphasis is on their aesthetic value.

In 1992, the World Heritage Committee revised the criteria for cultural sites in such a way as to allow three types of cultural landscape to be considered for inscription on the world heritage list. These are:

1 Landscapes designed or created intentionally by men.
2 Organically evolved landscapes (relict or continuing).
3 Associative landscapes, defined in terms of their powerful religious, artistic or cultural associations.

Source: Droste *et al.* 1995; Phillips 1998.

research: landscape policy is dynamic and designation systems can be changed (Box 16.1). Landscape evaluation research needs to develop a raft of methods that can assist future policy revision. Once designated, the practical problems of managing protected landscapes raise the very issues that landscape evaluation research addresses: managers need to clarify the purpose of designation and identify the qualities for which the landscape is valued before effective management prescriptions can be found. Evaluation research has found a practical application at this later stage in the planning process. Landscape managers are also required to justify the social and economic costs of management. This has led policy makers to perceive a need to express landscape value in monetary terms.

The applied researcher's job is first to translate all these policy needs into researchable questions, and then after the research is complete, to repackage and present the results to the practitioner in a way that is understandable, politically realistic and usable. The simplification that may occur at this stage does not mean that the research itself is any less rigorous than theoretical work, rather the reverse. If important resource allocation decisions are made on the basis of the results, it must be as rigorous as possible. The extent to which policy-related and applied research is actually used in practice is a function not only of the researcher's presentation skills but also of the policy maker's perception of the need for research.

The next section illustrates the different ways in which researchers have analysed the nature of the problem from different theoretical standpoints, and how they have reduced the issues to researchable questions.

Landscape evaluation research into the visual qualities of landscapes

The policy makers question 'which landscapes are the most beautiful, best or most preferred' and 'can they be mapped', but these questions in themselves are not directly researchable. The use of 'expert judgement' of aesthetic quality was quickly abandoned as a foundation for decision making because of the implied subjectivity and lack of rigour. In its place, researchers concentrated on the more fundamental research questions and methodological problems that had to be resolved before researchers could offer policy makers useful conclusions. These were:

1 How do people perceive landscapes?
2 Can people's preferences for different landscapes be measured?
 if so
3 What quantifiable visual landscape features are associated with the most preferred landscapes?
4 Is it possible to measure preferences for specific elements of the landscape?
5 Is there a consensus of opinion as to which landscapes are the most preferred?
 if not
6 What factors explain the variation in people's visual evaluation of landscapes (such as personality, motivation, socio-economic factors and cultural background).

Geographers have adapted and applied theories and techniques drawn from environmental psychology to explore these issues (Zube *et al.* 1982; Uzzell 1991). The research was generally approached from a positivist standpoint, using quantitative methods. The general assumption of this type of research was that people's response to the visual characteristics of the landscape could be measured accurately and that landscape preferences reflected aesthetic quality.

This type of research originally took the policy makers' view that landscape beauty was a function of the landscape itself (and was therefore mappable), but the results soon shifted research towards seeking an understanding of the psychological make up, the functional requirements and the cultural context of the viewer of the landscape.

Many studies, including cross-cultural studies, suggest a consensus of preference for land with high relative relief and with green (or varied) vegetation. However, cross-cultural quantitative research has shown some cultural differences in preference patterns (e.g. Yu 1995). This indicated that preference (or landscape taste) for some landscapes or combinations of landscape elements are culture-specific, suggesting that perhaps some preferences may be learned or acquired.

Landscape evaluation research into the social, cultural and heritage values of landscapes

The outcome of the quantitative research into the visual evaluation of landscape shifted the nature of the research question in a subtle way. Researchers now focused on questions such as 'why and for what are landscapes valued?' and 'who values them?'

Social geographers have been responsible for developing and adapting different techniques to explore these issues. The social, cultural and heritage values of landscape have been investigated using qualitative methods adapted from sociology, anthropology and cultural studies. This type of research has addressed very broad issues, including:

1. What are the social, cultural and heritage values that make a landscape significant to an individual, to a social group and to a nation?
2. What features of the landscape denote these values and meanings?
3. Do different social and cultural groups value landscapes for different things and in different ways?
4. How are these values expressed?

The values of contemporary cultures have been explored using ethnographic methods and the analysis and interpretation of narrative. This includes the analysis of all forms of documentary, written, spoken, visual and other recorded evidence. This type of research is preoccupied not only with the meanings that people and cultures attach to and read into landscapes (Harrison *et al.* 1986) but also with the importance of these meanings to people's social and cultural identity. These values are not measurable in the same way as visual preferences.

Approaching the issue from a very different standpoint, post-modern geographers are grappling with the issue of how landscapes acquire particular social, cultural or aesthetic values, and how these learned values change with time. This is being done through the analysis of art, the media and historic documents, and other historic evidence that provides semiotic or symbolic connections between landscape and perception (Cosgrove 1990).

The practical dilemma that policy makers face is that although analysis of literary accounts of landscapes has shown that cultural landscape values have changed fundamentally over time in some cultures, the contemporary population normally shows a strong preference for the *status quo* and will resist any major landscape change. However, little applied research has yet addressed the issue of contemporary perceptions and evaluations of landscape change, and post-modern geography has not yet got to the stage whereby its theoretical approach can be translated into practice. Nevertheless, research that rests on post-modern approaches to theory does have potential relevance to policy making in that it acknowledges complexity, contradiction and difference. This is essential in a world that increasingly tries to accommodate cultural diversity in many areas of social and environmental policy.

Landscape evaluation research into the monetary value of landscape

The policy makers' question, 'is the landscape worth what it costs to manage it?' is to them a most urgent and relevant one. It is a completely different policy question to that of the visual, cultural or social value of landscape. To some it is an unresearchable question. Monetary 'worth' and non-monetary 'value' are two completely different concepts. Nevertheless, environmental economists have applied techniques such as contingent valuation (CVT) to landscape in order to give policy makers a direct answer (Price 1994), but without appearing to analyse the question into its researchable components, and without following other researchers' methodological efforts of clearly identifying what it is about a landscape that people are ascribing a (monetary) value to.

Summary of the nature of the problem

It is clear that research cannot give policy makers the clear, unambiguous answers that they seek. It is not the role of research to make value judgements or political decisions on behalf of the policy makers. Landscape evaluation research, however, can provide information about, and interpretations of, the general population's landscape values that can clarify the nature of the decisions that face policy makers and landscape managers (Sidaway 1990). Research can help practitioners to understand the political nature and practical implications of the decisions they do make.

CASE STUDIES

Evidence of how research has influenced policy is difficult to obtain by those outside the planning and management process. The case studies presented here are not necessarily 'best practice' in research terms, but they do illustrate the ways in which research appears to have influenced management at site level, and policy at strategic level.

Photo preference studies have frequently been used to aid practical site management. At Cannock Chase it was an integral part of the site management plan (Box 16.2). More recently, Karjalainen (1996) has used a similar methodology to investigate stakeholders' preferences for landscapes produced by different forms of forest management in Finland. The emphasis was on measuring respondents' preferences for the visual characteristics of the vegetation patterns produced by different methods of clear felling. Evidence of the practical application of this example is less direct, but it once again shows the potential for this type of research to feed into forest management at both site and forest level.

Landscape evaluation research has also had some effect on strategic policy making. The evolution of ethnographic and phenomenological research into the social and cultural values and meanings of landscapes has the capacity to influence management policies. It would appear that the work of Harrison *et al.* (1986) had an impact on the review of UK national countryside recreation policy in the 1980s, while at a regional level, Kirby (1993) highlights impending management problems. She demonstrates a divergence in values between residents and the strategic management authority for the South West New Zealand (Waahipounami) World Heritage Area (Plate 16.2). This divergence presents critical management choices, particularly in the way that the area is interpreted to visitors. More fundamentally, it points to the need for changes in the management style and structures. Researchers from this school have also informed and perhaps led the world debate on the criteria for designating cultural landscapes. The redrafted cultural criteria for world heritage properties was finally produced in 1992 (see Box 16.1).

This change led directly to the Tongariro National Park in New Zealand being designated as the first associative cultural landscape under the new guidelines. The park itself was established in 1887, when the Maori donated the three sacred mountain peaks to the Crown for conservation purposes. To date, the government has prioritised the management of the natural qualities and the recreational use of the area. However, the newly affirmed recognition of Tongariro as an associative cultural landscape provides the government with the opportunity to take greater account of the spiritual meaning of the place in its practical management strategies. The World Heritage Committee insisted that the National Park Management Plan and related processes should reflect and involve Maori concerns more than they have in the past (Department of Conservation 1990). But it still remains to be seen whether and how the spiritual values of Tongariro impact on its management (Kirby 1997).

Uluru-Kata Tjuta National Park in Australia may be perceived by non-aboriginals as a unique natural feature, and Ayers Rock is indeed an icon of Australia. It was listed as a world heritage natural site in 1986. However, the evolving world debate on the social and cultural values of land-

Box 16.2 Cannock Chase Country Park, Staffordshire, England

The preparation of the Cannock Chase Country Park management plan was an exercise in applied research (Rodgers *et al.* 1982). A landscape preference study was used to clarify some critical management choices. The park is intensively used for recreation, but much of it is also designated as a site of special scientific interest (SSSI), showing ecological characteristics transitional between upland and lowland heath with fragments of ancient oak woodland and mire communities. It is a cultural landscape in that the open heathland is a result of woodland clearance, sheep grazing and other forms of land use over the last few hundred years (see Plate 16.1). However, these practices had ceased by the 1940s, and the landscape has measurably changed since then, with birch, hawthorn scrub and seedlings from adjacent coniferous plantations invading the heathland. Controlling this vegetational change would be very expensive as grazing was thought not to be a practical option.

Plate 16.1 Cannock Chase Country Park.

The research team assumed that the open heath was the most valued landscape for recreation, but evidence of user preferences was needed before an expensive management regime was recommended. A photo preference study was therefore undertaken to measure user preferences for different types of vegetation as recreation environments (Burton 1982). The study was designed as rigorously as possible to ensure that users were responding to the different vegetation types and not to any other aspect of the picture of the landscape.

Unexpectedly, the results indicated that users very strongly preferred the ancient woodland; only a minority ranked the open heathland vegetation as the most preferred. The research teams' response was to put far more research effort into investigating why the ancient woodland was not regenerating and on proposals for the management of its recreational use, while in the final plan the heathland management proposals were justified on ecological grounds rather than on their landscape value.

scapes led to a very significant change in its status ten years later (Phillips 1998). The realisation that the surrounding landscape was a product of a continuing regime of traditional (aboriginal) fire management, and the formal acknowledgement of the spiritual significance of Uluru to the Aborigines (Layton and Titchen 1995) led to the rescheduling of the site on the world heritage list as an associative cultural landscape (Plate 16.3). The implications of this change for tourism management are potentially profound. Uluru is one of Australia's most important international tourist attractions and is seen as an awesome natural feature that tourists wish to climb. The traditional owners (Aborigines) see it as a place of immense spiritual power that they do not climb. Again, it remains to be seen how or whether the reclassification of the park will be reflected in its management planning.

Whatever the objectives of landscape management, justifying the costs of management is an ever-present strategic policy issue. CVT has been applied to landscape in order to measure the monetary 'worth' of landscapes (Box 16.3). In the UK, the technique has been used in a project in the Yorkshire Dales National Park.

The extent to which the Yorkshire Dales research, and its extension to the Norfolk Broads (Bateman *et al.* 1994), has influenced UK strategic policy is not clear, as the results in fact appear to

Plate 16.2 Landscapes of the south west New Zealand World Heritage area, South Island, New Zealand.

Plate 16.3 Kata Tjuta in Uluru National Park, Northern Territory, Australia.

justify the *status quo*. The importance of this particular research may not concern its immediate impact on short-term financial decisions. It may have more to do with attitudes to the management of landscape change. If people do prefer the *status quo*, and in this case if the monetary value of the existing landscape justifies current expenditure, perhaps research attention will be directed to analysing the process by which preferences evolve, how current expenditure does effect landscape change and how people become accustomed to the changing landscape in which they live.

CONCLUSIONS

Visual and functional values of landscape can be demonstrated using environmental psychology methods; such research can be of practical use to

Box 16.3 Yorkshire Dales National Park, UK

The upland limestone country of the Yorkshire Dales National Park (UK) is in IUCN terms a category 5 protected cultural landscape (see Plate 16.4). Its field pattern was established during the 1760–1820 enclosures. The landscape depends on traditional sheep rearing, which is now under threat. Research (O'Riordan *et al.* 1992) was designed to assess user and resident preferences for pictures representing eight different versions of the Dales landscape that could be created in response to different types of change in the farming and land management systems. A parallel study by Willis and Garrod (1993) attempted to assess the monetary value put on these future landscapes. The authors did not seek separate assessments of the ecological, social or heritage values but attempted to obtain one overall assessment that took all these attributes into account. About half of the respondents chose landscapes depicting the existing situation as the scene they liked best. Most of the others chose the 'conserved' or 'planned' landscapes, options that were visually most similar to today's landscape. The willingness to pay for today's landscape was around £24 for all respondents, but the minority who preferred 'conserved', 'sporting' or 'wild' landscapes valued them far more highly (at around £34). Residents valued landscapes consistently lower than visitors, this discrepancy being particularly marked for the 'abandoned' landscape. The results show that although the overwhelming consensus is to preserve the *status quo*, there are some important differences in preferences and values among minority groups, who might be important players in the pragmatic process of landscape management.

However, the study uses the data in a different way. The results are aggregated to demonstrate that 'today's' landscape generates total 'benefits' four times higher than the cost of maintaining it. In contrast, the costs of the 'conserved', 'planned' and 'sporting' landscapes are far higher than the benefits they generate.

The conclusions thus support the *status quo*.

Plate 16.4 Agricultural landscapes of the Yorkshire Dales.

guide designation policies, to set management objectives and to prioritise resource allocation.

However, cultural geographers would warn that this approach may only be valid if all the stakeholders are drawn from one homogeneous cultural group and therefore are likely to share common cultural values. Social geography research highlights the role of landscape in the expression of cultural identity: in a multicultural state the issue of whose landscape is valued most highly is intensely political.

Techniques that seek to assign monetary values to landscape must also be entirely culture-specific, and strictly limited to Western capitalist cultures. Even in Western cultures, the assumptions on which the technique is based need to be far more closely analysed before being built into any policy-making process, not least the assumption that people are capable of assigning money values to environmental goods in a rational manner. The clarification of which aspect of landscape is being valued (aesthetic, ecological, cultural, functional, etc.) is also crucial, as the practical management prescriptions in each case can be so radically different.

Ethnographic methods have a critical role in the exploration of landscapes' cultural and heritage values. This type of research has tended to replace the concept of 'landscape' by that of 'place'. This may reaffirm moves towards more integrated approaches to management that acknowledge links between social, economic, cultural, aesthetic and ecological planning.

The analysis of the way in which people respond to landscape change may be the biggest research challenge in the future. Processes of climate change may necessitate profound changes in land management practices, even in traditional cultures: in tandem with the effects of globalisation, this may accelerate landscape change in many parts of the world. Longitudinal descriptive surveys have recorded and monitored landscape change, particularly that generated by changes in the agricultural and forestry economies. Very considerable methodological problems face researchers attempting to investigate people's perception and evaluation of contemporary landscape change. This is a crucial area for developing research, as the management of the speed of landscape change (not its direction) may in the end be the most important issue facing practitioners, rather than the maintenance of any particular aesthetic quality or cultural characteristic.

GUIDE TO FURTHER READING

Uzzell (1991) and Zube *et al.* (1982) provide excellent overviews of the range of environmental psychology approaches, and both provide extensive references to further case studies.

Cosgrove (1990) offers a similar analysis of the social geographer's approach, while Price (1994) gives a comprehensive review of the economic evaluation techniques with a very full bibliography. Sidaway (1990) debates some policy applications of landscape research in the UK context, while Burgess (1996) comments on recent trends in landscape research.

Up-to-date reports of research developments are most readily accessed in the journal *Landscape Research*, while *Landscape Planning* and *Landscape and Urban Planning* frequently publish applied research.

Related issues such as landscape mapping can be followed up in *Landscape Research* Vol. 19 No. 3, 1994, and landscape ecology in Selman (1993).

REFERENCES

Bateman, I., Willis, K. and Garrod, G. (1994) Consistency between contingent valuation estimates. A comparison of two studies of UK national parks. *Regional Studies* 28, 457–74.

Brabyn, L. (1996) Landscape classification using GIS and national digital databases. *Landscape Research* 21(3), 277–87.

Burgess, J. (1996) The future for landscape research. *Landscape Research* 21(1), 5–12.

Burton, R.C.J. (1982) *Visitor–public preferences for vegetation types*. Technical Report No 7. Cannock Chase Country Park Plan, Countryside Commission.

Cosgrove, D. (1990) Landscape studies in geography and cognate fields of the humanities and social science. *Landscape Research* 15(3), 1–6.

Department of Conservation (1990) *Tongariro National Park Management Plan Vol. 1 Objectives and Policies*. Turangi, New Zealand. Department of Conservation.

Droste, B., Plachter, H. and Rossler, M. (1995) *Cultural Landscapes of Universal Value*. New York: Gustav Fischer Verlag.

Harrison, C., Limb, M. and Burgess, J. (1986) Recreation 2000: views of the country from the city. *Landscape Research* 11(2), 19–24.

Jeurry Blankson, E. and Green, B. (1991) Use of landscape classification as an essential prerequisite to landscape evaluation. *Landscape and Urban Planning* 21, 149–62.

Karjalainen, E. (1996) Scenic preferences concerning clear fell areas in Finland. *Landscape Research* 21(2), 159–73.

Kirby, V.G. (1993) Landscape, heritage and identity. Stories from the West Coast. In C.M. Hall and S. McArthur (eds) *Heritage Management in New Zealand and Australia*. Auckland: OUP, 119–29.

Kirby, V.G. (1997) *Heritage in Place*. Unpublished Ph.D. thesis. University of Canterbury, Christchurch, New Zealand.

Layton, R. and Titchen, S. (1995) Uluru: an outstanding Australian Aboriginal cultural landscape. In van Droste, B., Plachter, H. and Rossler, M. (eds) *Cultural Landscapes of Universal Value*, New York: Gustav Fischer Verlag, 174–81.

Neveh, Z. and Leiberman, A. (1989) *Landscape*

Ecology: Theory and Application. New York: Springer Verlag.

O'Riordan, T., Wood, C. and Shadrake, A. (1992) *Interpreting Landscape Futures in the Yorkshire Dales National Park*. Yorkshire Dales National Park, Grassington.

Phillips, A. (1998) The nature of cultural landscapes – a nature conservation perspective. *Landscape Research* 23(1), 21–38.

Price, C. (1994) Appendix: Literature review. *Landscape Research* 19(1), 38–55.

Rodgers, H.B., Burton, R.C.J. and Shimwell, D.W. (1982) *Cannock Chase: the Preparation of a Country Park Management Plan*. CCP 154, Countryside Commission, Cheltenham.

Selman, P. (1993) Landscape ecology and countryside planning: vision, theory and practice. *Journal of Rural Studies* 9(1), 1–21.

Sidaway, R. (1990) Contemporary attitudes to landscape and implications for policy: a research agenda. *Landscape Research* 15(2), 2–6.

Uzzell, D.L. (1991) Environmental psychological perspectives on landscape. *Landscape Research* 16(1), 3–10.

Willis, K.G. and Garrod, G.D. (1993) Valuing landscape: a contingent valuation approach. *Journal of Environmental Management* 37, 1–22.

Yu, K. (1995) Cultural variations in landscape preference: comparisons among Chinese sub-groups and Western design experts. *Landscape and Urban Planning* 32, 107–26.

Zube, E.H., Sell, J.L. and Taylor, J.G. (1982) Landscape perception: research, application and theory. *Landscape Planning* 9, 1–33.

17

Environmental impact assessment

John Blunden

INTRODUCTION – THE BASICS OF THE ASSESSMENT PROCESS

The process of environmental impact assessment (EIA) was introduced for the first time in the United States in 1969 under the National Environmental Policy Act for all major federal activities. Since then, there has been an ever-widening acceptance, particularly by the industrialised nations of the world, of the view that environmental effects likely to be caused by a proposed development are material considerations within any planning decision-making process. The influence of the US federal measures led to the rapid incorporation of EIA into state and local statutes across that country and then by the government of Canada in 1973. Many other developed countries followed including, Australia at commonwealth level (1974), Japan (1984) and New Zealand (1991). Although a number of its member countries, such as France and Ireland, had embraced EIAs as early as 1976, followed by the Netherlands (1981), the Council of Environmental Ministers of the European Communities did not adopt a Directive on EIAs for certain types of development until 1985. Their implementation became mandatory in 1988 (Montz and Dixon 1993; Sanchez 1993; Geraghty 1996). As for developing countries, while many of the 121 sovereign states that might be so categorised had, by the 1990s, at least considered EIA legislation, only nineteen had put in place the necessary administrative, institutional and procedural frameworks for the implementation of EIA systems, only six of which were successfully operational (Ebisemiju 1993).

Sometimes there is a single well-defined catalyst for action in the decision to adopt EIA, as was the situation in Austria in the mid-1980s when Hainburg, the proposed site of a hydro-electric power plant on the Danube, became a symbol of environmental and citizens' activism (Davy 1995). Elsewhere, the process has been more incremental, especially where member states of a federation are concerned. This was certainly the case in Australia (Wood 1993). But always the concept of EIA has evolved in response to real needs, and wherever it is used it is in real-world situations. It is not an intellectual exercise practised by academics, nor is EIA designed to provide a passive record of environmental change. Its sole objective is that of making available environmental information on which informed decision making may take place in relation to projects both public and private (Beattie 1995).

EIA has had a number of definitions in the last three decades, many of which are founded on the objectives and experience of their authors, whether they be institutions, government agencies or individual researchers conducting an examination of the practice of EIA, and whether they be located in the developed or developing countries (Sankoh 1996). However, the United Nations Development Programme (UNDP) in 1992 accepted a simple definition of EIA from an authoritative source (Clark *et al.* 1980) that has widespread currency (Table 17.1). The UNDP has also usefully summarised what it sees as the common key activities involved in the preparation of EIAs, along with the techniques used in the presentation of the assessment (see Table 17.1).

However, the determination in 1985 of the European Commission (EC), that EIAs should become the practice in all member countries, led, in the preamble to Directive 85/337, to it both accepting and embellishing the 1980 definition. It did so by affirming that the EIA process involves a systematic approach that embraces both a structured methodology and a formal set of procedures, adding that the investigation that constitutes the EIA implies the preparation of a report – an environmental impact statement (EIS) – which in itself 'provides the basis for consultation, participation and decision making' (Wood *et al.* 1991). The systemic nature of the EIA process, wherever it is applied, ideally involves both the developer and the relevant planning authority in an iterative process where it is the feedback loops that will help to minimise impacts, improve attempts to mitigate environmental damage and assist in general project design (Figure 17.1). Ultimately, such an approach must also affect the quality of decision making that follows a full consideration of the EIS, a document for which the developer bears the full responsibility.

EIA – PROBLEMS OF APPLICATION AND INTERPRETATION

Subverting the EIA process

Having stressed the procedural strength of EIA in the provision of a clear-cut method of evaluation, the initial acceptance of the need for EIA, as well as its eventual application, can have political dimensions, sometimes amounting to the subversion of its intent. When the suggestion was made that the EU should embrace EIAs, some member states did not view the prospect with much enthusiasm. France, particularly, feared both national and transnational problems if EIAs were applied to the large number of nuclear

Table 17.1 Environmental impact assessment.

1 *Definitions:* EIA is an analytical process that systematically examines the possible consequences of the implementation of projects, policies and programmes. Its main objective is to provide decision makers with an account of the implications of alternative courses of action before a decision is made and then to alter, if necessary, the final project design.

2 *Activities:* most EIAs involve the:
- identification of the impact
- measurement of the impact
- interpretation of the significance of the impact
- display of the results of assessment
- development of ameliorative measures to eliminate or minimise adverse impacts
- identification of appropriate monitoring schemes

3 *Techniques:* typically the following techniques have been applied for EIA:
- mapping and overlay charts
- checklists of potential environmental impacts
- matrices of project actions and environmental impacts
- flow charts, based on systems analysis
- other models of economic environmental interaction

Source: UNDP 1992.

Figure 17.1 EIA as a systemic process.

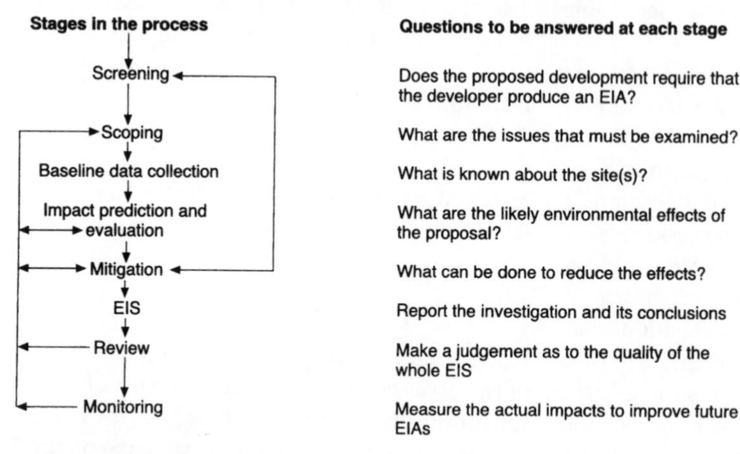

Source: Allen 1996.

power plants it was planning to build, many of which were to be located at the periphery of the state. At the same time, the UK took the view that EIA was an unnecessary and burdensome addition to its sophisticated land-use planning system, which had evolved effectively since the passing of the Town and Country Planning Act in 1947. It therefore took several years and more than twenty internal drafts before the proposal to the EC was formulated in 1985. Even then, it had to be pushed onto the agenda at a period of high unemployment throughout the Community on the back of the promise that environmental protection might hold out in terms of the creation of new jobs.

Once the desire to have an EIA programme has been accepted, the politicisation of its realisation can also become apparent. While it is clear that 'the scope, timing and content of EIAs everywhere in the world are invariably influenced largely by a variety of administrative and legislative measures' (Sankoh 1996), it is also evident that some countries can take matters very much further in their apparent subversion of the process. Here it may be said that if no checks are in place, 'EIA is open to capture by powerful government interests' (Horberry 1984). In Nigeria, for example, the guidelines on its fourth National Development Plan (1981–1985) contained a directive that 'feasibility studies for all projects both private and government shall be accompanied by an environmental impact statement'. Whether this was a general expression of government concern about the need to prevent environmental degradation, or intended to assuage aid agencies, or to pay lip service to international environmental conventions, it was certainly not backed by anything resembling an EIA system. Not surprisingly, ten years after the plan was published the government had to admit that environmental issues 'had been neglected or not given enough attention . . . in actions designed to increase the productivity of the society and to meet essential needs' (FEPA 1989).

In general, however, EIA can expose developing countries such as Nigeria to public scrutiny and debate, a largely unwelcome procedure for dictatorships no matter how interested such countries may be in environmental protection either notionally or in practice. Moreover, the EIA process is also open to subversion by powerful government interests where even evidence of ecological damage can be manipulated. As Ebisemiju (1993) has commented, 'unscientific and unprofessional practices thrive best in socio-political systems in which corruption and dictatorship are hallmarks'. These criticisms are less likely to be valid, however, among industrialised countries of long standing, where public scrutiny and a much more open decision-making process is in evidence. Nevertheless, examples can be found of attempts to subvert the EIA process.

For the EU, the enactment of its Directive has allowed member states a considerable degree of latitude in the way in which the systematic approach of the EIA is allowed to work out in practice in each member country in the name of subsidiarity. The government of the UK, because of its ideological objection to EIA, therefore allowed four major national infrastructural developments – the Twyford Down extension of the M3, the M11 link road, the East London River Crossing and the Channel Tunnel Rail Link – to provoke direct and unprecedented political confrontation with the EC in Brussels. All had infringed Directive 85/337, since preparatory work had been carried out on these but none had been subject to a formal EIA. When steps were taken by way of enforcement procedure to prevent further work on these projects, the EC was ignored by the British government. Unfortunately, from an environmental point of view, this head-on collision between government and Commission came at a time when the rejection of the Maastricht Treaty by the Danes (2 June 1992) meant that the EU needed the support of the UK government. By the end of the same month, in order to achieve this support, all proceedings relating to these projects were dropped, with the Commission using the flimsiest of pretexts to justify its *volte-face*. It is perhaps unfortunate that these changes of mind were possible 'without giving full explanations to complainants or to Parliament and that in the absence of the possibility of judicial review neither complainants nor Court can challenge the reasoning' (Kunzlik 1995).

Performance variables

Institutional and procedural arrangements – the developing world

In terms of the operationalisation of the EIA process, an overwhelming majority of the small number of developing countries with EIA systems have used the US federal model, whereby specific laws are enacted to make EIA mandatory and a clear system of procedures for environmental assessment laid down, albeit modified to suit local conditions. While this may well reflect a government's desire to enforce consideration of environmental issues in the decision-making process, the efficiency of the institutional arrangements that this implies have been questioned (Sankoh 1996; Ibaara 1987; Szelely 1987) and the suggestion made that these work only where the responsible environmental agency is placed within the office of president, prime minister or some other high-profile ministry such as those dedicated to national economic planning and budgetary control. Unfortunately, most agencies in developing countries are subsidiary functions of one ministry or another and thus have low status in the bureaucracy, and lack funding, trained staff and the status necessary to enforce compliance with environmental laws and regulations. In such a general situation of functional decentralisation, these problems are particularly manifest in the inability of the environment agency to muster the right degree of inter-organisational coordination and cooperation between the many sectoral agencies and tiers of government that have some responsibility for one or another aspects of the environment. Without this, it is especially difficult for the EIA to achieve its primary objective of incorporating environmental considerations into project planning, design and implementation through the disclosure of environmental effects and public scrutiny (Ebisemiju 1993).

However, as if to compound the chances of not achieving such a goal, in most developing countries the EIA is conducted as a separate exercise divorced from the technical and economic aspects of project planning and design, often appearing as an afterthought. This implies that there is little prospect of considering alternatives, with the EIA being used as a perfunctory endorsement of public or private actions rather than to influence decisions. Thus, although Thailand and the Philippines have been cited as having the most elaborate EIA systems in the developing world, their EISs have been described as 'nothing but a collection of unsynthesised biophysical data irrelevant to the choice among real alternatives' (Roque 1985). In spite of a considerable number of EIAs submitted to environmental agencies in Southeast Asia, few are able to have much impact in reducing the environmental consequences of the projects concerned.

Brazil, Malaysia and Mexico, however, do have systems that integrate the EIA into the project cycle. Apart from the advice that proponents of a scheme receive regarding the integration of the economic, technical and environmental elements of their project, it is mandatory for them to show that any conditions attached to the review report are complied with and that measures to be taken to alleviate or prevent the adverse impact on the environment are being incorporated into the design, construction and operation of the prescribed activity. In Mexico, the EIS must be approved before a project's final design is produced, while Brazil effectively internalises EIA in the project cycle by incorporating it into its long-standing three-stage project-licensing system. There are also built-in mechanisms for continuous monitoring of every stage of project implementation for compliance with recommendations contained in previous licences.

However, as far as the developing world is concerned, these three countries are quite exceptional in terms of the way their EIAs function.

Institutional and procedural arrangements – the developed world

Compared with the developing world, the implementation of the EIA process by the developed countries is generally of a higher standard. This is hardly surprising, given their advanced state of economic development, their political stability and the substantial period of time in which they have had the opportunity to embrace the original US federal initiative and to experience its operation. However, there are exceptions. For example, the Canadian federal recognition of EIA in 1973 led to the provision by government of a structure for its own system that was seen to be flawed because it allowed the initiators the responsibility to screen and to assess their own project environmentally (Cooper 1990). Moreover, the original legalisation did not provide for an enforcement mechanism. It also failed to clarify, among other things, the types of project that must be assessed, the content of an acceptable environmental assessment, and the nature of the role of the public in the process (FEARO 1987). Although the Canadian Environmental Assessment Act of 1995 was supposed to remedy such deficiencies, it continues to support self-assessment and apparently does little to clarify with precision what procedures should be followed in the EIA process. This is especially unfortunate since the federal approach fails to provide a yardstick against which the provinces, each with their very different approaches to EIA, might start to move towards common national standards (Delicaet 1995).

Perhaps more unexpected, however, are the results of a survey that considered NEPA's effectiveness after nearly three decades of its application to US federal projects. This drew attention to problems related to EIA practice rather than process, commenting not merely on the fact that the consideration of EIA in project planning and decision making was not early enough but also on the lack of post-EIS follow-up in monitoring, in the implementation of mitigating measures, in ecosystem management and in environmental auditing. The survey also remarked upon an insufficient consideration of both biophysical and socio-economic factors in an integrated mode in the EIA process. Although it is possible to allow that these issues, according to the survey's authors, particularly represent common concerns in the worldwide practice of EIA (Canter and Clark 1997), it is nevertheless unfortunate that they remain flaws in the US federal system so long after the establishment of NEPA. A lack of political will during the last two decades undoubtedly explains such inertia.

While such failings do remain a recurring phenomena and can, indeed, be recognised in countries of the developed world, there are nevertheless examples of best procedures and practices. Indeed, many of the states of the Commonwealth of Australia show an impressive record of the successful development of EIA since its adoption. In summarising the present position in Western Australia, where much has been done as a result of its new Environmental Protection Act of 1986, it has been suggested that in its 'numerous public participation and appeal provisions, the referral system, the various types of EIA report and recommendations and the strong links to action monitoring, together with the annual report on the EIA system, it provides a model worthy of wide-spread imitation' (Wood and Bailey 1994).

In Victoria, the first of the individual states to enact the relevant legislation, in 1978, the effectiveness of the system has, indeed, evolved over the years so that it has now become highly pro-active at all stages in the EIA process, culminating in the general use of inquiry panels and consultative committees. Moreover, it has succeeded in getting proponents and decision makers to consider environmental effects earlier in their planning stage. Its one remaining weakness is the lack of *mandatory* monitoring in the project cycle (Wood 1993).

CASE STUDIES

The three case studies that follow are designed to illuminate what has so far been outlined in the preceding narrative section, which has largely concentrated on the nature of the structures in place for EIA. Here the concern is much more with illustrating the practical difficulties that can arise in the execution of EIAs, which in turn can deeply affect the outcomes of the process.

As suggested on page 250 and as Box 17.1 affirms, what Brazil eventually put in place bears all the hallmarks of a system that, for the developing world, is well above the norm in terms of its apparent capacity to mediate the conflicting interests that are apparent in any project that has environmental implications. Apart from the fundamental exogenous problem that the pervading ethos in Brazil is for economic development to take priority over environmental

Box 17.1 Environmental impact assessment in Brazil

For many years, the government of Brazil had pursued the idea that rapid economic growth was of prime importance, to the extent that at the International Conference on the Environment at Stockholm in 1972, it suggested that environmentalism was an imperialist plot designed to thwart the interests of developing countries. However, it was ultimately unable to resist the demands of international financing agencies that their support for major projects in Brazil would be dependent on careful consideration of their environmental implications. The National Environmental Policy Law of 1981, suitably amended in 1989 and 1990, therefore established that potentially polluting activities would require an EIA, the criteria for which are defined by the National Environmental Council (CONSEMA).

The supporting information required for an EIA are:

1 an environmental diagnosis of the project's area of influence with a complete description and analysis of the physical, biological, cultural, historical and socio-economic environmental resources and their interactions prior to the implementation of the project;
2 an analysis of the project's likely environmental impacts based on its magnitude, temporality, reversibility and cumulative and synergistic properties in a cost–benefit analysis, along with a specification of technical or site alternatives;
3 a description of mitigating activities for negative impacts, including a comparative analysis of control equipment to be deployed and the systems to be used for waste treatment;
4 a statement as to who is responsible, whether public or private, for the programme of environmental monitoring, which is mandatory.

The conditions pertaining to (1) may require considerably more detail concerning environmental issues if the proposed project is deemed to be in an area of great sensitivity. Copies of the EIA must be sent to all government units related to the project so that it may be evaluated. Finally, the EIS that emerges is required to be written in accessible language.

Public access to information in the EIA is a constitutional right. Details of it are required to be published in local or regional newspapers and displayed in documentation centres for public environmental policy administration.

Source: Fowler and Dias de Aguiar 1993.

considerations, there are other reasons why EIA fails to fulfil its promise.

The principal problem is one of data inputs regarding the ecology of Brazil. Frequently it is not available, and where it does exist it is often out of date. Moreover, what information of value there is tends to be scattered across so many institutions that its acquisition within the time-scales available for the completion of an EIA is impossible.

Where public participation is concerned, the territory is familiar enough in so far as it is a problem that pervades systems around the world. Here, while its existence is not in doubt, it fails to

Box 17.2 EIA – Differences in attitude and practice: Norway and Estonia compared

ESTONIA

As early as 1978, the USSR Council of Ministers decided that all large projects must undergo expert review in terms of their impact on the environment. After the dissolution of the Soviet Union in 1991, Estonia was able to choose its own environmental strategies and decided to introduce instruments developed in the West, including EIA. Compared with most other post-Soviet states, Estonia made careful preparations before the formal introduction of EIA. The enabling laws were established in 1992, but it was in 1994 that the key player in the EIA adoption process, the Ministry of the Environment, produced an order that dealt with the procedural matters of EIA and gave practical advice on EIA report preparation. Since it is an aspiration of Estonia to join the EU, it is its intention to bring Estonian EIA practice into line with the EU Directives on the subject.

NORWAY

The basis for EIA in Norway when formally introduced in 1990 was the Planning and Building Act. This has long served as a means of land-use planning and the granting of building permits. Preparation for EIA, however, began in the late 1970s, but the process became protracted as a result of debates over the distribution of competencies to supervise the needs of EIA among the ministries. In order to achieve a level of integration between EIA and project planning, the responsibilities for EIA were ultimately located in the development ministries rather than in the Ministry of the Environment, whose role remains a strategic one. Because of the entry of Norway into the European Economic Area in 1994, Norway harmonised its EIA system with the EU in 1995.

The chief differences between the two countries in their approach to EIA are summarised below. They may be said to broadly reflect the situation in other Nordic states as against those of the ex-Communist Baltic states, which also include Latvia and Lithuania.

Baltic and Nordic EIA systems compared

Baltic EIA system	*Nordic EIA system*
Technical	Political
Pursuance of objectivity	Multidisciplinary approach
Staffed by engineers and natural scientists	Staffed by social scientists, natural scientists, lawyers and engineers
EIA sector homogeneous (closed) and distinct; responsibility concentrated in the EIA offices of the Ministry and in regional environmental departments; 'ecological expertise'	EIA sector heterogeneous (open); EIA tasks dispersed in various ministries, by developers, NGOs, and sectoral and local authorities
Tied to environmental/pollution control and permit procedures	Tied to land-use planning
Focus on EIA report	Focus on EIA process
Focus on effects	Focus on impacts
EIA late in project planning	EIA early in project planning

Source: Holm-Hansen 1997.

permeate all aspects of the EIA process, with the public having difficulty in making known those concerns that they believe the EIA should address. However, while it is not obligatory for EIAs to be defended publicly, it is possible for citizens to petition for it.

Finally, although the idea of an interdisciplinary approach to EIAs is mandated by the legislation, there has been a notable reluctance on the part of natural scientists and social scientists to work together, something that training programmes for those who service EIAs has as yet failed to address (Fowler and Dias de Aguiar 1993).

The lack of an interdisciplinary approach to the realisation of EIAs is the main theme in the next case study, but here the discussion is put into a comparative context to illustrate the problems that the countries of Eastern Europe, because of their Communist heritage, have in bringing their practice into line with their Western European neighbours.

Estonia represents one of the more advanced of the ex-Communist countries now seeking to conform with the EU approach to EIA as a prelude to joining the Union in the near future. Norway has moved in the same direction but is motivated by its membership of the European Economic Area. Both had two decades of experience of involvement with attempts to reconcile economic development with environmental concerns, and although they are each moving towards a common approach, their separate socio-economic histories mean that notable differences still remain. In this respect, it is Estonia's socialist heritage that plays an important role, as it does in other Eastern European countries. There, engineers still remain the backbone of the educated classes. Political development – in which the citizenry are encouraged to take part in decision-making processes and in the public debate – remain new features of life, as does the development of the social or the human sciences. Thus any problem is reduced to that in which solutions are largely technical ones. This permeates thinking in the sphere of environmental protection in a way that is perhaps stronger than in many other spheres, because it is an area where technicians, engineers and natural scientists feel most at home. But the contrast here with Norway is particularly strong because of its multi-disciplinary approach to such matters. As the tabular comparison in the case study shows, herein lie perhaps the major differences in the approach to EIA in the two countries.

As for the EU approach to EIA (Box 17.3), the decision taken to provide a basic system and to remit the implementation of its component parts to the individual member states under the banner of subsidiarity has, in retrospect, left much to be desired. Indeed, the Commission itself has recognised that its implementation had been very uneven with, at worst, some examples of the flagrant disregard of its intent. Spain, for example, blatantly reduced many attempts to profile the impact of a project on the environment merely to an account of that project's economic benefits (Pardo 1998).

It is not surprising, therefore, that the Commission should have sought, in due course, to review the process as set out in its original Directive in order to bring it into line with best practice worldwide. But this was made more compelling by the new imperative of sustainable development that followed from the Rio Earth Summit in 1992. What has been achieved as a result of the review, however, has not been without considerable compromise, necessitated by attempts to seek common ground among the member states, some of which, not least the UK, have been largely hostile to its intentions. Thus, while the key changes described in Box 17.3 do offer very positive improvements, the Amendment Directive of 1997 falls short of what might be recognised from the previous section as what was really needed. An important failure has been the inability of the Commission to insist on formal scoping, together with post-project monitoring and enhanced public participation. These should be common to any system put in place by advanced economies such as those of the EU as part of a desire to produce EIAs that offer something to the quest for sustainable development.

Box 17.3 The EU and EIA initiatives

Five years after the mandatory implementation of EC Directive 85/337, the EC saw fit to undertake its review. This highlighted a number of problems and difficulties experienced by member states, which led to the EIA Amendment Directive 97/11/EC due to be applied by all member states from March 1999.

The key changes wrought by this Directive are as follows:

1 *Project screening* This has been carried out according to two schedules. Annex I is a project list that covers all those types of development for which an EIA is mandatory. Annex II, however, is a list of development types that may require an EIA, and here the review had highlighted disparities in the ways in which member states had implemented it. Some had used thresholds and/or criteria (set at high or low levels), others had used a case-by-case approach, while some, such as the UK, used a combination of both, where thresholds/criteria were indicative only. The EC, therefore, introduced a set of selection criteria that member states must take into account when deciding how they determine which Annex II projects shall be subject to EIA. But to ensure that the most damaging projects are always subject to an EIA, a number of categories have been moved to Annex I.
2 *Scoping* While the Commission wanted the process to be a formal one that would entail consultation with the competent authority, at the insistence of the UK and Germany, it still remains open to the developer to make a decision on this. Formal scoping in public, is, though, suggested as best practice.
3 *Project information provided by the developer* It is now a requirement that an outline of alternative sites for the development is provided, together with reasons for the final choice 'taking into account the environmental effects'.
4 *Consultation* On the completion of an EIS the environmental authorities must now be consulted regarding its content, while details of the request for consent and information gathered during the EIA should be made available to the public and time given for them to respond.
5 *Transboundary effects* It is now mandatory that information from an EIA be made available to another member state if it is likely to be affected by a proposed project.
6 *Decision making* The competent authority making the decision about a project must provide the main reasons for it, together with an account of the considerations on which the decision is based.

Source: Sheate 1997.

CONCLUSION – SELECTED AREAS OF EIA RESEARCH

Given the problems that are sometimes evident in the EIA process, such as a lack of trained staff to undertake the relevant procedures, an inferior biophysical/socio-economic database and a less than adequate public participation, it is clear that 'expert systems' can offer considerable potential in terms of their resolution. Expert systems attempt to simulate the means by which a human expert tackles a real-world problem using a set of rules, heuristics and inferences programmed into a computer system. Indeed, as a problem solving device an expert system interprets information and reasons towards a conclusion obtaining the same results that the human expert would arrive at if presented with a comparable task. The component parts of the expert system and the means by which the knowledge base is amassed, then addressed and driven through the reasoning process (the neural network) to provide the relevant outcomes to the user have been described elsewhere in terms of their applications to the field of geographical research (Blunden *et al.* 1998). Suffice it to say here that this approach has been applied to EIA in a handful of pioneering experiments in North America and Europe, where they have been used to help environmental groups at public inquiries and non-experts to critique EISs that have already been prepared (Geraghty 1993).

Expert systems in the EIA context can be adapted to varying EIA assessment regimes and planning systems and have the advantage that their knowledge store is easily updated or revised as circumstances or techniques change. They also have potential in conjunction with geographical information systems and, through interface with ecological or environmental models, to produce highly sophisticated graphics. However, the use of expert systems to provide a more flexible and readily available source of expertise in this rapidly growing field is wide open to further research investigation, and applied geographers with a broad knowledge base across the environmental and the social sciences are well placed to contribute to its development.

Another area of concern for those wishing to support the effective use of the EIA in the planning decision-making process, one that is frequently in question worldwide, is that of public participation. For this to be worthwhile in terms of providing confident support for the legitimacy of the final outcome, local knowledge, both lay and expert, will need to have been examined. As part of this undertaking, affected parties will need to have explained their own views in a discourse with others of differing opinions and thus will have had an equal chance of influencing the conclusions ultimately reached. Ideally, a resolution will have been achieved to which all parties can give their support. However, in spite of the importance of this area of concern, research on the best way to achieve such ends seems less than adequate. While it may not seem the obvious topic for the applied geographer, this author has managed to demonstrate how participation can be handled effectively in the interests of achieving an environmental conservation policy, with which a wide range of actors with considerable diversity of lifestyles and interests could agree and ultimately accept as fair and reasonable. In this case, the need was to gain general acceptance for a policy for the Broadlands of East Anglia (Blunden 1985). Others have pursued similar work in connection with the siting of a waste disposal facility in the canton of Aargau in Switzerland (Webler *et al.* 1995), but both exercises have identified common characteristics of what can be described as social learning processes for those involved. These include face-to-face small groups meeting and working regularly over several months; opportunities for all participants to explain and justify their perceived needs; creating an atmosphere that encourages participants to discuss, criticise or challenge statements made by other group members; providing access to expert witnesses; and being able to undertake field visits. But this is not to say that more does not need to be done. These examples can only be early contributions to the development of more definitive models of effective participation processes.

FURTHER READING

Two major texts, not otherwise referred to in this chapter, are important for those wishing to read more deeply about EIA. *Environmental Impact Assessment – Theory and Practice*, (Routledge, London, 1988) edited by P. Wathern remains the original definitive work on EIA and, although reprinted in 1998, it has not been updated. After his introductory essay, the editor divides the book into sections with contributions from a range of authors on the mechanics of EIA; the efficiency of EIA; the practice of EIA around the world; and EIA as both art and science. A rather more up-to-date and, as the title suggests, forwarding-looking text *Environmental Impact Assessment: Cutting Edge for the Twenty-First Century* (Cambridge University Press, 1995), comes from A. Gilpin. This has rather more the style of a handbook about it, dealing, as it does, with procedures and methodologies for carrying out EIA. It also contains sections on practice and legislation in most continents, apart from Africa, as well as reviewing EIA practice by international agencies. But unlike this chapter, it extends its coverage to deal with strategic environmental assessment (SEA), i.e. the interrelationship between the environmental aspect of projects, programmes and plans, usually within a regional context. In addition to these books, three journals provide an invaluable source of material on contemporary work in the field. *Environmental Impact Assessment Review* is pre-eminent, but the *Journal of Environmental Management* and *European Environmental Law Review* have occasional articles of considerable interest.

REFERENCES

Allen, R. (1996) What is Environmental Impact Assessment? *Rural Wales* autumn, 8–9.

Beattie, R.B. (1995) Everything you already knew about EIA (but don't often admit). *Environmental Impact Assessment Review* 15(2), 109–14.

Blunden, J.R. (1985) Conflict management in rural resource planning. In *Problems of Constancy and Change – the complementarity of systems approaches to complexity*, 31st Annual Meeting of the International Society for General Systems Research, Budapest.

Blunden, J.R., Pryce, W.T.R. and Dreyer, P. (1997) Classification of rural areas in the European context: an exploration of a typology using neural network applications. *Regional Studies* 31, 3.

Canter, L. and Clark, R. (1997) NEPA effectiveness – a survey of academics. *Environmental Impact Assessment Review* 17(2), 313–27.

Clark, B., Bisset, R. and Wathern, P. (1980) *Environmental Impact Assessment: A Bibliography with Abstracts*. London: Mansell.

Cooper, K. (1990) Environmentalists reject federal environmental assessment bill. *Intervenor* 15(6), 1–44.

Davy, B. (1995) The Australian Environmental Impact Assessment Act. *Environmental Impact Assessment Review* 15(4), 361–75.

Delicaet, A. (1995) The New Canadian Assessment Act: a comparison with the environmental assessment review process. *Environmental Impact Assessment Review* 15(6), 497–505.

Ebisemiju, F.S. (1993) Environmental impact assessment: making it work in developing countries. *Journal of Environmental Management* 38, 247–73.

EC Directive 85/337 *The Assessment of the Effects of Certain Public and Private Projects on the Environment*, Brussels.

Federal Environmental Assessment Review Office (1987) *Reforming Federal Environmental Assessment*. Ottawa, Ontario, Ministry of Supply and Services, Canada.

Federal Environmental Protection Agency (1989) *National Policy on the Environment*, Lagos.

Fowler, G.G. and Dias de Aguiar, A.M. (1993) Environmental impact in Brazil, *Environmental Impact Assessment Review* 13(3), 169–76.

Geraghty, P.J. (1993) Environmental assessment and the application of expert systems: an overview. *Journal of Environmental Management* 39, 27–38.

Geraghty, P.J. (1996) Environmental impact assessment in Ireland following the adoption of the European Directive. *Environmental Impact Assessment Review*, 16(3), 189–211.

Horberry, J. (1984) *Status and Application of Environmental Impact Assessment for Development*, Gland, IUCN.

Holm-Hansen, J. (1997) Environmental impact assessment in Estonia and Norway. *Environmental Impact Assessment Review* 17(6), 449–63.

Ibaara, A.B. (1987) Reflections on the incorporation of an environmental dimension into the institutional framework and operation of the public sector in Latin America and the Caribbean. In *Conference on the Environment* 55–76, Washington DC, Inter-American Development Bank.

Kunzlik, P. (1995) EIA: the British cases. *Environmental Law Review* December, 336–44.

Montz, B.E. and Dixon, J.E. (1993) From law to practice: EIA in New Zealand. *Environmental Impact Assessment Review* 13(2), 89–105.

Pardo, M. (1998) Environmental impact assessment: myth or reality? Lessons from Spain. *Environmental Impact Assessment Review* 17(2), 123–41.

Roque, R. (1985) Environmental impact assessment in the Association of South-east Asian Nations. *Environmental Impact Assessment Review* 5(3), 257–64.

Sanchez, L.E. (1993) Environmental impact assessment in France. *Environmental Impact Assessment Review* 13(4), 255–65.

Sankoh, O.A. (1996) Making environmental impact assessment convincible to developing countries. *Journal of Environmental Management* 42, 185–9.

Sheate, W.R. (1997) The Environmental Impact Assessment Directive 97/11/EC – A small step forward? *European Environmental Law Review* August/September, 235–43.

Szelely, F. (1987) Strategies to strengthen environmental quality in the IDB development project cycle. In *Conference on the Environment* 77–102, Washington DC, Inter-American Bank.

Webler, T., Kastenholz, H. and Renn, O. (1995) Public participation in impact assessment: a social learning perspective. *Environmental Impact Assessment Review* 15(5), 443–63.

Wood, C.M. (1993) Environmental impact assessment in Victoria: Australian discretion rules EA! *Journal of Environmental Management* 39(4), 281–95.

Wood, C.M., Lee, N. and Jones, C.E. (1991) Environmental statements in the UK: the initial experience. *Project Appraisal* 6, 187–94.

Wood, C. and Bailey, J. (1994) Predominance and independence in environmental impact assessment: the Western Australian model. *Environmental Impact Assessment Review* 14(1), 37–59.

18

Countryside recreation management

Guy Robinson

The impacts of recreation and tourism, memorably termed the 'fourth wave' (Dower 1965: 123), have transformed many rural areas in recent decades, in some cases becoming predominant within the economy and contributing significantly to social change. The demand for rural land to be used for recreational purposes has added to pressures upon the countryside to fulfil multiple roles, thereby adding to the complexity of rural planning and land management. In their research on countryside recreation management, geographers have analysed the outcomes of existing management plans as well as contributing in both theoretical and practical form to the ongoing debate regarding the nature and use of the countryside. This chapter will outline some of the main avenues of geographical enquiry into the management of countryside recreation, with special reference to formulation of management plans, issues relating to access, and the relationship between recreational provision and social change.

RURAL RECREATION AND TOURISM

Two basic types of recreation are usually recognised:

1 *Formal.* This takes place on managed sites and is often associated with profit-seeking organisations. Management may involve provision of special areas, zoning or rationing demand by entrance charges, a membership fee or imposing maximum numbers;
2 *Informal.* The countryside provides a backdrop to a range of activities, including recreational driving, walking and general sightseeing.

Various general characteristics within society in the latter half of the twentieth century have produced increased opportunities for both types of leisure activity, notably greater affluence, increased personal mobility, and reduced and/or more flexible working arrangements. The growth in private ownership of cars and improvements in transport links between urban and rural areas have helped to direct a substantial proportion of this leisure towards the countryside, with urban residents attracted by the aesthetic qualities of the setting. This has produced both greater participation in traditional non-consuming rural pursuits (e.g. walking, nature study, sightseeing) and new activities that may utilise a specific rural resource (e.g. mountain biking, windsurfing) (Butler 1998). Many of the latter owe their increased popularity to a combination of greater affluence, ease of accessibility of rural areas to urban residents and technological developments that have been applied to sporting/leisure activity (Mieczkowski 1990). Some activities have also been relocated to the countryside to take advantage of cheaper greenfield sites, the pleasant surroundings and ease of access from multiple urban centres, e.g. golf courses, theme parks.

Another factor promoting increased rural recreation has been growing public concern for the environment and 'green' issues, promoting activities such as bird watching and nature study in general. The growth of rural-based ecotourism is part of this trend. A wider appreciation of the attractions of rural locales can be seen in the growth of rural-based second-home ownership (Cherry 1993), hobby farming (Gasson 1988), vacation renting of properties in the countryside, e.g. the French *gites* system (Campagne *et al.* 1990), and urban–rural migration in general (Champion 1989).

Concern for fitness and health among the more affluent in society has encouraged some of the more 'active' leisure pursuits, e.g. horse riding, downhill skiing, water skiing. Moreover, greater scope for recreational pursuits has been brought about by processes of rural restructuring, which have released land and buildings from agricultural use or encouraged farmers to diversify their enterprises. The latter has promoted numerous leisure activities, with farming activities themselves becoming tourist attractions (Bramwell 1994; Opperman 1996), e.g. wineries (Hall and Macionis 1998), children's farms, pick-your-own schemes, nature trails, horse riding and farm-based accommodation (Benjamin 1994; Evans and Ilbery 1992). This has contributed to the public's changing perception of rural areas: from regarding them as predominantly fulfilling a primary production purpose to areas combining production and recreation within attractive settings. Thus there is no longer a single type of rural space but rather a multiplicity of overlapping social spaces, including recreational space (Mormont 1987; 1990).

Different forms of recreation may often be concentrated in the same locality, creating a well-recognised set of problems for countryside managers: traffic congestion, pressure on available services, environmental damage. Furthermore, it is common for conflicting demands to be placed upon rural space by the various leisure pursuits; for example, contrast the different needs of ramblers seeking quiet relaxation while walking, horse riders, mountain-bike riders, hang gliders, powerboat enthusiasts, orienteers and bird watchers. Sometimes, provision of facilities for one activity may incidentally promote increased development of other pursuits, as in the case of chairlifts for skiers being used by walkers in the summer months (Sidaway 1993).

MANAGING RECREATION IN THE COUNTRYSIDE

The conflicting demands of the various forms of recreation provide a complex set of management issues, especially when added to the need to reconcile recreational demands with those of competing land uses, notably conservation and maintenance of landscape quality. Two critical characteristics of countryside recreation with respect to its management are the fragmented and diverse nature of both demand and supply, and the dynamic and variable claims made on its resource base (Sharpley and Sharpley 1997: 113). These features complicate the process of management and planning (summarised in Table 18.1) and contribute to the general lack of development policies for rural recreation/tourism (Jenkins *et al.* 1998). Indeed, there has been a surprising absence of positive planning for the provision and integration of recreation and tourism within broader policies for rural development. Most planning has been a response to pressures and demands and has often dealt with containment instead of positive promotion. This has tended to mean a neglect of wider economic and social implications. Pigram (1993), for example, refers to a 'policy implementation gap' between the recognition of potential benefits from tourism and recreation and their practical incorporation into rural development plans. Indeed, some plans see rural recreation/tourism as a challenge or threat to the established modes of productive resource use and to maintenance of the character of rural society (Sharpley 1994).

Various 'solutions' have been proposed with respect to the need for more effective management of rural resources for recreational purposes, with geographers making a significant contribution to the debate. For example, Glyptis (1991) advocates a fourfold approach in order to combine the use of the countryside for recreational and conservation purposes:

1 production of comprehensive resource appraisals to identify sites and key areas of significance for conservation or recreation;

Table 18.1 The rural tourism planning and management process.

Stages	*Processes*
1 Setting objectives	Determination of objectives Comparison of objectives Integration with broader plans and objectives
2 Surveys	Quantitative data (patterns and trends) Qualitative data Socio-economic, political, environmental factors
3 Analysis and synthesis	Evaluation of survey data Combined analysis of all data Projection of requirements, capacities
4 Proposals	Preparation of alternative plans Selection of most appropriate policies
5 Implementation and mangement	Continuous process of: • implementation • monitoring • evaluation

Source: Sharpley and Sharpley 1997, p. 116.

2 evaluation of demand, including allowance for minority groups and minority activities;
3 formation of partnerships between interest groups wishing to share the same resources (in order to avoid polarisation of interests);
4 development of an understanding of public interests and facilitation of access to resources to a wide section of the public through provision and dissemination of information (see Box 18.1).

Increasingly, it is being acknowledged that there is an interdependence between recreational use of rural areas and the physical and socio-cultural attributes of those areas (Dernoi 1991; Kariel and Kariel 1982). The attributes represent a resource base for tourism and recreation, but maintenance of the attributes is supported by the presence of recreationists. Therefore management has to strike a balance between satisfying the needs of recreationists and those of local rural communities, and preserving the inherent qualities of the rural environment (Butler 1991). Attaining this balance is often rendered more difficult by the fragmented nature of management, with various different 'managers' having responsibilities for different aspects of recreation in the countryside. Farmers, landowners, gamekeepers, conservationists, planners, sporting administrators, community organisations, local authorities, site owners and commercial interests all play a part in managing rural recreation. Hence, the formulation and operation of systematic plans can be extremely complex. However, it is this complexity that has attracted the attention of geographers, who have contributed both to studies of existing plans and plan formulation.

RESEARCH BY GEOGRAPHERS ON RURAL RECREATION

A multiplicity of problems associated with the management of rural recreation has attracted geographers' attention, forming the basis for a growing volume of research since the 1960s,

Box 18.1 Approaches to planning for leisure

Through a focus on the four activities in Glyptis' list (see main text), management can strive to integrate the multiplicity of demands made by recreation, access, amenity and conservation, although the task of satisfying diverse interest groups is rendered more difficult by the multiplicity of interests within tourism and recreation itself. Because of its informality and spontaneity, much countryside recreation has been deemed by government to require little positive planning and management (Glyptis 1991). Nevertheless, these critical characteristics mean that relatively small planning measures can have a disproportionate effect. The basic range of measures that have been utilised is as follows:

Approach	*Content*
1 Standards	Planning based on *per capita* specifications of levels of provision laid down by some authoritative body. Usually based on demand estimates.
2 Gross demand	Estimation of broad demand levels based on existing national or regional participation surveys. This is the most basic of demand estimation approaches but can be varied to consider local socio-demographic conditions.
3 Spatial approaches	Localised demand estimation incorporating consideration of facility catchment areas. This extends the gross demand approach when considering the question of facility location.
4 Hierarchies of facilities	Recognises that different types and scales offacility have different catchment areas. Especially relevant for planning new communities and for facilities involving spectator audiences.
5 Grid or matrix approach	Examines impacts of all of an authority's leisure services on all social groups via impact evaluation.
6 Organic approach	Strategy development based on assessment of existing service provision and spatial gaps in demand. It is incremental rather than comprehensive and is common within the private sector.
7 Community development approach	Planning and policy development based on community consultation.
8 Issues approach	Plans based on initial identification of 'key issues' rather than comprehensive needs/demand assessment. Corresponds to SWOT (strengths, weaknesses, opportunities, threats) analysis. Most common for *ad hoc*, one-off projects.

Source: Based on Veal (1993: pp. 92–3).

when concerted attention was first directed towards this topic (Coppock and Duffield 1975; Simmons 1975). Geographers played an important part in assessing the characteristics of sites/areas supplying recreation and in initial attempts to measure demand for recreation (Lavery 1971; Wall 1972), with more sophisticated methods being developed over time (Rodgers 1993; Veal 1987: 125–56), including assessments of the monetary cost of recreation (Bateman 1995).

The pioneering Clawson method placed a cost on the visit to a recreational site and explained levels of use in terms of that cost (Clawson 1981). This involved examining the origins of and distances travelled by visitors to a site and calculating an assumed cost for that trip. However, such estimates are prone to error, and it may not be travel costs that are regarded by recreationists as the most important ones (Field and Macgregor 1987: 179–80). Related work on recreational carrying capacities was also pioneered in the 1970s (Brotherton 1973). Subsequently, analysis has moved on to a deeper consideration of provision and access. For example, Harrison (1991) examined the way in which the interests of landowners, private property rights and efforts by planners to provide controlled access to key sites have combined to reinforce the attraction of 'honeypot' sites to the general public.

Empirical studies have provided an understanding of patterns of activity and consequent demands on land and water resources, and established a baseline for future work, across a continuum from description to explanation to prediction to policy formulation (Owens 1984; Smith 1983). This can be compared with Morgan's (1991) fivefold classification of survey, analysis, plan, monitoring and review. A continuing emphasis on demand and site surveys has revealed that the pursuit of recreational activities in the countryside is highly concentrated: spatially, temporally, and by activity, mode of transport, class, age and family type (Gilg 1996: 221–2).

Concern for the environmental effects of recreation upon 'fragile' areas such as national parks or wilderness has been manifest more recently (e.g. Mieczkowski 1995). There are numerous studies reporting substantial environmental damage at 'honeypot' sites in national parks and sites that attract large numbers of visitors, such as Stonehenge or Ayers Rock. However, the growing numbers of rural tourists and recreationists are spreading human impacts to areas previously spared such depredations, and creating new management problems (Bromley 1994). Until recently, little of this environmental concern was directed at the wider, more densely settled countryside, although government and academics have now shown greater awareness of the broader scale of such impacts (e.g. Countryside Commission 1991; 1994; 1995; Croall 1995).

There are conflicting views, though, of the severity of impacts of recreation upon the environment. For example, studies monitoring the impact of recreational use upon the countryside in the 1970s and 1980s contributed to growing official views that impacts were far less damaging than had been widely believed (Sidaway 1990). Furthermore, comparisons with other land uses revealed much recreational activity to be more desirable, for example when compared with the spread of improved pasture, afforestation and urban sprawl. However, clashes between 'active' and 'passive' forms of recreation have been widely recognised as problematic for managers.

From the early 1980s, economic and environmental foci have been accompanied by work on cultural and social impacts in destination areas (e.g. Bouquet and Winter 1987). Valuable work on the perceptions and views of users of the countryside has also been performed by geographers, generally recognising differences in the characteristics of users of different sites. For example, work by Cloke and Park (1982) revealed clear behavioural differences between people using two contrasting sites. An open moorland site (but accessible by road) attracted people who wanted to roam and 'to do their own thing'. In contrast, a formal country park site attracted those who preferred the security of an evidently 'managed' environment. This division cut across social class, and it also revealed the way in which management can influence recreational behaviour. For example, restrictions on car access and judicious use of signposts have been shown to promote activities associated with walking and enjoying rural peace and quiet (e.g. Countryside Commission 1976).

Another focus of recent research has been on the development of possible policies for tourism in rural areas (e.g. Luloff *et al.* 1994) and of policy implications (e.g. Clarke 1993). Some of this work has linked tourism to the sustainable development of the countryside (Lane 1994; Priestley *et al.* 1996). For example, research on the impacts of mass tourism and growing recreational demands in Alpine Europe prompted the emergence of suggestions for different types of management, using concepts such as alternative, responsible, soft, appropriate or green tourism (Hunter and Green 1995; Smith and Eadington 1992). These are now part of planning for sustainable development 'to balance demand and capacity so that conflicts are minimised and the countryside is used to its full potential without deterioration of the resource base' (Pigram 1983: 171). The management issue that this work raises is how to encourage tourism and recreation development that can not only help to maintain the rural economy but also preserve the environment that attracts the tourists. This issue can be compared with the three overriding goals of management of rural recreation that are generally recognised (Gilg 1998):

1 ensuring rational development of resources on a sustainable basis,
2 providing equitable development between different activities and groups within society;
3 reconciling consumption and production.

The remainder of this chapter considers examples of how geographers have contributed to the analysis of key issues in the management of rural recreation. Three issues are singled out for special attention: access and land zoning, commodification of the countryside and management via advertising, and reconciling competing land uses within national parks.

CASE STUDIES

Access to land for recreational use

Harrison's (1991) analysis of the UK's town and country planning system and the national parks system concluded that by the late 1960s it had failed both to protect the countryside from destructive forces and to provide sufficient access to meet the demands of recreationists. Recognition of these deficiencies then led policy makers to three principal policy responses (Curry 1985):

1 to improve site access, as best illustrated in the country parks, 40 per cent of which are in the urban fringe;
2 to provide small amounts of targeted funding to ease conflicts between conservation and recreation, e.g. upgrading footpaths, improving signage;
3 to continue to restrict wider access.

The last point reinforces the fact that the growing demand for rural recreation has not been matched in most developed countries by a similar increase in the supply of publicly accessible lands or routeways within the countryside (Millward 1993). Indeed, some loss of access has occurred through urbanisation, forestry development and the spread of modern farming methods. The question of the rights of the general public to have access to rural land for recreational pursuits continues to be a contentious issue in many countries, but especially in England and Wales (Countryside Commission 1988), where 'legal systems give enormous weight to private property rights in contrast to the very limited access they give to individuals' (Gilg 1996: 191). At the opposite extreme is Sweden, where there is legal right of access to all land, whether publicly or privately owned, and Norway, where the same situation applies subject to certain conditions laid down in legislation. In many countries, legal restrictions have placed significant constraints on access to rivers for fishing, and to land for field sports as well as casual rambling. These constraints can apply even in national parks (Watkins 1996) and despite the existence of statutory provisions (Curry 1994).

Growing demands for access have provoked unfavourable reactions from landowners as well as litigation and legislative responses. As a result, ease of access for the general public in certain areas has been reduced (Groome and Tarrant 1985). Related to this trend have been changing attitudes to some traditional rural pursuits, notably hunting. Both with hunting in North America using guns and fox hunting in Britain using hounds, protests by animal rights groups and changing public sentiments have closed off certain lands to hunters or imposed legal restraints. Elsewhere, sheer pressure of numbers of recreationists is causing attempts by landowners to restrict or control access to their land. However, although much has been written about the problems of restricted access in individual countries, there has been only limited work on comparative studies (Jenkins and Prin 1998).

In England, certain groups have gained preferential access to the countryside via both public and private provision of resources. In the private sector, the controls exerted by landowners have meant that rights of access have frequently been granted to small specialist groups, e.g. shooting rights, fishermen purchasing special licences. Harrison (1991) foresees a bigger role for public sector provision of recreational facilities but acknowledges the difficulty in constructing a publicly acceptable policy combining conservation and recreation. In addressing this issue of policy formation, Curry (1994) advocates a higher priority for recreation planning, although he recognises five areas of difficulty in attaining this higher priority, with accompanying solutions, as shown in Table 18.2 (see also Curry 1996).

In practical terms, among the most frequently used management tools for regulating access to land for recreational purposes has been employment of some form of land zoning and systems for concentrating activities in selected areas and corridors (Groome 1993; Table 18.3). This concentration, often through explicit use of 'honeypots', has been seen as a key method for protecting 'fragile' environments or the wider countryside. Even in areas specifically designated for recreational purposes, such as the UK's country parks or Ontario's recreation parks (Killan 1993), zoning has usually been employed in order to restrict recreational use in part of the park (Box 18.2).

In some cases, zoning may be enforced via charging for access or use of facilities (Broom 1991; Table 18.4). A pricing mechanism can fulfil four roles (McCallum and Adams 1980):

1 raising revenue;
2 allocating demand between recreation and non-recreation expenditure;

Table 18.2 Problems and solutions for recreation planning in the UK.

	Current difficulties	*Possible solutions*
1	The fragmented nature of the organisational structure for rural recreation has restricted implementation of comprehensive policies and plans	Realign the Countryside Commission to make it a Countryside Recreation Commission and expand farm-based recreation
2	Confusion of responsibilities between the public and private sectors	Increased market-orientation for public sector facilities; greater emphasis on charging for access and treating information as promotion; more use of powers to increase access and treat access rights as commodities
3	Rural recreation has been dominated by the middle-class, with other groups reluctant to participate	Replace attitude that the scientific elite knows best and that protection of the environment always transcends recreational needs of a broader spectrum of society; avoid focus on restricted 'honeypots'
4	Policies on recreation conceptualise it as a threat when compared with agriculture and forestry	Promote recreation as a potentially friendly land use
5	Recreation policies are usually residual to other rural policies	Use continuing ability of farming to produce more from less land to treat recreation facilities as market commodities, using demand as the key arbiter of what facilities to provide

Source: Based on Curry 1994; Gilg 1996, p.238.

Table 18.3 Access mechanisms and ideology.

Access mechanism	*Ideology*
De facto access	Non-market
rights of way	freedom to roam
rights of navigation	public rights over private land and water
rights of open access	
Access agreements	Market
management agreements	public intervention in the market
public ownership	
property rights	private rights of property
sporting rights	
permissive access	

Source: Groome 1993, p. 159.

3 allocating demand between alternative recreation facilities;
4 achieving certain objectives within the provision of countryside recreation. This can include differential pricing to divert users from one facility to another, resource protection and variable pricing to smooth random patterns of use.

Various combinations of these roles have been applied widely in North America, usually in conjunction with additional financial support from general taxes, specific bond issues or taxes and grant-in-aid (Robinson 1990: 262–5).

Rural recreation and commodification of the countryside

Harrison (1991) argues that the demands for wider access to the countryside by the urban populace have been largely ignored because of the 'drawbridge' mentality of new rural residents. These are the so-called service class or middle-class urbanites who have migrated to the countryside seeking a romanticised rural idyll, which must not be disturbed by urban-based recreationists (Cloke and Thrift 1987). The service class has developed certain attributes of taste that have been influential and widely adopted by other consumers of rural spaces, e.g. a reverence for the pastoral idyll, an acceptance of certain cultural symbols such as old houses, antiques, health foods and real ale, and an enjoyment of outdoor pursuits such as jogging, cycling, fly-fishing, windsurfing and mountaineering (Urry 1988: 41).

These traits of the service class are part of a 'decentring' of identity in which people lead more eclectic lives, unshackled by the legacy of tradition or collective expectation, responding freely to the marketplace. This eclecticism, a characteristic of post-modernist society, has developed new forms of leisure consumption, often associated with the 'spectacle' of theme parks, medieval 'fayres', pop festivals and 'living' museums, and coexisting with more traditional forms of rural leisure (Getz 1991; Janiskee and Drews 1998; Urry 1991). A crucial aspect of the new leisure consumption is the attachment of commercial value to items previously largely ignored by recreationists or not offered for sale by entrepreneurs, a process termed 'commodification' (Hopkins 1998; see Box 18.3).

There has been increased commodification of the countryside, giving rise to a series of new markets for countryside commodities, including the crafting, packaging and marketing of 'pay-as-you-enter' national parks and theme parks (see Plate 18.1), craft and food outlets and 'leisure experience' activities. Increasingly, via deregulation and privatisation, leisure commodities are

Box 18.2 The recreation opportunities spectrum

Zoning has frequently been implemented following a formal assessment of a potential recreational site. Indeed, site assessment has been a particularly popular avenue of research in North America, for example using the recreation opportunities spectrum (ROS) to combine the requirements of visitors with the constraints and characteristics of a site (Figure 18.1). The spectrum, employed to allocate uses in a designated area via a zoning system, ranges across the various combinations of physical, biological, social and managerial conditions. It recognises that each situation offers a range of opportunities for recreation, but with visitor numbers likely to increase as wilderness gives way to some form of development. This development and visitor influx then has to be managed, usually by placing restrictions on visitors' movements (van Oosterzee 1984). The ROS can help to identify the diversity of opportunities, but it does have limitations, especially as it plays down the importance of biophysical features, which are the natural attractions that are often the main draw for visitors (Hammitt and Cole 1987).

Figure 18.1 The recreation opportunities spectrum.

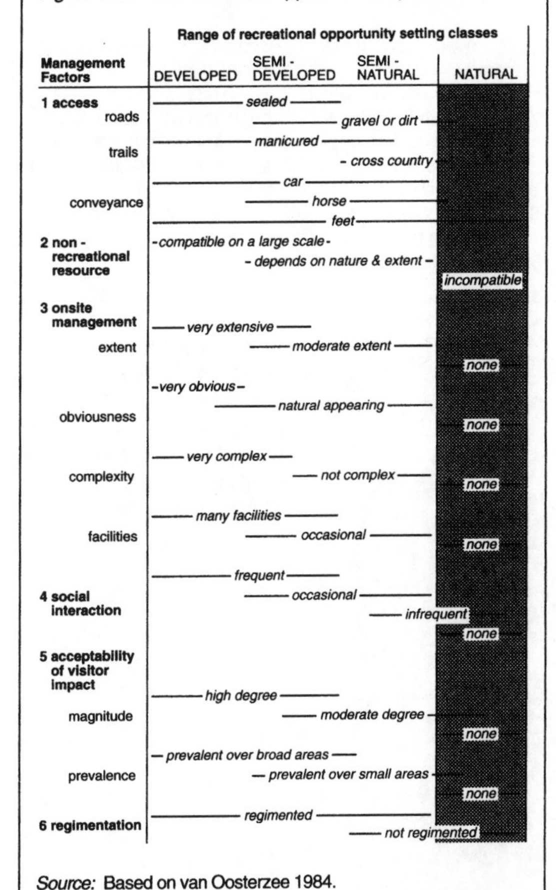

Source: Based on van Oosterzee 1984.

being sold by the private sector rather than the public sector. Thus Cloke (1993: 58) contends that 'the informal and free forms of outdoor recreation in the countryside are gradually being replaced for many by a more formal, attraction-based, day out in which the countryside experience is packaged and paid for.' He quotes the example of Powergen's Rheidol hydro-electric scheme in Dyfed, Wales, promoted as a 'great day out' and offering (for the price of admission):

- a visit to Rheidol Power Station Information Centre with its exhibition, video room, souvenir shop and refreshments;
- a tour of the power station and its fascinating fish farm;
- a picnic at the lakeside picnic area;
- a drive around the scenic upland reservoirs;
- fishing for trout in one of the lakes;
- enjoyment of a romantic view of the floodlit Felin Newydd weir.

This is one example of the growing importance of the marketing of rural recreation as a form of countryside management, with the repeated use of key icons or symbols in the advertising brochures: endeavouring to attract custom by using words such as landscape, nature, history, family orientation, craft and country 'fayre' (Cloke 1992; 1993; Cloke and Goodwin 1992). This is the effective means by which commercial recreation/tourist enterprises are attracting the urban population to spend time (and money) in the countryside. Its advance has been furthered by the process of farm diversification, stimulated by falling incomes from traditional farming activities, as farmers offer paying tourists the chance to spend time on farms, lured by attractions such as a working farm, themed park experience, waterfowl centres, rare breeds, farm museums, country sports, woodland parks, butterfly farms and shire horse centres (see Plate 18.2).

Marketing has increasingly become a management tool, by selling the 'right' areas and sites to visitors. This involves increasing the provision of information made available to recreationists while targeting certain types of

Table 18.4 Potential income streams from recreational facilities.

Method	*Desired outcome*
Charging for admission	Direct income through admission fee
Indirect charging for particular facilities, e.g. car parks	Obvious income from visitors using the countryside for walking, climbing and fishing as well as visiting specific sites
Selling goods or services, e.g. retailing activities	Provides money for wider site/facilities management
Voluntary donations, membership subscription	Can pay attendants; provides money for maintenance
Levies on complementary products	Portion of purchase price set aside of a good like a publication
Tourism taxes, e.g. road tolls	Can be used to meet high management costs

Source: Based on Groome 1993, pp. 140-1.

Box 18.3 Multi-authority collaboration in the Dartmoor Area Tourism Initiative (DATI)

The DATI strategy in the Dartmoor National Park incorporates the following (Greenwood, 1994):

1 increased emphasis upon marketing and information provision to attract visitors to key sites and activities;
2 expanded use of public transport;
3 focus on farm diversification to provide more farm-based accommodation and farm-based recreation;
4 developed more 'green' tourism via visitor information and education, and a programme directed at local businesses and communities;
5 improved public relations, especially by informing local communities of the park's aims and plans;
6 improved advisory and information role;
7 adopted an interpretation strategy both within and outside the park to help to disperse visitors from certain honeypots and to encourage visits to certain attractions and participation in certain activities;
8 alleviated erosion and damage to key sites by providing alternative routes and initiating restoration;
9 increased monitoring activities.

Plate 18.1 An example of a 'packaged, themed experience' at Shantytown, near Greymouth, New Zealand, where a nineteenth century goldrush community has been re-created.

Plate 18.2 The shire horse centre near Plymouth, Devon, where farmland has been converted to a themed visitor-pays attraction.

information at particular people. Analysis of this development has enabled geographers to apply ideas formulated within social theory to concrete examples of recreational marketing (Hall 1992; Urry 1995). The outcome is a growing understanding of the multiple roles that the countryside fulfils and of the increasing importance for entrepreneurs of their ability to package and sell a rural image to recreationists.

Managing recreation in national parks in England and Wales

In many countries, one of the principal foci of recreational and tourist pursuits in rural areas is the national parks system. For example, over 100 million visitor-days are spent each year in the national parks in England and Wales, contributing up to £900 million annually to local economies. A key factor in management practice has been whether national parks are essentially wilderness areas, as in North America (Blacksell 1993) and most developing countries, or contain human settlements and landscapes incorporating commercial agriculture and forestry, as in many European examples.

In the UK, it has been harder to develop coherent management strategies in the much-visited national parks than in countries where parks systems have been managed by the public sector for a long time and using common policies and guidelines (Butler 1998: 225). In national parks in England and Wales, the landscape is the product of long-term human impact and management. It needs continuation of careful management to maintain heather moorland, sheep and deer grazing, grouse shooting and a diversity of recreational activities. A summary of the typical conflicts and interactions faced by managers of the land is shown in Figure 18.2. However, this management is widely shared: by farmers, gamekeepers, landowners and their agents and tenants, as well as parks and local authorities. This division complicates many key management issues, as demonstrated in Statham's (1993) analysis of the North York Moors National Park. Here the national park authority can use planning Acts to control building development, but this does not apply to control of most agricultural and forestry land. Hence farmers have converted moorland to improved pasture (the area under heather moor has fallen by one-quarter since the park was designated in 1952) and the Forestry Commission has put large areas under conifers. National parks usually encompass the jurisdiction of several local authorities, so that multi-authority collaboration,

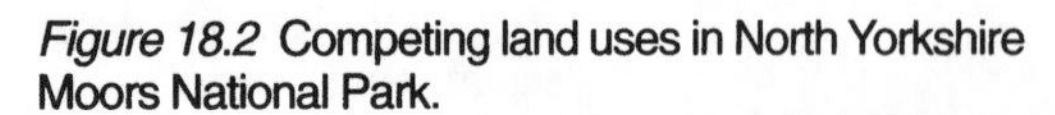

Figure 18.2 Competing land uses in North Yorkshire Moors National Park.

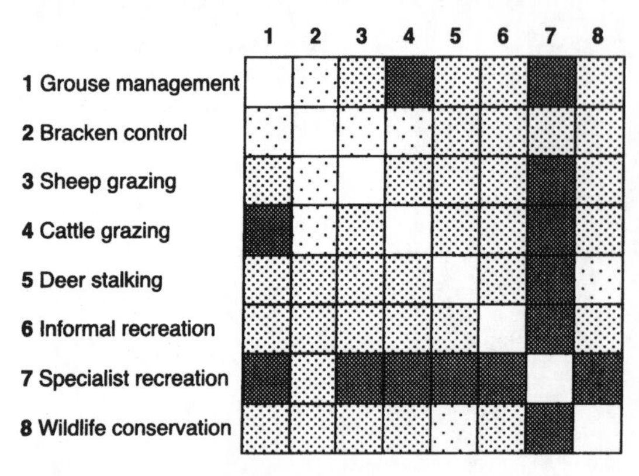

Source: Based on Statham 1993.

as in the Dartmoor Area Tourism Initiative (DATI), has been one way forward (Greenwood 1994; see Box 18.2).

One of the prime management strategies in ensuring the provision of attractive, 'unspoilt' countryside for recreation has been to pay landowners compensation for maintaining designated environmental features (Swales 1994). Management agreements have been used to influence tree planting, heather management, maintenance of stone walls and barns, and the conservation of important wildlife habitats. They have been operated by national parks authorities but also widely adopted elsewhere, as in the case of the UK's system of sites of special scientific interest (SSSIs), or for managing the countryside in an environmentally friendly fashion, e.g. the EU's environmentally sensitive areas (ESAs) scheme (Robinson 1994) and the countryside stewardship scheme in England and Wales (Morris and Potter 1995). Geographers have contributed both to monitoring the effects of these policies and to policy decisions on the designation of areas for conservation, notably in the ESAs (Whitby 1994).

Mather's (1993) work on farmers with SSSIs on their land revealed that the designation had not affected use and management of land on two-thirds of the farms surveyed, and that only one-fifth of farmers had actually been restricted in their farming activities by the presence of an SSSI. Nevertheless, other surveys show that damage to SSSIs continues, with 10 per cent at serious risk and with the Nature Conservancy Council in England and Wales spending 15 per cent of its 1991 budget on management agreements (Splash and Simpson 1994). Compulsory purchase of threatened sites or use of payments as nature conservation gain under the Town and Country Planning Acts may be more effective (Boucher and Whatmore 1993).

National parks authorities have compulsory powers to require public access to land in the parks, but over large areas access is often difficult, especially on privately owned farmland. The solution adopted in the North York Moors has been investment in promotion and management of routes giving a variety of recreational experiences in different park landscapes. A national trail, the Cleveland Way, is being complemented by a network of 'regional' and local routes. The nature of these landscapes is being modified by management agreements and direct purchase of land by the park authority. The example shown in Figure 18.3 demonstrates how former moorland, now converted largely to enclosed arable land, is being developed into broadleaved parkland landscape open for public access, and combined with areas of heath or common land. Conflicts between different types of recreation are still prevalent, though, despite zoning of areas for noisier pursuits. Erosion by motor-bike scrambling is a particular problem.

CONCLUSION

Management of tourism and recreation in rural areas has frequently been regarded as a mechanism for minimising visitor pressure while maximising economic benefit to rural communities. Much research effort has been directed at different aspects of this management issue,

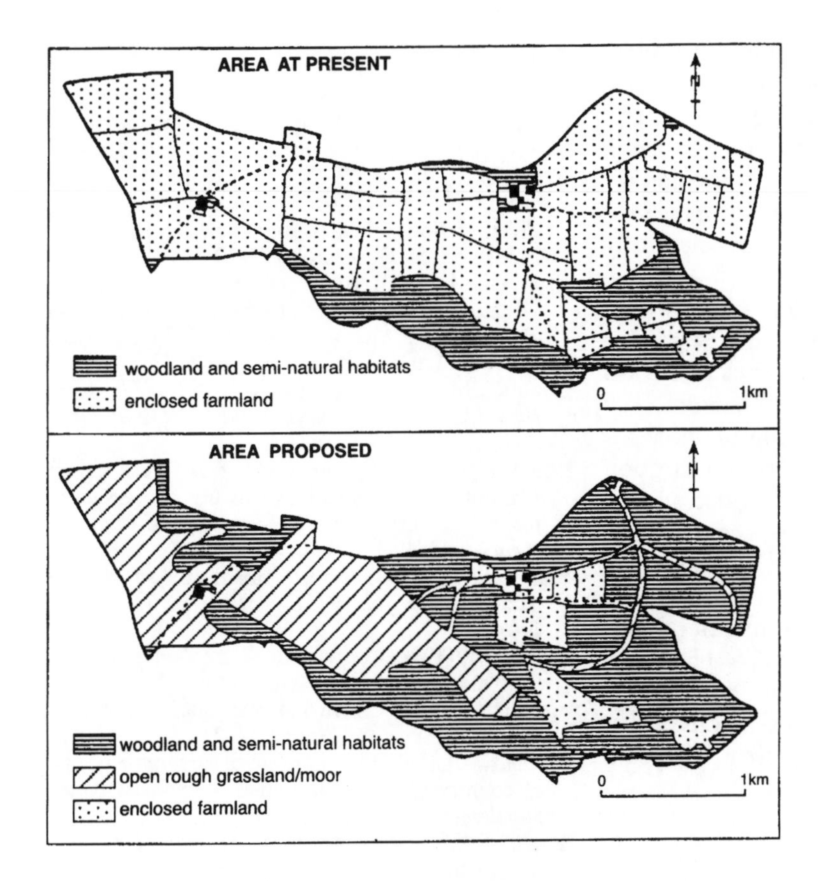

Figure 18.3 New landscapes, Murton Grange, North Yorkshire Moors National Park.

Source: Based on Statham 1993.

from how to assess demand to how to control access and to the multifaceted problem of reconciling different recreational uses of the same area of land. As in many other areas of human activity, an additional element within this research in recent years has been the issue of whether recreation and tourism in rural areas can be managed in a sustainable fashion. The dimension of sustainability seems destined to feature very prominently in applied research on management issues in future and may redress its previous neglect in government policies relating to tourism and recreation. It may also offer new ways of reconciling competing economic, social and cultural objectives in rural areas (Murdoch 1993). However, for this to occur there will need to be much more work on how sustainability can be measured in a practical fashion in particular situations and on how the various conflicts between different uses are to be resolved at a local level.

One of the key problems facing researchers on rural issues in general is the diversity of roles that the countryside fulfils in modern society. The days when rurality in the developed world could simply be equated with primary production have disappeared as 'rural' has taken on a more complex meaning. While agriculture and forestry still dominate rural land use, recreational, tourist, sporting, military, conservation and other commercial interests have assumed a far greater importance. In particular, the public's increased pursuit of leisure activities in rural areas has given it a voice on rural issues that has often conflicted with that of long-term agricultural interests. Yet an overriding problem is that the public hold ambivalent, ambiguous and contradictory views of what role they wish the countryside to fulfil and how they wish it to be managed with respect to recreational provision (McNaghten 1995). This makes it extremely hard to plan for multiple use, although Gilg (1996: 239) argues that planning is still ill-informed about the public's attitudes and motivations for visiting the countryside. He urges that more research be pursued in this area to overcome the fact that provision for recreation frequently meets the needs of the providers rather than being what people really want.

During the last three decades, work by geographers on the management of rural recreation and tourism has grown from a trickle to form a highly varied and substantial literature. The varied nature of the research contribution has mirrored the diverse nature of rural recreation itself. Nevertheless, themes highlighted in this chapter, notably relating to access, demand, marketing and conflicts with conservation interests, are likely to continue as significant elements in the research agenda. The changing nature of rurality has added to the scope for geographical research, and this continually evolving character, together with the potential new framework(s) offered by considerations of sustainability, present a dynamic prospect for future work.

GUIDE TO FURTHER READING

There are a number of books dealing with the topic of rural recreation and associated management issues, notably Curry (1994), Glyptis (1991) and Groome (1993). These focus primarily on the UK, as does Bromley (1994) in a handbook for recreation managers. Sharpley (1993; 1994) includes a wider range of examples, as do Sharpley and Sharpley (1997) who focus on rural tourism. Useful international collections of essays are Glyptis (1993) and Bouquet and Winter (1987) (on rural recreation), Butler *et al.* (1998) (on rural tourism) and Pearce and Butler (1993) (on tourism in general). A good introduction to rural planning in the UK is provided by Gilg (1996), and see Ilbery (1998) on general change in the countryside. Issues relating to access to land for recreation are considered by Harrison (1991) and Watkins (1996), to sustainability by Hunter and Green (1995) and Priestley *et al.* (1996), and to environmental issues by Mieczkowski (1995).

REFERENCES

Bateman, I. (1995) Environmental and economic appraisal. In T. O'Riordan (ed.) *Environmental Science for Environmental Management*, Harlow: Longman, 45–66.

Benjamin, C. (1994) The growing importance of diversfication activities for French farm households. *Journal of Rural Studies* 10, 331–42.

Blacksell, M. (1993) Wilderness and landscape in the United States. In S. Glyptis (ed.) *Leisure and the Environment: Essays in Honour of Professor J.A. Patmore*, London: Belhaven, 266–76.

Bouquet, M. and Winter, M. (eds) (1987) *Who from their Labours Rest: Conflict and Practice in Rural Tourism*. Aldershot: Gower.

Boucher, S. and Whatmore, S. (1993) Green gains? Planning by agreement and nature conservation. *Journal of Environmental Planning and Management* 36, 33–49.

Bramwell, B. (1994) Rural tourism and sustainable rural tourism. *Journal of Sustainable Tourism* 2, 1–6,

Bromley, P. (1994) *Countryside Recreation: A Handbook for Managers*. London: E & F.N. Spon.

Broom, G. (1991) Environmental management of countryside visitors. *Ecos* 12, 14–21.

Brotherton, I. (1973) The concept of carrying capacity of countryside recreation areas. *Recreation News Supplement* 9, 6–10.

Butler, R.W. (1991) Tourism, environment, and sustainable development. *Environmental Conservation* 18, 201–9.

Butler, R.W. (1998) Rural recreation and tourism. In B.W. Ilbery (ed.) *The Geography of Rural Change*, Harlow: Longman, 211–32.

Butler, R.W., Hall, C.M. and Jenkins, J.M. (eds) (1998) *Tourism and Recreation in Rural Areas*. Chichester: John Wiley & Sons.

Campagne, P., Carrere, G. and Valceschini, E. (1990) Three agricultural regions of France: three types of pluriactivity. *Journal of Rural Studies* 4, 415–22.

Champion, A.G. (ed.) (1989) *Counterurbanisation: The Changing Face of Population Deconcentration*. London: Edward Arnold.

Cherry, G.E. (1993) Changing social attitudes towards leisure and the countryside, 1890–1990. In S. Glyptis (ed.) *Leisure and the Environment: Essays in Honour of Professor J.A. Patmore*, London: Belhaven, 22–32.

Clarke, J. (ed.) (1993) *Nature in Question: An Anthology of Ideas and Arguments*. London, Earthscan.

Clawson, M. (1981) *Methods for Measuring the Demand for the Value of Outdoor Recreation*, 10th reprint, Washington DC: Resources for the Future.

Cloke, P.J. (1992) The countryside: development, conservation and an increasingly marketable commodity. In P.J. Cloke (ed.) *Policy and Change in Thatcher's Britain*, Oxford: Pergamon, 269–96.

Cloke, P.J. (1993) The countryside as commodity: new rural spaces for leisure. In S. Glyptis (ed.) *Leisure and the Environment: Essays in Honour of Professor J.A. Patmore*, London: Belhaven, 53–67.

Cloke, P.J. and Goodwin, M. (1992) Conceptualising countryside change: from post-Fordism to rural structured coherence. *Transactions of the Institute of British Geographers*, new series 17, 321–36.

Cloke, P.J. and Park, C.C. (1982) Country parks in national parks: a case study of Craig-y-Nos in the Brecon Beacons. *Journal of Environmental Management* 12, 173–85.

Cloke, P.J. and Thrift, N.J. (1987) Intra-class conflict in rural areas. *Journal of Rural Studies* 3, 321–33.

Coppock, J.T. and Duffield, B.S. (1975) *Recreation in the countryside*. London: Macmillan.

Countryside Commission (1976) *Tarn Hows: An Approach to the Management of a Popular Beauty Spot*. Cheltenham: Countryside Commission.

Countryside Commission (1988) *Changing the Rights-of-way Network*. Manchester: Countryside Commission.

Countryside Commission (1991) *Visitors to the Countryside*. Cheltenham: Countryside Commission.

Countryside Commission (1994) *Managing Public Access*. Cheltenham: Countryside Commission.

Countryside Commission (1995) *The Environmental Impact of Leisure Activities in the English Countryside*. Cheltenham: Countryside Commission.

Croall, J. (1995) *Preserve or Destroy: Tourism and the Environment*. London: Calouste Gulbenkian Foundation.

Curry, N. (1985) Countryside recreation sites policy: a review. *Town Planning Review* 56, 70–89.

Curry, N. (1994) *Countryside Recreation: Access and Land Use Planning*. Aldershot: E. & F.N. Spon.

Curry, N. (1996) Access: policy directions for the late 1990s. In C. Watkins (ed.) *Rights of Way: Policy, Culture and Management*, London: Pinter, 47–62.

Dernoi, L. (1991) About rural and farm tourism. *Tourism Recreation Research* 16, 3–6.

Dower, M. (1965) The fourth wave – the challenge of leisure. *Architects Journal* 20 January, 123–90.

Evans, N.J. and Ilbery, B.W. (1992) Farm-based accommodation and the restructuring of agriculture: evidence from three English counties. *Journal of Rural Studies* 8, 85–96.

Field, B. and Macgregor, B. (1987) *Forecasting Techniques for Urban and Regional Planning*. London: Hutchinson.

Gasson, R.M. (1988) *The Economics of Part-time Farming*. Harlow: Longman.

Getz, D. (1991) *Festivals, Special Events and Tourism*. New York: Van Nostrand Reinhold.

Gilg, A.W. (1996) *Countryside Planning*, 2nd edition. London: Routledge.

Gilg, A.W. (1998) Policies and planning mechanisms: managing change in rural areas. In B.W. Ilbery (ed.) *The Geography of Rural Change*, Harlow: Longman, 189–210.

Glyptis, S. (1991) *Countryside Recreation*. Harlow: Longman.

Glyptis, S. (ed.) (1993) *Leisure and the Environment: Essays in Honour of Professor J.A. Patmore*. London: Belhaven.

Greenwood, J. (1994) Dartmoor Area Tourism Initiative – a case study of visitor management in and around a national park. In A.V. Seaton (ed.) *Tourism: The State of The Art*, Chichester: John Wiley & Son, 682–90.

Groome, D. (1993) *Planning and Rural Recreation in Britain*. Aldershot: Avebury.

Groome, D. and Tarrant, C. (1985) Countryside recreation: achieving access for all. *Countryside Planning Yearbook* 6, 72–100.

Hall, C.M. (1992) *Hallmark Tourist Events: Impacts, Management and Planning*. New York: Halstead Press.

Hall, C.M. and Macionis, N. (1998) Wine tourism in Australia and New Zealand. In R.W. Butler, C.M. Hall and J.M. Jenkins (eds) *Tourism and Recreation in Rural Areas*, Chichester: John Wiley & Sons, 197–224.

Hammitt, W.E. and Cole, D.N. (1987) *Wildland Recreation Ecology and Management*. New York: John Wiley & Sons.

Harrison, C. (1991) *Countryside Recreation in a Changing Society*. London: TMS Partnership, University College.

Hopkins, J. (1998) Commodifying the countryside: marketing myths of rurality. In R.W. Butler, C.M. Hall and J.M. Jenkins (eds) *Tourism and Recreation in Rural Areas*, Chichester: John Wiley & Sons, 139–56.

Hunter, C. and Green, H. (1995) *Tourism and the Environment: A Sustainable Relationship?* London: Routledge.

Ilbery, B.W. (ed.) (1998) *The Geography of Rural Change*. Harlow: Longman.

Janiskee, R.L. and Drews, P.L. (1998) Rural festivals and community reimaging. In R.W. Butler, C.M. Hall and J.M. Jenkins (eds) *Tourism and Recreation in Rural Areas*, Chichester: John Wiley & Sons, 157–75.

Jenkins, J.M., Hall, C.M. and Troughton, M.J. (1998) The restructuring of rural economies: rural tourism and recreation. In R.W. Butler, C.M. Hall and

J.M. Jenkins (eds) *Tourism and Recreation in Rural Areas*, Chichester: John Wiley & Sons, 43–68.

Jenkins, J.M. and Prin, E. (1998) Rural landholder attitudes: the case of public recreational access. In R.W. Butler, C.M. Hall and J.M. Jenkins (eds) *Tourism and Recreation in Rural Areas*, Chichester: John Wiley & Sons, 179–96.

Kariel, H.G. and Kariel, P.E. (1982) Socio-cultural impacts of tourism: an example from the Austrian Alps. *Geografiska Annaler* 64B, 1–16.

Killan, G. (1993) *Protected Places: A History of Ontartio's Provincial Parks System*. Toronto: Dundurn Press.

Lane, B. (1994) Sustainable rural tourism strategies: a tool for development and conservation. *Journal of Sustainable Tourism* 2, 102–11.

Lavery, P. (1971) *Recreation Geography*. Newton Abbot: David & Charles.

Luloff, A.E., Bridges, J.C., Graefe, A.R., Saylor, M., Martin, K. and Gitelson, R. (1994) Assessing rural tourism efforts in the United States. *Annals of Tourism Research* 21, 46–64.

McCallum, J.D. and Adams, J.G.L. (1980) Charging for countryside recreation: a review with implications for Scotland. *Transactions of the Institute of British Geographers* new series, 5, 350–68.

McNaghten, P. (1995) Public attitudes to countryside leisure: a case study on ambivalence. *Journal of Rural Studies* 11, 135–47.

Mather, A. (1993) Protected areas in the periphery: conservation and controversy in northern Scotland. *Journal of Rural Studies* 9, 371–84.

Mieczkowski, Z. (1990) *World Trends in Tourism and Recreation*. New York: Peter Lang.

Mieczkowski, Z. (1995) *Environmental Issues of Tourism and Recreation*. Lanham, Md.: University Press of America.

Millward, H. (1993) Public access in the West European countryside: a comparative survey. *Journal of Rural Studies* 9, 39–51.

Morgan, G. (1991) *A Strategic Approach to the Planning and Management of Parks and Open Spaces*. Basildon: Institute of Leisure Management.

Mormont, M. (1987) Tourism and rural change: the symbolic impact. In M. Bouquet and M. Winter (eds) *Who From Their Labours Rest: Conflict and Practice in Rural Tourism*, Aldershot: Gower, 35–44.

Mormont, M. (1990) Who is rural? or how to be rural: towards a sociology of the rural. In T.K. Marsden, P. Lowe and S. Whatmore (eds) *Rural Restructuring*, London: David Fulton, 21–44.

Morris, C. and Potter, C. (1995) Recruiting the new conservationists: farmers' adoption of agri-environmental schemes in the UK. *Journal of Rural Studies* 11, 51–63.

Murdoch, J. (1993) Sustainable rural development: towards a research agenda. *Geoforum* 24, 225–41.

Opperman, M. (1996) Rural tourism in southern Germany. *Annals of Tourism Research* 23, 86–102.

Owens, S. (1984) Rural leisure and recreation research: a retrospective evaluation. *Progress in Human Geography* 8, 157–88.

Pearce, D. and Butler, R.W. (eds) (1993) *Tourism Research: Critiques and Challenges*. London: Routledge.

Pigram, J.J. (1983) *Outdoor Recreation and Resource Management*. Beckenham: Croom Helm.

Pigram, J.J. (1993) Planning for tourism in rural areas: bridging the policy implementation gap. In D. Pearce and R.W. Butler (eds) *Tourism Research: Critiques and Challenges*, London: Routledge, 156–74.

Priestley, G.K., Edwards, J.A. and Coccossis, H. (1996) *Sustainable Tourism? European Experiences*. Wallingford: CAB International.

Robinson, G.M. (1990) *Conflict and change in the countryside: rural economy, society and economy in the Developed World*. London and New York: Belhaven Press.

Robinson, G.M. (1994) The greening of agricultural policy: Scotland's environmentally sensitive areas (ESAs). *Journal of Environmental Planning and Management* 37, 215–26.

Rodgers, H.B. (1993) Estimating local leisure demand in the context of a regional planning strategy. In S. Glyptis (ed.) *Leisure and the Environment: Essays in Honour of Professor J.A. Patmore*, London: Belhaven, 116–30.

Sharpley, R. (1993) *Tourism and leisure in the countryside*. Huntingdon: ELM.

Sharpley, R. (1994) *Tourism, tourists and society*. Huntingdon: ELM.

Sharpley, R. and Sharpley, J. (1997) *Rural Tourism: An Introduction*. London: International Thomson Business Press.

Sidaway, R. (1990) Contemporary attitudes to landscape and implications for policy: a research agenda. *Landscape Research* 15(2), 2–6.

Sidaway, R. (1993) Sport, recreation and nature conservation: developing good conservation practice. In S. Glyptis (ed.) *Leisure and the Environment: Essays in Honour of Professor J.A. Patmore*, London: Belhaven, 163–73.

Simmons, I.G. (1975) *Rural Recreation in the Industrial World*. London: Edward Arnold.

Smith, S. (1983) *Recreation Geography*. London: Longman.

Smith, V. and Eadington, W. (1992) *Tourism Alternatives: Potentials and Problems in the Development of Tourism*. Philadelphia: University of Pennsylvania Press.

Splash, C. and Simpson, I. (1994) Utilitarian and rights-based alternatives for protecting sites of special scientific interest. *Journal of Agricultural Economics* 45, 15–26.

Statham, D.C. (1993) Managing the wilder countryside. In S. Glyptis (ed.) *Leisure and the Environment: Essays in Honour of Professor J.A. Patmore*, London: Belhaven, 236–52.

Swales, V. (1994) Incentives for countryside management. *Ecos* 15(3/4), 52–7.

Urry, J. (1988) Cultural change and contemporary holiday-making. *Theory, Culture and Society* 5, 35–55.

Urry, J. (1991) *The Tourist Gaze*. London: Sage.

Urry, J. (1995) *Consuming Places*. New York: Routledge.

van Oosterzee, P. (1984) The recreation opportunity spectrum: its use and misuse. *Australian Geographer* 16, 97–104.

Veal, A.J. (1987) *Leisure and the Future*. London: Allen & Unwin.

Veal, A.J. (1993) Planning for leisure: past, present and future. In S. Glyptis (ed.) *Leisure and the Environment: Essays in Honour of Professor J.A. Patmore*, London: Belhaven, 85–95.

Wall, G. (1972) Socio-economic variations in pleasure trip patterns: the case of Hull car-owners. *Transactions of the Institute of British Geographers* 59, 45–58.

Watkins, C. (ed.) (1996) *Rights of Way: Policy, Culture and Management*. London: Pinter.

Whitby, M. (ed.) (1994) *Incentives for Countryside Management: The Case of Environmentally Sensitive Areas*. Wallingford: CAB International.

19

The de-intensification of European agriculture

Brian Ilbery

CONTEXTUAL SETTING

European agriculture has undergone substantial restructuring in the post-war period and, while both Western and Eastern Europe experienced forms of agricultural intensification between 1950 and the 1980s, the direction of change has since been quite different. In Eastern Europe, this has been based on a return to private farming. The transition has not been easy, and many structural problems still confront the agricultural sector, not least the re-creation of landed property rights and the development of an efficient market system of agricultural production (Repassy and Symes 1993; Ilbery 1998). Controlling agricultural output has thus not been a priority, which is in contrast to Western Europe, where the emphasis since the mid-1980s has been on a post-productivist farming system. The objective has been to de-intensify agricultural production through extensification, diversification and farming in more environmentally beneficial ways (Ilbery 1992; Battershill and Gilg 1996; Evans and Morris 1997). This chapter therefore focuses on the applied characteristics of, and problems associated with, the de-intensification of agriculture in Western Europe.

Government policy, enacted through the Common Agricultural Policy (CAP), has been the main catalyst of change in European agriculture. Prior to the mid-1980s, a *productivist* ethos based on the principles of efficiency and rationality was engendered through high levels of government support for farming. A system of guaranteed prices stimulated farmers to maximise production, irrespective of market demand. As a consequence, agricultural systems became more intensive and specialised, and farming became more spatially concentrated in 'core' farming regions such as the Po valley, Paris basin and East Anglia (Bowler 1985a and b).

Each of the three dimensions of productivist agriculture – intensification, specialisation and concentration – was accompanied by what Bowler (1985a) described as secondary consequences (Table 19.1). For example, rising indebtedness and declining farm incomes occurred as farmers became trapped on a 'technological treadmill' (Ward 1993). Second, overproduction of many agricultural products increased as both efficient and inefficient farmers were encouraged to intensify production. Third, farmers took an exploitative rather than conserving attitude towards their natural resource base, creating a number of environmental disbenefits. These included the pollution of air, soil and water courses, the removal of hedgerows and woodlands, the drainage of wetlands, and the ploughing of moorland and herb-rich permanent grasslands. Finally, productivist agriculture polarised farm-size structures and further exaggerated spatial inequalities in farm types and farm incomes. Regions became overspecialised on particular crops or livestock, as for example in the production of table wine in the Languedoc region of Mediterranean France, where attempts to de-specialise and diversify agriculture were only partially successful (Jones 1989).

Table 19.1 Secondary consequences of productivist agriculture.

Structural dimension	*Secondary consequences*
Intensification	Development of supply (requisites) cooperatives Rising agricultural indebtedness Increasing energy intensity and dependence on fossil fuels Overproduction for the domestic market Destruction of environment and agro-ecosystems
Specialisation	Food consumed outside region where it was produced Increased risk of system failure Changing composition of the workforce Structural rigidity in farm production
Concentration	Development of marketing cooperatives New social relations in rural communities Inability of young to enter farming Polarisation of the farm-size structure Corporate ownership of land Increasing inequalities in farm incomes between farm sizes, types and locations State agricultural policies favouring large farms and certain regions

Source: Bowler 1985a.

Since the mid-1980s, the established model of agricultural productivism has been under challenge, and the European Union (EU) has placed considerable emphasis on the de-intensification of farm production. Alternative discourses associated with reduced price support, environmental protection, sustainability, food quality and integrated rural development have arisen both from within and from outside the agricultural community. The aim of this chapter, therefore, is to examine some of the challenges associated with the transition towards post-productivist farming systems in Western Europe. More specifically, the chapter first conceptualises *post-productivism* and outlines how government policy is now responding to, rather than stimulating, agricultural change. It then discusses some of the key applied aspects of agricultural de-intensification, drawing on case study evidence to highlight the spatially uneven impact of post-productivism in Europe. The chapter concludes by questioning whether the trend towards more extensive farming systems is sustainable.

CONCEPTUALISING POST-PRODUCTIVIST AGRICULTURE

The movement away from a predominantly productivist ethos in agriculture has been conceptualised as the post-productivist transition (PPT) (Shucksmith 1993; Lowe *et al.* 1993; Ilbery and Bowler 1998). While the exact nature of the PPT has still to be defined in developed market economies, it is associated with a number of known characteristics (Table 19.2). More specifically, Bowler and Ilbery (1997) conceptualised the PPT in terms of three bi-polar dimensions of change:

- intensification to extensification
- specialisation to diversification
- concentration to dispersion

The first two have been actively encouraged through reforms of the CAP, but greater difficulty has been experienced in dispersing agriculture away from its concentrated pattern in 'core' areas. Indeed, the productivist ethos is well

Table 19.2 Characteristics of the post-productivist transition.

- A de-intensification of agricultural production and movement towards food quality rather than quantity
- The progressive withdrawal of state subsidies for agriculture and the decoupling of farm incomes from the volume of food produced
- The production of food within an increasingly competitive and international market
- A growing environmental regulation of agriculture through a range of agri-environmental programmes
- The creation of more sustainable farming systems

entrenched in policy circles, so much so that the two farming systems (productivist and post-productivist) are likely to coexist in the future. This may comprise an intensive system of farming, which emphasises food *quantity*, and a more extensive system, which espouses sustainability and food *quality*. It is probable that the two farming systems will become further spatially differentiated, with the prosperous agricultural regions in Europe producing for the mass food market and the more marginal agricultural areas providing quality food products for niche markets.

During the PPT, the CAP has been responding to, rather than stimulating, change. Initially, this took the form of production control measures, through for example milk quotas and arable set-aside (Briggs and Kerrell 1992; Naylor 1993). However, these had limited impact because farm subsidies were still coupled to the amount of food produced. It was not until the so-called MacSharry reforms of the CAP in 1992 that the idea of decoupling farm incomes from the volume of food production was taken seriously (Robinson and Ilbery 1993). This began through a system of *income aid*, in the form of arable area payments (AAPs) and voluntary agri-environmental programmes (AEPs).

Nevertheless, the movement towards agricultural de-intensification in the EU has been slow, and in 1995 the equivalent of 49 per cent of farm incomes still came in the form of subsidies; this compared with 15 per cent in the USA and 3 per cent in New Zealand. Such a high level of government support reflects deeply embedded attitudes, which make a move away from agricultural productivism politically difficult. Despite this, mounting macro-scale pressures, including the internationalisation of the food supply system and the greening of agricultural policy, are being placed on the CAP to decouple farm incomes completely from government economic subsidies.

These pressures have gained momentum through the 1992 Rio Earth Summit, the 1993 GATT agreement on world agricultural trade, and the 1996 Federal Agricultural Improvement and Reform Act (FAIR) in the USA. The FAIR programme has replaced subsidies through deficiency and set-aside payments with a seven-year system of decoupled payments, where farmers are not obliged to produce particular crops or any crop at all in order to receive income aid (Harvey 1996). Thus income is not tied to production (as it is in the EU's AAPs). Farmers are still able to sell their farm produce, but income from this is dependent on market, not guaranteed, prices. These ideas, especially the move towards market orientation, are almost certain to influence the next round of the WTO (formerly GATT) negotiations beginning in 1999 (Ritson and Harvey 1997). Indeed, the European Commission has already responded to the pressures through publication of its Agenda 2000 proposals, which advocates further major cuts in guaranteed prices paid to farmers, more decoupled income aid (but with an upper ceiling) and the channelling of support to the poorest farmers in marginal agricultural areas.

Clearly, farm households are having to adjust

to policy change and the PPT; it is at the farm level where the consequences of policy reform are 'played out'. Bowler (1992) and Bowler *et al.*. (1996) referred to these adjustments as 'pathways of farm business development'. While one possible pathway for some farmers is the continuation of the productivist model of industrialised farming, most options involve a redistribution of farm resources into different types of agricultural de-intensification. This may involve an *extensification* of production through either the maintenance/re-creation of traditional farming systems or the adoption of AEPs. It might also involve a *diversification* of the income base into different types of agricultural (e.g. non-conventional crops and livestock enterprises, woodlands) and/or structural (e.g. farm tourism, direct marketing and processing of food) diversification. Finally, it may involve the re-localisation and thus the *dispersion* of the agro-food system in which locally produced quality products, with real authenticity of geographical origin and traceability, can act as niche markets (Marsden 1996). This is one possible way in which marginal agricultural areas can exploit the increasing demand for local and wholesome food products. These three applied aspects of the PPT are not mutually exclusive and could be interlinked through developments in, for example, organic farming (Park and Lohr 1996). Nevertheless, for the rest of this chapter, each dimension is examined in isolation, drawing on case study evidence where appropriate.

Extensification: agri-environmental programmes

Although initially conceived as an attempt to reduce agricultural production by paying farmers to de-intensify their farming systems, the main thrust of extensification in European agriculture has come through the adoption of different agri-environmental programmes (AEPs). AEPs can be interpreted as a political compromise between the demands of the farm lobby and the calls for change by environmental groups (Potter 1998). Whereas the former group demanded payments for environmental management in the EU, the latter group argued that payments to farmers (subsidies) should be withdrawn if specified conservation conditions are not met (the concept of *cross-compliance*).

The first signs of incorporating environmental dimensions into EU agricultural policy came in a Commission 'Green Paper' in 1985 (CEC 1985). This recommended the withdrawal of land along environmentally strategic buffer zones, within ecological corridors along field boundaries, and around water bodies. The Green Paper suggested the designation of environmentally sensitive areas (ESAs), within which farmers would receive an annual premium to introduce or maintain farming practices that were compatible with the protection of the environment and natural resources. These ideas were incorporated into the new Structures Regulation of 1985 (797/85), and EU funding for ESAs was confirmed in 1986. Indeed, continuous policy developments between 1985 and 1992, including extensification (Regulation 1760/87) and arable set-aside (Regulation 1094/88), were based on the principle of financial compensation for reducing agricultural output.

Although the major reforms of the CAP in 1992 were economically driven (Ritson and Harvey1997), Regulation 2078/92 stated that member states must implement a package of 'accompanying measures', to include AEPs. Each country had to have approved an agri-environmental package by 1993. The schemes were to be voluntary over five years, and farmers would be financially compensated for loss of income if they abided by one or more of the following:

- substantial reduction in the use of fertilisers and/or the introduction/continuation of organic farming methods;
- change to more extensive forms of crop/livestock production;
- use of other farming practices that are beneficial to the environment and natural resource protection;
- upkeep of abandoned land;

- long-term set-aside of agricultural land for environmental reasons;
- land management for public access and leisure.

Half of the eligible expenditure for the AEPs was to come from the EU budget, and the Regulation was open to interpretation by individual states. In the UK, for example, the 1993 package included more ESAs, access to ESAs and set-aside land, new nitrate measures, an organic farming scheme, a moorland scheme, and a habitat improvement scheme through long-term set-aside (Potter 1993). The principle of compensating farmers for their projected loss of income suggests that 'AEPs have amounted to little more than a continuation and extension of existing programmes such as ESAs' (Winter 1996: p. 255).

There has been limited research on the applied and geographical consequences of AEPs in Europe (Whitby 1996; Evans and Morris 1997). International comparisons are difficult, because different schemes exist in different countries. Even within one scheme, such as ESAs in the United Kingdom, prescriptions vary between designated areas. The focus of research, therefore, has been on uptake rates of specific schemes in particular regions or countries. For example, Wilson (1995; 1996; 1997) examined uptake and farmers' attitudes towards the MEKA programme in Baden-Württemberg (Germany) and the ESA scheme in Wales. Similarly, Wilson *et al.* (1996) and Curry and Stucki (1997) analysed AEPs in Switzerland (see Box 19.1). However, the real environmental benefits of AEPs are far from clear. Many farmers are 'passive' adopters who enter the schemes for financial rather than environmental reasons; as Morris and Potter (1995) suggest, there is a need to examine the longer-term impact of AEPs, well after the schemes have finished.

Indeed, academic research has pointed to a number of weaknesses of AEPs, including:

1. They are effectively 'bolted on' to a productivist-oriented agricultural policy. Because of a lack of cross compliance, where farmers receive economic payments (e.g. subsidies or income aid) only if they satisfy environmental prescriptions, AEPs do not automatically lead to the production of environmental goods.
2. Most AEPs focus on inputs, such as application rates of inorganic fertilisers, rather than outputs. Thus farmers get paid for satisfying certain conditions rather than for results achieved (in terms of environmental conservation).

Box 19.1 Agri-environmental programmes (AEPs) in Switzerland

Since the mid-1980s, Swiss agricultural policy has shifted away from price supports and towards both direct payments to farmers and specific objectives for the environment, ecology and welfare of rural communities (Curry and Stucki 1997). Unlike the CAP reforms of 1992, changes to the Swiss Federal Agricultural Law in 1992 involved a full reappraisal of agricultural policy. This covers all agricultural land and is invoked through cross-compliance. Two elements now dominate agricultural policy: first, the decoupling of farm incomes from price supports; and second, direct payments to farmers. Under the first element, farmers receive baseline payments (price and marketing safeguards) for the production of local and high-quality food products that are associated with environmentally sustainable farming methods. Under the second element, three types of payment are available. The first is compensation for reduced subsidies, but farmers receive these only if they satisfy the management practices of environmental sensitivity and animal welfare. Second, social payments are available to small family farms to help to retain people in rural areas; and third, farmers receive payments for entering voluntary AEPs.

Under the voluntary AEPs, the amount of income aid available to farmers varies according to different 'levels' of ecological farming. Thus the smallest payments are made for biological diversity (extensive grassland cultivation) and the highest for organic farming. The different schemes are administered by the twenty-six cantons, which have some flexibility to vary the prescriptions of AEPs according to local conditions (Wilson *et al.* 1996). After one year, 20 per cent of the 75,000 Swiss holdings had been accepted into an AEP, and by 1995 40 per cent of the entered agricultural area was involved in 'integrated production' or 'organic' farming.

In Switzerland, therefore, ecological farming has been proposed as the dominant farming system; it is a single scheme for the whole country, with real cross-compliance, and is not 'bolted on' to an essentially economically driven policy that is designed to maintain farm subsidies.

3 They are often spatially targeted, with little control on production and environmental management outside the designated areas, and this reduces their overall effectiveness.
4 It is questionable whether voluntary schemes can initiate land-use change on farms, especially if farmers do not have to enter the whole farm into an AEP.
5 AEPs are not really sustainable, in the sense that they lead to a significant reduction in both the energy and agrochemical dependencies of modern farming and the dependency of farmers on state financial support (Evans and Morris 1997).
6 There is a general lack of finance for AEPs, accounting for less than 5 per cent of the total agricultural budget of the EU.

AEPs will continue to have little impact as long as farmers continue to receive subsidies for producing food in designated and non-designated areas. Potter (1998) raises the question of whether policy makers should continue to 'green' existing agricultural policy or remove all support for agricultural production and provide farmers with income aid for environmental conservation.

Diversification: other gainful activities

Reforms of the CAP and especially cuts in guaranteed prices paid to farmers have seen the 'once peripheral option of supporting farm income diversification become more central' (Fuller 1990: p. 67). Defined as the generation, by farm household members, of income from either on- and/or off-farm sources in addition to that obtained from primary agriculture, other gainful activities (OGAs) represent a major pathway of farm business development. In a recent study of different pathways in the northern Pennines in England, Bowler *et al.* (1996) found that 33 per cent of farm households had off-farm OGAs and 29 per cent had on-farm OGAs. Nevertheless, the study identified a 'resistance' to diversification, which was often considered only when traditional farming was unable to address the income needs of farm households.

There can be little doubt that the incidence of OGAs in Europe has been increasing during the PPT. Indeed, a major research project conducted by the Arkleton Trust in twenty-four regions of Western Europe in the late 1980s indicated that nearly 60 per cent of farm households were pluriactive (Fuller 1990). A large number of OGAs are off-farm (over 50 per cent of all households compared with less than 10 per cent with on-farm OGAs), but this varies considerably between the twenty-four regions. So, while 81 and 72 per cent of farm households in Freyung–Grafenau (Germany) and West Bothnia (Sweden), respectively, have off-farm OGAs, this falls to 33 and 27 per cent in the west of Ireland and Picardie (France), respectively (Box 19.2). This pattern is further complicated by differences in which family member(s) participate in off-farm work. However, in all regions farm operators are more prevalent than spouses in off-farm work; in only five regions (in Spain, Italy, Austria and Ireland) do other family members (including spouses) outnumber farm operators.

Not surprisingly, therefore, the Arkleton project found that one-third of farm households obtain over 50 per cent of their income from off-farm sources (Plate 19.1). This again varies between the regions (Figure 19.1), from a high of 71 per cent in Freyung–Grafenau to a low of just 10 per cent in Picardie. The research found that just 17 per cent of farm households derive 100 per cent of their income from farming, whereas 43 per cent obtain less than 30 per cent in this way. Fuller (1990) concluded that patterns of pluriactivity (OGAs) in Western Europe are quite complex, reflecting 'the interplay of farming, household and labour market characteristics, as well as cultural factors' (p. 368).

It is the interaction of many 'external' and 'internal' factors that shapes geographical patterns of pluriactivity in Europe. For example, Efstratoglou-Todoulou (1990) identified a relationship between regional socio-economic conditions (e.g. local labour markets, unemployment, tourist activities) and rates of pluriactivity in Greece. In contrast, Edmond and Crabtree (1994) found that OGAs were more important in

Box 19.2 Other gainful activities (OGAs) in France

The amount of farm household income coming from on- and off-farm OGAs in France increased from 15 per cent in 1956 to 42 per cent in 1988 (Benjamin 1994). However, there are wide variations between the regions, from 25 to 52 per cent, and Campagne *et al.* (1990) related these to different levels of agricultural modernisation. They examined the different types of pluriactivity practised in three contrasting agricultural regions:

1 *Picardie*, an area of large-scale arable production in northern France.
2 *Languedoc*, a region in southern France with a long history of combining wine production with other work.
3 *Savoy valleys*, a physically marginal area for agriculture in southeast France.

Campagne *et al.* described Picardie as a zone of 'business pluriactivity' because household members are using agricultural resources to increase non-agricultural activities. Many spouses (especially wives) work outside agriculture and there is a culture of accumulation among the mainly family-based farms in this prosperous agricultural area. In contrast, the generation of mainly off-farm income in Languedoc, especially by the spouse and children, is vital to the maintenance and modernisation of the farm business. The authors referred to this as the 'pluriactivity of maintaining farming'. Finally, in the Savoy Valleys, where there is a progressive abandonment of marginal farming, farm households are having to search for a diverse range of on- and off-farm income-generating activities. Not surprisingly, this was described as 'pluriactivity for survival'.

Explanations for such patterns would need to take account of both 'external' and 'internal' factors. For example, Benjamin (1994) indicated that CAP reforms have increased wives' participation in off-farm work. However, she also showed that the age and education of farm wives have a greater positive effect on off-farm participation, especially among women aged between 26 and 42 who do not have a high school diploma.

Plate 19.1 Diversifying the farm business into alternative (non-agricultural) enterprises.

Figure 19.1 Off-farm OGAs in selected regions of Western Europe.

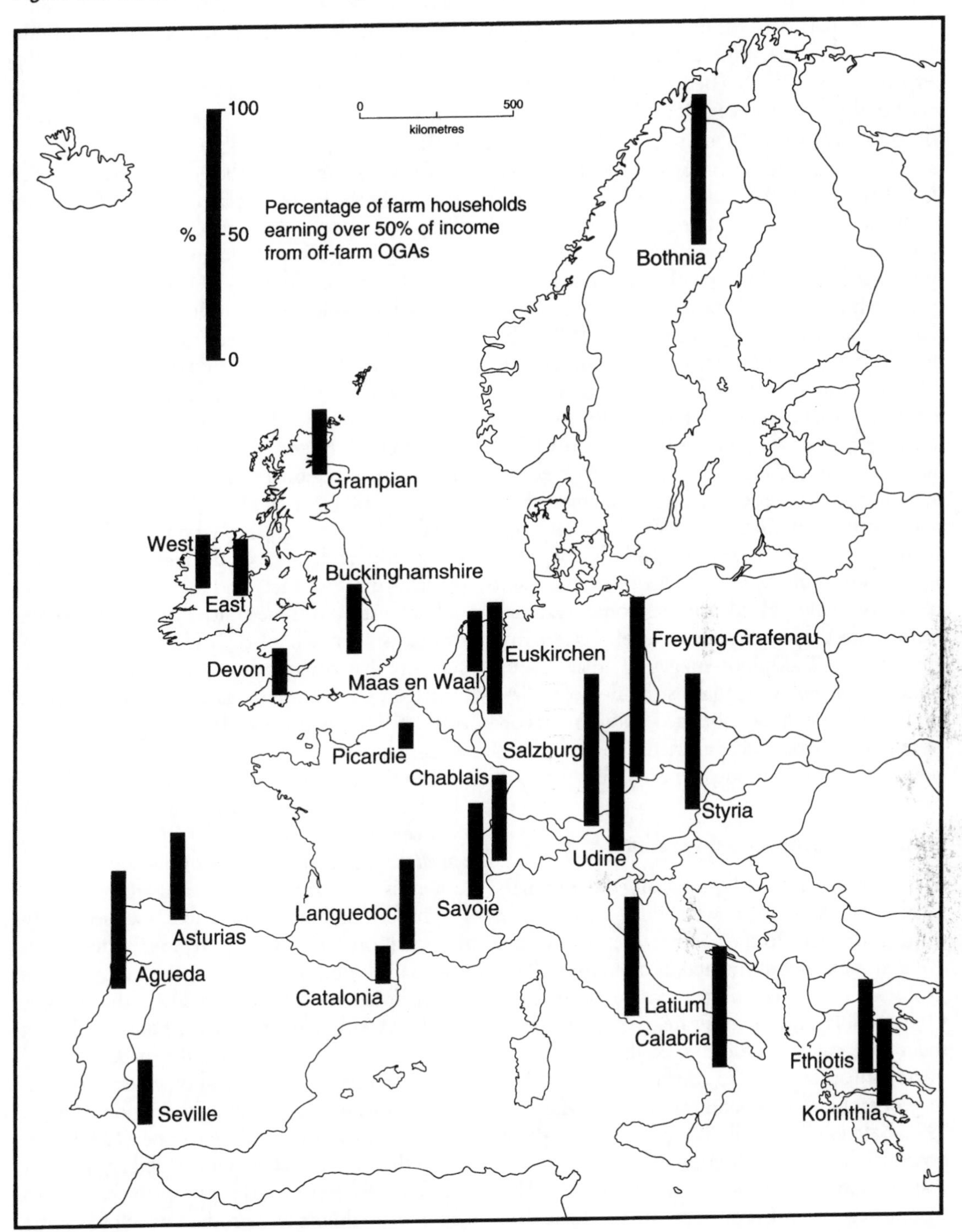

Source: Based on data derived from Fuller 1990.

the least densely populated areas, reflecting social and cultural rather than economic factors. The increasing participation of women in the labour force is one such example that helps to explain the expansion of OGAs (Benjamin 1994). Similar relationships have been found between such 'internal' factors as farm size, family life cycle, succession, age and education and the adoption of OGAs (Gasson 1987; Ilbery and Bowler 1993).

In a rare study of regional patterns of pluriactivity, Efstratoglou-Todoulou (1990) hypothesised that adoption of OGAs would be positively related to off-farm opportunities ('pull' factors) and negatively related to favourable farming conditions ('push' factors). Applying his ideas to Greece, he found that pull factors were stronger in less favoured areas, with low farm incomes and low farm opportunities; here OGAs were a *necessity*. In contrast, push factors exerted a significant inverse effect on OGAs in areas where agricultural structures and farm incomes were higher; in such areas, farm households have alternative opportunities and so OGAs are the result of *choice*. Although identifying some 'external' factors affecting regional patterns of pluriactivity, Efstratoglou-Todoulou suggested that a full understanding of the spatial distribution of OGAs in Europe could be obtained only by incorporating 'internal' characteristics of the farm household into the modelling exercise.

Dispersion: speciality food products

Many farmers in the EU, especially those in marginal agricultural areas, will find it increasingly difficult to adjust to the continual reforms of the CAP and the environmental, health and welfare problems associated with intensive agriculture. One possible adjustment strategy is the development of locally produced speciality food products (SFPs; see Box 19.3), with real authenticity of geographical origin and traceability (Marsden 1996; Battershill and Gilg 1996). This offers some potential for a relocalisation of the agro-food system and thus agricultural dispersion. The lagging regions in Europe are among those that could benefit from the increasing demands for SFPs, especially if they are tied to a regional image and notions of sustainability and environmental friendliness. Indeed, the Committee of the Regions (1996) described SFPs as a possible trump card for the regions and one that could engender endogenous and bottom-up rural development (Bryden 1994).

There is currently little geographical research that links SFPs to specific places (Ilbery and Kneafsey 1998). However, there is growing interest in quality and regional imagery. The Scottish Food Strategy Group (1993: p. 3) defined a quality food product as 'one which is differentiated in a positive manner by reason of one or more factors from the standard product, is recognised as such by the consumer, and can therefore command a market benefit if it is effectively marketed.' The key to success is to link quality to a regional image through marketing and the promotion of place. As some consumers, notably from within the 'service class', are on a quest for authenticity, through for example quality rural products and services, private and public institutions need to develop marketing techniques to 'sell places'. Urry (1995: p. 163) suggests that places can be 'substantially reconstructed', so that the focus shifts away from the sales of what is produced to the production of what will sell.

Despite the lack of work on SFPs, Moran (1993) provided a lead when he made a direct link between product and place in the French wine *appellation* system. The basic philosophy of this system is that wine is an expression of the geographical individuality of places. Consequently, the creation of a strong regional identity is essential if wines are to sell at the best prices in international markets. Moran quotes the example of Chateauneuf-du-Pape, where the *appellation* laws (rather than any scientific measurement) created an image of high-quality wines; here is an example of a product creating the regional image. More recently, Bell and Valentine (1997) suggested that 'we are where we eat' and argued that the link between product and place can be so strong that 'almost any product which has some tie to place – no matter how invented this may be

Box 19.3 Speciality food products (SFPs) in England

The speciality food and drinks sector in Britain employs over 20,000 people and has an annual turnover of nearly £3 billion. It is dominated by small and medium enterprises (SMEs) using traditional recipes and/or innovative ideas to make high-quality products. The focus is on using wholesome ingredients, and many SMEs cater for niche markets. Concerns over agricultural intensification have increased the opportunity to sell SFPs outside their regions of origin. However, they rarely have the resources or skills to find outlets in national and international markets. This is the justification for speciality food groups.

A national network of regional and county speciality food groups exists in Britain; fifteen regional and county groups cover England and Wales (Figure 19.2), and the Scottish and Northern Ireland groups are managed by Sottish Enterprise and A Taste of Ulster, respectively. Local and county groups in England often lacked the necessary critical mass to succeed. In 1991, therefore, the Ministry of Agriculture, Fisheries and Food launched a six-year grant scheme for regional food groups; there are currently six, and a seventh (Heart of England) was launched in early 1998. Their development is coordinated and managed by Food from Britain.

Over 60 per cent of the speciality food and drink producers employ five or less employees and sell a majority of their products through either their own shops or local caterers (although mail orders and the multiple retailers are other important outlets). A wide range of SFPs are produced in the different regions of England, from farmhouse cheeses (Plate 19.2) and smoked/cured meats to salad dressings and special occasion cakes and biscuits. Taste of the West, for example, has nearly 200 producer members in its seven counties, over seventy of which are concentrated in Devon. Here the emphasis is very much on dairy and speciality drink products, followed by meats, bakery products and preserves. Most producers are strongly attached to their regions, and there is now a need to use regional imagery to further develop the link between product and place.

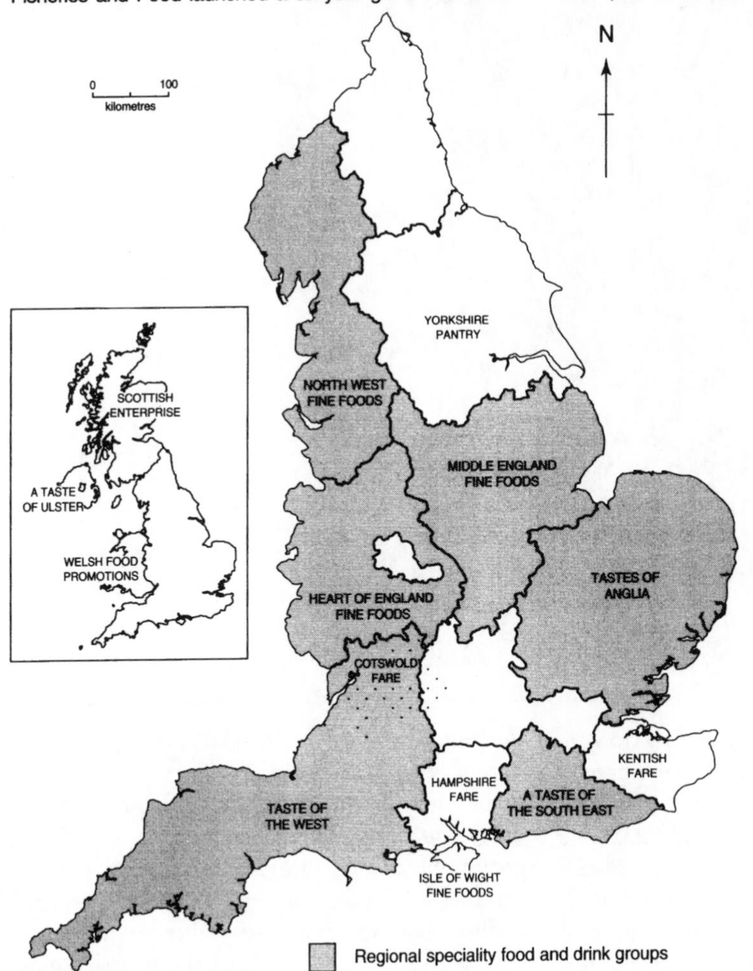

Figure 19.2 Regional and county speciality food groups in England.

Box 19.3 continued

Plate 19.2 Speciality food products in Britain: farmhouse cheeses.

– can be sold as embodying that place' (p. 155).

The European Commission has encouraged the production of SFPs from specific regions through Regulations 2081/92 and 2082/92. While the latter introduced *certificates of special character* for quality products produced with local raw materials and/or a traditional mode of production, the former introduced *protected designations of origin* (PDOs) and *protected geographical indications* (PGIs). These are European quality marks for products from a specific region where quality is due exclusively to a particular geographical environment (PDO) or where a product from a specific region possesses a specific quality, but not necessarily due to its natural environment (PGI). Both of these designations are available to producers working in cooperative groups, but their uptake across Europe still awaits detailed research.

In response to an OECD (1995) report on niche markets and the statement by the Committee of the Regions (1996), a major European project, funded by the European Commission, is attempting to help public institutions to develop strategies, policies and structures to enable the successful marketing of SFPs in the lagging regions of the EU. The RIPPLE project is attempting to link product and place by conducting research on regional imagery and marketing in relation to the creation of SFPs from twelve lagging rural areas (Ilbery and Kneafsey 1998). Investigating a range of quality products and services, RIPPLE will involve surveys of both producers and consumers of SFPs, as well as the public and private agencies concerned with their promotion and successful marketing. In the future, the production of SFPs in the peripheral regions of Europe may contribute to the process of agricultural dispersion.

CONCLUSION

This chapter has focused on the applied dimensions of, and problems relating to, the de-intensification of agriculture in Western Europe. It has highlighted the duality between 'productivist' and 'post-productivist' farming systems and examined the uneven spatial development associated with the PPT. Different regions are dominated by different pathways of farm business development, and de-intensification is being encouraged through a combination of extensification (AEPs), diversification (OGAs) and dispersion (SFPs).

The PPT is being regulated by the state, which will need to reform the CAP further in response to the growing trends towards market orientation and sustainability. While the former is pressing for farmers to receive market prices for their products without economic subsidies, the latter is seeking a more environmentally sound system of farming that is less dependent on major energy and agrochemical inputs. Indeed, at the core of the PPT is the concept of sustainable agriculture. Although this can be interpreted in environmental, socio-economic or productive senses (Brklacich *et al.* 1990), there are two main models of sustainable agriculture: idealist and instrumental (Bowler 1992). While the former adopts an 'alternative agriculture' perspective and argues that 'no' or 'low' growth modes are the only long-term option for agriculture, the latter is more conventional and sees sustainability as a contextual process rather than a set of specific prescriptions. The instrumentalist model is thus less rigorous and advocates an extensive, diversified and conservation-oriented system of farming. This contrasts with the organic, biodynamic and ecological systems of farming put forward by the idealist school of thought. In Western Europe, only a minority of farmers are pursuing the idealist model. The instrumentalist model is more popular with farmers and is being encouraged through such state regulation as limits on fertiliser application, imposition of minimum standards of pesticide residues in food, constraints on types and rates of application of agrochemicals, and subsidies to farm under lower input – lower output systems.

In the longer term, it could be that the only payments made to farmers under the CAP will be for environmental conservation. Indeed, the future focus of policy will be broader than agriculture, incorporating rural diversification and integrated rural development. The latter needs to be multisectoral, sustainable and based on local needs through the concept of subsidiarity.

GUIDE TO FURTHER READING

Bowler, I. (1985) *Agriculture under the Common Agricultural Policy: A Geography*. Manchester University Press, Manchester. Examines the main elements of the CAP and the spatial dimensions of 'productivist' agriculture in the EU up to the early 1980s.

Evans, N. and Morris, C. (1997) Towards a geography of agri-environmental policies in England and Wales. *Geoforum* 28, 189–204. Provides a review of the development of AEPs at national, regional and local geographical scales.

Ilbery, B. and Bowler, I. (1998) From agricultural productivism to post-productivism. In Ilbery, B. (ed.) *The Geography of Rural Change*, Longman, London, 57–84. Offers a detailed account of both theoretical and empirical aspects of 'productivist' and 'post-productivist' farming systems.

Robinson, G. and Ilbery, B. (1993) Reforming the CAP: beyond MacSharry. In A. Gilg (ed.) *Progress in Rural Policy and Planning*, Volume 3, Belhaven Press, London, 197–207. Gives an insight into agricultural policy development and reforms of the CAP in the EU.

Winter, M. (1996) *Rural Politics: Policies for Agriculture, Forestry and the Environment*. Routledge, London. Is a study of the evolution and content of policies affecting the countryside, especially in Britain.

REFERENCES

Battershill, M. and Gilg, A. (1996) Traditional farming and agro-environmental policy in southwest England: back to the future. *Geoforum* 27, 133–47.

Bell, D. and Valentine, G. (1997) *Consuming Geographies: We Are Where We Eat*. Routledge, London.

Benjamin, C. (1994) The growing importance of diversification activities for French farm households. *Journal of Rural Studies* 10, 331–42.

Bowler, I. (1985a) Some consequences of the industrialisation of agriculture in the European Community. In M. Healey and B. Ilbery (eds) *The Industrialisation of the Countryside*, GeoBooks, Norwich, 75–98.

Bowler, I. (1985b) *Agriculture under the Common Agricultural Policy: A Geography*. Manchester University Press, Manchester.

Bowler, I. (1992) Sustainable agriculture as an alternative path of farm business development. In I. Bowler, C. Bryant and D. Nellis (eds) *Rural Systems in Transition: Agriculture and Environment*, CAB International, Wallingford, 237–53.

Bowler, I. and Ilbery, B. (1997) The regional consequences for agriculture of changes to the Common Agricultural Policy. In C. Laurent and I. Bowler (eds) *CAP and the Regions: Building a Multidisciplinary Framework for the Analysis of the EU Agricultural Space*, INRA, Versailles, 105–16.

Bowler, I., Clark, G., Crockett, A., Ilbery, B. and Shaw, A. (1996) The development of alternative farm enterprises: a case study of family labour farms in the north Pennines of England. *Journal of Rural Studies* 12, 285–95.

Briggs, D. and Kerrell, E. (1992) Patterns and implications of policy-induced agricultural adjustments in the European Community. In A. Gilg (ed.) *Restructuring the Countryside: Environmental Policy in Practice*, Avebury, Aldershot, 85–102.

Brklacich, M., Bryant, C. and Smit, B. (1990) Review and appraisal of concepts of sustainable food production systems. *Environmental Management* 15, 1–14.

Bryden, J. (1994) Prospects for rural areas in an enlarged Europe. *Journal of Rural Studies* 10, 387–94.

Campagne, P., Carrère, G. and Valceschini, E. (1990) Three types of agricultural region: three types of pluriactivity. *Journal of Rural Studies* 6, 415–22.

Commission of the European Communities (1985) *Perspectives for the Common Agricultural Policy*. CEC, Brussels.

Committee of the Regions (1996) *Promoting and Protecting Local Products*. Opinion of the Committee of the Regions, CDR 54/96.

Curry, N. and Stucki, E. (1997) Swiss agricultural policy and the environment: an example for the rest of Europe to follow? *Journal of Environmental Planning and Management* 40, 465–82.

Edmond, H. and Crabtree, R. (1994) Regional variation in Scottish pluriactivity: the socio-economic contexts for different types of non-farming activity. *Scottish Geographical Magazine* 110, 76–84.

Efstratoglou-Todoulou, S. (1990) Pluriactivity in different socio-economic context: a test of the push-pull hypothesis in Greek farming. *Journal of Rural Studies* 6, 407–13.

Evans, N. and Morris, C. (1997) Towards a geography of agri-environmental policies in England and Wales. *Geoforum* 28, 189–204.

Fuller, A. (1990) From part-time farming to pluriactivity: a decade of change in rural Europe. *Journal of Rural Studies* 6, 361–73.

Gasson, R. (1987) The nature and extent of part-time farming in England and Wales. *Journal of Agricultural Economics* 38, 167–91.

Harvey, D. (1996) The US Farm Bill – 'FAIR' or 'FOUL'? In D. Colman (ed.) *The American Farm Bill: Implications for CAP Reform*, Manchester University Press, 82–93.

Ilbery, B. (1992) Agricultural policy and land diversion in the European Community. In A. Gilg (ed.) *Progress in Rural Policy and Planning*, Volume 2, Belhaven Press, London, 153–66.

Ilbery, B. (1998) The challenge of agricultural

restructuring in the European Union. In D. Pinder (ed.) *The New Europe: Economy, Society and Environment*, Wiley, London, 341–57.

Ilbery, B. and Bowler, I. (1993) The Farm Diversification Grant Scheme: adoption and non-adoption in England and Wales. *Environment and Planning* C 11, 161–70.

Ilbery, B. and Bowler, I. (1998) From agricultural productivism to post-productivism. In B. Ilbery (ed). *The Geography of Rural Change*, Longman, London, 57–84.

Ilbery, B. and Kneafsey, M. (1998) Product and place: promoting quality products and services in the lagging rural regions of the European Union. *European Urban and Regional Studies* 5, 329–41.

Jones, A. (1989) Reform of the European Community's table wine sector: agricultural despecialisation in the Languedoc. *Geography* 74, 29–37.

Lowe, P., Murdoch, J., Marsden, T., Munton, R. and Flynn, A. (1993) Regulating the new rural spaces: the uneven development of land. *Journal of Rural Studies* 9, 205–22.

Marsden, T. (1996) Rural geography trend report: the social and political bases of rural restructuring. *Progress in Human Geography* 20, 246–58.

Moran, W. (1993) The wine appellation as territory in France and California. *Annals of the Association of American Geographers* 83, 694–717.

Morris, C. and Potter, C. (1995) Recruiting the new conservationists: adoption of agri-environmental schemes in the UK. *Journal of Rural Studies* 11, 51–63.

Naylor, E. (1993) Milk quotas and changing patterns of dairying in France. *Journal of Rural Studies* 9, 53–63.

OECD (1995) *Niche Markets as a Rural Development Strategy*. OECD, Paris.

Park, T. and Lohr, L. (1996) Supply and demand factors for organic produce. *American Journal of Agricultural Economics* 78, 647–55.

Potter, C. (1993) Pieces in a jigsaw: a critique of the new agri-environment measures. *Ecos* 14, 52–4.

Potter, C. (1998) Conserving nature: agri-environmental policy, development and change. In B. Ilbery (ed.) *The Geography of Rural Change*, Longman, London, 85–105.

Repassy, H. and Symes, D. (1993) Perspectives on agrarian reform in east–central Europe. *Sociologia Ruralis* 33, 81–91.

Ritson, C. and Harvey, D. (eds) (1997) *The Common Agricultural Policy*, second edition. CAB International, Wallingford.

Robinson, G. and Ilbery, B. (1993) Reforming the CAP: beyond MacSharry. In A. Gilg (ed.) *Progress in Rural Policy and Planning*, Volume 3, Belhaven Press, London, 197–207.

Scottish Food Strategy Group (1993) *Scotland Means Quality*. SFSG, October.

Shucksmith, M. (1993) Farm household behaviour and the transition to post-productivism. *Journal of Agricultural Economics* 44, 466–78.

Urry, J. (1995) *Consuming Places*. Routledge, London.

Ward, N. (1993) The agricultural treadmill and the rural environment in the post-productivist era. *Sociologia Ruralis* 33, 348–64.

Whitby, M. (ed.) (1996) *The European Environment and CAP Reform: Policies and Prospects for Conservation*. CAB International, Wallingford.

Wilson, G. (1995) German agri-environmental schemes: II The MEKA programme in Baden-Württemberg. *Journal of Rural Studies* 11, 149–59.

Wilson, G. (1996) Farmer environmental attitudes and ESA participation. *Geoforum* 27, 115–31.

Wilson, G. (1997) Factors influencing farmer participation in the Environmentally Sensitive Areas Scheme. *Journal of Environmental Management* 50, 67–93.

Wilson, G., Lezzi, M. and Egli, C. (1996) Agri-environmental schemes in Switzerland. *European Urban and Regional Studies* 3, 205–24.

Winter, M. (1996) *Rural Politics: Policies for Agriculture, Forestry and the Environment*. Routledge, London.

20

Wetlands conservation

Max Wade and Elena Lopez-Gunn

SETTING THE SCENE

Wetlands represent only 6 per cent of the Earth's surface, but it is believed that in 1900 this percentage might have been twice as much (Barbier *et al.* 1994). Wetlands include a wide array of habitats, ranging from fens and marshes to mangrove forests and rice paddies, and are considered one of the most threatened landscapes in the world (Gardiner 1994). A simple definition is 'land with soils that are permanently flooded' (Williams 1990: p. 1). The Ramsar Convention, an international treaty to conserve wetlands, defines wetlands as: 'areas of marsh, fen, peatland or water, whether natural or artificial, permanent or temporary, with water that is static or flowing, fresh, brackish or salt, including areas of marine water the depth of which at low tide does not exceed six metres.' This encompasses a wide range of habitats, the main types being shown in Box 20.1 based on just one recognised classification (Gleich 1993). According to the Ramsar classification, there are marine, coastal, inland and man-made types, subdivided into thirty categories of natural wetland and nine human-made ones, such as reservoirs, barrages, and gravel pits (Dugan 1993). The International Wetlands Research Bureau has established a wetland database. In relation to natural wetlands, the World Conservation Monitoring Centre (Groombridge 1992) summarises the extent of different types relative to latitude (Figure 20.1). Surveys have established the extent of wetlands, both past and present, in different areas of the world, leading to the compilation of inventories of wetland sites, particularly for plants, birds and mammals, and investigations into physical, chemical and biological processes. Collectively, these have developed a real insight into wetland ecology.

Wetlands epitomise the problem of trying to classify ecosystems and habitats. While mangroves, for example, meet the criteria set for wetlands, they also meet those for both forest and coastal systems (Groombridge 1992), indicating the integral role that wetlands play in the broader

Box 20.1 A classification of wetland habitats

Marsh A frequently or continually inundated wetland characterised by emergent herbaceous vegetation adapted to saturated soil conditions. In European terminology, a marsh has a mineral soil substrate and does not accumulate peat.

Swamp A wetland dominated by trees or shrubs (US definition). In Europe, a forested fen would easily be called a swamp. In some areas, wetlands dominated by reed grass are also called swamps.

Fens A peat-accumulating wetland that receives some drainage from surrounding mineral soil and usually supports marsh-like vegetation.

Bogs A peat-accumulating wetland that has no significant inflows or outflows and supports acidophilic mosses, particularly spaghnum.

Peatland A generic term for any wetland that accumulates partly decayed plant matter.

Source: Gleick 1993.

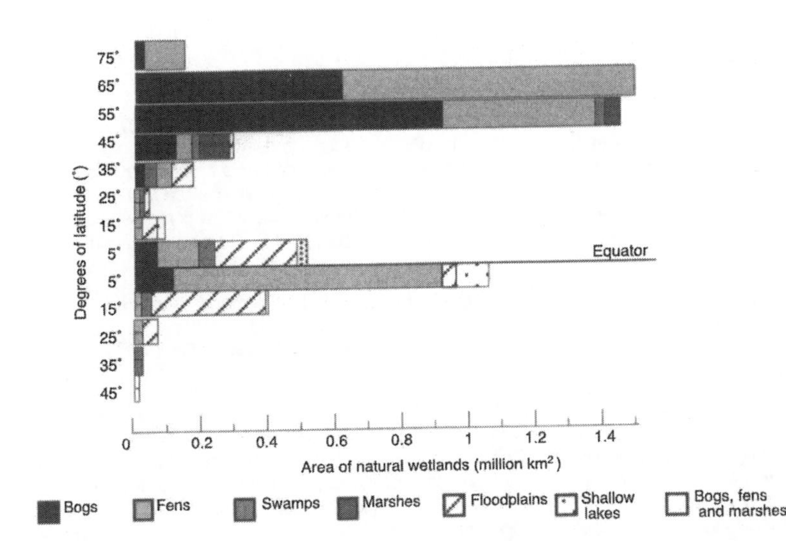

Figure 20.1 Latitudinal distribution of natural freshwater wetlands.

Source: Groombridge 1992.

ecology of countries and regions. Another problem encountered in assessing wetlands is delineation of their boundaries, typically an ecotone between the wetland and either aquatic or terrestrial habitats (Committee on Characterisation of Wetlands 1995).

The human perception of wetlands has always been ambivalent. Misunderstandings over their ecology and functioning lead to their perception as a hazardous wasteland, or an area to be drained for agriculture and other arguably more productive land uses (Box 20.2). Historically, wetlands were considered hazardous, marginal waterlogged lands, harbouring disease. Malaria, dengue fever, filiariasis and yellow fever are all tropical diseases associated with wetlands (Dugan 1993). However, local people often respect and understand wetlands as a resource and are dependent on them (Box 20.2), while more recently, others are happy to enjoy them as tourists or in the comfort of their own home via the medium of television. This ambivalence has resulted in conflict and a substantial loss of wetlands due to, e.g. drainage, and degradation of many of the remaining wetlands due to, e.g. pollution or over-abstraction. More recently, research has begun to show the fundamental role that wetlands play locally, regionally and globally, highlighting the need for geographers to apply knowledge and skills in order to resolve not only environmental problems but also social and economic issues. This requires inputs from historical, physical, environmental and human geographers on an interdisciplinary basis.

WHY ARE WETLANDS IMPORTANT?

Wetlands are one of the most productive ecosystems in the world. Recently, more emphasis has been placed on understanding and valuing wetlands and their functions, and on the need to achieve sustainable management. The conflict over wetlands is in large measure a failing of the current socio-economic systems to recognise their value. Different ways of categorising wetland functions and values have been put forward (Williams 1990). As a system, the total economic value (TEV) of wetlands has often been underestimated. This TEV includes *direct use values* of products such as fish and fuelwood and services, such as recreation and transport; *indirect use values* (or functional values) such as flood control and storm protection provided by e.g. mangroves; *option values*, which could be discovered in the future; and *intrinsic values*, the value of the wetland 'of its own right' with its attributes (Barbier *et al.* 1994). Wetlands have been likened to a sponge and filter (functions), a larder or hardware store (products) and an

Box 20.2 Views of wetlands: past and present

Romney Marsh, England, in 1576 was depicted as: 'evil in winter, grievous in summer and never good' and of the Fens, England, in 1629: 'The Air nebulous, grosse and full of rotten harres; the water putred and muddy, yea full of loathsome vermine; the earth sprung, unfast and boggie' (*Source:* William Lambarde, archivist to Elizabeth I, as cited by Purseglove 1988: p. 25).

A contemporary description of the Everglades, Florida, USA included: 'aquatic flowers, of every variety and hue, are to be seen on every side, in pleasant contrast with the pale green of the sawgrass' (*Source:* Smith 1847).

An account of the land use of grazing marsh wetlands at Hatfield Chase, England, in the seventeenth century: 'The region was based on an agricultural economy which incorporated hunting and fishing. Furthermore the agriculture was mainly pastoral. . . . The main livestock included cattle, sheep and horses, in that order of importance. . . . The peat fens were grazed and for each village it was usual for such summer grazing to be unstinted. Stock was often brought into the area from other districts. . . . Many grasslands were flooded from November to May and during this time alluvium ("natural warp") was deposited, improving fertility. In Epworth Manor the local population was permitted to catch white fish on Wednesdays and Fridays' (*Source:* Cory 1985: p. 35).

'These fish, called *binni*, were barbel. . . . Madan [a local tribe] poisoned fish in the winter and in the spring before the water in the Marshes began to rise. They used datura which they bought from the local merchants and mixed into pellets with flour and chicken droppings or inserted into freshwater shrimps. The datura stupefied the fish, which rose to the surface and were easily collected.' – Local exploitation of the marshes of southern Iraq (*Source:* Thesiger 1964: p. 91).

'The very existence of Dutch wetlands is at stake. The sustainability of wetlands should therefore be the prime ecological research priority' (*Source:* Best *et al.* 1993: pp. 318–19).

'Wetlands are among the world's most important environmental resources, yet remain some of the least understood and most seriously abused assets. . . . A major challenge to scientists, economists, decision-makers, managers, users and the conservation community, is to bridge the gap between socio-economics and ecology' (*Source:* Maltby 1989: p. 46).

'Over the past ten years there has been a major increase in concern for wetlands worldwide. Several aid agencies now include wetland management among their environmental priorities. The US Treasury has adopted voting standards which instruct the US Executive Directors to the multilateral banks, and the Administrator of USAID, not to support projects which will destroy or degrade wetlands' (*Source:* Dugan 1994: p. 11).

earthly paradise (attributes). Many wetlands have a number of functions, yield more than one product and can have a range of attributes, hence even among those who recognise their value, there can be competition between those exploiting the different resources (Table 20.1).

THE NATURE OF THE PROBLEM

Historical geographers have described how the value of wetlands has been ignored and have explained how consequently their conservation has received low priority. It is now accepted that wetland loss is a result of natural and man-made

Table 20.1 A categorisation of the value of wetlands.

Functions	*Products*	*Attributes*
Flood control	Fisheries	Biological diversity
Sediment accretion and deposition	Game	Culture and heritage
Ground water recharge	Forage	
Ground water discharge	Timber	
Water purification	Water	
Storage of organic matter		
Food-chain support/cycling		
Water transport		
Tourism/recreation		

causes. For example, natural causes include sea-level rise, drought, hurricanes and other storms, erosion, and biotic effects. However, the main cause has been human action, either direct or indirect. The reclamation of wetlands has been an expensive engineering operation that, in the short term, increased the economic value of the wetland area, outweighing the costs of drainage and/or flood alleviation. Equally, dam construction typically causes the loss of wetland habitat due to impoundment of the floodplain. Other human actions directly responsible for the loss and degradation of wetlands have included conversion for aquaculture, mining of wetlands for peat, coal, gravel, phosphate and other materials, and groundwater abstraction. Recently, indirect human action has also been recognised as a cause of wetland loss and degradation: discharges of pesticides, herbicides and nutrients; hydrological alterations by canals, roads and other structures; and subsidence due to extraction of groundwater, oil, gas and other minerals have all damaged wetland sites.

Historical geographers have described this wetland loss. For example, in Europe, most of the loss occurred before the modern era, largely due to agricultural drainage enterprises, many masterminded by Dutch engineers. This experience and skill was exported not only to countries like England but further afield, for example to the peatlands of Indonesia. The pressure for wetland alteration and reclamation is now increasing in the tropics and developing nations. The rate of loss cannot be quantified in most countries but is relatively well documented in the USA (Figure 20.2).

The increased perception and recognition of the importance of wetlands at global, regional and local levels has led to a conservation backlash, trying to halt wetland loss and degradation worldwide. The problem facing applied geographers is first to take stock of the remaining wetland resource and monitor its extent and quality. This should also involve establishing a value for any given wetland in order to ensure that decision making that might impact on it directly or indirectly is soundly based. The various wetland processes must be properly researched, leading to tried and tested management in the form of conservation, restoration and creation. Lastly, the success of this management should be appraised, thereby validating the management and providing useful feedback by way of action research. These various stages can be viewed from different levels:

- *International* e.g. the Ramsar convention confers protection on wetland sites of global importance, virtually worldwide (Box 20.3).
- *Regional/national* e.g. developing a policy for water resource management in the Mekong basin; the Med–Wet programme financed by the European Union.

Figure 20.2 Percentage of wetland area lost in the USA between the 1780s and 1980s.

Source: Dahl 1991.

- *Local* e.g. establishing a plan for the sustainable exploitation of the fishery in a given wetland; management plans and conservation objectives for specific sites.

The advent of aerial photography, remote sensing and spatial analysis techniques, such as GIS, are making an important contribution to wetland conservation. For example, wetland boundaries are sometimes delineated through the use of aerial photography as an alternative or complementary tool to field data (National Research Council 1995). Meanwhile, satellite remote-sensing data can help to identify wetland hydrology, particularly useful in the context of developing countries (Haack 1996). In the case of GIS, the analysis of spatial information can be used for the management and study of wetlands, for example by looking at land-use change or water quality data. It helps to set the wetland in its surrounding landscape and human pressures. The application of aerial photography and remote sensing has identified the importance of technology and the need for personnel trained in such methods and with a field knowledge of wetlands (Committee on Characterisation of Wetlands 1995).

There is also a need to quantify wetland values, not just over the short term but on into the future. While quantities can be put on certain components of wetlands, for example the weight of fish caught per year, the number of geese overwintering and mass of nitrogen stored in a given wetland, it is harder to turn these into monetary values as the basis for decision making. This is a current area of research, which has focused on aspects such as:

1 Evaluating the global and regional role of wetlands. The commercial and environmental value of coastal marsh in Georgia, USA, has been calculated at $50,000–125,000 per hectare, and the 'life support' value of saltmarsh (based on the conversion of solar energy) has similarly been put at $212,500 per hectare (Odum, in Maltby 1986).
2 Calculating the value of harvested resources. The market value of the fish caught from a wetland can be calculated for a given year or season, as can the value of any associated industry and employment.
3 Evaluation systems that seek to compare natural wetlands with human economic systems. This approach uses the measure of willingness to pay to achieve monetary value (Mitsch and Gosselink 1993).

Not surprisingly, these different methodologies typically give different values for the same wetland. The lack of consistent and accepted methodologies for comparing wetlands with conventional economic goods and services limits the usefulness of the estimates that have been made, and there is a need for much more research by human geographers and economists.

Box 20.3 The Ramsar Convention (http:/iucn.org/themes/ramsar/)

The Convention on Wetlands of International Importance especially as Waterfowl Habitat (usually referred to as the Ramsar Convention after the place of its ratification in Iran in 1971) is one of the most important instruments for conserving wetlands of international importance. This international treaty laid the basis for international cooperation in conserving wetlands and by 1991, more than sixty countries had signed up to the Ramsar Convention.

The convention requires the signatories:

1 To designate wetlands of international importance for inclusion in a list of so-called Ramsar sites.
2 To maintain the ecological character of their listed Ramsar sites.
3 To organise their planning so as to achieve the wise use of all of the wetlands on their territory.
4 To designate wetlands as nature reserves.

There are more than 500 wetland sites on the Ramsar list covering in excess of 30 million hectares of wetland habitat. To be considered a wetland of international importance, a site must (*Source:* Gleick 1993: p. 287):

1 Support a significant population of waterfowl, threatened species, or peculiar fauna or flora.
2 Be a regionally representative example of a type of wetland or an exemplar of a biological or hydrogeomorphic process.
3 Be physically and administratively capable of benefiting from protection and management measures.

MANAGEMENT: CONSERVATION, RESTORATION AND CREATION OF WETLANDS

Historical analysis reveals a long relationship between communities and wetlands throughout civilisation. For example:

- Conflict arose in the seventeenth century over the draining of Hatfield Chase, England, resulting in violence to the Dutch engineers and their workers (Purseglove 1988).
- The Marsh Arabs of southern Iraq have built up a complete way of life based on wetland conservation (Thesiger 1964); evidence has been found of the restoration of systems that used silt deposited in floodplains that had been destroyed by unusually heavy flooding.
- Medieval excavations of peat for fuel undertaken in various parts of Europe created what have become important wetland sites, such as the Norfolk Broads (George 1992).

Many traditional societies have developed complex systems to regulate access to resources. These can in many instances provide the basis for multiple use under today's conditions. Yet where control over all natural resources is vested in agencies of central government, often based hundreds of kilometres away, such locally based management is often severely hampered. In designing and establishing planning and management frameworks for sustainable conservation and use of wetland resources, special care needs to be taken to ensure that these are pursued within an appropriate institutional or government policy (Box 20.4).

Only rarely are the main components of a wetland managed in an effectively integrated manner; rather, emphasis tends to be upon maximising benefit from a single product. The critical need today is to recognise the interlinkages and benefits to be obtained from integrated management of resources such as fish, trees, water and wildlife. This introduces a new dimension to wetland conservation, the requirement for integration of institutions such as departments of fisheries, forestry, water resources and tourism. Such integration is required beyond the wetland site itself in the form of planning and management of the catchment or coastal zone within which the wetland lies. For example, productivity in most wetlands depends upon the flow of water and nutrients into them. Consideration should also be given to the downstream benefits of wetland conservation such as flood control and maintenance of water quality, emphasising further the central role that wetlands can play in regulating the hydrological and biogeochemical cycles. In addition to researching into institutional change, human geographers need to explore the accompanying need for the development of appropriate policies.

National economic and agricultural policies frequently determine the rate at which wetlands are lost. For example, artificially high prices paid for a crop such as winter wheat, available under the Common Agricultural Policy until the early 1990s, made it profitable to drain lowland wet grazing meadows (see also Box 20.4). The rapid rise in demand for land for urban and industrial development generates an economic momentum, which renders invalid many conservation arguments based upon the multiple values of these natural systems. Thus, private developers continue to invest in drainage and conversion in the expectation that, even if aquaculture fails, the

Box 20.4 Conversion of natural floodplain to agricultural land, northern Nigeria

Despite efforts by the states of Borno, Kano and Bauchi, northern Nigeria, to conserve the natural floodplain of the Hadejia River system, agricultural and economic policies are driving conversion to agricultural land. In response to falling oil prices and the need to save hard currency, the federal government banned all wheat imports from January 1987. At the same time, a 50 per cent subsidy was offered on fertilisers and equipment for wheat cultivation, while the producer price for domestic wheat rose by 1000 per cent between 1986 and 1989. By the 1988–9 growing season in Kano alone, 30,000 ha had been converted for wheat cultivation. While this wheat boom will generate a profit for individual farmers, the benefits will be short-lived. The sandy soils are predicted to degrade rapidly under irrigated wheat cultivation, thus compromising long-term options for rural development in the region.

Source: Kimmage 1991.

land is worth more as wasteland for housing than as wetland (see also Box 20.5 for an example of 'best practice').

The importance of excessive water abstraction for commercial agriculture can be seen in the case of the Tablas de Daimiel wetland, in Spain. Tablas de Daimiel, a national park in central Spain, is one of the few floodplain wetlands remaining in the country (Casado *et al.* 1992) (Plate 20.1). However, its hydrological and ecological functioning no longer operates as a natural system. Tablas de Daimiel is located in the Spanish central plateau, in the semi-arid region of Castille–La Mancha, in the province of Ciudad Real. It was designated a Ramsar site in 1982 and a Special Protection Area (SPA) under the Birds Directive and a candidate special area of conservation (cSAC) under the Habitats Directive. In 1973, it was declared a national park to mitigate plans to drain this area for irrigation. In the same year, the first wells were legalised in the area to irrigate corn and barley, thus substituting traditional, extensive dryland Mediterranean agriculture of olives, vines and wheat.

Tablas de Daimiel National Park has a designated area of 2000 ha, out of a total wetland area of 8600 ha of a very complex endorrheic hydro-geological system. It results from a confluence of surface waters from the Ciguela River (saline water) and the Guadiana River (fresh water), small seasonal streams and groundwater from Aquifer 23, the key hydrological feature in the Upper Guadiana basin (Cirujano *et al.* 1996; Llamas 1988). During high groundwater levels, the wetland is a groundwater discharge area; during low groundwater levels, it is a groundwater recharge area.

From the moment that the park was declared, in 1973, there was an increase from 30,000 irrigated hectares to 130,000 ha in 1989 in the area surrounding the park. It is estimated that the aquifer has renewable resources at an average of 335–400 Mm^3/yr, yet the net abstraction for irrigation has been 520–600 Mm^3/yr. The groundwater levels over the last thirty years have dropped by as much as a metre a year, compounded by a drought between 1991 and 1995. As a result, from 1984 the park ceased to be

Box 20.5 An institutional framework recommended for wetland conservation in Zimbabwe

These recommendations were put forward at a seminar on the wetlands of Zimbabwe as the basis for developing an institutional framework. The existing institutions were weak in the country and not supportive of wetland management. There was seen to be a need to set up institutions catering for wetlands and involving the following:

1 Development of a National Conservation Strategy with the objective of setting up an effective National Environmental Action Plan with a section on wetlands, objectives of which are:
   - identification of wetlands and threats to their existence;
   - development of a wetland inventory;
   - zonation of wetlands and site protection;
   - delineation of wetland boundaries;
   - utilisation of wetlands for pasture development, cropping, fishing and freshwater habitats;
   - a study of human settlements and the benefits they derive from wetland utilisation;
   - as they are the resource users, local communities, their traditions, cultures and needs, should be involved in policy formulation;
   - provision of incentives to local people to promote long-term conservation, such as being co-researchers;
   - sustainable use of natural resources;
   - conservation of biodiversity;
   - legislative review;
   - manpower development and training;
   - education in wetland issues and environmental awareness.
2 Strengthening of existing institutions.
3 Setting up of a training and research institute.
4 Incorporation of other institutes, such as the Department of Physical Planning and city councils, into policy making on wetlands issues.
5 Setting up information centres for wetlands.
6 Coordination of activities at inter-ministerial and inter-departmental levels.
7 Cooperation with NGOs on wetlands issues.
8 Instigation of siltation and pollution abatement and prevention of dumping of toxic wastes.
9 Monitoring of siltation and pollution, rehabilitation of rivers, and conservation of catchment areas and riverine environments.
10 Establishment of an environmental impact assessment monitoring unit.
11 Establishment of social impact assessments of wetland conservation or utilisation.

Source: Katerere 1994 – see also Maltby 1989.

Plate 20.1 Tablas de Daimiel, July 1998.

the natural overflow of the aquifer and for a period was desiccated except for a small area. The main water source for the park is now the Ciguela River, which implies a higher saline content for the park's waters. At present, both water quantity and quality are key issues for the conservation of the Tablas.

In terms of water quantity, the aquifer was officially declared overexploited. This meant an immediate stop on new water abstractions, the setting up of 'Aquifer User Communities' and the preparation of a plan to control water abstractions. This plan identified the short-, medium- and long-term strategy for Tablas de Daimiel Park:

- The short-term strategy was based on setting up pumps to pump water directly into the park, approximately 18 Mm^3/yr from Aquifer 23. This would allow the flooding of 600 to 1200 ha. However, salinity has affected the efficiency of the pumping stations due to corrosion, and only 400 ha were flooded in the dry summer period.
- The mid-term strategy included plans to transfer water via the Tagus–Segura transfer, from the Tagus catchment, along 150 km of the Ciguela river bed, to the park. This meant a transfer over three years of 60 Mm^3 (never more than 30 Mm^3/yr) for environmental purposes. However, the Ciguela River is saline, and 20 Mm^3/yr cannot be compared to the aquifer contribution before the 1970s of 200 Mm^3/yr. Additionally, much of this transfer was illegally abstracted by farmers for irrigation. In the first transfer, in 1985, approximately 75 per cent of the water transferred made it to the park, by 1994 the quantity arriving in the park had declined to 40 per cent (Table 20.2).
- The long-term strategy was centred on the construction of two dams to store 212 Mm^3/yr. These dams would collect water that could be used to flood the park at regular intervals. Yet the estimated external environmental and social costs are very high.

In the case of water quality, progressive water salination of the park is reflected by a slow invasion of halophytic plants, such as *Ruppia maritima.* Eutrophic water could be the result of diffuse pollution from the use of fertilisers (Montes and Bifani 1993). Surface waters, in view of aquifer overexploitation, are the only lifelines of the park. Yet there are inherent dangers as, for example pollution from *alpechines*, a waste product of olive production, pesticides and uncontrolled sewage discharges into the Ciguela.

At present, none of the strategies directly tackles the reason for overexploitation, i.e. abstractions for irrigation from Aquifer 23 in the perimeter of the park. The only initiative has been a plan to compensate farmers not to irrigate, financed under the European Union's agri-environment directive. In 1993 and for a period of five years, until 1997 (now extended to 2003),

Table 20.2 Water transferred from the Tagus–Segura and received in the Tablas de Daimiel National Park.

Year	*Transferred water (Mm^3)*	*Water received in the park (%)*
1988	12–90	75
1989	13–33	75
1990	15–78	75
1991	17–72	68
1992	6.6	40
1993	0	–
1994	15	–

compensation was paid to farmers to reduce pumping in the area. Yet this programme has not encouraged changes of behaviour or alternatives but rather the same practices, but with less intensity. It is a lost opportunity and goes against the environmental policy principle of 'pumper/ resource user pays'.

To summarise, the Tablas de Daimiel case is an example of unsustainable regional development, of short-term economic and social development at the expense of long-term environmental damage. Some think that restoration is now impossible; at a recent Ramsar meeting in Seville, consideration was given to removing the Tablas de Daimiel from the Ramsar site list, since it was a 'dry' wetland. It appears that slowly all parties involved are realising that the depletion of the aquifer damages all stakeholders. Ironically, climate change may be responsible for three unusually wet years after five years of drought; this rainfall has temporarily reduced tension (and abstractions) in the area. It has also allowed dialogue to restart, to identify future strategies for achieving sustainable rural development and the survival of the Tablas de Daimiel wetland.

This case study illustrates well the problems facing those responsible for the conservation of wetlands into the twenty-first century:

- the prevention of further damage and loss of wetland habitat;
- conserving remaining wetlands;
- rehabilitating damaged wetlands; and
- creating new wetlands.

The challenge to geographers is developing management systems and practices that allow the sustainable use of wetlands. This is increasingly concerned with the human dimension, for example the socio-economic aspects related to wetlands, reasons why communities are led, through the current economic system which emphasises short-term benefit, and unsustainable management. Also how, on a regional or global scale, wetland functions are affected by policies such as the Common Agricultural Policy or international trade, e.g. of fisheries.

An example of sensitive management is the case of Wicken Fen in England, located in the East Anglian fenland. This fenland covered approximately 3380 km^2 in the seventeenth century, of which the southern part, peat-based Black Fens, occupied 1480 km^2 (Butlin 1990; Newson 1994). The main attempt to drain the Cambridgeshire fenland was the general drainage of the seventeenth century, which was conducted by Sir Cornelius Vermuyden, a Dutch engineer, in two phases between 1630 and 1653. The investment in the drainage came from the Earl of Bedford and thirteen other 'adventurers', who as payment, obtained 38,500 ha of drained land from the local landowners (Williams 1990). This fenland drainage system forms the largest ditch network in Britain. Between 1637 and 1954, there was a reduction in area of the East Anglian fens from 3380 km^2 to 10 km^2.

Wicken Fen[1] is one of the oldest nature reserves in Britain, and since 1899 it has been owned and managed by the National Trust as 'a remnant of a once extensive landscape' (Gilman 1994: p. 24). Wicken Fen is a Ramsar site, a National Nature Reserve, a Site of Special Scientific Interest and an cSAC, famous mainly as an entomological and botanical reserve, e.g. for the fen violet (*Viola persicifolia*). It is located approximately 15 km northeast of Cambridge and in the southeastern edge of the fenland basin, in the East Cambridgeshire district. The geology is sedge peat over Gault clay (Rowell 1986).

Wicken Fen was a summer dry/winter wet fen during the 1630s and was not intended to be drained sufficiently to be winter dry (Rowell 1986). It had various traditional uses: sedge for thatching, some peat cutting for fuel and 'litter' (common reed *Phragmites australis* and purple moor-grass *Molinia caerulea*) for animal bedding (Godwin 1978). The underlying clay was used for local brick making (Gilman 1994). Wild crops of reed and sedge are still harvested under the present management, to the benefit of 30,000

[1] Students are advised to visit the Wicken Fen web site on: http://www.demon.co.uk/ecoln/wicken_fen/

visitors a year, who can experience the 'lost landscape' of the peat fens.

The reserve is about 305 ha and consists of four sections: Wicken Sedge Fen; Adventurer's Fen, St Edmund's Fen and, only secured in 1992, Priory Farm. The most important section – Wicken Sedge Fen (103 ha) – is bounded by clay banks to the north and west, on the east by the rising land of Wicken Ridge and to the south by a broad, man-made watercourse, Wicken Lode. It stands like a reverse island, three meters above the surrounding farmland (Plate 20.2) and is kept wet by pumping water into it, and through water-proofing around its perimeter (Friday 1997).

The main problems for Wicken Fen's conservation have been, on the one hand, falling water levels and, on the other, encroachment by scrub. There is a hydrological gradient between Wicken Sedge Fen and the farmland to the north and east. This drained, shrunken land due to peat wastage acts like a sponge for the water in the reserve through gravity flow (Purseglove 1988; Friday 1997). 'Fens would quickly turn into woodland if not continuously wet. Most of Wicken Fen has thus become a wood in the last 50 years' (Rackham 1986: p. 381). In addition to falling water levels, a compounding factor was 'lack of management' in that sedge and reed were not harvested after 1920. However, in 1961 a management plan was drawn up to arrest the fen's decline and to restore its former habitats. This new management has been based, for example, on reopening old ditches, excavating new ditches, cutting down trees and reintroducing sedge harvesting. In 1982, the fen violet reappeared – it had last been seen in 1916 (*ibid.*).

Plate 20.2 Wicken Fen.

In 1942, Eric Ennion wrote *Adventurer's Fen* as a requiem to a disappearing landscape, drained as part of the 'dig for victory campaign' and requisitioned for cultivation. Now half a century later, and looking forward to the next century 'the last crops of potatoes, linseed and sugarbeet have been lifted from Priory Farm land and the combined forces of men, machinery and sheep are beginning the transformation of arable black peatland back to fenland' (Friday 1997: xiii). Therefore, it seems that time is going backwards in this reserve, and agricultural land is being reverted, given back to nature, a true example of wetland restoration.

CONCLUSION

Applied geographical research has a particularly important contribution to make to the development of effective programmes for wetland conservation. Actions are being taken to develop and implement conservation programmes such as those undertaken by the Ramsar bureau. However, a first, crucial, step is an increase in the understanding and awareness of the value of wetlands, their rate of loss, and the social and economic impact of these losses. This includes:

1 *Valuing wetlands.* Studies have been undertaken for some wetlands in Europe and North America, with a gradual increase in studies in the tropics. This has increased initial awareness of the importance of wetland ecosystems, but it needs to be expanded, coupled with targeted efforts to increase the capacity of training institutions to provide instruction in wetland evaluation.
2 *Quantifying the benefits of wetland conservation.* Studies in wetland economics need to go beyond the analysis of wetland value to

examine the broader economic impact of wetland conservation and management and, in particular, need to demonstrate that carefully designed investments in wetland conservation can make a significant contribution to local and national economies.

3 *Documenting wetland loss.* Lessons learned in assessing and analysing wetland loss in countries such as the USA can help geographers in other countries to identify the critical data needed to achieve meaningful status reports, an essential foundation for the development of national awareness programmes and management and conservation policies.

Future investments in wetland management need to be based on the best possible understanding of the capacity of wetland ecosystems to sustain different forms of use and of the way in which future changes in human population, development policy and climate will impact upon wetland ecosystems. At the same time, lessons from traditional systems of wetland management have much to contribute to modern-day management. To meet these needs, five areas of research require special attention.

1. *Resource analysis.* The assessment of the capacity of a wetland to sustain different uses requires analyses of water, soil, flora and fauna, and an understanding of problems such as overgrazing and loss of forest resources. Effective solutions must be based on a good understanding of ecosystem functioning, which of itself necessitates more collaborative research between institutions and between different disciplines.
2. *Socio-economic studies.* Wetland degradation is often due to mismanagement by rural communities well aware of many of the consequences of their actions but through factors such as poverty obliged to pursue non-sustainable practices. Management needs to be based on socio-economic studies that provide an understanding of the changing rural economy. Similarly, studies of the structures and mechanisms through which resource use is administered should focus on ways to provide incentives to people to manage resources more effectively.
3. *Climate change.* Wetland management needs to plan for predicted trends in climate such as an unprecedented and rapid rise in sea level and flooding of many coastal wetland systems, the increase frequency of droughts, and changes in the distribution of species over large parts of the world.
4. *Population growth and pressure.* Substantial effort needs to be made in examining the impacts of increasing human population upon wetland resources, particularly in developing countries, and in identifying mechanisms that might be used to reduce these impacts.
5. *Restoration and creation of wetlands.* Many countries have initiated wetland restoration and creation programmes as a response to loss and degradation. The scientific basis for this is at an early stage, and research is needed to assess both the requirements for restoration of specific wetlands and the successes and failures of completed projects.

Applied geographers, with expertise in environmental management and sensitivity to impacts of human and physical processes, are thus well placed to contribute to discussion and practical implications of wetland conservation.

GUIDE TO FURTHER READING

Finlayson, M. and Moser, M. (eds) (1991) *Wetlands.* International Waterfowl and Wetlands Research Bureau, Facts of File, Oxford/New York. Documents the status of the world's major wetlands through a series of well-illustrated wetland directories.

Maltby, E. (1986) *Waterlogged Wealth. Why Waste the World's Wet Places?* London: Earthscan. Although published in the mid-1980s, this well-illustrated text usefully describes the value of wetlands and the threats against them in what is a very readable style.

Williams, M. (ed.) (1990) *Wetlands: A Threatened Landscape.* Institute of British Geographers Special Publication 25. Oxford: Blackwell. A comprehensive appraisal of the world's wetlands from both physical and human perspectives focusing on the nature of wetlands, the effects of human impacts and strategies for their management.

REFERENCES

Barbier, B., Burgess, J.C. and Folke, C. (1994) *Paradise Lost? The Ecological Economics of Biodiversity*. Beijer International Institute for Ecological Economics, The Royal Swedish Academy of Sciences, London: Earthscan 116–31.

Best, E.P.H., Verhoeven, J.T.A. and Wolff, W.J. (1993) The ecology of the Netherlands' wetlands: characteristics, threats, prospects and perspectives for ecological research. *Hydrobiologia* 265, 305–20.

Butlin, R.A. (1990) Drainage and land use in the fenlands and fen-edge of northeast Cambridgeshire in the 17th and 18th centuries. In D. Cosgrove and G. Petts (eds) *Water, Engineering and Landscape*. London: Belhaven, 54–76.

Casado, S., Florin, M., Molla, S. and Montes, C. (1992) Current status of Spanish wetlands. In M. Finlayson, T. Hollis and T. Davis (eds) *Managing Mediterranean Wetlands and their Birds*, Proceedings of an IWRB International Symposium, Grado, Italy, February 1991; Slimbridge: IWRB Special Publication No. 20, 56–8.

Cirujano, S., Casado, C., Bernues, M. and Camargo, J.A. (1996) Ecological study of las Tablas de Daimiel National Park (Ciudad Real, Central Spain): differences in water physico-chemistry and vegetation between 1974 and 1989. *Biological Conservation* 75, 211–15.

Committee on Characterisation of Wetlands (1995) *Wetlands: Characteristics and Boundaries*. National Research Council. Washington: National Academy Press.

Cory, V. (1985) *Hatfield and Axholme. An Historical Review*. Ely: Providence Press.

Dahl, T. (1991) *Wetland Losses in the United States, 1780s–1980s*. Report to Congress, US Department of the Interior, Fish and Wildlife Service, Washington DC.

Dugan, P.J. (ed.) (1993) *Wetland Conservation: A Review of Current Issues and Required Action*. Gland: International Union for the Conservation of Nature and Natural Resources.

Dugan, P.J. (1994) Integrated management of wetlands: a perspective from IUCN's global programme. In T. Matiza and S.A. Crafter (eds) *Wetlands Ecology and Priorities for Conservation in Zimbabwe*, Proceedings of a seminar on wetlands ecology and priorities for conservation in Zimbabwe, Harare Kentucky Airport Hotel, 13–15 January 1992, 11–20.

Friday, J. (ed.) (1997) *Wicken Fen: The Making of a Wetland Nature Reserve*. Colchester: Harley Books.

Gardiner, J. (1994) Paper 5: Pressures on wetlands, Wetlands Management International Conference, 2–3 June. London: The Institution of Civil Engineers, 1–23.

George, M. (1992) *The Land Use, Ecology and Conservation of Broadland*. Chichester: Packard.

Gilman, K. (1994) *Hydrology and Wetland Conservation*. Institute of Hydrology; Chichester: Wiley.

Gleick, P.H. (ed.) (1993) Section F. Water and ecosystems. In P.H. Gleick (ed.) *Water in Crisis: a Guide to the World's Fresh Water Resources*; Pacific Institute for Studies in Development, Environment and Security; Stockholm Environment Institute; New York: Oxford University Press, 287–320.

Godwin, H. (1978) *Fenland: Its Ancient Past and Uncertain Future*. Cambridge: Cambridge University Press.

Groombridge, B. (ed.) (1992) *Global Biodiversity: Status of the Earth's Living Resources*. A report compiled by the World Conservation Monitoring Centre. London: Chapman & Hall.

Haack, B. (1996) Monitoring wetland changes with remote sensing: an East African example. *Environmental Management* 20(3), 411–19.

Katerere, D. (1994) Policy, institutional framework and wetlands management in Zimbabwe. In T. Matiza and S.A. Crafter (eds) *Wetlands Ecology and Priorities for Conservation in Zimbabwe*, proceedings of a seminar on wetlands ecology and priorities for conservation in Zimbabwe, Harare Kentucky Airport Hotel, 13–15 January 1992, 129–36.

Kimmage, K. (1991) Small-scale irrigation initiatives in Nigeria: the problems of equity and sustainability. *Applied Geography* 11, 5–20.

Llamas, R. (1988) Conflicts between wetland conservation and groundwater exploitation: two case histories in Spain. *Environmental Geology and Water Science* 11(3), 241–51.

Maltby, E. (1989) Wetland management goals: wise use and conservation. In M. Marchand and H.A. Udo de Haes (eds) *The People's Role in Wetland Management*, Reproduktieafdeling Biologie, Leiden University, 46–5.

Mitsch, W.J. and Gosselink, J.G. (1993) *Wetlands*. New York: Van Nostrand Reinhold.

Montes, C. and Bifani, P. (1993) Spain. In K. Turner and T. Jones (eds) *Wetlands: Market and Intervention Failures Four Case Studies*, London: Earthscan, 144–95.

Newson, M. (1994) *Hydrology and the river environment*. Oxford: Clarendon.

National Research Council (1995) Maps, imaging and modelling in the assessment of wetlands. In NRC (ed.) *Wetlands: Characteristics and Boundaries*, Washington: National Academy Press, 190–206.

Purseglove, J. (1988) *Taming the Flood.* Oxford: Oxford University Press.

Rackham, O. (1986) Marshes, fens, rivers and the sea. In O. Rackham (ed.) *The History of the Countryside: The Classic History of Britain's Landscape, Flora and Fauna*, London: Dent.

Rowell, T.A. (1986) The history of drainage of Wicken Fen, Cambridgeshire, England, and its relevance to conservation. *Biological Conservation* 35, 111–42.

Smith, B. (1847) *Everglades of Florida: Acts, Reports and Other Papers, State and National, Relating to the Everglades of the State of Florida and their Reclamation.* Senate Document 89, 62nd Congress.

Thesiger, W. (1964) *The Marsh Arabs.* London: Longman.

21

Land-use conflict at the urban fringe

Gordon Clark

INTRODUCTION AND BACKGROUND

What and where is the urban fringe? A precise definition and map are not possible, but generally the urban fringe means those areas just beyond the built-up part of a city, although still close enough to the city to be subject to intense development pressures (for a discussion of definitions, see Bourne and Simmons 1978: pp. 18–41). The fringe is not a line on a map; it is a zone of radially diminishing urban-style activities. It is the existence of a fringe that prevents one being able to distinguish the urban from the rural, since the fringe has features of both. Yet it is more than an amalgam of the two; the fringe is a distinctive place with features of its own. It is, above all, a place of heightened land-use conflict, uncertainty and profit potential, hence its interest to geographers.

Arguably every section of every city was once on the fringe of the built-up area at some point in the city's history – St Martin-in-the-Fields, now in central London, was once precisely that. How land evolved when on the fringe will have left a stamp on the built form of the area that will have endured long after the fringe has been swallowed up by the expanding city. Equally, how a society has dealt with its urban fringes tells us a great deal about how that society works, the values it holds to be important and how these have evolved.

The earliest geographical models of cities did not recognise the concept of an urban fringe. The city met the countryside and there was no transition between them, each being distinctive in terms of economic and social structures and culture. This approach was replaced by what might be called 'stage' or 'gradient' models. Some of these identified broad categories of area from the urban and peri-urban out to the rural and very rural. Others built on the work of von Thünen and the classic models of urban structure by Burgess and Hoyt (for a review of these models see Johnson 1972; Northam 1975). In these models, a continuous gradient runs from the city centre to the deepest countryside, with an inexorable decline from the former to the latter in land values, profits per unit area, and the density of building and population (Figure 21.1). In these models, the urban fringe is an area where land values rise over time as more productive and intensive land uses replace, for example, agriculture. This is shown in Figure 21.1 by the rise in land values at point UF. A revision of the model by Sinclair (1967) suggested that, although development value rose as one approached the city centre, agricultural value fell because vandalism and the high probability of urban building increasingly reduced farming's profitabilities towards the city's edge.

The twin themes of 'urban influence on the countryside' and 'the transition from rural to urban' have inspired much research at the urban fringe by geographers. This has sought to examine urban influence and urban transition as *processes*. What happened and how did it occur? There were studies at the urban fringe focusing on the intensification of farming, the increase in building, recreational development, and how the land market operates (for reviews see Pacione 1984; Gilg 1985; Mather 1986; Robinson 1990).

Another strand of research has been concerned

Figure 21.1 Land values around a city and the effect of urban expansion on land values.

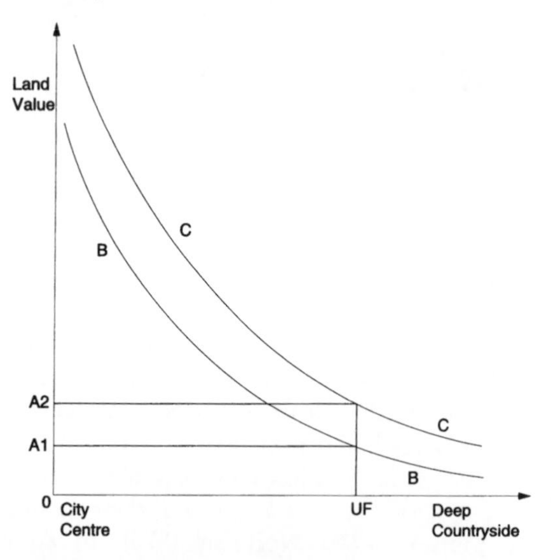

Notes: B = original bid-rent curve; C = new bid-rent curve after urban expansion; UF = arbitrary point at urban fringe where land values rise from A1 to A2 after urban expansion.

with planning at the urban fringe. If the transition from rural to urban is not to be left to the free market, how might public interests be formulated and brought to bear on the development process? What would be the consequences and side-effects of land-use planning on the pace and form of urban growth? Another dimension is to examine the positions taken by pressure groups as they contest the process of development.

The urban fringe is not a homogeneous type of area – its land uses and history vary from city to city and culture to culture. Nor is it a wholly unique type of area, since the development pressures faced as open fields are encroached on by the advancing suburbs can also be found in the cores of cities when major sites become available for redevelopment (e.g. former port areas). Yet the edge of the city exerts a particular fascination for so many groups. For the city dweller, it is where they first experience open countryside, farms and nature. For the farmer, the nearness of the city may hinder farming yet offer the prospect of better access to customers and speculative gains by selling land to a housebuilder. For the impoverished rural migrant, the urban fringe's caravan parks, slums or squatter settlements may be as near as they can afford to get to housing that allows them access to urban jobs. The urban fringe has a distinctive image and its own blend of land uses, issues and problems.

THE NATURE OF THE PROBLEM: WHETHER AND HOW TO MANAGE THE URBAN FRINGE

Much geographical research on urban fringe areas has focused on whether and how to plan for the transition from rural to urban, and the consequences of trying to plan the process.

The issue of 'whether to plan' focuses attention on free-market cities, where private enterprise is allowed a largely free rein. Los Angeles is a good example (see the case study in Box 21.1) and is typical of the kind of urban-fringe management

Box 21.1 Los Angeles

Los Angeles is the second largest city in the USA (with a statistical area population of 11.4 million), yet it has a population density less than half that of London. Los Angeles has relied on the rapid outward expansion of its suburbs and those of formerly separate cities to create a huge, low-density, car-based metropolitan area. Little weight was given during its outward expansion to factors such as conserving the beauty of the landscape, protecting farmland, providing recreational open space, limiting the physical or population size of the city, protecting the central business district or minimising the cost of providing transport or infrastructure. Land was sold for building wherever willing buyers and sellers could agree terms, sometimes leapfrogging open fields and across jurisdictional boundaries before later development filled in the gaps. In general, strong centralised planning powers at federal or state level were politically unwelcome, and any infringement of the right to develop one's land as one wished was resisted. In the last twenty years, there have been some moves to ensure a more orderly progression of urban expansion. However, the complexity (by European standards) of local and state government and of planning and regulatory agencies has led to a confusing picture. Power, influence and insider knowledge are key weapons for landowners, authorities, developers and pressure groups alike in the highly negotiable question of where to develop next on the fringe of Los Angeles and hence who will reap the profits. Uncertainty and competition are the key features of the geography of the fringe of Los Angeles.

Sources: Marchand 1986; Davis 1990.

in many US cities. The emphasis remains on the right of the individual to profit from the development of his/her land. Open competition for sites is likely to reduce land prices (and hence the cost of development) and to balance the demand for, and supply of, building land. However, the resulting newly formed urban area may lack non-profit activities such as public parks and habitats; it may not economise on land loss from farming; and it may not result in a compact city, because of fragmented built-up areas. The profits from land development will accrue to individuals, and speculative gains and losses may be frequent, since the pattern and timing of development will be hard to predict in detail. Without developing a full planning system, one could attempt to achieve some of these public objectives by, for example, a taxation system that favours retaining land in farming (as in California, Maryland and New York state), the public purchase of development rights (as in Vermont, but an expensive option) and voluntary conservation agreements (as in Wisconsin).

If the benefits of the free market are felt to be outweighed by its demerits, the idea of a British-style planning system deserves attention. The key features are that the right to develop land is separated from its ownership, and development rights are vested in the government. A similar result comes from the Dutch system whereby land to be developed must be sold by the original owner to a public agency, which can then sell it to the developer. Either way, the state controls the process and can affect the rate and direction of land transfers. The (British) Town and Country Planning Act 1932 allowed those whose development proposals had been refused by local government to claim compensation for lost income; this effectively nullified the planning process, as the rapid suburbanisation of the 1930s testifies. Equally, one could tax the unearned fortuitous gains made by those selling land for development on the grounds that society should share the financial rewards from land-use changes made possible by society through its planning mechanism. Such a tax may work if it has all-party support in parliament; if not, the hope of its repeal may intensify the short-term shortage of development land and inflate land prices even further. State controls make it easier to include 'uneconomic' land uses such as public-access woodlands, examples of this being found around London, Amsterdam and Paris. The trade-off is between the greater certainties of the planned development process and the need to regulate the oligopoly of development land that it creates.

Clearly a planning system needs a plan, or rather it needs two or more tiers of local, regional and national plans, that sets out what society wants to achieve and what the private land market will not provide. In Britain (see the London case study in Box 21.2), the goals included smaller continuous built-up areas, more compact cities, less loss of farmland and more land for public access. Interestingly, the benefits of compactness (first espoused by the vilification of 'urban sprawl' in the 1930s) have been rediscovered by the 'sustainable city' ideal in the 1990s (Haughton and Hunter 1994).

The success of the plan depended on its credibility. The more exceptions and deviations from the plan, the less credible and effective it would be. The approved London Green Belt was in fact preserved with only minor exceptions. Yet flexibility is also vital. To achieve this without fatally undermining the plan, the inner limit of the Green Belt was initially beyond the then built-up area, so allowing for some further expansion before reaching the Green Belt. Flexibility also came from the use of 'proposed' and 'interim' green belts. Until these were approved, a more flexible view of development could be taken in these areas.

Flexibility also arose from the meaning given to 'development'. What are the permissible uses of an urban fringe? Green Belt designation prohibited 'urban' developments (e.g. housing and industry) but did not of itself develop positive features of the countryside. Control over agriculture was limited (although slightly greater than elsewhere in the UK), since farmers could diversify into non-agricultural and recreational activities. In smaller British cities, often without a green belt, the city edge was the preferred location for

Box 21.2 London

London is the largest city in the UK (population 6.7 million in 1991) and, until twenty years ago, one of the largest cities in the world. Was it too big, in area or population? In the inter-war period, there was concern over the rapid outward expansion of London's suburbs, focusing on the loss of farmland and the difficulty of providing public transport for the suburbs. The idea of a green belt was first enacted in 1938, and land was bought to protect open countryside. The Town and Country Planning Act 1947 provided the powers for a circular belt of land roughly 10–15 kilometres wide, beyond the city, where there was a very strong presumption against development. The Green Belt was both a policy in its own right (stopping London's expansion) and a component in a wider plan of urban management that also included the redevelopment of slums within London and moving people to new and expanded towns beyond the Green Belt.

The Green Belt has survived for fifty years, added to in some places and eroded a little in others. It did not aim to enhance the countryside; that was left to separate policies concerned with conservation and the protection of landscapes. Nor was the exclusion of development absolute; much of the orbital motorway around London (the M25) was built through the Green Belt. Other national planning interests did sometimes take precedence over the green belt function, yet it did successfully stop almost all urban development within its area.

The management of the urban fringe around London was characterised by a coordinated hierarchy of regional and local plans, very detailed land-use planning using published criteria, and a system of inquiries and appeals for aggrieved parties. The whole system was staffed by a cadre of professionally trained planners. It was part of a UK system that expressed national ideals about how development should proceed. However, in practice an acceptable compromise had to be found between national needs (e.g. more houses), regional requirements (how many more in the south east), and local priorities (whether to encourage growth in particular localities). This planning system was designed to confine the political system and politicians to issues of policy (e.g. the future of the Green Belt) and to separate them from direct involvement with individual land-use decisions, except within the framework of published plans and criteria or the final adjudication of the most contentious cases.

The system created a high degree of certainty about what development would be permitted, confined speculative land purchase to small areas, and established a regulated oligopoly of suppliers of development land. It was inflationary of land and house prices, since it created an artificial shortage of building land. The London Green Belt has become a symbol of the British planning system, and this helped its survival even during the deregulatory and pro-growth period of the 1980s.

Sources: Munton 1982; Elson 1986; Simmie 1994.

space-extensive activities (such as supermarkets), bad neighbours (e.g. abattoirs) and for activities generating traffic (new hospitals or leisure complexes) (see Figure 21.2 and Plate 21.1). The urban-edge location, where public transport might be limited, tended to promote further car usage, while simultaneously the 'sustainable city' ideal promoted more compact cities, less fringe development and less use of cars. The case study of Lancaster (Box 21.3) shows these features clearly.

The city edge is the first part of the countryside that urban dwellers meet, so there will be conflicts over what city folk would like their most adjacent countryside to be and what the fringe-area residents and landowners want from it. Conflicts may arise over access to farmland for walking, the tipping of rubbish and vandalism. City people may object to some farming practices (e.g. the smell of slurry or traffic from a farm shop). These are not issues unique to the urban fringe, but they are more common and intense there because

Figure 21.2 Urban edge land uses, Lancaster.

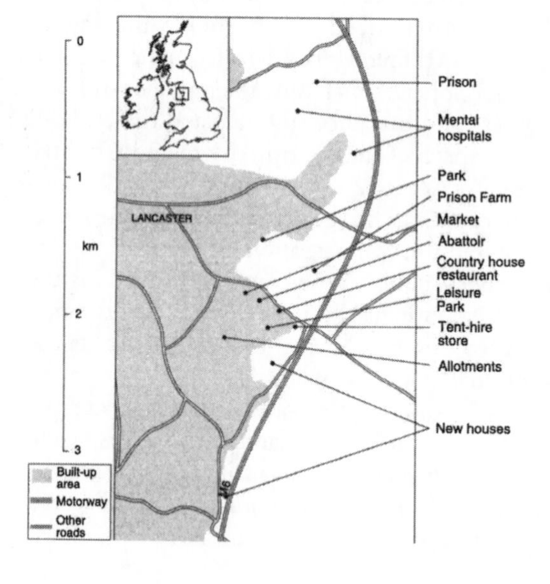

Plate 21.1 New retail site on the edge of Lancaster and Morecambe – a typical car-based land use on the urban fringe.

urban and rural values and expectations for the accessible countryside come into the sharpest conflict and proximity.

The changes in economic activity and social structure on the fringe are usually pervasive, even when 'formal' urbanisation (such as extensive housebuilding) does not occur. There may be more part-time farming (as farmers take urban work and urban workers buy a farm as a hobby). The use of farmland may reflect the closeness of the city through car boot sales, pick-your-own crops and horse livery. The appearance of the countryside may become important, and there will be pressures from urban people to buy houses (and to have more houses built) in the surrounding towns and villages. This can give rise to the 'urban village'. Newcomers restore old properties in pseudo-traditional styles and take over village institutions. They may use the village shop less but support the village school against closure. They may be uninterested in the fate of the bus service but keen to oppose the development of employment in the village and new housebuilding. Longer-term residents of the

Box 21.3 Lancaster

Lancaster is a small city in northwest England with a population of about 50,000, and it is typical of many similarly sized towns. It is growing slowly but without the economic dynamism of a capital city or a boom town. Its direction of expansion is limited by physical and development cost barriers such as a river and a motorway (see Figure 21.2). The expansion of the city has been guided by three principles:

- Expansion should be adjacent to the existing built-up area, so avoiding leapfrogging and ribbon development, and minimising the cost of infrastructure (see Plate 21.2).
- The urban fringe is an acceptable location (even the ideal one) for activities that are either too land-extensive to fit into a mediaeval urban core (e.g. a university, supermarket and sports centre), or generate a lot of traffic (a hotel and leisure park), or are un-neighbourly land uses for a built-up area (e.g. kennels, a prison and slaughterhouse) or are site-oriented (a water treatment works and microwave towers above the town).
- Development densities are fairly high so as to keep the city compact, minimise the loss of farmland and reduce house prices, since land is expensive. The hope of development profits is concentrated on the city edge and inflates land prices there.

Some smaller and picturesque villages around Lancaster have been 'taken over' by incomers, who work in Lancaster and who oppose further village expansion. Other villages with more mixed populations take a more relaxed view of some types of development.

The map of a small part of eastern Lancaster's fringe (see Figure 21.2) demonstrates clearly the mixture of public- and private-sector land uses, which are either drawn into the fringe from the countryside or are pushed out to the fringe from the town. Allotments and farms link to the open fields of the countryside, and the country house restaurant has similar resonances. Un-neighbourly land uses (e.g. the abattoir and perhaps the mental hospitals) and space-extensive ones (e.g. the prison) compete for space with activities generating a lot of traffic (e.g. the livestock market and leisure park).

Plate 21.2 Allotments at the edge of Lancaster surrounded by three generations of private housing and a leisure park (right) (see also Figure 21.2).

village may be more favourably disposed to these. The two groups (rather simplified here in their views) will contest how the parish or village should evolve (for reviews see Pacione 1984; Champion 1989; Robinson 1990).

For larger cities, the pressures may be more focused on major developments such as airports. The rapid growth of air travel has encouraged the building of new airport terminals and runways and even complete new airports. Since airports service metropolitan regions and must be accessible to travellers, employees and suppliers, they should be as near to the city as possible; but noise issues will push them deeper into the countryside. The environmental impact of airports is intense, and their negative effects on openness and quietness are overwhelming. The adjacent residents (probably escapees from the noisy city) are likely to be against such development. It is little surprise that airport proposals have attracted some of the bitterest conflicts at the urban fringe (e.g. Tokyo, London, Frankfurt, Manchester). The rise of environmentalism has intensified the opposition, as environmental groups have formed alliances with local people against airports. Opposition similar to that against airports has also arisen against motorways at the urban fringe and retail and entertainment complexes.

Whereas these developments tend to be close to the inner edge of the urban fringe, its outer edge witnesses conflicts over second and holiday homes. Beyond the realm of daily commuting, there may still be urban residential pressures from retired people and the owners of second homes. The retired may be unwelcome, since they reduce the vitality and workforce in the area. On the other hand, they may have the capital to improve their houses (so helping the local building trade) and have the personal interest and free time to revive social and community organisations. Second-home owners may be viewed less favourably. Their sporadic residence in the area may be characterised as exploitative – putting little back into the community and denying a house to a full-time resident. The balance of effects varies geographically – in some areas they may take over derelict houses that no one else wants (as in parts of Spain, Greece and France). In other places, they may displace people. If the second-home owners and local people are culturally distinct (different social groups or nationalities), then the conflict could expand from resentment to open hostility (as in parts of Sardinia and Wales) (for reviews see Champion 1989; Robinson 1990).

So far, this chapter has focused on the fringes of cities in the developed world. Those in the less developed world are no less varied and interesting (Drakakis-Smith 1980; Gilbert and Gugler 1992). The specific issues are usually rather different, but the principles are the same. The city edge is the place where many of the poorest migrants to the city will arrive from the countryside. Housing and health conditions on the fringe will often be very poor, with a severe lack of infrastructure. Yet the city may offer a better future than the countryside could. Issues of houses flooding and the physical safety of fringe sites are more important than in the developed world. The special conditions of apartheid gave South African cities an unusual structure, with townships at their fringe by law for ideological reasons. Economic forces keep the squatters of Nairobi and Lima in a similarly marginal position. The effects of the lack of a planning system and the way that political groups try to gain advantage from the poverty (and the hope of escaping it) are examined in the Lagos case study (Box 21.4).

CASE STUDIES: THE KEY ISSUES

This section highlights the key issues that this discussion and the case studies have revealed.

1 The way the urban fringe works (in the private and public sectors) reflects the wider economy and society of the country. If nationally individuals' rights, planning or corruption are prominent features of public life, then they will also be driving forces in how the urban fringe works.
2 Since these national traits will change over time and be more dominant in some areas than others, so the exact characteristics of the fringe

Box 21.4 Lagos

Lagos is the largest city in sub-Saharan Africa and rapidly growing in population. The 5–6 million people in Lagos state generate 20 per cent of Nigeria's GNP, and the city has become a magnet for Nigerian and foreign migrants. The city is characterised as thriving, expensive, corrupt, chaotic and exciting. Given its high cost of living, poor in-migrants congregate in slums and squatter settlements (though many fewer of the latter than in Latin American and East African cities), many being on the edge of the city. The way that urban-fringe land is transferred to urban uses mirrors other aspects of Nigerian society. There is a planning system based on the Town and Country Planning Ordinance of 1946, which in turn is based on a UK planning act of 1932, the latter being so ineffective that it was replaced in the UK but still has influence in Nigeria. Planning is very limited, a low national priority and poorly staffed. The Land Use Decree (1978) and the Land Use Act (1980) have not helped (as was hoped) in bringing forward land for development and curtailing land hoarding and speculation. There is a general lack of resources for infrastructure throughout the country but particularly on the fringe, and this has been exacerbated by the devaluation of the Nigerian currency and the shortage of skilled building workers. Hence there is a lack of water supplies, rubbish collection, roads, electricity and drains; overcrowding and ill-health are notable features. There is no national sense of the need to limit the growth of Lagos or to preserve peri-urban farmland and landscapes. Yet the release of land for housing is not random, since housing the very poor very inadequately can still be a profitable activity. Land release is controlled quite directly by politicians and is marked by clientism and patronage, and by the exchange of money, in which senses it reflects common practice in life and business in Nigeria. Indeed, the hope of development (as much as actual development) consolidates local leaders' powers. Development tends to be fragmented and ribbon-like along the roads. The rate of development is inadequate to meet the needs of the current urban fringe population, let alone the new in-migrants. There are groups for whom it is in their financial and political interests to ensure that some under-provision at the fringe continues.

Source: Piel 1991; Taylor 1993.

will vary between cities and will not be fully predictable.

3. Who is to profit from land development at the city edge – the landowner, the politician, the state or some combination of these? The answer to this question will influence how the urban fringe evolves.
4. Is the urban fringe to be contained or managed in a formal public way? If it is, there will have to be some kind of planning system. A balance needs to be struck between private rights and public interests in the planning system. How should planning operate and what effects will it have?
5. Are there acceptable and unacceptable uses of the urban fringe? If there are, how do you allocate land uses to meet these judgements? And on what basis do you form this judgement?
6. What is to be the balance between local and national requirements for the urban fringe?
7. The provision of transport facilities and the level of personal mobility greatly affect the pressures for development at the fringe.

The resolution of these issues will set the parameters within which each city will regulate its fringe. Hence, despite the general principles, every urban fringe will be different and the geographer will have to unravel how each city reached its current state and how it might evolve in the future.

CONCLUSIONS: A PROSPECTIVE VIEW OF USEFUL APPLIED GEOGRAPHICAL RESEARCH

When it comes to land-use conflicts at the urban fringe, how can the geographer be 'useful', and what should applied geographical research comprise? The applied geographer has four possible roles, which are not mutually exclusive but are distinct. He/she can be a gatherer of information, an interpreter of situations, a forecaster of events or an advocate for a cause:

- The *gatherer of information* is the geographer who collects details on, for example, land-use and landscape changes, the evolving social composition of villages, and the rate of migration. Without such spatially referenced detail, informed debate and planning will be impossible. The geographer's task, perhaps using geographical information systems and survey techniques, is to inform all who are concerned exactly what is happening at the urban fringe.
- The geographer who is an *interpreter* uses such

information to provide an explanation of the processes that are unfolding at the fringe, often basing the interpretation on a theoretical understanding of issues such as planning and the workings of the political economy. He/she may also evaluate the effectiveness of previous policies.

- The *forecaster* uses *ex-post* evaluation and interpretation of events to forecast either how the situation will evolve or how policy should change in order to achieve a given position in the future that is different from the *status quo*.
- The final role, that of *advocate*, sees the geographer leaving the fairly safe world of the 'expert' and becoming an advocate for a particular position in which he/she believes.

Clearly, the skills of the information gatherer, interpreter and forecaster can be turned to good use when combined with political *savoir faire* to create the effective geographer-advocate.

The urban fringe is a dynamic and exciting place in large and small cities in the developed and less developed world. The stakes are high there, and land-use changes are often controversial, not least because their effects will be long-lasting. Consensus on whether and how to manage the urban fringe is as far away as ever. There is plenty of scope for the geographer to do good applied research on the fringes of cities around the world.

GUIDE TO FURTHER READING

Johnson, J.H. (ed.) (1974) *Suburban Growth*. London: Wiley. Although dated now, this book gives a broad background to urban-fringe issues.

Bryant, C.R., Russwurm, L.H. and McLellan, A.G. (1982) *The City's Countryside*. Harlow: Longman. Good international coverage with a focus on planning issues.

Burtenshaw, D., Bateman, M. and Ashworth, G.J. (1986) *The European City*. London: David Fulton. Best coverage of European cities, setting the fringe in the context of the whole urban area.

Gilbert, A. and Gugler, J. (1992) *Cities, Poverty and Development*. Oxford: Oxford University Press. Covers the less developed world and emphasises the 'buffer' role of the fringe.

Short, J.R. (1996) *The Urban Order*. Oxford: Blackwell. Less directly on the fringe, but the ideas are readily transferable.

REFERENCES

Bourne, L.S. and Simmons, J.W. (1978) *Systems of Cities*. Oxford: Oxford University Press.

Champion, A. (1989) *Counterurbanisation*. London: Edward Arnold.

Davis, M. (1990) *City of Quartz*. London: Verso.

Drakakis-Smith, D. (1980) *Urbanisation, Housing and the Development Process*. New York: St Martin's Press.

Elson, M. (1986) *Green Belts*. London: Heinemann.

Gilbert, A. and Gugler, J. (1992) *Cities, poverty and development*. Oxford: Oxford University Press.

Gilg, A. (1985) *An Introduction to Rural Geography*. London: Edward Arnold.

Haughton, G. and Hunter, C. (1994) *Sustainable Cities*. London: Jessica Kingsley.

Johnson, J. H. (1972) *Urban Geography*, 2nd edition. Oxford: Pergamon.

Marchand, B. (1986) *The Emergence of Los Angeles*. London: Pion.

Mather, A. (1986) *Land Use*. Harlow: Longman.

Munton, R. (1982) *London's Green Belt*. London: Allen & Unwin.

Northam, R. (1975) *Urban Geography*. New York: Wiley.

Pacione, M. (1984) *Rural Geography*. London: Harper & Row.

Piel, M. (1991) *Lagos*. London: Belhaven.

Robinson, G. (1990) *Conflict and change in the countryside*. London: Belhaven.

Simmie, J. (1994) *Planning London*. London: UCL Press.

Sinclair, R. (1967) Von Thünen and urban sprawl. *Annals of the Association of American Geographers* 57, 72–87.

Taylor, R. (1993) *Urban Development in Nigeria*. Aldershot: Avebury.

22

Derelict and vacant land

Philip Kivell

INTRODUCTION

Derelict land became an obvious fact of life in many older industrial districts of Europe and North America in the economically depressed years of the 1920s and 1930s, but it did not attract systematic attention from geographers and planners until after the Second World War. The pioneering work of Beaver (1946) in Britain drew attention to the economic and environmental consequences of dereliction, as well as to the successes of some early reclamation efforts, in localities such as the Black Country, the northeast of England and South Wales. It was the combination of attention from geographers, the development of new mining and industrial technologies, the process of industrial restructuring and the establishment of a comprehensive planning system that placed the problem on the post-war political agenda. At first, the problem was connected with heavy manufacturing and mining industries and was largely confined to those localities in which these were located in Britain, Belgium, northern France and Germany, and to related problems of large-scale strip-mining in the northeast USA, Poland and Germany.

Throughout the 1950s and 1960s, with economic growth and new industrial investment, concern for the problems of derelict land remained largely confined to those with specialised professional interests in planning and mining, although comprehensive studies by Oxenham (1966) and Barr (1969), and later by Wallwork (1974), did much to publicise both the problem and some of its solutions. Gradually, however, large-scale reclamation efforts started, prompted by a concern for economic and environmental improvement, the need for more public open space, technical advances in land reclamation, and the shock of 144 lives lost through the collapse of a coal spoil tip at Aberfan in South Wales in 1966.

By the 1980s, the problem had gained more widespread urgency. Industrial change was still at the core of land dereliction as Britain and other early manufacturing nations lost their competitive advantages in the new global markets. A massive restructuring of production capacity followed, prompted by new styles and techniques of manufacturing, widespread mergers and closures, new forms and locations of investment at national and international scales, and new patterns of land use. But it was not just industry that was restructuring and abandoning its old sites; the same process was happening to docks (and the cities that had grown up around them), utilities and power sources, military installations, and public institutions, including hospitals. The land-use requirements of modern society were being transformed. Sometimes individual sites were abandoned (often in a severely damaged state), but in other cases whole localities and communities became effectively redundant. Nowhere was this more marked than in Eastern Europe, where the opening up of borders and the pressing economic reorganisation after 1990 revealed dereliction and contamination on a massive scale created by chemical works, lignite power stations and steel plants in the former East Germany, in the Czech Republic and in the Don basin.

No longer was dereliction a narrowly defined

problem of industrial closures. In effect, the fundamental economic and settlement geographies of the older industrial nations were being recast; the use of, and demand for, land was changing profoundly, and as a consequence derelict, vacant and otherwise waste land was being created in many localities faster than it could be dealt with. Large-scale movements, especially out of the larger cities, left many inner urban areas bereft of investment and with an accumulating record of derelict and vacant land. This was an intrinsic part of the 'inner city problem', which, in Britain, was identified as one of the most severe social and planning issues of the era.

THE NATURE AND SCALE OF THE PROBLEM

Problems occur in trying to define the different types of land embraced by this issue. Derelict land in Britain is normally taken to be 'land so damaged by industrial or other development that it is incapable of beneficial use without treatment'. However, this excludes categories such as land that is 'naturally derelict' (e.g neglected farmland and marshes), land that is damaged but subject to restoration conditions, land still in sporadic use, and vacant land. Vacant land is particularly problematical. Although there is no statutory definition, it was clear from the late 1970s onwards that vacant (not necessarily derelict) land was a serious problem in many cities (Civic Trust 1977; Burrows 1978), a theme that was developed by Coleman (1982) and later by Chisholm and Kivell (1987) and Adams *et al.* (1988) in the context of inner city redevelopment.

Subsequently, the related problem of contaminated land became the focus of much concern, and there was some legislative confusion in the early 1990s. In Britain, contaminated land was initially defined in terms of previous uses and activities of a contaminative nature, but this was changed in the Environment Act of 1995 to land that appears to a local authority to be in such a condition – because of substances it contains – that water pollution or significant harm is being, or is likely to be, caused. 'Harm' refers to the health of living organisms and the ecological system of which they form part, to humans and to property. Plans to establish public registers of contaminated land were shelved for fear that this might discourage development of brownfield sites and cause land values to fall.

In broad terms, derelict, vacant or contaminated land can present some or all of the following problems:

- *A waste of a valuable resource*. Allowing land to remain unused may be both economically and morally unjustifiable, especially where development continues to take place on greenfield sites.
- *An eyesore*. Because of its topography and the abandoned installations that often remain, derelict and vacant land is invariably ugly. This in turn can lead to further neglect or misuse.
- *A disincentive to development*. Derelict sites blight large areas, deter new investment and degrade the wider environment for local communities.
- *A danger*. This can include hidden shafts and voids, flooded pits, unstable tips and a variety of toxins with different levels of toxicity and longevity.

Related to these general issues are many specific problems that may deter or delay the redevelopment of individual sites. These can be summarised as follows:

- *Land prices*. Developers will normally base the price they are willing to bid for a site upon the value of the completed project minus their development costs. Frequently, the price thus calculated is below the price at which a vendor is willing to sell, and sometimes the price is a negative one. Government agencies have applied a variety of subsidies and reclamation grants to encourage development.
- *Cost of reclamation*. Technically, most things are possible; land can be reshaped and stabilised, new drains can be laid, shafts can be capped and toxins can be dealt with, but all of these procedures are costly.

- *Ownership.* The owners of vacant sites are sometimes reluctant to sell them, and the complexities and fragmentation of land ownership commonly make it difficult to assemble plots for comprehensive redevelopment.
- *Location.* Sites may be derelict or vacant because whole industries have collapsed and whole areas have been abandoned. These locations are often in the wrong place for modern investment and fail to convey the correct prestige and place image.
- *Access.* Many derelict sites were originally served by canal or railway and are surrounded by dense housing and other development. They do not permit easy access by large road vehicles; nor are they convenient for motorway connections.

Another important aspect of the problem is to attempt to estimate its scale. In Britain, this is possible through a series of estimates and surveys, but most of the figures have various important shortcomings. Elsewhere, such surveys are non-existent or, as in the case of France, lacking in precision (Couch 1989, Dechosal 1992; see Box 22.1).

For contaminated land, the shifting definitions and lack of systematic surveys make quantitative estimates particularly difficult. In Wales, where long-established industries have caused particular problems, one desk study (Environmental Advisory Unit 1984) identified over 700 sites totalling nearly 3800 hectares that were likely to be contaminated. For England, the Department of the Environment suggested to the House of

Box 22.1 Dereliction and reclamation in northeast France

Compared with Britain, France has had a different timing of industrial growth and restructuring, a different industrial composition and less pressure upon development land. For these reasons, the problems of dereliction in France were not evident as early, nor were they quite as widespread. However, from the late 1970s, the issue of *friches industrielles* became more pressing, notably in the smokestack industrial areas of the northeast, and a major report in 1986 (Lacaze 1986) formally drew attention to the issue.

The problem was greatest in the Nord–Pas-de-Calais region, which, with only 2.3 per cent of the overall area of France, possessed around 12,000 hectares of dereliction, over half of the estimated national total (Dechosel 1992). The causes of this concentrated dereliction were linked to the decline of older industries, notably coal and steel in the *bassin minier* around Bethune, Lens, Douai and Valenciennes, and to a lesser extent the textile industry around Lille, Roubaix and Tourcoing. Although concern originally focused upon the usual economic and community issues, there is now fear about pollution of the chalk groundwater resources.

Other pockets of dereliction occur, again mainly from industrial decline, in regions including Lorraine and in the vicinity of Lyon and St Etienne. Generally, however, French cities are less affected than their British counterparts. This is due to the different industrial histories alluded to above, but also to differences in housing policies and traditions of urban living, which mean that French cities have retained a greater residential and commercial vitality.

In some parts of France, state money and assistance for land purchase and reclamation has been available for many years. An inter-ministerial group for redeveloping coal-mining areas was established in 1972, and much progress has been made through localised bodies such as the *Établissement Public de la Métropole Lorraine*, but it was only from the middle of the last decade that programmes became more general. Between 1984 and 1988, approximately 2 billion francs of public money was allocated to land reclamation, over half of which came from local authorities, mainly in the form of low-cost loans. Where reclamation projects are too large for local authorities to handle on their own, there is a mechanism through the *Contrat de Plan* whereby the government and region act together. In mining areas, where demand for development land may be modest, reclamation for leisure projects and public open space is common, but in the main cities more elaborate projects are often pursued.

In Paris, there are some direct parallels with British urban experience as a range of utilities, industries, railway and dock installations (river and canal) have declined. Over 1000 hectares of 'derelict' land has been identified in the Île-de-France region (Chaix 1989), although there have been subsequent reductions. A good example is to be seen in the area of Paris known as La Villette, where a site of 55 hectares, just inside the *peripherique*, was becoming derelict in the early 1980s with the closure of a complex of canal basins, warehouses, abattoirs and associated facilities. Initially, many of the classic derelict land problems arose (e.g. multiple and confused ownership, poor access, difficult ground conditions, etc.), but an imaginative redevelopment scheme, backed jointly by the city, the state and private investors, quickly rescued the situation. There now exists on this site the largest centre in Europe devoted to popular science and technology. It boasts exhibition halls devoted to science and industry, a programme of displays and games, a park, and a centre devoted to music in the form of teaching, concerts and a museum. A fuller appreciation can be gained by a real or virtual visit (http.//www.cite-sciences.fr/).

Not only has this project made imaginative use of a once derelict site but it is also one example of a type of urban planning that has been used to re-image cities across Europe and North America.

Commons Environmental Select Committee (1990) that 27,000 hectares may be contaminated, although the same committee heard other evidence that suggested a total of 50,000 hectares. In a wider context, estimates placed before the committee implied that approximately 185,000 hectares of land in ten EC member countries were contaminated. In Belgium, geographers have been instrumental in helping to assess the risks and compile inventories of contamination (Miller 1994). In England and Wales, the Environmental Agency has more recently lowered estimates to between 5000 and 20,000 sites likely to need assessment under the regime introduced by the 1995 Environment Act (ENDS 1996).

The extent of vacant land is also difficult to quantify, despite the fact that it has been at the heart of urban policy throughout the 1980s and 1990s. From 1980 to 1996, a land register recorded vacant land in the public sector. This revealed that in February 1987, for example, 40,235 hectares of vacant land was held by public sector bodies. Independent analysts suggested that this figure was a gross underestimate and that the real total of vacant and derelict land for England alone might be as high as 210,000 hectares, with between 5 and 10 per cent of land lying vacant in many inner city areas (Chisholm and Kivell 1987). A survey of vacant land in 1990 (DoE 1992) formally identified a total of 49,080 hectares but suggested that a truer estimate would be 60,000 hectares in urban districts alone (for comparison, the total area of Bristol is around 11,000 hectares). By the early 1990s, vacancy was being tackled by more aggressive land disposal and development policies pursued by both the public sector and the newly privatised companies.

For derelict land, the figures initially look rather more comprehensive and reliable. Local authorities in England have been required to collect information regularly, and the results of surveys carried out in 1974, 1982, 1988 and in 1993, have been published (DoE 1995). For Scotland, the only comparable figures come from a survey of conditions in 1990 (Scottish Office 1992). Accepting the figures from these surveys as the best that are available, it is possible to examine the national pattern of dereliction. In 1993, a total of 39,600 hectares was derelict in England and, from 1990 figures, Scotland possessed a further 8297 hectares. There was an urban bias to the pattern, although in England this was only marginal at 52 per cent compared with the 80 per cent urban share in Scotland (see Plate 22.1) (Kivell and Lockhart 1996).

Table 22.1 gives a breakdown of the English

Plate 22.1 The Scottish Conference and Exhibition Centre and the Moat House Hotel in Glasgow, on former derelict land caused by the closure of Queen's Dock. Land in the foreground, also previously dock land, has remained derelict since the Glasgow Garden Festival in 1988, although there are currently plans to develop it as a Science Centre using a grant from the Millennium Commission.

Table 22.1 Derelict land by type in England, April 1993.

Type of dereliction	*Hectares*	*%*
Spoil heaps	9191	23
Colliery	4109	10
Metalliferous	3003	8
Other	2079	5
Mining subsidence	674	2
Excavations and pits	5807	15
Military dereliction	3275	8
Railway land	5615	14
General industrial dereliction	9749	25
Other forms of dereliction	5289	13
Total	39,600	100

Source: DoE 1995.

total. This indicates that mineral extraction in various forms has caused the largest amount of dereliction, accounting for approximately 40 per cent of the total, a proportion that rises to nearly 55 per cent in Scotland. The second largest category, that of general industrial dereliction, accounts for a further quarter in England. Overall, the distribution of dereliction – Figures 22.1 and 22.2 – reflects Britain's industrial past. Industrial decay and closure has often been accompanied by a loss of population, a deterioration of housing, a collapse of community confidence and the closure of utilities and services, leading in turn to cumulative processes of dereliction. The local authorities with the highest densities of dereliction were, unsurprisingly, urban: in Newham, Barking and Dagenham, Greenwich, Sandwell, Stoke-on-Trent, Salford, Lincoln, Liverpool and Bury, over 4 per cent of the authority's total area was derelict in 1993.

Although individual sites may remain unused for many years, derelict land is not a static phenomenon. New dereliction is created when businesses collapse or activities are abandoned, but at the same time land reclamation and redevelopment removes sites from the record. The problem for Britain has been that the overall stock has remained disappointingly stable. Between the surveys of 1988 and 1993, for example, nearly 9500 hectares of reclamation was offset by 8600 hectares of 'new' dereliction, giving a net reduction of just 900 hectares (equivalent to 2.2 per cent). Broadly speaking, dereliction caused by mining activity decreased, as a result of reclamation, preventive legislation and improved engineering techniques, but there were increases in the categories of military land (as a result of the 'peace dividend'), industrial dereliction and 'other forms' (including former landfill sites and commercial and residential premises).

THE REUSE OF DERELICT LAND

Much of the study of derelict, vacant and contaminated land, and the collection of survey information, is designed to encourage its productive reuse and prevent its further occurrence. Most countries in northern Europe have planning policies and financial packages to deal with damaged industrial land (DoE 1989), and most involve a degree of public/private sector collaboration. Germany, for example, has had an active programme since 1979 through the *Grundstucksfond-Ruhr*, but across the European Union as a whole the procedures are very variable. Despite differences in approach, there are often similarities in the outcome of major restoration schemes across Europe. For example, the Festival Park development of shops, leisure facilities and landscaping undertaken in the late 1980s on a former steel mill site in Stoke-on-Trent (Plate 22.2) has a direct, albeit larger, counterpart in the reclamation of the steel mill at Oberhausen in the Ruhr, completed in 1996 to form the Centro development. Similarly, the redevelopment of the London docks has many parallels in the enormous reconstruction of the dockland area along the estuary of the River Tagus in Lisbon.

Successive British governments have pursued policies to prevent dereliction occurring, by imposing restoration conditions upon permissions granted for mineral extraction, and to encourage the reclamation of existing dereliction through a regime of grants and subsidies. Limited

Figure 22.1 Derelict land in England in 1993, by type of dereliction and by county.

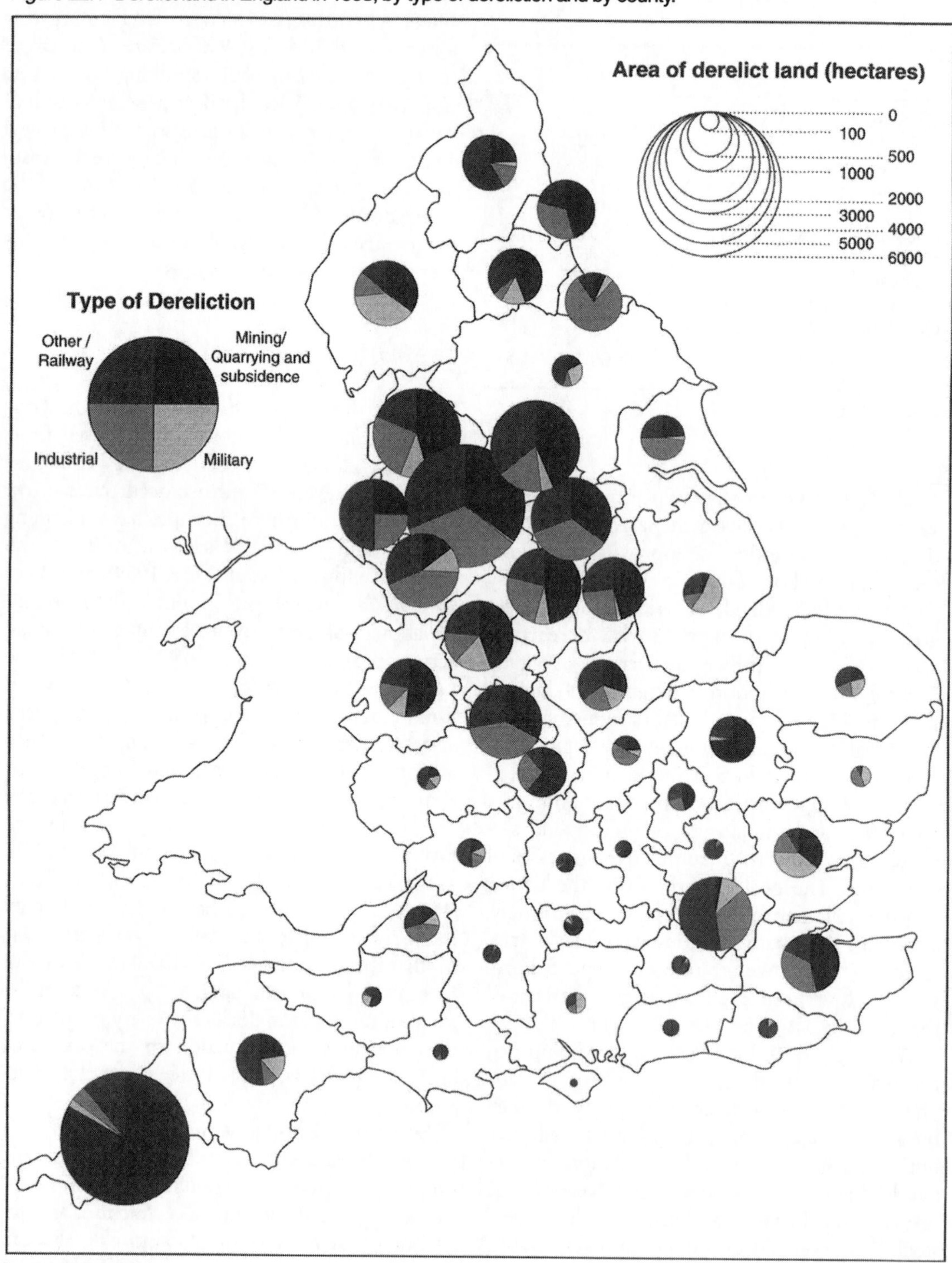

Source: DoE 1995.

Figure 22.2 The regional distribution of derelict and vacant land in Scotland, 1990.

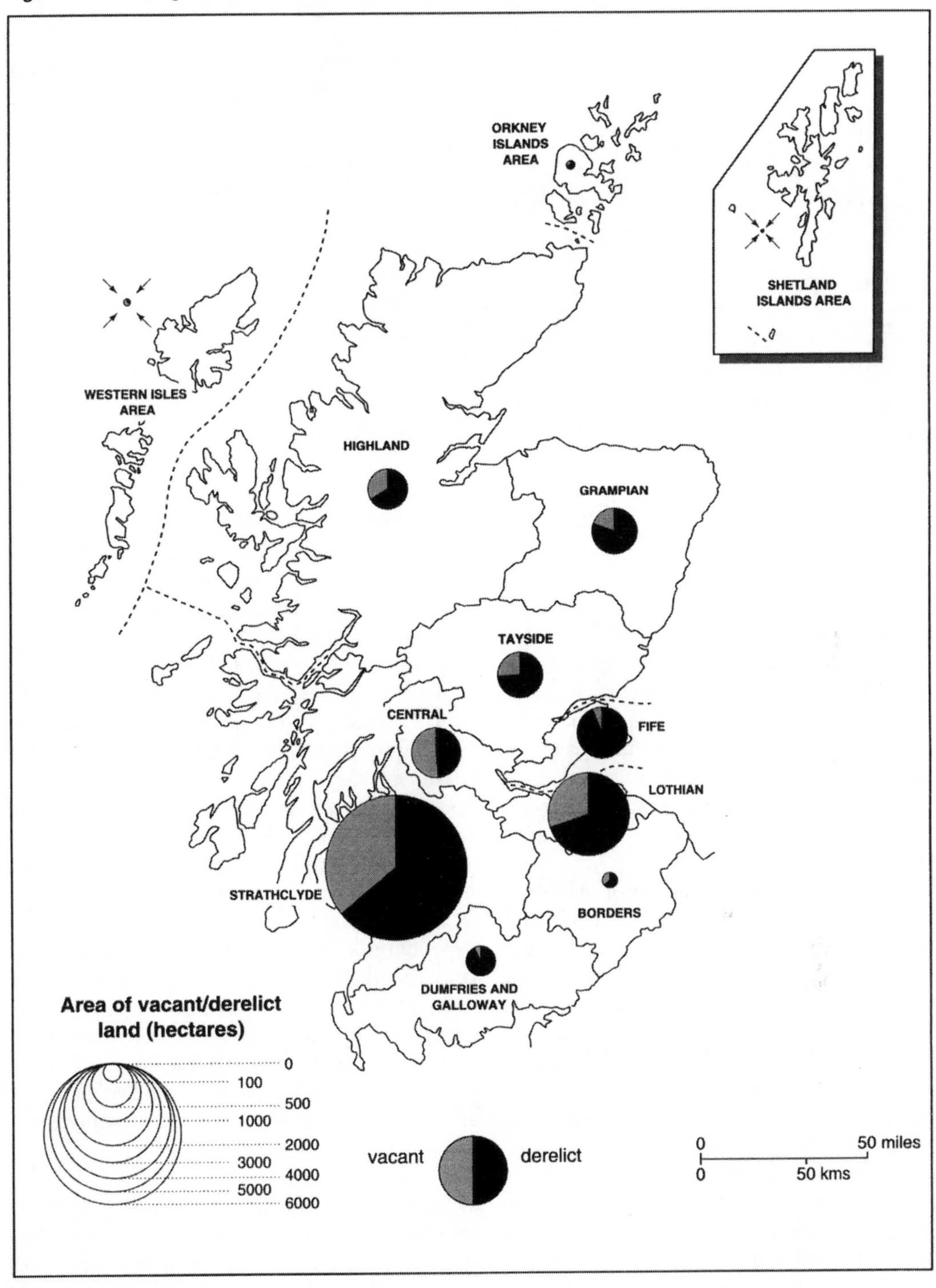

Source: Scottish Office 1992.

Plate 22.2 A view over Festival Park in Stoke-on-Trent. After many years of industrial closures a site of 67 ha was left with a mixture of derelict industrial buildings old mineshafts, tar lagoons and waste deposits. A Garden Festival in 1986 paved the way for the present-day mixed development of retail, office and leisure uses.

subsidies have been available to selected local authorities in Britain since the 1950s, but a comprehensive scheme was established by the Derelict Land Grant in 1982, which was later extended to the private sector. During the 1980s, finance for reclamation also came from city grants, the Urban Programme and urban development corporations, but in 1997 this was simplified; the Land Reclamation Programme continued under the auspices of English Partnerships, and most other aspects of the urban programme were rolled together into the Single Regeneration Budget.

Although the UK was an innovator in dealing with derelict land, its approach to contaminated land has been somewhat slower. The United States, through its Superfund Programme, has been tackling the problem since 1980, and in the Netherlands a comprehensive approach has been applied since 1983. Across the EU, the practice varies widely (Christie and Teeuw 1996). Most countries have registers of contaminated land, and most have laws relating to hazardous waste, water quality or planning that impinges upon contaminated land issues. In Britain, legislation requiring local authorities to identify and remedy contaminated land was not formalised until the 1990 Environmental Protection Act and did not really become operational until modified by section 57 of the 1995 Environment Act. The principles were endorsed by the government in 1997 (DoETR 1997a), although the practicalities are still under review.

Even though the total stock of dereliction remains resistantly high, substantial areas have been restored and brought back into productive use. Figures from the 1988 survey showed that between 1982 and 1993 approximately 23,485 hectares was restored in England, over half by local authorities with the help of derelict land grants, with urban development corporations making a substantial contribution to the balance.

The use to which reclaimed land is put depends upon local need, the location and type of land, the quality of the restoration, and government policy at the time. Original reclamation efforts were largely aimed at environmental improvements in some of the most blighted industrial areas, although in the development-led urban regeneration ethos of the 1980s there was pressure to use reclaimed sites for 'hard' end uses, that is housing, industry or commerce. Of the 9485 hectares reclaimed in the period 1988–93, 89 per cent was in productive use on 1 April 1993, with an approximately even split between hard end uses (44 per cent) and 'soft' uses (56 per cent),

Box 22.2 Birmingham Heartlands

The redevelopment in the Birmingham Heartlands area illustrates several themes concerning industrial decline and dereliction, the choice of regeneration policies and the changing patterns of land use and activities of a mainly 'post-industrial' form.

Situated northeast of Birmingham city centre, and astride the elbow formed by the M6 and the Aston Expressway, the Heartlands district was, by the early 1980s, showing multiple signs of inner city decay. It covered over 950 hectares and embraced 13,000 people in an area of extensive economic, social and environmental disadvantage that had once been at the heart of the West Midland's industrial economy. Between 1979 and 1989, 10,000 jobs were lost as factories, railway installations, gas, electricity and other utilities either closed or shrank in size. With these closures, the problem of derelict and vacant land increased, reaching a peak of 300 hectares in 1989. Despite severe handicaps, it was clear that the area possessed great potential: it is located in the centre of the country, with a very large surrounding population; it is adjacent to major motorway corridors; and it is close to Birmingham International Airport and to other elements of Birmingham's contemporary development, such as the National Exhibition Centre.

Initially, Birmingham City Council was wary of regenerating the area through an urban development corporation (UDC) similar to those being adopted elsewhere. It was felt that to do so might deter investors by unduly emphasising the 'problem' nature of the area, and it would certainly rob the city council of control over regeneration strategies. Instead, a Heartlands Development Agency was established in 1988 as a partnership between the private sector and Birmingham City Council. This was run as a private company but importantly, and unlike the situation within urban development corporations, the local authority retained planning control.

Ambitious plans for the redevelopment of the area were started, and significant progress was made; however, it was realised that there were advantages to be gained from UDC status, particularly in terms of more straightforward decision-making procedures and more generous levels of funding. Accordingly, the Birmingham Heartlands Development Corporation was set up in 1992, although it consciously maintained close links with the city council. The designation of the UDC made available an extra £50 million of government money, as well as additional grants from the European Regional Fund.

Over the past ten years, a substantial transformation has been wrought in the area (Figure 22.3). This consists of a number of major projects in which support for the remaining major industries, including Jaguar cars, LDV vehicles, GEC Alsthom's railway works and the SP and Dunlop Tyre companies has featured prominently, but the plans also acknowledge that the traditional manufacturing character of the area must be complemented by new activities. Land reclamation and redevelopment figure prominently in these

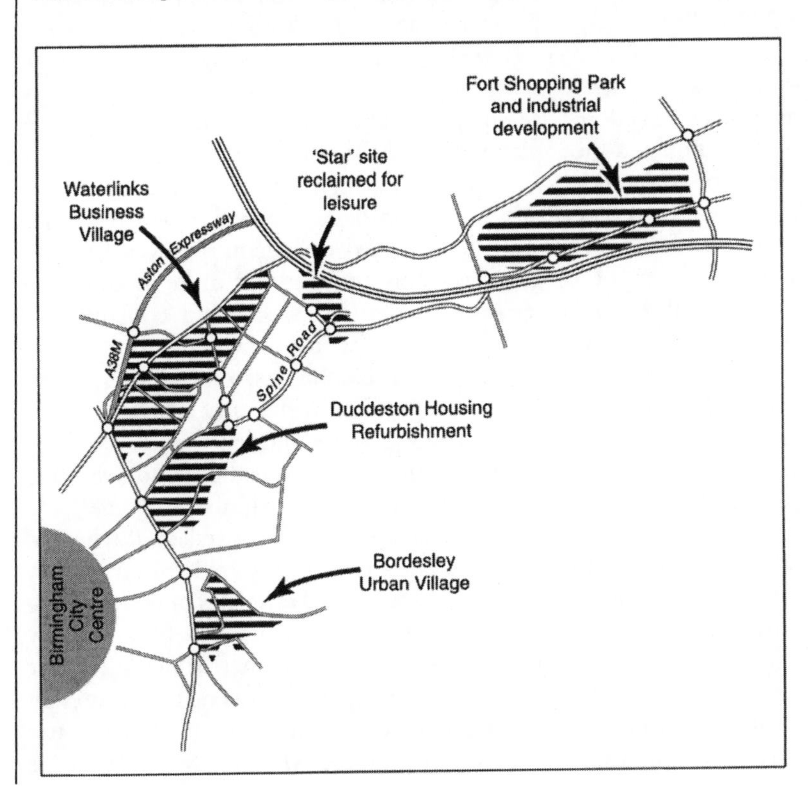

Figure 22.3 New developments in Birmingham Heartlands.

Box 22.2 continued

projects. One large example is the reclamation of land alongside the Birmingham and Fazeley Canal to create the Waterlinks Business Village, with 750 new jobs in offices and clean industries. Alongside another of the area's canals, the Grand Union, 40 hectares of derelict land has been reclaimed to build Bordesley, an urban village with over 1000 houses, a school and other community facilities

Other major developments on derelict or vacant land in the area include the development of the Fort Shopping Park and the 'Star' site close to the M6/A38M motorway interchange. This latter site of 14 hectares, previously the location of a power station, has now been reclaimed with the help of a £6.8 million grant from the development corporation and is set to become a leisure and entertainment centre with one of the largest cinema complexes in Europe. A key element in making all these developments possible has been the construction of a new 6 km dual-carriageway spine road through the area.

The Birmingham Heartlands Development Corporation has been involved in many activities, but one of its central and largest concerns has been to bring land and buildings into effective use. In the four financial years 1993/94–1996/97 it reclaimed more than 115 hectares of land, and the acquisition and reclamation of land have accounted for between half and three-quarters of its budget in most years. In the space of a generation, this part of Birmingham has been transformed from a mix of traditional manufacturing industries and utilities in decline, through various stages of dereliction and decay to a situation today where new activities and new environments are creating a post-industrial future for the area. It is clear that problems remain: deprivation and social disadvantage have not been entirely overcome; parts of the environment are still unattractive; and the area is most unlikely to provide as many jobs as it did in its heyday, but it is equally clear that a major turnaround has been achieved, and investment is now being attracted back into what was effectively an economic no-go area just ten years ago.

which included recreation, public open space and agriculture/forestry.

CONCLUSION AND PROSPECTIVE VIEW

Questions relating to the use and misuse of land are particularly pressing in the United Kingdom in view of the density of the population and the changing land-use needs that must be accommodated upon a small surface area. Over the years, the geographical connections and contributions to this debate have been substantial. An important starting point was the publication of *Applied Geography* (Stamp 1960), a work that combined the themes of land use and abuse referred to in this chapter, and the present book's broader theme of applied geography.

Geography has contributed greatly to an understanding of the spatial pattern of dereliction and the way in which this relates to contemporary land needs. Not only are there regional concentrations of derelict, vacant or contaminated land but there are also marked differences between urban and rural conditions and between inner and outer urban localities. Such land cannot be seen in isolation, for it is clearly one part of a complex interrelationship between changing economic and social forces and resulting land use patterns.

A further point of contact for geographers comes through the role of planners. Many planners started out as geographers (and a few *vice versa*), and for the most part they speak the same language and use some of the same tools. Many geographers have worked closely with planners and have influenced the formulation of land policy through debate or through commissioned, and highly applied, research. A final point to be noted is the contribution that geographical tools and methods make to an understanding of derelict, vacant and contaminated land. In particular, the techniques of GIS are increasingly providing powerful and flexible ways of collecting, storing and analysing information relating to land quality.

The debates about the end use of derelict and vacant land have recently shifted as planners and politicians grapple with the problem of where to accommodate Britain's projected growth of new housing. Controversial estimates of new housing needs for the next twenty years have recently been raised from 4.25 million to 5 million, a figure that is approximately equivalent to the present housing stock of Greater London, plus Greater Manchester, plus the whole of Wales. There is pressure from many quarters to place as much as possible of this upon reclaimed and vacant land, or 'brown-field' sites, in preference to taking green-field sites. In reality, this latter has been happening to a significant degree in recent years. In 1992, half of all new urban development was on land previously used for urban purposes (i.e. recycled land), and a further 8 per cent was on land in urban localities that had not previously been developed (DoETR 1997b). The scope for major increases in this proportion may be limited by the locational mismatch between the concentrations of derelict/vacant land in the Midlands and north of England and the greater demand for new housing in the south, but it is clear that derelict and vacant land is going to be central to this debate.

GUIDE TO FURTHER READING

A comprehensive overview of the early years of derelict land study appears in K. Wallwork's *Derelict Land* (1974, Newton Abbott: David & Charles). E.M. Bridges, *Surveying Derelict Land* (1987, Oxford: Clarendon Press), reviewed an interdisciplinary methodology for the assessment of derelict and contaminated land, and *Inner City Waste Land* by M. Chisholm and P.T. Kivell (1987, London: Institute of Economic Affairs) assessed failures in land development in a way that has renewed relevance for the present debate about land for new housing.

The Department of the Environment produced surveys of derelict land in 1974, 1982, 1988 and most recently in 1993, and its report, *Assessment of the Effectiveness of Derelict Land Grant* (1994, London: HMSO), gives useful details of land reclamation, including a number of case studies. Derelict land in a wider European context was dealt with by a collection of papers in French, mainly by geographers, entitled *La Problematique des friches industrielles* (1994, Strasbourg: Centre Européen du Developpement Regional).

Up-to-date information, including government press releases, can be found on the World Wide Web. Useful sites are:

The Department of the Environment: (*http://www.detr.gov.uk/detrhome.htm*)

The Environment Agency: (*http://www.environment-agency.gov.uk/home.html*)

The European Commission DG XI: (*http://www.europa.eu.int/en/comm/dg11/dg11home.html*)

REFERENCES

Adams, C.D., Baum, A.E. and MacGregor, B.D. (1988) The availability of land for inner city redevelopment: a case study of Manchester. *Urban Studies* 25, 62–76.

Barr, J. (1969) *Derelict Britain*. Harmondsworth: Penguin Books.

Beaver, S.H. (1946) *Report on Derelict Land in the Black Country*. London: Ministry of Town and Country Planning.

Burrows, J. (1978) Vacant land – a continuing crisis. *The Planner* January, 7–9.

Chaix, R. (1989) Friches industrielles et réaffectations en Île-de-France, Evolution 1985–1988. *Hommes et Terres du Nord* 4, 320–24.

Chisholm, M. and Kivell, P.T. (1987) *Inner City Wasteland*. London: Institute of Economic Affairs.

Christie, S. and Teeuw, R. (1996) European perspectives on contaminated land. *European Environment* 6, 85–94.

Civic Trust (1977) *Urban Wasteland – A Report on Land Lying Dormant in Cities, Towns and Villages in Britain*. London.

Coleman, A. (1982) Dead space in the dying inner city. *International Journal of Environmental Studies* 19(2), 103–7.

Couch, C. (1989) Vacant and derelict land in France. *Land Development Studies* 6, 183–99.

Dechosel, L. (1992) Land reclamation in France and England. *Working Paper No.2*, Policy Research Centre, Sheffield Business School.

DoE (1989) *A Review of Derelict Land Policy*. London: HMSO.

DoE (1992) *The National Survey of Vacant Land in Urban Areas of England, 1990*. London: HMSO.

DoE (1995) *Survey of Derelict Land in England 1993*, Volume 1, Report; and Volume 2, Reference Tables, London: HMSO.

DoETR (1997a) *Contaminated Land Review Completed*. Press Release No. 539, 22 December.

DoETR (1997b) *Land Use Change in England No. 12*. Statistical Bulletin.

ENDS (1996) MPs urge changes to contaminated land proposals, *ENDS Report* 263, p. 22.

Environmental Advisory Unit (1984) *Survey of Contaminated Land in Wales*. Cardiff.

House of Commons Environmental Select Committee (1990) First Report on Contaminated Land. *House of Commons Paper No. 170* (three volumes), London: HMSO.

Kivell, P.T. and Lockhart, D.G. (1996) Derelict and vacant land in Scotland. *Scottish Geographical Magazine* 112(3), 177–80.

Lacaze, J.P. (1986) *Les Grandes Friches Industrielles*. Paris: DATAR.

Miller, J. (1994) La contamination des friches industrielles. In *La Problematique des Friches Industrielles*, Strasbourg: Centre Européen du Developpement, 139–49.
Oxenham, J.R. (1966) *Reclaiming Derelict Land*. London: Faber & Faber.
Scottish Office (1992) *Scottish Vacant and Derelict Land Survey, 1990*. Edinburgh: Environment Department.
Stamp, L.D. (1960) *Applied Geography*. Harmondsworth: Penguin Books.
Wallwork, K.L. (1974) *Derelict Land*. Newton Abbot: David & Charles.

23

Sustainable tourism

Lesley France

INTRODUCTION

> who could have foreseen all this three summers ago; when their yachts first dropped anchor here; when the first village houses were bought and converted ... the first property acquired and developed? ... it is time to weigh anchor again and seek remoter islands and farther shores and pray for another three years reprieve.
>
> (Fermor 1983: 120)

Originally published in 1966 with reference to the coasts of southern Spain, this graphic description of the spread of Christaller's (1964) 'pleasure periphery' hints at some of the negative impacts of tourism in destination areas. These impacts became more acute and more widespread during the following thirty years and included ecological damage, the loss of traditional values and societies, and the operation of the economically and environmentally disastrous 'resort cycle' (Butler 1980; Lane 1990). Essentially, they were an outcome of the post-war growth in the numbers of tourists resulting from increased leisure and paid holidays, greater disposable income, and cheaper and easier travel for many in the urban industrial areas of Northwest Europe and North America (see Box 23.1).

Widespread discussion in the 1970s by a range of people from futurologists to academics and church-based groups took place about the negative effects of the 'unfettered growth in mass tourism' (Lane 1990). France, Germany and Switzerland led the search for alternative forms of sustainable tourism (defined in the following section) with the economist and sociologist Pierre Laine, the theologian and holiday psychologist Paul Rieger and Professor Jost Krippendorf (Krippendorf 1987).

Academics from a variety of disciplines published tourism studies in the late 1970s and 1980s. This work was drawn from the following fields: anthropology (Smith 1977), which focused on notions of authenticity both in artefacts sold to tourists and in non-material customs such as rituals and dances; sociology (O'Grady 1981), which centred around changes resulting from contacts between hosts and guests e.g. in language, religious practices, prostitution; economics (Vaughan and Long 1982), within which the nature and extent of employment, foreign exchange earnings, linkages with other sectors of the economy and dependency were among the factors considered; ecology (Stroud 1983; Pawson *et al.* 1984), which examined the effects of air and water pollution and the destruction of flora and fauna; and geography (Pearce 1987; Shaw and Williams 1994), which concentrated on the spatial aspects of tourism activity. Much detailed research was collected on tourism within general texts (Mathieson and Wall 1982; Lea 1988; Pearce 1989). These also often proposed planning and policy measures to alleviate the problems arising from negative tourism impacts. Such measures – such as the imposition of quotas on visitors or cruise ships, land-use planning restrictions, conservation work, employment regulations – tried to establish a viable alternative approach to tourism that is less destructive for the

Box 23.1 The rise and decline of a resort: Torremolinos, Spain

The 'self-destruct' theory (Shaw and Williams 1994) of tourism highlights the rise and decline of a destination in a cyclical manner (Figure 23.1; see also Butler 1980). These ideas can be clearly illustrated with reference to the resort of Torremolinos on the Costa del Sol in Spain.

Figure 23.1 Growth and decline of a tourist destination.

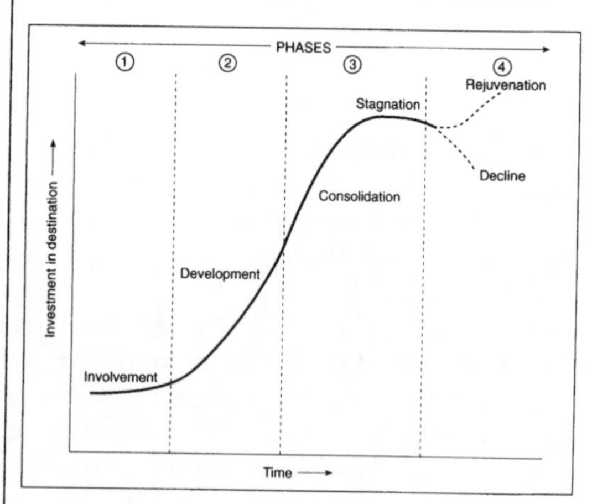

Sources: Butler 1980; Shaw and Williams 1994.

PHASES

1 *Involvement.* Initially, tourism acts as a catalyst for local community initiatives that provide facilities for adventurous visitors, often drawn from an elite market. In the 1950s, Torremolinos typified this situation, with fashionable visitors who enjoyed the resort's distinctive characteristics.
2 *Development.* Dynamic growth leads to large numbers of new, often middle-class visitors. Control passes out of local hands as multinational companies move in and, as development proceeds, so the character of the area begins to change. Tourism structures like hotels were frequently Spanish-owned in Torremolinos, but foreign tour operators like Clarkson and Thomson from the UK and Neckermann from West Germany began to control the nature and direction of tourism in the boom years of the 1960s.
3 *Consolidation and stagnation.* Mass tourism becomes established and the leaders of fashion, who once popularised the destination, have now moved on to more exclusive venues. Social and environmental degradation become apparent as peak visitor numbers are reached. Many of the luxury hotels in Torremolinos were downgraded in the late 1960s and early 1970s. By the 1980s, 3-star hotels served a 'cheap and cheerful' market. A concern over the standards of public behaviour of tourists in bars and nightclubs, and increasing arrests, highlighted the deteriorating image of the resort. Pollution of the sea became a problem.
4 *Decline and/or rejuvenation.* The destination either sinks under its problems as tourists leave, or attempts to revitalise tourism. Torremolinos has tried to meet a continued demand for budget holidays from the North European market through an increase in self-catering apartments and an extension of the season into the winter months, aimed specifically at the elderly. More rigorous policing has improved standards of tourist behaviour in public. Attempts have also been made to upgrade the physical environment by such means as improved sewerage schemes to reduce seawater pollution, the rebuilding of the promenade in the Carihuela area, and the refurbishment of public areas, e.g. through tree planting.

Sources: Barke and France 1996, Butler 1980, Shaw and Williams 1994.

Plate 23.1 Torremolinos in the late 1950s: Phase 1 – Involvement.

Plate 23.2 Torremolinos in about 1990: Phase 3 – Stagnation.

host society, economy and environment yet still provides a satisfying experience for tourists. Frequently based on natural or cultural resources such as climate, scenery, wildlife, historic monuments, and local customs and ceremonies, the forms of tourism that this approach encouraged are given a variety of labels: green, responsible, alternative, soft (Krippendorf 1987; Wheeller 1991). In practice, they included activities like walking holidays, wildlife safaris and culture-based trips. As part of this movement increased, popular interest in the environment led to the emergence of ecotourism as a major thrust within the field of sustainable tourism development (Cater 1994). There were, however, those (Wheeller 1991) who insisted that alternative tourism of the types described could never fulfil mass demand for tourism, which inevitably involved very large numbers of people. New types of tourism that cater for small numbers can only stand alongside and not replace a more sustainable approach to all forms of tourism (Muller 1994). Perhaps by its nature tourism can never achieve a full measure of sustainability, but it can move towards a lower-impact situation in which more benefit accrues to local people, tourists gain a higher degree of satisfaction and the host environment is less threatened than in more traditional forms of mass tourism (France 1997). These ideas will be explored and possible solutions outlined within the chapter.

SUSTAINABLE DEVELOPMENT TO SUSTAINABLE TOURISM

While sustainable tourism is a pragmatic outcome of the need to respond to the negative effects of the industry in destination areas, as a philosophy it is rooted in sustainable development.

The 1987 definition of sustainable development by the World Commission on Environment and Development Report (usually known as the Brundtland Report) is most widely accepted. It states that to be sustainable, development must 'meet the needs of the present without compromising the ability of future generations to meet their own needs' (WCED 1987: 43).

Further refinements and a more obvious application to tourism emerged with the United Nations Conference on Environment and Development, held in Rio de Janeiro, Brazil, in 1992. In its action plan, Agenda 21, it proposed a number of routes forward to achieve more sustainable development. Although tourism is scarcely mentioned in the documents resulting from the conference, among the issues that are especially appropriate for tourism are those designed to change consumption patterns, combat poverty, provide socio-culturally sensitive and environmentally sound programmes, empower groups and communities, and engender economic benefits.

Recognition of the importance of Agenda 21 by those involved in tourism occurred in 1995, when the First World Conference on Sustainable Tourism was held in Lanzarote, the Canary Islands. Discussion here led to the publication of a Charter for Sustainable Tourism and a Plan of Action that formally attempted to apply Agenda 21 to tourism (de Avila 1996). Along with many others, participants at the conference wrestled with definitions of sustainable tourism and with efforts to describe and illustrate examples of good practice.

DEFINITION OF SUSTAINABLE TOURISM AND ASSOCIATED DIFFICULTIES

While much dispute exists about the nature of sustainable tourism, there is general agreement that those characteristics listed in Table 23.1 typify sustainable tourism approaches (Lane 1990; Cater 1994; Muller 1994). Academics, pressure groups and practitioners (Lane 1990; Eber 1992; Elliott 1997) often suggest that any benign and sustainable tourism development should possess as many of these characteristics as possible. Ideally all the actors – the host community, economy and environment; the tourists; and the industry – should have the same focus and achieve equal satisfaction. Unfortunately, a situation of conflict exists because the aims of these

Table 23.1 Characteristics of sustainable tourism and approaches to their achievement.

Characteristic	*Nature of characteristic*	*Approaches to successful achievement*
Scale	Appropriate, with numbers of tourists and size of development limited, so avoiding overconsumption and waste.	Apply carrying capacity limits to: • environment, economy and society of the host area, • needs of holiday makers, • demands of the industry.
Growth	Controlled and slow.	Apply mandatory planning and policy measures to host areas and voluntary/mandatory restrictions to and by the international industry.
Style of development	Appropriate to the environment, economy, culture and society of host areas, maintaining their diversity and distinctiveness.	Apply planning and policy measures in host areas, voluntary regulation by the industry and educate the holiday maker. Encourage the maintenance of local culture.
Degree of local involvement	Reduce dependency on metropolitan areas/institutions.	Encourage local ownership, management and employment through regulations, education, training and the provision of opportunities for the acquisition of local experience. Consult stakeholders and the public.
Integration with other sectors of the economy	Reduce leakages and increase local benefits.	Provide encouragement through administrative mechanisms like cooperatives. Provide education and training.
Repeat visitors	Retain/increase visitor numbers and types of visitor over time.	Plan and manage the tourist environment to ensure careful marketing to maintain host economies, societies and environment; provide ongoing profits for the industry; provide high satisfaction levels for visitors.

Sources: Lane 1990; Eber 1992; Squire 1996.

actors are not necessarily the same (Table 23.2), and sustainable tourism is not perceived as a single, uniform phenomenon.

Within the industry, issues of cost and the profit motive dominate. The achievement of sustainable tourism is low on its list of priorities other than as a desirable model of 'political correctness' or a useful marketing tool. Yet tourism is a market-led industry, dominated by multinational companies that play a powerful role in manipulating consumer demand. Ultimately, it is that demand, i.e. the tourists themselves, that determines the nature and extent of international tourism activity. Multinational companies provide a mechanism for organising tourism. They transport tourists to their holiday destinations where they arrange accommodation and other services, like visits to attractions. Therefore they have the power to influence, or even dictate, the form of tourism and the size of the

Table 23.2 Aims and conflicts of actors/interest groups.

Actors/interest groups	*Aims*	*Potential conflicts*
Host community	Income; employment; meeting other people	Increased income for hosts means higher costs for tourists and possible lower profits for the industry. Too many visitors can turn 'euphoria' into 'antagonism' (see Doxey model – Burns and Holden (1995).
Host economy	Increased GDP and employment.	Danger of dependency on multinational companies and consequent leakages; other sectors of the economy losing labour.
Host environment	Conservation and restrictions on visitor numbers	Fewer visitors reduces income and employment opportunities for hosts and results in higher operating costs for the industry but can create greater visitor satisfaction.
Tourists	Value for money/low cost; satisfying experience; education.	Can conflict with aims of hosts and the industry for higher income/more profits.
Tourism industry	Profit; repeat visits by satisfied clients.	Can conflict with hosts' aims for income and employment generation and conservation; also with tourists' aims of low costs.

Source: France 1997.

industry to host governments and local communities.

It is precisely within these local communities and tourist-receiving nations that the negative effects associated with tourism are most apparent and are at their most acute. Sustainable approaches to tourism are therefore high on the agenda in host areas, where there is least power to generate them, and are least demanded by consumers and the industry, where control is focused.

A further problem lies in a lack of consensus in the definition of sustainable tourism. This arises from varying perceptions of tourism by those involved (see Box 23.2) and goes hand in hand with a changing vision of sustainable approaches (Hjalagar 1996). Acceptable attitudes and

Box 23.2 Perspectives on tourism

There are major difficulties in reconciling the perspectives of various tourism interest groups, as each has its own interpretation of any given situation. A report on all-inclusive resorts by McNeill (1997) clearly illustrates this problem. On the one hand, a spokeswoman for the pressure group Tourism Concern claimed that 'all-inclusive resorts deny the local economy the opportunity to become involved in tourism', while a director of one of the Caribbean's leading all-inclusive operators countered with the assertion that they employ many local staff, buy 'huge quantities of food and drink from local farmers' and 'alone provide 10 per cent of Jamaica's foreign currency earnings'. Tour operators believe that such all-inclusives offer the best value for money for holiday makers – an important motive in the choice of destination and type of holiday (Callaghan *et al.* 1994) – and therefore in providing tourist satisfaction. Although operators suggest that all-inclusive clients are more likely to spend money on local sightseeing and souvenirs, thereby bringing benefits to the host community, professional tourism experts from the Tourism Society believe that all-inclusive holiday makers may enjoy a convenient and high-quality experience but lose their independence and often fail to venture outside their resort. This conflict of views sheds little light on the real level of sustainability within a sector of the tourism industry that is growing in terms of both the number of establishments and their geographical spread.

Sources: Callaghan *et al.* 1994; McNeill 1997.

activities vary over time and from place to place as a result of fashion, education, the psychology of visitors and the costs of developing more sustainable features.

Some form of compromise is therefore inevitable. Perhaps it is realistic to envisage a spectrum along which the degree of sustainability of each actor can be assessed. The aim for all should be to move from a situation in which damage, conflict and dissatisfaction are high towards a more benign situation with higher benefits and lower costs.

TYPES OF TOURISM AND SUSTAINABILITY

Since it is large-scale traditional mass tourism that has been accused of having so many adverse impacts (Mathieson and Wall 1982; O'Grady 1990), small-scale developments are often used as exemplars for the promotion of sustainable approaches (O'Grady 1990; Bird 1995). However, more important than scale alone is the capacity of resources to absorb visitor numbers. Blackpool, Torremolinos or Miami Beach can absorb much larger numbers of people than the trekking trails around Mount Everest, the area of outstanding natural beauty in the north Pennines in England, or in a small game park in East Africa, before capacity levels are exceeded and costs begin to outweigh benefits. It is partly a question of the relative resilience of the natural environment and indigenous culture that determines capacity levels and partly a question of the level of crowding that reduces the appeal of a destination to tourists. That appeal also depends on the characteristics of tourists at different types of destination, as well as the physical ability of particular destinations to absorb visitors while still retaining the illusion that relatively few tourists are present.

Some types of tourism, superficially at least, reveal more sustainable characteristics than others. This is highlighted in the theoretical contrasts that exist between traditional large-scale mass package tourism and ecotourism (see Table 23.3).

Table 23.3 Theoretical characteristics of mass package tourism and ecotourism.

Mass package tourism	*Ecotourism*
Large scale	Small scale
Large numbers	Small numbers
Imported cultural environment, e.g. food, language	Acceptance of local culture by visitors, e.g. food, language
Commercialised	Relatively little commercialisation
Artificial attractions an integral part of the holiday experience	Reliance on natural and local cultural features as attractions
Multinational companies dominate the industry	Much local ownership and management
Foreigners dominate decision making	A high level of decision making by host governments, companies and individuals
Environmental degradation common	Relatively pristine environment
Inadequate planning and management	Careful and effective planning and management

Sources: Burns and Holden 1995; Cater 1994; Shaw and Williams 1994.

Mass tourism – can it become more sustainable?

Traditionally, this type of tourism caters for those who prefer a risk-free environment, often within a familiar setting, where the cultural components have been imported, e.g. language, beer, 'tea like mother makes', fish and chips. Such people usually prefer to spend their holiday among large numbers of like-minded companions, who require English-speaking guides and hotel staff, and entertainment similar to that found at home. Demands of this nature have often provoked conflict with host communities and environments (among others, see Mathieson and Wall 1982; Pearce 1989). Nevertheless, to some extent this pattern provides a degree of sustainability for the industry and for the market as a whole, although some sectors of the market decline as wealthy and fashionable tourists quickly move away to remoter, less commercial venues as their satisfaction levels fall.

A pragmatic acceptance of these changes does not inevitably lead to an abandonment of the search for a more sustainable approach to tourism. The creation of tourist ghettos can satisfy a considerable proportion of existing demand and focus it into areas and resorts whose environment has been degraded in the process and the lives, customs and economy of the local people irrevocably altered. However, concentrating visitors into tourist ghettos does not necessarily involve a continued deterioration of the physical and human environment after the initial changes. Many seaside holiday resorts, like Blackpool and Scarborough, are over 100 years old. Their construction changed the physical environment, but subsequent modifications have led to minimal structural changes. Local authorities, even in 'notorious' resorts like Torremolinos, have attempted to improve the image of the destination by measures such as planting trees, cleaning and upgrading buildings, promenades, street lighting and monuments, removing rubbish, and improving sewerage systems.

Many of the resorts of the Spanish Costas, including Torremolinos, have continued to provide a focus for tourists over a long period of time. They were developed at an early stage in international mass tourism during the 1960s, when rising incomes, increased leisure time and holidays with pay, technological advances in transport, and a desire to travel fuelled a tourism boom in Europe. In spite of the loss of their fashionable status as the 'jet set' visitors – who often create the image of a subsequently popular venue – moved on to new, less frequented destinations, the total numbers of tourists to the Spanish Costas have not fallen substantially. Instead, the nature of the industry there has changed as luxury hotel and restaurant provision has been replaced or outnumbered by lower-class hotels, self-catering units, cheap cafes and supermarkets (Barke and France 1996).

In a similar way, there is to be a revitalisation of Butlin's holiday camps in Britain (Walsh 1997), both as a response to changing consumer demand and in an attempt to revive the flagging profits of the company. By providing a focus for holiday demand, these revitalised camps should help to ease pressure on more vulnerable locations elsewhere. Similar ventures, like Center Parcs and a range of theme parks, provide popular short-break destinations in areas close to centres of demand. They are also based on robust artificial rather than less resilient natural attractions.

The perception of mass tourism solely as a provider of ghettos that focus visitors away from less resilient areas is perhaps a little defeatist. It *is* possible for this traditional form of tourism to become gradually less destructive and begin to move towards achieving higher levels of sustainability, even in less developed countries. The use of more local food and menus, thereby improving links with local agriculture and encouraging the employment of local chefs, is one step in this direction. As such, it begins to introduce the concept of local participation and empowerment through the increased employment of local people, especially in skilled and managerial positions, which leads to a rise in local ownership and more involvement in decision making. These are examples of 'green viruses', which are described by Muller (1994) and illustrated in Box 23.3 in relation to the West

Box 23.3 Green viruses in the Caribbean

Local businesses and syndicates own many tourism enterprises, even within the smaller Caribbean Islands. Typical are the local ownership and management of small budget hotels, like the Yellow Bird, which lie along the south coast of Barbados; the dominance of local ownership (70 per cent) in Dominica; hotels on St Lucia, which range from those offered by multinational tour operators, such as the Green Parrot, to the independently marketed Anse la Raye; and Morne Fendue Guesthouse on Grenada, which is praised in widely available guidebooks (Henderson 1994). Local ownership and management extend beyond the accommodation sector into transport and local tour provision on islands like Barbados, Dominica, Grenada and Jamaica. Associated with these positive moves is the increased employment of local people, especially in skilled and managerial positions, on all the islands as education, training and experience have spread.

The degree of penetration of local food into the tourism sector, through the increased use of both ingredients and menus, is a further reflection of the rising level and nature of local participation. A decline in imported food has occurred on islands like Barbados (Momsen 1994), where local agriculture has begun to provide a greater proportion of produce in hotels and restaurants. Most of the chicken and pork is reared locally, as are increasing amounts of vegetables and flowers for tourist establishments.

The conservation and restoration of attractions for visitors, such as Old San Juan in Puerto Rico; Brimstone Hill on St Kitts; the upgrading of neglected botanical gardens on Nevis; cleaning the sea off Grenada; the creation of a marine park at Montego Bay, Jamaica; demonstrate that links between tourism and the environment can be positive. Conservation and tourism can not only be interdependent – the economic benefits of tourism create a strong motive for the existence of protected areas – but can also provide recreation provision for local people. The latter is one of the aims of the refurbishment of the Salt Pond near Speightstown in Barbados (Stancliffe 1997). Some new construction projects also begin to address environmental issues. The marina development at Port St Charles in Barbados may have transformed the coastline, thereby altering the natural environment, but there have also been attempts here to incorporate a range of measures that should safeguard the area in the future. These include the construction of an efficient sewerage and effluent system to avoid seawater contamination and measures to protect nesting turtles (Miller and Miller 1997). Such efforts are being acknowledged and encouraged. Thirteen hotels have been identified across the Caribbean, from Antigua to Jamaica, by the International Hotels Environment Initiative and the Caribbean Hotel Association as having good environmental practice. This includes such elements as staff training; monitoring energy consumption; waste management; control of hazardous chemicals; links with local communities; keeping buildings in local styles; and purchasing policies (Elliott 1997).

Sources: Elliott 1997; Henderson 1994; Miller and Miller 1997; Momsen 1994; Weaver 1991.

Indies. Changes like this, which are incremental in nature, increase the benefits of tourism in host areas. Many of the initiatives are private-sector-driven, although they are often carried out with the approval and/or support of government.

Governments can also act directly, via mandatory planning and policy measures, to achieve greater levels of sustainability for one or more of the interest groups involved in tourism. Bermuda and Bhutan, for example, have both restricted visitor numbers in order to sustain the industry in the long term and control its nature so as to reap maximum profits while minimising further adverse impacts on local people. Kenya and the USA have imposed visitor quotas to some national parks in order to try to protect the environment, thereby providing a more enjoyable experience for visitors.

As an example, the tourist industry in New Zealand clearly illustrates that, although there are still problems, such as the size of the market, remoteness from main markets, a poor awareness of existing plans by local people, and noise in remote areas, efforts to date do move towards achieving a greater level of sustainability in an industry vital to the economy. Built around the environment, the range of approaches used attempts to introduce good practice (see Table 23.4). In particular, as elsewhere in the world, planning at national, regional and local levels exists, or is encouraged, alongside experiments with visitor management techniques (Human 1997).

Is ecotourism sustainable?

Academics often assume (Mowforth 1993) that alternative *types* of tourism, like ecotourism, have many sustainable characteristics. It is described as small-scale, carefully planned, locally owned and managed, closely integrated with other sectors of the economy, such as agriculture, and supporting

Table 23.4 Examples of good tourism practice in New Zealand.

Practice	*Examples*
Some companies work with the community and have environmental programmes	Fiordland Travel
Historic buildings are conserved	Art Deco buildings in Napier
Department of Conservation has a visitor strategy for the natural areas it manages	
Visitor services provide education and are easily booked	Heritage trails, visitor centre information network
Environmentally friendly transport reduces energy uses and pollution	Shuttle buses to attractions

Source: Human 1997.

the indigenous culture and environment.

However, ecotourists are not a single homogeneous group. Table 23.5 illustrates the range of tourist types who consider themselves ecotourists. In practice, once ecotourism, initially involving 'rough' or 'specialist' types of tourist, begins to emerge in previously remote and underdeveloped areas, these become more widely known and more fashionable, and visitor numbers increase rapidly. Overseas developers move in, and the adverse effects of the industry begin to outweigh its benefits. Negative impacts are quickly apparent because of the relatively fragile nature of the environments in which ecotourism often occurs (Hailes 1991). Indeed ecotourism, on occasions, can become a precursor to traditional forms of mass tourism. 'Smooth' ecotourists who undertake safaris would typify this situation, in which small-scale expeditions in East Africa during the early post-war period have been replaced in national parks like Masi Mara and Amboseli by package tours run by multinational operators. Such tourists might be more appropriately described as mass tourists rather than those who seek ecotourism as a variant of alternative tourism. The large-scale safari enterprises in which they participate have led to increased damage to the flora and fauna of these

Table 23.5 Types of ecotourist.

Types	*Travel group size*	*Arrangements*	*Mode at destination*	*Accommodation*	*Food*
Rough	Individuals and small groups	Independent	Local buses, foot	Cheap, locally owned	Local
Smooth	Groups	Tours	Taxis	3–5-star hotels, multinationally owned	Luxury restaurants. Much use of imported food and beverages
Specialist	Individual or specialised group	Often independent	Wide range used	Wide range used	Wide range used

Source: After Mowforth 1993.

Box 23.4 Ecotourism in Belize

Ecotourism in Belize emerged in the 1980s and 1990s in response to a demand from those living in advanced industrial countries for new and more remote holiday destinations that offered relatively untouched natural environments and cultures. Belize had the resources to fulfil this demand, including a spectacular barrier reef, scenic cays, tropical rain forest, and a Mayan cultural history.

During the first phase of development, existing dwellings were often enlarged to provide tourist accommodation. These locally owned and managed establishments added diversity to a healthy fishing economy. However, by the mid–late 1980s, foreign capital was introduced to build resort-style hotels in areas like Ambergris Cay. Tourism growth was dramatic as visitor numbers more than doubled from 1985–1990, and tourism became an important revenue earner.

At this stage, a discrepancy began to emerge between the aims and image of tourism as projected by the government, and the reality of the situation on the ground. Official sources promoted the concept of ecotourism, and at the Rio Earth Summit in 1992 Belize was praised for pursuing 'conservation and therefore ecotourism' (Godfrey, quoted in Munt and Higinio 1993: 13). The government claimed that it was emphasising sustainable tourism through measures such as the establishment of reserves to maintain attractions, e.g. Hoi Chen Marine Reserve, and efforts to encourage the involvement of local communities, e.g. the Community Baboon Sanctuary at Bermudian Landing.

Yet in practice there are high levels of foreign ownership, foreign exchange leakages and environmental degradation. Examples include the building of all-inclusive hotels, golf courses and polo fields, which involve loss of income for local people and the creation of structures alien to indigenous culture and society. Three-quarters of the land set aside for these and similar facilities in the early 1990s was to be transferred to foreign developers, although protests delayed this move. Even the Belizean Tourism Industry Association was largely composed of expatriates.

These contradictions question whether a truly alternative form of tourism has emerged in Belize, or whether this so-called ecotourism is merely a precursor to more traditional forms of mass tourism activity.

Sources: Pearce 1989; Munt and Higinio 1993.

national parks (Lea 1988) and have trivialised the society and culture of people like the Maasi (Olerokonga 1992). Similarly, journeys of exploration by travellers in southern Spain, like Laurie Lee (1971) and Penelope Chetwode (1985) in the early 1960s, helped to open up inland Andalusia, where mass operators like Thomson now run coach tours. One of the most striking examples of the way in which a 'theoretically correct' ecotourism can become a precursor to a more damaging, less sustainable yet more fashionable and popular form of ecotourism is that of Belize (see Box 23.4).

Nevertheless, examples of long-established, relatively successful forms of ecotourism do exist. One of these, which attracts 'rough' and sometimes 'specialist' ecotourists, is associated with wilderness holidays in the United States. Small numbers of backpackers travelling on foot or by canoe are allowed into federally owned and protected forest wilderness areas, or the back country in some national parks, on a strict quota system. Quotas are set at levels that avoid damage to the pristine environment and also allow visitors to enjoy an experience of isolation in an untamed wilderness setting, such as parts of Yosemite National Park. No facilities are provided within the wilderness areas, although the local population near entry and exit points gain some economic benefit from people who holiday within these areas. However, numbers are too low to swamp such communities, thereby avoiding the perils of negative social and cultural impacts.

CONCLUSION

One of the world's leading industries, tourism is based on the annual temporary migration of millions of people. Traditionally, they travel from major urban industrial parts of the developed world, where demand is focused, to more peripheral areas of supply. As incomes rise and political affiliations change, so new markets open up, like those in Eastern Europe. The effect of this temporary migration upon destinations is substantial and often damaging to the economy, society, culture and environment. This impact can be traumatic, as the popularity of destinations rises and falls in a cyclical manner (Butler 1980, and illustrated in Box 23.1) that is controlled by fashion and by the marketing policies of the multinational companies that dominate the tourism industry.

Attempts to reduce negative impacts have led to a search for more sustainable approaches towards tourism. But real-world examples show that, how ever desirable the concept, it is extremely difficult to develop a completely sustainable approach in practice. In addition, the idea of establishing new, sustainable types of tourism to stand alongside, or even replace, more traditional mass forms is unrealistic (Wheeller 1991). Nevertheless, some success has been achieved in gradually introducing a range of more sustainable measures to *all* types of tourism activity. These should be pursued in order to contain or even reduce damage to host areas, to increase visitor satisfaction and to achieve at least some of the aims of the industry.

So the introduction of sustainable approaches is a serious and urgent problem. As an interdisciplinary subject, its resolution will require inputs from a variety of fields of study. Nevertheless, geographers have an important role to play in researching and applying their knowledge and skills to many of the controversial and compelling issues that arise in the pursuit of more sustainable tourism activity. The distillation and application of best practice gathered from a wide variety of locations would be a useful starting point on this quest.

GUIDE TO FURTHER READING

Burns, P.M. and Holden, A. (1995) *Tourism. A New Perspective*. Hemel Hempstead: Prentice-Hall. A useful survey of tourism, including both its impacts and potential solutions, like sustainable tourism and planning.

Cater, E. (1994) Ecotourism in the third world: problems and prospects for sustainability. In E. Cater and G. Lowman (eds) *Ecotourism. A Sustainable Option?* Chichester: Wiley, 69–86. A thorough survey of ecotourism, with a wide range of case studies.

France, L. (ed.) (1997) *The Earthscan Reader in Sustainable Tourism*. London: Earthscan. An outline of sustainable tourism, with a wide-ranging collection of relevant readings.

Shaw, G. and Williams, A.M. (1994) *Critical Issues in Tourism. A Geographical Perspective*. Oxford: Blackwell. A good geographical approach to tourism issues, including impacts and mass tourism.

REFERENCES

Barke, M. and France, L. (1996) The Costa del Sol. In M.T. Newton (ed.) *Tourism in Spain. Critical Issues*, Wallingford: CAB International, 265–308.

Bird, C. (1995) Communal Land, Communal Problems. *In Focus* Summer (16), 7–8.

Burns, P.M. and Holden, A. (1995) *Tourism. A New Perspective*. Hemel Hempstead: Prentice-Hall.

Butler, R.W. (1980) The concept of a tourist area cycle of evolution: implications for management of resources. *Canadian Geographer* 24(1), 5–12.

Callaghan, P., Long, P. and Robinson, M. (eds) (1994) *Travel and Tourism*, second edition. Sunderland: Centre for Travel and Tourism, and Business Education Publishers.

Cater, E. (1994) Ecotourism in the third world: problems and prospects for sustainability. In E. Cater and G. Lowman (eds) *Ecotourism. A Sustainable Option?* Chichester: Wiley, 69–86.

Chetwode, P. (1985) *Two Middle-aged Ladies in Andalusia*. London: Century Publishing.

Christaller, W. (1964) Some considerations of tourism location in Europe. *Papers and Proceedings of Regional Science Association* 12, 95–105.

de Avila, A.L. (1996) First World Conference on Sustainable Tourism. *One Europe Magazine* update on Internet: January 24 1996 – Patrick.

Eber, S. (ed.) (1992) *Beyond the Green Horizon. Principles for Sustainable Tourism*. A discussion paper commissioned from Tourism Concern by WWF UK, Surrey: WWF UK.

Elliott, H. (1997) Hotels pass the green test. *The Times* Thursday, 5 June, 43.

Fermor, P.L. (1983) *Roumeli. Travels in Northern Greece*. Penguin: Harmondsworth.

France, L. (ed.) (1997) *The Earthscan Reader in Sustainable Tourism*. London: Earthscan.

Hailes, J. (1991) Ecotourism: a load of rubbish? *The Independent on Sunday* 28 April, 39.

Henderson, J. (1994) *The South-Eastern Caribbean*. London: Cadogan.

Hjalager, A.M. (1996) Tourism and the environment: The innovation connection. *The Journal of Sustainable Tourism* 4(4), 201–17.

Human, B. (1997) Sustainable tourism in New Zealand. *Tourism. The Journal of the Tourism Society* 94, Autumn, 14.

Krippendorf, J. (1987) *The Holiday Makers*. Oxford:

Butterworth Heinemann.

Lane, B. (1990) Sustaining host areas, holiday makers and operators alike. Paper to the Sustainable Tourism Development Conference, Queen Margaret College, November.

Lea, J. (1988) *Tourism and Development in the Third World*. London: Routledge.

Lee, L. (1971) *As I Walked out one Midsummer Morning*. Penguin: Harmondsworth.

McNeill, L. (1997) All-in holidays cause upset. *The Times* Thursday, 28 August, 21.

Mathieson, A. and Wall, G. (1982) *Tourism. Economic, Physical and Social Impacts*. London: Longman.

Miller, K. and Miller, S. (1997) *The Ins and Outs of Barbados*. St James, Barbados: Miller Publishing Co.

Momsen, J. (1994) Tourism, gender and development in the Caribbean. In V. Kinnaird and D. Hall (eds) *Tourism. A Gender Analysis*, Chichester: Wiley, 106–20.

Mowforth, M. (1993) In search of an eco-tourist. *In Focus* Autumn 9, 2–3.

Muller, H. (1994) The thorny path to sustainable tourism. *Journal of Sustainable Tourism* 2(3), 131–6.

Munt, I. and Higinio, E. (1993) Belize – eco-tourism gone awry. *In Focus* Autumn 9, 12–13.

O'Grady, R. (1981) *Third World Stopover*. Geneva: World Council of Churches.

O'Grady, R. (ed.) (1990) *The Challenge of Tourism*. Bangkok: Ecumenical Coalition on Third World Tourism.

Olerokonga, T. (1992) What about the Maasi? *In Focus* Summer 4, 6–7.

Pawson, I.G. *et al.* (1984) Growth of tourism in Nepal's Everest region: impact on the physical environment and structure of human settlements. *Mountain Research and Development* 4(3), 237–46.

Pearce, D. (1987) *Tourism Today. A Geographical Analysis*. Harlow: Longman.

Pearce, D. (1989) *Tourist Development*, second edition Harlow: Longman.

Shaw, G. and Williams, A.M. (1994) *Critical Issues in Tourism. A Geographical Perspective*. Oxford: Blackwell.

Smith, V. (ed.) (1977) *Hosts and Guests: An Anthropology of Tourism*. Philadelphia: University of Pennsylvania Press.

Squire, S.J. (1996) Literary tourism and sustainable tourism? Promoting Anne of Green Gables in Prince Edward Island. *The Journal of Sustainable Tourism* 4(3), 119–34.

Stancliffe. A. (1995) *Agenda 21 and tourism*. Unpublished mimeo.

Stroud, H.B. (1983) Environmental problems associated with large recreational subdivisions *Professional Geographer* 35(3), 303–13.

Vaughan, R. and Long, J. (1982) Tourism as a generator of employment: a preliminary appraisal of the position in Great Britain. *Journal of Travel Research* 21(2), 27–31.

Walsh, D. (1997) Noddy recruited by Redcoats in £139m Butlin's revamp. *The Times* Thursday, 4 September, 10.

Weaver, D. (1991) Alternative to mass tourism in Dominica. *Annals of Tourism Research* 18, 414–32.

Wheeller, B. (1991) Tourism's troubled times: responsible tourism is not the answer. *Tourism Management* June 12(2), 91–6.

World Commission on Environment and Development (1987) *Our Common Future*. Oxford: Oxford University Press.

24

Townscape conservation

Peter Larkham

> A historic city is essentially a product of the time and place of those who shape it and it is also a link between the past, the present and the future.
>
> (Ashworth and Tunbridge 1990: p. 28)

INTRODUCTION

The urban landscape, or 'townscape', in the sense of the cumulative layering in the majority of settlement locations of elements belonging to different historical and cultural periods, is one of the most common human experiences. It is difficult not to perceive, to interpret and to use this richness as important everyday occurrences, whether at the macro-scale of ready visual evidence, or in response to more subtle cues. These are familiar experiences of the majority of the population, certainly of Westernised industrialised countries, and for the occupants of the world's fastest-growing cities, in the developing world.

The production and maintenance of this physical fabric of settlements absorb a large amount of the wealth of the Western world in particular, and have done so for centuries, giving rise to the historic compositeness of the townscape. The landscape of historical settlements – most particularly urban ones, but the same is often true of smaller places, even rural villages – has rightly been described as a palimpsest. Strong cases have been made for the social, cultural and psychological significance of the townscape. Many studies have shown the need, in these terms, for the preservation of historical townscapes – at least in outward appearance. Yet this leads to tension and conflict. For there is also a widespread agreement that settlements must change, or they will stagnate. Adaptation of the townscape is necessary, but this is hard to achieve without some wastage of the investment of previous societies.

Urban geographers, in particular, have long investigated these phenomena. Townscape conservation is a rich area of study in applied geography (*cf.* Conzen 1975). This has led geographers to explore related fields of architectural and urban design; environmental perception and linkages between environment and behaviour; town planning, and in particular the development of related law, guidance and practice; development economics; and social and cultural relations. This is a complex field, defying attempts to simplify theory or practice.

Townscapes can be understood, using the approach of urban morphology, as complexes of street patterns, which are extremely conservative, changing very infrequently; plot patterns, rather more subject to change; and building structures, changing yet more frequently. Changes should be understood through identifying and examining the actors (individuals, institutions, corporate bodies, etc.) and processes (particularly planning and legal systems) involved. Together, these 'people and processes' represent a microcosm of the society and culture shaping a settlement at any one point.

ISSUES IN APPLIED GEOGRAPHY

This perspective can be applied to townscape

conservation in a number of ways but has been found to be relevant to the inclusivity of disciplines mentioned earlier, to all cultural and national circumstances thus far explored, and to all historical periods.

Problems in identifying areas for conservation

The identification of individual buildings or monuments, and of areas, that might merit conservation is a historico-geographical activity, and there are many examples of such work in that literature. Two key problems arise.

Areas and their boundaries

The delineation of area boundaries, in particular, draws on geographical concepts of 'area', 'character', locality/proximity, and identity. Yet, since conservation is a political activity (part of the planning process at local or national level), actual boundaries and designations do not always correlate to these geographical concepts. The boundary of Bradford's Little Germany area (Box 24.1), for example, was, for convenience, the new ring road.

Townscape conservation is not static; society's concepts of what it is acceptable to retain, and the values placed upon these monuments and areas, change. In the UK, this can be seen with the acceptance of particular architectural and morphological periods as conservation-worthy, with the consequent foundation of related pressure groups:

- Ancient Monuments Society (1921)
- Georgian Group (1937)
- Victorian Society (1958)
- Thirties Society (1980)
- Twentieth Century Society (renaming of Thirties Society) (1992).

The types of area seen as suitable for designation

Box 24.1 Industrial heritage problems: Little Germany, Bradford, UK

Bradford, 13 km west of Leeds, is an industrial city suffering from economic decline, a problem clearly visible in the physical fabric of older quarters of the city centre. The industrialising of the woollen industry in the nineteenth century demanded large new warehouses, many of which were built close together on an 8 ha site adjoining the town centre between 1860 and 1874: with the strong German connection, this district became known as 'Little Germany' (Figure 24.1).

A comprehensive redevelopment scheme began in the city centre from the late 1950s through to the 1970s, although several key late nineteenth-century public buildings were retained. This scheme did not encroach upon Little Germany, although it, together with vague plans for an inner ring road, blighted part of the area. The main perceived threat to the area was from commercial developers purchasing buildings and demolishing them to provide surface-level car parks – future development sites. The changing nature of the textile industry, and the decline of manufacturing industry in general in this part of the country in the late twentieth century, led to the redundancy of many of these large, bricks and stone-built, five- or six-storey buildings. No suitably extensive uses were available, and the area became run-down and neglected. Yet it is well positioned, immediately adjoining the city centre, although the recent completion of the inner ring road has isolated Little Germany from other quarters of the industrial town.

Conservation area designation, and the collapse of the property market in 1973, helped to ward off threats from commercial developers. The area was designated as 'outstanding' in the mid-1970s, which allowed applications for grant aid from central government. The local planning authority (LPA) was then concerned to retain the confidence of the remaining occupiers and to prevent further neglect and demolition. By the early 1980s, fifty-five of the eighty-eight buildings in the area had been listed: England's highest concentration of protected Victorian industrial structures. By the same time, however, about 50 per cent of Little Germany's floor area was vacant – although most buildings were in use as the vacant space was concentrated into the largest buildings.

In March 1982, much of Little Germany was declared a commercial improvement area 'to improve the appearance of old industrial and commercial properties in the inner cities for those who work, live, visit and pass through them . . . to help improve the image of the city and district'. In 1985, the LPA and the English Tourist Board felt that a more pro-active approach, going further than traditional physical planning, was needed. They commissioned URBED to produce a strategic direction for the revitalisation of Little Germany. A new public open space was created in Festival Square, and a Little Germany Festival was inaugurated in 1986. Further initiatives in the late 1980s included a considerable extension of the environmental improvements, particularly to pavement and street surfaces, and including new street furniture. Central government grant aid for the repair and restoration of various properties was offered: the LPA's commercial improvement area and sites and premises schemes both gave priority to Little Germany in this period.

In 1989, the Council accepted that Little Germany had become 'a major asset in the marketing and presentation of the district'. However, having encouraged private sector involvement through grants and carrying out environmental works, and having completed several key projects, the LPA reassessed its priorities and decided to complete existing commitments to public works but to withdraw from further high-profile activities in the longer term to encourage the area's self-sufficiency.

This case shows several features, principally revolving around problems of large concentrations of industrial heritage buildings in a small area; of the need for large grants to make conversions financially viable; and of the competing demands for limited public funding. Stone cleaning and paving improvements changed the area's appearance; although there were some public protests that the very process of cleaning had 'removed some of the city's industrial heritage'. Some of the early work in regenerating confidence, particularly by the local business association, had little impact as companies tended to look after their own interests, particularly in the recession of the late 1980s and 1990s. Property speculation, forcing prices up and encouraging site acquisition for financial investment reasons rather than refurbishment and reuse, also had an adverse impact.

Although it is recognised that regeneration work in this area has not been completed, attention and finance has moved elsewhere in the city. New government grant regimes are targeted elsewhere, as is English Heritage's partnership scheme for conservation area finance. Nevertheless, the injection of significant amounts of public money into this area over a decade has led directly to visible improvements to buildings and streets; and to increased occupancy rates by a diversity of new commercial and residential owners and tenants. The area is marketed as part of the city's heritage and has its own cultural attractions.

Note: This case was researched for the EU *Compendium of Spatial Planning Systems* and draws heavily on unpublished material from Bradford MBC and English Heritage. See also Falk (1993) and Tiesdell *et al.* (1996).

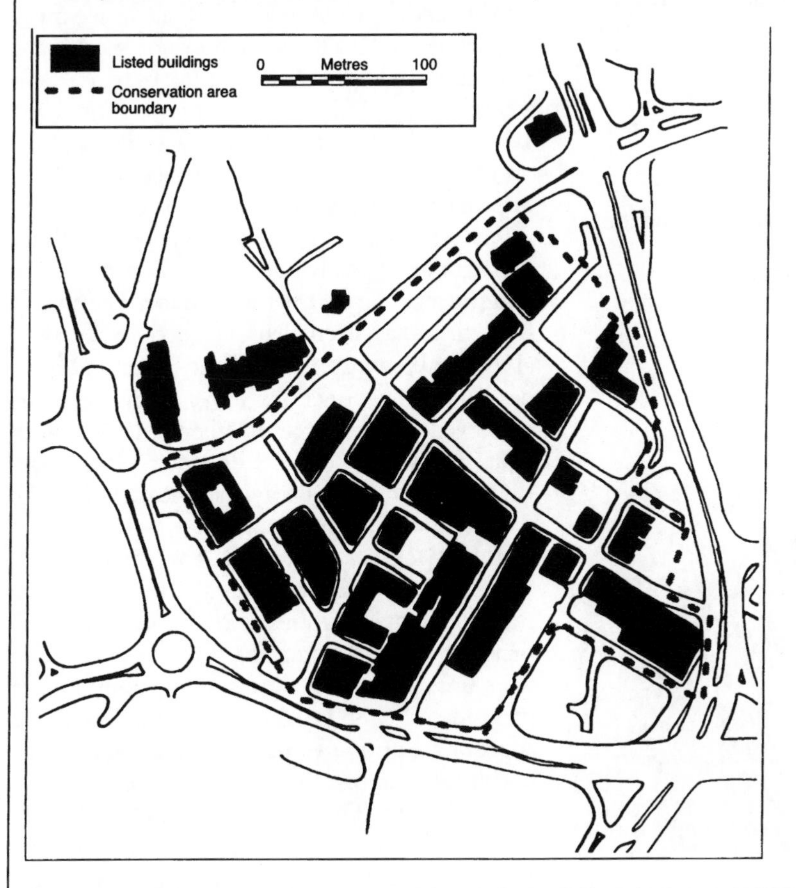

Figure 24.1 Little Germany conservation area, Bradford.

Source: Redrawn from information supplied by Bradford Metropolitan Borough Council.

have also changed over time. Originally, most societies concentrated on historic (i.e. mediaeval and earlier) town centres. Industrial areas became recognised following the rising academic and lay interest in industrial archaeology and architecture; and the much larger scale of industrial buildings and areas brings new problems for the funding of townscape conservation and the finding of new uses for redundant structures (Box 24.1). Suburbs are now increasingly popular designations in both the UK and USA; Ames (1999) relates this development to the change of direction in architectural history away from elite or 'high' styles towards vernacular architecture. But conservation of such familiar townscapes causes much controversy (Plate 24.1).

The geographical and temporal patterns of conservation area designations, and their potential causes, from national-level statute and government advice to local politics and events, can be reviewed (Larkham 1996). Knowing the trends in the types of area being designated is also useful. More, however, needs to be done on the social, economic and physical impacts of designation (*cf.* Gale 1991).

Plate 24.1 Inter-war speculative semi-detached suburbia conserved: Hall Green, Birmingham, designated in 1988 and featured in *The Times*.

Character

An enduring element within townscape conservation is consideration of the 'character' of the area to be conserved (see Plate 24.2). In the UK, this is enshrined in the statutory definition of the conservation area as one 'of special historical or architectural interest, the character or appearance of which it is desirable to preserve or enhance' (1967 Civic Amenities Act). This wording is significant in explicitly separating 'character' from simple 'appearance'.

The examination of 'character' lends itself to geographical approaches. Straightforward issues of physical form may be examined, for example using the tools and concepts of urban morphology. Other relevant factors include the history and development of the area, which might in part be revealed through morphological analysis, and its past, present and future uses. The relevance of use and activity patterns should not be underestimated: increasing car dependence and the need for on- and off-street parking has affected the character of many suburban conservation areas, and Ludlow has changed with the conversion of many Georgian town houses to antique retailing for the tourist trade. The change over time has been demonstrated by Buswell (1984) in Newcastle upon Tyne, as central business and retail functions move away from the preserved historic core: if other functions with equal investment powers do not replace them, then the core will decay, notwithstanding the conservation designation.

Box 24.2 gives current UK government guidance on area character. But the problem with traditional investigations of character is twofold: first that, in many countries, planning authorities were unsystematic in their investigations and, in many cases, simply made designations without appropriate prior investigations. This may lead to problems in subsequent planning and management. Only within the last decade in the UK has the need for suitably detailed character assessment become accepted. Second, the traditional methods of description implied 'that character can be identified, even perhaps measured and

Plate 24.2 Bamberg, Germany: market square. Character relates as much to uses as to physical form. Here, the lively market use and the complete built form of the square draw attention away from the unsympathetic Herte shopfront (left).

quantified; and that whatever constitutes character is readily identifiable to, and agreed by, different groups in society' (Larkham and Jones 1993: p. 399). Environmental and behavioural psychology shows that different groups in society have very different reactions to familiar and historic townscapes. In reviewing this literature, Hubbard (1993) examines the dangers of basing conservation purely on architectural and/or historical criteria, since such approaches largely ignore the key role that townscapes play in maintaining cultural identities. The challenge, then, is to integrate 'traditional' historico-geographical approaches to area character with explorations of users' and residents' perceptions, and with culture.

Box 24.2 Defining 'character' as applied to UK conservation areas: extract from current guidance

The distinctiveness of a place may come from much more than its appearance. It may draw on other senses and experiences, such as sounds, smells, local environmental conditions or historical associations, for example those connected with particular crafts or famous people. The qualities of a place might change from daytime to night. Such elements of character can be identified, but not directly protected or controlled. By defining and protecting the tangible, such as buildings and the spaces formed between them (streets, squares, paths, yards, and gardens), the activities and uses that make up the special character of a place can be sustained. Effective physical conservation measures should be rooted in firm land use policies in an adopted development plan.

Most of the buildings in a conservation area will help to shape its character in one way or another. The extent to which their contribution is a positive one depends not just on their public face, but on their integrity as historic structures and the impact they have in three dimensions, perhaps in an interesting roofscape or skyline. Back elevations can be important, as can side views from alleys and yards.

In a large conservation area, or one where its development spans a considerable period, the character may vary greatly within its boundary. For example, a small market town may have a medieval core, focused on a market place or church, then a Georgian phase of development of grander houses and formal streets, followed by the railway, and eventually by modern housing at the edges and on gap sites. Where the character is composite in this way and the phases of growth are clear, it will often be worth analysing them separately.

Elsewhere, rebuilding may have taken place many times over the same sites, resulting in overlays of building forms which are often contained within an ancient framework. The richness of an area today may thereby reflect the build-up of successive historic periods.

(English Heritage 1997: pp. 2–3; Relevance to the approach of urban morphology is evident)

Capacity

'The fundamental planning problem facing historic cities is the tension between the need to conserve the physical fabric of the city (both its core and its setting) and the demands of the activities currently taking place within it or attracted to it' (Ove Arup *et al.* 1994: p. 6). How much capacity for continued growth do historic centres have, particularly in the face of continued pressures for business and retail use in the central business district (CBD), suburban expansion and intensification, and recreation and tourism – and all of the transport needs that such uses generate?

Pioneering research in Chester (*ibid.*: p. 14) has developed a methodology to explore these issues for historic centres. It explores the concept of 'carrying capacity' in terms of

- *physical capacity*: the amount of space available for an activity;
- *ecological capacity*: the ability of the space to absorb the uses; and
- *perceptual* (or *behavioural*) *capacity*: estimates of capacity in terms of personal satisfaction.

The approach first carries out a range of quantitative technical assessments of key indicators, including emission levels, traffic flows, noise levels and pedestrian density, and comparing them to tolerance thresholds. Second, a range of perception studies of local residents and organisations, using a variety of techniques, gains qualitative data on what is liked or not liked about the city. Combining both approaches produces a capacity framework. This may then be tested against a range of scenarios showing the different long-term ways in which the city could develop and function, resulting in planning and management guidelines. However, as Strange (1997) concludes, despite widespread interest the Chester example is the only such study to date; and the wider applicability of the capacity concept, and its use in generating sustainable development policies, is not yet completely clear.

A closely related theme, developed more in the USA, is that of 'growth management'. State and local ordinances may stipulate that settlements identify 'urban growth boundaries', outside which no new urban development is permitted. Not only does this help to protect 'the wider historic landscape' from continued sprawl and 'edge cities', but this tactic concentrates the tax dollar within the city and thus permits expenditure on more traditional preservationist concerns.

Culture and conflict

Culture is central to issues of conservation and heritage. Cultural processes through time produce the urban landscapes that we choose to conserve – itself a cultural phenomenon. In this sense, townscape conservation reflects many of the concerns of contemporary cultural geography in that culture is seen as a major, rather than residual, factor in social (especially urban) life; and at issue is the nature of relationships between the cultural, economic and political processes leading to the production, use, representation and modification of townscapes. But one key problem is that cultures change through time: war, conquest, disease, exploitation, as well as periods of rapid cultural evolution such as the Industrial Revolution, are difficult at the time and result in a range of heritages and heritage choices. If 'history is written by the victor' then so, too, is heritage selected by and interpreted for the victor, conqueror, exploiter or survivor. Box 24.3 examines conservation in a post-colonial city, where there was much scope for conflict in the selection and interpretation of heritage – although, in the event, these conflicts have been minimised.

Townscape conservation and perception

Too often, 'arguments for the social, psychological and aesthetic significance of the conserved townscape are taken for granted and rarely addressed in any explicit manner. . . . As such, conservation policy is shackled by the stigma of subjectivity and is open to accusations of elitism' (Hubbard 1993: p. 361). Studies of those who actively participate in conservation, through membership of organisations of activities, overwhelmingly show them to be the

Box 24.3 Colonial heritage problems in Stone Town, Zanzibar

The city of Zanzibar has a complex urban form resulting from an extreme diversity of cultural origins. Early wooden shelters on irregular plots were replaced by coral limestone houses after Sultan Seyyid Said moved his administrative headquarters here in 1832. A range of imposing public and private buildings were constructed as the town grew on the fruits of the slave trade, together with trade in gold, ivory and spices, in the nineteenth century (Figure 24.2). The major cultural influences were Arab, Indian traders, Europeans and particularly the British following the closure of the slave market in 1873. The native Swahili and mixed-blood islanders, together with descendants of African slaves, were hardly represented in the Stone Town but occupied the neighbouring Ngambo district. Thus the Old Town contained no relics of the Swahili culture.

An African revolution in 1964 overthrew the sultan and resulted in the flight of the Arab and Indian population, severe economic decline, and the abandonment and decay of many of the Stone Town's buildings. Some were confiscated by the new government and used to rehouse people from the temporary slum buildings of the Ngambo. But this increased population density had neither the skills to maintain stone buildings nor the money to hire those skills; the government was also unable to invest in the area. Buildings collapsed as joists rotted. Despite their cultural and colonial overtones, some of the public buildings were occupied by the new administration.

After fifteen years, the government became concerned at

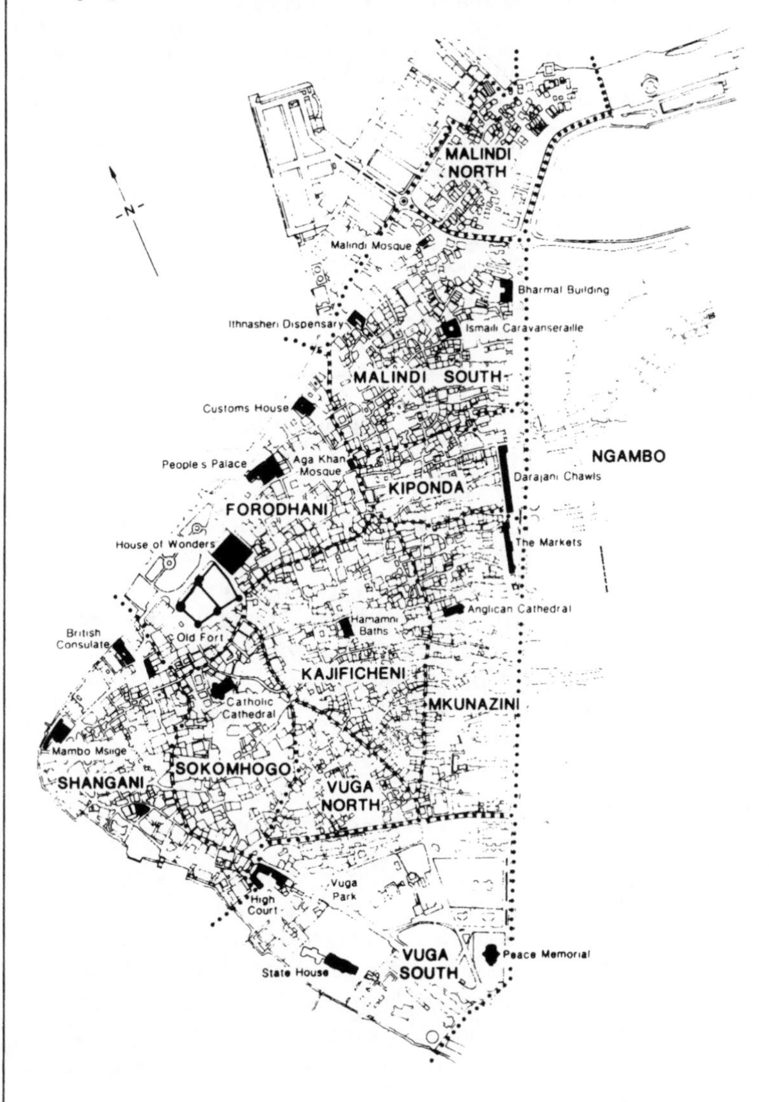

Figure 24.2 Zanzibar, showing key conserved buildings in black.

Source: McQuillan 1990, figure 18.3

Box 24.3 continued

the loss of housing stock and, with some United Nations support, decided

> to halt the collapse of housing, revitalize the local economy and so preserve the urban patrimony of the Stone Town as a testament to the diverse origins of the population of Zanzibar, notwithstanding the recent painful memories which many of the structures evoked.
> (McQuillan 1990: p. 405)

Only six structures – all public buildings – had been protected in 1979. Concerted preservation action was, in practice, almost a by-product of solutions to the housing problem. The collapse of ill-maintained property was seen as a needless waste of resources; the United Nations' involvement arose from concern (outside Zanzibar itself) over the loss of one of the most distinctive urban heritages in sub-Saharan Africa.

A comprehensive redevelopment plan, with the creation of a historic district and an independent authority responsible for preservation and rehabilitation, was devised (LaNier and McQuillan 1983). As part of its innovative mixture of preservation and housing improvement, residents of government-owned buildings in danger of collapse could purchase them for a small sum (approximately 25 per cent of market value), provided that the property was renovated to agreed standards specific to each building within two years of purchase. By 1990, over 100 buildings had been transferred; the government had been relieved of the problems of tenancy and maintenance, and had derived some income, which was directed into repair of other historic property.

However, funding is a major problem, and overseas aid has been responsible for major support to restore landmark buildings: with grants of $400,000 from the United Nations Development Programme and $600,000 from the European Community (McQuillan 1990: p. 411). The rate of building collapse has declined dramatically, and in 1988 the Stone Town was officially designated as a conservation area.

Despite this apparent success, Marks (1996) shows significant contradictions and ambiguities, not least that tourism (and the need for tourist-generated overseas income) can 'simultaneously preserve individual notable buildings while destroying many of the town's fragile social and cultural networks and much of the urban fabric they inhabit . . . resulting in sporadic gentrification of the Stone Town and marginalisation of the poor'.

Thus the restoration of individual buildings and public spaces has sanitised the past, with the vision being that of a 'white-washed paradise . . . a caricature of Zanzibar culture' where there is no space for informal trade. 'Privately and publicly owned "monuments" . . . are rehabilitated into a series of tourist preserves, where a reified notion of "tradition" and "culture" is put on display' (Marks 1990: p. 274). Informal and illegal encroachments upon public and private land increase as the demand for space rises, sanctioned by some high-level politicians. The character and appearance of the Stone Town are changing significantly and quickly.

Sources: LaNier and McQuillan 1983; McQuillan 1990; Marks 1996.

wealthy, articulate, educated, middle classes (e.g. Lowe and Goyder 1983). They, of course, work for a particular selection of conservation, particularly in the third world and former colonies, where, until the last decade or two, it has been relatively rare to see townscapes of, or the contribution of, the indigenous population preserved. Their view has been that the indigents were not civilised, not urbanised – or not there. They have particular perceptions of what should be conserved, and how.

For the most part, among the active elite or the passive majority (who nevertheless consume commodified heritage), there is support for the past, for the qualities of past places, and for their replication in conserved modern townscapes. But, as Hubbard (1993) again reminds us, different groups in society do have different values and attitudes, and these can be revealed with sufficiently sensitive techniques. Education and age appear to be significant variables, and there is a distinct gap between built environment professionals and the lay public – for example in the attitude towards historical authenticity, as reflected in façadism and pastiche replication. Strange (1997) argues that the public are sensitive to authenticity, perceive losses to the non-renewable conserved townscape resource, and would not choose to visit a replication: arguably, the popularity of the National Trust's Uppark House, rebuilt after its 1989 fire, may disprove this.

Conservation and remaking places

Heritage has become increasingly used in the marketing of products and, especially relevant in the current context, places (see Plate 24.3). Both case studies illustrate this use, in widely differing cultures and locations. Heritage images are widely used in place promotion, even for places with a short, or contested, history; and a sub-set of

Plate 24.3 Łód'z, Poland. Main street pedestrianised in 1993. Removal of original surfaces and tramway, new street furniture. Place marketing here is drawing away from the 'heritage' feel.

'tourist-historic' cities has been identified (Ashworth and Tunbridge 1990). Regeneration efforts in a wide range of neglected urban quarters across the world have used heritage as a key component of the place-marketing and revitalisation strategies, and this can clearly be seen in Bradford's Little Germany, Birmingham's jewellery quarter (see Box 24.1; Tiesdell *et al.* 1996) and Cape Town's waterfront district. Yet criticism surrounds the selectivity of the heritage, which excludes aspects of local heritage deemed 'unsalable' to tourists or investors (Kearns and Philo 1993), and its sanitization, for example in pseudo-historicist street furniture and enhancement schemes (Booth 1993). But such is the competitive nature of contemporary place-marketing that such questioning is seen as unwelcome, even traitorous.

In the UK and other Western countries, de-industrialisation has led to a growing reliance on service sector industries, of which heritage tourism is very significant. In many countries in the developing world, heritage tourism is also a great supporter of the economy; but while the first world commodifies its heritage, often for internal consumption, the third world relies on producing attractions for overseas tourists.

CONCLUSION

In his early study of townscape conservation as applied geography, Conzen (1975: p. 83) concluded by asking three questions: 'What is the social purpose of conservation? What are the dangers threatening the conservation of historical townscapes in Europe? What is the general nature of conservation?' All three of these questions are now much better understood through the great volume of work published since that date – much of it undertaken by geographers, and much of it closely connected with the evolving concerns of the discipline of geography. Nevertheless, there is still an evident lack of a widely accepted ethic or philosophy of townscape conservation, and this is a key area for future development.

Much more is known of the history and development of conservation thought, legislation and practice (for example, see Delafons (1997) for an insider's view of the UK system). Quantitative information on numbers and types of preserved monuments and areas is being supplemented by qualitative data on perceptions, reactions and uses. Most tellingly, the concept of 'conservation', having developed from 'preservation', has itself now developed into 'heritage' (Ashworth 1994). Studies have suggested that, although the heritage concept could be argued to have popularised conservation, its ideologies have clearly restricted choice and freedom. The power of heritage selection and promotion is vested in powerful elites, whether multinational corporations or municipal authorities. Little effective consultation with local groups takes place. The commodification of heritage is a further part of the impact of the capitalist system on the built environment.

Where next? First, developments in technology continue apace and may be harnessed in the study of the conserved townscape. For example, recent developments using GIS and computer analysis have considerably assisted the 'layering', representation and analysis of such morphological, historical and archaeological data (Koster 1998). Second, cultural issues of conservation and heritage are likely to become even more significant, particularly involving groups dispossessed of their heritage, marginalised in heritage planning, or otherwise involved in dissonant or contested heritage problems. Third, a development of the previous point, conservation of places 'in the periphery', outside the North Atlantic axis, where most work of the last two decades has focused, will become more prominent (*cf.* Shaw and Jones 1997). Fourth, conservation worldwide will become more sharply focused on the issues of resource efficiency, growth management and capacity. In short, these are elements of what has become *the* contemporary key word in planning, 'sustainability'. Awareness of finite world resources, and the implications for continued unfettered development, will impinge upon conservationist concerns, and not always to the good. The 'compact city' concept, for example, may mean large-scale urban redevelopment at much higher densities in order to both minimise resources wasted in travel and productive land occupied with urban sprawl; but the consequences for valued urban landscapes may be severe. Fifth, academic concepts such as Conzen's more sophisticated concept of 'historicity' (1975) could be used to refine practice-based and theoretical approaches to issues such as area character.

In short, the future for geographical views of townscape conservation is likely to be one of increasing work, addressing the 'elitist' criticisms levelled at conservationism in recent years and the cynicism of the heritage industry, through broadening views on peripheral conservation, cultural contributions to difficult heritages, and the rising relevance of sustainability.

GUIDE TO FURTHER READING

There are many excellent texts that focus on single structures, localities or countries, or that adhere to narrow views of preservation or conservation. Those listed here take wider, more inclusive perspectives, developing themes introduced in the chapter.

Ashworth, G.J. and Tunbridge, J.E. (1990) *The Tourist-Historic City*. London: Belhaven. Developing the concept of the 'tourist-historic city', bringing together issues of place-identity and marketing, planning, and developing key ideas of 'heritage'.

Ashworth, G.J. and Larkham, P.J. (eds) (1994) *Building a New Heritage: Tourism, Culture and Identity in the New Europe*. London: Routledge. Chapters deal with a range of applications of conservation and heritage, particularly to issues of European nationalism (historical and as re-emerging in the late twentieth century), dissonant heritage, the role of architecture and design, tourism, and the consumption of conservation and heritage.

Graham, B.J., Ashworth, G.J. and Tunbridge, J.E. (1999) *A Geography of Heritage: Power, Culture and Economy*. London: Arnold. This is a significant geographical overview of the heritage/conservation field, focusing on the nature of heritage, the political, social and economic contexts, and the use of heritage in place management.

Larkham, P.J. (1996) *Conservation and the City*. London: Routledge. A recent morphological perspective on conservation and townscape change, drawing explicit links with architectural, planning and urban design practice, and beginning to develop an international dimension.

Shaw, B.J. and Jones, R. (eds) (1997) *Contested Urban Heritage: Voices from the Periphery*. Aldershot: Ashgate. Useful collection focusing on contestation, heritage and culture; and in a geographically neglected peripheral area, away from the usual North Atlantic axis.

Whitehand, J.W.R. (1987) *The Changing Face of Cities*. Institute of British Geographers Special Publication 21, Oxford: Blackwell.

Whitehand, J.W.R. (1992) *The Making of the Urban Landscape*. Institute of British Geographers Special Publication 26, Oxford: Blackwell. Taken together, these books provide an overview of urban morphological approaches to studying the changing urban landscape and concepts of management – of which conservation is a part. Case studies are UK-based but include town centres, residential areas and institutional developments.

REFERENCES

Ames, D.L. (1999) Understanding suburbs as historic landscapes through preservation. In R. Harris and P.J. Larkham (eds), *Changing Suburbs: Foundation, Form and Function* London: Spon, 222–38.

Ashworth, G.J. (1994) From history to heritage: from heritage to identity: in search of concepts and models. In G.J. Ashworth and P.J. Larkham (eds)

Building a New Heritage: Tourism, Culture and Identity in the New Europe, London: Routledge, 13–30.

Ashworth, G.J. and Tunbridge, J.E. (1990) *The Tourist-Historic City*. London: Belhaven.

Booth, E. (1993) Enhancement in conservation areas. *The Planner* 79(4), 22–3.

Buswell, R.J. (1984) Reconciling the past with the present: conservation policy in Newcastle upon Tyne. *Cities* 1(4), 500–14.

Conzen, M.R.G. (1975) Geography and townscape conservation. In Anglo-German symposium in applied geography, Giessen-Würzburg-München, 1973, *Giessener Geographische Schriften*, 95–102 (reprinted in J.W.R. Whitehand (ed.) (1981) *The Urban Landscape: Historical Development and Management. Papers by M.R.G. Conzen*, London: Academic Press).

Delafons, J. (1997) *Politics and Preservation*. London: Spon.

English Heritage (1997) *Conservation Area Appraisals*. London: English Heritage.

Falk, N. (1993) Regeneration and sustainable development. In J.N. Berry, W.S. McGreal and W.G. Deddis (eds) *Urban Regeneration: Property Investment and Development*, London: Spon, 161–74.

Gale, D.E. (1991) The impacts of historic district designation. *Journal of the American Planning Association* 57(3), 325–40.

Hubbard, P. (1993) The value of conservation: a critical review of behavioural research. *Town Planning Review* 64(4), 359–73.

Kearns, G. and Philo, C. (eds) (1993) *Selling Places: The City as Cultural Capital, Past and Present*. Oxford: Pergamon.

Koster, E. (1998) Urban morphology and computers. *Urban Morphology* 2(1), 3–7.

LaNier, R. and McQuillan, D. (1983) *The Stone Town of Zanzibar: A Strategy for Integrated Development*. Nairobi.

Larkham, P.J. (1996) Designating conservation areas: patterns in time and space. *Journal of Urban Design* 1(3), 315–27.

Larkham, P.J. and Jones, A.N. (1993) The character of conservation areas in Great Britain. *Town Planning Review* 64(4), 395–414.

Lowe, P.D. and Goyder, J. (1983) *Environmental Pressure Groups in Politics*. London: Allen & Unwin.

Marks, R. (1996) Conservation and community: the contradictions and ambiguities of tourism in the Stone Town of Zanzibar. *Habitat International* 20(2), 265–78.

McQuillan, A. (1990) Preservation planning in post-colonial cities. In T.R. Slater (ed.) *The Built Form of Western Cities*, Leicester: Leicester University Press, 394–414.

Ove Arup & Partners, M. Breheny, Donald W. Insall & Associates and DTZ Debenham Thorpe (1994) *Environmental Capacity: A Methodology for Historic Cities*. London: English Heritage.

Shaw, B.J. and Jones, R. (eds) (1997) *Contested Urban Heritage: Voices from the Periphery*. Aldershot: Ashgate.

Strange, I. (1997) Planning for change, conserving the past: towards sustainable development policy in historic cities? *Cities* 14(4), 227–33.

Tiesdell, S., Oc, T. and Heath, T. (1996) *Revitalizing Historic Urban Quarters*. Oxford: Butterworth.

Part III

Challenges of the human environment

25

Urbanisation and counterurbanisation

Tony Champion

Changes in the distribution of population constitute a primary focus for geographical investigation, both theoretical and applied. The single most important dimension at global level continues to be the process of urbanisation, with the proportion of the world's population living in urban places rapidly approaching the 50 per cent mark (UNCHR 1996). At the same time, particularly in more urbanised countries but also in some parts of the developing world, there is clear evidence of the largest cities losing population to smaller urban centres, as well as of a wider dispersal process that has produced a rural population turnaround, sometimes referred to as counterurbanisation (Champion 1989). While academics are still locked in argument about the significance of the latter for the future evolution of settlement patterns, there is no doubt that these centrifugal population shifts can have just as impressive an impact on people and places as is already well documented for the urbanisation process, nor any doubt that, while these impacts are generally positive in their effects on human welfare, they may also generate problems, which policy makers attempt to tackle. This chapter examines the main problems caused by both urbanisation and counterurbanisation and gives examples of ways in which research on the nature and causes of these processes can help towards curbing their less desirable consequences.

BACKGROUND

Before going into detail about the problems caused by these two types of population shift, it is essential to provide some background to what they involve on the ground, for as with many words ending in '-isation', things are usually more complex than they seem at first glance. At its most basic, urbanisation can be defined as a process of population concentration, the main result of which is an increase in the proportion of the population living in urban places. Additionally, however, it is associated with the faster growth of the larger urban places or, in more technical terms, a positive correlation between growth rate and settlement size. Meanwhile, counterurbanisation represents – for most people, including the originator of the term (Berry 1976) – the direct antithesis of urbanisation and thus a process of population deconcentration, yet it is rarely associated with a diminution in the proportion of people considered urban and is instead more commonly seen in terms of a redistribution from larger urban places to smaller ones, together with the outward expansion of individual urban centres into the surrounding countryside (for a review of the problems of studying counterurbanisation, see Champion 1998a).

Leading on from this, a second aspect needing clarification concerns the direct causes of these population shifts. If asked about this, most people would point to migration as being the key element in raising the level of urbanisation, and indeed this was undoubtedly true in the nineteenth century, when death rates in the industrial cities were very high and urban growth was possible only because of strong inward migration from rural areas or overseas. In theory, however, there are two other sources of urbanisation: surpluses of births over deaths and reclassification of places from rural to urban. It is estimated that in recent years, almost three-fifths of the world's urban population growth has been the result of natural increase alone, with a significant part of the remainder being caused by rural settlements achieving urban status or being absorbed into expanding cities (McGee and Griffiths 1994).

Third, in relation to the migration element, it must be stressed that population redistribution is merely the net effect of both rural–urban and urban–rural movements. Even in the nineteenth century, there was evidence of strong two-way movements between places, leading Ravenstein to develop his rule that for every migration stream, there would be a counterstream, while in the present-day developing world context this phenomenon has become so intense that it is commonly referred to as 'circulation'. The same is true in situations of overall counterurbanisation, where the numbers of people moving down the urban hierarchy into less urban areas may be only marginally in excess of those moving up it into larger centres. The significance of this wider picture of population turnover looms much greater when there are differences in the characteristics of in-migrants and out-migrants. This is commonly the case in terms of age structure, with older people and families with children tending to form the rump of the counterurbanisers and with more school leavers and young adults moving to the larger cities (see Boyle *et al.* 1998 for a review of migration).

Nevertheless, the majority of studies that have looked at the problems associated with urbanisation and counterurbanisation have focused on the effects of net migratory gains on the destination areas and of net losses on the source areas, so this is where the main emphasis will be put in the rest of this chapter. What must be borne in mind, however, is that, first, the net migration picture is merely the result of much larger and more complex patterns of population turnover; and second, there are other processes at work in producing these population shifts besides migration.

IMPACTS OF URBANISATION

The centripetal movement of population from rural areas into much higher-density urban concentrations carries implications for both the zones of departure and the areas of reception. It needs to be stressed that, on the whole, the impacts are positive in nature. For one thing, these shifts are usually good for the national economy in that workers are switching from rural production activities – with generally low average productivity in a traditional urbanising society (especially in areas that have been 'overpopulated' in relation to available resources) – into factory-based and other urban activities characterised by higher output per person. The faster growth of larger cities than smaller urban centres reinforces this process, since it is these larger cities that gain greatest benefit from agglomeration economies. This movement is also good for most of the individuals involved, since they will normally be gaining from the higher wages that derive from that greater productivity and from access to a wider range of facilities and amenities than are normally available to those living in villages and small country towns. After all, although a proportion of migrants may be lured to cities by false expectations, in most cases the migrants find themselves better off in the cities or at least, if they reckon their moves to have been misguided, have the chance of returning to their rural origins.

Nevertheless, the urbanisation process is also associated with a range of negative effects for both the departure zones and the reception areas, which reduce the welfare of the residents of these areas and which tend to become more severe over time unless checked by appropriate countermeasures. As regards the rural areas, the main problem is that net out-migration will reduce the population to below the level that it would otherwise have been and may in the end lead to absolute population decline. While in the short term this may lead to the achievement of a more favourable balance between population levels and the local resource base, thereby reducing under-employment and pressures for the subdivision of family land holdings, it also lowers the ability of the local area to support community facilities. This has the effect of reducing the attractiveness of life in the countryside at the same time as the growing cities are providing an increasing range of amenities, including higher-order consumer goods, with the result that the pull of the city becomes that much greater and the incentive for further rural–urban migration stronger.

The impact of migration on the rural source areas is made more acute because of the selective nature of the process. This has several facets. Out-migrants are usually young adults for whom the countryside holds few job opportunities and is unable to compete with the 'city lights', with the result that the next generation of children is lost to the city and the rural population becomes progressively older. The rural exodus also tends to be skewed more heavily towards women than men, because rural production is primarily the domain of male labour, whereas women can gain access to a much wider range of jobs in cities, not only in domestic and other services, but also in some branches of manufacturing like textiles. This gender selectivity reduces the 'marriage market' for the men that stay behind in the countryside, with the result that fewer of these have families or else eventually succumb to the temptations of the city. Third, while rural out-migration can be bipolar in social terms and involve the landless and destitute as well as the more gifted and enterprising, it is normally the latter that dominate the outflows, whether moving to get further education or seeking work. This will tend to enhance the human capital of the city at the expense of the countryside – a process that will be reinforced by the return to their original community of migrants who have failed to achieve what they expected in the city.

Turning to the impacts on the reception areas, it can be deduced from what has just been said that urban areas are clear beneficiaries of the process, in that they gain from the arrival of young, enterprising and adaptable workers and from the 'multiplier effects' of their consumer demands and in due course of their biological reproduction. On the other hand, there are some important 'downside' effects, notably on existing urban residents and on the overall functional efficiency of the city. Put bluntly, the arrival of new residents in the short-term increases the competition for work and for the scarce resources of space, housing, food and other everyday needs, while in the longer term the city grows larger, developing at higher densities and expanding in area. The newcomers do not incur the full costs of their migration decisions, since the additional congestion is felt by all and the provision of the extra infrastructure (roads, drainage, schools and so on) is largely financed from taxes and charges levied on the whole community.

Further diseconomies arise in situations where the newcomers are significantly different in their characteristics from the longer-established urban population. In fact, it is usually the case that migrants from rural areas will arrive with few resources, given that they come from essentially poorer areas and that any family wealth will be invested in the land, let alone the fact that most will be at the start of their working lives. As a result, initially they can afford only the cheapest housing, which leads to the emergence of 'slum housing' areas either through the recycling of older housing or through the construction of 'shanty towns'. This social polarisation is reinforced if the newcomers are also distinctive in their cultural attributes, not merely being associated with a rural lifestyle but being drawn from different racial stock or indeed from different national origins. The 'ghettos' of cities in the USA resulted from the influx of black Americans from the rural South as well as from the immigration of the rural poor from southern and eastern Europe and more recently from Latin America and Pacific Asia. Perhaps the most extreme examples of poor living conditions in the city are those associated with people intending fairly temporary residence there, such as those engaged in some form of seasonal movement or 'circular migration' and those whose aim is to remain in the city only long enough to acquire enough capital to set up in business back home.

The two case studies illustrate well the nature, and indeed the complexity, of the urbanisation process and its impacts, as well as indicating some of the solutions that have been put forward to deal with the problems caused. Box 25.1 provides an example of the traditional process of rural–urban population movement: that of India, with its very large absolute increases in urban population in recent decades. Box 25.2 focuses on the phenomenon of the 'guest workers' recruited to northwest Europe in the third quarter of the twentieth century – a movement that was originally designed to be short-term but did not turn out that way, serving to demonstrate the potential for migration to be a 'self-feeding' process.

IMPACTS OF COUNTERURBANISATION

The centrifugal movement of population is a long-established phenomenon at local scale, where it is traditionally referred to as suburbanisation because of its domination by housing. Progressively over time, these shifts have been involving a wider range of urban functions and taking place over longer distances, as daily personal mobility has grown and urban centres have expanded to embrace their previously rural hinterlands. As a result, suburban centres and 'edge cities' have increasingly been

Box 25.1 Urbanisation and rural–urban migration in India

Although India still has a relatively low degree of urbanisation, with only just over one-quarter (26.1 per cent) of its population living in urban areas in 1991, its urban population has been growing rapidly in both absolute and relative terms over the past three decades, with the larger cities constituting a progressively larger share of the total (Table 25.1). Rural–urban migration has accounted for only about one-fifth of this growth, the remainder being due to the rapid rise in the number of settlements classified as towns and to the relatively high rate of natural increase in urban areas. Moreover, the rates of urbanisation and rural–urban migration both fell somewhat in the 1980s compared with the previous decade.

Table 25.1 Urban population growth in India, 1961–91.

Census year	*Number of towns*	*Total urban population (million)*	*Change from previous census*		*Proportion in towns with 100,000 and over (%)*
			million	*% p.a.*	
1961	2270	77.6	18.2	2.34	51.4
1971	2476	107.0	29.4	3.26	57.2
1981	3245	156.2	49.2	3.86	60.4
1991	3609	212.9	56.7	3.15	65.2

Source: 1991 Census of India, after Mohan 1996.

Indian cities share most of the characteristics and problems of rapidly growing cities in the developing world, particularly in relation to housing. Census data indicate that the quality of shelter *per capita* has declined over the last thirty years as measured by indices of overcrowding. Typically, the poorest of urban dwellers, as many as 30–50 per cent in most cities, live in dwellings that have been constructed by themselves or with the help of neighbours, friends and other locals, usually without formal design and often using waste materials. There is an inadequate supply of sites for development, leading to squatting. Much depends on the availability of public funding, but the main local revenue sources for urban authorities (property taxes and a tax on goods entering urban areas) have proved difficult to administer.

At the same time, there are some positive signs, notably the relatively low level of rural–urban migration – equivalent to only about 1 per cent of the rural population annually – and also the fact that this appears to be contributing as much to the growth of medium-sized and small towns as to the larger cities. Rural areas are characterised by a significant degree of surplus labour, so current levels of migration are not likely to have an adverse effect on production capacity. With so much scope for further urbanisation remaining, however, the main challenges are to accelerate the pace of economic transformation into non-agricultural activities and to reinforce the trend towards a more balanced distribution of growth across the urban system.

Sources: Mohan 1996; Papola 1997; Visaria 1997.

Box 25.2 'Guest workers' in Western Europe

The 'guest worker' system was established in the early postwar period by the more industrialised countries of northwest Europe (especially West Germany and France) to combat labour shortages arising from renewed economic growth and the low fertility of the interwar years. Normally through bilateral agreements with countries with labour surpluses (especially southern Europe but also Turkey and North Africa), workers were recruited on relatively short-term contracts, usually of no more than one or two years, with the expectation that afterwards they would return to their origins, taking with them the money that they had saved and the skills that they had acquired. The system proved highly successful in terms of the numbers involved, with annual flows of labour migrants to West Germany averaging 0.4 million in the 1960s and reaching a peak of 0.7 million in 1970; and flows to France averaging 0.25 million in the later 1960s. By the end of 1973, foreign workers made up 12 and 10 per cent, respectively, of the total labour forces of these two countries and the proportion was even higher in Switzerland, at 30 per cent. Active recruitment terminated soon after this, mainly because of economic downturn in these countries occurring at the same time as their rate of indigenous labour supply was again increasing, but also because the 'guest worker' system seemed to be taking on a dynamic of its own and having some unintended impacts.

The main problems with the system arose because the migrants' stays tended not to be as temporary as originally anticipated, chiefly because it took them longer to earn the money that they had hoped for. In due course, they were joined by family and others from their home area in a process known as 'chain migration'. In the end, it was not uncommon for whole communities to become transplanted, including shopkeepers and in some cases even the village priest, as the continued exodus of young men progressively undermined local farming and social life in the rural areas that formed the main source areas. The cumulative nature of this migration, referred to as a 'self-feeding' process by Böhning (1972), was reinforced by the way in which the migrants soon came to dominate whole sectors of the economies of their host countries, such as metal manufacturing, textiles, and hotels and catering, making these sectors less 'respectable' for the native population and increasing their dependence on the immigrant groups.

The system therefore produced some extreme examples of the effects of urbanisation. At the rural end, it threatened the long-term survival of settlement in areas that not long before had been overpopulated, prompting policy responses in the form of trying to diversify the local economy away from dependence on farming and to stabilise the service sector base by concentrating new public investment into key settlements. At the big-city destinations in northwest Europe, the biggest challenge concerned the very poor living conditions of the immigrant workers – perhaps not so serious an issue in the early stages when only young men were arriving for temporary work, but a major concern when they stayed longer and formed families. A quarter of a century on from the end of the 'guest worker' system, there remains a clear legacy in the form of high concentrations of immigrant stock in the poorest housing and most deprived neighbourhoods.

Sources: Böhning 1972; Salt 1976; Salt and Clout 1976; Ogden 1993; Blotevogel *et al.* 1993.

challenging the original central business districts (Hartshorn and Muller 1986; Garreau 1991). It is only since the 1960s, however, that records have shown a large-scale net exodus from these wider metropolitan regions into smaller urban regions and rural areas that lie beyond the primary commutersheds of the major cities, the process commonly termed counterurbanisation (see, for instance, Berry 1976; Fielding 1982; Champion 1989). Suburbanisation, or 'local urban decentralization', and counterurbanisation, or 'urban deconcentration', carry largely similar implications for the older urban cores, or 'central cities' in American parlance, which both of them are denuding of residents and activities, but their impacts on the reception areas are differentiated to a greater extent, mainly because of contrasts in the character of areas affected but also because of some differences in the types of people involved.

Urban decline and inner-city problems are nowadays high on the policy agenda of most developed countries. As with urbanisation, many of the changes induced by population deconcentration are positive in nature, being associated with long-term social trends that most people embrace wholeheartedly, for instance rising real incomes, greater personal mobility, a widespread desire for living in relatively new low-density settlements and the economic advantages of home ownership. Other irreversible trends aggravating urban decline are the shrinking share of the workforce in manufacturing and the expanding use of electronic communications technology. Yet, in the words of Downs (1994: p. 60):

> The same forces that have successfully produced the suburban American dream of single-family homes, two cars in every garage,

> and a better life have left many of the poor behind in central-city locations. Poverty breeds deterioration and despair, which feed on themselves in the form of crime, ignorance, and poor health. And so a downward spiral of life perpetutates itself.

This vicious circle of decline has proved extremely difficult to break. Certainly, the big-city authorities appear unable to redress these problems by themselves, because of lack of resources. As noted by Eversley (1972), in an early analysis of the inner-city problem in Britain, the fiscal resources available to city governments shrink as their wealthier residents and most rapidly expanding businesses depart, while at the same time their *per capita* needs for public sector services increase because of their progressively rising proportions of very poor households. The latter is further aggravated if cities continue to act as reception areas for low-income people arriving from backward rural areas and other countries, as has been very widely the case in Western Europe and North America in recent decades (see previous section).

At the same time, the process of receiving the city exodus in the suburbs and beyond is not without its difficulties, just as the process of urbanisation has proved not totally beneficial to the cities involved. As regards the suburbs, the problems are principally those of congestion and costs. As the development pressures build up, land becomes more scarce, building tends to take place at higher densities and newcomers get less housing for their money than earlier suburbanites. The latter, however, also face disadvantages in due course, as they lose the green space and low local taxes that helped to persuade them to move in the first place. With further building, they find their homes further from open land, and the taxes rise as the local powers need extra funds to provide new schools, roads, drains and community facilities, while road congestion increases and makes the commute back to the city-based job more stressful. Little wonder that the earlier cohort of suburbanisers attempt to 'raise the drawbridge' behind them by opposing new development schemes in NIMBY (not in my back yard) fashion and by campaigning for no-go areas (e.g. green belts) and exclusionary zoning (see, for instance, Murdoch and Marsden 1994) in an effort to keep further growth to a small trickle of the most wealthy!

Similar issues arise from the counterurbanisation that can be prompted by restrictions on suburban growth as well as arise from more deep-seated forces, but the impacts of these longer-distance moves from cities tend to be more complex and diverse. In the first place, as with the suburbs, any influx pushes up land and property prices, a rise made all the sharper by the fact that previously these areas will have been languishing in economic terms and have had house prices attuned to what low-income rural workers could afford. Second, as the reception areas are primarily smaller urban centres and more remote areas that have traditionally been little affected by metropolitan influences, the arrival of 'city folk' can administer a major social and cultural jolt to the existing community, captured well in Pahl's (1966) phrase *urbs in rure*. Third, unlike the suburbs, which attract primarily younger families, the counterurbanisation process involves a much greater proportion of older people, including retirees, reinforcing the top-heavy age structure already resulting from the departure of young adults and in due course increasing the burden on health care and social services. Many of these areas are also affected by their attractiveness for city-based second-home owners, removing housing from the local market without any gain in permanent residents. Despite the overall boost given to the local economy, the outcome may well be an acceleration in the out-migration of less well-off local people, particularly where restrictions on new building (for instance, in national parks and other protected areas) focus the extra pressures on the existing stock of housing.

Ideas for solutions abound, but as can be seen from the case studies, they have so far proved to have been of only limited effect. Given the deep-rooted nature of the urban deconcentration process, it is perhaps not surprising that the most

Box 25.3 Suburbanisation and central city decline in the USA

America is now very much a 'suburban' nation, with its 1990 Census showing almost half (46.2 per cent) of total population living in non-central parts of metropolitan areas and less than one third (31.3 per cent) remaining in central cities. Over the previous decade, America's suburbs had increased their number of residents by 1.4 per cent a year, while the central cities had grown annually by only 0.6 per cent – and this almost entirely due to the contribution of the southern and western states. In the northeast, central cities were already in decline in the 1960s and lost over 10 per cent of their population in the 1970s before recovering to close to a zero growth balance in the 1980s. Central cities in the mid-west recorded a 9 per cent loss in the 1970s, as the USA's former manufacturing heartland switched to 'rust belt', and were still declining overall in the 1980s.

The impacts on the central cities have been huge. Alarm bells began ringing in the 1960s when several city authorities, notably New York, became effectively bankrupt as a result of borrowing money to finance current expenditure on services. Rising local taxes and deteriorating local services merely served to accelerate the flight of better-off residents and more footloose firms into the burgeoning suburbs, leaving behind the less dynamic economic sectors and least wealthy people, notably blacks and recent overseas immigrants. Even in the more stable 1980s, when New York City's population grew by 3.5 per cent, its white, non-Hispanic population fell by 11.5 per cent and the proportion of its total residents accounting for the 'minority population' rose to over 60 per cent in 1990. Similar patterns of white population loss and high minority shares were recorded by a number of other cities (Table 25.2). Meanwhile, across America in 1990 the central cities accounted for three-quarters of people living in 'extreme poverty neighbourhoods' and 91 per cent of the nation's population living in 'underclass neighbourhoods' (Downs 1994).

Table 25.2 Population change in selected US cities, 1960–90.

City	*1990 population (thousands)*	*Population change (%)*			*White population change 1980–90 (%)*	*Minority share of total population 1990 (%)*
		1960–70	*1970–80*	*1980–90*		
New York	7323	1.4	–10.4	3.5	–11.5	60.5
Chicago	2784	–4.7	–10.7	–7.4	–17.2	62.7
Philadelphia	1586	–3.1	–13.5	–6.1	–12.3	48.4
Detroit	1028	–8.5	–19.2	–14.6	–47.7	79.7
Baltimore	736	–2.8	–12.5	–6.4	–15.8	61.6
Cleveland	506	–14.3	–23.6	11.9	–19.0	52.5
Pittsburgh	370	–14.1	–18.5	–12.8	–14.9	28.5
Cincinnati	364	–9.8	–15.0	–5.5	–11.2	39.9

Source: US Census; Downs 1994.
Note: Data relate to central cities only. Cities ranked by 1990 population size. Hispanics included in the minority population.

Though intervening in these market-forces outcomes is considered un-American by many people, a variety of strategies have been experimented with over the years. The most basic need is for extra public funds to compensate for the shrinking tax base, which can be achieved either by directly redistributing local tax revenues from wealthy suburban municipalities to central cities or more commonly by funding regeneration programmes through state and federal governments. Beyond this, in a review of American urban policy, Downs (*ibid.*) identifies four key problems facing deprived inner-city neighbourhoods: crime and insecurity, children raised in poverty, poor education and poor worker integration into the mainstream labour force. These can best be tackled through four types of strategy: area development (addressed particularly at policing, education and job creation); personal development (focusing on parents and children at home and in schools); household mobility (especially facilitating the migration of poorer families into better-off areas); and worker mobility (via better child-care facilities and easier commuting to suburban jobs).

Sources: 1990 US Census; Fainstein *et al.* 1992; Downs 1994.

impressive achievements have occurred where policies have sought to channel or moderate the main trends rather than reverse them. The new towns programme was one of the great success stories of postwar planning in Britain, although it proved inadequate at coping with the total

Box 25.4 Counterurbanisation and rural change in the UK

Counterurbanisation in Britain dates back to the 1960s, when for the first time the areas situated well away from metropolitan influence began to grow faster than the main conurbations and their dependent regions. Population growth in rural Britain was particularly strong in the late 1960s and early 1970s but has continued over the past two decades, with net out-migration from the main metropolitan areas to the rest of the UK averaging about 90,000 people a year, a rate of 0.5 per cent. The main contributor to this urban exodus in both absolute and relative terms is Inner London, while the main beneficiary is the most remote rural category of local authority districts (Figure 25.1). Indeed, the growth of the latter is powered entirely by migration, as these areas are now experiencing a surplus of deaths over births – a product of the above-average age of the in-migrants combined with the continuing 'urbanisation' of school leavers. What these statistics fail to show is the growth of temporary residents and visitors, as it is not easy to monitor the occupancy of second homes and holiday lets or to gauge the volume of day trips and overnight stays.

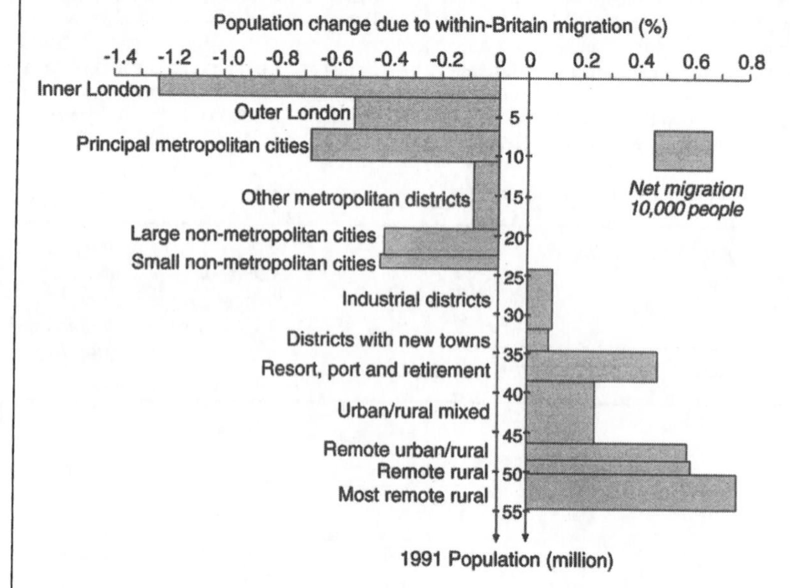

Figure 25.1 Net within-Britain migration, 1990–1, by district types.

Source: Calculated from 1991 Census SMS and LBS/SAS (ESRC/JISC purchase), Crown copyright.
Note: 'Metropolitan' includes the Central Clydeside conurbation. Areas of bars are proportional to volume of net migration.

While this 'rural renaissance' is encouraging in economic terms after decades of depopulation, the benefits are not as great as the pure population statistics might suggest and also need to be set against some important negative impacts. As a significant proportion of the people moving into the more remote areas are of older working age or are already retired, their arrival does little to boost the demand for places in local schools threatened by closure because of the departure of young people. Their strong purchasing power, helped by selling a family-size home in metropolitan Britain, raises property prices, thereby making it more difficult for local children to remain in the area when they want to set up home. Moreover, these post-family-age newcomers are not big consumers of everyday goods and, anyway, tend to do most of their buying by car on outings to supermarkets in nearby towns rather than patronise the more expensive local stores. To the extent that they are still gainfully occupied, a significant proportion are self-employed in freelance work, with little local multiplier effect, or in tourist-related shop and accommodation ventures, which often are of limited success and duration. Later, as the newcomers age and become less mobile, they place extra demands on already stretched public transport, social support and health-care facilities (Gant and Smith 1991).

Solutions to these challenges have tended to be limited in both variety and effectiveness, not helped by a general reluctance to engage in social engineering and impose limits on personal freedom. In the Lake District, an attempt to restrict the sale of new housing to local people only initially caused rapid inflation in prices of existing houses and was eventually declared illegal by central government (Shucksmith 1991). Efforts have been made to curb the inflow of elderly people to the 'costa geriatrica' of southwest England by refusing applications for permission to build new bungalows and convert seaside hotels into nursing homes, but this approach has often proved controversial locally (Phillips and Vincent 1987). Perhaps the most effective solutions lie outside the rural areas themselves, namely in improving the attractiveness of the main source areas and persuading more older people to stay put there, closer to urban amenities and to the support of family and friends.

Sources: Champion 1994; Gant and Smith 1991; Shucksmith 1991; Phillips and Vincent 1987.

volume of urban out-migration and in the end was accused of starving the cities of industrial investment (Aldridge 1979). The green belts helped to limit the physical spread of the conurbations but prompted the leapfrogging of population and housebuilding into more distant towns and deeper countryside (Hall *et al.* 1973). Land development restrictions and landscape protection measures have increased congestion in the cities and further enhanced the appeal of the 'rural idyll', somewhat paradoxically magnifying the benefits to be gained from urban–rural migration (Champion 1998b). Meanwhile, neither the much publicised waterfront renewal schemes, such as in New York, Baltimore and London Docklands (Hoyle *et al.* 1988), nor the wider 'reurbanisation' process (Bourne 1996) have generally been sufficient to produce a significant long-term reversal in the urban exodus – 'islands of renewal in a sea of decay', according to Berry (1985). On the other hand, the last two decades have seen some stabilisation of large-city populations, partly as a result of growth in financial services and of 1960s' 'baby boomers' reaching adulthood and moving to the 'city lights', but also because of higher immigration from overseas and periodic recessions in the building industry (Frey 1993; Champion 1994; Downs 1994).

CONCLUDING OBSERVATIONS

Both urbanisation and counterurbanisation, as defined for the purposes of this chapter, constitute fundamental processes of population redistribution that are taking place in response to deep-seated societal changes and, in their turn, also have major impacts on people and places. They are complex even when being analysed directly in demographic accounting terms, because these geographical shifts in population are produced not only by migration but also by trends in births and deaths and by changes in which settlement systems are conceptualised and delineated. Their complexity becomes infinitely greater when attempts are made to study the factors that influence people's behaviour, partly because of the huge gap that exists between these two terms as essentially academic constructs and the everyday reality of individuals making decisions about choosing where to live. It is therefore perhaps not surprising to find that, as demonstrated in this chapter, most of the literature explicitly concerned with urbanisation and counterurbanisation is descriptive and analytical, striving to make sense of developing tendencies and thus helping to provide a better-informed context for discussions in the policy arena.

It is also important to recognise the general reluctance of democratic governments to become involved in what might be seen as 'social engineering'. There are few contemporary examples of direct government intervention into people's decisions on where to live within countries (e.g. China), unlike the actions taken to prevent migration between states. Policies on internal population distribution are invariably indirect in nature, coming partly in the form of exhortation and sometimes financial inducements but most commonly being implemented through attempts to modify the wider planning environment. Most of the examples of policy intervention given in this chapter comprise measures directed at patterns of economic growth, social welfare and physical development, using a combination of 'stick' (e.g. restrictions on new building) and 'carrot' (e.g. subsidies to developers and employers). In research terms, these aspects have tended to be of secondary interest to the population geographers who have dominated the study of urbanisation and counterurbanisation trends over the past quarter of a century. But the widespread incidence and increasing intensity of social, economic and environmental problems arising from urbanisation and counterurbanisation commend the field to applied geographers, whose problem-oriented perspective can offer an invaluable complement to the process-oriented perspective of the demographer.

GUIDE TO FURTHER READING

Champion, A.G. (1998) Studying counter-urbanisation and the rural population turnaround. In

P. Boyle and K. Halfacree (eds) *Migration into Rural Areas: Theories and Issues*, Chichester: John Wiley & Sons, 21–40. An outline history of the study of counterurbanisation, highlighting the preoccupation of population geographers with conceptualising, measuring and explaining the phenomenon.

Downs, A. (1994) *New Visions for Metropolitan America*. Washington, DC: The Brookings Institute. A clear analysis of the relations between cities and suburbs in the USA, followed by an evaluation of past policies and recommendations for the future.

Hall, P., Thomas, R., Gracey, H. and Drewett, R. (1973) *The Containment of Urban England*. London: Allen & Unwin, 2 vols. Still the most detailed and comprehensive account of the impact of town and country planning measures designed to restrict urban sprawl in Britain.

Jones, G.W. and Visaria, P. (eds) *Urbanisation in Large Developing Countries: China, Indonesia, Brazil and India*. Oxford: Clarendon Press. A collection of well-researched essays on the urbanisation process in the developing world, including details of national urban development strategies and assessments of their effectiveness in combating the problems caused by rapid urban population growth.

Murdoch, J. and Marsden, T. (1994) *Reconstituting Rurality: Class, Community and Power in the Development Process.* London: UCL Press. The results of a detailed case study in England showing how counterurbanisers have altered the social and political complexion of the countryside and attempted to realise their notions of the 'rural idyll' and resist migration and development pressures that might lead to a further reshaping of their settlements.

UNCHR (1996) *An Urbanising World: Global Report on Human Settlements 1996*. Oxford: Oxford University Press. A massive compilation prepared for the 1996 (Habitat II) Conference organised by the United Nations Centre for Human Settlements, covering global and regional perspectives on population and urbanisation, social and environmental conditions and trends, developments in key problem areas (housing, land, infrastructure, governance) and policy responses in settlement planning and environmental protection.

REFERENCES

Aldridge, M. (1979) *The British New Towns: A Programme without a Policy*. London: Routledge & Kegan Paul.

Berry, B.J.L. (ed.) (1976) *Urbanisation and Counter-urbanisation*. Beverly Hills, Calif.: Sage.

Berry, B.J.L. (1985) Islands of renewal in seas of decay. In P. Petersen (ed.) *The Urban Reality*, Washington, DC: The Brookings Institute, 69–96.

Blotevogel, H.H., Jung, U.M. and Wood, G. (1993) From itinerant worker to immigrant? The geography of guestworkers in Germany. In R. King (ed.) *Mass Migrations in Europe: The Legacy and the Future*, London: Belhaven, 83–100.

Böhning, W.R. (1972) *The Migration of Workers in the United Kingdom and the European Community*. London: Oxford University Press.

Bourne, L.S. (1996) Reurbanisation, uneven urban development and the debate on new urban forms. *Urban Geography* 17, 690–713.

Boyle, P., Halfacree, K. and Robinson, V. (1998) *Exploring Contemporary Migration*. Harlow: Longman.

Champion, A.G. (ed.) (1989) *Counterurbanisation: The Changing Pace and Nature of Population Deconcentration*. London: Arnold.

Champion, A.G. (1994) Population change and migration in Britain since 1981: evidence for continuing deconcentration. *Environment and Planning A* 26, 1501–20.

Champion, A.G. (1998a) Studying counter-urbanisation and the rural population turnaround. In P. Boyle and K. Halfacree (eds) *Migration into Rural Areas: Theories and Issues*, Chichester: John Wiley & Sons, 21–40.

Champion, A.G. (1998b) *Urban Exodus*. London: Council for the Protection of Rural England.

Champion, A.G. and Atkins, D.J. (1996) The counter-urbanisation cascade: an analysis of the 1991 census special migration statistics for Great Britain *Seminar Paper No. 66*, Department of Geography, University of Newcastle upon Tyne.

Downs, A. (1994) *New Visions for Metropolitan America*. Washington, DC: The Brookings Institute.

Eversley, D.E.C. (1972) Rising costs and static incomes: some economic consequences of regional planning in London. *Urban Studies* 9, 347–68.

Fainstein, S., Gordon, I. and Harloe, M. (1992) *Divided Cities: New York and London in the Contemporary World*. Oxford: Blackwell.

Fielding, A.J. (1982) Counterurbanisation in Western Europe. *Progress in Planning* 17, 1–52.

Frey, W.H. (1993) The urban revival in the United States. *Urban Studies* 30, 741–74.

Gant, R. and Smith, J. (1991) The elderly and disabled in rural areas: travel patterns in the north Cotswolds. In T. Champion and C. Watkins (eds) *People in the Countryside: Studies of Social Change in Rural Britain*, London: Paul Chapman, 108–24.

Garreau, J. (1991) *Edge City: Life on the New Frontier*. New York: Doubleday.

Hall, P., Thomas, R., Gracey, H. and Drewett, R. (1973) *The Containment of Urban England*. London: Allen & Unwin, 2 vols.

Hartshorn, T.A. and Muller, P.O. (1986) *Suburban Business Centers: Employment Implications*. Washington, DC: US Department of Commerce, Economic Development Administration.

Hoyle, B.S., Pinder, D.A. and Husain, M.S. (eds) (1988) *Revitalising the Waterfront: International Dimensions of Dockland Redevelopment*. London: Belhaven.

McGee, T. and Griffiths, C.J. (1994) Global urbanisation: towards the twenty-first century. In *Popuation Distribution and Migration*, New York: United Nations Population Division, 55–74.

Mohan, R. (1996) Urbanisation in India: patterns and emerging policy issues. In J. Gugler (ed.) *The Urban Transformation of the Developing World*, Oxford: Oxford University Press, 93–132.

Murdoch, J. and Marsden, T. (1994) *Reconstituting Rurality: Class, Community and Power in the Development Process*. London: UCL Press.

Ogden, P. (1993) The legacy of migration: some evidence from France. In R. King (ed.) *Mass Migrations in Europe: The Legacy and the Future*, London: Belhaven, 101–17.

Pahl, R.E. (1966) *Urbs in Rure*. London: Weidenfeld & Nicolson.

Papola, T.S. (1997) Extent and implications of rural–urban migration in India. In G.W. Jones and P. Visaria (eds) *Urbanisation in Large Developing Countries: China, Indonesia, Brazil and India*, Oxford: Clarendon Press, 315–20.

Phillips, D. and Vincent, J. (1987) Spatial concentration of residential homes for the elderly: planning reponses and dilemmas. *Transactions of the Institute of British Geographers New Series* 12, 73–83.

Salt, J. (1976) International labour migration: the geographical pattern of demand. In J. Salt and H. Clout (eds) *Migration in Post-war Europe: Geographical Essays*, London: Oxford University Press, 80–125.

Salt, J. and Clout, H. (1976) International labour migration: the sources of supply. In J. Salt and H. Clout (eds) *Migration in Post-war Europe: Geographical Essays*, London: Oxford University Press, 126–67.

Shucksmith, M. (1991) Still no homes for locals? Affordable housing and planning controls in rural areas. In T. Champion and C. Watkins (eds) *People in the Countryside: Studies of Social Change in Rural Britain*, London: Paul Chapman, 53–66.

UNCHR (1996) *An Urbanising World: Global Report on Human Settlements 1996*. Oxford: Oxford University Press.

Visaria, P. (1997) Urbanisation in India: an overview. In G.W. Jones and P. Visaria (eds) *Urbanisation in Large Developing Countries: China, Indonesia, Brazil and India*, Oxford: Clarendon Press, 266–88.

26

Boundary disputes

Gerald Blake

INTRODUCTION

Since the Second World War, the international community has formally declared itself in favour of the stability of boundaries between states. The Charter of the United Nations set the scene in 1945 by recognising that the sovereignty of a properly constituted state is absolute and exclusive, and that states must respect the territorial integrity of one another. The principle of the inviolability of boundaries has been confirmed on a number of subsequent occasions. In 1964, member states of the Organisation of African Unity agreed to respect the borders that they had inherited from colonial times. In 1975 in the Helsinki Final Act, the Conference on Security and Cooperation in Europe affirmed the same principle, while in 1991 the Commonwealth of Independent States (the majority of the states of the former Soviet Union) also agreed to accept their inherited boundaries. States in Latin America had set the trend in the nineteenth century when they became independent of Spain and Portugal, a principle known as *Uti possidetis*. These worthy declarations, alas, have not rid the world of boundary disputes. On the contrary, Africa, Europe, the former Soviet Union and Latin America have all witnessed a large number of boundary disputes, and some spectacular changes to the political map. The truth is that the world political map is changing all the time, and will continue to change in future.

Boundary disputes must therefore be seen in the context of an evolving political mosaic. In many cases, disputes result in territorial adjustments, but the political map also changes in response to other powerful processes. Goertz and Diehl (1992) undertook an analysis of territorial changes worldwide from 1816 to 1980. Of 770 territorial transfers, 42.3 per cent were by cession, 15.6 per cent by conquest, 15.5 per cent on independence and 14.5 per cent by annexation. Other causes were secession, unification and mandates. Since 1980, there have been large-scale changes to the world map, particularly in the wake of the break-up of the Soviet Union. Altogether, twenty-two new states have emerged, and more than fifty new land boundaries. A series of world maps in Foucher (1988) tracing the evolution of international boundaries since 1800 strikingly reveal how the political arrangement of space can change through time. The world map of 100 years ago is scarcely recognisable; it is doubtful whether today's world map will be recognisable a century hence. The popular perception is that the map we know is permanent, a kind of finished product. In reality, it is a snapshot of geopolitical history.

THE DIMENSIONS OF THE PROBLEM

There are 191 independent states and seventy dependent territories in the world today. Most of the surviving dependencies are islands. Seven states retain territorial claims to Antarctica, but these boundaries are not considered here. There are 308 land boundaries between sovereign states (Biger 1995), although some authorities find a few more. Out of the 191 independent states, 148 are coastal states with the right to delimit their offshore areas. These coastal states, together with the island dependencies mentioned above, will generate approximately 420 international maritime boundaries. Maritime boundary delimitation only began in earnest less than fifty years ago, and by 1998 only about 150 (35 per cent) had been formally agreed. At the present rate of about five agreements per annum (Charney and Alexander 1993; 1998), the offshore international boundary map will take fifty-four years to complete.

How many of the world's boundaries are disputed? Attempts to draw up lists of disputed boundaries around the world always differ in certain respects, and the totals rarely tally. There are problems of definition and information. Paul Huth (1996) has written a most valuable analysis, covering land and island disputes (not maritime disputes) from 1950 to 1990. Using a strict definition of 'dispute', Huth identified 129 cases, 116 of which he categorised as 'boundary' rather than 'territorial' disputes. During the period 1950–90, 33 per cent of borders were at some point in dispute (*ibid.*: p. 34). This coincides with best estimates of the proportion of land boundaries in dispute in 1998. Since the Soviet Union broke up in 1989, it has been difficult to estimate the number of true boundary disputes there. Kolossov *et al.* (1992: pp. 42–50) counted 168 ethno-territorial conflicts, a quarter of which involved boundary changes on state borders. As to maritime boundaries, there are 270 undelimited. In a significant proportion of cases, no attempt is yet being made to delimit boundaries, and therefore no dispute has emerged. There are, however, a number of ongoing maritime boundary disputes, including at least thirty associated with disputed island sovereignty, and about thirty other delimitation disputes (McDorman and Chircop 1991: pp. 344–86). Thus approximately 22 per cent of the undelimited maritime boundaries are already known to be in dispute, and the signs are that there will be a lot more. Many of these should be resolved peacefully, but among them are nasty potential flashpoints, in the Aegean and South China Seas, for example.

By most reckoning, this amounts to a large number of boundary disputes, but we should not be surprised. The land boundary system was largely the creation of Europe, usually superimposed on the underlying geography, and boundary lines of no width were an alien concept in many parts of the world. Many boundaries came about almost by accident, and the wonder is that they have survived at all. But whatever the causes of disputes may be, they should be taken seriously; one-third of the wars fought since 1945 had territorial questions as a major contributory factor. At a time when globalisation and the 'borderless world' are topics of debate among social scientists, it is tempting to regard border troubles as of diminishing significance. Unfortunately, the evidence is to quite the contrary.

THE CHARACTERISTICS OF INTERNATIONAL BOUNDARIES

Land boundaries

The land boundaries between states appear on our world maps as thin lines, usually depicted by the same colour and symbol throughout. In reality, state boundaries differ greatly in their origins, age, permeability and degree of conformity to the human and physical geography. Some boundaries have been formally agreed and marked out on the ground. Others have never been properly delimited or demarcated and continue to carry the seeds of potential conflict. The borderlands on each side of the boundary may be hostile, or may enjoy a high level of integration. Generalisations about the world's 300-plus land boundaries are therefore difficult, but one universal characteristic is their sensitivity. International boundaries mark the absolute limits of state sovereignty, of the identity of its peoples, the extent of its legal and administrative system, resources, and security arrangements. The boundary marks the interface between neighbouring states and is often used as the setting for symbolic acts of friendship or hostility.

State boundaries extend vertically into the air

to an undefined height, and vertically into the Earth, where they define ownership of resources, to unlimited depth. Although many boundaries are marked by complex systems of barbed wire and other obstacles sometimes several metres wide, the boundary itself is a line of no width. Where two boundaries meet, the result is a precise point. With the availability of global positioning systems (GPS), boundaries can be fixed with great accuracy, and territories allocated with precision. Strictly speaking, there are therefore few frontiers (or zones of transition) between states as boundaries are delimited and demarcated more accurately. Geographers like to distinguish between frontiers and boundaries, but the terms are commonly regarded as synonymous, not least by the eminent boundary scholar S.W. Boggs (1940).

Air boundaries

The 1944 Convention on International Civil Aviation and the 1982 UN Convention on the Law of the Sea recognise absolute state sovereignty over the land area of the state and its territorial sea, and over the airspace above. The upper limit of state control of airspace has never been defined, but for all practical purposes it is the limit of powered flight. Aircraft of other states must seek permission to overfly the land territory or the territorial sea of another state. In normal circumstances this is granted, but certain airlines such as El Al (Israel) and South African Airways were once excluded from the airspace of a number of states for political reasons. States guard their airspace jealously for the sake of security and safety, and a number of serious incidents have occurred in contested airspace. Aircraft have the right of transit passage through straits used for international navigation, even though the strait may fall within the territorial sea of coastal states (UN Convention on the Law of the Sea (UNCLOS) 1982, Article 38).

Maritime boundaries

Unlike land boundaries, which generally may be visibly marked, maritime boundaries are not visible, nor do they affect the everyday lives of people to such an extent. There are three kinds of maritime boundary. First, *adjacent boundaries*, which start from the land boundary terminus, 160 of which reach the coast worldwide. Second, *opposite boundaries* are those between opposite states in enclosed and semi-enclosed seas. Third, there are boundaries between coastal state jurisdiction and the international seabed beyond. This is primarily determined by a 200-nautical-mile limit (1 nautical mile = 1.15 statute miles or 1.852 km) measured from the baseline along the coast, but some twenty-two states are also entitled to claim continental shelf beyond 200 miles in accordance with Article 76 of UNCLOS 1982 (Prescott 1996: pp. 51–82).

The legal functions of maritime boundaries are more varied than for land boundaries, which define absolute state sovereignty. Within the *territorial sea* to a maximum distance of twelve nautical miles state sovereignty is absolute, except that ships of other states are entitled to innocent passage (UNCLOS 1982, Article 17). Beyond twelve miles to a distance of 200 nautical miles is the *exclusive economic zone* (or EEZ), in which the coastal state enjoys exclusive rights to living and non-living resources (UNCLOS 1982, Article 56). States also have the exclusive right to the resources of their continental shelves, including notably hydrocarbons, minerals and bottom-dwelling marine life (UNCLOS 1982, Article 77). In theory, therefore, a coastal state may have maritime boundary agreements with neighbouring states in respect of territorial sea and/or continental shelf and EEZ. In practice, most modern maritime boundary agreements are 'all-purpose' boundaries, rather confusingly referred to as 'single maritime boundaries'.

In spite of their complexity, maritime boundaries have one advantage over land boundaries in that UNCLOS offers some guidelines for boundary delimitation. For territorial sea boundaries, there is a presumption that the line will be equidistant between the parties' coasts (UNCLOS 1982, Article 15) unless they agree otherwise or some special circumstance justifies an alternative alignment. Continental shelf and EEZ boundaries, on the other hand, are to be drawn 'by agreement on the basis of international law . . . in order to achieve an equitable solution' (UNCLOS 1982, Articles 74 and 83), which may or may not mean an equidistance line.

WHY BOUNDARY DISPUTES OCCUR ON LAND

There is a considerable volume of literature on the causes and consequences of land boundary disputes, to which geographers have made a significant contribution, particularly in the reporting of individual cases. Although the subject has been explored by scholars for a century or more, there is still no useful theory as to why boundary disputes occur. Part of the problem is their sheer diversity through time and across space. Another difficulty is distinguishing between the underlying and the immediate causes of a dispute. Furthermore, in many cases tension at the state boundary is symptomatic of the relationship between neighbours, and the boundary is not the real substance of their quarrel: 'a boundary, like the human skin, may have diseases of its own or may reflect the illnesses of the body' (Jones 1945: p. 3). These factors make simple explanations of boundary disputes both difficult and potentially misleading. Most disputes comprise a number of ingredients, and it is not always clear which are the most important.

Positional disputes

When the precise location of a boundary is in doubt, it is known as a positional problem. Such disputes can occur for a number of reasons. River boundaries are notorious for positional disputes, especially in areas where the course of the river or the main navigation channel shifts with natural physical processes. The Rio Grande (Mexico–United States) proved to be a most inconvenient international boundary because of marked changes in its course during the nineteenth century. A modern example is the Congo River (Angola–Democratic Republic of Congo, formerly Zaire), where the international boundary follows the navigation channel under an agreement of 1891. Figure 26.1 shows the extent of the changes in the navigation channel between 1960 and 1977, raising questions as to the ownership of islands that find themselves on one side and then the other side of the boundary. Many other positional questions are raised by river boundaries, including finding the midpoint, the fastest stream, or the deepest water, all of which have been used to locate international boundaries. Positional disputes also occur in a variety of other circumstances, especially where boundary demarcation has not taken place. Problems can also arise after demarcation if boundary markers are removed or destroyed. In recent years, the use of GPS has facilitated the resolution of such disputes, but it has also revealed a number of incorrectly located boundaries, resulting in low-level disputes.

Territorial disputes

The distinction between positional and territorial disputes is primarily a matter of scale. Because so much more is at stake, the risk of serious conflict is greater. States may lay claim to the territory of another state for many reasons. A major category of motives is *geopolitical*. Territory may be needed to ensure access to the sea, or to give control of a strategic location, or to deprive a hostile neighbour of a vital road or rail link. Campaigns have been fought in the Middle East and in southeast Europe for these reasons in the past thirty years. Territorial disputes also occur quite commonly because peoples of the same ethnic background are divided by the international boundary. Figure 26.2 illustrates typical cases where *divided ethnic groups* seek independence, or wish to secede to a neighbouring state. There are perhaps between 6000 and 9000 'suppressed nationalisms' in the world today, many of which are in or near borderlands, and they are likely to be more influential in defining the shape of tomorrow's world political map (Griggs and Hocknell 1995: pp. 49–58). Disputes that feature ethnic groups often depend heavily on history as part of the case

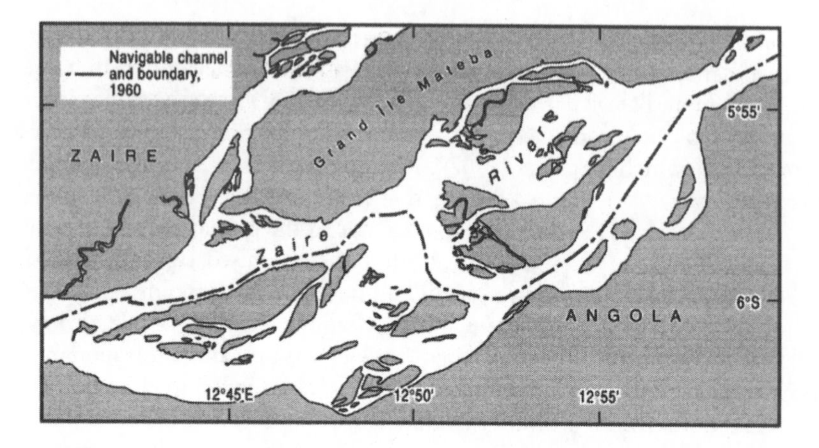

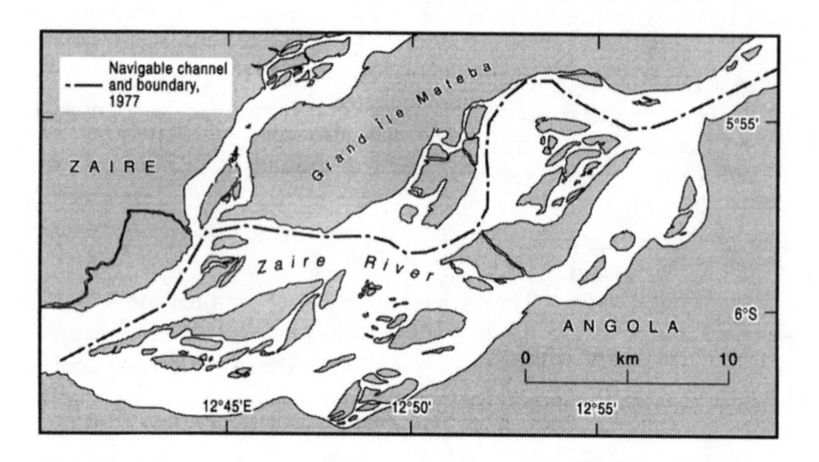

Figure 26.1 The Zaire–Angola boundary in the Zaire River, 1960 (top) and 1977 (bottom).

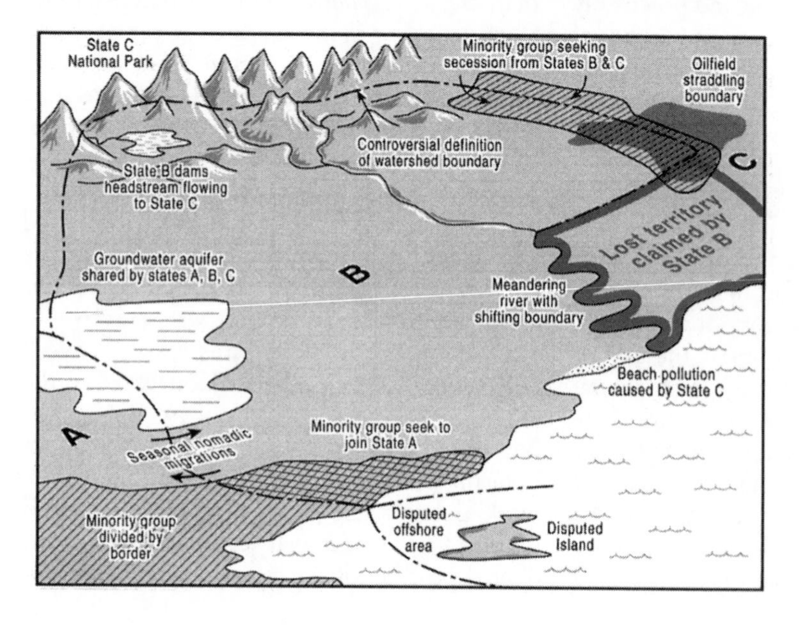

Figure 26.2 Some geographical causes of international stress along state boundaries.

for territorial change. The concern of governments to acquire or protect *resources* also helps to explain territorial ambition. As the demand for scarce resources of oil, gas, fish and fresh water has increased, states have turned to more marginal areas, and territorial disputes may be the result. A final, and crucial, element in many territorial disputes is human *territoriality*. In some territorial conflicts, it is difficult to detect any other motive than the intense desire by one party to protect what it believes to be its own, and the desire of another, fired by nationalist fervour, to extend its domain. Passions can run high where even small areas of land are in question.

Functional disputes

Stress can be caused between states by the consequences of the international boundary rather than its location. Functional disputes are very common and cover a wide range of activities. Fortunately, they are commonly managed peacefully by the parties and no serious dispute results. Many busy boundaries maintain joint commissions or groups of some kind to trouble-shoot as problems arise and are very successful in keeping the peace. A large category of grievances concerns unauthorised border crossings by smugglers, bandits, migrant workers and refugees, with the allegation that state A is doing too little to prevent incursion into state B. Other functional disputes include illegal transboundary fishing, or accusations that a state is taking out more than its fair share of water or oil from a straddling reservoir. The examples are legion, and the closer one examines any particular boundary, the more functional problems become evident. Figure 26.1 illustrates a number, including a nature conservation area in state C that is not matched in state B, making borderland species protection and biodiversity programmes ineffective. Pollution of rivers, lakes and beaches from across the international boundary is also a common grievance, and the consequences of dam building and the extraction of water by upstream states in shared river basins is another. All this underlines the growing need for training in boundary management skills, and continuing exchange of ideas about best practice (Blake *et al.* 1995).

Technical problems

As with charts (see below), the misuse and misunderstanding of topographic maps can create border problems between states. Some of the major pitfalls are discussed by Adler (1995), Rushworth (1997) and Blake (1995). Common problems are concerned with misunderstanding scale, and geodetic datums, misuse of coordinates, and failure to understand the limitations of maps.

WHY BOUNDARY DISPUTES OCCUR AT SEA

Maritime boundary questions are less inclined to create serious international friction than land boundaries, because emotional and historic ties to the seabed are naturally much weaker. Islands are an exception. Rival claims to offshore islands can engender powerful nationalist emotions, and the media are fascinated by such disputes. Many tiny islands with romantic names have hit the headlines in recent years, such as Imia/Kardak, Sipadan and Ligitan, Diaoyo Tai/Senkaku and the South Sandwich Islands. A large number of these disputed islands have little intrinsic value, but their ownership can give title to surrounding seabed resources. Smith and Thomas (1998) identified thirty-two disputed islands or groups of islands worldwide involving forty-four states. The list is likely to lengthen as more states turn to maritime boundary delimitation in future. The process of offshore boundary delimitation is likely to encounter a number of other problems, which are outlined below.

Questions of sovereignty

Sovereignty over islands and mainland territory must clearly be resolved before maritime boundary drawing can begin. Thus in the current (1998) dispute between Eritrea and Yemen over their Red Sea boundary, the Court of Arbitration will first decide who owns the Hanish Islands and then address the boundary problem. Similarly, if there is uncertainty about the location of the land boundary terminus at the coast, the adjacent maritime boundary cannot be agreed. In the Red Sea, there is no Egypt–Sudan maritime boundary because of a longstanding land boundary dispute, and offshore oil exploration has been abandoned. The world's most dazzling island sovereignty question concerns the Spratly Islands, scattered over a vast area in the South China Sea. All the 500 or so features are claimed by China, Taiwan and Vietnam, thirty-three by the Philippines and twelve by Malaysia, while Brunei claims an area of seabed, but no islands.

Definition of geographical terms

Islands can cause headaches for negotiators in two other respects. First, UNCLOS 1982 defines an island as 'a naturally formed area of land surrounded by water which is above water at high tide' (Article 21.7), whereas 'rocks which cannot sustain human habitation or economic life of their own shall have no exclusive economic zone or continental shelf' (Article 121.3). Not surprisingly, there are arguments about what is an island and what is a 'rock'. Genuine islands give states the right to surrounding EEZ and seabed, whereas rocks give only limited advantages in territorial waters (up to 12 miles offshore) and no advantage elsewhere. Second, negotiators must agree on what weight to give to islands when delimiting a boundary. Depending on size and location, islands may be given the same effect as mainland, or half effect, partial effect, or no effect at all. Many overlapping maritime claims occur because states give more weight to their own islands than to those of their neighbour.

Excessive baseline claims

Another common cause of offshore disputes is the adoption of excessive baselines. Coastal states are permitted to use straight baselines along coastlines that are 'deeply indented or cut into, or if there is a fringe of islands along the coast in its immediate vicinity' (UNCLOS 1982, Article 7.1). Unfortunately, these terms have never been defined, nor has the requirement 'not to depart to any appreciable extent from the general direction of the coast' (Article 7.3). Because offshore limits are measured from either the low water mark or straight baselines, states are often tempted to make excessive claims inconsistent with the Convention. Geographers have proposed some sensible guidelines for straight baseline evaluation, but they have never been formally adopted (US Department of State 1987). Victor Prescott (1988) has made a significant contribution to the exposure of illegal baselines, while the United States has lodged objections to almost half the seventy straight baseline claims declared (Roach and Smith 1994: pp. 77–81).

Technical problems

Both land and maritime boundary disputes can arise for technical reasons, sometimes because those involved in negotiations have insufficient training or have not consulted technical experts. Accessible publications on technical aspects of boundary delimitation are published by the International Boundaries Research Unit in Durham (Adler 1995; Beazley 1994). The International Hydrographic Organisation has published the standard work on the subject (IHO 1990). Some examples of what can go wrong are:

- *Misunderstanding nautical charts*, especially the properties of the Mercator projection. Scale increases with latitude, boundary lengths are difficult to measure, and a straight line drawn on such a chart does not represent the shortest distance between two points (i.e. the geodesic line).
- *Matching geodetic datums*. The regular geometrical shape approximating the shape of the Earth used for practical mapping and surveying is known as the *ellipsoid*. In the past, (especially) national surveys used ellipsoids of different shapes and sizes. Coordinates derived from one system have to be matched with those from another. The geodetic datum used is often not stated in boundary

agreements, and confusion results.

- *Tidal datums*. States adopt different criteria for the measurement of low water marks, which can affect the definition of low-tide elevations, and the selection of basepoints for baseline delimitation.

BOUNDARY DISPUTES AND DISPUTE RESOLUTION

Unresolved disputes

Four unresolved disputes (two land and two maritime) are briefly outlined in this section to illustrate why certain boundary disputes can be difficult to resolve. Where inter-state relationships are poor, perhaps for historical or ideological reasons, negotiations may be impossible or extremely difficult, as China–India and Greece–Turkey illustrate. Where the territory or seabed in question is of high economic potential (Cambodia–Thailand, Saudi Arabia–Yemen, for example) or is regarded as strategically critical (China–India, Greece–Turkey), compromise is unlikely. Significantly, too, in all the chosen examples there are perplexing legal, historical and geotechnical questions, which require time, money and expertise to answer. By contrast, Boxes 26.1 to 26.4 illustrate four disputes that have been successfully resolved, although there were extremely tricky issues to settle. The most

Box 26.1 France–United Kingdom continental shelf

(Resolved by a Court of Arbitration, June 1977.)

France and the United Kingdom (UK) could not agree on a continental shelf boundary in the western approaches to the English Channel, and in 1975 they asked a Court of Arbitration for a solution. The chief problems were how far to take into account the UK Channel Islands lying close to the French coast, Eddystone Rock and the Scilly Isles (both UK features), and the French isle of Ushant (Figure 26.3). The court's decision was announced in June 1997. The Channel Islands were enclaved within twelve mile territorial seas, and median lines with the French mainland to south and east, while Eddystone was recognised as a legitimate basepoint. Because of their distance beyond UK territorial waters, the Scillies were given only half effect. This was achieved by drawing a line following the French and UK mainlands, and another treating the islands as mainland. A line was drawn to bisect these two equidistant lines (Francalanci and Scovazzi 1994: pp. 238–9).

In March 1978, the court made adjustments to the line as a result of UK complaints that the court line was a loxodrome, or straight line, on a Mercator chart that took no account of the curvature of the Earth. At its western extremity, it lay four nautical miles north of the correct position and was accordingly adjusted in favour of the UK (Jagota 1985: pp. 140–55).

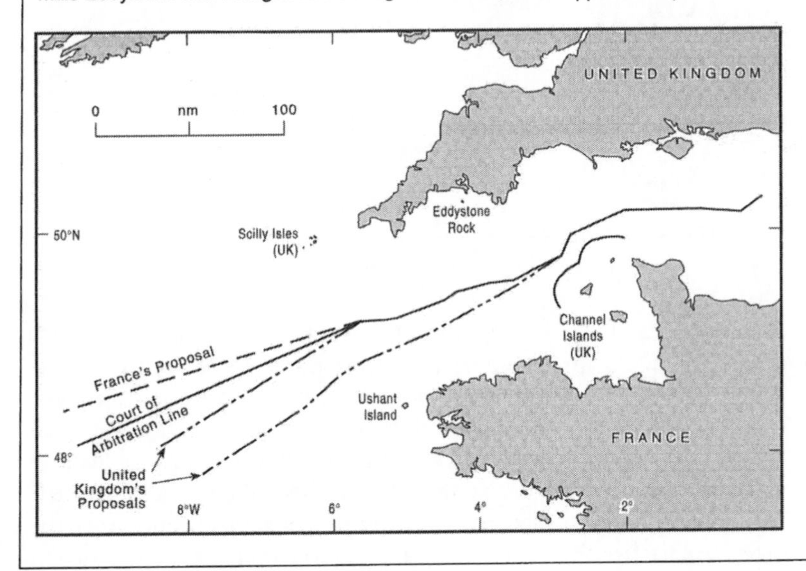

Figure 26.3 The France–United Kingdom continental shelf boundary (resolved 1977, 1978).

Box 26.2 Canada–United States Gulf of Maine maritime boundary

(Resolved by a Chamber of the International Court of Justice, October 1984.)

The Gulf of Maine is geographically complex and of considerable economic importance because of the rich fisheries on George's Bank. Canada and the USA tried very hard to delimit a maritime boundary there but after five fruitless years of negotiation, in 1981 they asked a Chamber of the ICJ to delimit a single continental shelf and fisheries boundary. It was a formidable task. The Canadians claimed a median line between mainland coasts, giving them a large part of George's Bank. The USA argued for a line perpendicular to the general direction of the coast and for control of the whole of George's Bank (Figure 26.4). More scientific, legal, economic, social and historical data were assembled to support the arguments of the parties than for any similar case. It took nearly four years for the evidence to be prepared and the case heard. In the event, geographical considerations were of paramount importance.

The court rendered its judgement in October 1984, proposing a line in three sectors. A–B was delimited by drawing the bisector of an angle determined by perpendiculars to the general direction of the adjacent coasts. B–C started as an equidistance line between opposite coasts but was adjusted in the ratio of 1.32:1 in favour of the USA to reflect the greater length of the US coast in this sector. C–D was determined as a perpendicular to the closing line of the gulf, to a distance of 200 nautical miles offshore. The sector between the land boundary and point A remains undelimited because of a sovereignty dispute over Machias Seal Island (Charney and Alexander 1993: pp. 401–16).

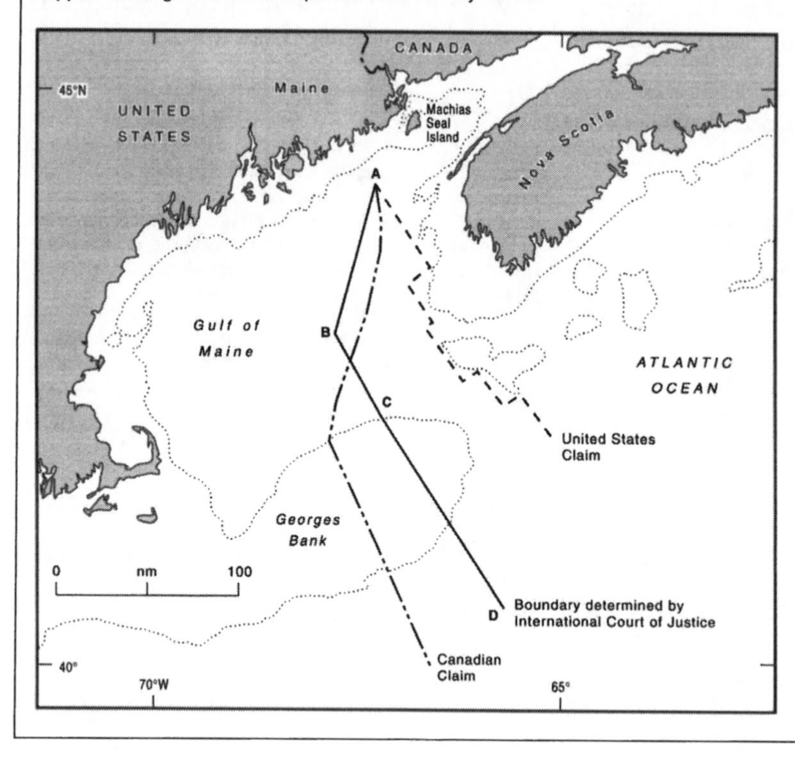

Figure 26.4 The Canada–United States maritime boundary in the Gulf of Maine (resolved 1984).

important aspect, however, was the political will of the parties to reach a peaceful agreement.

China–India

The 2500 mile (4000 km) China–India boundary (Figure 26.7) involves 32,000 square miles (83,000 km²) of contested territory. Apart from size, the dispute has some potent ingredients for a dangerous confrontation, including a remote mountain environment, geostrategic objectives, ideological differences, and the legacy of British imperial involvement. Although the problems go back at least to the beginning of the century, Chinese claims were not pressed until after the Communist Revolution in 1949. Half a dozen

Box 26.3 Chad–Libya boundary in the Sahara

(Resolved by the International Court of Justice, February 1994.)

The boundary dispute between Chad and Libya came to a head in June 1973, when Libya's troops occupied 11,000 km² of northern Chad known as the Aouzou Strip. There were exaggerated reports at the time of uranium, iron ore and other riches in the region to help to explain Libya's action. Libya relied politically on a 1935 boundary agreement between the French and Italians by which the area was ceded to Italian-occupied Libya. In return, Italian claims to parts of French-occupied Tunisia were dropped. Neither state ratified the agreement, however, partly because the Second World War broke out in 1939, and it seems to have been forgotten until the 1970s. The dispute was referred to the ICJ in 1990. In its submission, Libya surprised the court by invoking Ottoman history and the influence of the Senoussi Order to claim Chad as far south as 15° north (Figure 26.5). After three and a half years, the ICJ gave its judgement in February 1994. The Chad–Libya boundary had been confirmed by a Treaty of Friendship and Good Neighbourliness concluded by the parties in August 1955, which recognised boundaries in place at the time of Libyan independence in 1950. Franco-British conventions of 1898 and 1919 had clearly established the boundary, and the Aouzou region was south of the boundary with Chad. In due course, Libya withdrew from the Aouzou Strip in accordance with the findings of the court (Blake 1994b).

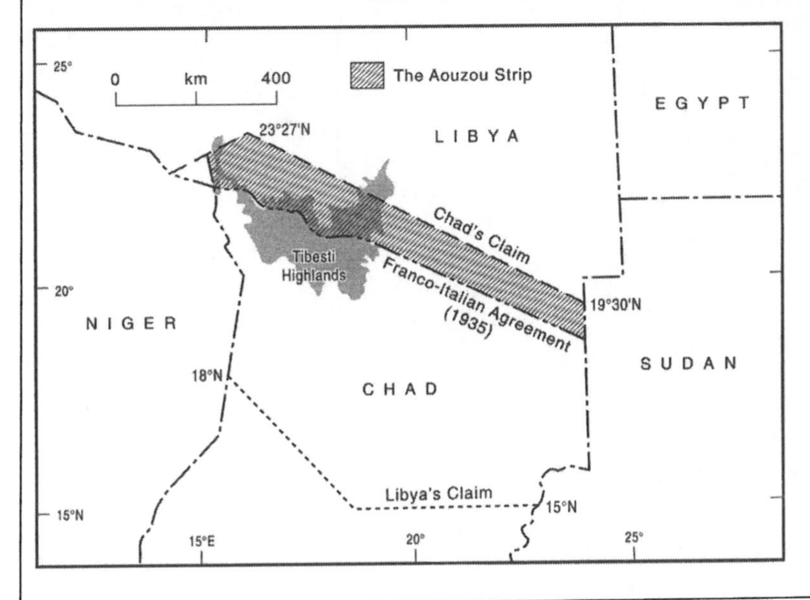

Figure 26.5 The Chad–Libya boundary (resolved 1994).

areas are claimed by China, the largest being Aksai Chin in the west and Arunchal Pradesh in the east. China took military action on several occasions in the 1950s and 1960s, culminating in a border war in 1962 in which the Chinese gained control of the strategic Aksai Chin region. Aksai Chin now provides a vital road link between Tibet and China's Sinkiang province (Allcock *et al.* 1992: pp. 428–39).

Much of the China–India problem turns on the validity of the McMahon line, which India regards as marking its northernmost limits, but China does not. Sir Henry McMahon proposed the line at the Simla Conference (1913–14), convened to discuss the status of Tibet. It seems doubtful whether the McMahon line is a legitimate claim, but it has never been put to the test. Several efforts have been made to resolve the border problem in bilateral talks, the most encouraging of which resulted in reopening the border in 1991 after being closed for thirty years, and subsequent border troop reduction agreements.

Box 26.4 Argentina–Chile boundary in the southern Andes

(Resolved by British arbitration, 1902, 1966 and 1994.)

In 1855, Argentina and Chile formally agreed to accept that the boundaries they inherited from Spain in 1810 (a doctrine known as *Uti possidetis*). The precise alignment of the inherited boundaries was not clear, however, and brought the countries close to conflict on several occasions. An agreement signed in Buenos Aires in July 1881 was an attempt to settle the Andes boundary, which 'shall run . . . over the highest summits of the said Cordilleras which divide the waters' (Boggs 1940: pp. 85–93). Closer examination revealed that the watershed and the highest peaks do not coincide, because of headward erosion by the steeper and stronger streams flowing westwards to the Chilean coast. In some areas, the watershed was far to the east of the highest peaks; thus Chile favoured a watershed boundary and Argentina the highest peaks (Figure 26.6). The parties asked the British monarch to arbitrate in 1896, and after exhaustive field surveys a compromise line was proposed by a British commission in 1902, which largely avoids the watershed or the highest peaks. Argentina gained 15,450 square miles and Chile 20,850 square miles of formerly disputed territory. Ambiguities were subsequently revealed in two relatively short sectors around Palena and Laguna del Desierto, which were resolved by British-led commissions in 1966 and 1994, respectively.

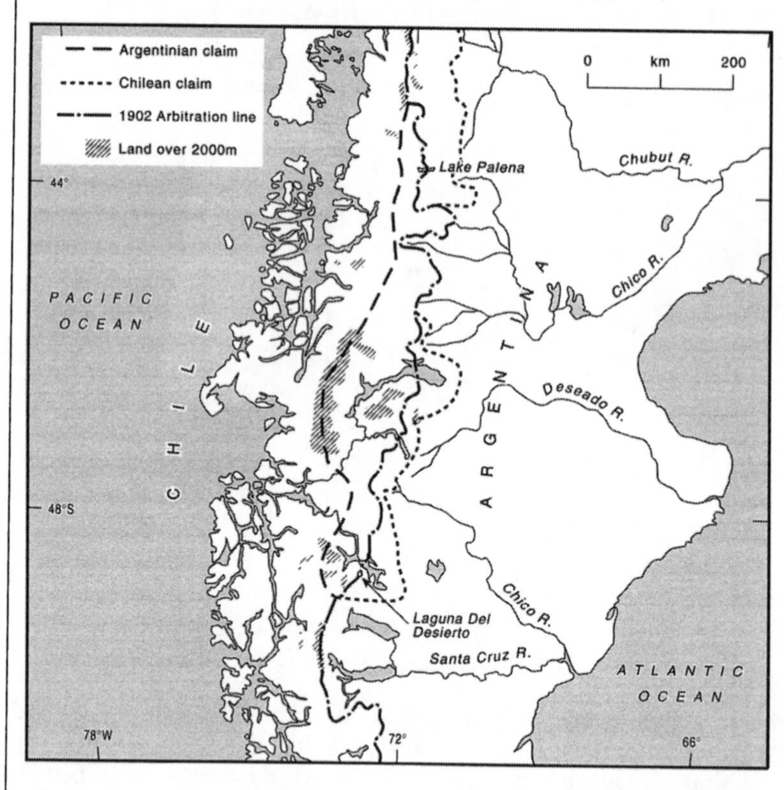

Figure 26.6 The Argentine–Chile boundary in the southern Andes (resolved 1902, 1996, 1994).

Saudi Arabia–Yemen

Saudi Arabia and the Republic of Yemen are engaged in a major territorial dispute, which the parties have tried to resolve in high-level talks since 1993. Political relations between the two states have rarely been good, in spite of a degree of economic interdependence. Yemen was formed in 1990 by the unification of the Yemen Arab Republic (North Yemen) and the People's Democratic Republic of Yemen (South Yemen), which had been ruled by Britain as the Aden Protectorate until 1967. Saudi Arabia and North Yemen agreed on part of their boundary by the Treaty of Taif (1934), which runs mostly through populous mountain regions, but no boundary

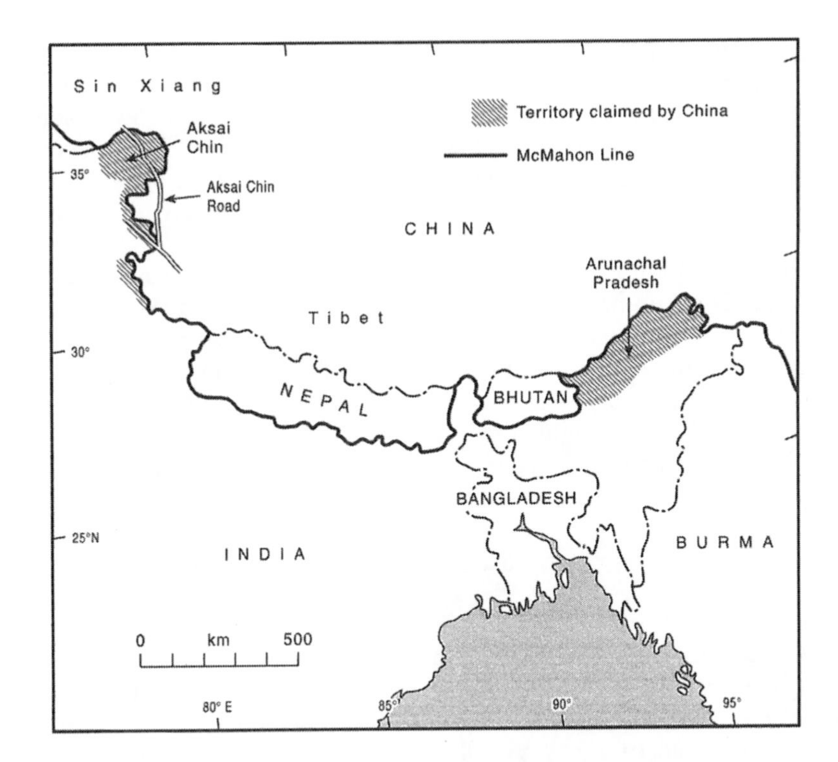

Figure 26.7 The China–India boundary dispute.

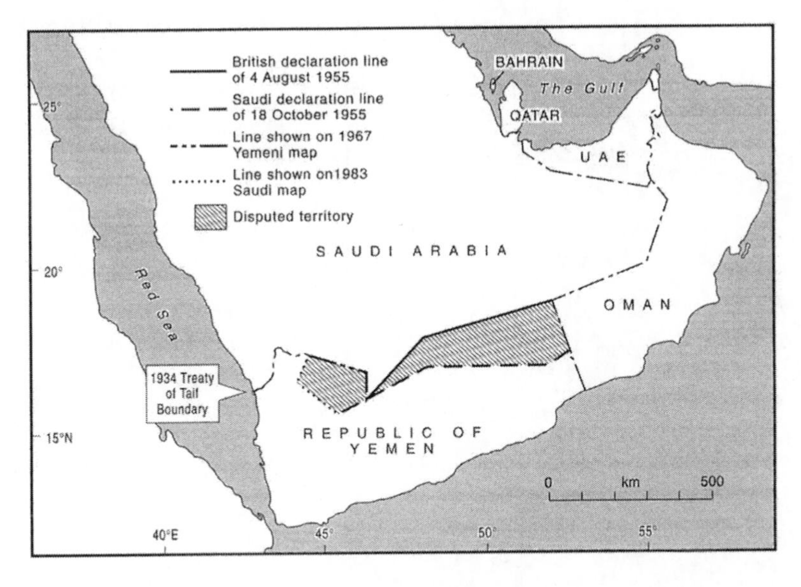

Figure 26.8 The Saudi Arabian–Yemen boundary dispute.

was ever agreed east of the Taif line. Here the border region is sparsely populated, gravelly or sandy desert. Neither party seems ready to compromise, not least because of the discovery of large oil deposits, especially in the central section.

A complex series of claims have been made in the contested region since the early years of the century. They are all straight lines although tribal allegiances to Saudi Arabia or Yemen are allegedly the basis of the claims. Figure 26.8

shows the substantial difference between their claims. The two states have been effectively operating a border zone that is much narrower and lies somewhat between the two extremes. Oil concessions have been made by both states close to the contested boundary, and in 1993 the Saudi government warned foreign oil companies operating on behalf of Yemen that they were inside Saudi territory (Schofield 1994: p. 26). Another problem revealed in discussions was that the Treaty of Taif boundary will have to be resurveyed and demarcated. Since 1934 it has been neglected, and the pillars lost, stolen or destroyed, a classic demonstration of the need to manage agreed international boundaries once they are in place.

Greece–Turkey

Centuries of rivalry between Christian Greece and Moslem Turkey help to explain the bitterness of their dispute over maritime boundary delimitation in the Aegean Sea (Figure 26.9). The Turkish invasion of Cyprus in 1974 and subsequent division of the island into Greek Cypriot and Turkish Cypriot areas has deepened the degree of animosity. Tension has remained high in the Aegean since Turkey attempted oil exploration in contested waters in 1976. On that occasion and again in 1986, Greece and Turkey threatened to go to war, and the dispute remains potentially dangerous in 1998.

Following the defeat of the Ottoman Turkish empire in the First World War, the Treaty of Lausanne (1923) deprived Turkey of all but a handful of islands within three miles of its mainland coast. Greece, on the other hand, possesses hundreds of islands and islets, some of which are very close to the Turkish coast. Although island sovereignty is a grievance in Turkey, it is not the basis of the Turkish quarrel with Greece. According to the 1982 UN Convention on the Law of the Sea, the Greek islands are entitled to up to twelve nautical miles of territorial sea, and to the resources of the adjacent continental shelf. At present, Greece claims only six nautical miles of territorial sea, but that means that almost 44 per cent of the Aegean already falls into Greek territorial waters; with a twelve mile limit, it would exceed 71 per cent

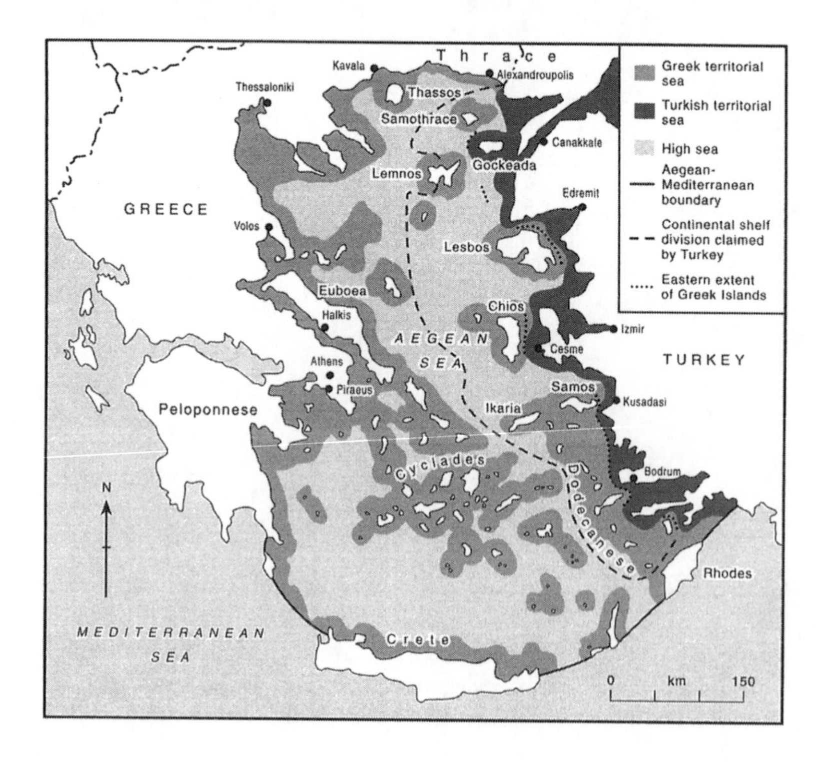

Figure 26.9 The Greece–Turkey dispute in the Aegean Sea.

(Allcock *et al.* 1992: p. 102). Turkey sees the Aegean as geographically a special case; geologically, the Aegean is regarded as the 'natural prolongation' of Anatolia, and Turkey should therefore enjoy continental shelf rights up to a median line between the mainlands. Turkey would gladly allow a six mile territorial sea for the Greek islands, but it demands access to the seabed and its resources between the islands to the median line. Turkey is also determined to maintain freedom of movement for ships and aircraft in the Aegean, and would regard a Greek extension of territorial waters to twelve miles as a *casus beli*. There have been spasmodic talks about these fundamental questions, but no progress has been made. In 1975, Greece and Turkey agreed to place the matter before the International Court of Justice – but could not even agree on the terms of reference.

Cambodia–Thailand

The prospect of hydrocarbons in the Gulf of Thailand (Figure 26.10) began a rush of maritime boundary claims by the coastal states of Thailand, Cambodia, Malaysia and Vietnam in the 1970s. These claims were not always well considered and were complicated by the shape of the gulf, coastal lengths, and the size and location of islands. Today, over 30 per cent of the Gulf of Thailand is subject to overlapping claims. The

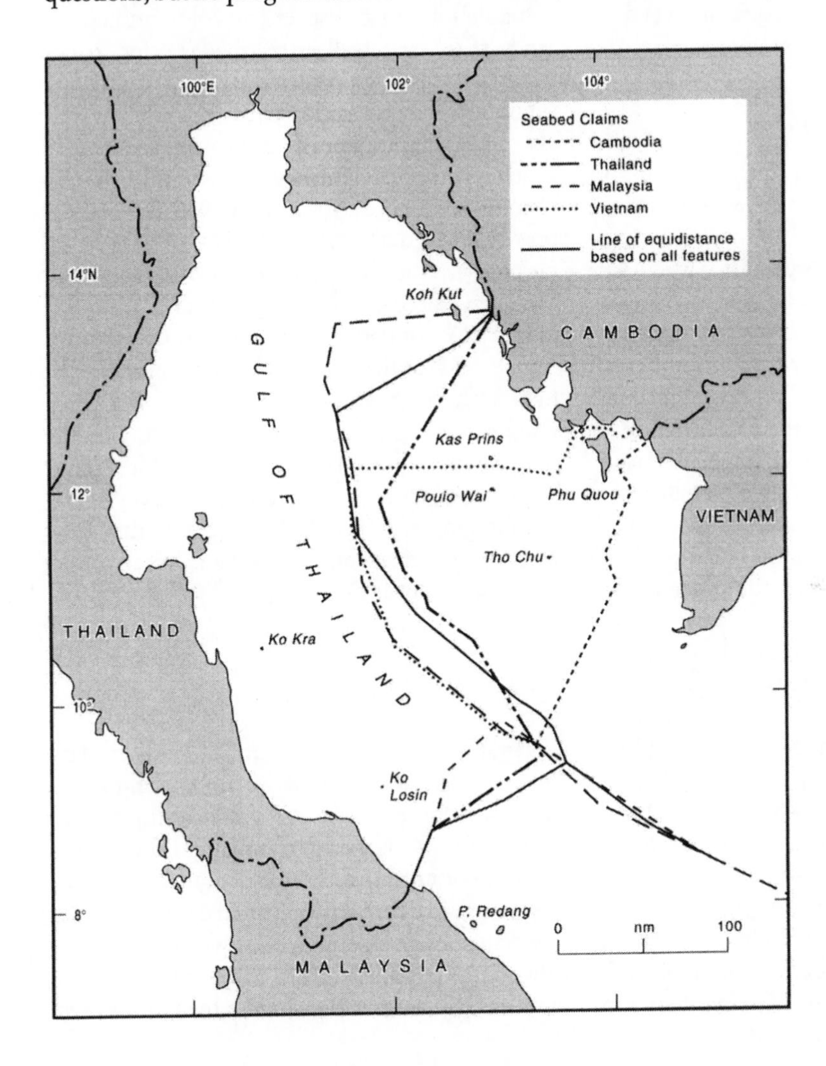

Figure 26.10 The Cambodia–Thailand seabed dispute in the Gulf of Thailand.

Source: Adapted from Prescott 1998 by permission of MIMA.

Cambodia–Thailand case illustrates the kind of problem. Cambodia's line westwards from the land boundary cuts across the Thai island of Koh Kut. Thailand, on the other hand, gives maximum advantage to Koh Kut while conveniently discounting Cambodia's offshore islands. An equidistance line taking into account all features gives yet another result. The seabed contested between the Cambodian and Thai claims is 19,900 km^2. Gas and possible oil deposits have been discovered near the boundary, and the difference in potential hydrocarbon revenues could be colossal. Talks have been convened on a number of occasions in the 1990s to find a solution, without success. Eventually, the two states might propose a joint development zone modelled on the 1992 Thailand–Malaysia agreement.

Boundary dispute resolution

The most common means of dispute resolution is through bilateral *negotiation*, which has the merit of immediacy and economy, and states retain control over the process. In particularly tough cases, a third party may be called in to assist through *mediation* or *conciliation*. Failing this, the dispute may be taken to *arbitration*, either by a tribunal established for the purpose or by an established authority such as the Organisation of African Unity. Some of the most problematic cases may be settled by costly *litigation* before the ICJ or the International Tribunal for the Law of the Sea (ITLOS). Many disputes are resolved through negotiations, although the process may be time-consuming and hard. Paul Huth's invaluable analysis of 116 land boundary disputes between 1945 and 1990 suggests that fifty-seven disputes were resolved, fewer than a dozen of which involved arbitration or litigation (Huth 1996: pp. 195–239). Although nearly half the disputes were settled, it was no easy matter. In fifty-five cases, there had been armed confrontations, sometimes repeatedly, while over 80 per cent had dragged on for more than ten years, one-third of them for over thirty years (Huth 1996: p. 31). Overall, approximately one in three of the world's land boundaries were contested to some degree in the 1945–1990 period, and the proportion remains much the same in 1998.

There is a good chance that more and more boundary disputes will be resolved peacefully in future, for a number of reasons. First, there are proven mechanisms to assist with dispute resolution, including notably the ICJ, and ITLOS in Hamburg. Second, there are strong economic incentives to avoid conflict, especially where oil company investment is sought to explore for hydrocarbons. Third, modern technology such as GPS and geographical information systems (GIS) are facilitating more accurate and rapid boundary delimitation. Fourth, more alternatives to absolute state sovereignty are emerging as valuable interim measures to avoid conflict. On land, there are well-established buffer zones, international zones, protected areas, demilitarised zones, no-fly zones and neutral zones. There are also a growing number of transboundary collaborative ventures requiring the surrender of an element of sovereignty in particular activities in borderlands (Blake 1994a). Most significant of all perhaps are the sixteen joint development zones at sea, several of which were established following failure of boundary negotiations in the spirit of Article 83.3 of the 1982 UN Convention on the Law of the Sea requiring states to enter into 'provisional arrangements of a practical nature' in the absence of a boundary agreement. Most have been remarkably successful, and they provide encouraging evidence that states are finding new ways to organise political space, which offers hope for the future.

CONCLUSION

Geographers have been very much involved with international land boundaries for a century or more, sometimes with considerable distinction. In the past, they took part in delimitation and demarcation commissions, wrote academic commentaries, and drew and interpreted maps. Indeed, few activities lent themselves so well to the application of geographical skills. More recently, geographers have been in demand for their knowledge of maps and cartography, and the application of GPS and GIS in the preparation of cases and as expert witnesses. As described above, geographical training is also extremely valuable in maritime boundary delimitations. For the future, there may be a shift of emphasis from delimitation questions to functional problems, from the boundary line to borderlands. Geographers can contribute much to understanding the dynamics of borders and borderlands, to ensure that borderlands develop as regions of opportunity and not zones of deprivation. None of this implies that the geographical community is either for or against boundaries as we have known them; as long as they are there, we wish them to be managed as peacefully as possible.

GUIDE TO FURTHER READING

Malcolm Anderson (1996) *Frontiers: Territory and State Formation in the Modern World*. Cambridge: Polity Press. A balanced and highly digestible introduction, including thirty pages on Asian and African disputes.

J.B. Allcock *et al.* (eds) (1992) *Border and Territorial Disputes*, 3rd edn. Harlow: Longman. An excellent and reliable summary of most of the key disputes worldwide.

Boundary and Security Bulletin. This quarterly journal published by the International Boundaries Research Unit in Durham monitors boundary disputes worldwide in its news section, and includes articles on boundary matters.

George Demko and William Wood (eds) (1994) *Reordering the World: Geopolitical Perspectives on the 21st Century*. Boulder, Colo.: Westview Press. A superb collection of essays providing a progressive geographical backdrop for boundary disputes.

S.B. Jones (1945) *Boundary-Making: A Handbook for Statesmen, Treaty Editors and Boundary Commissioners*. Washington: Carnegie Endowment. A classic work by a geographer, and a goldmine for border scholars. If you find one buy it!

J.R. Victor Prescott (1985) *The Maritime Political Boundaries of the World*. London: Methuen.

J.R. Victor Prescott (1987) *Political Frontiers and Boundaries*. London: Allen and Unwin. Two standard texts by the world's leading geographer on international boundaries.

REFERENCES

Adler, R. (1995) *Positioning and Mapping International Land Boundaries*. Boundary and Territory Briefing 2(1). Durham: International Boundaries Research Unit.

Beazley, P. (1994) *Technical Aspects of Maritime Boundary Delimitation*. Maritime Briefing 1(2). Durham: International Boundaries Research Unit.

Biger, G. (ed.) (1995) *The Encyclopedia of International Boundaries*. New York: Facts on File.

Blake, G.H. (1994a) International transboundary ventures. In W.A. Gallusser (ed.) *Political Boundaries and Coexistence*. Berne: Peter Lang, 359–71.

Blake, G.H. (1994b) The International Court of Justice Ruling on the Chad–Libya dispute. *Boundary and Security Bulletin* 2(1), 80–3.

Blake, G.H. (1995) The mapping of international boundaries. *Society of University Cartographers Bulletin* 28(2), 1–7.

Blake, G.H., Hildesley, W.J., Pratt, M.A., Ridley, R.J. and Schofield, C.H. (eds) (1995) *The Peaceful Management of Transboundary Resources*. Dordrecht: Martinus Nijhoff.

Boggs, S.W. (1940) *International Boundaries: A Study of Boundary Functions and Problems*. New York: Columbia University Press.

Charney, J.A. and Alexander, L.M. (eds) (1993; 1998) *International Maritime Boundaries* Vols 1 and 2, 1993; Vol. 3, 1998. Dordrecht: Martinus Nijhoff.

Foucher, M. (1988) *Fronts et frontières*. Paris: Fayard.

Francalanci, G. and Scovazzi, T. (eds) (1994) *Lines in the Sea*. Dordrecht: Martinus Nijhoff.

Glassner, M.I. (1993) *Political Geography*. New York: John Wiley.

Gallusser, W.A. (ed.) (1994) *Political Boundaries and Coexistence*. Proceedings of the IGU Symposium Basle, 24–27 May 1994. Berne: Peter Lang.

Goertz, G. and Diehl, P.F. (1992) *Territorial Changes and International Conflict*. London: Routledge.

Griggs, R. and Hocknell, P. (1995) Fourth World Faultlines and the Remaking of 'International' boundaries. *Boundary and Security Bulletin* 3(3), 49–58.

Huth, P.K. (1996) *Standing Your Ground: Territorial Disputes and International Conflict*. Ann Arbor: University of Michigan Press.

International Hydrographic Organisation (IHO) (1990) *Manual on Technical Aspects of the UN Convention on the Law of the Sea*. Monaco: International Hydrographic Bureau.

Jagota, S.P. (1985) *Maritime Boundary*. Dordrecht: Martinus Nijhoff.

Kolossov, V.A., Glezer, O. and Petrov, N. (1992) *Ethno-Territorial Conflicts and Boundaries in the*

Former Soviet Union. Boundary and Territory Briefing. Durham: International Boundaries Research Unit, pp. 1–51.

Luard, E. (ed.) (1970) *International Regulation of Frontier Disputes*. New York: Praeger.

McDorman, T.L. and Chircop, A. (1991) The resolution of maritime boundary disputes. In E. Gold (ed.) *Maritime Affairs: A World Handbook*, 2nd edn, Harlow: Longman.

Prescott, J.R.V. (1988) *The Gulf of Thailand*. Kuala Lumpur: The Maritime Institute of Malaysia.

Prescott, J.R.V. (1996) Contributions of the United Nations to solving boundary and territorial disputes, 1945–1995. *Political Geography* 15(3/4), 287–318.

Prescott, J.R.V. (1998) National rights to hydrocarbon resources of the continental margin beyond 200 nautical miles. In G.H. Blake, M. Pratt, C. Schofield and J.A. Brown (eds) *Boundaries and Energy: Problems and Prospects*, London: Kluwer Law International, 51–82.

Roach, J.A. and Smith, R.W. (1994) *United States Responses to Excessive Maritime Claims*, 2nd edn, The Hague: Martinus Nijhoff.

Rushworth, D. (1997) Mapping in support of frontier arbitration. *Boundary and Security Bulletin* 5(3), 55–60.

Schofield, R.N. (ed.) (1994) *Territorial Foundations of the Gulf States*. London: UCL Press.

Smith, R.W. and Thomas, B.L. (1998) *Island Disputes and the Law of the Sea: An Examination of Sovereignty and Delimitation Disputes*. Maritime Briefing 2(4). Durham: International Boundaries Research Units 1–27.

Tägil, S. *et al.* (1977) *Studying Boundary Conflicts*. Lund Studies in International History, Lund: Essette Studium.

United Nations (1982) *Convention on the Law of the Sea*. New York: United Nations Publications.

US Department of State (1987) *Developing Standard Guidelines for Evaluation of Straight Baselines*. Limits in the Seas No. 106. Washington: Bureau of Oceans and International Environmental and Scientific Affairs.

27

Political spaces and representation within the state

Ron Johnston

> regional geographers may perhaps be trying to put boundaries that do not exist around areas that do not matter.
>
> (Kimble 1951: 159)

'Region' is one of the commonest words in the geographical lexicon, adopted by adherents to a range of different philosophies within the discipline as a key concept with which their area of study can be identified. Much effort has been expended on the definition of regions, a great deal of it *ad hoc*. To some, defining and describing regions is the highest form of the geographer's art (Hart 1982), whereas to others, like Kimble (1951), regions are largely irrelevant, geographers' constructions of reality rather than reality itself: those constructions may become reality, however, as illustrated in this chapter.

Two major types of region have been identified. *Formal regions* are relatively homogeneous areas on one or more predetermined characteristics – whether physical (such as climatic regions), human-made (social areas, say), or both (landscapes). Their definition involves determining the salient criteria, mapping those over the selected area and defining the boundaries around the separate regions, either subjectively, using a single identifier (such as the number of frost-free days per annum), or by statistical procedures based on the analysis of variance, in which the units within each region are more like each other than they are like the units in adjacent regions. The result is a mosaic of areas, with each relatively homogeneous internally. *Functional regions* are defined on a more limited range of characteristics, usually flows: the goal is to define areas dominated by a particular flow pattern – usually focused on a node (hence the alternative term 'nodal region'). The outcome is a set of regions (which may not each comprise single contiguous blocks of territory), each focused on a particular core – such as the hinterland of a shopping centre or market town and the commuter-shed of a factory or industrial estate.

From Kimble on, some geographers have argued that both types of region are largely irrelevant in the contemporary world, because of the growing interconnectedness of life, frequently expressed in relatively vague concepts such as 'globalisation' and 'the global village'. Against this, it is argued that regions are crucial elements of the structuring of economic, social, cultural and political life, for three main reasons:

1. Regions (or places, or localities) provide the contexts within which most people are socialised, particularly although not only during the early years of their lives. We learn to be people in contexts that are both culturally and territorially defined, and from those among whom most of our daily interactions take place. Local cultures are spatially constrained – at a variety of scales – and their structuration (i.e. their creation and continual recreation) is the basic cause of the complex mosaic of cultural regions that comprises the contemporary world.
2. Although information technology allows the rapid movement of ideas and abstract

commodities (such as money) around the world almost instantaneously, and virtual reality will allow people to co-locate even when they are physically apart, nevertheless many constraints to human movement limit the areas within which regular journeys – such as the journey to work and the journey to shop – can be made. Severe limits to where a person can move to from a set base within a defined period will remain crucial for many years, imposing important constraints on the organisation of economic, cultural, social and political activity: functional regions, however much they overlap, are a key component of the spatial structures within which lives are necessarily organised.

3 Territoriality is a key strategy for programmes of control (Sack 1986). Bounded spaces can be defined physically, and power over people exercised within them relatively easily – compared with the alternative of exerting power and influence over people, wherever they are. Such areas provide both refuges and prospects – safe havens, retreats from which unwanted potential trespassers can be excluded, and yet also outlooks on the 'world outside'. Surveillance is feasible, allowing control to be exercised. For these and a range of other reasons, states, which are institutionalised apparatus for the exercise of power, influence and control, are all associated with bounded territories – and since it is widely accepted that states are necessary to the operation of complex societies, it follows that a division of the world into bounded territories within which their power strategies can be exercised is a necessary component of spatial structure.

Regions, both formal and functional, are thus key elements in the organisation of many aspects of society, therefore, from the individual right up to the nation-state and its emerging successor, the international regime.

REGIONS IN CONTEMPORARY SOCIETY

Given this general argument, regional definition is clearly a necessary component of how people, communities and nation-states are structured. People want to live in communities with similar others – hence the processes of separation that mark the creation of distinct residential districts in urban areas. Their residents (often identified as communities) also want to be represented separately in relevant democratic forums and decision-making arenas: they want to ensure not only that their separate points of view are heard but also that they are allocated sufficient representation to ensure that they are influential. Thus regions, or communities of interest, are central elements to the structuring of daily lifeworlds and their political representation, at a variety of scales.

Regions are also important to the state's internal structuring. Territoriality is a key strategy in the exercise of state power and is as important to their internal as it is to their external relations. Most states are too extensive for all aspects of administration to be undertaken from a single point. Central control needs to be allied with local application, allowing easier surveillance at local scales (whether in the maintenance of law and order, as with policing, or in the implementation of policies, such as tax collection); easier appreciation of local needs and requirements, which can be catered for separately from that undertaken in other areas; and the ability to involve local people in the governance of their home areas by making it locally accountable. Thus all states have a local state apparatus, with various degrees of independence from the central state; most local state apparatus comprises a complex, overlapping system of administrative and governmental areas, including all-purpose (within constitutional limits) local governments, for example single-purpose local governments, and *ad hoc* quangos serving areas defined for that purpose alone.

The political impulse

Regions, broadly defined as separate territorial segments of space, are crucial elements in the structuration of many aspects of social organisation. It should follow that as the study of regions is a central element of the geographical discipline, then the definition of regions for a variety of purposes should be based on geographical appreciation and methods.

A major caveat has to be inserted at this stage of the argument, however, one that is particularly relevant to regions that are part of the state apparatus. As pure scientists, geographers adopt objective stances to their subject matter – so that in defining regions they operate the sequence outlined above (determining the criteria then delimiting the boundaries): the resulting regional definitions are answerable only to the pre-determined criteria. But as applied scientists, geographers invited to define regions are often asked to delimit areas that fit a particular need, which may not easily be integrated with their credentials of objectivity and neutrality. People wanting a region to be defined usually require it to serve a particular purpose that fits their political agenda; the geographers employed to advise on the boundary-drawing process may feel under pressure to compromise their 'scientific credentials' in order to produce what is wanted by the political bodies who commissioned the regionalisation – or their own political agenda may lead them to adopt such compromises.

This difficulty is exacerbated because even 'objective' scientific methods for regionalisation call for subjectivity in determining and applying criteria (on which Johnston (1968) is still relevant). Very few regions are clearly delimited on the ground, especially where human decision making is involved: few residential areas are entirely homogeneous in their population characteristics, for example; many people prefer to shop at centres other than those patronised by the majority of their neighbours, or to commute to workplaces well outside the area that constrains most of their fellow residents; and most agricultural regions include some farms whose managers operate different regimes from their neighbours, even though the objective conditions in which they operate are the same. In other words, the definition of regions is a fuzzy activity: there is rarely a single right answer.

Given this combination of political needs and a multitude of acceptable answers, we can appreciate that although regional definition is a central geographical activity, relatively few geographers have been involved in the political task of regional boundary drawing: the task is too sensitive to leave to 'objective scientists'. This argument is developed with a number of brief case studies.

DEFINING REGIONS: THE GEOGRAPHER'S DILEMMA

Consider the problem of defining regions on a crucial criterion, their residents' racial–ethnic status. Very few areas are racially homogeneous, certainly very few areas of any considerable spatial extent within a continental land mass (Mikesell and Murphy 1991): over time, various social and other processes produce intermixing, unless strong policies of ghettoisation (i.e. regionalisation with strong surveillance mechanisms) are imposed – as in South Africa's notorious *apartheid* policy, which separated the races at the individual scale (banning inter-racial sexual conduct and marriage), at the scale of individual buildings and vehicles (post offices had separate counters for blacks and whites, and trains had separate carriages), in residential areas (blacks could live in defined townships only), in workplaces (occupations were reserved to either blacks or whites), and by 'nation' of residence (all blacks were allocated to a homeland, of which they were citizens, and they could enter 'white South Africa' only with passes).

If the Republic of South Africa for forty years of its recent history is a paradigm example of explicit political regionalisation based on homogeneous areas, the Balkans is a clear example of the counter case – a racially very mixed area, where defining homogeneous regions based on population characteristics is fraught with difficulties. This was apparent when the modern system of nation-states was being created in the area under the Treaty of Versailles, which redrew the world political map after the First World War. A Serbian geographer (Johan Cvijic) was involved in this unenviable task, but his work was strongly influenced by political considerations reflecting both his own opinions and those of the people who employed him.

One of the areas under dispute between several countries was Macedonia, where 'the Albanian, Greek, Bulgarian and Serbian linguistic provinces meet and overlap, and where in addition exclaves of Romanian and Turkish speech are found' (Wilkinson 1951: 3). In the early years of the twentieth century, Serbia laid claim to the area, as part of its expansionist policies, and one of its main tools in this was cartographic evidence that the residents were part of the Serbian linguistic province. Several of those maps, some of which influenced diplomats from other countries, especially after the First World War, were drawn by Cvijic. Wilkinson presents five of them (Figure 27.1), on which basis he accuses Cvijic of 'gross inconsistency' (p. 176):

> Even scientists of the highest personal integrity were guilty of the practice of misrepresentation, excusing themselves on the ground that the end justified the means. In some cases, notably that of J. Cvijic, an unmerited, perhaps unconscious rationalization of false distributions was prompted by the irresistible spirit of patriotism of the period.

As Figure 27.1 shows, Cvijic extended the Serb area southwards by adopting the concept of Macedo-Slavs, a group previously regarded as Bulgarian: this idea was not widely accepted in 1908 but gained greater support later (according to Wilkinson, after 1918, 'the popularity of his ethnographic map knew no bounds' in western Europe: p. 182): his 1913 map was 'designed to support Serbia's plan for a reorganisation of the Western Balkans' (*ibid.*: 180) and had a major impact on the final process of boundary drawing there, with the consequence that 'the Slavs of Serbian Macedonia were denied any freedom of self-expression, and for all practical purposes were held to be Serbian in culture and in national outlook' (p. 235).

Defining such a political map may advance one 'national interest', but in many cases it may do no more than contain inter-community strife while surveillance and power can be exercised. This occurred with the 1919 creation of the Federal Republic of Yugoslavia, one of whose component states – Bosnia-Herzegovina – contained a complex mosaic of regions (some far from homogeneous) occupied by members of three separate ethnic communities: Croats, Serbs, and Muslims. While surveillance was strict their differences were contained, but collapse of the Yugoslav state in the 1980s and the Serbs' growing hegemonic project stimulated war, plus ethnic cleansing strategies whereby the dominant group in an area excluded the others to create an 'ethnically pure' region. Ending that strife and creating a new political map was a major problem, which increasingly it seemed could not be resolved locally, and international mediation was introduced.

Two politicians involved in that tortuous mediation process for several years were David Owen, a former UK Foreign Secretary and leader of the Social Democratic Party, and Cyrus Vance, a former US Secretary of State. The racial composition of the country's *opstina* before 'ethnic cleansing' began is shown in Figure 27.2. Owen and Vance discussed five options for restructuring the state apparatus and chose 'a centralized federal state with significant functions carried out by between four and ten regions' (Owen 1996: 65) as 'the best compromise . . . since much of the predicted intercommunal friction could be kept from the central government by giving the provinces competence over the most divisive issues, e.g. police, education, health and culture, while depriving them of the right to be a state within a state'. Defining those provinces was a key regionalisation task: they decided on ten, which were largely groups of contiguous *opstina* although with 'corridors' in some places dividing them (Figure 27.3). This plan failed for a range of political reasons (on which see Owen 1996) and the ethnic cleansing continued, resulting in a very different ethnic map (Figure 27.4), which formed the basis for a final division into two separate republics within a federal state after the Dayton accords were signed in November 1995 (Figure 27.5). One of those republics, the Muslim–Croat Federation, was itself divided into ten cantons to

Figure 27.1 Cvijic's five ethnographic maps of the Macedonia area.

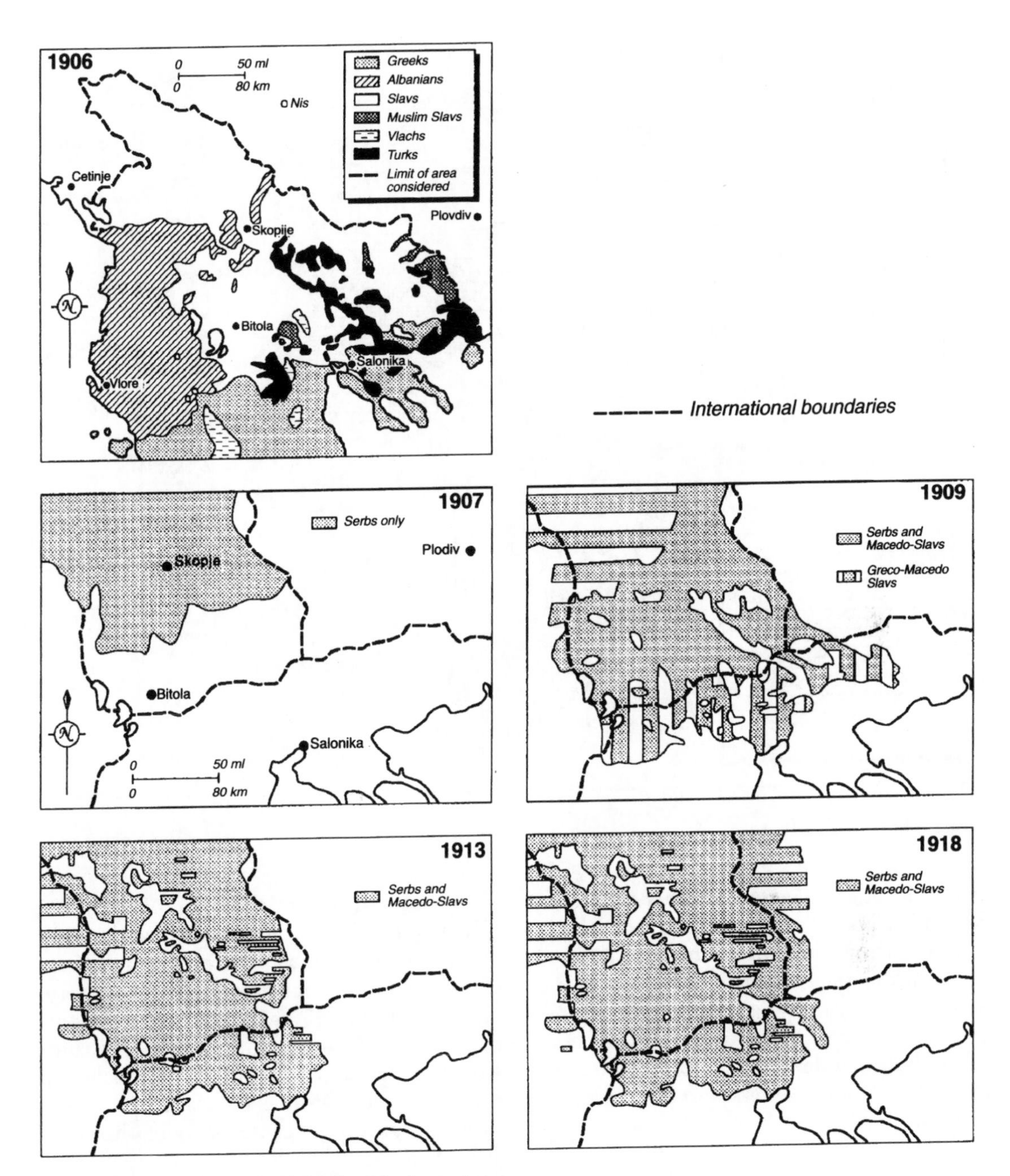

Source: Wilkinson 1951: reprinted with the author's permission.

Figure 27.2 The ethnic map of Bosnia-Herzegovina before *c.*1990.

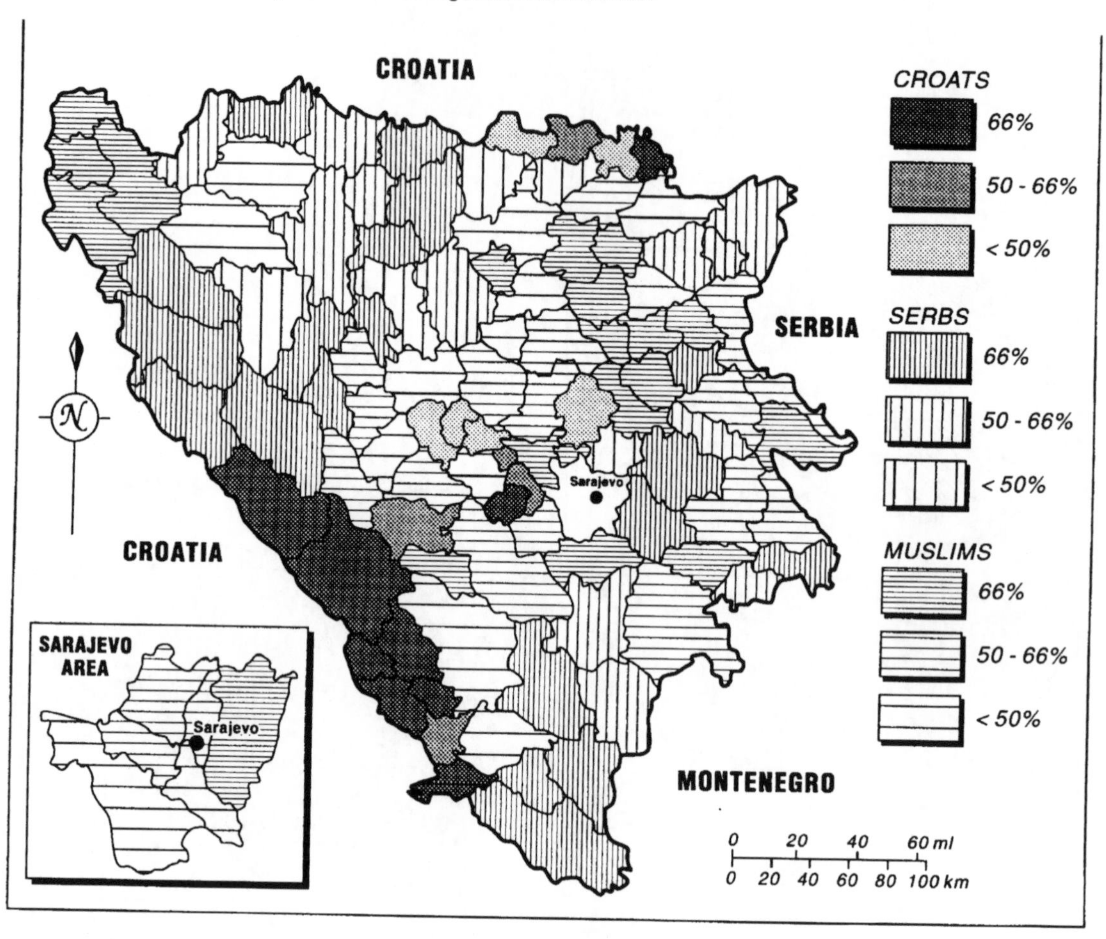

Source: Owen 1996.

reflect the ethnic diversity, whereas the other – Republica Sprska – was over 98 per cent Serb.

The extremes involved in political regionalisation within states exemplified by *apartheid* in South Africa and ethnic cleansing in Bosnia-Herzegovina are not repeated in many other places, but such deviant cases bring into sharp focus the problems common to most regionalisations.

Constituencies as regions

Government of most liberal democracies involves legislative assemblies popularly elected from territorial constituencies. In some, like the UK and the USA, those constituencies each return a single member to represent a separate (almost invariably contiguous) block of territory: there are currently 659 in the UK House of Commons and 435 in the US House of Representatives. Because of the criteria on which they are based, the constituency maps have to be redrawn regularly: a new set of regions has to be defined.

Those criteria are embodied in four concepts of fairness, which underpin many 'theories' of democratic representation.

Figure 27.3 The February 1993 Vance–Owen plan for dividing Bosnia-Herzegovina.

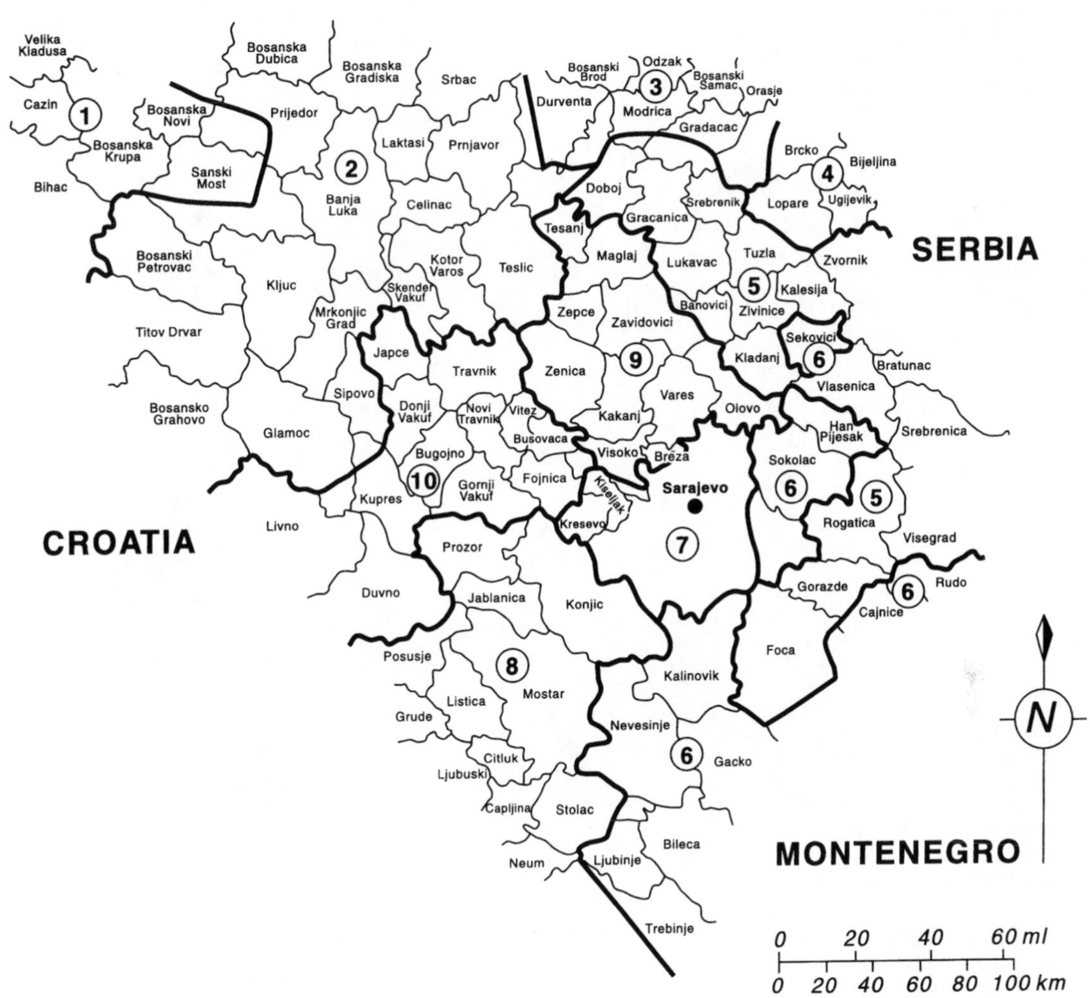

Source: Owen 1996.

- *Fairness to individuals.* All citizens should have equal power, with each person's vote worth the same as everybody else's: if that is not the case, then unequal power to influence the outcome of an election results and some preferences carry less weight than others. To meet this criterion, constituencies should have equal-sized electorates.
- *Fairness to communities.* A state's civil society is not an aggregation of atomised individuals but comprises separate geographically defined communities with their own cultures and interests. Each should be represented, otherwise power is not fairly distributed among society's major interest groups: if those communities are spatially delimited, the fairness criterion can be met by making community and constituency boundaries coterminous.
- *Fairness to minorities.* Few national societies are culturally homogeneous; most contain one

Figure 27.4 The ethnic map of Bosnia-Herzegovina in October 1995.

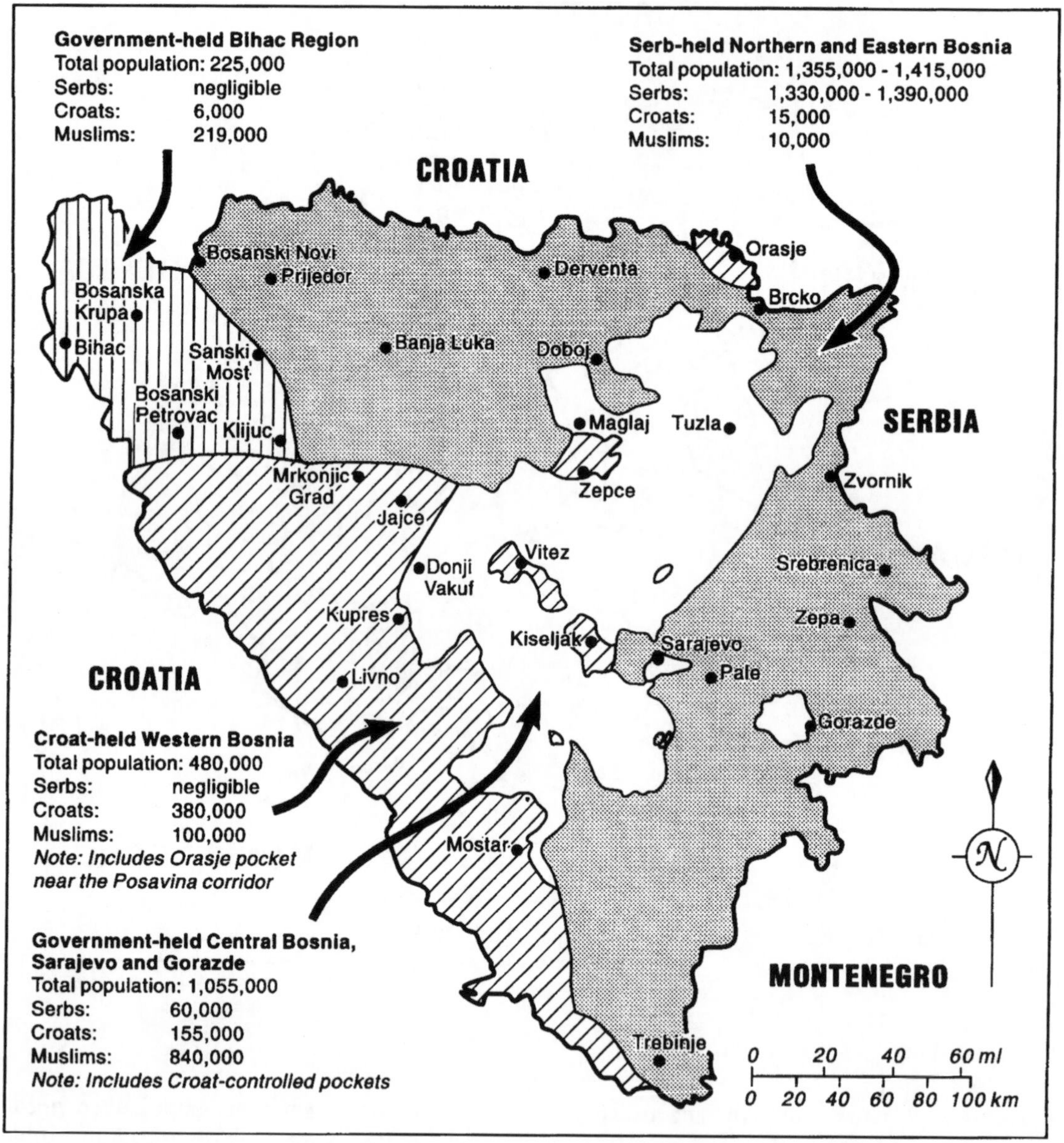

Government-held Bihac Region
Total population: 225,000
Serbs: negligible
Croats: 6,000
Muslims: 219,000

Serb-held Northern and Eastern Bosnia
Total population: 1,355,000 - 1,415,000
Serbs: 1,330,000 - 1,390,000
Croats: 15,000
Muslims: 10,000

CROATIA

Bosanski Novi
Prijedor
Derventa
Orasje
Brcko
Bosanska Krupa
Bihac
Banja Luka
Doboj
Sanski Most
Bosanski Petrovac
Kljuc
Maglaj
Tuzla
SERBIA
Mrkonjic Grad
Jajce
Zepce
Zvornik
Donji Vakuf
Vitez
Srebrenica
Kupres
Zepa
Kiseljak
Sarajevo
Pale
CROATIA
Livno
Gorazde

Croat-held Western Bosnia
Total population: 480,000
Serbs: negligible
Croats: 380,000
Muslims: 100,000
Note: Includes Orasje pocket near the Posavina corridor

Mostar

Government-held Central Bosnia, Sarajevo and Gorazde
Total population: 1,055,000
Serbs: 60,000
Croats: 155,000
Muslims: 840,000
Note: Includes Croat-controlled pockets

MONTENEGRO
Trebinje
0 20 40 60 ml
0 20 40 60 80 100 km

Source: Owen 1996.

or more separate minorities defined by criteria such as ethnicity, language and religion. National cohesion and stability depend on these minorities being fully integrated and equally treated within the society, frequently raising questions regarding either positive discrimination or affirmative action. The design of electoral systems may involve ensuring that significant minority groups form a majority of the electorate in a proportion of constituencies, consistent with their relative size.

- *Fairness to political parties.* Political parties are

Figure 27.5 The changed ethnic map of Bosnia-Herzegovina and the division after the Dayton Accord.

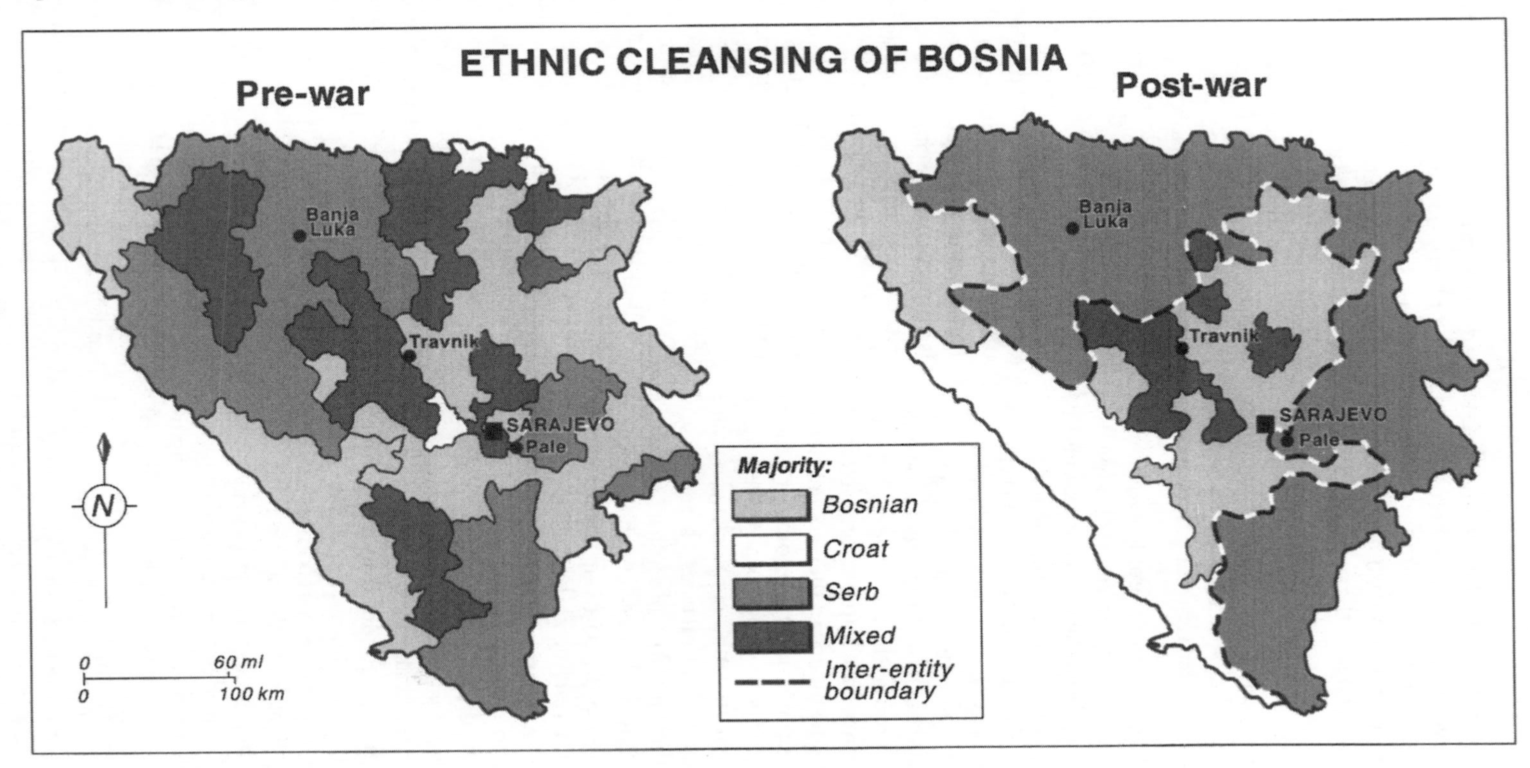

Source: *The Economist*, 24 January 1998.

at the centre of contemporary liberal democracies for two main reasons: they provide stability for a government within a legislature, and continuity of support among a portion of the electorate. They are the focus of electorate mobilisation and government organisation, and therefore, it is argued, should achieve a level of representation in the legislature consistent with their electoral support – hence the term proportional representation.

Only the first two of these are implemented in the UK, through the rules for redistribution that the independent Boundary Commissions are required to apply under the *Parliamentary Constituencies Act 1986*. (The first criterion is violated by the rules, however, since Scotland and Wales are over-represented relative to England and Northern Ireland: such over-representation is not explicitly linked to the third criterion.) For each administrative area, the Commissions determine the number of constituencies they are entitled to and then draw up provisional recommendations for their composition using local government electoral wards as the building blocks. They are required to make constituencies as equal as is practicable in their number of electors (meeting the fairness to individuals criterion) and to respect local community ties (i.e. to group together areas with common interests, according to the second criterion). Those recommendations are subject to public consultation when the major participants are the political parties, which want to ensure constituencies are created to their electoral advantage; they may present alternatives to an assistant commissioner at a public local inquiry, whose report is used by the commission to determine whether to alter its provisional recommendations before making final recommendations to Parliament. (For a full discussion, see Rossiter *et al.* 1998.) Electoral considerations are crucial in the debates over alternative constituency configurations, therefore, as exemplified by the City of Sheffield (see Box 27.1).

In the USA, there are no legal conditions defining how constituencies should be determined, and separate procedures were developed in the various states. In the 1960s, however, the Supreme Court upheld cases brought by plaintiffs who claimed that congressional districts violated the equal protection clauses of the American Constitution because they had unequal electorates: malapportionment was outlawed, and districts with equal electorates were mandatory. Subsequent judgements made this a very stringent demand, culminating in a case that rejected a recommended redistricting of New Jersey – in which the congressional districts varied in their populations by less than 1 per cent around an average of more than 600,000 – because plaintiffs showed that even greater equality was possible.

Fairness to individuals dominates US redistricting as a consequence of these judgements. Fairness to communities has not been claimed (in large part because States are equally represented in the Senate, irrespective of their size). Fairness to minorities became important after the 1960s civil rights movement and was interpreted in the states covered by the *Voting Rights Act* as requiring each to have a proportion of its Congressional Districts with a black majority equivalent to the black proportion of the State's population (i.e. if 40 per cent of a State's population is black and it has ten Congressional Districts, then four of them should have a black majority – what is known as the minority-majority requirement): recent decisions suggest that this requirement should not be implemented if to do so involves race as the predominant factor in the redistricting plan. Some interpret this as being the case if minority-majority Districts are oddly-shaped but the task of choosing a set of Districts is extremely large and race may predominate even though there are no oddly shaped districts – as Box 27.2 shows.

The fourth criterion – fairness to parties – has been introduced in some States, where the redistricting procedure was vested in the party controlling the State legislature and resulted in an 'unfair' outcome for the other party (two parties – Republican and Democratic – predominate in American politics). Legal complaints against such gerrymandering led to the appointment of independent individuals or bodies to produce

Box 27.1 Redistricting Sheffield

The size of the task faced by the UK Boundary Commissions is illustrated by the city of Sheffield. During its Third Periodic Review of all constituencies, the Boundary Commission for England determined that Sheffield should have six parliamentary constituencies, which would be combinations of city council wards (the commission always uses wards as the building blocks for creating constituencies: the wards, and their estimated Labour vote in 1979, are in Figure 27.6A). Those six should be contiguous blocks of territory, and each should have an electorate as close as practicable to 65,753 – the national electoral quota. The Commission's provisional recommendations (Figure 27.6B) allowed up to 10 per cent variation around this norm.

How many ways could six contiguous constituencies have been created out of the twenty-nine wards, within a 12 per cent variation around the average? A computer algorithm

A

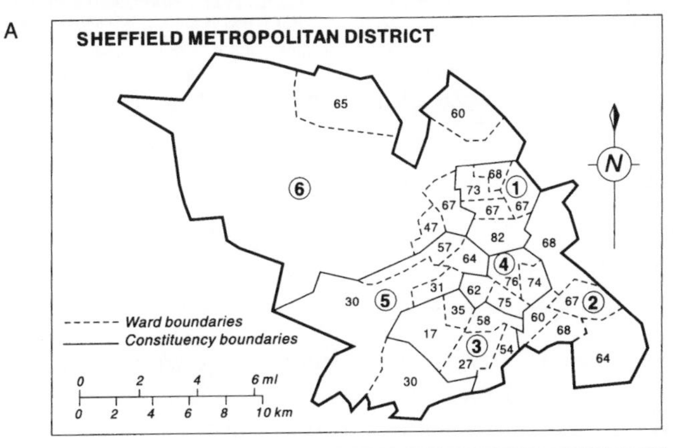

B

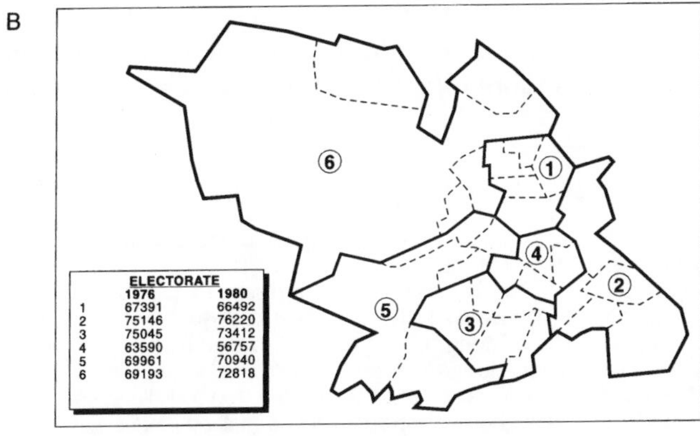

	ELECTORATE 1976	1980
1	67391	66492
2	75146	76220
3	75045	73412
4	63590	56757
5	69961	70940
6	69193	72818

C

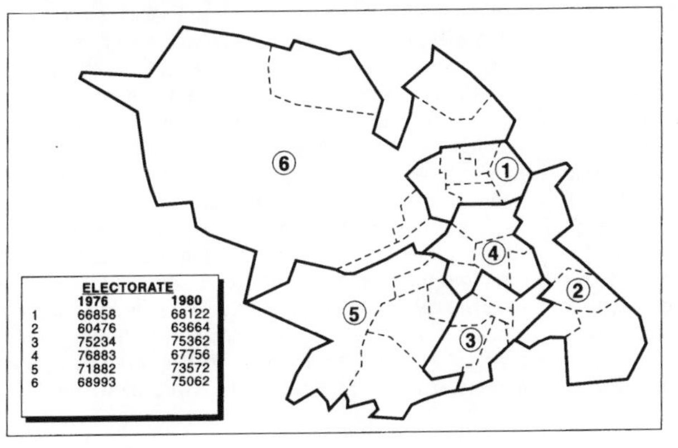

	ELECTORATE 1976	1980
1	66858	68122
2	60476	63664
3	75234	75362
4	76883	67756
5	71882	73572
6	68993	75062

Figure 27.6 Redistricting Sheffield: (A) the wards and their political complexion (estimated Labour support); (B) the Boundary Commission's provisional recommendations for six constituencies; (C) the Boundary Commission's final recommendations.

Source: Johnston and Rossiter 1982.

Box 27. 1 continued

identified 13,317 (Johnston and Rossiter 1982), indicating that the commission had a major task to identify that which it felt best met the various criteria it was supposed to meet.

An important feature of the provisionally recommended six constituencies is that, if how Sheffield's electors voted at the most recent (1980) local government elections was a reasonable guide to how they would vote in a general election, then four of those six were likely to be won by Labour and one by the Conservatives, compared with five for Labour and only two for their opponents in the most recent (1979) general election fought in the old constituencies. Statistically, this was something of a surprise since, as the following table shows, most of the possible 'solutions' would have retained the *status quo* of a 5:1 Labour victory.

Seats won	*Solutions*
Labour 6: Conservative 0	697
Labour 5: Conservative 1	12,327
Labour 4: Conservative 2	2,913

Clearly, the Labour Party wished to get this changed and presented an alternative scenario at the local inquiry, whereas the Conservatives sought to convince the Assistant Commissioner to recommend that the provisional recommendations be retained as the final version. In addition, two geographers presented independent evidence promoting a set of constituencies that were more equal in their electorates than that provisionally recommended, by 1980 electoral data if not those for 1976, which the Commission was using. This last formed the basis for the Assistant Commissioner's recommendation for a different set of constituencies (Figure 27.6C), which was accepted by the Commission (and in which Labour won five seats at the next general election).

more neutral configurations. In 1971, geographer Dick Morrill was appointed to produce a plan for Washington state without any knowledge of what the two parties had proposed. Table 27.1 shows the outcome for the 102-seat lower house of the Washington State legislature. In 1970, each party had won fifty-one seats, with slightly more of the Republicans' than the Democrats' being marginal. Each party's plans would have increased its number of fairly safe and safe seats (especially the former: no point in piling up very large majorities) and reduced the numbers of marginals

Box 27.2 Redistricting Mississippi

The population of the State of Mississippi was around 40 per cent black in the 1960s, but none of the State's five congressional districts defined in 1966 had a black majority, the main area of black population (the Delta region) being split between four of the five districts. This 'disenfranchisement' of the blacks was continued in the State's districting plans after the 1970 and 1980 censuses, but the latter was challenged under the *Voting Rights Act* in 1981 and a court created a black majority district in the Delta region; this was largely retained after the 1990 census, and the district has been won by a black candidate since 1986.

What is the likelihood of a districting plan for Mississippi including no black majority district, given the size of the black population and its concentration in one part of the State? This question was addressed using a similar computer method to that employed by Johnston and Rossiter (1982) in Sheffield. The 'building blocks' used in American redistricting are much smaller than the wards used in the UK, and although the equal size criterion is much more rigorously applied, the number of possible districting plans is very large – even if other constraints, such as 'shapeliness' and not splitting counties, are also applied. The procedure allowed only a 1 per cent maximum deviation around the average population for the five districts, and generated 100,000 separate plans. In half of these, at least one district had a black majority, and sixty-eight had two.

The implication is that a districting plan for Mississippi without a black majority would be suspect, with the boundaries having been drawn to prevent a black representative being elected from the State (a classic example of a negative gerrymander). But none of the 100,000 plans generated through the computer-intensive method produced a district with as large a black population as District 2 in Figure 27.7, suggesting that blacks were 'packed in' to it; a larger black majority district had been created by careful cartography than was likely to occur by chance. Given that the adopted plan splits more than three times as many counties as do plans created by the 'County-conscious algorithms', the results suggest that race was given undue precedence over preservation of the integrity of political subdivisions.

In this case study, computer programs have been used, not to generate a solution to a regionalisation problem, therefore, but rather to provide a datum against which to evaluate regionalisations. By creating a large frequency distribution – a sample of all possible regionalisations in that context – an 'objective' means of testing whether a proposed regionalisation is 'unusual' has been provided.

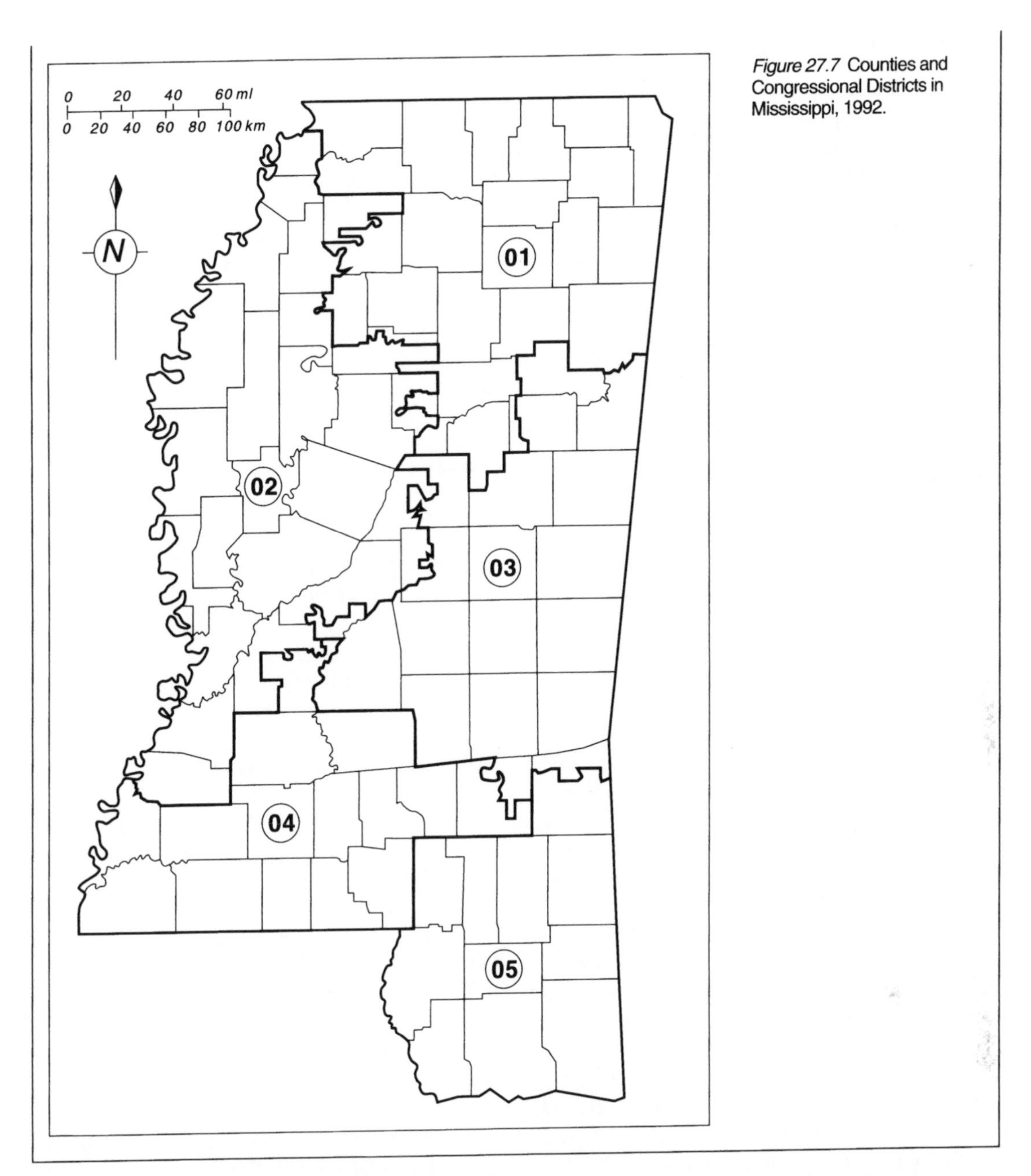

Figure 27.7 Counties and Congressional Districts in Mississippi, 1992.

and those won by its opponent. Morrill's neutral court plan 'reproduced' the 1970 election almost exactly, however, with many more marginals than either party wanted and a much more even division of the safer seats, and this was reflected in the 1972 election result, except that most of the marginals went to the Democrats.

In this case, the independent redistricter was used as a check on the partisan ambitions of the parties. This potential use of social science cartographic skills to evaluate plans has recently been extended much further with regard to the third criterion, which became increasingly important in the redistricting process from the 1970s on

Table 27.1 Redistricting the lower house of Washington's State legislature.

	1970 election	*Party plans*		*Court plan*	*1972 election*
		Democrat	*Republican*		
Democrat					
Safe	15	10	17	16	16
Fairly Safe	28	40	21	25	27
Marginal	5	12	12	19	7
Marginal	5				3
Fairly Safe	25	20	36	24	24
Safe	14	16	12	14	14
Republican					

Source: Morrill 1973.

because of concerns that blacks and other minorities were under-represented. Box 27.2 illustrates this, as yet unrealised, potential.

CONCLUSIONS

The easiest problems to tackle in most situations are those to which there is a single, right solution: the task is to determine that solution, and then implement it. Unfortunately, most problems in applied geography – and certainly so in the area discussed in this chapter – have no single right solution. Instead, there is a multiplicity of solutions, some at least of which will be favoured by different interest groups. The task of people allocated such problems is thus not solution but resolution: they must not only identify feasible solutions but recommend either that which is favoured by the interest group that makes the strongest case for its preferred outcome or that which seems to provide most for everybody (i.e. is the least worst outcome for all, even if not the best for anyone).

Resolution rather than solution is clearly the case with the application of principles of regionalisation to the definition of political spaces and representation within the state. There is a large range of possible outcomes to each problem, because of the 'messiness' of the geographical mosaic onto which the regionalisation is to be imposed and range of interest groups (most of them political in the widest usage of that term), which differ in their preferred outcomes.

GUIDE TO FURTHER READING

There is a massive literature on regions and regional geography: the most recent general survey is provided by Claval (1998), whereas a more critical stance is covered in several of the essays in Johnston *et al.* (1990). The classic work on territoriality as a spatial strategy – in effect, a form of practical regionalisation – is by Sack (1986). On the drawing of constituency boundaries in the United Kingdom, the standard work is Rossiter *et al.* (1999).

REFERENCES

Claval, P. (1998) *An Introduction to Regional Geography*. Oxford: Blackwell.

Freeman, T.W. (1967) *The Geographer's Craft*. Manchester: Manchester University Press.

Hart, J.F. (1982) The highest form of the geographer's art. *Annals of the Association of American Geographers* 72, 1–29.

Johnston, R.J. (1968) Choice in classification: the subjectivity of objective methods. *Annals of the Association of American Geographers* 58, 575–689.

Johnston, R.J., Hauer, J. and Hoekveld, G.A., (ed.) (1990) *Regional Geography: Current Developments and Future Prospects.* London: Routledge.

Johnston, R.J. and Rossiter, D.J. (1982) Constituency building, political representation and electoral bias in urban England. In D.T. Herbert and R.J. Johnston. *Geography and the Urban Environment, Volume 5*. Chichester: John Wiley, 113–56.

Kimble, G.H.T. (1951) The inadequacy of the regional concept. In L.D. Stamp and S.W. Wooldridge (ed.) *London Essays in Geography: Rodwell Jones Memorial Volume*. London: Longmans Green, 151–74.

Mikesell, M.W. and Murphy, A.B. (1991) A framework for comparative study of minority-group aspirations. *Annals of the Association of American Geographers* 81, 581–604.

Morrill, R.L. (1973) Ideal and reality in reapportionment. *Annals of the Association of American Geographers* 63, 463–77.

Owen, D. (1996) *Balkan Odyssey*. London: Indigo.

Rossiter, D.J., Johnston, R.J. and Pattie, C.J. (1999) *The Boundary Commissions: Redrawing the UK's Map of Parliamentary Constituencies*. Manchester: Manchester University Press.

Sack, R.D. (1986) *Human Territoriality: Its Theory and History*. Cambridge: Cambridge University Press.

Vasovic, M. (1980) Jovan Cvijic, 1865–1927. In T.W. Freeman and P. Pinchemel (eds) *Geographers: Biobibliographic Studies, Volume 4*. London: Mansell, 25–32.

Wilkinson, H.R. (1951) *Maps and Politics: A Review of the Ethnographic Cartography of Macedonia*. Liverpool: University of Liverpool Press.

28

Housing problems in the developed world

Keith Jacobs

INTRODUCTION

It is paradoxical that in spite of economic growth and technological innovation in the developed countries of the world, many people still experience serious housing problems over the course of their lives. These problems are most prevalent for working people unable to secure regular employment and those households who, through ill-health or old age, are reliant on state welfare for their income. Poor-quality housing has important implications. Inadequate housing can undermine good health, impede educational attainment and jeopardise an individual's employment prospects. At a social level, poor housing has a detrimental impact on the environment, the economy and neighbourhood communities.

Although many people might have housing problems, this chapter confines itself to looking at some of the housing problems that poor people experience in the developed world and the steps undertaken by governments to address the most pressing difficulties. The first part is primarily theoretical and seeks both to untangle the nature of these housing problems and to understand the reasons why governments intervene. The second part of the chapter develops this discussion by providing three case studies to illustrate specific measures adopted by central government, and statutory and local agencies to address acute housing problems. The case studies examine council housing renewal in *London*; street homelessness in *New York*; and finally, efforts to tackle social exclusion in *Toulouse*. The conclusion explores possible future government interventions, anticipating the types of problem that are likely to occur and the potential policy responses in developing countries.

DEFINING HOUSING PROBLEMS

How then should we view housing problems? Usually, the housing problems experienced by those on limited incomes are viewed as an inevitable outcome of an increasing reliance on market mechanisms to allocate resources. Under a market system, individuals with the most resources will secure the best-quality homes and those with least resources will end up living in the poorest housing. At a fundamental level, many of the poor's housing problems in developing countries can be traced back to the difficulties that individuals have in securing *affordable* housing. However, other factors also need to be considered, especially if we are to understand why certain housing issues become problems. In particular, there is a need to explore how ideology and power conflicts impact on both the definition of a problem and subsequent government intervention.

To explore these issues, it is necessary to adopt a critical approach, recognising how government and powerful interest groups promote particular issues as *problems* that need tackling in specific ways. A useful contribution has been advanced by Kemeny (1992). He argues that it is important to understand *how* powerful groups are able to successfully *define* certain issues as a *problem* that requires resolution. What becomes a problem is, to a considerable extent, contingent on the ways in which interest groups compete with each other to impose a particular definition and exclude others. In this respect, problems are socially constructed, as policy makers seek to establish a dominant policy agenda in response to changing economic and social conditions and their own needs.

Such a perspective is very different to those approaches to the subject that maintain that problems are simply a reflection of underlying realities. Instead, a social constructionist perspective emphasises the dynamic aspects of problems and how definitions articulated by policy makers can change in a short space of time. For example, system-built tower blocks are now viewed in the United Kingdom as inadequate, even though in earlier periods they were seen as high-quality accommodation (Plate 28.1). Homelessness is another problem subject to changing definitions and policy responses. In some developed countries, for example the United States, homelessness is not seen as a failure of economic policy but as a result of individual choice. This redefinition of homelessness has taken place in spite of the attempts by pressure groups to coerce government to undertake more radical policy responses. The measures now being adopted in many developed countries reflect the relative weakness of those least able to influence the political agenda or define their concerns as a problem meriting substantive policy intervention.

The merit of the 'social constructionist' approach adopted by Kemeny is three-fold. First, it acknowledges that housing problems are not capable of ostensible definition. Second, seeing housing policies as the outcome of competing claims can help us to understand why so many policies are often contradictory and rarely directed towards one consistent, unified aim. Housing priorities are ultimately subordinate to government's overall ideological concerns, even if such interventions impact detrimentally on those reliant on social housing. Third, such a perspective establishes a linkage between housing problems and decisions in other areas of social and economic policy. A clear example of the construction of housing problems can be illustrated by the way in which housing allowances to claimants in some developed countries have been eroded when interest groups have not been able to protect their entitlements. For example, in the economic recession of the 1930s in the United States and in

Plate 28.1 Tower block demolition, Clapton Park Estate, Hackney, London. Only one tower block remains, the rest having been demolished to make space to build new homes for rent.

the post-war years in the UK, the governments of both countries embarked on major house-building programmes, partly in response to demands made upon them by trade unions and organised labour. In the 1980s and 1990s, the pressures on governments to resource adequately public sector housing were sublimated by other interest group claims. In particular, the competing demand that tax levels should be controlled enticed governments to make savings by curbing spending on housing and welfare projects. As Bramley (1997: p. 95) has written, 'an endemic condition of modern welfare states, is where the potential demand for and cost of providing an ever-widening set of services and benefits constantly outruns the ability and willingness of the economy, in the form of workers/taxpayers, to finance it'.

HOUSING POLICY INTERVENTION

It is important that we should be critical of approaches that seek to portray housing policy as the State's response to a problematic environment: principally overcrowding, poor conditions and most recently homelessness. Common to these approaches is the assumption that there is a general unified housing market composed of owner occupiers, social housing and private rented accommodation. The analysis usually entails a chronological examination of the consequences of government policies on specific public and private tenure divisions construing government as enacting policy primarily to ameliorate the problems experienced by consumers of housing. The evidence to justify this formulation is often a manifesto or a declaration of intent by a government minister. Housing policy, seen simply in terms of initiatives that governments undertake, is not an appropriate model with which to interpret the intervention undertaken by governments.

It should be apparent from the preceding discussion that problems are to a large extent contingent on definition and political pressures on governments to instigate solutions. Of course, such a perspective is at odds with the ways in which governments state their policy objectives, as these are usually couched in terms of meeting housing need. These pronouncements are often for public consumption, for an important aspect of contemporary housing policy intervention is its symbolic impact (Edelman 1988). Interventions are widely publicised to convey an impression that the government is taking effective action. To conclude this first section, it is worth reiterating the benefit of viewing government housing policy interventions as responses to these processes rather than simply the measures of a benign state seeking to ameliorate the problems of housing. This perspective will help us to understand the political context in each of the following three case studies.

CASE STUDIES

The primary purpose of the three case studies is to discuss the practical policy measures that have been undertaken and to ascertain how governments have responded to the pressures to intervene in the areas of most acute housing stress. Over recent years, as a consequence of criticism and mounting concern about the impact of poor housing and poverty, many governments have responded with specific policies targeted in areas of acute housing stress. Many of these current policies, as the case studies show, attempt to offset failures arising from overall fiscal policies that have impacted detrimentally on those groups with limited incomes. However, it is apparent that the levels of resources and intervention are not sufficient to redress the entrenched inequalities that exist or offset the overall impact of economic and fiscal policies that have continued to exacerbate the gulf between rich and poor.

Housing renewal in Hackney, East London

In the UK, those unable to afford to own their own home or secure adequate housing in the private sector are reliant on social housing organisations to provide accommodation. Over the last twenty years, the overall investment in social housing has fallen dramatically. It has been estimated by the Chartered Institute of Housing (1997) that the additional expenditure required to tackle the backlog of council housing disrepair currently stands at £20 billion. The main reason why council housing has fallen into disrepair stems from the public's apathy to public expenditure cuts in housing. As Hills and Mullings (1990: p. 144) observed,

> Housing as an issue has moved down the political agenda – the public reaction to the substantial reduction in public spending ... has been much more muted than the reaction against constraints on a health budget whose cost has grown in real terms. Unlike education or the National Health Service, the extent of the problem of council housing disrepair has not become an important political issue, and rarely features in the election campaigns of political parties.

Neither the housing profession nor tenant groups has been able to protect spending on council house renewal at a time of fiscal restraint. As Bramley (1997: p. 395) recognised,

> housing expenditure comprised a large element of capital investment, and it is well known that capital expenditure is easier to cut in a crisis than current expenditure. Such cuts defer future benefits, while reducing both present borrowing requirements and future debt interest commitments. They do not entail withdrawing benefits from current service recipients or sacking public sector employees.

Through the 1980s and 1990s, investment in housing maintenance has borne the brunt of expenditure cuts. While overall public expenditure on services increased in real terms from £191.2 billion in 1994/5 (an increase of 34.6 per cent), housing expenditure has been cut from £11.7 billion in 1980/1 to £5.4 billion in 1994/5, a decrease of 53.9 per cent (Wilcox 1995: p. 88). The cumulative impact of these expenditure cuts has been accentuated by government legislation encouraging council tenants to buy their own property. In practice, this meant that the most desirable properties have been sold, while flats on unpopular estates remain under the control of local authorities.

However, there are examples of individual estates that have benefited from investment; for example the Holly Street Estate in Hackney (see Box 28.1). The approach adopted by recent governments is that the problems on council estates cannot be addressed successfully solely by physical renewal of the properties and that a more holistic approach is required. As a consequence, government-funded initiatives now entail an integrated approach, including investment in the stock, housing management initiatives, tenure diversification and employment projects. In

Box 28.1 The Holly Street Estate, Hackney, London

The Holly Street Estate in Hackney, built in the early 1970s, consisted of a complex system of nineteen system-built, five story blocks and four nineteen-storey tower blocks. In the early 1990s, 30 per cent of those potentially economically active were out of work. As many as 80 per cent of those residing on the estate had applied for a transfer, and voids and squatted properties amounted to 25 per cent of all properties. Between 1990 and 2003, the estate will have received government resources to the tune of £200 million. The modernisation, when complete, will provide a mixture of different tenures, including new housing association flats and private properties for sale.

In contrast to earlier initiatives to tackle housing problems, the renewal of Holly Street has entailed a partnership with a range of government and private sector agencies. Initiatives such as this have, over a short space of time, become prototypes for future housing developments. In addition, considerable effort is undertaken by housing officers to engender tenant participation to encourage tenants to play a role in the development of the estate. The holistic approach now being adopted can be contrasted with earlier interventions by local authorities and government, where there was a tendency to pursue housing policies in isolation from other social policies.

In some respects, it is too early to undertake an assessment of the effectiveness of the intervention taking place on the Holly Street Estate. Clearly, the renewal of the estate, including new street properties to replace the system-built housing and tower blocks that have been demolished, will provide a significantly better environment than before.

Britain, resources for this type of intervention are provided by the government through the Department of the Environment, Transport and the Regions Single Regeneration Budget programme.

However, whether the housing renewal strategies now being undertaken in Hackney and in many other areas of the UK will successfully regenerate these localities is another question. In the UK, along with other countries in the developed world, social housing is widely seen as an inferior tenure to owner-occupied housing, and as such many better-off residents continue to move out of council property as soon as their personal economic circumstances permit. This process known as residualisation in effect undermines efforts to tackle the social and economic problems that exist on many council estates. In addition, it seems unlikely that enough resources have been set aside by the government to enable many other estates to benefit from increased investment. In fact, public expenditure on Single Regeneration Budget programmes will contract by 29 per cent in real terms between 1994/5 and 1998/9 (Centre for Urban and Regional Studies 1995: p. 20).

For these reasons, it can be argued that the interventions taking place are insufficient to prevent this process of residualisation. Indeed the tenants who will be living on estates such as Holly Street are, according to Page (1993: p. 30),

> Even more economically disadvantaged than those housed previously – although younger and more likely to be economically active, their incomes are lower and they are less likely to have a job, more likely to be unemployed and more likely to be wholly dependent on state benefits or pensions.

The policy intervention taking place on the Holly Street Estate provides an example of the holistic approach to housing renewal. In contrast to earlier periods, the objective is to target resources on a specific area and to integrate housing policies with economic and social measures. Although the quality of individual council tenants lives will improve as a result of the physical renewal, the overall status of public sector housing will continue to deteriorate because of the residualisation process. Until steps are taken to restore its status as a form of tenure attractive to those who are able to exercise choice and not just those in greatest need, the economic outlook of localities like Holly Street will remain forlorn (Plate 28.2).

Tackling street homelessness in New York

The second example of a contemporary housing problem concerns the rise of homelessness in the

Plate 28.2 The Holly Street Estate, Hackney, London. before the current £200 million redevelopment programme, as many as 80 per cent of the tenants had requested a move off the estate.

USA and the efforts of government agencies to tackle the issue. Homelessness is so debilitating that individuals without a home are highly susceptible to a range of other problems and have great difficulty in gaining access to health care and other services. How homelessness is measured is of course subject to competing pressures. US government agencies usually insist on a strict definition of rooflessness when making estimates, while pressure groups and charities tend to argue for a broadly based definition to include those individuals living in substandard, overcrowded accommodation (see Daly 1996: p. 7 for a discussion of competing definitions). As in other countries in the developed world, there has been a large increase in the number of homeless people in the USA. In the 1980s, estimates of the homeless population ranged from 250,000–300,000 (US Department of Housing and Urban Development 1984) to 3 million (Homes and Synder 1982). In the 1990s, estimates now range from 840,000 to 5,000,000 (Link *et al.* 1994; Takahashi 1996). In New York, agencies working on behalf of homeless groups estimate the number of homeless to be between 70,000 and 90,000, half of whom live on the street and the remainder in public or private emergency sheltered accommodation (National Coalition for the Homeless 1989). Approximately one-fifth of this number are parents and children, one-third young adults aged between 16 and 21 years and the remainder single people, 80 per cent of whom are men. As many as 90 per cent of the residents in sheltered accommodation are from ethnic minority groups, even though as a proportion of the city population, ethnic minority groups constitute 40 per cent (Cohen 1994; National Coalition for the Homeless 1989).

As discussed in the first part of the chapter, homelessness represents the outcome of a complex set of economic, social and political factors, all of which impact upon the availability and location of housing. In most of the principal cities of the USA, the supply of homes has been affected by abandonment of properties in the 1970s and recent gentrification processes in which professionals moved to the inner city (Daly 1996: p. 21). So, for example, in New York city between 1970 and 1985, 109,000 single-room occupancy units were either demolished or converted into flats for sale or rent (many of which were let to people on low incomes, including those in receipt of welfare support). Over the same period, the number of single households has increased enormously, forcing rents up and literally pricing many individuals out of the market. The problems have been compounded by economic restructuring (leading to unemployment and lower-paid jobs), reductions in welfare support (as many as 1 million of New York city's population is in receipt of welfare allowances) and the curtailment of new public sector housing. The result has been a relative fall in the incomes of the poor and a long waiting list (200,000) for public housing (Marcuse 1990). In addition, it has been suggested (Kearns and Smith 1993) that the deinstitutionalisation of mental health care has added to the number of homeless people along with demographic changes and changes to family structures.

The policy response to the increasing incidence of homelessness in New York reflects the issue's low political priority. There is no constitutional right to permanent housing, and as Cohen (1994: p. 772) argues, 'there have only been piecemeal programmes for the homeless'. In New York city, $500 million is spent each year by the city and state authorities on programmes to assist the homeless (New York City Commission on the Homeless 1992). This includes funds set aside for soup kitchens, emergency hostels and drop-in centres. As in other countries, voluntary organisations have played an increasingly important role, especially for individuals with mental health problems. In recent years, following the McKinney Homelessness Assistance Act (Interagency Council on the Homeless 1990; 1994), there is a requirement for city authorities to establish an implementation plan (Comprehensive Homeless Assistance Plan). However, this assistance has made only a limited impact (Berman and West 1997) and, as any visitor to New York can testify, there remain an enormous number of destitute individuals living on the city's streets.

It is important to explore why the amount of resources in what is one of the wealthiest countries in the world is so clearly insufficient to tackle the problems of homelessness. One possible explanation is that homelessness is seen as individual 'fecklessness' and not an issue that the government should address. For many Americans, homelessness is not seen as an issue that merits substantial government intervention. Other political priorities are deemed to be more important, in particular the commitment to reduce federal taxes for those in work. The culture of 'individualism', which is so prevalent in the USA, also has an effect. There is longstanding hostility to 'big' government, and increases in welfare intervention are not generally supported by the majority of the population. As Takahashi (1996: p. 291) has observed 'policy responses have become increasingly punitive . . . at the local level, there have been expanding efforts both to criminalize homelessness through anti-camping ordinances and to prevent homeless persons from entering and staying in specific jurisdictions'.

Social exclusion in Toulouse: the establishment of *contract de ville.*

As much as 90 per cent of social housing in France is managed by *Habitations à Loyer Modéré* (HLM) (housing at moderate rents). There are as many as 1000 HLM organisations managing a stock of 4.6 million dwellings. Financial support and lending facilities for HLM are provided by the *Caisse des Dépôts Consignations*, a national savings and investment bank. Along similar lines to other countries, social housing estates in France have increasingly housed the poorest groups in society. This process of residualisation accelerated in the 1970s. As Power (1993: p. 60) writes, 'HLMs came to house disproportionate numbers of large households, of one parent families, of immigrants, and of French citizens from overseas provinces.' This was accentuated by 'better-off' residents moving into owner occupation. The changing economic and social conditions on many of the large estates in conjunction with disrepair intensified the problems.

In response to the physical, social and economic problems on large French housing estates (*Grand ensembles*), the government developed a number of initiatives (see Tuppen 1995). For example, in 1981 a national body entitled Commission Nationale pour le développement des Quartiers with a budget of 721 million francs (1985 prices) was set up to improve not only the physical fabric but also the social and recreational facilities of residents. In 1983, two other joint initiatives were established: Mission Banlieus 89 and the Council for Preventing Delinquency. This was followed shortly by another important initiative, Guidelines for the City. This placed a requirement on all urban communes to develop a *programme local de l'Habitat*, a local housing strategy focusing on those who are badly housed or deprived. In 1991, the national HLM organisation established a job creation scheme with government for tenants living on some of the most deprived estates.

The most important recent initiative that the government in France has undertaken is the *contract de ville* programme, intended to revitalise areas of deprived social housing (see Box 28.2). The contract entails a partnership primarily between central and local government and HLM social

Box 28.2 Contract de ville in Toulouse

Toulouse, France's fifth largest city, is one locality that has already benefited from *contract de ville*. The population is currently 3 million, unemployment stands at 19.7% (compared to a national average of 10.8%) and the proportion of social housing units is over 55% (compared to a national rate of 14.6 %). In addition, many immigrants and ethnic minorities have settled in Toulouse (comprising 18.3% of the total population). Among the contract's key objectives under the Right to Housing and Right to City Life budgets are provision of low-rent HLM housing units; special allocations arrangements to improve the social and economic composition of the estate; refurbishment of 4400 system-built HLM properties; an emergency reception service for homeless people; sites for travellers; and 200 new units of emergency temporary housing. As much as 1.67 billion francs (£175.86 million) has been set aside for these initiatives.

housing organisations. Overall, there are 214 contracts in existence, fifty-eight of which are in the Paris region (*Île de France*). An additional 4 billion francs is provided by regional councils. Contracts last five years (1994–8) and are funded by central government. The key aims are to enhance public service provision, renovate poor housing (120,000 units a year), support economic regeneration and prevent crime. The amount of resources set aside by the government is 9.5 billion francs (approximately 1 billion) to be spent between 1994 and 1998 (LHU 1997). The implementation of the contract is the responsibility of the senior *chef de project* and a steering committee, on which serve elected representatives and the local prefect. The actual implementation of the contract is the task of officers.

The policies now being pursued in France are in certain respects similar to the initiatives being undertaken in the UK and other developed countries. Both illustrate the pressures on government to target resources selectively and at the same time integrate housing measures alongside economic initiatives. However, as is the case in Britain, it is doubtful whether these initiatives can reverse the accelerating social problems now emerging in developed countries and the increasing divide between the 'well-off' and those who are deemed to be socially excluded. In addition, there is increasing concern that the complex bureaucracy established by *contract de ville* means that implementation is now contingent on a wider network of agencies than was the case previously, making it much more difficult to undertake quick and effective decision making or attract private sector investment (Tuppen 1995: p. 371).

CONCLUSION

This chapter began with a discussion about how problems of housing are defined and what are deemed the appropriate policy responses. It was argued that a reliance on a narrow economic formulation as either one of limited supply and excess demand or as an enduring facet of industrial society is best supplemented with an analysis of other considerations such as politics and ideology if we are to understand the nature of contemporary housing problems. The three case studies provide an illustration of the way in which pressures on government to act may not be as substantial as other competing pressures to control public expenditure and stimulate growth. Although considerable publicity and energy is devoted to the new initiatives, this should not conceal the stark underlying reality that social housing problems are slipping from the mainstream political agenda of most governments in the developed world.

However, in spite of this pessimistic assertion, there are some positive aspects to the way that specific problems are being addressed. In both France and the UK, efforts to integrate housing policies within wider economic and political concerns reverse long-standing policies that saw government tiers working in isolation. Yet in both countries it is doubtful whether the resources provided are sufficient to address the scale of the problem. In New York, the response to homelessness reflects the low status accorded to social welfare. It is no surprise that in this environment government rhetoric has not been matched by spending commitments.

In any assessment of specific policy responses, it is important to recognise that throughout the developing world there is little enthusiasm to commit large-scale resources to tackle the problems that exist for the poorest members of society. If we judge housing policies by the level of resources set aside by governments and not statements of intent, then the only conclusion that can be meaningfully drawn is that the problems of housing have marginal importance in the policy agenda framed by governments throughout the developed world. Ultimately, this can be traced back to the capacity of sophisticated electorates to resist policies that promote increased taxation as a means of alleviating inequality (Galbraith 1992) and the willingness of countries to adopt market solutions irrespective of the impact that these policies have on the poorest and most vulnerable residents. While predicting the future is always difficult, it would appear from current housing policies that even though the widening gulf between rich and poor citizens may remain an issue for all governments, it will not be a sufficiently high priority to alter government's overall economic and political priorities of controlling public expenditure and maintaining low taxation. In practice, this will mean that many of the housing problems described in the case studies will continue to afflict the poorest members of society. Applied geographers and other social scientists can play an emancipatory role by undertaking research that reveals the real causes of the housing problem.

GUIDE TO FURTHER READING

The key international journal on housing policy is *Housing Studies*. Journals also worth consulting include *Policy and Politics* and *Urban Studies*. Recent books that successfully combine a theoretical analysis with a critique of housing policy include Kemeny (1992), Malpass and Means (1993) and Malpass and Murie (1996). The most outstanding book published in recent years is Michael Harloe's (1995) *The People's Home*, a detailed critique of social housing provision in Europe and the USA. Those interested in comparative research should consult Ball *et al.* (1992), Barlow and Duncan (1994), McCrone and Stevens (1995) and Kleinman (1996). The most recent book in this area by Doling (1997) has the added advantage of examining housing in Australia and Japan as well as in North America and Europe. In connection with the specific concerns dealt with in the case studies, three valuable books are Ambrose (1994) on UK housing renewal policy; Daly (1996) on homelessness in the United States; and the consequences of social exclusion on housing estates in Europe by Power (1997). Finally, those readers interested in a broad-ranging discussion of welfare and wider ideological processes should consult the work of Esping-Andersen (1990).

REFERENCES

Ambrose, P. (1994) *Urban Process and Power.* London: Routledge.

Ball, M., Harloe, M. and Martens, M. (1988) *Housing and Social Change in Europe and the USA.* London: Routledge.

Barlow, J. and Duncan, S. (1994) *Success and Failure in Housing Provision: European Systems Compared.* Oxford: Pergamon.

Berman, E. and West, J. (1997) Municipal responses to homelessness: a national survey of 'Preparedness'. *Journal of Urban Affairs* 19(3), 303–18.

Bramley, G. (1997) Housing Policy: a case of terminal decline. *Policy and Politics* 25(4), 387–407.

Centre for Urban and Regional Studies (1995) *The Single Regeneration Budget: The Stocktake* Birmingham: CURS.

Chartered Institute of Housing (1997) Housing and The Comprehensive Spending Review. *CIH Briefing*, Coventry: CIH.

Cohen, C.I. (1994) Down and out in New York and London: a cross-national comparison of homelessness. *Hospital and Community Psychiatry* 45(8), 769–76.

Daly, G. (1996) *Homeless.* London: Routledge.

Doling, J. (1997) *Comparative Housing Policy: Government and Housing in Advanced Industrialized Countries.* Basingstoke: Macmillan.

Edelman, M. (1988) *Constructing the Political Spectacle.* Chicago: University of Chicago Press.

Esping-Andersen, G. (1990) *The Three Worlds of Welfare Capitalism.* Cambridge: Polity Press.

Galbraith, J.K. (1992) *The Culture of Contentment.* London: Sinclair Stevenson.

Harloe, M. (1995) *The People's Home: Social Rented Housing in Europe and America.* Oxford: Blackwell.

Hills, J. and Mullings, B. (1990) Housing: a decent home for all at a price within their means? In J. Hills (ed.) *The State of Welfare: the Welfare State in Britain since 1974.* Oxford: Clarendon Press, 135–205.

Homes, M.E. and Synder, M. (1982) *Homelessness in America: A Forced March to Nowhere.* Washington, DC: Community for Creative Nonviolence.

Interagency Council on the Homeless (1990) *1990 Annual Report.* Washington, DC: Interagency Council.

Interagency Council on the Homeless (1994) *Priority Home! The Federal Plan to Break the Cycle of Homelessness.* Washington, DC: Interagency Council.

Kearns, R.A. and Smith, C.J. (1993) Housing stressors and mental health among marginalized urban populations. *Area* 25, 274–87.

Kemeny, J. (1992) *Housing and Social Theory.* London: Routledge.

Kleinman, M. (1996) *Housing, Welfare and the State: A Comparative Analysis of Britain France and Germany.* Cheltenham: Edward Arnold.

Link, B.G., Susser, E., Stueve, A. and Phelan, J. (1994) Lifetime and five-year prevalence of homelessness in the United States. *American Journal of Public*

Health 84, 1907–12.
London Housing Unit (1997) *The French Connection: Regeneration Lessons from Contracts de Ville*. London: London Housing Unit.
Malpass, P. and Means, R. (eds) (1993) *Implementing Housing Policy*. Bletchley: Open University Press.
Malpass, P. and Murie, A. (1996) *Housing Policy and Practice*, 4th edition. London: Routledge.
Marcuse, P. (1990) Homelessness and housing policy. In C. Caton (ed.) *Homeless in America*, New York: Oxford University Press, 138–59.
McCrone, G. and Stephens, M. (1995) *Housing Policy in Britain and Europe*. London: UCL Press.
National Coalition for the Homeless (1989) *One Hundred Thousand and Counting. Homelessness in New York State*. Albany, NY: National Coalition for the Homeless, February.
New York City Commission on the Homeless (1992) *The Way Home: A New Direction in Social Policy*. New York: Office of the Mayor.
Page, D. (1993) *Building for Communities: A Study of New Housing Association Estates*. York: Joseph Rowntree Foundation.
Power, A. (1993) *Hovels to High Rise: State Housing in Europe Since 1850*. London: Routledge.
Power, A. (1997) *Estates on the Edge: The Social Consequences of Mass Housing in Northern Europe*. Basingstoke: Macmillan.
Takahashi, L.M. (1996) A decade of understanding homelessness in the USA: from characterisation to representation. *Progress in Human Geography* 20(3), 291–310.
Tuppen, J. (1995) After les minguettes: 'problem housing estates in France'. *European Urban and Regional Studies* 2(4), 367–71.
US Department of Housing and Urban Development (1984) *The extent of homelessness in America: a report to the Secretary on the homeless and emergency shelters*. Washington, DC: US Department of Housing and Urban Development.
Wilcox, T. (1995) *The 1995/6 Housing Finance Review* York: Joseph Rowntree Foundation.

29

The geography of poverty and deprivation

Michael Pacione

Despite the economic and social advances of the post-war era, seen in increased life expectancy, a reduction in the proportion of the world's population facing hunger and life-threatening deprivation, and increased access to health-care and education, poverty remains a major problem for a significant proportion of the population in most countries in the contemporary world. At the heart of this dilemma is the uneven distribution of the world's resources. Uneven development is an inherent characteristic of capitalism that stems from the propensity of capital to flow to locations that offer the greatest potential return. The differential use of space by capital in pursuit of profit creates a mosaic of inequality at all geographic scales from global to local. Consequently, at any one time certain countries, regions, cities and localities will be in the throes of decline as a result of the retreat of capital investment, while others will be experiencing the impact of capital inflows.

Applied human geographers have focused particular attention on the conditions of poverty and deprivation experienced by those people and places at the disadvantaged end of the quality of life spectrum. This chapter reviews the major dimensions of applied research into the geography of poverty and deprivation. The discussion first identifies the nature of poverty and deprivation before considering the question of its measurement. The extent and incidence of poverty is then examined with reference to research undertaken at a variety of scales from the global to local level in a variety of settings throughout the world. In the final section, we examine the value of an applied geographical perspective for the identification and amelioration of the multiple problems of poverty and deprivation.

THE ANATOMY OF POVERTY

Poverty implies deprivation or human needs that are not met. It is generally understood to arise from a lack of income or assets, which means that people are unable to meet basic physical needs such as an adequate diet and decent housing. The poor are, in many instances, also unable to attain health-care when sick or injured and, outside the welfare states of the North, have no means of subsistence when unemployed, ill, disabled or too old to work. Other 'higher-order' needs that many would incorporate in any definition of poverty include self-esteem and access to civil and political rights (Townsend 1993).

The causes of poverty are complex. In the countries of the developing world, many problems are associated with economic stagnation and/or debt crisis, and with the difficulties of structural adjustment. In most of the 'transition countries' of Eastern Europe and the former Soviet Union, problems of poverty are linked to the collapse of communism, although in many countries social progress had already slowed in the years prior to these changes. In several of the wealthiest countries, the increase in poverty during the 1980s was associated less with economic stagnation and more with changes in the labour market, including a growth in long-term unemployment. Other contributing factors were related to political strategies that reduced expenditure on social welfare. In many Western states, these trends are reflected by a growing inequality in income distribution. (Goodman *et al.* 1997).

A key factor in the debate over the nature and extent of poverty and deprivation is the distinction made between absolute and relative poverty. The absolutist or subsistence definition of poverty contends that a family would be considered to be living in poverty if its 'total earnings are insufficient to obtain the minimum necessaries for the maintenance of merely physical efficiency' (Rowntree 1901: p. 186). This notion of a minimum level of subsistence and the related concept of a poverty line exerted a strong influence on the development of social welfare legislation in post-war Britain. The system of National Assistance benefits introduced following the Beveridge Report of 1942 was based on calculations of the amount required to satisfy the basic needs of food, clothing and housing plus a small amount for other expenses. A similar concept of a safety net for particularly vulnerable social groups underlies the Medicare and Medicaid programmes in the USA. In general, most estimates of the scale of poverty within and between nations employ a *per capita* level of income as a definitive poverty line.

If, however, we accept that needs are culturally determined rather than biologically fixed then poverty is more accurately seen as a relative phenomenon. The broader definition inherent in the concept of relative poverty includes job security, work satisfaction, fringe benefits (such as pension rights) plus various components of the 'social wage', including the use of public property and services as well as satisfaction of higher-order needs such as status, power and self-esteem. In essence, the absolutist perspective carries with it the implication that poverty can be eliminated in an economically advanced society, while the relativist view accepts that the poor are always with us.

As Figure 29.1 indicates, poverty and deprivation are related concepts. Poverty is a central element in the multi-dimensional problem of deprivation whereby individual difficulties reinforce one another to produce a situation of absolute disadvantage for those affected. The root cause of deprivation is economic and stems from two sources. The first arises due to the low wages earned by those employed in declining traditional industries or engaged, often on a part-time basis, in newer service-based industries in post-industrial societies, or by the mass of the self-employed in the informal sector of third world economies. The second cause is the unemployment experienced by those marginal to the job market such as single parents, the elderly, disabled and, increasingly, never-employed school leavers. Significantly, applied geographical research has demonstrated that the complex of poverty-related problems shown in Figure 29.1 exhibits marked spatial concentration. This patterning serves to accentuate the effects of poverty and deprivation for the residents of particular localities. The phenomenon of socio-spatial segregation is seen at its starkest in the squatter settlements of the third world (Plate 29.1) but is to be found to a varying degree in most of the cities of the modern world (Box 29.1).

In declining older industrial regions of the North neighbourhood unemployment rates of three times the national average are common, with male unemployment frequently in excess

Figure 29.1 The anatomy of multiple deprivation.

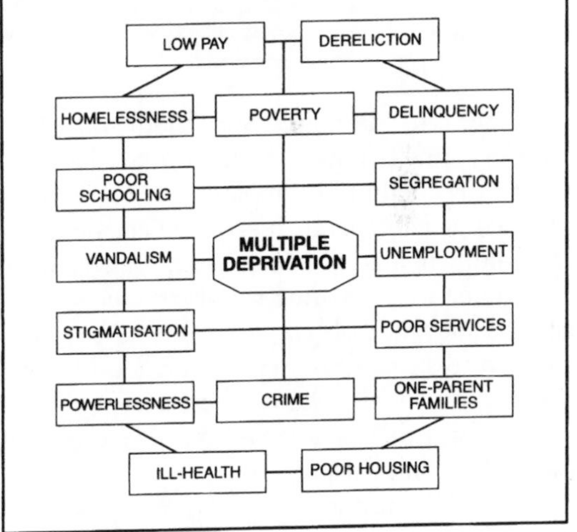

Source: Pacione 1997b.

Plate 29.1 The McDonald's Farm squatter camp in Soweto, South Africa (*photograph: David Drakakis-Smith*).

Box 29.1 The Chheetpur squatter settlement in Allahabad, India

Allahabad is one of the five major cities in the Indian state of Uttar Pradesh. In-migration has played a significant role in the growth of the city's population, which increased by 500 per cent between 1981–91. The economic base of the city is weak, and the informal sector accounts for over 40 per cent of the total workforce. Urban infrastructure is unable to cope with the increasing population, and a quarter of the inhabitants live in slums and squatter settlements.

The squatter settlement of Chheetpur occupies a hectare of flood-prone land along a railway line in the south of the city. The majority of residents are members of the lowest castes. Living conditions are inhuman. One- or two-room mud houses shelter families of five–seven members, including adult males, females and children, with each room measuring not more than 12 m^2. None of the houses has a separate kitchen or lavatory, and defecation occurs outside along the railway track or other open spaces or in open drainage lines at the side of the street. Animals and humans live together, and the low level of sanitation spreads diseases, which find ready victims in the malnourished Chheetpurians. Despite the location of the settlement adjacent to several schools and colleges, two-thirds of the squatters are illiterate and the struggle for daily survival means that investment in the future receives a low priority. Despite the general poverty of the environment, even within the area socio-spatial divisions are evident between higher-caste poor and the poorest of the poor. In this and other squatter settlements throughout the third world, the problems stemming from absolute and relative poverty and deprivation are endemic.

Source: H. Misra 1994.

of 40 per cent (Donnison and Middleton 1987; Convery 1997). Within such regions, the shift from heavy industrial employment to service-oriented activities, and the consequent demand for a different kind of labour force, has served to undermine long-standing social structures built around full-time male employment and has contributed to social stress within families. Dependence upon social welfare, where available, and lack of disposable income lowers self-esteem and can lead to clinical depression.

Poverty also restricts diet and accentuates poor health. Malnutrition is a major factor underlying the high infant mortality rates in the deprived environments occupied by the majority of population in the South. Even in the developed world, however, infant mortality rates are often significantly higher in poor areas (Benzeval *et al.* 1995). Children brought up in such environments are more likely to be exposed to criminal subcultures and to suffer educational disadvantage (Pacione 1997a). The physical environment in deprived areas is typically bleak. The ghettos of American cities and parts of deprived local authority estates in the UK exhibit extensive areas of dereliction, little landscaping, and shopping and leisure facilities that reflect the poverty of the area. The extreme degradation that characterises the poor areas of third world cities frequently poses a threat to the very survival of inhabitants (Main and Williams 1994). The residents of poverty areas are often also victims of stigmatisation, which operates as an additional obstacle to obtaining employment or credit facilities. Many deprived areas are socially and physically isolated from the mainstream, and those who are able to move away do so, leaving behind a residual population with limited control over their quality of life.

MEASURING THE EXTENT OF POVERTY

The measurement of poverty and deprivation is one of the most direct means of monitoring socio-spatial variations in quality of life, and of assessing the performance of public agencies charged with improving the well-being of the poor. Applied researchers must be aware, however, that, in practice, these key objectives are often impeded by the non-availability of accurate and comparable statistics. Accordingly, researchers must make explicit the assumptions and limitations of the data employed and must be aware of the fact that in some situations what is measurable and measured becomes what is real, standardising the diverse and excluding the divergent and different (Chambers 1994).

As indicated, most estimates of the scale of poverty employ an income level as the poverty threshold. However, as the WHO (1992) reported, poverty defined solely by level of personal income cannot cover health, literacy or access to public goods or common property resources. To overcome the deficiencies of a simple income-based measure of poverty, researchers may either select a single indicator of more direct relevance to the problem, or employ a suite of social, economic and demographic variables to produce a multivariate measure of the condition. One of the most valuable single indicators of social conditions in a country is life expectancy at birth (Figure 29.2). This measures

Figure 29.2 Life expectancy at birth, 1990.

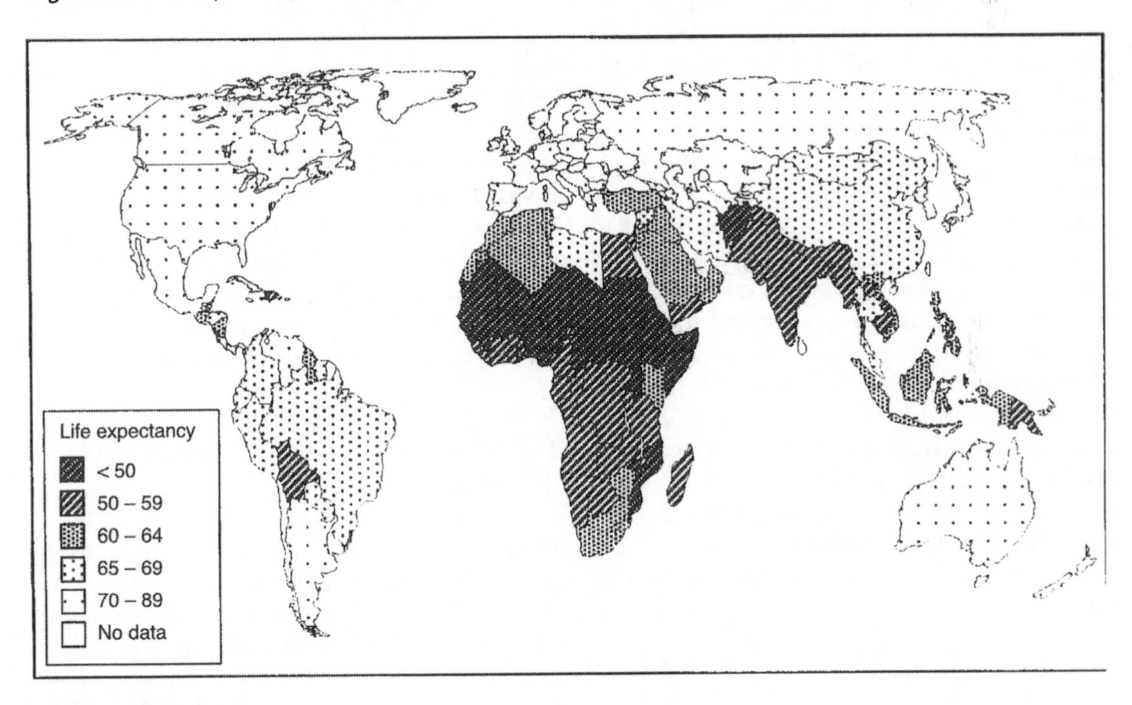

Source: Findlay 1994.

the extent to which prevailing economic, social and political factors make it possible for citizens to avoid premature death. This indicator has been found to exhibit a strong correlation with multivariate measures of living standards or overall well-being (Partha and Weale 1992) and may therefore be regarded as a sound indicator of the extent of poverty and deprivation. Clearly, it is not possible for a country to exhibit a high average life expectancy if most of the population do not enjoy access to good quality housing, safe water supplies, adequate sanitation and health care provision. Even prosperous countries cannot rank among those with highest life expectancies if a significant proportion of their population lack the income, living conditions and access to services that protect against premature death.

While life expectancy provides a useful guide to national levels of poverty, use of an indicator of average life expectancy for a country can disguise major socio-spatial variations in well-being *within* nations. Two of the most important factors found to explain intra-national variations in poverty refer to the extent of inequality, and the level of basic service provision. Thus, in a survey of twenty-two low-income countries, Anand and Ravallion (1998) found that one-third of the increase in life expectancy over the study period was due to reduced poverty and two-thirds to increased spending on social services. These results suggest that the significance of economic growth in expanding life expectancy is tied to the ways in which the benefits are *distributed* among the population. The association between life expectancy and levels of income inequality has also been demonstrated in twelve countries of the European Union (Wilkenson 1992).

Of particular significance for applied geographical research into poverty and deprivation is the fact that these data place strong emphasis on the need for analysis at the intra-national scale. Accordingly, in the following sections we examine the conduct and value of such investigation at three different levels.

URBAN AND RURAL POVERTY

Although socio-spatial variations in the incidence of poverty and deprivation exist in all countries in the context of the developing world, particular attention has been focused on differential levels of living between urban and rural areas, with most studies indicating a higher incidence of poverty in the latter. This conclusion is supported by the data presented in Table 29.1, which is based on a nationally defined poverty line related to the income needed to satisfy basic minimum needs in each country. A note of caution is required, however, not least since the use of a single income-based poverty line across both urban and rural areas may underestimate the extent of urban poverty by failing to take account of the higher costs of urban living. When allowance is made for differences in living costs between urban and rural areas, the scale of urban poverty generally increases (Feres and Leon 1990).

Nothwithstanding the difficulties of comparative analysis, the evidence from Table 29.1 indicates severe poverty in rural areas. A study of rural poverty in 114 countries in the South found that 940 million or 36 per cent of the population had incomes below the poverty line (Jazairy *et al.* 1992), while in the USA Lyson (1989) found that the rural and black-belt labour market areas of the American south have remained poorer and more underdeveloped than their urban counterparts. There is, however, little to be gained by seeking to demonstrate the relative severity of poverty in urban and rural areas. As Table 29.1 indicates, in both contexts, the scale of the problem is critical and demands urgent remedial action.

REGIONAL VARIATIONS IN POVERTY AND DEPRIVATION

The geographical and statistical overlap between individual indicators of poverty and deprivation stimulated the development of composite multivariate territorial social indicators as a means of revealing the differential patterns of quality of life between regions. Early studies in the field include

Table 29.1 The extent of absolute poverty in selected countries.

Country or region	*Proportion of the population below the poverty line*			
	In urban areas	*In rural areas*	*In whole nation*	*Date*
Africa	29.0	58.0	49.0	1985
Botswana	30.0	64.0	55.0	1985/86
Côte d'Ivoire	30.0	26.0	28.0	1980/86
Egypt	34.0	33.7	33.8	1984
Gambia	63.8	57.7		1989
Ghana			59.5	1985
Morocco	28.0	32.0		1985
Mozambique	40.0	70.0	55.0	1980/89
Swaziland	45.0	50.0	49.0	1980
Tunisia	7.3	5.7	6.7	1990
Uganda	25.0	33.0	32.0	1989/90
Zambia	40.0		80.0	1993
Asia (excluding China)	34.0	47.0	43.0	1985
Bangladesh	58.2	72.3		1985/86
China	0.4	11.5	8.6	1990
India	37.1	38.7		1988
Indonesia	20.1	16.4	17.4	1987
Korea, Republic of	4.6	4.4	4.5	1984
Malaysia	8.3	22.4	17.3	1987
Nepal	19.2	43.1	42.6	1984/85
Pakistan	25.0	31.0		1984/85
Philippines	40.0	54.1	49.5	1988
Sri Lanka	27.6	45.7	39.4	1985/86
Europe				
France			16.0	*c.*1990
Germany			10.0	*c.*1990
Hungary			15.4	1991
Ireland			19.0	*c.*1990
Italy			15.0	*c.*1990
Poland			22.7	1987
Spain			19.0	*c.*1990
United Kingdom			18.0	*c.*1990
Latin America	32.0	45.0		1985
Argentina	14.6	19.7	15.5	1986
Brazil	37.7	65.9	45.3	1987
Colombia	40.2	44.5	41.6	1986
Costa Rica	11.6	32.7	23.4	1990
El Salvador	61.4			1990
Guatemala	61.4	85.4	76.3	1989
Haiti	65.0	80.0	76.0	1986
Honduras	73.9	80.2	77.5	1990
Mexico	30.2	50.5	29.9	1984
Panama	29.7	51.9	41.0	1986
Peru	44.5	63.8	51.8	1986
Uruguay	19.3	28.7	20.4	1986
Venezuela	24.8	42.2	26.6	1986
North America				
Canada			15.0	*c.*1990
United States of America			13.0	*c.*1990

Source: United Nations Centre for Human Settlements 1996.
Notes: These are all estimates based on data from a household budget, income or expenditure survey and are based on the concept of an 'absolute poverty line' expressed in monetary terms. The figures for different countries are not necessarily comparable, since different assumptions will have been made for setting the poverty line. (NB comparisons between these countries should be avoided, as different criteria were used to set poverty lines).

Wilson's (1969) analysis of inter-state variations in quality of life in the USA and Gordon and Whittaker's (1972) study of prosperity for local areas of southwest England. Smith (1973) employed a set of forty-seven social variables relating to income, health, housing, education, social order and social belonging to create a map of social well-being for the coterminous states of the USA, which highlighted the relative poverty of the southern states (Figure 29.3).

More recently, analysis of a number of deprivation-related indicators from the 1991 census illuminated the contrasting geographies of poverty and affluence among local authority districts in Britain. Compare, for example, the position of Glasgow district and the spatially contiguous suburban district of Bearsden and Milngavie in Tables 29.2 and 29.3. Green (1994) has also shown that in terms of unemployment and inactivity rates the gap between the best and worst wards in the UK increased between 1981 and 1991 to accentuate an already significant degree of polarisation. Significantly, the socio-spatial incidence of poverty exhibited a high level of consistency over the decade. A similar conclusion was demonstrated at the conurbation level by Pacione's (1995a; b) analysis of multiple deprivation in Strathclyde, which revealed the spatial and temporal consistency of deprivation in specific localities. As well as illuminating regional variations in well-being, these studies also provided further support for the examination of socio-spatial variations in poverty and deprivation at the intra-urban level.

POVERTY AND DEPRIVATION WITHIN THE CITY

The problems of poverty and deprivation experienced by those people and places marginal to the capitalist development process have intensified over recent decades. A substantial proportion of the disadvantaged live in towns and cities. In the countries of the North, large areas of many of the older industrial cities have been economically and socially devastated by the effects of global economic restructuring, de-industrialisation, and ineffective urban economic policy (Pacione 1990a). For many of the residents of cities in the UK and USA, the nature and extent of multiple deprivation represents a contemporary urban crisis (Box 29.2) (Pacione 1997b).

In order to understand the incidence and impact of poverty and deprivation on disadvantaged communities, it is necessary to complement national- and regional-level analyses by working at the lowest possible spatial scale; a scale that draws closest to the context of people's daily lives. In general, the finer the spatial scale of analysis the more detailed and more policy-relevant are the research findings. Consequently,

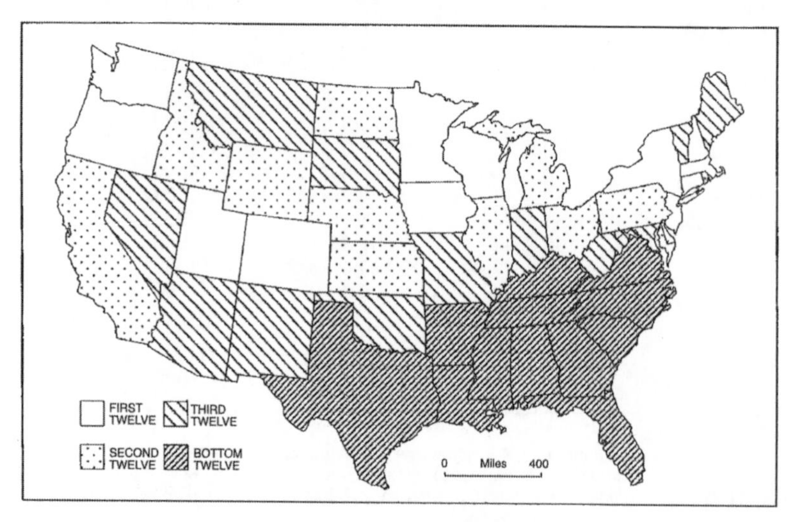

Figure 29.3 Social well-being in the USA.

Source: Smith 1973.

Table 29.2 Most deprived local authority districts in Britain, 1991.

Unemployment (%)			*Inactivity rate (20–59)(%)*			*Households owning no car (%)*		
1	Hackney	22.5	1	Rhondda	32.3	1	Glasgow City	65.6
2	Knowsley	22.1	2	Easington	31.2	2	Hackney	61.7
3	Tower Hamlets	21.8	3	Merthyr Tydfil	30.4	3	Tower Hamlets	61.6
4	Liverpool	21.1	4	Afan	30.2	4	Islington	59.9
5	Newham	19.3	5	Rhymney Valley	29.3	5	Clydebank	58.7
6	Glasgow City	19.1	6	Blaenau Gwent	29.3	6	Southwark	58.0
7	Manchester	18.7	7	Cynon Valley	29.2	7	Westminster	57.7
8	Southwark	18.2	8	Knowsley	28.2	8	Liverpool	57.0
9	Haringey	17.7	9	Cumnock and Doon Valley	28.0	9	Manchester	56.6
10	Lambeth	17.1	10	Glasgow City	27.8	10	Camden	55.8

Source: Hills 1995.

Table 29.3 Most affluent local authority districts in Britain, 1991.

Households from social classes 1 and 2 (%)			*Adults with high-level qualifications (%)*			*Households with two or more cars (%)*		
1	City of London	71.4	1	City of London	35.7	1	Surrey Heath	51.0
2	Richmond upon Thames	63.3	2	Bearsden and Milngavie	31.4	2	Hart	50.5
3	Bearsden and Milngavie	63.0	3	Richmond upon Thames	29.2	3	Wokingham	48.5
4	Eastwood	62.3	4	Cambridge	27.9	4	Chiltern	47.7
5	Elmbridge	60.5	5	Kensington and Chelsea	27.6	5	South Buckinghamshire	47.3
6	Kensington and Chelsea	60.3	6	Eastwood	26.5	6	Uttlesford	44.4
7	Wokingham	59.8	7	Oxford	26.2	7	Tandridge	43.1
8	Chiltern	59.7	8	Camden	25.8	8	Wycombe	42.8
9	St Albans	59.5	9	St Albans	25.7	9	Mole Valley	42.7
10	Mole Valley	57.3	10	Chiltern	24.0	10	East Hampshire	42.3

Source: Hills 1995.

over recent decades applied geographers have focused particular attention on local variations in poverty and deprivation within cities.

A striking early example of intra-urban variation in well-being is provided by Bunge and Bordessa's (1975) map of infant mortality in Detroit, which revealed that an American born in the central areas of the city had the same chance of survival as an infant born in several third world countries (Figure 29.4). The skewed distribution of life chances within the modern Western city was also revealed in more recent evidence from Glasgow, which showed infant mortality rates in the disadvantaged council estate of Easterhouse to be five times higher than those in the nearby middle-class suburb of Bishopbriggs (Pacione 1993).

Applied geographical research into poverty at the intra-urban scale has generally used the census tract as the basic frame of reference. Smith (1994) employed data on median family income and racial composition to explore the differential incidence of poverty in Tampa (Figure 29.5). The marked socio-spatial concentration of poverty in

Box 29.2 Experiences of poverty

We're getting low now. We've got food for today, but there's none for tomorrow. Last week we ran out of money for the electric meter. . . . I just have one meal a day, sometimes I don't eat at all. . . . I've had to get the little ones out of nappies – I couldn't afford them any more. . . . We're going steadily downhill.

(Terry – a resident of inner London)

The soup kitchen opens at noon. The volunteers who prepared the food have now been joined by an assortment of students, 'lunch ladies' and people here in a less than voluntary capacity working a community service sentence. They feed up to 150 people a day, rising from about seventy at the beginning of the month to a peak at the end when times are hardest. About four-fifths of those served are black, the vast majority are men aged twenty-five to forty. There are a small number of drug addicts, but many are decently dressed, and some are at work in minimum-wage jobs. Many are regulars, recognised and greeted by the servers.

(Feeding the poor in a city in the Southern United States)

My day begins at four in the morning especially when my companero is on first shift. I make his breakfast. I then make about 100 saltenas (pies) which I sell in the street to pay for necessities that my husband's wage doesn't cover. Then the kids that go to school in the morning get ready while I wash the clothes I left soaking overnight. From what we earn between us we can eat and dress, but food is expensive: 28 pesos for a kilo of meat, 4 pesos for carrots, 6 pesos for onions . . . Considering that my husband earns 28 pesos a day working in the mine that's hardly enough, is it?

(Domitila – the wife of a Bolivian tin miner)

Sources: Harrison 1985; Weild 1992; Smith 1994.

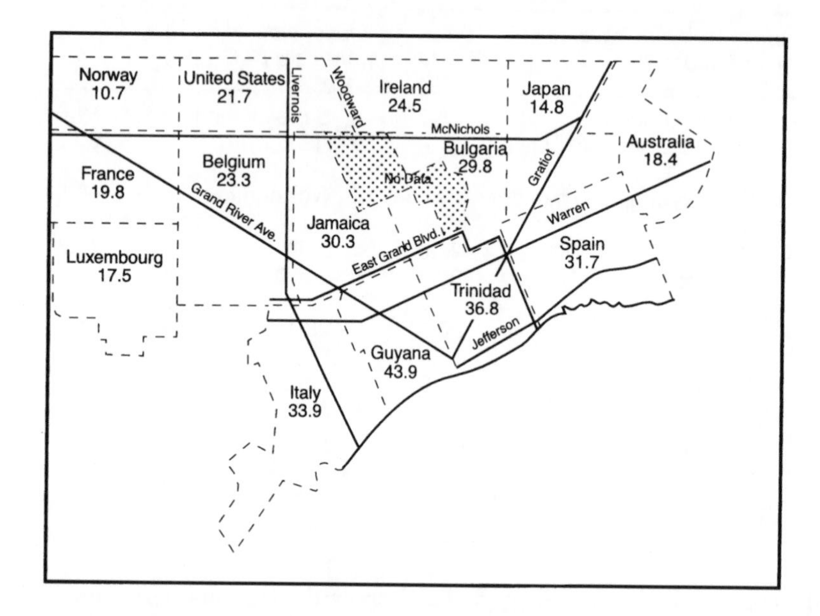

Figure 29.4 Infant mortality in Detroit.

Source: Bunge and Bordessa 1975.

the inner-city black residential areas was reproduced in other cities of the southeastern USA, lending support to the existence of an excluded urban underclass within advanced urban society (Mingione 1996).

The value of local-level applied research can be illustrated with reference to the geography of poverty and deprivation in the city of Glasgow, Scotland. In this study, data on sixty-four census-based social, economic and demographic indicators for each of the 5374 census areas in the city were subjected to principal component analysis. The first component, accounting for 15 per cent of the variance, was clearly identifiable as an indicator of multiple deprivation with high positive loadings on variables relating to unemployment, overcrowded housing, single parents, car-less households, long-term illness and local authority rented housing. Calculation of component scores provided a measure of deprivation for each of the

Figure 29.5 Variations in social well-being in Tampa.

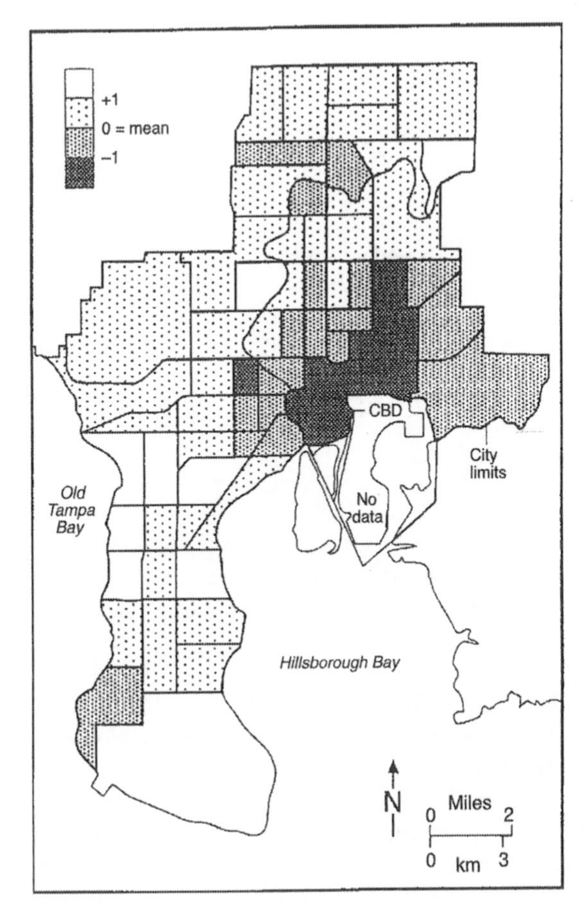

Source: Smith 1973.

census tracts in the city. The spatial expression of the multiple deprivation component is shown in Figure 29.6. This identified major concentrations of deprivation in inter-war inner suburban areas such as Possilpark, Garngad, Haghill and Blackhill; in isolated pockets south of the river in Govan, Gorbals and Pollokshaws; parts of the Maryhill Corridor and the Glasgow Eastern Areas Renewal (GEAR) area; and in the post-war peripheral council estates of Drumchapel, Castlemilk and Easterhouse.

Comparison of the results of this analysis with earlier studies of poverty and deprivation in the city indicated a significant change in the distribution of the problem in Glasgow over recent decades. In 1971, a higher proportion of deprived areas were located in the inner city, particularly in the East End and the Maryhill Corridor (Pacione 1989). Over the intervening period, the traditional inner tenement housing, which previously exhibited severe deprivation, recorded a relative improvement in status largely as a result of massive clearance and redevelopment by the local authority, combined in some areas, like GEAR, with modernisation and new building aided by housing associations and private developers. This process involved the large-scale relocation of residents in a general process of decentralisation. The inner areas now contain a much reduced and ageing population living in improved accommodation. Conversely, the outer estates exhibit a younger demographic structure and, while the housing is generally well provided with basic amenities, overcrowding is widespread. Serious social problems such as unemployment and a high proportion of single-parent families are also present. These spatial changes in the incidence of deprivation have been accompanied by a redistribution in terms of housing tenure. Whereas in 1971 a high proportion of deprived areas included older and frequently private rented properties (notably in the East End, Maryhill, Springburn and Govan), by 1991, deprivation had become increasingly concentrated in the public sector.

The geographical incidence of multiple deprivation in 1991 also sheds light on the effectiveness of the city's system of priority planning areas. While significant improvements have been made to living conditions in the two inner areas of Maryhill and GEAR, comparatively little progress was achieved in the outer estates. In addition, the geographical boundaries of the priority areas are open to question, with major concentrations of multiply deprived households (e.g. in Possilpark and Haghill) excluded from the official priority designations. In general, however, as Figure 29.6 reveals, the most significant trend over the past two decades has been the increasing concentration of deprivation in areas of council housing. This phenomenon represents a major

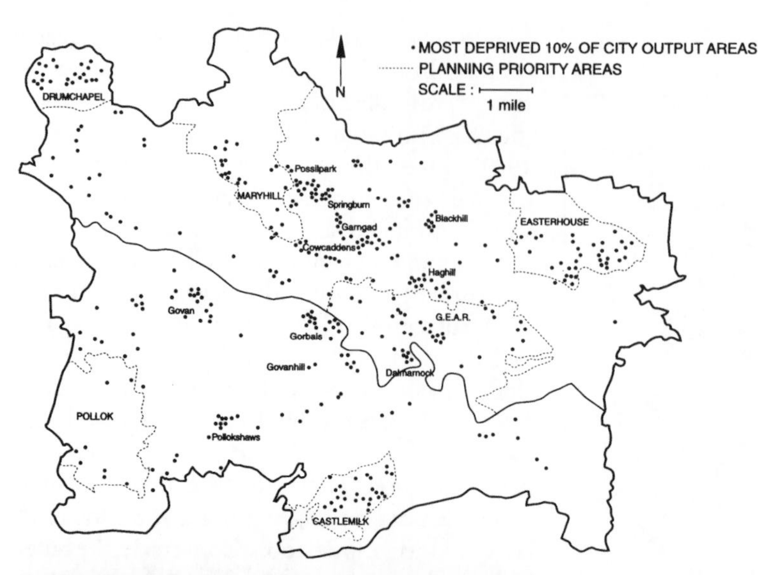

Figure 29.6 The distribution of multiple deprivation in Glasgow, 1991.

Source: Pacione 1995b.

Plate 29.2 A modern equivalent of 'bread and circuses'? The proliferation of satellite television in a deprived council estate in Glasgow (*photograph: the author*).

challenge for contemporary urban planners and policy makers.

THE VALUE OF AN APPLIED GEOGRAPHICAL PERSPECTIVE

While the principal underlying causes of local concentrations of poverty and deprivation are structural, stemming from the effects of global economic restructuring and state social and economic policy, the impacts of these processes are felt by people living within particular local contexts. Study of the difficulties experienced by these localities, which, although remaining meaningful places to their inhabitants, are not considered profitable spaces by capital, affords valuable insight into the nature and incidence of poverty and deprivation in contemporary society and informs a critique of state policy.

Critics of geographical analyses of socio-spatial variations in poverty, however, have sought to dismiss area-based research on the grounds that it does not offer an explanation of the underlying causes of revealed patterns. This viewpoint reflects the radical critique of the empirical–statistical analyses that dominated applied human geography throughout the 1970s. Such critics are now pushing at an open door. The claim that uncritical spatial analysis lends itself to preservation of the injustice present in the *status quo* has been widely accepted by applied human geographers and has informed subsequent policy-oriented analyses of poverty and deprivation in which the identification of pattern is used to advance a critique of current policy aimed at alleviating multiple deprivation (Pacione 1990b; 1992).

Proponents of geographical analyses maintain that an area-based perspective can be justified on several grounds. Some, such as Donnison (1974), have argued for a degree of 'area effect' in accentuating if not actually causing deprivation. Structural forces notwithstanding people live locally and experience the prosperity and problems of their own localities. Furthermore, as we have seen, at all geographical scales poverty and deprivation have been shown to exhibit strong spatial concentration. Second, the identification of spatial patterns is an essential starting point in understanding the local incidence of social disadvantage. As Brunhes (1920) noted more than seventy-five years ago,

> the study of poverty should mean not simply statistics but an attempt at precise localisation. Since to fix the topographical distribution of poverty is a means of knowing it more exactly, it is doubtless also a means of relieving it and curing it in a less abstract and more efficacious manner.

As Pacione (1995c) demonstrated, the application of area-based indicators at the national and regional levels provides a basis for local survey analyses that can facilitate a more informed allocation of available resources targeted at specific places and populations. Third, as Rivlin (1971: p. 146) pointed out, 'to do better we must have some way of distinguishing better from worse.' Analysis of the nature, intensity and distribution of multiple deprivation permits comparisons both spatially within countries, regions and cities and over time (Pacione 1986), and facilitates monitoring of the effectiveness of remedial strategies. Fourth, while the long-term ideal may remain a fundamental political–economic restructuring to tackle the roots of inequality in society, area-based policies of positive discrimination formulated on the basis of applied geographical research can provide more immediate benefits that enable some people to improve some aspects of their quality of life. For most applied human geographers, to seek to reduce the difficulties facing deprived people and places represents a realistic assessment that both long-term restructuring and short-term ameliorative action are required. To do nothing in the short term in the hope of the capitalist system being brought to an end by its inherent contradictions is tantamount to allowing the excesses of the system to continue unchecked. This is of little benefit to those currently burdened by the effects of poverty and deprivation.

An effective assault on the problems of poverty and deprivation requires appropriate action at several scales, comprising a combination of 'people policies' operating over the longer term at the structural level with the aim of achieving redistribution of society's wealth, and more immediate 'place policies' to improve the current position of the disadvantaged. More specifically, at the structural level a primary requirement is recognition of the fact that the disadvantaged position of the poor is linked inextricably with the privileged position of the wealthy. This requires enaction of a policy that addresses the fundamental question of the distribution of society's wealth. In the shorter term, easing central controls over local authorities in accordance with the principle of subsidiarity could enable the lower tier more latitude to undertake activities of direct benefit to local disadvantaged groups. Local government itself must seek to devise more effective strategies to attract private sector investment into deprived areas, via either persuasion (e.g. tax relief) or direction (via linked development and leverage schemes), with the particular mix depending on the strength of the local economy. Local authorities must also foster the empowerment of local communities to capitalise on the human resources of deprived areas. Ideally, this will involve devolution of financial and political power to the neighbourhood level, with greater participation of residents in budget formulation and plan preparation for their area. Applied geographers are well placed to contribute to these objectives by undertaking research into the causes and consequences of poverty, and by initiating critical appraisal of policies aimed at alleviating the economic, social and environmental problems that continue to impact on the well-being of a significant proportion of the world's population.

GUIDE TO FURTHER READING

The nature and definition of poverty in an international context is discussed in P. Townsend (1993) *The International Analysis of Poverty*, New York: Harvester Wheatsheaf. Poverty in the particular context of the third world is examined in United Nations Centre for Human Setlements (1996) *An Urbanising World*, Oxford: OUP; while K. McFate, R. Lawson and W. Wilson (1995) *Poverty, Inequality and the Future of Social Policy*, New York: Russell Sage; and A. Walker and C. Walker (1997) *Britain Divided: The Growth of Social Exclusion in the 1980s and 1990s*, London: Child Poverty Action Group, consider the problems of poverty in advanced capitalist societies. Particular examples of the application of an applied geographical approach to the analysis of poverty and deprivation are provided by M. Pacione (1995) 'The geography of multiple deprivation in Scotland', *Applied Geography* 15(2), 115–33; M. Pacione (1995) 'The geography of deprivation in rural Scotland', Transactions of the Institute of British Geographers 20, 173–92; and in Part II of D. Smith (1994) *Geography and Social Justice*, Oxford: Blackwell.

REFERENCES

Allen, T. and Thomas, A. (1992) *Poverty and Development in the 1990s*. Oxford: OUP.

Anand, S. and Ravallion, M. (1998) Human development in poor countries. *Journal of Economic Perspectives* 7(1), 133–50.

Benzeval, M. *et al.* (1995) *Tackling Inequalities in Health*. London: King's Fund.

Beveridge, W. (1942) Social insurance and allied services. *Cmnd 6404*. London: HMSO.

Brunhes, J. (1920) *Human Geography*. London: G. Harrap.

Bunge, W. and Bordessa, R. (1975) The Canadian alternative. Geographical expeditions and urban change. *Geographical Monograph No. 2*, Toronto: York University.

Chambers, R. (1994) Poverty and livelihoods: where reality counts? *Environment and Urbanisation* 7(1), 7–14.

Convery, P. (1997) Unemployment. In A. Walker and C. Walker (eds) *Britain Divided: The Growth of Social Exclusion in the 1980s and 1990s*, London: CPAG, 170–97.

Donnison, D. (1974) Policies for priority areas. *Journal of Social Policy* 3, 127–35.

Donnison, D. and Middleton, A. (1987) *Regenerating the Inner City*. London: Routledge & Kegan Paul.

Feres, J. and Leon, A. (1990) The magnitude of poverty in Latin America. *CEPAL Review* 41, 133–51.

Findlay, A. (1994) Life expectancy. In T. Unwin (ed.) *Atlas of World Development*, Chichester: Wiley, 102–4.

Goodman, A., Johnson, P. and Webb, S. (1997) *Inequality in the United Kingdom*. London: Institute for Fiscal Studies.

Gordon, I. and Whittaker, R. (1972) Indicators of local prosperity in the south-west region. *Regional Studies* 6, 229–313.

Green, A. (1994) *The Geography of Poverty and Wealth*. Coventry: University of Warwick.

Harrison, P. (1985) *Inside the Inner City*. London: Pelican.

Hills, F. (1995) *Inquiry into Income and Wealth*, Vol. 2. York: Joseph Rowntree Foundation.

Jazairy, I., Mohiuddin, A. and Panuccio, T. (1992) *The State of World Rural Poverty*. London: IT Publications.

Lyson, T. (1989) *Two Sides to the Sunbelt: The Growing Divergence Between the Rural and Urban South*. New York: Praeger.

Main, H. and Williams, J. (1994) *Environment and Housing in Third World Cities*. Chichester: Wiley.

Mingione, E. (1996) *Urban Poverty and the Underclass*. Oxford: Blackwell.

Misra, H. (1994) Housing and environment in an Indian city. In H. Mair and S. Williams (eds) *Environment and Housing in Third World Cities*, Chichester: Wiley, 191–206.

Pacione, M. (1986) The changing pattern of deprivation in Glasgow. *Scottish Geographical Magazine* 102, 97–109.

Pacione, M. (1989) The urban crisis: poverty and deprivation in the Scottish city. *Scottish Geographical Magazine* 105, 101–15.

Pacione, M. (1990a) What about people? A critical analysis of urban policy in the UK. *Geography* 75, 193–202.

Pacione, M. (1990b) The ecclesiastical community of interest as a response to urban poverty and deprivation. *Transactions of the Institute of British Geographers* 15, 193–204.

Pacione, M. (1992) Citizenship, partnership and the popular restructuring of urban space. *Urban Geography* 13, 405–21.

Pacione, M. (1993) The geography of the urban crisis: some evidence from Glasgow. *Scottish Geographical Magazine* 109(2), 87–95.

Pacione, M. (1995a) The geography of multiple deprivation in the Clydeside conurbation. *Tijdschrift voor Economische en Sociale Geografie* 86(5), 407–25.

Pacione, M. (1995b) The geography of multiple depri-

vation in Scotland. *Applied Geography* 15(2), 115–33.
Pacione, M. (1995c) The geography of deprivation in rural Scotland. *Transactions of the Institute of British Geographers* 20, 173–92.
Pacione, M. (1997a) The geography of educational disadvantage in Glasgow. *Applied Geography* 17(3), 169–92.
Pacione, M. (1997b) Urban restructuring and the reproduction of inequality in Britain's cities. In M. Pacione (ed.) *Britain's Cities: Geographies of Division in Urban Britain*, London: Routledge, 7–60.
Partha, D. and Weale, M. (1992) On measuring the quality of life. *World Development* 20(1), 119–31.
Rivlin, A. (1971) *Systematic Thinking for Social Action*. Washington, DC: Brookings Institute.
Rowntree, S. (1901) *Poverty*. London: Macmillan.
Smith, D. (1973) *Human Geography: A Welfare Approach*. London: Arnold.
Smith, D. (1994) *Geography and Social Justice*. Oxford: Blackwell.
Townsend, P. (1993) *The International Analysis of Poverty*. New York: Harvester Wheatsheaf.
United Nations Centre for Human Settlements (1996) *An Urbanising World*. Oxford: OUP.
Weild, D. (1992) Unemployment and making a living. In T. Allen and A. Thomas (eds) *Poverty and Development in the 1990s*, Oxford: OUP, 55–77.
WHO (1992) *Our Planet, Our Health*. Geneva: WHO.
Wilkenson, R. (1992) Income distribution and life expectancy. *British Medical Journal* 304, 165–8.
Wilson, J. (1969) *Quality of Life in the United States*. Kansas City: Midwest Research Institute.

30

Segregation and discrimination

David Herbert

INTRODUCTION

Segregation and discrimination are key words in the lexicon of social geography. Whereas segregation is the more widely used word and concept, it has integral links with the process of discrimination and also with other key processes such as assimilation and prejudice. Over the longer history of social geography, studies have been dominated by the influence of race and ethnicity, but now include gender, sexuality, impairment and age. Race has dominated, but one thesis is that race and class are closely intertwined, and a key function of discrimination and segregation is to deny access to greater wealth and status. There are *de facto* separate residential areas as the products of discrimination and segregation. The ability of the suburb to maintain and enhance its separateness and distinctive character is as much a testimony to the power of these processes as is the persistence of the impoverished ghetto. In the social geography of the city, this mosaic of residential areas with its visible symbols of power and prestige on the one hand and disadvantage and poverty, on the other, offers evidence of discrimination and segregation as key social, economic and political processes.

Discrimination is defined variously as 'to set up or observe a difference', 'to treat differentially', especially on the grounds of sex, race or religion and is a set of values from which actions may flow. Banton (1994) argued that discrimination is an individual action but that since members of the same group are treated in similar ways, it is typically a social pattern of aggregate behaviour. Again, the sets of attitudes tend to be transmitted from one generation to another and are difficult to dispel or even to modify. The effect of discrimination may be to create or increase inequalities between classes of persons and make discrimination more frequent (*ibid.*: 8).

Segregation is the more common theme in social geography, probably because it has meaning both as a process and as an outcome or condition. To segregate is defined as 'isolating', 'putting apart from the rest', 'the separation of one particular class of people from another on grounds such as race. Segregation has stronger behavioural imperatives than discrimination; it is more an action or activity that underpins the actuality of separate and different geographical spaces. The main such space is residential, but segregation also finds expression in education with segregated schools and in the workplace, reflecting real divisions within society.

Discrimination and segregation are common processes that underpin most of society, but recent geographies have tended to focus on the exceptional rather than the broad bases, and one aim in this discussion will be to maintain the kind of balance that the theme deserves. The chapter therefore begins with a summary of the proven significance of discrimination and segregation in the understanding of race and class and residential areas. It will take the opportunity to examine the value of new approaches, including those of cultural geographies, to our understanding of these well-studied schemes. Second, it will examine school segregation, which has been a powerful theme in the United States, especially during the second half of the twentieth century. Third, the more recent focus on the exceptional groups such as gypsies and the sexually deviant will be considered, and then, finally, attention is given to the new forms of financial exclusion being practised by organisations able to influence the flows of funds to individuals and groups.

DISCRIMINATION AND SEGREGATION IN RESIDENTIAL AREAS

When Burgess (1925) proposed his concentric model and Hoyt (1939) his sector theory of neighbourhood change, they were both identifying the facts of residential separation and segregated areas. Neither focused strongly on the meaning and significance of the processes of discrimination and segregation that produced these patterns. Their key assumptions were related to the economics of land use and the power of the bid-rent curve. The social area analysts used segregation in more explicit ways, and the ethnic and class 'dimensions' were central features of the type of social area analysis developed both by Shevky and Bell (1955) and by the legion of factorial ecologists (see Herbert and Thomas 1997). Whereas the thrust of this form of social geography was to identify and classify residential

Table 30.1 Measures of segregation: indices of dissimilarity in British cities, 1991 (from white population).

	Bangladeshi	*Pakistani*	*Black Caribbean*
London	65	54	49
Birmingham	80	74	54
Bradford	75	63	47
Oldham	79	76	44
Leeds	82	64	69
Sheffield	72	71	49
Leicester	81	56	38
Luton	70	65	24

Source: After Peach 1996.
Note: The index of dissimilarity has a range from 0 to 100, where 100 equals totally segregated.

Table 30.2 Measures of segregation: indices of dissimilarity in US cities, 1970–90 (black–white segregation).

	1970	*1980*	*1990*
Boston	81.2	77.6	68.2
Chicago	91.9	87.8	85.8
Detroit	88.4	86.7	87.6
Los Angeles	91.0	81.1	73.1
New York	81.0	82.0	82.2
Atlanta	82.1	78.5	67.8
Dallas/Fort Worth	86.9	77.1	63.1
Miami	85.1	77.8	71.8
Washington DC	81.1	70.1	66.1

Source: After Massey and Denton 1993.
Note: The index of dissimilarity has a range from 0 to 100, where 100 equals totally segregated.

areas in the city, the separate strand of social segregation researchers, best exemplified by the 'dissimilarists' (Peach 1996; Taeuber 1988), was concerned exclusively with racial segregation and ethnic areas. From the studies of the social area analysts came consistent evidence of the dimensions that led to residential separation in cities – social class, ethnic differences, stage in family life cycles, migrant status, housing conditions. From the dissimilarists came clear statements on the extent of racial segregation and its persistence over time. Consistently, black Americans have the highest levels of segregation (Massey and Denton 1993); in British cities, Peach (1996) found evidence for change, with Bangladeshis possessing the highest levels of segregation, although conditions nowhere resembled a ghetto.

THE ETHNIC GHETTO

The term 'ghetto' is emotive but draws together many of the significant features of discrimination and segregation. In the literature of the social sciences, ghetto is a racial concept. Peach (1996) noted that the two defining conditions for a ghetto – being dominated by a single ethnic or racial group and containing most members of that group – were only really met in ethnic minority areas. Boal (1978) provided an early attempt by a geographer to give specific meaning to the term 'ghetto' with his typology of colony, enclave and ghetto. Boal recognised the diversity of ethnic areas and the varying contributions of choices and constraints that led to their formation. The status of ethnic minorities relates to their migration histories and to the social distance that separates them from the charter group or host society. Jackson and Penrose (1993) argued that terms such as 'minority', 'race' and 'ethnic' were socially constructed, the products of specific historical and geographical forces. Assimilation is a key process, and whereas behavioural assimilation, whereby members of a minority group acquire the values and mores of the host society, is both achievable and within the control of the minority group, structural assimilation, or acceptance into occupational, educational and housing markets, is much more difficult. Discrimination becomes active as minority group members are prevented from achieving structural assimilation, and the outcome is segregation and disadvantage.

Mason (1995) argued that studies of minority groups often stress the differences rather than the diversity. Difference implies a 'norm' from which some groups deviate and can be 'rescued' by assimilation; diversity accepts differences with no necessary imperative of integration. Geographers have recognised the overarching significance of this process of discrimination, although Boal (1978; 1987) and others have shown that some choice mechanisms are also at work. Boal classified these as defensive, avoidance, preservation and resistance and argued that they could be applied, for example, to some Asian communities in Britain and to Chinese immigrants in American cities (Plate 30.1). Preservation is probably the most important of these choice mechanisms and suggests that minority groups may want to remain segregated in order to preserve their language, religion and culture, which might otherwise be quickly lost in a new society. Choice mechanisms are of much less significance where discrimination is greatest and the facts of segregation are accompanied by disadvantage, as summarised by the huge disparities between parts of black inner cities and exclusive white suburbs in the United States.

Black minority groups have typically suffered the worst discrimination and have experienced the highest levels of residential segregation (Figure 30.1). Despite the establishment of civil rights legislation and the signs of progress and change, de Vise (1994) stated that *de facto* residential segregation in Chicago remained very high. Massey and Denton (1993) argued that few

Plate 30.1 An ethnic area: New York's Chinatown. Chinatowns are among the most persistent segregated areas and involve choice as well as constraint. Such areas allow the preservation of language, religion and cultural values.

Figure 30.1 Johannesburg, 1985; segregated city.

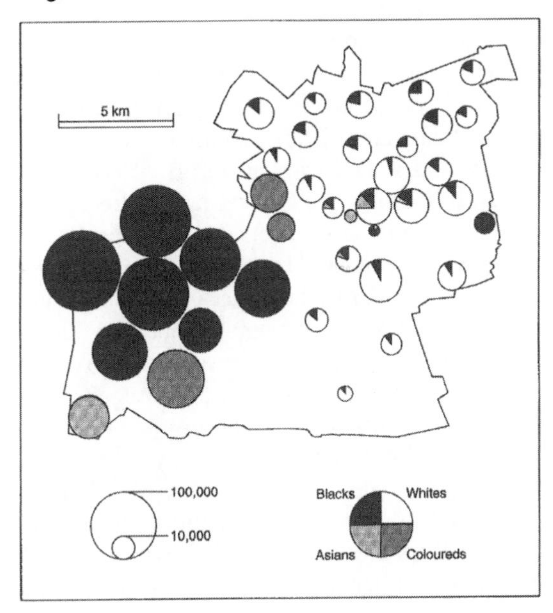

Source: After Christopher 1994.

Americans realised the depth of black segregation or the degree to which it was maintained by ongoing institutional arrangements and individual actions, and Aponte (1991) argued that substantial black inner city populations were still 'hopelessly mired in poverty'. Wilson (1987) was a catalyst for studies of change among the American black population. His central argument was that residualisation, marginalisation and exclusion *within* black residential areas had led to an underclass of the 'truly disadvantaged'. As stable families, those with jobs and social aspirations, and community leaders moved out, they left welfare-dependent areas with many lone-parent families, high unemployment, and a high incidence of crime and drugs. Perhaps the most significant loss was that of a sense of place, of belonging and of community, which had held many of these people together.

> Because of racial segregation, a significant share of black America is condemned to experience a social environment where poverty and joblessness are the norm, where a majority of children are born out of wedlock, where most families are on welfare, where educational failure prevails, and where social and physical deterioration abound.
>
> (Massey and Denton 1993: 2)

For American black populations, these ghetto-forming processes are well understood and spring almost entirely from prejudice and discrimination. For Massey and Denton (1993), the black ghetto was constructed through a series of well-defined practices and policies designed by whites to contain the growing black urban populations. There are other ethnic minorities groups where the processes are less straightforward, as Dennis (1997) showed in his analysis of the emergence of Jewish areas in Toronto. Established Jewish families played a role and:

> Ghetto-ization was not just a consequence of poverty or a need for a cultural identity but also a strategy on the part of self-styled community leaders to limit the impact of large-scale immigration on mainstream Toronto society.
>
> (*ibid.*: 378)

SEGREGATED SCHOOLS

Racial segregation in the United States and elsewhere is not confined to residential areas. Education is another example of the impact of discrimination, and it is only in the second half of the twentieth century that the American justice system has made substantive moves to end the dual system of segregated schools. From the 1950s to the 1970s, the issue of school desegregation dominated the politics of education. Studies such as that by Lowry (1973) in Mississippi documented the slow process of change from completely separate schools with wide disparities to some measure of integrated state education. Devices such as 'separate but equal' and 'freedom of choice' were used by local school boards to hinder the federally driven process of desegregation. Key decisions in the courts, such as Brown versus Topeka School Board in 1954 and

1955, Alexander versus Holmes County in 1969, and Swann versus Mecklenburg in 1971, removed the legal bases of segregation. Fiss (1977) argued that one problem for the law makers was that of distinguishing between segregation as a process or activity whereby students were assigned to particular schools on the basis of race, and segregation as a demographic pattern with whites in one school and blacks in another. In the Brown decision, the court felt no need to make the distinction and simply held the dual school system to be unlawful. The issue of whether racial assignment *per se* was illegal was settled in the Swann case, which judged that racial assignment was unlawful if it was used to achieve segregation. On the other hand, racial assignment intended to reduce or eliminate segregation was lawful. Segregation as a demographic pattern was judged to be unlawful because it led to inequalities in schooling and stigmatised black children.

Since the 1970s, the effectiveness of policies to reduce levels of school segregation has diminished for several reasons. First, white parents have been removing their children from state schools. This 'white flight' has involved families with the means to afford school fees, and the effect has been to maintain segregation and widen disparities. Second, some judicial decisions have strengthened the powers of school boards, and many of the actions designed to produce integration, such as busing, have been abandoned. Third, the absence of any significant changes in residential segregation means that school populations reflect the residential areas in which they are placed and remain segregated. Clark (1984) showed that the changing ethnic compositions of Los Angeles' schools were directly related to demographic changes in their neighbourhood catchment areas. His sample set reflected the effects of white flight, outward white migration and the large influx of Hispanic families.

THE OTHER MINORITIES

Geographies of minority groups have focused on the larger minorities such as American blacks and British Asians, who aspire to accommodation within mainstream society. There are other minorities who arguably suffer even higher levels of discrimination and segregation. Indigenous Indian populations in both Canada and the United States are still linked with territorial reservations and rank lowest in terms of economic status. In Europe, the gypsies have suffered discrimination in many societies and remain on the margins of acceptability. Sibley (1992) argued that the boundaries of society are continually being redrawn to distinguish between those who belong and those who are excluded. Gypsies commonly belong to this excluded category with characteristics that make them not just different but deviant. Because they have a way of life that is regarded as negative and inferior, gypsies become legitimised targets for discrimination. Their economy revolves around domestic scrap, low-level repairs and services and appears to fit their designation on the margins of society. Gypsies occupy marginal places that by association become labelled as unsafe and undesirable. Gypsy sites have to be provided by law in the United Kingdom, but they are inevitably contested places as local residents object to their proximity. Images of lack of cleanliness, dubious work practices, violence and antagonism are associated with gypsies. Those who control nearby space will seek to exclude gypsies.

Another extreme case in the landscapes of exclusion has been the fate of the mentally ill. Over long periods, they were institutionalised and isolated by forces motivated more by fear and the need to contain rather than by welfare considerations. When the process of de-institutionalisation that brought many mentally ill people back into communities took place in the 1960s, it was driven by economics rather than by any changing attitude towards the well-being of the group. The experience of de-institutionalisation has served to confirm the prevalence of discrimination against the mentally ill and to expose the myth of community care. Conflict arose from both the assignment problem and the attempt to match mentally ill people to treatment settings and the siting problem, or the fitting of type of facility to community. Unless close

Plate 30.2 A rescue mission in Vancouver's skid row. The run-down areas of low-cost rooming houses in North American cities are inhabited by the real have-nots in society. Many suffer from diseases such as mental illness and alcoholism.

family was available, both the community-based facilities and the mentally ill were rejected. New forms of spatial segregation arose through neighbourhood resistance to facilities, from planners' tendency to locate after-care facilities in those inner city areas that showed least resistance, and from the informal filtering or drift of mentally ill people towards transient rented areas. The mentally ill remain a class of outsiders, clustering in inner city areas and strongly over-represented among the homeless, the low-cost boarding houses (Plate 30.2) and the prison population. As Sibley (1992) argued, outsiders are those groups that do not fit into dominant models of society and are seen as 'polluting'. Such groups disturb the homogeneity of a locality, and the common reaction of a hostile community will be to expel them and purify spaces.

There are other minorities, distinguished by their sexuality, that have become visible in Western cities and rank as outsiders. Cities have always had districts associated with the 'sex trade', which carry euphemisms such as 'vice areas' and 'red-light districts'. Generally, the attitude of society is that if such activities are to be tolerated, they should be confined to specific areas where they can be controlled and monitored; districts such as London's Soho and the red-light district of Amsterdam are outcomes of this process. The professionals of the sex industry are confined to such districts for the practice of their trade, although not necessarily for their residences. In the moral geographies of the city, vice areas are at the lowest point of the scale. A different kind of moral geography has emerged with the greater willingness of sections of society with different sexual proclivities to become visible. Gays and lesbians form minority groups that are regarded as deviant by the dominant society.

Forest (1995) studied the West Hollywood district of Los Angeles as an example of a place with a gay identity. His focus was on portrayals of the gay community in the press and, in particular, the attempts by the gay press to link sexual meanings to particular places and thus represent gay minorities in ways similar to ethnic minorities. This concept of diversity and the rights of minorities to occupy specific spaces combats the older image of exclusion or the need to confine 'perverts' and moral failures to excluded places. This perspective suggests that place has a key role in allowing minorities to resist domination: there is some shift from constraint and exclusion to choice and recognition. Places where gays and lesbians are accepted become places where they are empowered, and the whole process of 'coming out' is enabled in environments of this kind. Gay territories play significant parts in the evolution of gay identities and subcultures. In West Hollywood, there are symbols of gay identity that conform to the characteristics of many of its inhabitants; place plays a fundamental role in the creation of a 'normative ideal'. Valentine (1993a; 1993b) studied the space behaviour of lesbians in British cities and revealed the difficulties faced by this minority group in a society dominated by a different form of sexuality.

LANDSCAPES OF PRIVILEGE

The most commonly cited examples of segregation and discrimination focus on the disadvantaged, but at the other ends of the social status spectrum are the landscapes of privilege, where the wealthy establish their residential spaces. Places such as Beverly Hills, Hampstead and Nob Hill also serve as icons of segregation and discrimination. Hoyt's (1939) sector model was strongly influenced by his identification of high-grade residential areas, and Firey (1947) provided a classical analysis of the way in which residents of 'old Boston' worked to maintain the character and status of Beacon Hill as it came under threat from city centre expansion. He spoke of the sentiment and symbolism associated with the area and the cultural motives of the families who strove to preserve it. In a later study, Domosh (1992) studied the comparable high-status area of Back Bay in Boston. She argued that by the late nineteenth century Boston's upper classes were remarkably cohesive and the Boston Association had been established to protect their interests. The Back Bay residential development embodied the ideologies of the elite and allowed them to distance themselves from new immigrants. Whereas Beacon Hill belonged to 'old Boston', Back Bay provided residences for the new moneyed upper-income groups. Both places were outcomes of the felt need for the wealthy to segregate and to place distance between themselves and the lower classes.

From new cultural geographers there have emerged studies of the ways in which specific residential areas have acquired and retained segregated high status, often in the face of significant difficulties. Shaughnessy Heights in Vancouver is an elite landscape with strong historical roots and a modern reproduction (Duncan 1992). Originally, Shaughnessy Heights was a high-status development featuring English-style country houses, which carried the dual connotation of a link to the gentry and to Englishness. In response to threats to its status and character, residents successfully organised a preservation movement and were helped by a wide cross-section of the urban population that recognised its symbolic significance to the city as a whole.

> Shaughnessy Heights is not simply a cultural production (a material landscape) interpenetrated by political and economic structures; it is also a cultural (re-)production in that it reproduces the meaning of belonging to an Anglophile elite in a west Canadian city.
>
> (*ibid.*: 50)

FINANCIAL EXCLUSIONS

A key economic mechanism that stems from discrimination and produces forms of segregation is related to the control of financial credit. The term 'financial exclusion' has been used to describe a situation in which large numbers of people are denied access to credit and the benefit of financial services. Some people, such as those on very low incomes, will be 'outside' the financial system by choice (Ford and Rowlinson 1996), but for many their inability to achieve inclusion in financial services is an increasing disadvantage (Box 30.1). The broadest correlate of financial exclusion is social class, but there is evidence of the relevance of other factors such as race, age and gender.

Some of the older forms of financial exclusion reveal both racial and social class dimensions and has led directly to spatial segregation. Leyshon and Thrift (1997: 229) argued that the geography of income and wealth shaped access to the financial system.

> The rich get richer and the poor get poorer as lending flows towards wealthy areas, with low credit risks, and in high volumes per capita as credit available is often a multiple of income. In poorer areas there may be a downward spiral of decline as residents of such areas find it hard to sell or buy property and businesses are unable to obtain credit.
>
> (Dymski and Veitch 1992)

The impact of financial exclusion policies in the housing market has been evident for some time. Building societies in Britain practised a policy of 'red-lining' by which they refused loans in specific areas of perceived high risk. A study of inner city Birmingham (CDP 1974) showed that only 7 per cent of owner-occupiers held a mortgage at the prevailing rate of interest, and low-income families experienced a general round of housing

Box 30.1 Financial exclusion, discrimination and segregation

Who are the financially excluded?	Mainly low-income, high-risk.
Form of discrimination	Restricted access to financial services. Refusal of credit/loan facilities. Closure of banks in targeted areas.
Sources of credit unavailable	Personal loans from banks. Mortgages from building societies. Credit cards and store cards. Hire-purchase terms from companies.

GROUPS WITH NO CURRENT BANK ACCOUNT

	%		%
All adults	31	Full-time housewives	63
Women	35	Unemployed	52
Social class E	59	On housing benefit	60
Social class D	44	On income support	58
Income < £3000	35	Council tenant	47
Lone parents	52	Run-down area	55

CASE STUDIES OF EXCLUSION/COMMODIFICATION

1 Client with overdraft and other debts sought advice at an interview with his bank manager. He was charged £75.00 for the interview.
2 An unemployed client wishing to open a bank account was refused both an account and an explanation.
3 An applicant with debts and who lived on benefits was told by a bank that 'it was only interested in quality clients'.

GEOGRAPHICAL IMPLICATIONS

1 Deepens the disparities between wealthy areas and the financially excluded groups living in poorer areas.
2 Poorer areas experience a withdrawal of financial services.

Source: After Leyshon and Thrift 1997.

discrimination. The continuing decay of the inner city was strongly driven by the unwillingness of financial institutions to invest in these areas. Restrictive practices in the housing market have ameliorated with civil rights and race relations legislation and with the re-regulation of the financial system, which led to mortgage availability under the 'right to buy' policy. This flexibility ended in the 1990s with the retreat of the financial system to its traditional middle-class heart land (Leyshon and Thrift 1997) and the redirection of credit away from poorer and towards richer social groups as a strategy of risk avoidance.

Financial exclusion has ramifications well outside the particular case of the housing market. Bank accounts, cheque cards, credit cards, non-cash transactions and short-term loans have become essential parts of society. Without access to at least some of these facilities, people are placed at significant disadvantage. In Britain, credit transfers or crossed cheques are used to pay, for example, benefits and pensions but about 35 per cent of this group do not have a bank account. Credit cards, store cards and hire purchase are not available to low-income groups, yet these are increasingly common as modes of transaction. There is evidence for more aggressive banking practice involving, for example, the re-introduction of charges for those with low balances, the closing of accounts with low balances and few transactions, and branch bank closures in areas where the volume of business is judged to be unsatisfactory. Christopherson (1993) argued that there was a relationship between an underlying geography of income and class and the pattern of financial service branch closure. The evidence from the United States is of withdrawal of financial services from poorer communities, principally the African-American and Hispanic inner city areas. Davis (1990) referred to this as contributing to the 'spatial apartheid' of the American urban system (Box 30.2). No study of segregation should

Box 30.2 The two worlds of Los Angeles

THE EXCLUDED

Unemployment among Black youths in Los Angeles county remained at 45 per cent through the late 1980s. A survey of ghetto housing projects revealed 120 employed out of 1060 households in Nickerson Gardens.

(p. 305)

The deteriorating labour market position of young Black men is the main reason for the counter-economy of drugs and crime. Forty per cent of children in Los Angeles county lived at or below the official poverty line.

(p. 306)

Aside from 230 Black and Latino gangs, there are over 80 Asian. Gangs are now much more interested in drug-sales territories than traditional turfs.

(p. 316)

The Californian educational system is in steep decline. Public schools in veritable 'children's ghettoes' are over-burdened. Racial isolation has assumed an overlay of class isolation.

(p. 307)

THE INCLUDED

The carefully manicured lawns of Los Angeles' Westside sprout forests of ominous little signs warning 'Armed Response'. Even richer neighbourhoods in the canyons and hillsides isolate themselves behind walls guarded by gun-toting private police and state-of-the-art electronic surveillance.

(p. 223)

Where the itineraries of the Downtown powerbrokers unavoidably intersect with the habitats of the homeless or the working poor, extra ordinary design precautions are taken to ensure the physical separation of the different humanities.

(p. 234)

We live in 'fortress cities' brutally divided between the 'fortified cells' of affluent society and 'places of terror' where police battle the criminalized poor.

(p. 224)

Source: After Davis 1992.

underestimate the roles played by the major financial institutions and their managers.

CONCLUSION

Despite the long record of legislation designed to reduce or eliminate the effects of discrimination and segregation, these show a remarkable persistence in many parts of the world. There have been changes of major significance such as civil rights legislation in the United States, race relations acts in the United Kingdom and the transition of South Africa from an apartheid state. Legislation will always be needed to provide the legalistic framework within which policies can be developed and practices improved, but the main battle is for hearts and minds. Whereas major studies, such as those on American ethnic minorities (Wilson 1987; Massey and Denton 1993) confirm the continuing existence of segregation and disparities as the norms of urban life, there are signs of progress, which include more access to jobs, education and housing for groups with long histories of segregation. Residential segregation on racial lines persists and underlies many other forms of separation in schools, community activities, workplaces and financial markets. In the foreseeable future, it is unlikely that these patterns of residential segregation will change, and the need to direct resources, power and involvement to underprivileged communities remains a high priority. In the United States, the 1977 Community Reinvestment Act significantly empowered local communities in their fight to retain access to financial services; credit unions and community development banks help to fill the gap for lower-income households. Leyshon and Thrift (1997) make a case for 'financial citizenship' in Britain, with more responsible banks and alternative financial services for the low-paid.

Desegregation is a more elusive goal than greater equality, fewer disparities and segregation resting much less on discrimination and much more on choice and preference. Perhaps the most worrying trend in Western societies is the accentuating nature of social exclusion in its many forms. Reforms allow more people to move 'inside' society and to take part in its processes, but outside is the increasingly sharply defined group of the truly disadvantaged, the real poor with their sub-culture of differences. An understanding of the causes and consequences of these processes is a priority for applied social geography.

GUIDE TO FURTHER READING

Herbert, D.T. and Thomas, C.J. (1997) *Cities in Space: City as Place*. London: Fulton. Comprehensive text with detail on many forms of spatial segregation.

Leyshon, A. and Thrift, N. (1997) *Money/Space: Geographies of Monetary Transformations*. London: Routledge. Detailed commentary on the geographies of financial exclusion.

Massey, D.S. and Denton, N.A. (1993) *American Apartheid: Segregation and the Making of the Underclass*. Cambridge, Mass.: Harvard University Press. Analysis of the effects of racial discrimination and segregation in the United States over time.

Anderson, K. and Gale, F. (eds) (1992) *Inventing Places: Studies in Cultural Geography*. Melbourne: Longman-Cheshire. Excellent range of case studies exemplifying the new cultural geography.

REFERENCES

Aponte, R. (1991) Urban Hispanic poverty, disaggregations and explanations. *Social Problems* 38, 516–28.

Banton, M. (1994) *Discrimination*. Buckingham: Open University Press.

Boal, F.W. (1978) Ethnic residential segregation. In D.T. Herbert and R.J. Johnston (eds) *Social Areas in Cities*, London: Wiley, 57–95.

Boal, F.W. (1987) Segregation. In M. Pacione (ed.) *Social Geography: Progress and Prospect*, London: Croom Helm, 90–128.

Burgess, E.W. (1925) The growth of the city. In R.E. Park, E.W. Burgess and R.D. McKenzie (eds) *The City*, Chicago: Chicago University Press, 47–62.

Christopherson, S. (1993) Market rules and territorial outcomes: the case of the United States. *International Journal of Urban and Regional Research* 17, 214–32.

Clark, W.A.V. (1984) Judicial intervention, busing and local residential change. In D.T. Herbert and R.J. Johnston (eds) *Geography and the Urban Environment* 6, 245–81.

Community Development Project (1974) *Inter-Project Report*. London: HMSO.

Davis, M. (1990) *City of Quartz: Excavating the Future in Los Angeles*. London: Vintage.

Dennis, R. (1997) Property and propriety: Jewish landlords in early twentieth century Toronto. *Transactions, Institute of British Geographers* 22, 377–97.

de Vise, P. (1994) Integration in Chicago forty years after Brown. *Urban Geography* 15, 454–69.

Domosh, M. (1992) Controlling urban form: the development of Boston's Back Bay. *Journal of Historical Geography* 18, 288–306.

Duncan, J. (1992) Elite landscapes as cultural (re)-productions: the case of Shaughnessy Heights. In K. Anderson and F. Gale (eds) *Inventing Places: Studies in Cultural Geography*, Melbourne: Longman-Cheshire, 37–51.

Dymski, G. and Veitch, J. (1992) *Race and the Financial Dynamics of Urban Growth: LA as Fay Wray*. Working Paper 92-21, Department of Economics, University of California, Riverside.

Firey, W.E. (1947) *Land Use in Central Boston*. Cambridge, Mass.: Harvard University Press.

Fiss, O.M. (1977) School desegregation: the uncertain path of the law. In M. Cohen, T. Nagel and T. Scanlon (eds) *Equality and Preferential Treatment*, Princeton, NJ: Princeton University Press, 155–91.

Ford, J. and Rowlinson, K. (1996) Low-income households and credit: exclusion, preference and inclusion. *Environment and Planning A* 28, 1345–60.

Forest, B. (1995) West Hollywood as symbol: the significance of place in the construction of a gay identity. *Society and Space* 13, 133–57.

Hoyt, H. (1939) *The Structure and Growth of Residential Neighbourhoods in American Cities*. Washington, DC: Federal Housing Administration.

Jackson, P. and Penrose, J. (eds) (1993) *Constructions of Race, Place and Nation*. London: UCL Press.

Lowry, M. (1973) Schools in transition. *Annals, Association of American Geographers* 63, 167–80.

Leyshon, A. and Thrift, N. (1997) *Money/Space: Geographies of Monetary Transformation*. Routledge: London.

Massey, D.S. and Denton, N.A. (1993) *American Apartheid: Segregation and the Making of the Underclass*. Cambridge, Mass.: Harvard University Press.

Peach, G.C.K. (1996) Does Britain have ghettos? *Transactions, Institute of British Geographers* 21, 216–35.

Mason, D. (1995) *Race and Ethnicity in Modern Britain*. Oxford: Oxford University Press.

Shevky, E. and Bell, W. (1955) *Social Area Analysis*. Stanford: University of California Press.

Sibley, D. (1992) Outsiders in society and space. In K. Anderson and F. Gale (eds) *Inventing Places: Studies in Cultural Geography*, Melbourne: Longman-Cheshire, 107–22.

Taueber, A.F. (1988) A practitioner's perspective on the index of dissimilarity. *American Sociological Review* 41, 884–9.

Valentine, G. (1993a) Negotiating and managing multiple sexual identities; lesbian time–space strategies. *Transactions, Institute of British Geographers* 18, 237–48.

Valentine, G. (1993b) (Hetero)-sexing space: lesbian perceptions and experiences of everyday spaces. *Society and Space* 11, 395–413.

Wilson, W.J. (1987) *The Truly Disadvantaged: The Inner City, the Underclass and Public Policy*. Chicago: University of Chicago Press.

31

Socio-spatial variations in health

Matthew Smallman-Raynor and David Phillips

INTRODUCTION

The term 'health' (stemming from the Old English word *hael*, or 'whole') means different things to different people (Kiple 1993: pp. 45–110). In modern Western medicine, for example, a 'healthy' person or place is often judged according to the absence (or otherwise) of a medically defined disease or disorder. The charter of the World Health Organization (WHO) favours a broader definition of health as a 'state of complete physical, mental and social well-being and not merely the absence of disease or infirmity' (WHO: 1988: p. 1). At a more abstract level, health can be defined according to the unidirectional nature of time; as Hudson (1993) notes, unless other factors intervene, our genetic programmes are inexorably geared towards disease and death. Other medical systems have placed yet further interpretations on health. In ancient Greek medicine, for example, health was viewed in terms of a balance in the bodily humours (blood, phlegm, and yellow and black bile), while notions of balance and harmony also underpinned conceptions of health in ancient Chinese and other Asian medical systems (Shigehisa 1993).

There are numerous and varied examples of the ways in which health – however defined – can vary between individuals, groups of people and, equally importantly, between places (Vågerö 1991; 1995; Vågerö and Illsley 1993; West 1991; Wilkinson 1987). Moreover, health variations are apparent at every geographical scale, from the continents and macro-regions of the planet to the districts and sub-districts of a single city. Seminal studies by G. Melvyn Howe (1986) on the global and world regional incidence of cancers, Gerald F. Pyle (1971) on national and local patterns of heart disease and stroke in the United States, and John A. Giggs (1973; 1988; 1990) on schizophrenia, affective psychoses and substance abuse in the city districts of Nottingham, England, are illustrative of the breadth and geographical range of the problem.

Today, efforts to improve health status and to erase (or at least to substantially reduce) variations in the well-being of people and places lie at the heart of much global health policy. Spurred by the WHO's *Global Strategy for Health for All by the Year 2000*, national governments and international agencies have launched a plethora of research initiatives to identify and monitor health inequalities (WHO 1994). The socio-spatial dimensions of the research issue have secured an important and growing role for medical and health geographers. In particular, geographers have brought an increasing methodological sophistication in spatial analysis and statistical modelling (Cliff and Haggett 1988; Cliff, *et al.* 1998; Thomas 1992), geographical information systems (Openshaw 1990; de Leper *et al.* 1995; Bailey and Gatrel 1995; Gatrell and Bailey 1996) and, most recently, qualitative techniques (Litva and Eyles 1995; Eyles 1997) to bear on the problem. At the same time, a traditional concern of medical geography with the spatial and environmental parameters of infectious and parasitic diseases (classic diseases such as cholera, malaria, measles and tuberculosis, but also 'newly emergent' conditions such as the acquired immunodeficiency syndrome or AIDS) has broadened to include a spectrum of acute, chronic and degenerative conditions. These include life-threatening ailments such as cerebrovascular and cardiovascular disease (Learmonth 1988; Phillips and Verhasselt 1994; Iyun *et al.* 1995), possible autoimmune diseases such as multiple sclerosis (Foster 1992), and potentially disabling conditions that are likely to have both genetic and environmental underpinnings, such as eczema, asthma and hay fever (McNally *et al.* 1998). In part, these developments reflect a growing recognition that environmental factors may aggravate or trigger some health conditions, and specialists in fields as diverse as dermatology, genetics, oncology and toxicology have increasingly turned to medico-health geography for clues to the environmental links in disease causation (Howe and Loraine 1980; Bentham 1994).

THE NATURE OF THE PROBLEM

Why does people's health appear to vary so markedly according to who they are and where they live? The evidence is stark, but explanations are often complex. Early responses to poor public health involved the implementation of a range of sanitary measures to improve physical health. That approach, when combined with improved nutrition, education, housing and other welfare measures, resulted in a much reduced mortality from infectious diseases in the cities of late nineteenth-century Europe and North America (Jones and Moon 1987; Jones and Curtis 1997; Cliff *et al.* 1998). Further reductions in mortality from infectious diseases, including the remarkable feat of smallpox eradication, were achieved by vaccination and other medico-scientific interventions in the twentieth century (Fenner *et al.* 1988). By contrast, the twentieth century has witnessed relatively few 'magic bullet' successes in the control of chronic, degenerative and mental illnesses, many of which display distinct social and spatial patterns. So, with a few exceptions, areas with a preponderance of people who are unemployed, or who are engaged in manual and semi-skilled jobs, are generally at an inflated risk for illnesses such as heart disease, certain types of cancer, cerebrovascular disease and some infectious conditions (Luoto *et al.* 1994; Lamont *et al.* 1997). Likewise, mental ill-health shows some socio-economic variations as well as marked spatial differences (Smith and Giggs 1988). However, the cause-and-effect issue is far from resolved. For example, do people with acute mental illnesses cluster in the less salubrious areas of cities due to some precipitating environmental factor? Alternatively, does the illness itself precipitate downward mobility in the housing market?

While socio-spatial variations in health are apparent at all geographical scales, it is much more difficult to reach adequate explanations for those patterns. As described more fully below, spatial patterns of ill-health and mortality often correlate closely with measures of social class, income and/or deprivation. But even at the finest levels of analysis (say, the districts of a single city), such associations rarely provide satisfactory explanations for health variations. In some instances, the explanations are evident: an overcrowded slum with an unreliable supply of drinking water, inadequate sanitation and sewerage systems is likely to provide an environment ripe for the rapid spread of an infectious disease such as cholera. But why are there similar socio-spatial variations in ailments (including many chronic, degenerative and mental illnesses) that have no apparent link to unsanitary conditions? This is much more difficult to explain, and a range of examples are to be seen in Iyun *et al.* (1995) and Harpham and Tanner (1995), among other sources.

The determinants of health

Although explanations of health variations are often elusive, exponents of Western medicine have reached a broad consensus on the nexus of factors that impinge on human health (Learmonth 1988; WHO 1992; Tarlov 1996).

These factors are commonly referred to as the 'determinants of health' and fall into four broad categories:

1 *Human biological determinants*, encompassing factors that are internal to the human body (for example, genetic composition, ageing and gender).
2 *Environmental determinants*, encompassing environmental factors that are external to the human body. These factors can be further classified according to the physical environment (for example, climate and altitude), the social environment (for example, housing and population density), and the biological environment (for example, the presence and persistence of disease-causing micro-organisms).
3 *Lifestyle determinants*, encompassing personal behaviours that can threaten health (for example, personal hygiene, smoking, substance abuse and diet).
4 *Health-care system determinants*, encompassing the resources devoted to health care and medicine in a population.

Current thought on the aetiology of many chronic, degenerative and mental illnesses has implicated a range of human biological, environmental and lifestyle factors in disease expression (see Kiple 1993; McNally *et al.* 1998). For example, factors implicated in the aetiology of cardiovascular disease have included nervous stress, hypertension, cigarette smoking, physical inactivity, obesity, genetic predisposition and the existence of other diseases such as diabetes mellitus. It seems likely that some complex and, perhaps individualised, combinations of predisposing and precipitating factors ensure that some people, groups and places experience higher or lower rates of certain diseases. It is also apparent that, even in countries with high-grade systems of medical care, there may be spatial variations in the efficacy of programmes aimed at the early detection and successful treatment of life-threatening illnesses (Expert Advisory Group on Cancer 1995).

The epidemiological (health) transition

Environments, lifestyles and health-care systems alter with the processes of social and economic development. As development rarely (if ever) progresses at an even rate in a given geographical area, it follows that socio-spatial variations in health are intrinsically dynamic phenomena. One framework for analysing this dynamism is the concept of epidemiological transition (Omran 1971) and the broader concept of health transition (Caldwell *et al.* 1990). The concept of epidemiological transition was widely publicised by Abdel Omran in 1971 and since then has been the subject of empirical examination in many countries of both the developed and developing world (Frenk *et al.* 1989; 1996; Phillips 1990; 1994; Phillips and Verhasselt 1994). Indeed, such has been the impact of the concept that the wider subject area has spawned its own journal since 1991, *Health Transition Review*.

In essence, the epidemiological transition envisages a process by which the mortality profile (and, by implication, the health profile) of a human population progresses through three distinct stages: Stage 1, a period with a preponderance of Old World epidemics and pandemics of infection and famine; Stage 2, a period of receding pandemics; and Stage 3, a period in which chronic, degenerative and human-induced ailments predominate. Each stage is associated with a decreasing death rate, a decreasing birth rate, an increasing life expectancy and demographic ageing. More recently, and for developed countries at least, Olshansky and Ault (1986) have posited a Stage 4 of the transition. This fourth stage is associated with advances in the medical treatment of chronic and degenerative diseases, thereby giving rise to increased survivorship (but potentially worsening overall health status) in middle-aged and elderly populations.

While the linear model of epidemiological transition provides a conceptual framework for studying the evolution of mortality patterns in particular, and health patterns more generally, it is obviously a simplification of a complex reality.

The transition is widely affected in its timing and extent by socio-economic, scientific and infrastructural developments. The cities of many developing countries, for example, are experiencing a combination of high death rates from both infectious and degenerative diseases – a phenomenon that has been referred to as 'delayed transition' (Frenk *et al.* 1989; 1996). In this instance, the better-off residents have rapidly exhibited the disease profiles typical of Stage 3 of the transition, but their poorer neighbours often stand in double jeopardy from infectious and chronic/degenerative diseases (Phillips 1993; 1994). Moreover, for some countries of Latin America at least, Frenk *et al.* (1996) identify the re-emergence and intensification of diseases such as dengue, malaria and some sexually transmitted diseases as evidence of a 'counter-transition'.

CASE STUDIES

Case study 1. Health variations and contemporary research in the United Kingdom

Socio-spatial variations in health are a major issue influencing the policy and health practice of many countries (Power 1994; Marmot 1996). In the United Kingdom, for example, inequalities of the type described in Box 31.1 have formed the backdrop for a series of government health initiatives, including the 1980 *Black Report* (DHSS 1980), the 1992 *Health of the Nation* White Paper (Secretary of State for Health 1992) and, most recently, the 1998 *Our Healthier Nation* Green Paper (Secretary of State for Health 1998). These various initiatives have included the promotion of multidisciplinary research programmes, one of the most prominent of which is the *Health Variations Programme* (HVP) funded by the Economic and Social Research Council (ESRC). The HVP was launched in 1997 and seeks to further an understanding of the social processes that underpin health variations. In the programme's first newsletter, Graham (1998) has identified seven key issues for research:

1 *Life-course influences*, concerned with the one-off and cumulative effects on health at different life stages, and throughout life, and which later contribute to patterns of inequality.
2 *Area effects*, concerned with the extent to which the socio-economic characteristics of places have an effect on people's health, over and above the effect of their socio-economic background.
3 *Income dynamics*, concerned with the health effects of short- and long-term dependence on low income.
4 *Psycho-social processes*, concerned with the health effects of psychological and social difficulties (for example, low self-esteem, depression and lack of social support).
5 *Policy impact*, concerned with the impact of social welfare policies on health inequalities;
6 *Ethnicity*, concerned with the extent to which patterns of health vary between and within ethnic groups.
7 *Gender and age differences*, concerned with the extent to which patterns of health vary by gender and age group.

As described in Box 31.1, socio-spatial variations in health indicators such as longevity are persisting and, indeed, even widening in some parts of the United Kingdom. One priority of the HVP research agenda is to set explanations for these variations within a practical policy framework. This priority raises a number of operational issues, over and above any explanations for the variations. For example, to what extent are health variations sensitive to the provision of health and welfare services, and which aspects of health variation are most effectively targeted by health interventions? To what extent do health service reorganisations (past, present and future) impinge on health variations? How should interventions to reduce variations be monitored, and what are the appropriate time scales on which to judge policy success? Answers to these, and similar, questions are paramount to any effective and concerted action on health inequalities.

Box 31.1 Spatial patterns of longevity and deprivation in England

Set against the policy framework of the *Health of the Nation* White Paper, Soni Raleigh and Kiri (1997) have examined the spatial association of life expectancy and socio-economic deprivation in the 105 District Health Authorities (DHAs) of England. As illustrated by a map of male life expectancy at birth (LEB) for the years 1992–4 (Figure 31.1), the analysis reveals that male longevity declines along a south–north gradient. So, with the exception of parts of Greater London, DHAs to the south, east and west of England generally display a high LEB (≥ 74.5 years). From here, life expectancies decline to intermediate levels in the Midlands, reaching their lowest values (LEB ≤ 73.4 years) in some northern districts. Life expectancies for females mimic the same basic spatial pattern, albeit at relatively higher values of LEB.

Aspects of the spatial association of life expectancy and socio-economic deprivation in the DHAs of England are examined graphically in Figure 31.2. The horizontal axis of each graph plots the 105 DHAs according to a seven-category ranking of a standard index of socio-economic deprivation (a Jarman Index, formed in Figure 31.2 such that category 1 represents the least deprived areas and category 7 the most deprived) against, on the vertical axis, various indices of life expectancy. Trend lines (linear regression lines fitted by ordinary least squares) and Pearson's *r* correlation coefficients are shown to assist in the interpretation of the graphs.

Figure 31.2 identifies three main features of the spatial association of life expectancy and socio-economic deprivation in England:

1 *Life expectancy is inversely associated with the level of deprivation in an area* For each deprivation category and DHA, Figure 31.2A plots estimates of LEB for males (lower line trace) and females (upper line trace) in the period 1992–4. While there is considerable variation in gender-specific life expectancy within a given deprivation category, the overall trends are for life expectancy to fall as the level of deprivation in an area increases. These visual associations are confirmed by the statistically significant

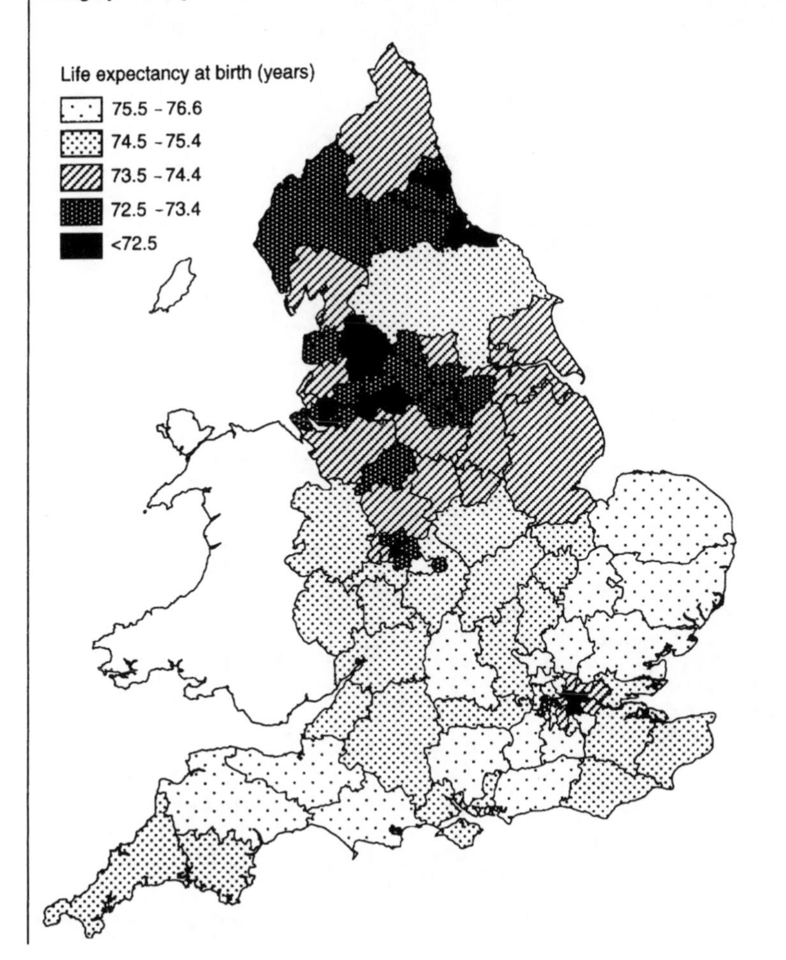

Figure 31.1 Life expectancy at birth (LEB) for males in the district health authorities (DHAs) of England, 1992–4.

Source: Drawn from data in Soni Raleigh and Kiri 1997: Table 1, pp. 654–5.

Box 31.1 continued

and negative correlation coefficients for both males ($r = -0.77$; $p < 0.01$ in a two-tailed test) and females ($r = -0.56$; $p < 0.01$ in a two-tailed test).

2 *Gender differentials in life expectancy are directly associated with the level of deprivation in an area* Figure 31.2B plots the (female – male =) gender difference in the LEB. Again, there is considerable variation within a given deprivation category. However, the overall tendency is for gender differentials to widen as the level of deprivation in an area increases ($r = 0.77$; $p < 0.01$ in a two-tailed test).

3 *Relatively affluent areas made the most pronounced gains in life expectancy during the late 1980s and early 1990s* For each deprivation category and DHA, Figure 31.2C plots the percentage change (computed as an annual average) in LEB between the periods 1984–6 and 1992–4. For males (upper line trace) and females (lower line trace) in all districts, the LEB for 1992–4 was higher than 1984–6, marking a systematic gain in life expectancy during the observation period. But, Figure 31.2C shows that the level of improvement was inversely associated with the level of deprivation for both males ($r = -0.47$; $p < 0.01$ in a two-tailed test) and females ($r = -0.30$; $p < 0.01$ in a two-tailed test). By implication, the gap in life expectancy between relatively affluent and relatively deprived areas widened during the late 1980s and early 1990s.

These findings are consistent with a growing literature on the spatial association of mortality differentials and socio-economic deprivation in the United Kingdom (see, for example, Eames *et al.* 1993; Morris *et al.* 1996; Watt and Ecob 1992; Wilkinson 1987). However, a lively debate surrounds the nature and persistence of the association (DHSS 1980; Blane 1985; Illsley and Le Grande 1993; Macintyre 1998), and this impinges on the policy implications of the research. On the one hand, the association could be an artefact of the available data, while the problems inherent in ecological studies (the modifiable areal unit problem and the so-called ecological fallacy) cannot be ignored. Alternatively, spatial associations may reflect patterns of social mobility as they relate to health status, while Illsley and Le Grande (1993) argue that health-related behaviour and ethnicity (rather than deprivation *per se*) underpin regional variations in mortality.

Clearly, much research and, more importantly, well-informed interpretation of research findings is still required to elucidate the nature of the association between ill-health, longevity and the socio-economic environment. As described in the main text, it is precisely these issues that research initiatives such as the ESRC Health Variations Programme seek to address. Not only has work of the type outlined by Soni Raliegh and Kiri served as a catalyst for such investigations, but it also provides a baseline for monitoring the success of future policy initiatives.

Figure 31.2 Associations between life expectancy at birth (LEB) and deprivation category for the District Health Authorities (DHAs) of England. For each graph, the horizontal axis plots the 105 DHAs according to a seven-category ranking of Jarman deprivation scores (category 1 = least deprived); (A) LEB and deprivation category for males (triangular symbols, lower line trace) and females (circular symbols, upper line trace), 1992–4; (B) gender (female - male =) difference in LEB and deprivation category, 1992–4; (C) average annual change in LEB and deprivation category for males (triangular symbols, upper line trace) and females (circular symbols, lower line trace). Average annual change is expressed as a percentage and has been computed for the period between 1984–6 and 1992–4.

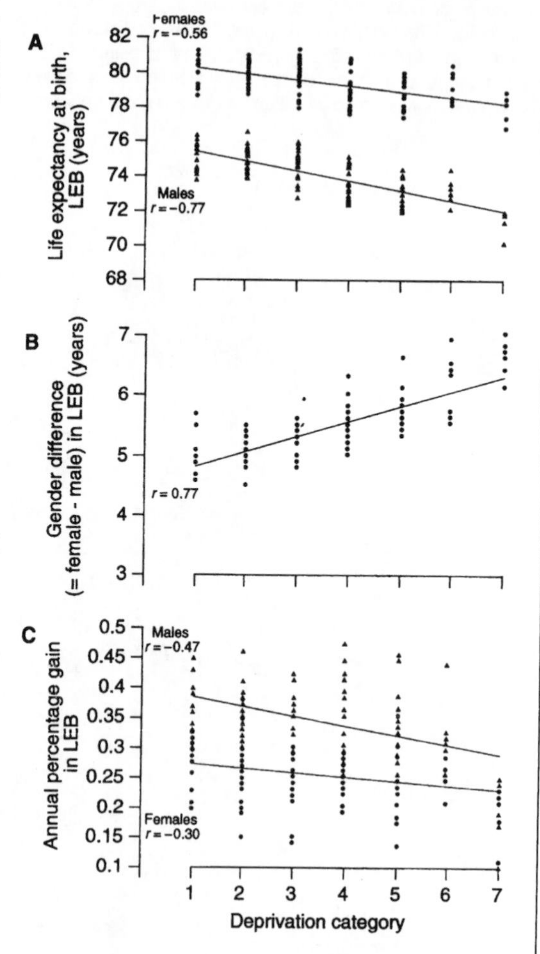

Source: Data from Soni Raleigh and Kiri 1997: Tables 1, p. 651 and Table 2, pp. 654–5.

Plate 31.1 Queen's Medical Centre and University Hospital, Nottingham – a vast, modern, teaching hospital situated close to Nottingham city centre and a major provider of health services in Trent Health Authority Region.

Plate 31.2 A suburban general practitioner's surgery within walking distance of the Queen's Medical Centre.

Case study 2: socio-spatial patterns of AIDS

Nowhere are the complexities of health variations more apparent than at the level of applied health intervention – the districts of a single city. In this case study, we draw on the work of Rodrick Wallace and colleagues (Wallace 1988; 1990; 1991; Wallace and Wallace 1993; 1997a; 1997b; Wallace *et al.* 1994; 1997) to illustrate how techniques of spatial analysis, when combined with empirical evidence from disciplines such as sociology and criminology, can be used to isolate the circumstances that have fuelled a local-level epidemic.

In his paper *A Synergism of Plagues*, Wallace (1988: p. 1) describes the Bronx area of New York City (see Figure 31.3 for location) as 'symbol of a systematic catastrophe in American cities.' At the heart of this catastrophe lie the profound health consequences of officially, if covertly, sanctioned urban decay and its social sequelae. These sequelae include the disintegration of community networks and the social dislocation of minority groups, the deterioration of public order, and the rise in a gamut of behavioural pathologies such as substance abuse and drug-related prostitution, violence and murder. Box 31.2 illustrates how this nexus of social and physical decay has shaped the geography of one of the most prominent public health problems in the Bronx, namely the

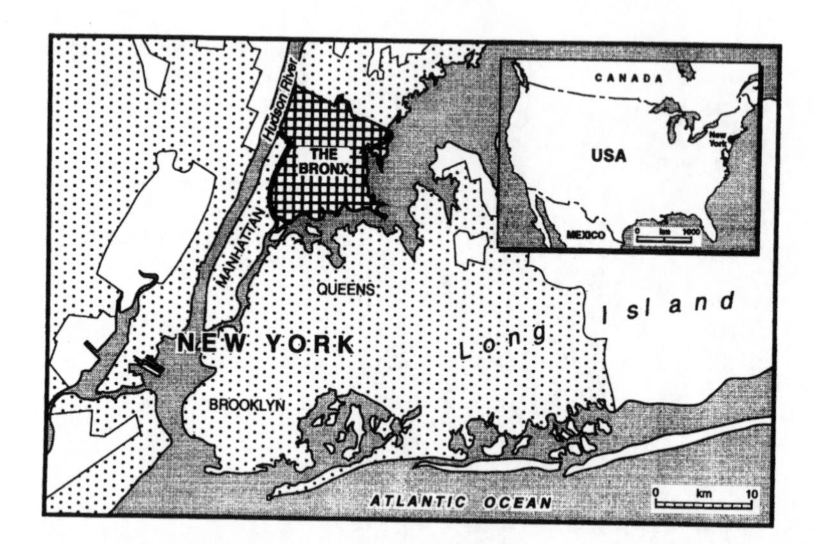

Figure 31.3 Location map of the Bronx, New York City.

epidemic of AIDS and its causative agent, the human immunodeficiency virus (HIV).

Remedial responses to the spread of HIV/AIDS in the cities of North America, and elsewhere, are largely dependent on the integrity of local social networks. These networks underpin the operation of outreach services and provide the conduits for an efficient propagation of HIV-related information and education. When these networks have been shredded through the processes of population relocation, as in the case of the Bronx (Box 31.2) and areas of other US cities, the success of HIV/AIDS control programmes is crucially dependent on a wider set of urban policies that promote geographical stability and the reknitting of personal, family and community networks.

CONCLUSION

In his report for the year 1995, the Director-General of the World Health Organization observed that 'The world's most ruthless killer and the greatest cause of suffering on earth is listed in the latest edition of WHO's International Classification of Diseases under the code Z59.5. It stands for extreme poverty' (WHO 1995: p. 1). Consistent with this observation, the geographical and allied literature is replete with examples of the manner in which health, disease and mortality vary spatially with levels of absolute and relative poverty, deprivation and other socio-economic indicators. Precisely what these variations mean, however, remains open to considerable debate. More often than not, it seems that *complexes* of factors are likely to underlie many socio-spatial variations (for examples, see WHO 1992; McNally *et al.* 1998). Clarification of these factors, and the formulation of appropriate policy responses to erase the related health variations, are set to occupy medical geographers, health scientists and others well into the third millennium. But the new millennium will undoubtedly bring fresh health challenges, and a number of intrinsically geographical factors can be expected to impinge on health variations. These factors are likely to include: continued and rapid demographic growth; population migration and urban growth; increased geographical interaction and the telescoping of time–space; land-use change and agricultural colonisation; and climate change and global warming. Geographers, by their training, are uniquely placed to study the health outcomes of these phenomena.

In this chapter, two case studies have illustrated very different aspects of socio-spatial variations in health. The case study in Box 31.1, set in England, confirms many of the expected socio-economic

Box 31.2 Social disintegration and the geography of AIDS in the Bronx, New York

The Bronx, with a land area of 109 km^2 and a population of about 1.2 million, is a borough of economic extremes. Some of the most affluent districts of New York City are located in the northwest of the borough. By contrast, the districts in the west and southwest ('South Bronx') are populated largely by ethnic minorities and rank among the most impoverished ghetto areas of the United States. That same area is also the focus of a major epidemic of intravenous (IV) drug use and, with it, an intense epidemic of HIV/AIDS. Thus, as early as the mid-1980s, the sharing of injecting equipment had resulted in the dissemination of HIV to 10–20 per cent of all young adult males in the South Bronx (Drucker and Vermund 1989). To give an impression of the geography of this early spread process, Figure 31.4A plots those Bronx Health Areas ranked 1–9 (solid shading) and 10–18 (cross-hatched shading) in terms of AIDS deaths during the period 1980–85. The overall pattern is dominated by the southern and western health areas, which constitute the South Bronx, with secondary foci in the central and northern Health Areas.

Socio-spatial variations in health can be directly influenced by public policies, inadvertently or deliberately. In the instance of the Bronx, Wallace (1988; 1990) argues that the spatial patterns of IV drug use during the early 1980s, and the resulting spread of AIDS to its 1985 pattern (Figure 31.4A), were inextricably linked to a covert policy of so-called 'planned shrinkage'. In order to clear impoverished areas of New York City, essential municipal services were withdrawn or dramatically reduced in parts of the South Bronx during the 1970s. This resulted in a mass relocation of the South Bronx population as the housing infrastructure began to deteriorate. To give an impression of this relocation, Figure 31.4B maps those Health Areas ranked 1–20 in terms of badly overcrowded housing in 1970. Similarly, map C shows the equivalent pattern for 1980. Together, the maps indicate a 'contagious-like' relocation of ghetto areas from the central area of the South Bronx to neighbouring areas in the west and east.

A large body of evidence in sociology, criminology, psychology and the health sciences (see, for example, Wallace 1988; 1990) relates the dislocation of social networks to a nexus of public disorder, violence, drug use, prostitution and disease. Given that some of these behaviours are associated with HIV transmission, the forced displacement and community dislocation of populations to the west and east of the South Bronx may have actively fuelled the spread of HIV in these areas. To assess this hypothesis, Wallace (1990) has examined the manner in which two measures of HIV-related behaviour altered during the course of planned shrinkage. The two measures, which have been assessed for each of the Health Areas of the Bronx, are (1) the average number of liver cirrhosis deaths (denoted y_1); and (2) the average number of intentional violent deaths (y_2). Both variables serve as markers of the geographical distribution of drug use in the Bronx; liver cirrhosis can be a direct health outcome of drug use, while the association between violent crime and drug use is well established.

To determine whether the geographies of these HIV-related behaviours were driven by social dislocation,

Figure 31.4 AIDS and ghetto relocation in the Bronx, New York City: (A) distribution of AIDS deaths, 1980–85; (B) distribution of badly overcrowded housing, 1970; (C) distribution of badly overcrowded housing, 1980. All maps have been formed by ranking (1 = highest) the respective variables across the 62 health areas; following Wallace, health areas ranked 1–18 (map A) and 1–20 (maps B and C) are indicated.

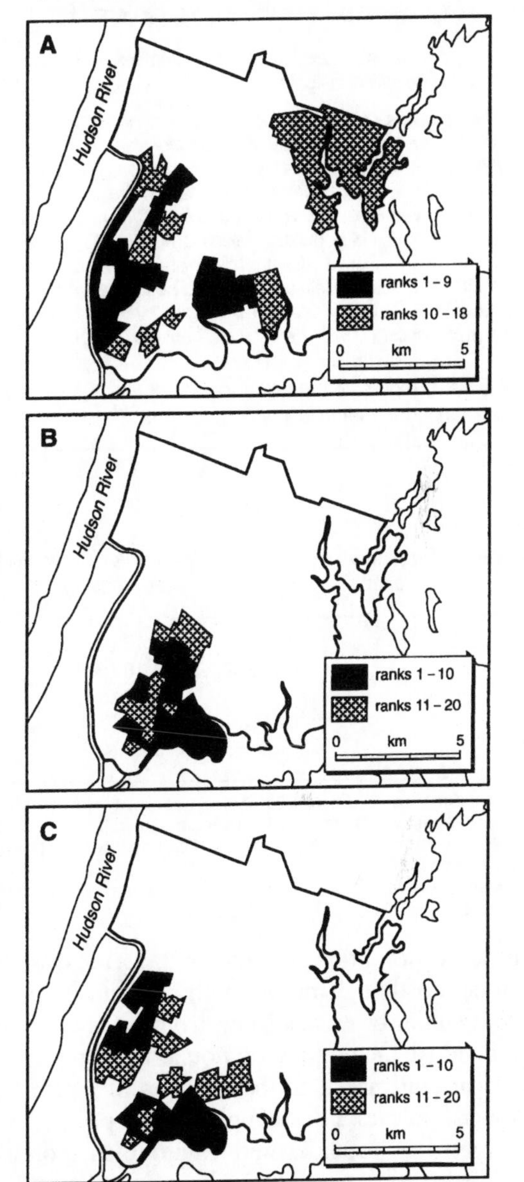

Source: Maps redrawn from Wallace 1988: Figure 11, p. 16 and Figure 14, pp. 20–1.

Box 31.2 continued

Wallace defined two indices to characterise that dislocation in a given Health Area. These indices are formally defined elsewhere (*ibid.*) but, in brief, the poverty index (denoted x_1) is a community-wide measure of deprivation, while the community disintegration index (x_2) is a measure of existing social support networks. These indices served as the independent (x) variables in a series of least squares multiple regression models in which cirrhosis deaths (y_1) and intentional violent deaths (y_2) variously formed the dependent (y) variables.

The regression results, which are summarised in Table 31.1, yield two critical insights:

1 *The process of population displacement served to strengthen the geographical relationship between HIV-related pathology and social dislocation.* Models 1 and 2 relate deaths from liver cirrhosis to the two independent variables for a period immediately prior to the main phase of population relocation (1970–73, model 1) and a time period after the main phase of relocation (1978–82, model 2). Between these two periods, accounted variability in cirrhosis deaths (as judged by R^2) doubled from 28 per cent (1970–73) to 57 per cent (1978–82).

2 *The process of population displacement was associated with an absolute rise in the mean level of HIV-related pathology.* Models 3 and 4, which relate intentional violent deaths in 1970 and 1980 to the independent variables, reveal an approximate stability in accounted variability. However, inspection of the Student's *t* statistics associated with the intercept coefficients indicates that this parameter made a significant contribution to the 1980 model but not the 1970 model.

Taken together, these results suggest that the social sequelae of population displacement in the South Bronx during the 1970s resulted in both (1) a geographical focusing and (2) an intensification of two correlates of substance abuse, the predominant route of HIV transmission in the borough. It appears more than coincidental that the same social sequelae should also be highly predictive of the geographical pattern of the borough's early AIDS cases (Table 31.1, model 5). An inference from these findings is that the historical spread of HIV in the Bronx was geographically focused and fuelled by the social dislocation engendered by an iniquitous policy of planned shrinkage.

Table 31.1 Regression results for tests of the relationship between HIV-related behaviour and population relocation in the health areas of the Bronx.

Model	*Dependent variable*	*Date*	*Intercept coefficient (t-statistic)*	*Slope coefficients for independent variables (t-statistic)*		
				x_1	x_2	R^2
1	y_1	1970–73	2.10 (2.06)*	0.30 (1.34)	1.83 (3.68)*	0.28
2	y_1	1978–82	2.18 (5.50)*	0.68 (3.52)*	1.17 (4.50)*	0.57
3	y_2	1970	0.12 (0.17)	0.02 (10.15)*	0.02 (4.83)*	0.78
4	y_2	1980	1.59 (3.07)*	0.02 (7.89)*	0.02 (5.47)*	0.79
5	y_3	1980–85	0.27 (0.28)	2.92 (6.40)*	2.07 (3.37)*	0.67

Source: Data from Wallace 1990: Tables 1, 2 and 4, pp. 806–10).
Notes: * Significant at the $p = 0.01$ level (one-tailed test).
Dependent variables: y_1 = average annual cirrhosis deaths; y_2 = average annual intentional violent deaths; y_3 = AIDS deaths.
Independent variables: x_1 = poverty index; x_2 = community disintegration index.

associations of poorer health and social disadvantage. Although the case study does illustrate the difficulties of extrapolating from an area to an individual, the findings do not, as they stand, give clear guidance as to the direction of health and social policies. For example, should policies be targeted at localities and environments or individuals and households? The case study in Box 31.2, set in the Bronx district of New York City, illustrates the sophistication of spatial and statistical analysis now used in much health geography research. The study also illustrates how a socio-spatial analysis of recent historical data can give guidance to future public health and urban redevelopment policies in large urban settings. Such policies are likely to become ever more pressing in the mega-cities of the world.

GUIDE TO FURTHER READING

The monumental *Cambridge World History of Human Disease*, edited by Kenneth F. Kiple and published by Cambridge University Press (Cambridge) in 1993, provides a comprehensive introduction to the global history and geography of health and disease. A useful overview of some contemporary themes and approaches in medical geography is given in a 1996 special issue of *Social Science and Medicine* 42(6), 787–964, entitled *Research in Medical Geography*. For a recently updated edition of his classic account of health, disease and mortality in Great Britain, see Howe, G.M. (1997) *People, Environment, Disease and Death: A Medical Geography of Britain Throughout the Ages*, Cardiff: University of Wales Press. One of the more accessible introductory texts on the spatial and environmental associations of many infectious and some noncommunicable diseases is Learmonth, A. (1988) *Disease Ecology: An Introduction*, Oxford: Blackwell. Medical geographers have developed a wide range of techniques for the spatial analysis of health and disease data, many of which are outlined in Cliff, A.D., and Haggett, P. (1988) *Atlas of Disease Distributions: Analytic Approaches to Epidemiological Data*, Oxford: Blackwell Reference. The fruitful collaboration of medical geographers and public health officials is illustrated in Gordon, A. and Womersley, J. (1997) 'The use of mapping in public health and planning health services', *Journal of Public Health Medicine* 19(2), 139–47. Useful introductions to the terminology and methods of epidemiology are provided by Last, J.M. (1995) *A Dictionary of Epidemiology*, New York: Oxford University Press, and Beaglehole, R., Bonita, R. and Kjellstrom, T. (1993) *Basic Epidemiology*, Geneva: World Health Organization.

REFERENCES

Bailey, T.C. and Gatrell, A. (1995) *Interactive Spatial Data Analysis*. Harlow: Longman Scientific.

Bentham, G. (1994) Global environmental change and health. In D.R. Phillips and Y. Verhasselt (eds) *Health and Development*, London: Routledge, 33–49.

Blane, D. (1985) An assessment of the Black Report's explanation of health inequalities. *Sociology of Health and Illness* 7(3), 423–45.

Caldwell, J.C., Findley, S., Caldwell, P., Santow, G., Cosford, W., Braid, J. and Broers-Freeman, D. (1990) *What We Know About Health Transition: The Cultural, Social and Behavioural Determinants of Health*, 2 Volumes. Canberra: Australian National University Press.

Cliff, A.D. and Haggett, P. (1988) *Atlas of Disease Distributions: Analytic Approaches to Epidemiological Data*. Oxford: Blackwell Reference.

Cliff, A.D., Haggett, P. and Smallman-Raynor, M.R. (1998) *Deciphering Global Epidemics: Analytical Approaches to the Disease Records of World Cities, 1888–1912*. Cambridge: Cambridge University Press.

de Leper, M.J.C., Scholtern, H.J. and Stern, R.M. (1995) *The Added Value of Geographical Information Systems in Public and Environmental Health*. Dordrecht: Kluwer Academic.

DHSS (1980) *Inequalities in Health: Report of a Research Working Group*. London: Department of Health and Social Security.

Drucker, E. and Vermund, S.H. (1989) Estimating population prevalence of human immunodeficiency virus infection in urban areas with high rates of intravenous drug abuse: a model of the Bronx in 1988. *American Journal of Epidemiology* 130(1), 133–42.

Eames, M., Ben-Shlomo, Y. and Marmot, M.G. (1993) Social deprivation and premature mortality: regional comparison across England. *British Medical Journal* 307(6912), 1097–102.

Eyles, J. (1997) Environmental health research: setting an agenda by spinning our wheels or climbing the mountain. *Health and Place* 3(1), 1–13.

Expert Advisory Group on Cancer (1995) *A Policy Framework for Commissioning Cancer Services*. London: Department of Health and the Welsh Office.

Fenner, F., Henderson, D.A., Arita, I., Jezek, Z. and Ladnyi, I.D. (1988) *Smallpox and its Eradication*. Geneva: World Health Organization.

Foster, H.D. (1992) *Health, Disease and Environment*. London: Belhaven Press.

Frenk, J., Bobadilla, J.L. and Lozano, R. (1996) The epidemiological transition in Latin America. In I.M. Timaeus, J. Chackiel and L. Ruzieka (eds) *Adult Mortality in Latin America*, Oxford: Clarendon Press, 123–39.

Frenk, J., Bobadilla, J.L., Sepulveda, J. and Cervantes, M.L. (1989) Health transition for middle-income countries: new challenges for health care. *Health Policy and Planning* 4(1), 29–39.

Gatrell, A. and Bailey, T.C. (1996) Interactive spatial data analysis in medical geography. *Social Science and Medicine* 42(6), 843–55.

Giggs, J.A. (1973) The distribution of schizophrenics in Nottingham. *Transactions of the Institute of British Geographers* 59, 55–76.

Giggs, J.A. (1988) The spatial ecology of mental illness. In C.J. Smith and J.A. Giggs (eds) *Location and Stigma: Contemporary Perspectives on Mental*

Health and Mental Health Care, London: Unwin Hyman, 103–33.

Giggs, J.A. (1990) Drug abuse and urban ecological structure: the Nottingham case. In R.W. Thomas (ed.) *Spatial Epidemiology*, London: Pion, 218–36.

Graham, H. (1998) *Health Variations Programme. Health Variations; Newsletter of the ESRC Health Variations Programme* 1 (January), 2–3.

Harpham, T. and Tanner, M. (1995) *Urban Health in Developing Countries*. London: Earthscan.

Howe, G.M. (1986) *Global Geocancerology: A World Geography of Human Cancers*. Edinburgh: Churchill Livingstone.

Howe, G.M. and Loraine, J.A. (1980) *Environmental Medicine*, second edition. London: William Heinemann.

Hudson, R.P. (1993) Concepts of disease in the West. In K.F. Kiple (ed.) *Cambridge World History of Human Disease*, Cambridge: Cambridge University Press, 45–52.

Illsley, R. and Le Grande, J. (1993) Regional inequalities in mortality. *Journal of Epidemiology and Community Health* 47(6), 444–9.

Iyun, B.F., Verhasselt, Y. and Hellen, J.A. (1995) *The Health of Nations: Medicine, Disease and Development in the Third World*. Aldershot: Avebury.

Jones, I. and Curtis, S. (1997) Health. In M. Pacione (ed.) *Britain's Cities*, London: Routledge, 218–43.

Jones, K. and Moon, G. (1987) *Health, Disease and Society: An Introduction to Medical Geography*. London: Routledge & Kegan Paul.

Kiple, K.F. (ed.) (1993) *The Cambridge World History of Human Disease*. Cambridge: Cambridge University Press.

Lamont, D.W., Toal, F.M. and Crawford, M. (1997) Socioeconomic deprivation and health in Glasgow and the west of Scotland – a study of cancer incidence among male residents of hostels for the single homeless. *Journal of Epidemiology and Community Health* 51(6), 668–71.

Learmonth, A. (1988) *Disease Ecology: An Introduction*. Oxford: Blackwell.

Litva, A. and Eyles, J. (1995) Coming out: exposing social theory in medical geography. *Health and Place* 1(1), 5–15.

Luoto, R., Pekkanen, J., Uutela, A. and Tuomilehto, J. (1994) Cardiovascular risks and socioeconomic status: differences between men and women in Finland. *Journal of Epidemiology and Community Health* 48(4), 348–54.

Macintyre, S. (1998) Area inequalities in health. *Health Variations: Newsletter of the ESRC Health Variations Programme*, 1(January), 7–8.

McNally, N.J., Phillips, D.R. and Williams, H.C. (1998) The problem of atopic eczema: aetiological clues from the environment and lifestyles. *Social Science and Medicine* 46(6), 729–41.

Marmot, M. (1996) The social pattern of health and disease. In D. Blane, E. Brunner and R. Wilkinson (eds) *Health and Social Organisation: Towards a Health Policy for the Twenty-First Century*, London: Routledge, 42–67.

Morris, J.N., Blane, D.B. and White, I.R. (1996) Levels of mortality, education, and social conditions in the 107 local education authority areas of England. *Journal of Epidemiology and Community Health* 50(1), 15–17.

Olshansky, S.J. and Ault, A.B. (1986) The fourth stage of epidemiological transition: the age of delayed degenerative diseases. *Milbank Memorial Fund Quarterly* 64(3), 355–91.

Omran, A.R. (1971) The epidemiologic transition: a theory of the epidemiology of population change. *Milbank Memorial Fund Quarterly* 49(4), 509–38.

Openshaw, S. (1990) Automating the search for cancer clusters: a review of problems, progress and opportunities. In R.W. Thomas (ed.) *Spatial Epidemiology*, London: Pion, 48–78.

Phillips, D.R. (1990) *Health and Health Care in the Third World*. London: Longman.

Phillips, D.R. (1993) Urbanisation and human health. *Parasitology* 106 (supplement), S93–S107.

Phillips, D.R. (1994) Epidemiological transition: implications for health care provision. *Geografiska Annaler* 76B(2), 71–89.

Phillips, D.R. and Verhasselt, Y. (1994) *Health and Development*. London: Routledge.

Power, C. (1994) Health and social inequality in Europe. *British Medical Journal* 308(6937), 1153–6.

Pyle, G.F. (1971) *Heart Disease, Cancer and Stroke in Chicago*. Chicago: Department of Geography, University of Chicago.

Secretary of State for Health (1992) *The Health of the Nation: A Strategy for Health in England*. London: HMSO.

Secretary of State for Health (1998) *Our Healthier Nation*. London: HMSO.

Shigehisa, K. (1993) Concepts of disease in East Asia. In K.F. Kiple (ed.) *Cambridge World History of Human Disease*, Cambridge: Cambridge University Press, 52–9.

Smith, C.J. and Giggs, J.A. (1988) *Location and Stigma: Contemporary Perspectives on Mental Health and Mental Health Care*. London: Unwin Hyman.

Soni Raleigh, V. and Kiri, V.A. (1997) Life expectancy in England: variations and trends by gender, health authority, and level of deprivation. *Journal of Epidemiology and Community Health* 51(6), 649–58.

Tarlov, A.R. (1996) Social determinants of health: the

sociobiological transition. In D. Blane, E. Brunner and R. Wilkinson (eds) *Health and Social Organisation: Towards a Health Policy for the Twenty-First Century*, London: Routledge, 71–93.

Thomas, R. (1992) *Geomedical Systems: Intervention and Control*. London: Routledge.

Vågerö, D. (1991) Inequality in health – some theoretical and empirical problems. *Social Science and Medicine* 32(4), 367–71.

Vågerö, D. (1995) Health inequalities as policy issues – reflections on ethics, policy and public health. *Sociology of Health and Illness* 17(1), 1–19.

Vågerö, D. and Illsley, R. (1993) *Explaining the Difference. A Review of Theories about Health Inequalities*. Stockholm: Stockholm University.

Wallace, R. (1988) A synergism of plagues: 'planned shrinkage,' contagious housing destruction, and AIDS in the Bronx. *Environmental Research* 47(1), 1–33.

Wallace, R. (1990) Urban desertification, public health and public order: planned shrinkage, violent death, substance abuse and AIDS in the Bronx. *Social Science and Medicine* 31(7), 801–13.

Wallace, R. (1991) Travelling waves of HIV infection on a low dimensional socio-geographic network. *Social Science and Medicine* 32(7), 847–52.

Wallace, R., Fullilove, M., Fullilove, R., Gould, P. and Wallace, D. (1994) Will AIDS be contained within US minority urban populations? *Social Science and Medicine* 39(8), 1051–62.

Wallace, R. and Wallace, D. (1993) The coming crisis of public health in the suburbs. *Milbank Quarterly* 71(4), 543–64.

Wallace, R. and Wallace, D. (1997a) Resilience and persistence of the synergism of plagues: stochastic resonance and the ecology of disease, disorder and disinvestment in US urban neighbourhoods. *Environment and Planning A* 29(5), 789–804.

Wallace, R. and Wallace, D. (1997b) The destruction of US minority urban communities and the resurgence of tuberculosis: ecosystem dynamics of the white plague in the dedeveloping world. *Environment and Planning A* 29(2), 269–91.

Wallace, R., Wallace, D. and Andrews, H. (1997) AIDS, tuberculosis, violent crime, and low birthweight in eight US metropolitan areas: public policy, stochastic resonance, and the regional diffusion of inner-city markers. *Environment and Planning A* 29(3), 525–55.

Watt, G.C.M. and Ecob, R. (1992) Mortality in Glasgow and Edinburgh: a paradigm of inequality in health. *Journal of Epidemiology and Community Health* 46(5), 498–505.

West, P. (1991) Rethinking the health selection explanation for health inequalities. *Social Science and Medicine* 32(4), 373–84.

WHO (1988) *Priority Research for Health for All*. Copenhagen: World Health Organization Regional Office for Europe.

WHO (1992) *Our Planet, Our Health. Report of the WHO Commission on Health and the Environment*. Geneva: World Health Organization.

WHO (1994) *The Work of WHO 1992–1993: Biennial Report of the Director-General to the World Health Assembly and to the United Nations*. Geneva: World Health Organization.

WHO (1995) *The World Health Report 1995: Bridging the Gaps*. Geneva: World Health Organization.

Wilkinson, R. (1987) *Class and Health: Research and Longitudinal Data*. London: Tavistock.

32

Crime and fear of crime

Norman Davidson

In the last two decades, crime has become one of the most pervasive features of quality of life. 'Law and order' vies with health and the economy as the most salient concern of British citizens, and the picture is not radically different in most Western societies. The influence of this concern is such that many British police forces now see delivery of quality of life as or more important than catching criminals or preserving public order. At the same time, there have been significant shifts in public policy on law and order, from detection to prevention, from offender to victim, from imprisonment to community sentences, which both reflect and mediate public opinion. In this chapter, I will explore the meaning of these transformations from a geographical perspective and illustrate the increasing role for geographical skills in analysis of the problem and in the search for solutions.

The point of departure for this review is some observations about contemporary patterns of crime with resonances in classical conceptions of spatial concentration. The causes of crime are complex, and no single theory is sufficient to explain the wide variety of offences that are proscribed by law. In addition, there may be competing definitions of events by victims, police and prosecutors that confound simple prescriptions. Two current theories within criminology provide a focus for understanding. One is rational choice theory, which suggests that offender behaviour can be best understood in terms of choices made about costs and benefits on the information available at the time of the event. Offenders are thus seen as decision makers, and this may also be applied to victims and others involved in the event, whether directly or indirectly. Routine activities theory provides the framework for analysis. It suggests three essential elements for a crime to take place: a suitable target; a motivated offender; and the absence of capable guardians (protecting the victim) or an intimate handler (inhibiting the offender).

This 'criminal triangle' – victim, offender and situation – has implications for crime prevention, which will be reviewed below. It also has powerful links with observations of the uneven geographical distribution of the three elements. These are illustrated in Figure 32.1. Recent work on the incidence of crime has shown the extent to which victimisation is not a random event. Data from the British Crime Survey show that just 4 per cent of victims suffer 44 per cent of crime. Repeated victims are concentrated in *high-crime*

Figure 32.1 The concentration of crime and the links with place.

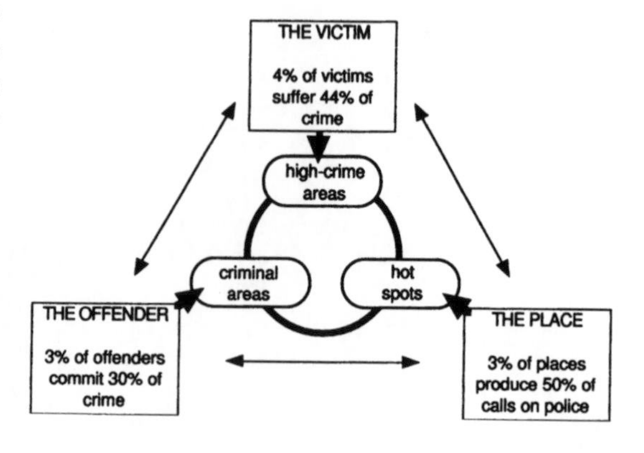

areas – such as the inner city and poorest estates – that have been recognised in geographical studies for many years. Concentration also exists for offenders. The Home Office Research and Statistics Department has indicated that 3 per cent of offenders commit 30 per cent of crime. The Audit Commission (1996) refined a Home Office study to suggest that 5 per cent of 14–17-year-old males commit 68 per cent of the offences of that age group. Again, there is a long tradition of identifying persistent offending with particular areas – *criminal areas* (Morris 1957). The third element in the triangle is the place where the crime is committed. Sherman (1995), in a study of Minneapolis, suggests that 3 per cent of places produce 50 per cent of calls on police time. These concentrations represent *hot spots of crime* such as pubs and clubs, transport facilities, hospitals and even troublesome households. The serial nature of crime – whether of victims, offending or location – is clearly a factor in its uneven geographical distribution, working alongside inequalities in economic, social and physical environments.

UNDERSTANDING CRIME RATES

In order to understand the processes that underpin patterns of crime, and indeed that influence the very definition of crime itself, it is necessary to see the criminal event as the subject of choices, that is of rational decisions made by a variety of people at different times. There are six fundamental decision-making areas:

- the offender's decision to offend;
- the victim's decision that an offence has been committed;
- the victim's decision to do something about the offence – normally to report it to the police;
- the police decision to record the offence;
- the police or Crown Prosecution Service's decision to prosecute the offender (including police cautions);
- the court's decision about guilt and sentence (in the English legal system, guilt and sentence decisions are made separately for more serious offences by jury and judge, respectively).

The British Crime Survey provides illustrations of the relative importance of these decisions (Figure 32.2). Official statistics – the published data used for media reports of crime trends – relate to the minor part of the stock of incidents regarded as crime by victims. The major part, about 70 per cent, constitutes the 'dark figure' of crime – incidents that are not reported to, or recorded by, the police. The data for 1991 also show that 7 per cent of crime is cleared up, just 3 per cent results in an offender being charged or cautioned, and about 0.3 per cent in an offender sentenced to immediate custody. The icebergs of crime reveal how attenuated crime becomes as the case is processed. The complexity of the relationship between official statistics and underlying victimisation levels is shown in Figure 32.3 for 1981 to 1995. During this period, victimisation has risen consistently, whereas the percentage of incidents reported rose during the 1980s and declined again in the 1990s. The decrease in the proportion of incidents recorded has steepened in the 1990s. The 'dark figure' of crime thus fell in the 1980s (due to more reporting) and rose again in the 1990s (less reporting and less recording).

GEOGRAPHIES OF CRIME

Just as crime is acknowledged as multi-causal, so geographical contributions to the explanation of the incidence of crime must make explicit the conditionality of understanding that can be achieved. First, explanations will necessarily be crime-specific: what may be concluded about burglary will be very different about shoplifting and again about sexual offences. Second, it can be argued that the spatial scale of any analysis will be reflected in the salience of different factors. The factors that differentiate countries in the incidence of burglary (laws, cultures) will not be the same as those differentiating neighbourhoods or houses in a street (factors influencing local concentrations of domestic burglary are highlighted

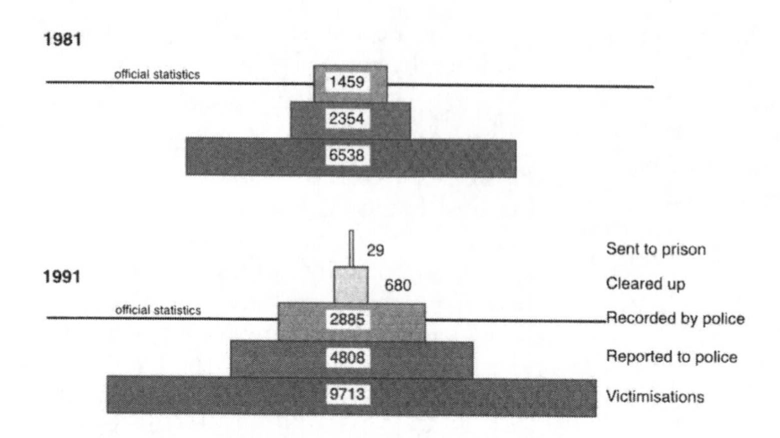

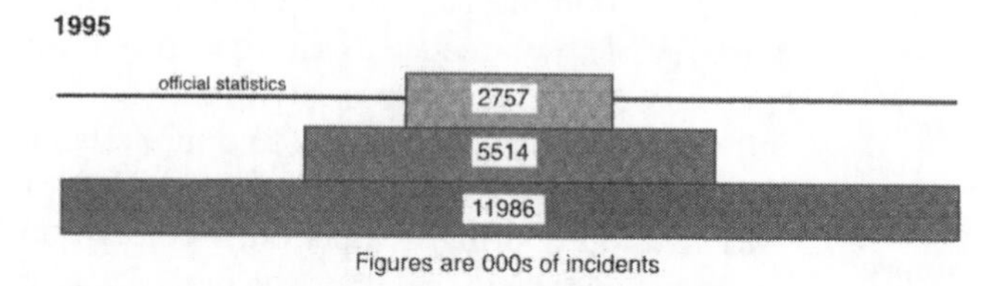

Figure 32.2 Icebergs of crime: trends in crime in England and Wales.

Sources: Mayhew *et al.* 1993; Mirrlees-Black *et al.* 1996; Hough and Tilley 1998.

Figure 32.3 Trends in reporting and recording crime: England and Wales, 1981–95.

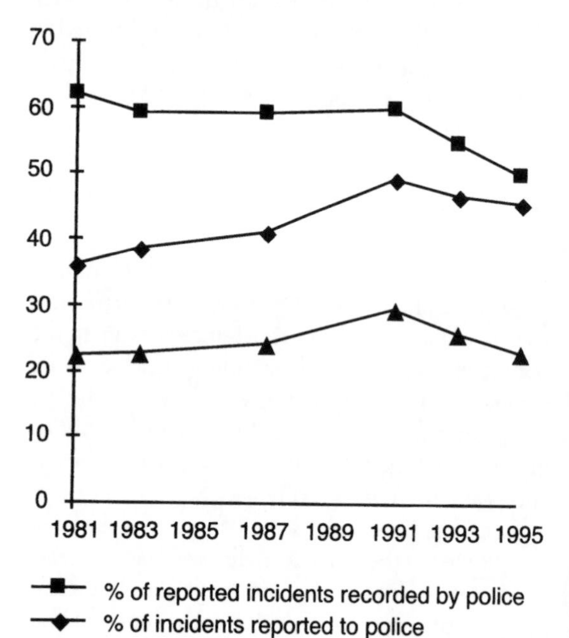

Source: Mirrlees-Black *et al.* 1996.

in Box 32.1). In the review that follows, the emphasis will be on the major categories of crime and on the meso- and micro-scales of analysis. This emphasis will do less justice to minority crimes, which should not be read as of less geographical interest.

Offending

This perspective on crime has the strongest historical antecedents. An emphasis on area can be dated right back to the beginnings of modern criminology in the nineteenth century. Guerry (1833) remarked on area variations in Metropolitan France, and later Mayhew (1864) described in detail the conditions of 'rookeries of crime' (neighbourhoods with high levels of criminal activity) in Victorian London. But the real development of a spatial perspective on crime came from the Chicago School of Ecology. Shaw and McKay (1942) painstakingly mapped the location of juvenile delinquency in Chicago and proposed the theory of social disorganisation to explain the pattern of delinquency areas that emerged. They noted that delinquency declined

Box 32.1 Case study: domestic burglary

Burglary is unique among crimes for the conjunction between victim's residence and the place of occurrence. It is one of the main sources of fear of crime, because of the strong emotions aroused by the violation of personal space. Burglary has been well researched, and its prevention is a key priority of criminal justice.

Six hypotheses found useful in explaining geographical concentrations of burglary:

1 *Affluence*
Wealthier households are, on the whole, more likely to be burgled, but this role needs heavy qualification – it is relative affluence rather than absolute, since visibility and accessibility are also important.

2 *Vulnerability*
There needs to be a basic distinction between the vulnerability of the house and the vulnerability of the neighbourhood, since burglars appear to make two separate spatial decisions – to go burgling (choice of area) and which house to burgle (while cruising the chosen area).

3 *Social cohesion*
This is important in the visual cues that are offered to potential offenders – signs of surveillance/occupancy – as well as in the mobilising of community protection, for example Neighbourhood Watch.

4 *Reputation*
Some areas will attract more burglary for the image they present. Well-lit, tidy, well-cared-for neighbourhoods attract less burglary than scruffy, run-down neighbourhoods. But highly visible symbols of wealth are also an attraction.

5 *Proximity*
Most burglars are young and opportunist. Proximity to areas of offender residence raises the risk of burglary. This factor has become less important, as the mobility of offenders has increased by stealing cars.

6 *Concentration*
Having already been burgled is a strong predictor of burglary patterns. The pattern of repeats is highly localised, so concentration serves to reinforce existing patterns.

Repeat burglary arises from a combination of:

1 offenders returning because they are familiar with how to get in, did not finish the job the first time round, think that insurance will have replaced the goods, etc.; and
2 the house is 'marked' as an attractive target due to affluence, location, ease of access, lack of occupancy, etc.

Sources: Davidson 1984; Bennett 1989; Hough and Tilley 1998.

with distance from the centre of Chicago and that the inner-city zones with the highest delinquency rates were characterised by physical decay and obsolescence, by high rates of in-migration and transience, and by family breakdown. Delinquency areas were found to persist over time despite rapid turnover of population. The focus on the neighbourhood is Shaw and McKay's most enduring legacy, together with a research methodology carefully rooted in the communities studied.

The concentration of offending in particular areas was central to Terence Morris's (1957) development of the concept of criminal areas based on a study of Croydon. He found that the pattern of delinquency was rather different from the Chicago model, with more emphasis on council housing estates well away from the urban core. In the 1960s and 1970s, the emphasis on offender areas waned somewhat, partly as a response to new statistical methodologies based on offence location, and partly with the re-emergence of interest in the victim. However, two contrasting studies continued in the ecological tradition. Gill's (1977) study of a small area of Liverpool followed the ethnographic tradition but yielded valuable new insights into the relationship between youth, crime and neighbourhoods. The Sheffield Urban Crime Study was on an altogether different scale. It covered the whole city and combined new statistical analyses with interviews and research on the processes whereby crime was identified and categorised. It started with the work of Baldwin and Bottoms (1976) on offence and offender patterns, continued with Mawby's (1979) study of policing, and has remained active in a series of papers concerned with the role of housing markets in crime, and the persistence of criminal areas through the notion of community crime careers.

The Sheffield studies identified a clear distinction between offender residence (the focus of criminal areas research) and offence areas (the basis for high-crime areas). Although offence and offender areas could overlap, for example in some inner-city areas and council housing estates, there was no necessary accordance. In the 1970s, new ideas began to emerge linking offence and offender location. One strand concentrated on the notion of journey-to-crime, showing the strong distance decay effects linking offender residence and the place of the crime, but a more fruitful framework has been developed by Paul and Patricia Brantingham in their work on the geometry of crime (Brantingham and Brantingham 1981). This suggested that crime would be concentrated where the offender's perception of opportunity (awareness space) overlapped with actual opportunities in the environment. Offenders' perceptions are skewed towards areas in which most of their normal legitimate activity takes place – home, school, work and leisure.

In the last decade, the dualism of offence and offender has been reaffirmed, although the slant has been less towards theoretical understanding of offender behaviour and more towards its pragmatic utility for crime prevention and policing. Developments in offender profiling explicitly aimed at managing risk, rather than pure rehabilitation, are beginning to show their potential. This is not just about psychology but about all aspects of the predictability of serial offending. Geographical profiling is one element in the crime analyst's toolkit, aided by mapping technologies provided by geographical information systems (GIS) to display patterns of offending, known associates and *modus operandi* against demographic, land-use or other information relevant to the distribution of potential targets. Equally, these techniques are used to identify hot spots of crime – clusters of offences in particular locations (Hirschfield *et al.* 1996; Sherman 1995) – using the power of modern computers to provide regular updates that highlight patterns as they happen rather than after the event.

Victims

Victims are the crucial element in the criminal process. Numerically, they make the bulk of decisions about crime, without which no incident would enter the criminal justice system. The introduction of victim surveys over the last twenty years means that we now know more about victims, who previously had been the forgotten item. Much of this knowledge relates to the risk of victimisation, where it is important to recognise the distinctions summed up by the simple equation:

incidence = prevalence × concentration

$$\text{i.e. } \frac{\text{crimes}}{\text{population}} = \frac{\text{victims}}{\text{population}} \times \frac{\text{crimes}}{\text{victims}}$$

where: incidence = risk of individual being victimised in a given period
prevalence = risk of individual becoming a victim in a given period
concentration = rate of repeated victimisation

So a high rate of victimisation can mean a few people repeatedly victimised (for example, domestic violence) or many people rarely victimised more than once (e.g. pickpocketing). The role of repeat victimisation has only recently been fully recognised. It applies to all the high-volume categories of crime – in 1995, 19 per cent of burglary victims suffered two or more victimisations, 28 per cent of car crime victims, and 37 per cent of contact crime victims (wounding, assault, mugging, snatch theft). We know that there is diffusion of repeat victimisation across crime types (being a victim of violence increases your risk of burglary, and *vice versa*), but as yet little is known about the geography of repeats.

The various sweeps of the British Crime Survey show a persistent association of the incidence of victimisation with certain types of neighbourhood. High risks tend to be associated with the poorest council estates, mixed inner metropolitan areas and high status non-family areas. Conversely, the lowest risks are associated with agricultural and retirement areas, modern high-income family housing, and affluent suburban areas. However, these are very much generalisations subject to considerable variation between types of victimisation. Area risks interact with other risks (demography, lifestyle and status) to produce complex patterns, for example the relatively rich in the relatively poor areas suffer most, and if they also have a lifestyle that involves going out and drinking, the risks are further augmented.

Environmental criminology also has a contribution to make to the geography of victimisation. Victim behaviour, particularly in public places, may be instrumental in increasing vulnerability, perhaps by simple exposure through engaging in certain activities at certain times in certain places, or by being careless or not taking precautions, or by being attractive through the visual cues offered. At another level, the nature of places may mediate victimisation, either through perceptual qualities such as privacy and familiarity or through functional qualities such as lighting, architecture or land use.

Reporting and recording

The victim's decision to report a crime is mediated primarily by the seriousness of the offence. There are important distinctions between reasons for reporting less serious offences (feelings of vulnerability, intrusion of private space, wanting something done, insurance) and reasons for not reporting serious offences (belief that nothing can be done, feelings of powerlessness, too much effort/cost, knowing offender). It is clear that victims do not make the decision to report in isolation – family and neighbourhood expectations may be just as important as personal feelings. Geographical factors that significantly increase the likelihood of reporting include intrusion into victim's own home, job-related offence, home ownership and council tenancy.

Details of police recording are not covered by the British Crime Survey, and relatively little work has been done to demonstrate the extent of variations in police practices. In the study by Sparks *et al.* (1977) of London, incidents were followed through directly from victimisation to police record, and the proportions recorded varied between the three areas chosen for analysis (see Davidson 1981: pp. 98–100 for a re-analysis of these data). An analysis of Nottinghamshire's high official crime rate concluded that the main reason was police recording practices (Farrington and Dowds 1985).

Detection and prosecution

The means by which offenders are identified and brought before the courts is another area where research has whetted geographical interest without providing up-to-date evidence. The Sheffield Urban Crime Project demonstrated two decades ago that suspects are brought forward in three ways: (1) by victims directly identifying who did it; (2) by indirect means (normally offenders in custody admitting to other offences); and (3) by the police (either by catching the offender in the act or by police work). Police discretion was most commonly used in high-offender areas, where attitudes to the police were likely to be negative (see Mawby 1979). Since then, there has been a gradual decline in detection ratios, although variation between police forces remains high. There also continue to be differences in the use of police cautions and in the proportions of offenders prosecuted.

Justice

By whatever criterion, spatial disparities in justice are considerable. Most justice systems endow sentencers with considerable powers of discretion, and this had led to a mosaic of sentencing patterns. Tarling (1979) notes that courts even within a single police force area could be inconsistent in their sentencing, and indeed were not overly concerned that it should be so. Some clues as to why this should be so are suggested in Hogarth's (1971) penetrating study of Toronto magistrates. He concludes that magistrates adopt a variety of sentencing models that emphasise different aspects of the justice process – punishment, rehabilitation, just deserts, etc. Among these and variable in its impact is the need to serve the community through the sentencing process. How this is achieved will reflect the perceived needs of the community as well as the predispositions of sentencers with considerable autonomy and powers of discretion. Variations in punishment in England and Wales are reviewed by Harvey and Pease (1989) and between European countries by Pease (1994). Geographical perspectives on the death sentence in the USA are mapped out by Harries (1993).

Fear of crime

It is ironic to leave discussion of fear of crime to last, as anxiety about crime is much more prevalent and may cost more than crime itself. Everyone, every day, employs crime-avoiding behaviour. Securing person and property has become so routine that victimisation comes as a shock. And like other aspects of crime, fear is unevenly distributed. Women fear more than men, older people fear more than young, inhabitants of inner cities fear more than rural dwellers. Studies of fear suggest that there is an important distinction to be made between general worries about crime and specific concerns about personal safety. Fear in a general sense is connected to perceptions of crime derived from the media (Plate 32.1), from indirect experience conveyed within the neighbourhood, and from environmental cues such as graffiti and vandalism. There is a growing body of evidence of the intimate links between fear of crime and other problems in the local area seen as symbolic or actual threats. These are loosely termed incivilities and include rubbish, drunks, noisy neighbours, loitering youths, even speeding cars and dogs not under control. Fear of crime at the neighbourhood level is profoundly connected with social disintegration and disadvantage, feelings of isolation, alienation and powerlessness, and the lack of local resources in coping with crime. These connections highlight the social perspective on fear that sees the impact of crime and fear located firmly with community life as well as individual perceptions. Where you live is therefore as important to fear as it is to crime itself. There is moreover a dynamic to the fear–community relationship. Fear of crime is not just a product of local conditions but is itself a determining factor in the decline of an area. So area regeneration is not just a matter of improving the physical environment; it needs to address the social disruption of community life.

Individual factors in fear of crime focus on vulnerability – people's feelings about their ability to protect themselves or cope with the consequences. Killias (1990) suggests that there are three key factors in vulnerability: exposure to risk, loss of control (ability to defend, protect, escape), and anticipation of serious consequences. Vulnerability is also mediated by lifestyle and

Plate 32.1 Media images of crime. Since most people do not become the victim of crime as it is still a rare event, the media, particularly local media, are important as a source of information about the risks of crime. Media images are often distorted by selective or partial reporting and can lead to contradictions between perceptions and realities of crime (*photograph: University of Hull*).

previous experience of crime to produce some complex patterns of fear differentiated by gender and age. Women and older people are generally more fearful but have lower actual risks of becoming a victim – paradoxes that have been explored in some detail (Pain 1993; Valentine 1989; Fattah and Sacco 1989).

Environmental factors are influential in personal fears through their role in vulnerability. Being out alone after dark will increase fear, as will an environment that sensitises vulnerability through visual or auditory cues. The way that people perceive danger and learn to cope with it is the key to understanding how individuals negotiate the consequences of their fear.

REDUCING CRIME AND FEAR OF CRIME

The rise in crime over the last four decades has led to increasing demands for effective solutions. In the public eye, this means more often than not more punishment or at least the threat of punishment. Yet a reading of the data in Figure 32.2 will question whether raising the 'ante' on crime (more jail sentences) is likely to have a significant effect on victimisation when so few crimes result in imprisonment for the offender. Similarly, increased policing has strictly limited potential for reducing crime directly by catching criminals. The key to crime reduction is preventing crime in the first place.

Crime prevention has a long and honourable history. With public safety, it was one of the twin pillars of the 'Peel Act' that introduced the police as a public service in Britain in 1830. From the latter part of the nineteenth century, its role waned under the pressure of the increasing professionalisation of the police so that by the 1960s it had become reduced to offering advice on the security of premises. The last two decades have brought crime prevention back to the forefront, much reinforced and with clear geographical orientations, which are summarised in Table 32.1. Hough and Tilley (1998) suggest a division of prevention into two basic categories. *Crime prevention* includes enforcement activities designed to deter or incapacitate offenders and situational measures aimed at foreclosing opportunities for the commission of offences. *Criminality prevention* covers social or community measures that inhibit the development of criminal motivation and the rehabilitation of offenders to reduce the risk of re-offending.

The range of activities covered illustrates the strength of the geographical basis of crime prevention. Closed-circuit television (CCTV) appears twice to distinguish its complementary roles in detecting crime and deterring offenders. Recent

Table 32.1 Crime prevention by geography.

	Hot spots	*High-crime areas*	*Criminal areas*
1 Enforcement	Visible patrols CCTV	Intelligence-led targeting Zero tolerance Security patrols	Witness protection Tenancy restrictions
2 Situational prevention	Target hardening CCTV Better design	Neighbourhood Watch Cocoon Watch Anti-vandalism Environmental design	Environmental regeneration
3 Community/social/ developmental prevention	Pub Watch	Community action	Diversion schemes Youth work Social regeneration
4 Rehabilitation	Curfews	Community-based restorative justice	Mentoring Drug treatment

Plate 32.2 Reclaiming the streets. Rising crime rates have forced businesses into protective measures that have made city streets less attractive as a social milieu. Getting people back into city centres is encouraged by community safety initiatives such as CCTV that reassure the public while increasing the risks for offenders (*photograph: Norman Davidson*).

developments in zero-tolerance policing are likely to be most effective in high-crime areas, where public support would be strong. Zero tolerance is less likely to be effective in criminal areas where the subculture of crime is pervasive. Enforcement there needs to protect the law-abiding while isolating, tackling or rehabilitating offenders, and dealing with the root causes of criminality through environmental and social regeneration. In high-crime areas, the emphasis would be on helping communities to help themselves by enabling more effective protection. There are new initiatives in restorative justice aimed at harnessing community resources to rehabilitate offenders by mediation and reparation. The problems of hot spots are being tackled by inhibiting offenders (curfews and restrictive orders) as well as by strengthening the role of capable guardians, for example by pubs and clubs

Box 32.2 The Garths Crime Reduction Scheme

The Garths Crime Reduction Scheme illustrates partnership working in crime prevention in a high-crime area of Kingston-upon-Hull. The partners included the Hull Community Safety Partnership, Humberside Police, Hull City Council, the University of Hull and Garths Enterprises Ltd. The project took place in 1996 and was funded by the Single Regeneration Budget Challenge Fund.

BACKGROUND

- The Garths is a 1970s estate of about 2200 mainly low-rise council housing in the Bransholme area of Hull.
- The estate suffered an extremely rapid rise in crime in the late 1980s and early 1990s. By 1994, the incidence of burglary was more than six times the national average.
- In 1994, there were 540 burglaries on the Garths. Crime pattern analysis established that one in five homes burgled was victimised two or more times, 50 per cent higher than chance occurrence. Repeat burglary was likely to be within three months and to take place in daylight. Neighbours were also more likely to be burgled.

AIMS AND OBJECTIVES

- To reduce crime and fear of crime by tackling the key crime of burglary, attending especially to the problem of repeats.
- To adopt a two-level approach to crime reduction. The first stage involved property marking to promote both crime prevention and detection. On subsequent burglary, members of the scheme would have installed a sophisticated alarm system with a priority response guaranteed by Humberside Police.
- The scheme's targets were:
  - to register 50 per cent of homes on the estate, the high proportion to ensure the effectiveness of property marking;
  - to reduce domestic burglary by 50 per cent in marked homes and 10 per cent elsewhere.

OUTCOMES

- Sixty per cent of homes on the Garths joined the scheme and had their property marked.
- Forty-seven per cent reduction in burglary and 65 per cent reduction in repeat burglary.
- Reduction in value of goods stolen over two years exceeds total cost of the scheme.
- Other crime on the estate also decreased by up to 32 per cent.
- All homes on the estate benefited from reduction in crime. Members felt safer in their own homes after scheme.
- Little evidence of displacement to other forms of crime or other nearby areas.

REASONS FOR SUCCESS

- Involvement of local people ensured a high take-up rate through confidence in the scheme.
- Strong partnership and careful planning ensured delivery of protection where needed.

Source: Davidson 1998.

employing trained door staff. This brief review does scant justice to huge efforts being made to find effective ways of reducing crime – efforts in which geographical skills have a vital place. Box 32.2 provides a case study in the reduction of residential burglary in Hull that was cited as an example of good practice in the White Paper for the 1997 Crime and Disorder Bill. Examples of good practice in American burglary reduction may be traced through Sherman *et al.* (1997).

CONCLUSIONS: REVIVING THE ROLE OF COMMUNITY

Crime and fear of crime remain an area of great geographical potential for student, researcher and practitioner alike. For the student, analysis of spatial patterns has thrown up a number of paradoxes (Box 32.3), which illustrate the complex nature of the relationships involved. While some are well researched, like the spatial paradox of women's fear, in others explanation remains elusive due to the dynamic impact of crime over space and time. An emerging and consistent focus of geographical perspectives on crime is community. In one sense, this is a reaffirmation of old-established themes – the ecological tradition of the Chicago School – but it also represents a much more contemporary emphasis on the social and cultural, in the ways in which identity is conceived and realised. The concept of community may lack clarity, but the importance of factors beyond the individual and his or her immediate surroundings is well attested across a wide range of issues relating to crime and fear. Whether it is street, neighbourhood or some less tangible spatial entity, the defining role of community is evident if not always readily unravelled. There remains much to be done, not least in attempting to explain the paradoxes, but also in grappling with new conceptions of spatial justice and equity that the recognition of differences in community reactions to crime demands.

For the practitioner, the future lies in the application of new technologies and new skills of geographical enquiry. GIS are coming of age in crime pattern analysis. No longer confined to retrospective application by highly skilled specialists, they are increasingly being used by front-line staff – police officers, probation officers, social workers – as a management tool, particularly in the assessment of risk. This sort of application is still in its infancy, and much remains to be done to make it more effective. Display mapping practices are still crude, but maps are beginning to appear as a means of communicating understanding and assessment in an increasing variety of professional contexts. The power of the map as a means of illustrating the connections between

Box 32.3 Spatial paradoxes in crime and fear of crime

1 *Even in high-crime areas, people tend to believe that crime is committed by someone else, somewhere else*

The displacement of fear, outsider theory, area stereotyping. (see Conklin 1975; Hindelang *et al.* 1978; Damer 1974)

2 *Local events increase fear, distance events do not*

Influence of media, cultivation hypothesis, concept of resonance.
(see Gerbner *et al.* 1979; Smith 1985; Williams and Dickenson 1993)

3 *Crime increases social cohesion in some areas, decreases it in others*

Durkheim's 'functionality of deviance'.
(see Conklin 1975)

4 *Women's fear is greatest in places where they are least at risk*

Explaining the 'irrationality' of fear; space and the patriarchy. (see Pain 1993; Valentine 1989)

5 *Neighbourhood Watch is most successful where it is least needed*

Contradictory definitions of success makes evaluation of crime prevention difficult.
(see Husain 1988; Bennett 1990; Rosenbaum 1987)

6 *The best-lit areas have the highest crime rates*

Long-term effects in crime prevention may be very different from short-term effects.
(see Atkins *et al.* 1991; Townshend 1997)

crime and the communities served by the practitioners is increasingly realised. Geographical profiling of crime is more than just mapping the pattern of events but involves the selection of relevant social, demographic and environmental variables that will enrich appreciation of the feasibility of solutions to the problem of crime. The challenge for geography is to move from understanding why there are spatial concentrations of crime to engaging in the development of crime prevention strategies, sensitive to local need, that will reduce the impact that crime has on the quality of individual and community life.

GUIDE TO FURTHER READING

John Conklin's (1975) study of Boston is an excellent starting point on crime and community and is still highly relevant. Harries (1980), Herbert (1982) and Davidson (1981) provide complementary wider reviews of the geography of crime. Brantingham and Brantingham (1984) is a comprehensive text covering temporal as well as spatial patterns and different levels of enquiry. Two more recent edited volumes have a much wider range of material – Evans and Herbert (1989) and Evans *et al.* (1992). For a contemporary review of environmental criminology, see Bottoms (1994). The best review of fear of crime is Hale (1996), which has a section on environmental influences. A straightforward introduction to crime prevention is offered by Hough and Tilley (1998), and a more comprehensive review by Gilling (1997).

REFERENCES

Atkins, S., Husain, S. and Storey, A. (1991) *The Influence of Street Lighting on Crime and Fear of Crime*. Police Research Group Crime Prevention Unit Paper 28, London: Home Office.

Audit Commission (1996) *Misspent Youth – Young People and Crime*. London: Audit Commission.

Baldwin, J. and Bottoms, A.E. (1976) *The Urban Criminal*. London: Tavistock.

Bennett, T. (1989) Burglar's choice of targets. In D.J. Evans and D.T. Herbert (eds) *The Geography of Crime*, London: Routledge, 176–92.

Bennett, T. (1990) *Evaluating Neighbourhood Watch*. Aldershot: Gower.

Bottoms, A.E. (1994) Environmental criminology. In M. Maguire, R. Morgan and R. Reiner (eds) *The Oxford Handbook of Criminology*, Oxford: Clarendon Press, 585–650.

Brantingham, P.J. and Brantingham, P.L. (1984) *Patterns in Crime*. New York: Macmillan.

Brantingham, P.L. and Brantingham, P.J. (1981) Notes on the geometry of crime. In P.L. Brantingham and P.J. Brantingham (eds) *Environmental Criminology*, Beverly Hills, Calif.: Sage, 27–54.

Conklin, J.E. (1975) *The Impact of Crime*. New York: Macmillan.

Damer, S. (1974) Wine alley: the sociology of a dreadful enclosure. *Sociological Review* 22, 221–48.

Davidson, N. (1998) *Garths Crime Reduction Scheme: Final Evaluation Report*. Hull: University of Hull, Centre for Criminology and Criminal Justice.

Davidson, R.N. (1981) *Crime and Environment*. London: Croom Helm.

Davidson, R.N. (1984) Burglary in the community: patterns of localisation in offender–victim relations. In R. Clarke and T. Hope (eds) *Coping with Burglary*, Boston: Kluwer–Nijhoff, 61–75.

Evans, D.J., Fyfe, N.R. and Herbert, D.T. (eds) (1992) *Crime, Policing and Place: Essays in Environmental Criminology*. London: Routledge.

Evans, D.J. and Herbert, D.T. (eds) (1989) *The Geography of Crime*. London: Routledge.

Farrington, D.P. and Dowds, E.A. (1985) Disentangling criminal behaviour and police reaction. In D.P. Farrington and J. Gunn (eds) *Reactions to Crime: The Police, Courts and Prisons*, Chichester: Wiley, 41–72.

Fattah, E.A. and Sacco, V.F. (1989) *Crime and Victimization of the Elderly*. New York: Springer-Verlag.

Gerbner, G., Gross, L.P., Morgan, M. and Signorelli, N. (1979) The demonstration of power: Violence Profile no. 10. *Journal of Communication* 29, 177–96.

Gill, O. (1977) *Luke Street: The Making of a Delinquency Area*. London: Macmillan.

Gilling, D. (1997) *Crime Prevention: Theory, Policy and Politics*. London: UCL Press.

Guerry, A.M. (1833) *Essai sur la statistique moral de la France avec cartes*. Paris: Crochard.

Hale, C. (1996) Fear of crime: a review of the literature. *International Review of Victimology* 4, 79–150.

Harries, K.D. (1980) *Crime and the Environment*. Springfield, Ill.: Charles C. Thomas.

Harries, K. (1993) Geography, homicide and execution: the US experience, 1930–1987. *Geoforum* 24(2), 205–13.

Harvey, L. and Pease, K. (1989) Variations in punishment in England and Wales. In D.J. Evans and D.T. Herbert (eds) *The Geography of Crime*, London: Routledge, 298–314.

Herbert, D.T. (1982) *The Geography of Urban Crime*. London: Longman.

Hindelang, M.J., Gottfredson, M.R. and Garofalo, J. (1978) *Victims of Personal Crime: An Empirical Foundation for the Theory of Personal Victimization*. Cambridge, Mass.: Balinger.

Hirschfield, A., Brown, P. and Bowers, K. (1996) Neighbourhood composition and crime 'hot spots': preliminary results from the Merseyside Crime and Disadvantage Study. *Focus on Police research and Development* 7, 38–41.

Hogarth, J. (1971) *Sentencing as a Human Process*. Toronto: University of Toronto Press.

Hough, M. and Tilley, N. (1998) *Getting the Grease to the Squeak: Research Lessons for Crime Prevention*. Crime Detection and Prevention Series Paper 85, London: Home Office.

Husain, S. (1988) *Neighbourhood Watch in England and Wales: A Locational Analysis*. Police Research Group Crime Prevention Unit Paper 12, London: Home Office.

Killias, M. (1990) Vulnerability: towards a better understanding of a key variable in the genesis of fear of crime. *Violence and Victims* 5, 97–108.

Mawby, R.I. (1979) *Policing the City*. Farnborough: Saxon House.

Mayhew, H. (1864) *London, Labour and the London Poor*. London: Griffin.

Mayhew, P., Maung, N. and Mirrlees-Black, C. (1993) *The 1992 British Crime Survey*. London: Home Office Research and Planning Unit.

Mirrlees-Black, C., Mayhew, P. and Percy, A. (1996) The 1966 British Crime Survey: England and Wales. *Home Office Statistical Bulletin* (12/96), 127–35.

Morris, T.P. (1957) *The Criminal Area*. London: Routledge & Kegan Paul.

Pain, R. (1993) Women's fear of sexual violence: explaining the spatial paradox. In H. Jones (ed.) *Crime and the Urban Environment*, Aldershot: Avebury, 55–68.

Pease, K. (1994) Cross-national imprisonment rates. *British Journal of Criminology* 34(Special), 116–30.

Rosenbaum, D.P. (1987) The theory and research behind neighbourhood watch: is it a sound fear and crime reduction strategy? *Crime and Delinquency* 33(1), 103–34.

Shaw, C.R. and McKay, H.D. (1942) *Juvenile Delinquency and Urban Areas*. Chicago: University of Chicago Press.

Sherman, L.W. (1995) Hot spots of crime and criminal careers of places. In J.E. Eck and D. Weisburd (eds) *Crime and Place*, Monsey, NY: Willow Tree Press, 35–62.

Sherman, L., Gottfredson, D., Mackenzie, D., Eck, J., Reuter, P. and Bushway, S. (1997) *Preventing Crime: What Works, What Doesn't, What's Promising: A Report to the United States Congress*. internet http//www.ncjrs.org/works/index.htm

Smith, S.J. (1985) News and dissemination of fear. In J. Burgess and J. Gold (eds) *Geography, the Media and Popular Culture*, London: Croom Helm, 229–53.

Sparks, R.F., Genn, H.G. and Dodd, D.J. (1977) *Surveying Victims*. London: Wiley.

Tarling, R. (1979) *Sentencing Practice in Magistrates' Courts*. Home Office Research Study No. 98, London: HMSO.

Townshend, T. (1997) Safer city centres: the role of public lighting. In T. Oc and S. Tiesdell (eds) *Safer City Centres – Reviving the Public Realm*, London: Paul Chapman, 119–29.

Valentine, G. (1989) The geography of women's fear. *Area* 21(4), 385–90.

Williams, P. and Dickenson, J. (1993) Fear of crime: read all about it? The relationship between newspaper crime and the reporting and fear of crime. *British Journal of Criminology* 33(1), 33–56.

33

Retail location analysis

Cliff Guy

INTRODUCTION

Shopping is one of the most commonplace human activities; shops and shopping centres characterise most settlements in the developed world. It is not surprising, therefore, that retail location has for many years been a prime concern of geographers. Following the development of central place theory in the 1920s, the spatial patterns formed by periodic markets and urban settlements became a subject for study. Much empirical research of a descriptive nature was carried out into these patterns, increasingly linked with studies of shopping travel (for reviews see, for example, Beaujeu-Garnier and Delobez 1979; Carter 1995). More recently, research into retail location has become informed by theories of consumer choice behaviour and strategic decision making by retail organisations (Guy 1980; Brown 1992a; Jones and Simmons 1990). There are signs now that the subject is forging renewed links with mainstream concerns in economic and cultural geography (Wrigley and Lowe 1996).

There has been a strong tradition of applied research in retail geography, in developing methods of analysis and forecasting that can be applied to enhance the profitability of commercial organisations in retailing. These methods, which first appeared in the United States in the 1920s, later became established as 'marketing geography' (for a historical review, see Thrall and del Valle 1997). They are summarised well in texts such as Davies (1976), Davies and Rogers (1984), Wrigley (1988), Berry and Parr (1988), and Jones and Simmons (1990). More recently, geographical techniques have been widely used in town planning practice to assess the impacts of proposed retail developments upon existing shopping provision (BDP Planning 1992; Bromley and Thomas 1993a).

This chapter has two broad purposes. First, it sets out some broad principles of marketing geography, relating these to the UK context. Second, it explores two areas of applied practice in retail geography: the forecasting of retail store or shopping centre sales; and the analysis of the impacts of major new shopping developments. Examples are given of ways in which geographical techniques can be used to assist these objectives, using British case studies. The economic and social contexts for these applications are also examined in some depth. The chapter ends with a brief examination of possibilities for further research, in the interests of integrating retail location analysis more firmly with current concerns in human geography.

ESSENTIAL CONCEPTS IN RETAIL LOCATION ANALYSIS

Retail location analysis is founded upon a series of interrelationships, which have strong geographical reference. This section summarises these interrelationships as a prelude to the case studies.

Catchment areas: where do a retail outlet's customers live?

The first of these relationships lies between retail outlets and the areas within which their consumers live. Underlying much of the methodology of retail locational analysis has been the notion that a retail outlet will possess a discrete *catchment area*, which includes a high proportion of the outlet's regular customers. In some cases, this area will be largely exclusive to the store in question, thus creating a local monopoly conferred by the outlet's geographical position in relation to its customers. Even where catchment areas overlap, as is more likely to occur, the outlet in question may have the largest *market share* of expenditure by the local population for those goods and services provided by the outlet. Analysts often define a *primary catchment area*, within which the outlet possesses the largest market share, and a *secondary catchment area*, within which other outlets dominate (see Figure 33.1A).

This 'traditional' model of retailer/consumer relationships is found in many geographical texts, usually with a warning that the use of cars for shopping in modern societies has to some extent undermined the simple relationships between local stores and local residents. Up to about the 1970s in Britain, it was broadly the case that convenience outlets and small centres served a local population, and that a set of such centres within an urban area would have largely non-overlapping catchment areas. At that time, most convenience shopping expenditure (i.e. on routine, regularly needed items such as food and groceries), was in fact made in the nearest shopping area to the home. Thus it was possible from empirical research to define such catchment areas (e.g. Potter 1982), and local centres themselves showed characteristics related to the socio-economic structure of their catchment areas (e.g. Davies 1968). This situation has partly broken down with the increasing use of cars in routine shopping, and increasing differentiation between retail outlets and shopping centres (see below). In Britain, most convenience expenditure now takes place in food superstores or large supermarkets

Figure 33.1 Retail catchment areas: (A) primary and secondary; (B) primary, for large and small stores.

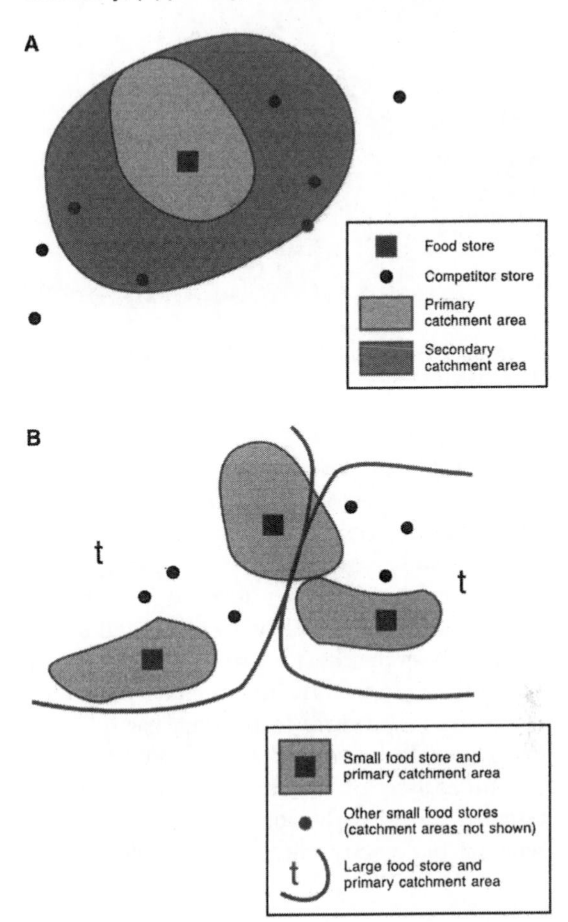

(Plate 33.1), and the small local centres (which generally still survive, somewhat reduced in importance) serve an ancillary function by supplying fresh foods or 'emergency' groceries. However, in forecasting the level of sales for a proposed large grocery store, analysts still use the concept of the catchment area: it is assumed to be likely that most customers will be drawn from an area that is closer to that store than to other similar large grocery stores. A large store's catchment area will include several catchment areas belonging to small local grocery stores or groups of shops (Figure 33.1B).

For *comparison shopping*, involving purchase of personal or fashion goods such as clothing,

Plate 33.1 A large modern food store: Morrison's, Morecambe.

catchment areas overlap to a greater extent, because convenience is a less important concept and competing centres differ more. In particular, shoppers will often neglect local shopping opportunities and travel further afield to a centre that offers a wide choice of styles, prices and so on. This has been the case for many years. However, recent trends in the UK have been for the largest town and city centres to prosper, partly at the expense of smaller town centres (Teale 1997).

The analysis of catchment areas remains an essential part of marketing geography. Knowing where an outlet's customers live, or are likely to live in the case of a proposed outlet, is important information for the retailer or marketer. Evaluation of a store's sales performance, decisions on whether to expand the store network and developments in direct marketing to consumers in their homes all rely upon an intimate knowledge of typical home locations. The applied science of *geodemographics* has become important in recent years as attempts are made to relate consumers typically associated with various types of retail outlet to the demographic and socio-economic characteristics of small residential areas (Longley and Clarke 1995; see also this volume, Chapter 42).

Retail attributes and consumer choice: how do people choose where to shop?

The second major relationship is the interaction between consumer choice of store and store attributes. These include store location but also the range of goods available, prices, standards of service, internal layout, provision of car parking, and quality of external environment. Consumer choice is typically based upon at least some of these attributes and is mediated through characteristics of the consumer her/himself and of the transport system and the methods of travel available to the consumer (see Box 33.1).

Many analysts have identified a basic trade-off

Box 33.1 Relationships between store size, type of consumer and shopping travel

It is suggested in this chapter that larger stores tend to have larger catchment areas, within which the catchment areas for smaller stores may nest (Figure 33.1B). One way of examining this topic empirically is to study the travel behaviour of different groups of shoppers in an urban environment.

In a study of shopping travel in Swansea, Bromley and Thomas (1993b) compared the choice of main food store between different groups of the sample population surveyed. They found that:

- Sixty-nine per cent of car-owning households used a superstore for their main food shopping, whereas only 31 per cent of families without cars did so.
- In four different areas surveyed, the car-less households consistently used the city centre and district and local centres more than did the car-owning households.
- The more elderly respondents, whether car-owning or not, relied more heavily on their local district centre or neighbourhood shops than did younger respondents.

A shopping diary survey carried out in Cardiff in 1982 obtained similar results. Households that had full use of a car for shopping spent on average just under 50 per cent of their food budget within 2 km of the home, whereas those without the use of a car spent about 75 per cent of their total food budget within this distance (Guy 1984: p. 58). These households tended to use small supermarkets located within walking distance or a short bus ride from their homes, whereas the full car users tended to visit larger supermarkets or superstores located further from home.

These and other surveys show that households that cannot easily reach or choose not to use superstores tend to make short trips to smaller local grocery stores. There is also evidence that superstore users often use the nearest superstore to the home. However, these stores are more widely dispersed than small stores and hence will have larger catchment areas.

between the 'attractiveness' or 'size' of a potential retail destination and the 'friction of distance' involved in travelling there. This notion underlies the family of spatial interaction approaches, including the so-called 'gravity' and 'retail potential' models. (Haynes and Fotheringham 1984; Fotheringham and O'Kelly 1989; Brown 1992b; Huff and Black 1997).

The retail potential model is probably the most widely used in the context of predicting aggregate consumer choice of shopping opportunity. It is based upon the following assumptions (Huff 1962; Lakshmanan and Hansen 1965):

- The larger in size the shopping opportunity, the more likely the shopper is to choose it. The relationship between choice probability and size may be linear or non-linear.
- The closer the shopping opportunity is to the shopper's home, the more likely the shopper is to choose it. The relationship between choice probability and distance (or travel cost) is non-linear.
- The likelihood of choosing any one shopping opportunity is affected by the size and relative location of all other shopping opportunities within the consumer's 'choice set'. In effect, the consumer, while considering the trade-off between size and distance for any one centre, simultaneously considers the trade-offs for all other centres in the choice set.

Although widely used in practice, this formulation has been extensively criticised. Three broad criticisms can be mentioned here:

- The model usually assumes that all consumers within the region concerned make identically structured trade-offs between shopping centre size and distance, and take competition into account in a similar way. Differences in choice of destination are thus simply a reflection of consumers' home locations. This assumption is clearly erroneous in a society in which mobility varies from one person to another.
- The model fails to take into account any determinants of consumer choice other than the size and distance factors mentioned. This is clearly incorrect in a society in which tastes and preferences vary from one person to another.
- The model assumes that all consumers within the region concerned have the same choice set. In practice, consumers will ignore many possible shopping opportunities, on grounds of ignorance, antipathy or problems of access.

These criticisms have led to the development of more complex 'choice models', which can incorporate several characteristics of shopping opportunities and consumers. Among these are the family of 'decompositional multi-attribute preference models' developed in the Netherlands by Timmermans (1984: see also Timmermans and Golledge 1990). These allow for competitive factors in a manner similar to the retail potential model, but they can simulate the effects of a wider range of determinants of shopping destination choice and more complex structures of comparison of destinations and trade-offs between attributes of those destinations.

Hierarchies of retail areas and shopping centres: how is retailing spatially organised?

The third set of basic relationships lies in the spatial disposition of retail opportunities within urban areas. Although some stores such as large food stores can function adequately in isolated sites, most store types benefit from clustering with other stores, either of the same broad type or of different types. Here we encounter some problems of definition. This author takes the view that a cluster of retail outlets that has grown gradually over time and is in multiple ownership should be called a 'retail area' (see Plate 33.2). A group of outlets built at the same time and under one ownership should be called a 'shopping centre'. The term 'town centre' defines the part of a town's central area that is devoted mainly to retailing and other consumer services: it may include both retail areas and shopping centres.

Many geographers have studied spatial patterns of retail and business areas within urban regions (for a review, see Brown 1992a: pp. 40–51;

Plate 33.2 A typical suburban 'retail area' in Cardiff.

Carter 1995: pp. 66–77). Central place theory, originally conceived as an explanation for the spatial distribution of settlements in southern Germany (Christaller 1966), was adapted by North American geographers in the 1950s and 1960s to the study of urban business location. Berry's (1963) study of Chicago was a classic in this respect, identifying several separate systems of retail and business provision. Berry's work was important in identifying a separate hierarchy of planned shopping centres, which had been built in the suburbs of Chicago from the 1930s onwards and were even then replacing the older 'traditional' retail and business areas (Guy 1998a).

In the UK, the distinction between systems of traditional retail areas and of car-oriented one-stop destinations such as *hypermarkets* and *retail warehouses* emerged during the 1970s (Guy 1994). The influence of car ownership was seen to be important: consumers who regularly use cars for shopping are much more likely to patronise off-centre stores with free car-parking facilities (e.g. Bromley and Thomas 1993b). Those without cars are nowadays associated with use of traditional retail areas, which are slowly declining in size and quality as their more affluent customers desert them for the off-centre stores (Thomas and Bromley 1995).

Studies of relationships between shopping destinations have become yet more complex in recent years because of the growth in importance of so-called *leisure shopping* outlets. These include factory outlet centres (Fernie and Fernie 1997), and speciality shopping centres such as the 'festival marketplaces,' which first appeared in the United States during the 1970s (Frieden and Sagalyn 1989). These centres form systems of retailing different from the traditional retail areas and the 'one-stop' centres referred to above. This is because they tend to attract families taking part in mixed shopping/leisure trips, often associated with holidays and tourism.

Figure 33.2 summarises ways in which various forms of retail development have become related to different types of shopping trip purpose.

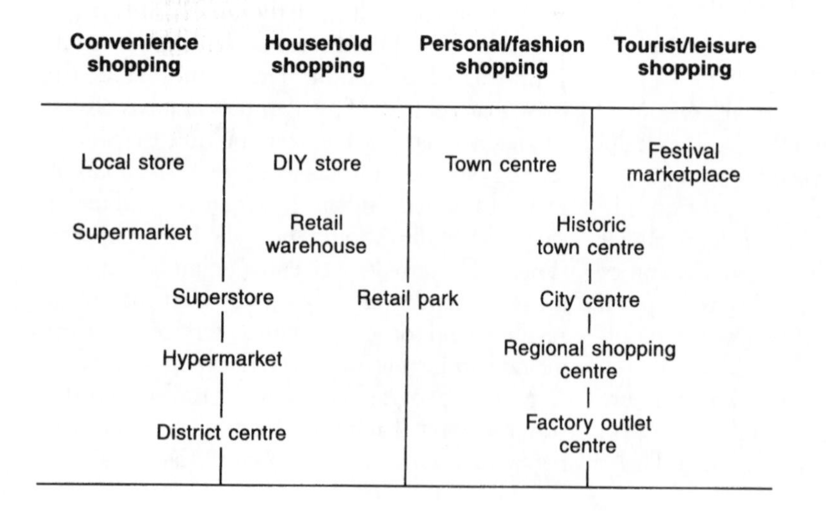

Figure 33.2 Relationships between shopping trip purpose and type of retail destination.

APPLICATION: SALES FORECASTING FOR RETAIL STORE LOCATION

A typical problem in retail planning practice is to determine the likely volume of sales to be derived from a new retail outlet of a given size at a given location. On such calculations may rest the ultimate profitability of the store as well as the amount that the company concerned is willing to pay for the site (Guy 1995).

The calculation of sales embodies the principles discussed in the previous section, including definition of a catchment area, recognition of competition from existing stores, and the influences of both the transport network and the socio-economic composition of the area surrounding the store site. Clarkson *et al.* (1996) show how these geographical principles underlie the range of methods typically used by major food retailers in the UK.

Broadly speaking, there are two ways of calculating store sales. The first is one of the *market area analysis* family of techniques, which rest upon the principles of traditional marketing geography. The process, which is described in more detail in Figures 33.3 and 33.4, may be summarised as follows:

- Define 'drive time' isochrones around the store, typically at 10 and 20 minutes distance. Relate these to residential areas and, using population census data, estimate the total population living within these drive time zones.
- Estimate the probable *market penetration* (or market share) of the new store within each of these zones, dividing each zone if necessary into smaller areal units such as census wards. Estimation uses simple rules based upon analogies with existing stores owned by the same company, and also on a knowledge of the locations of rival stores, existing and proposed.
- Calculate the amount of expenditure made at the store from within the 20-minute drive time zone by multiplying the population of each small area by its assumed expenditure per head on the goods likely to be sold in the new store, multiplied by the percentage market share as estimated above. This gives rise to estimates of *trade draw*.
- Adjust for trade drawn from outside the 20-minute isochrone, usually using a simple 'rule of thumb' such as an additional 10 or 20 per cent of the sales as derived above. This percentage accounts for 'random' visits by people living outside the catchment area and may be established through analogies with existing stores.

The second method involves use of a shopping model, typically a spatial interaction formulation

Figure 33.3 Estimating sales in a large food store using market area analytical methods.

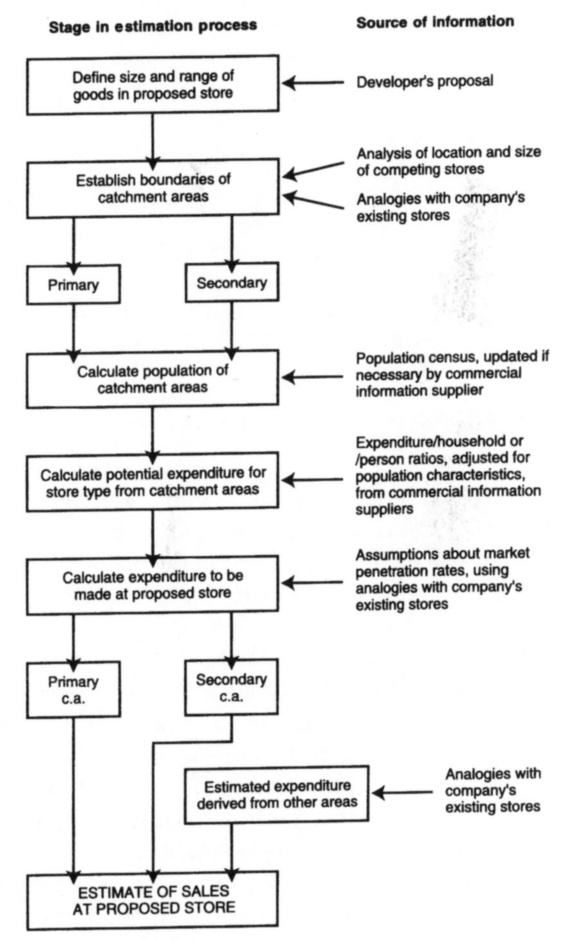

Figure 33.4 (A) Market penetration rates, and (B) trade draw pattern, for a large food store.

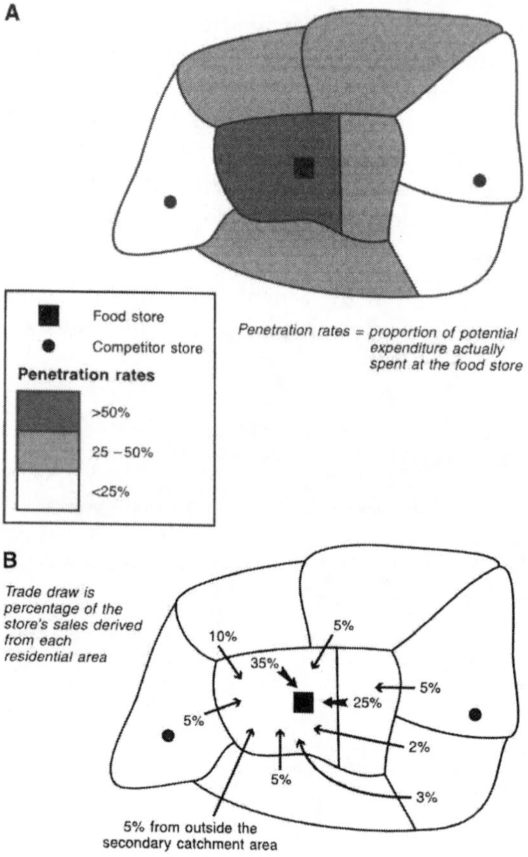

Source: Based on Doidge 1995, Figure 3.

Figure 33.5 Estimating sales in a new large food store using a spatial interaction model.

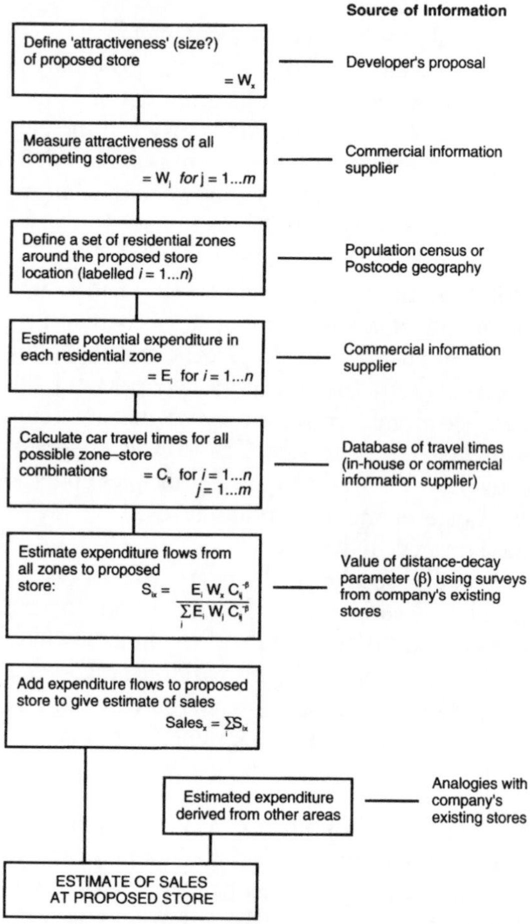

as discussed in the previous section. Penny and Broom (1988) describe the method used at that time by the food retailer Tesco in calculating the probable volume of sales for a proposed store. This is summarised in Figure 33.5. Many other companies, especially in North America, routinely use similar methods. As with market area analysis, these methods make heavy use of analysis of existing store sales and catchment population records. The recent tendency for food retailers to issue *loyalty cards* to customers is aimed partly at establishing a reliable and comprehensive database that associates expenditure patterns in particular stores with shoppers' residential addresses and information about their family composition (Clarkson *et al.* 1996).

Methods similar to these can be extended to deal with other, more complex, problems such as the forecasting of sales for a new shopping centre. Some retail companies carry out a performance evaluation for existing outlets through making sales forecasts using market analytical or modelling methods. These forecasts are then compared with actual sales (Davies 1977). An examination of the values of residuals between forecast and actual sales may be used either to improve the forecasting method or to identify strengths and weaknesses within the existing store portfolio.

A rather different problem, discussed mainly in the North American literature, is the identification of the *optimal site* for a new store: that which will produce the largest increase in overall store sales (not necessarily sales for the new store, since this might prosper only at a cost to existing stores). This problem is not discussed further here, due to the complexity of the methods involved (interested readers are referred to Drezner (1995) for a review of methods of optimising facility location across a whole network). In any case, in the UK context this type of problem is unlikely to arise in its pure form. Retail companies are usually highly constrained in their choice of store locations through land-use planning restrictions (Guy 1994).

APPLICATION: EVALUATION OF RETAIL IMPACTS

One of the most important issues in land-use planning over the last few years has been the growth in 'out-of-town' retailing (more correctly termed 'off-centre' retailing). This has involved the development of large purpose-built stores and shopping centres on suburban land, characterised by mass-market one-stop shopping and abundant free car-parking facilities (Guy 1994). The debate over the control of off-centre retailing has reflected public concern over the potential monopoly power of large retail and development companies, and a desire to protect the more fragile traditional retail environments. The latter are seen both as the home of the 'small businessman' and as well-used familiar places that give towns and cities their character and identity (Guy 1998b).

The impacts of such retail development are usually classified into economic, social and environmental (BDP Planning 1992). *Economic impact* constitutes the diversion of trade away from traditional retail areas and into the new stores and centres. In a situation where total retail expenditure within an area changes only slowly in the short term, new shopping opportunities are likely to take trade from established retail areas and shopping centres selling similar products. Thus new large food stores will have a noticeable trading impact upon smaller supermarkets and fresh food shops within the local area if these were previously the main source of food; but in the 1990s they are more likely to affect trading in other existing large food stores.

Social impact is related to the social profiles of both users and non-users of the new facilities: because some people find it difficult to gain access to the new facilities, the perceived gap between 'haves' and 'have-nots' becomes wider. In addition, the 'have-nots' suffer when their local retail areas fall victim to economic impacts.

This process gives rise to *environmental impact*, which occurs when economic and social impacts result in physical changes to traditional retail areas and town centres. These may include an increase in vacancy rates or the replacement of established shops of real value to the local community by transient 'twilight' uses. A 'downward spiral' of events can occur in serious cases, where the initial loss of trade caused by retail competition leads to several key retailers leaving the centre (Figure 33.6).

Partly for these reasons, government policy in the UK and several other Western European countries has increasingly focused upon maintaining and enhancing the *vitality* and *viability* of existing town centres and retail areas. Vitality 'is reflected in how busy a centre is at different times and in different parts', and viability 'refers to the

Figure 33.6 The 'downward spiral' model of environmental impact.

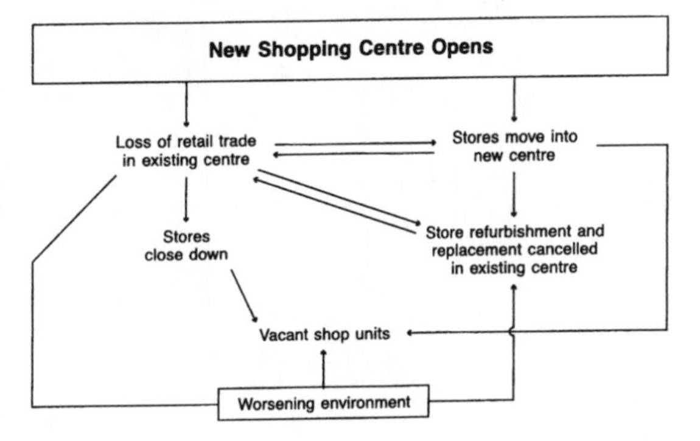

ability of the centre to attract continuing investment, not only to maintain the fabric, but also to allow for improvement and adaptation to changing needs' (URBED and Comedia 1994). The British government has since issued a list of indicators of vitality and viability, which local planning authorities and other town centre interests are expected to use in assessing the possible impacts of retail development (Department of the Environment 1996: Figure 2).

The method described in Figure 33.7 is typical of those used by planning consultants in the UK to investigate the probable impacts of proposed new retail facilities. For the sake of simplicity, the example taken is of a proposed large food store. The ultimate aim of the exercise is to predict the extent to which the proposed development will affect the vitality and viability of a nearby established town centre. Impacts upon other modern large food stores in the local area can also be assessed, but these are not usually held to be significant for planning purposes. The method clearly builds upon the principles of market area analysis already discussed. This leads to a prediction of trading impact: that the town centre will lose *x* per cent of its retail trade, for example. There is then a further stage of assessing impacts upon vitality and viability. This requires the exercise of informed judgement rather than simple application of some technical procedure.

Figure 33.7 Methodology for assessing the impacts of a new large food store on town centre vitality and viability.

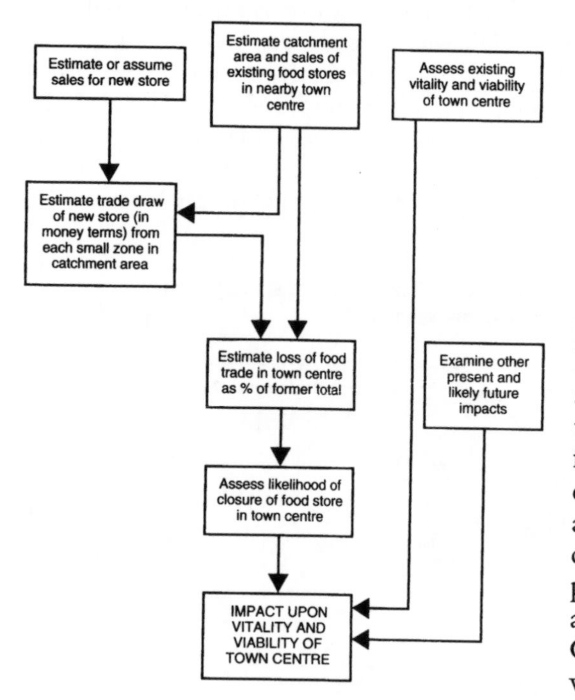

This type of impact assessment clearly makes use of geographical skills applied to a policy-making environment. Ultimately, however, the assessment is incomplete: it concentrates mainly on economic impact and to a limited extent on environmental impact. UK planning policy and practice pay little attention to the social impact of retail change, as one might expect in a situation where the distributional effects of social and economic change generally have no official place in planning law and procedures. However, this is an area where geographical research could make a substantial impact. This theme is developed further in the concluding section to this chapter.

Another area of research that needs to be applied more rigorously is the investigation of retail impacts *after* the store or centre in question has opened. The very extensive literature review carried out by BDP Planning (1992) for the Department of the Environment was able to identify only a few competent and thorough studies of retail impacts. More recently, the impacts of regional shopping centres such as the Metro Centre in Gateshead, Merry Hill in the West Midlands and Meadowhall in Sheffield have been investigated (Howard 1993; Howard and Davies 1993; Tym 1993; see Box 33.2). A study of long-term changes in grocery store location and size in Cardiff (Guy 1996b) suggested that trading impact is difficult to identify in isolation from the many other causes of retail growth and decline. In this particular case, the development of a few large new food stores coincided with the closure of several smaller supermarkets and independently owned food shops, but precise cause-and-effect sequences were hard to detect. Changes in the spatial provision of food shopping were better described as the outcome of a general

Box 33.2 Merry Hill regional shopping centre and its impacts

The Merry Hill shopping complex is located in the West Midlands, 10 miles west of Birmingham. It lies in an area formerly of heavy industry that was made an enterprise zone by the British government in the early 1980s in order to encourage urban regeneration. This entailed an absence of planning controls over certain types of retail development.

Retail development started in 1984 with a series of retail warehouses and a hypermarket. Following the success of these schemes, a regional shopping centre of over 1 million square feet was completed in 1989, anchored by a Debenhams department store and a large Marks & Spencer.

The town centre of Dudley, two miles away from Merry Hill, was severely affected. This process occurred in several stages, which match approximately the process shown in Figure 33.6:

- Some major retailers (e.g. Marks & Spencer) closed their stores and in effect moved them to Merry Hill.
- An immediate loss of trade was felt by the remaining stores, and some of these were closed. By 1992, over half of the seventy or so multiple retailers represented in Dudley town centre had left. According to a survey sponsored by the Department of the Environment and local authorities, over 70 per cent of Dudley's comparison goods trade was lost during this period (Tym 1993).
- The empty shops were reoccupied by low-quality discount and variety stores. This attracted a new type of clientele to the centre, mainly older and poorer customers. The centre has now 'repositioned' itself in the local economy as a discount shopping destination and is beginning to show a recovery in physical terms.

Merry Hill and Dudley were visited by the influential House of Commons Environment Committee in 1993, and its recommendations for tighter government control over off-centre development were based partly upon this case study.

process of concentration of trade in a few large companies, and the rationalisation of the store network.

CONCLUSIONS

Retail location analysis has been an important branch of applied geography for many years. First, it has provided an essential set of techniques that the retail and development sectors need in their attempts to maximise profitability, especially where this can be done through geographical growth, contraction or more efficient use of existing property. Particularly in North America, 'marketing geography' graduates are routinely recruited into the store location and assessment divisions of major retail and property development companies (Thrall and del Valle 1997). Second, retail location analysis has proved of value to both private and public sector agencies in the many battles waged in Western Europe over the control by governments of new retail development. The methods typically used here are grounded in geographical principles, although the practitioners involved in this work do not necessarily have specific geographical training.

In the last ten years or so, market area analysis has probably become more important than ever. Retail facilities in many parts of the developed world are reaching saturation point, first noted in a general slowdown of regional shopping centre development in the United States since the 1970s (Carlson 1991). In these circumstances, it becomes more important than ever to make locational decisions based upon sound methods and the best possible quality of information. Such methods can be carried out nowadays much more comprehensively and rapidly using desktop computers, detailed geodemographic information (see this volume, Chapter 42), and more sophisticated modelling techniques.

Market area analysis and its modern derivatives are still useful in carrying out basic procedures for retail location analysis, but are not always sufficient for this purpose in situations of complexity, for example where several different systems of retail development and consumption coexist in an urban region. Shopping models can to some extent cope with these situations, but they introduce theoretical and practical difficulties that restrict their use to a small number of practitioners.

Methods of analysis are also having to become more sensitive to environmental and political pressures. They need to be embedded in a deeper understanding of growth and change in retailing at local, national and international levels, and the outcomes for the retail environment, for consumers and the built and natural environment generally (Marsden *et al.* 1998).

A major deficiency in much of the retail locational literature is perhaps its detachment from mainstream economic and social theory. Both 'central place' methods and spatial interaction modelling have been criticised many times for their simplistic assumptions and their lack of a sound theoretical basis. The subject has also in recent years failed to follow mainstream geography into more subjective and experiential areas of investigation. If there is a need to develop a 'new' applied retail geography, the following avenues may offer ways of enlightening our understanding of the retail environment:

- A focus on retailer strategy, developing at national and international levels from theories of the firm and an understanding of methods of investment appraisal (for example, Wrigley 1996; Guy 1995; 1997). At local level, knowledge of geodemographics, property market features and local demand–supply relationships are needed in the search for the investigation of issues such as trading impact, sectoral growth and decline, and saturation (Langston *et al.* 1997; Guy 1996a). The research methods necessary include digesting company publications, analysts' reports, etc. and carrying out semi-structured interviews with leading experts and decision makers (as in Wrigley 1997).
- A focus on those consumers who have been disadvantaged and marginalised by retail change, the problems they face daily as consumers, and their coping strategies. This would relate to ongoing work on nutrition and health, household budgeting and the nature of 'food poverty' among the poorest sectors of the population (National Consumer Council 1992; Dobson *et al.* 1994). Research methods include participatory investigation and discussion in focus groups.
- A focus on cultural interpretations of retail spaces and consumption (Shields 1992; Wrigley and Lowe 1996; Miller *et al.* 1998). Such investigations are growing in popularity at present, although their immediate relevance to commercial pressures and public policy making appears to this author to be less strong when compared with the two other research areas noted above. Research methods include discussion in focus groups, and the interpretation of contemporary accounts, factual or fictional, on the experiences of retailing and consuming.

Retail locational analysis remains an important area of applied geography. Its basis lies in an understanding of relationships between retail outlets and consumers in space. That understanding can be enhanced if research in the three areas mentioned above continues to blossom. The danger here is that the conventional description and understanding of retail location and its relationships with patterns of consumption might be regarded as somehow less important than social or cultural interpretations. Marketing geography methods remain vital in providing an understanding of retail patterns in space and have considerable value as an area of applied research. All these approaches need to be maintained in geographical teaching and somehow brought together to enhance our understanding of retail geography.

GUIDE TO FURTHER READING

Bromley, R.D.F. and Thomas, C.J. (1993) *Retail Change: Contemporary Issues*. London: UCL Press. This edited collection of papers is particularly useful in exploring questions of retail impact and the advantages and disadvantages of 'out-of-town' shopping.

Brown, S. (1992) *Retail Location: A Micro-Scale Perspective*. Aldershot: Avebury. This book discusses retail location from geographical and economic perspectives, and it contains a massive bibliography.

Guy, C.M. (1994) *The Retail Development Process*. London: Routledge. This book describes both traditional and modern retail developments and relates these to property markets and land-use planning processes.

Jones, K. and Simmons, J. (1990) *The Retail*

Environment. London: Routledge. A thorough presentation of methods of retail location selection and appraisal, based mainly upon market area analysis: case studies are mainly North American.

O'Brien, L.G. and Harris, F. (1991) *Retailing: Shopping, Society, Space*. London: David Fulton. A wide-ranging exploration of retail issues of interest to geographers, with an emphasis on the UK.

Wrigley, N. and Lowe, M. (1996) *Retailing, Consumption and Capital: Towards the New Retail Geography*. Harlow: Longman. An edited collection of papers that attempts to transform the study of retail geography, emphasising structural and cultural approaches.

REFERENCES

Beaujeu-Garnier, J. and Delobez, A. (1979) *Geography of Marketing*. London: Longman.

BDP Planning and Oxford Institute of Retail Management (1992) *The Effects of Major Out of Town Retail Development: A Literature Review for the Department of the Environment*. London: HMSO.

Berry, B.J.L. (1963) *Commercial Structure and Commercial Blight*. Department of Geography, University of Chicago, Research Paper 85.

Berry, B.J.L. and Parr, J.B. (1988) *Market Centers and Retail Location: Theory and Applications*. Englewood Cliffs, NJ: Prentice-Hall.

Bromley, R.D.F. and Thomas, C.J. (eds) (1993a) *Retail Change: Contemporary Issues*. London: UCL Press.

Bromley, R.D.F. and Thomas, C.J. (1993b) The retail revolution, the carless shopper and disadvantage. *Transactions of the Institute of British Geographers* 18, 222–36.

Brown, S. (1992a) *Retail Location: A Micro-Scale Perspective*. Aldershot: Avebury.

Brown, S. (1992b) The wheel of retail gravitation? *Environment and Planning A* 24, 1409–29.

Carlson, H.J. (1991) The role of the shopping centre in US retailing. *International Journal of Retail and Distribution Management* 19(6), 13–20.

Carter, H. (1995) *The Study of Urban Geography*, 4th edition. London: Edward Arnold.

Christaller, W. (1966) *Central Places in Southern Germany*, (trans. C.W. Baskin). Englewood Cliffs, NJ: Prentice-Hall.

Clarkson, R.M., Clark-Hill, C.M. and Robinson, T. (1996) UK supermarket location assessment. *International Journal of Retail and Distribution Management* 24(6), 22–33.

Davies, R.L. (1968) Effects of consumer income differences on the business provisions of small shopping centres. *Urban Studies* 5, 144–64.

Davies, R.L. (1976) *Marketing Geography: With Special Reference to Retailing*. Corbridge: Retailing and Planning Associates.

Davies, R.L. (1977) Store location and store assessment research: the integration of some new and traditional techniques. *Transactions of the Institute of British Geographers* 2, 141–57.

Davies, R.L. and Rogers, D.S. (1984) *Store Location and Store Assessment Research*. Chichester: Wiley.

Department of the Environment (1996) *Planning Policy Guidance 6: Town Centres and Retail Developments*. London: HMSO.

Dobson, B., Beardsworth, A., Keil, T. and Walker, R. (1994) *Diet, Choice and Poverty: Social, Cultural and Nutritional Aspects of Food Consumption among Low-income Families*. London: Family Policy Studies Centre.

Doidge, R. (1995) How to improve financial returns. *Estates Gazette Shopping Centres Directory* 12–20.

Drezner, Z. (ed.) (1995) *Facility Location: A Survey of Applications and Models*. New York: Springer.

Fernie, J. and Fernie, S.I. (1997) The development of a US retail format in Europe: the case of factory outlet centres. *International Journal of Retail and Distribution Management* 25, 342–50.

Fotheringham, S. and O'Kelly, M. (1989) *Spatial Interaction Models: Formulations and Applications*. Dordrecht: Kluwer Academic.

Frieden B.J. and Sagalyn, L.B. (1989) *Downtown, Inc.: How America Rebuilds Cities*. Cambridge, Mass.: MIT Press.

Guy, C.M. (1980) *Retail Location and Retail Planning in Britain*. Farnborough: Gower.

Guy, C.M. (1984) Food and Grocery Shopping Behaviour in Cardiff. UWIST Department of Town Planning, *Papers in Planning Research* 86.

Guy, C.M. (1994) *The Retail Development Process*. London: Routledge.

Guy, C.M. (1995) Retail store development at the margin. *Journal of Retail and Consumer Services* 2, 25–32.

Guy, C.M. (1996a) Grocery store saturation: the debate continues. *International Journal of Retail and Distribution Management* 24(6), 3–10.

Guy, C.M. (1996b) Corporate strategies in food retailing and their local impacts: a case study of Cardiff. *Environment and Planning A* 28, 1575–602.

Guy, C.M. (1997) Fixed assets or sunk costs? An examination of retailers' land and property holdings in the UK. *Environment and Planning A* 29, 1449–64.

Guy, C.M. (1998a) Classifications of retail stores and shopping centres: some methodological issues.

GeoJournal, 45, 255–64.

Guy, C.M. (1998b) Controlling new retail spaces – the impress of planning policies in Western Europe. *Urban Studies*, 35, 953–79.

Haynes, K. and Fotheringham, A.S. (1984) *Gravity and Spatial Interaction Models*. Newbury Park, Calif.: Sage.

Howard, E.B. (1993) Assessing the impact of shopping centre development: the Meadowhall case. *Journal of Property Research* 10, 97–119.

Howard, E.B. and Davies, R.L. (1993) The impact of regional, out-of-town retail centres: the case of the Metro Centre. *Progress in Planning* 40(2), 1–49.

Huff, D.L. (1962) *Determination of Intraurban Retail Trade Areas*. Los Angeles: Graduate School of Management, University of California.

Huff, D.L. and Black, W.C. (1997) The Huff model in retrospect. *Applied Geographic Studies* 1, 83–93.

Jones, K. and Simmons, J. (1990) *The Retail Environment*. London: Routledge.

Lakshmanan, T.R. and Hansen, W.G. (1965) A retail market potential model. *Journal of the American Institute of Planners* 31, 134–43.

Langston, P., Clarke, G.P. and Clarke, D.B. (1997) Retail saturation, retail location and retail competition: an analysis of British food retailing. *Environment and Planning A* 29, 77–104.

Longley, P. and Clarke, G. (eds) (1995) *GIS for Business and Service Planning*. Cambridge: GeoInformation International.

Marsden, T., Harrison, M. and Flynn, A. (1998) Creating competitive space: exploring the social and political maintenance of retail power. *Environment and Planning A* 30, 481–98.

Miller, D., Jackson, P., Thrift, N., Holbrook, B. and Rowlands, M. (1998) *Shopping, Place and Identity*. London: Routledge.

National Consumer Council (1992) *Your Food: Whose Choice?* London: HMSO.

Penny, N.J. and Broom, D. (1988) The Tesco approach to store location. In N. Wrigley (ed.) *Store Choice, Store Location and Market Analysis*, London: Routledge.

Potter, R.B. (1982) *The Urban Retailing System: Location Cognition and Behaviour*. Aldershot: Gower.

Shields, R. (1992) *Lifestyle Shopping: The Subject of Consumption*. London: Routledge.

Teale, M. (1997) Big in retail is beautiful and getting bigger. *Centre Forward (Hillier Parker Retail)* summer issue, 8–9.

Thomas, C.J. and Bromley, R.D.F. (1995) Retail decline and the opportunities for commercial revitalisation of small shopping centres: a case study in South Wales. *Town Planning Review* 66, 431–52.

Thrall, G.I. and del Valle, J.C. (1997) Applied geography antecedents: marketing geography. *Applied Geographic Studies* 1, 207–14.

Timmermans, H. (1984) Decompositional multiattribute preference models in spatial choice analysis: a review of some recent developments. *Progress in Human Geography* 8, 189–221.

Timmermans, H. and Golledge, R.G. (1990) Applications of behavioural research on spatial problems II: preference and choice. *Progress in Human Geography* 14, 312–54.

Tym, Roger and Partners (1993) *Merry Hill Impact Study*. London: HMSO.

URBED and Comedia (1994) *Vital and Viable Town Centres: Meeting the Challenge*. London: HMSO.

Wrigley, N. (ed.) (1988) *Store Choice, Store Location and Market Analysis*. London: Routledge.

Wrigley, N. (1996) Sunk costs and corporate restructuring: British food retailing and the property crisis. In N. Wrigley and M. Lowe (eds) *Retailing, Consumption and Capital*, Harlow: Longman.

Wrigley, N. (1997) British food capital in the USA. *International Journal of Retail and Distribution Management* 25, 7–21 and 48–58.

Wrigley, N. and Lowe, M. (eds) (1996) *Retailing, Consumption and Capital: Towards the New Retail Geography*. Harlow: Longman.

34

Urban transport and traffic problems

Brian Turton

THE NATURE OF THE PROBLEM

The pressure of excessive flows of motorised traffic upon urban road networks is the principal component of the urban transport problem in most towns and cities of the Western and the developing world. The identification of the difficulties caused by vehicle movements in urban areas and the efforts made to apply effective solutions form the basis of this chapter, but urban transport also has a significant impact upon many aspects of the urban environment, such as air pollution and landscape conservation, which are discussed elsewhere in this volume.

The continually increasing rates of transport congestion and its detrimental impact upon the social and economic functions of towns and cities has stimulated the production of a wide range of problem-solving exercises, which constitute the urban transport planning process (Pass 1995) (see Figure 34.1). This varies in complexity and scope according to the urban area involved, and the efficacy of the planning process has increased with the number of disciplines that have contributed to the production of plans and policies (Dimitriou 1990a). In recent years, the environmental impact both of urban transport problems and of the solutions advanced has received particular attention (Banister 1994).

The earliest comprehensive urban transport plans date from the mid-twentieth century and were applied to large North American cities, where rapidly increasing volumes of motorised traffic were causing severe congestion. A demand-driven solution was commonly adopted, with the construction of new high-capacity urban expressways in city centres and suburbs (Muller 1995). By the early 1950s, car

Figure 34.1 Stages in the urban transportation planning process.

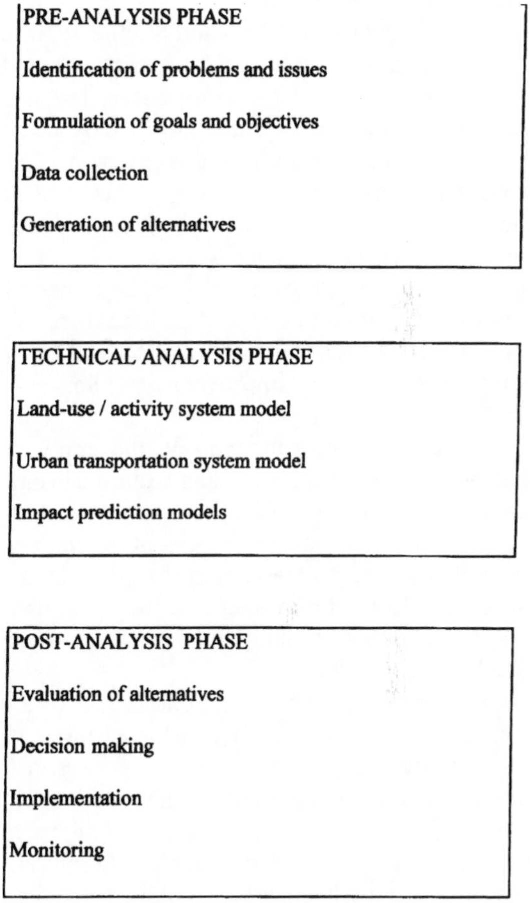

PRE-ANALYSIS PHASE
Identification of problems and issues
Formulation of goals and objectives
Data collection
Generation of alternatives

TECHNICAL ANALYSIS PHASE
Land-use / activity system model
Urban transportation system model
Impact prediction models

POST-ANALYSIS PHASE
Evaluation of alternatives
Decision making
Implementation
Monitoring

ownership and use in many Western European cities had also embarked upon a dramatic rise, and the initial response of urban planners was to adopt the North American approach and increase road capacity to meet the growing demand.

However, the physical patterns of many European towns and cities, often based upon medieval cores, hindered construction of the extensive urban highway networks advocated by American transport planners, and the results were often a compromise between the need to conserve the historic cores of inner cities and acceptance that private cars had become the dominant means of urban transport and required a more effective road system (Dunn 1981).

In Latin America, Africa and Southeast Asia, urbanisation in the last four decades has created some of the world's largest cities, but facilities for the movement of people and goods have failed to keep pace with physical expansion. Car ownership levels are significantly lower than in Western cities, and the principal transport problems are associated with inadequate public transport systems, the mixture of large volumes of motorised, animal-drawn and pedestrian traffic on unsuitable roads, and the lack of finance to support infrastructure improvements (Dimitriou 1990b).

In North America and Europe, the emphasis has now shifted from providing additional road capacity to devising policies for the restraint of private car use, accompanied by the upgrading of rail and bus undertakings to provide an alternative form of movement for the commuter and other users of facilities in central areas (Box 34.1).

Implementation of these policies has often generated strong opposition from the car owners' lobby, which enjoys the benefits of personal motorised travel within the city but is reluctant to accept that such travel is now exerting a negative impact upon the urban environment. However, promoting the advantages of public transport, coupled with programmes of car restraint, is now accepted by the transport planner as one of the most essential elements in future policy.

Contemporary patterns of urban transport, together with the strategies for planned improvements, must be seen in the context of private and public sector involvement. Public ownership and management of bus and rail undertakings has often been the basis of coordinated systems of transport designed to achieve an increased usage of such services. However, the recent trends towards deregulation and privatisation or liberalisation of urban bus and rail companies have replaced the environment of coordination with one of competition in an attempt to create more efficient services at lower costs to the consumer, but the effects of these changes have yet to be fully analysed. At the national level, government attitudes towards the public funding of urban transport improvement projects also have an impact upon local planning. In the United Kingdom, for example, the encouragement of joint funding initiatives for road and new rail building in urban areas was a feature of the early 1990s, but progress has been sporadic and very limited in extent. Given the very high costs of urban transport improvements, the availability of finance has always been a critical issue in both the Western and the developing world.

PATTERNS OF MOVEMENT IN URBAN AREAS

The various types of movement of people and goods within urban areas are achieved through the choice of appropriate modes of transport. Personal travel flows are the aggregate of individual trip-making decisions, each of which reflects the various needs of the urban population to gain access to a range of facilities and services. This accessibility is dependent upon the degree of personal mobility, which in turn is determined by the availablity of time, suitable transport, money and levels of physical mobility. Personal trips on a regular basis are dominated by journeys to work, an activity responsible for many of the more intractable transport problems. Journeys whose main purposes are education, shopping, social or leisure pursuits are also increasing in number and complexity, and the flexibility of

Box 34.1 San Francisco: reinvestment in urban rail to combat road congestion

San Francisco was the first major North American city to recognise that continued investment in urban freeways was not an effective answer to road congestion, and the opening in 1972 of a new rail rapid transit system initiated a period of public transport renewal and revival that has spread to other metropolitan centres in the United States and Europe. The city core of San Francisco–Oakland and the peripheral suburbs around the Bay are interconnected by a freeway network that suffers from increasing levels of congestion, particularly on the Bay Bridge, where cars and buses compete for space at peak hours. The Bay Area Rapid Transit (BART) is an urban rail network designed to attract commuters away from their cars, and it was estimated at the initial planning stage in 1967 that the new facility would carry 253,000 daily travellers (Figure 34.2). Two years after opening, however, the total was only 126,500, and it was not until 1993 that the forecast level of passengers was achieved. Difficulties in operating trains at high frequencies and coordinating bus services with BART timetables were partly responsible for the low usage in the first decade, but even after twenty-five years the effect of BART upon the city transportation system has been limited. At Bay Bridge, the peak congestion location, rail has replaced commuter bus services, but at the cost of encouraging more car traffic, and accessibility levels have only been substantially enhanced by BART in the central business district and in the East Bay suburbs. Experience of BART has shown that as more commuters use the system, with its suburban carparks, so highways become less congested and encourage other car-borne commuters to use them. The rail network has had little effect upon residential patterns, and whereas in 1970 about 40 per cent of the Bay area's 1.8 million workers lived in the BART-served zone, by 1990 only 33.8 per cent of the 3.02 million workers had elected to reside in this zone. Despite the failure of BART to make the expected impact upon San Francisco's travelling patterns, it can be regarded as the pioneer initiative in the current phase of urban public transport renewal, and it is significant in the contemporary context of environmental concern that the reduction of traffic-induced air pollution and energy, as represented by wasteful consumption of car fuels, were two of the major aims of the original BART project (Fielding 1995).

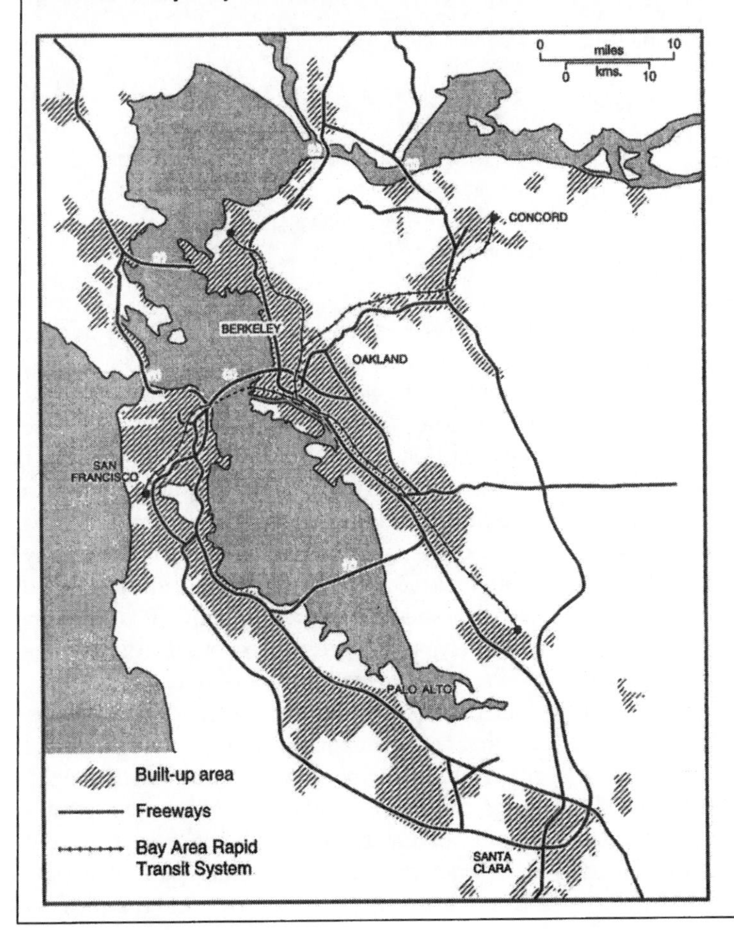

Figure 34.2 The San Francisco Bay Area Rapid Transit (BART) system.

travel conferred by the car has created a large group of multi-purpose trips carried out within urban areas.

Changes in land use, such as the trend towards establishing large retail complexes, and industrial and business parks on the periphery of urban areas, have altered the traditional centripetal/centrifugal nature of many personal flows and created new journey patterns, with the majority being made by car rather than by public transport, which is less able to cater for new directions of travel.

Freight traffic assumes a complex pattern, with the transfer of finished or semi-finished products between factories in towns and cities and the export and import of raw materials and products between urban centres. Scheduled and irregular goods distribution within urban areas involves deliveries to individual households, to scattered small retail locations and to larger shopping units or retailing complexes.

Current modal choice for personal and freight movements illustrates the conflict between the consumers' preferred means of travel and the transport planners' perception of the most efficient blend of modes, public and private, road or rail, for a particular city. The search for the most suitable methods of meeting demand with what are recognised as the most efficient transport modes is at the heart of the urban transport planning process, and a set of solutions must be found and applied so as to satisfy movement needs without seriously damaging the overall urban environment. The major problems are closely interrelated and can be identified to varying degrees of seriousness in all major towns and cities. Road congestion is a primary problem (Plate 34.1), caused by the overloading of urban highways by private cars, buses and commercial vehicles, and is most apparent during the journey-to-work periods. Pressure on inner-city car-parking facilities is closely linked to congestion, and plans for car restraint must involve both mobile and parked vehicles. In third world cities, the difficulties caused by congestion are exacerbated by the mixture of motorised and other traffic on the roads. Problems in these cities

Plate 34.1 Traffic congestion in central Glasgow (*photograph: P.T. Kivell*).

will become much more serious if the current forecasts for national car growth in developing countries are realised: in Malaysia, for example, the total number of cars is likely to rise sixfold between 1986 and 2025 (Button and Ngoe 1991). Congestion is also common on public transport systems, particularly in the morning and evening peak periods, when bus and rail vehicle capacity is often exceeded. This problem can occur even when these systems are operating at maximum frequency, and overcrowded buses in turn contribute to the overall pattern of road congestion.

Declining patronage of urban public transport is a direct consequence of the continuing increase in private car trips in towns and cities. In the period before road congestion reduced the attractiveness of car travel, municipal and privately owned rail and bus undertakings saw their passenger markets begin to decrease, and this process has continued through to the late 1990s in most urban areas. A cycle of decline is established, in which declining revenues are countered by reduced frequencies and increased fares, which in turn dissuade more passengers, who then make use of the car, either as drivers or passengers. Bus fleets and staff numbers are reduced in response to declining incomes, making it difficult to maintain a satisfactory service level throughout the day and thereby reducing the appeal of public transport even more.

In the third world there is a rising market for bus travel, but in many major urban areas the existing vehicle fleets and service patterns cannot meet demand, creating unacceptable levels of congestion on buses, and long waiting periods and hence unduly protracted journey times (Table 34.1).

In both the developed and developing worlds, many groups within the urban population are prevented by age, income or personal disability from using cars and continue to rely upon public transport systems that can no longer cater to their needs (White 1990). The road safety issue is particularly linked to the elderly, especially pedestrians, but it also affects cyclists and young children, whose opportunities for walking are becoming further eroded with the increase in motorised traffic.

The urban transport planning process is devoted to identifying problems, devising and evaluating a range of solutions, and implementing what are judged to be the most appropriate within cost constraints and predetermined time periods (Figure 34.1). Remedial action can take the form of traffic management, the upgrading of public transport, the introduction of 'green transport' strategies, transport coordination and policies designed to reduce the necessity for travel through the introduction of 'non-transport solutions'. The latter include extending the opening periods of essential services such as post offices, local government offices, health centres and solicitors into the evenings to enable urban workers to make use of them after conventional working hours. Each of these remedies can be applied at different times during the period of a transport plan and in different locations, but all are seen as part of the overall strategy for overcoming contemporary transport problems (Tolley 1997).

At its basic level, traffic management alleviates current congestion and increases road capacity by such measures as one-way systems, restrictions

Table 34.1 Private car and bus transport in major cities of the developing world.

City	*Population 1985 (000s)*	*Cars per 1000 population in 1980*	*Percentage increase in cars 1970–80*	*Buses per 1000 population in 1980*
Bangkok	6100	71	7.9	1.22
Bogota	4500	42	7.8	2.13
Bombay	10,100	21	6.1	0.36
Buenos Aires	10,900	53	10.0	1.20
Cairo	7,700	32	17.0	1.10
Calcutta	11,000	10	5.6	0.33
Hong Kong	5100	39	7.4	1.83
Jakarta	7900	33	9.8	0.72
Karachi	6700	35	8.4	2.32
Manila	7000	45	8.0	5.30
Mexico City	17,300	105	–	1.23
Rio de Janeiro	10,400	104	12.1	1.20
São Paulo	15,900	151	7.8	1.28
Seoul	10,300	15	11.7	1.55

Source: Dimitriou 1990.

Box 34.2 Harare: new initiatives in public passenger transport

The urban structure of Harare still retains a rigid division between low-density European housing areas and the high-density residential townships provided for the African population. The bulk of car travel in the city, accounting for about 14 per cent of all passenger trips, is carried out by white residents, with the low-income African workers relying upon either the bus or various forms of paratransit or walking for access to the industrial zones, the city centre or the suburbs, where many are employed as domestic staff (Armstrong-White 1993). Most of the African housing areas are on the western and southwestern periphery of the city, but over 280,000 also live in the town of Chitungwiza, 20 km to the south of the capital (Rakodi 1995).

Scheduled city bus services provided by the Zimbabwe United Passenger Company (ZUPCO) have been improved with the purchase of new vehicles and an increase in bus productivity and now account for 20 per cent of all trips. However, protracted delays during journey-to-work periods are still frequent, and the 'emergency taxi' is an essential alternative. These vehicles operate at government-controlled fares on prescribed routes between the suburbs and the city centre and account for 10 per cent of all trips.

In 1993 the public transport system was diversified with the introduction of 'commuter omnibuses', a private initiative sanctioned by the state as part of its economic structural adjustment programme. Most of these minibuses link low-income housing areas with the centre, and it was estimated in 1994 that commuter omnibus services, during their first three years of operation, had expanded to account for 16 per cent of all trips. The 'emergency taxis' have lost custom as a result and have concentrated upon shorter journeys (Maunder and Mbara 1995).

Various proposals for busways and a light rapid transit system for Harare have been made, but existing passenger flows are insufficient to justify the necessary investments, and low-income urban workers could not afford the travel costs. The Chitungwiza–Harare corridor, with its heavy commuter traffic, would appear to be a suitable candidate for a rail-based facility, but any available funds are better devoted to improvements of the ZUPCO bus services.

on parking on through roads, reversal of traffic flows on multi-lane highways and electronic co-ordination of light-controlled intersections within inner urban areas. Traffic management can also aim to alter modal choice by introducing strategies for car restraint in combination with measures to improve bus circulation. Vehicle restraint involves filtering schemes that permit only fully loaded cars into inner cities or a progressive increase in car-parking charges as the core is approached. Car drivers can also be discouraged from using inner urban roads by the imposition of tolls or special licences payable according to the length of time spent in city centres.

Buses can gain an advantage over cars by using bus-only lanes on major access roads, priority turns at intersections and by having exclusive rights in selected city centre streets.

Box 34.3 Singapore: achieving a compromise between public and private urban travel

Urban transport planners in this densely populated city-state of 3 million people face the major problem of providing for heavy volumes of commuters from the city and its satellite towns into a highly congested central business district. The current policy is based upon investment in high-capacity public rail and bus services, coupled with the restraint of private car travel within the inner city. Most of the population is now within five minutes access of a comprehensive bus network, with over 250 routes operated by one major and a group of smaller companies, which also provide contract services for employees and schools. Feeder and trunk routes, together with bus priority lanes in the inner city, ensure the most efficient use of the buses, and 80 per cent of services have intervals of ten minutes or less. Increasing demands for peak hour travel are met by the mass rapid transit network, the first two routes of which were opened between 1987 and 1992. Feeder bus services focus upon many of the stations, and extensions are planned so that by 2000 ridership could exceed 950,000 and the system could cater for 33 per cent of all public transport journeys (Turton and Knowles 1998).

Singapore pioneered fiscal measures of car restraint in urban centres with its Area Licensing Scheme, introduced in 1975. A restricted zone of the city centre can be entered at peak periods only by vehicles with a supplementary licence, but an electronic pricing scheme is now being introduced. This demand management initiative has reduced car traffic in the city by 20 per cent since 1975, although the city labour force has risen by one-third, and the average vehicle speed of 26 kph in peak periods is higher than those in London, New York and Hong Kong. In addition to car usage restraint, a vehicle quota system was introduced in 1990 in order to restrict car ownership levels, and both measures are seen by the government as part of the overall strategy for encouraging a greater use of the upgraded public transport facilities.

The public transport sector as a whole can receive more positive support by upgrading existing metro systems and by the introduction of light rail transit (Plate 34.2). Metro rail systems are generally confined to urban areas of over 500,000 and are currently operating in over ninety cities, with twenty-eight more planned or being built. Extensions to these systems are expensive, particularly where underground routes are necessary, but new lines are being built, especially where they are seen as contributing to urban revival and regeneration plans in addition to providing additional passenger capacity (Hall and Hass-Klau 1985).

Plate 34.2 Metrolink light rail rapid transit in central Manchester.

Light rail is being increasingly adopted as an alternative to a metro as it is cheaper, more easily adapted to city centre road networks and offers greater levels of access to passengers. Within city centres, at least 40 per cent of the light railway must be provided with an exclusive right-of-way to avoid delays from traffic congestion, but elsewhere light rail, buses and cars share the highway until the fringes of the core are reached, when light rail vehicles often make use of conventional rail tracks (Table 34.2). About 117 cities, mainly with populations of between 100,000 and 1

Table 34.2 Urban tramway, light rail, metro and suburban rail systems.

	Streetcars/trams	*Light rail*	*Suburban rail*	*Metro system*
Route characteristics:				
Route length from CBD (km)	Under 10	Under 20	Under 40	Under 24
Track position	On street	Over 40 % segregated	Segregated	Segregated
Access to CBD	At surface	At surface	At surface to CBD edge	Underground
Station spacing in suburbs	350 m	1 km	1–3 km	2 km
Station spacing in CBD	250 m	300 m	–	500 m–1 km
Maximum gradients (%)	10	8	3	3–4
Minimum radius of track (m)	15–25	25	200	300
Rolling stock:				
Number of carriages	1 or 2	2 or 4	Up to 12	Up to 8
Capacity of each carriage	50 seated, 75 standing	40 seated, 60 standing	60 seated, 120 standing	50 seated, 150 standing
Carriage access	Step	Step or platform	Platform	Platform
Performance:				
Average speed (km per hour)	10–20	30–40	45–60	30–40
Maximum speed (km per hour)	50 to 70	80	120	80
Typical peak headway (minutes)	2	4	3	2–5
Maximum number of passengers carried per hour	15,000	20,000	60,000	30,000

Source: Adapted from table in Knowles and Fairweather 1991.

million, now have light rail systems, which are currently regarded as one of the most effective means of upgrading urban public passenger transport (Turton and Knowles 1998).

Growing concern for the urban environment has stimulated attempts to reinstate and encourage walking and cycling as acceptable modes of transport. Both modes have been largely displaced by the car in Western cities, but urban transport planners believe that if safer facilities, such as pedestrian walkways and defined cycle tracks, can be provided then more trips can be made by non-motorised transport. In the United Kingdom, many of the 60 per cent of car journeys that are less than 8 km in length could be carried out by cycle, and walking could be used for a

Box 34.4 Manchester: the introduction of light rail transit to a British conurbation

Greater Manchester, with a population of 2.58 million, has adopted light rapid transit to help to solve problems of city centre road congestion and upgrade the public transport network in the city and surrounding towns and suburbs. The conurbation has a suburban rail network of 262 km, the largest of any English urban area outside London, but until recently the northern and southern components were unconnected, causing difficulties for passengers using both local and long-distance rail services passing through the city centre. Rail improvements have been made during a period of change in road-based transport policy, with rejection of the more ambitious highway schemes and the replacement, after 1986, of the period of public transport coordination by bus deregulation and privatisation.

During the 1980s, several city centre streets were pedestrianised, and bus-only access routes were introduced as part of the Central Manchester Traffic Plan. The two separate sections of the conurbation rail network were linked in 1988, and in 1992 the first part of the Metrolink light rail network was completed, making use of two existing suburban railways connecting Altrincham and Bury with the central termini and a short section along city centre streets and on reserved track (Figure 34.3). At peak periods, the Metrolink has a capacity of 2500 persons per hour in each direction, and in contrast to the BART system passenger flows on Metrolink have exceeded forecast levels, with an estimated 2.6 million car journeys having been captured. In particular, Metrolink has benefited from the existence of major rail/bus/car interchanges at its two outer termini of Altrincham and Bury. The network is to be extended to other parts of Greater Manchester and can be seen as one of the more successful transport initiatives in the post-deregulation period (Knowles 1996).

Metrolink has been operating during a decade of substantial change in bus transport, with the former PTA-owned Greater Manchester Transport's share of bus-kilometres declining from 97 per cent in 1985 to 66 per cent in 1991. It is still the dominant operator, but many private companies, including several with minibus fleets, now offer strong competition in the city centre, the suburbs and other towns such as Rochdale and Stockport, although service quality has declined and the number of bus passengers entering the city centre each day has fallen with increased use of the car.

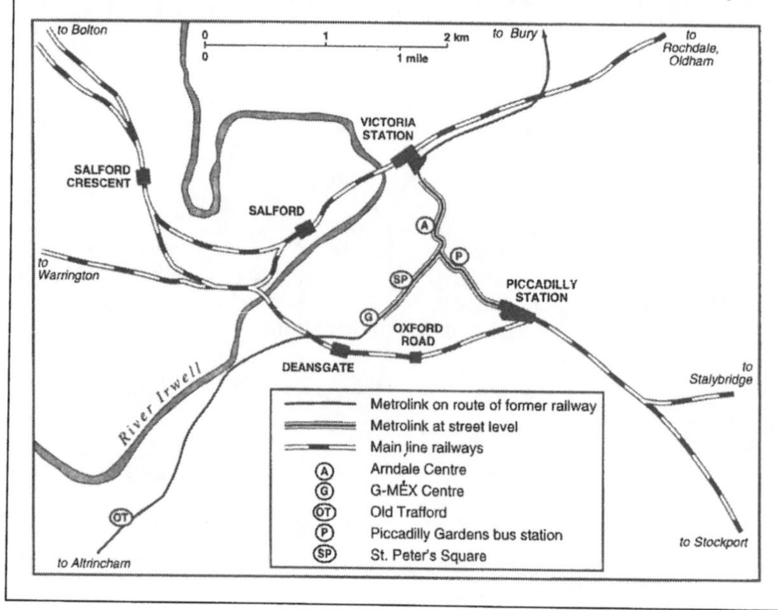

Figure 34.3 The Manchester Metrolink light rail system.

substantial share of the 60 per cent of trips by all modes that are less than 5 km. Many Dutch towns, and notably Delft, have promoted cycle routes, but much remains to be accomplished within this 'green transport' initiative (Tolley 1997).

Many of the more ambitious transport planning programmes have been implemented by municipal authorities that exercise administrative control over entire cities or conurbation areas, and also own and manage the public transport undertakings. This enables transport planning and policy to benefit from the coordination of services between rail and bus, thus ensuring that every effort can be made to provide urban travellers with an efficient and integrated transport system. Major cities in the United States and the United Kingdom have established such transport authorities, and investment has been coordinated to provide major facilities such as the Tyne and Wear metro and electrification of Glasgow's railways. However, transport coordination is subject to national political decisions, and in the United Kingdom it has been displaced since the early 1990s with the deregulation and privatisation of the conurbation transport authorities.

CONCLUSION

Recent decades have seen a variety of innovations directed towards solving some of the more serious problems in urban transport. It is likely that the continuing evaluation of these initiatives will be an essential part of future research activity in applied geography and related disciplines, especially as the investment of private capital in so many of these schemes will require an appraisal of economic and social return rates. New light rapid transit systems, for example, will require careful evaluation both as separate undertakings and in terms of their contribution towards alleviating citywide transport problems.

The continuation of efforts to integrate transport and land-use planning more effectively will depend substantially upon a geographical input, particularly in the context of measuring and analysing both personal transport requirements and traffic generation from industrial and commercial zones within urban areas. If the objective of reducing overall levels of trip generation within urban areas can be only partially achieved, then the pressure on transport resources will be lessened. Such pressures can also be reduced if the approach to transport problems involving 'non-transport' solutions can be further developed by extending the time periods in which facilities are available to the consumer. Shopping hours in many retail parks are already organised on a seven-day basis, and 'flexi-time' has been adopted in an increasing number of workplaces, but the overall impact upon urban traffic congestion is still slight. An extension of home-based working practices using electronic communications with head offices will also aid the reduction of journey-to-work flows, and what has been described as 'the death of distance' (Cairncross 1997) is gradually being achieved as the need for and amount of personal contact is reduced with increasing use of information technology.

Although the problems of traffic congestion and environmental pollution caused by the rising use of motor vehicles are still the primary concern of transport geographers and planners, the difficulties faced by groups within urban communities who will continue to rely upon public transport for most of their trips must also be addressed. These 'transport-deprived' sections of the urban population suffer from the continuing reduction in the quality of public bus and train services, particularly in areas where decentralisation of retail facilities has reduced their choice of shopping in town centres. In the growing cities of the third world, where income levels still make most workers dependent upon public transport, there is an urgent need for investment in bus and rail services to cater for peripherally located communities and decrease the time spent in daily travelling.

At the national policy level, the deregulation and privatisation of public urban transport have often eroded many of the advantages to the consumer that were offered by an integrated and coordinated transport system, where subsidies ensured an acceptable standard of services. Although competition within modes, particularly bus transport, has often improved conditions on the more profitable routes, other areas have suffered a decline in service levels. The application of new technologies such as light rapid transit has improved the efficiency of urban transport along some major routes, but the underlying problem of satisfying the demand for acceptable levels of movement within cities at all times and in all areas has still to be tackled. Many urban dwellers still see this demand as being met most conveniently by the continuing use of the car, but this practice, if left unchecked, will eventually result in unacceptable and chaotic levels of urban congestion that can be resolved only by a reduction in overall travel and a greater use of public transport for essential journeys. The long-term reorganisation of urban land-use patterns in order to achieve a decrease in both the number and length of personal trips will require a vital input from the applied geographer, whose spatial skills will also be necessary in the planning and designing of future transport systems, which will seek to minimise the use of the private car and maximise the role of public and non-motorised transport.

GUIDE TO FURTHER READING

Banister, D. (1994) *Transport Planning in the UK, USA and Europe*. London: E. & F.N. Spon. Deals with the development of public passenger transport planning, together with a comparative analysis of its application to four leading industrial nations and a discussion of issues to be resolved in the twenty-first century.

Dimitriou, H.D. (ed.) (1990) *Transport Planning for Third World Cities*. London: Routledge. Provides a comprehensive review of current urban transport planning practice in developing countries.

Hanson, S. (ed.) (1995) *The Geography of Urban Transportation*, 2nd edition. New York: Guilford Press. Discusses both the theory and practice of urban transportation in a North American context.

Tolley, R.S. (ed.) (1998) *The Greening of Urban Transport: Planning for Walking and Cycling in Western Cities*. London: Belhaven. Presents a world viewpoint on current policies and planning strategies for developing the 'green modes' in cities of Europe and North America.

White, P. (1995) *Public Transport: Its Planning, Management and Operation*, 3rd edition. London: UCL Press. Provides a practical guide to all aspects of public transport policy, planning, organisation and finance in British urban areas.

REFERENCES

Armstrong-White, A. (1993) *Public Transport in Third World Cities*. Transport Research Laboratory, London: HMSO.

Button, K.J. and Ngoe, N. (1991) *Vehicle Ownership and Use Forecasting in Low Income Countries*. Transport Research Laboratory, London: HMSO.

Cairncross, F. (1997) *The Death of Distance: How the Communications Revolution will Change our Lives*. London: Orion.

Dimitriou, H.T. (1990a) The urban transport planning process. In H. Dimitriou (ed.) *Transport Planning for Third World Cities*. London: Routledge, 144–83.

Dimitriou, H.T. (1990b) Transport problems of Third World cities. In H. Dimitriou (ed.) *Transport planning for Third World Cities*. London: Routledge, 50–84.

Dunn, J.A. (1981) *Miles To Go: European and American Transportation Policies*. Cambridge: Lexington.

Fielding, G.J. (1995) Transit in American cities. In S. Hanson (ed.) *The Geography of Urban Transportation*, 2nd edition, New York: Guilford, 287–304.

Hall, P. and Hass-Klau, C. (1985) *Can Rail Save the City? Rail Rapid Transit and Pedestrianisation in British and German Cities*. Aldershot: Gower.

Knowles, R.D. (1996) Transport impacts of Greater Manchester's Metrolink light rail system. *Journal of Transport Geography* 4(1), 1–15.

Knowles, R.D. and Fairweather, L. (1991) *The Impact of Rapid Transit*, Metrolink Study Working Paper 2, Department of Geography. Salford: University of Salford.

Maunder, D.C. and Mbara, T.C. (1995) *Initial Effects of Introducing Commuter Bus Services in Harare, Zimbabwe*. Crowthorne: Transport Research Laboratory.

Muller, P.O. (1995) Transportation and urban form. In S. Hanson (ed.) *The Geography of Urban Trans-*

portation, 2nd edition, New York: Guilford, 26–52.

Pass, E.I. (1995) The urban transportation planning process. In S. Hanson (ed.) *The Geography of Urban Transportation*, 2nd edition, New York: Guilford, 53–80.

Rakodi, C. (1995) *Harare*. Chichester: Wiley.

Rimmer, P. (1986) *Rikisha to Rapid Transit: Urban Public Transport and Policy in Southeast Asia*. Oxford: Pergamon.

Tolley, R.S. (1997) *The Greening of Urban Transport*, 2nd edition. London: Belhaven.

Turton, B.J. and Knowles, R. (1998) Urban transport problems and solutions. In B.S. Hoyle and R. Knowles (eds) *Modern Transport Geography*, 2nd edition. Chichester: Wiley, 192–221.

White, P.R. (1990) *Public Transport*, 3rd edition. London: UCL Press.

35

Rural accessibility and transport

Stephen Nutley

INTRODUCTION: THE PROBLEM

Living in the countryside confers advantages and disadvantages that are experienced in very unequal proportions by different groups of people. Those for whom the benefits outweigh the difficulties include people who have control over rural resources and people whose incomes are derived elsewhere and who may have moved into the countryside voluntarily to enjoy its environment and amenities. Those for whom the difficulties outweigh the benefits include people who are dependent on rural resources, such as working the land, but have no control over them. The latter might be identified with 'traditional' communities that have always lived in rural localities. Problems of rural areas range from the macro-scale to the extremely localised. The first type result from rural areas' subordination to external forces, their economic and political weakness, and their peripherality (Marsden *et al.* 1993; Hoggart *et al.* 1995; Ilbery 1998). It is the local problems, however, that bear upon the struggle for day-to-day living, especially for the more vulnerable social groups. Because of the nature of rural areas, the activities that people habitually undertake include many that involve making journeys to other places for normal everyday purposes. While the ability to make such trips may be taken for granted in the city, in rural environments the difficulties of doing the same are frequently so great as to cause hardship and isolation for many people.

While *accessibility* as a spatial concept is universal, it is made particularly acute as a social issue in rural areas due to the inherent characteristics of 'rurality' itself. These are a relatively low population density, a dispersed settlement pattern with low population totals at any point, a scattered pattern of small service outlets, a concentration of middle- and high-order facilities in widely separated urban nodes, and hence long and costly travel distances. In pursuit of a 'normal' lifestyle, people need to consume a range of goods and services, to get to work, to make shopping trips, to use medical, financial and information services, and to take part in social and recreational activities. Under rural conditions, only a small proportion of these needs will be achievable within walking distance of home (and even this makes certain assumptions about health and physical fitness), while a greater proportion will require some form of *transport*. An aggravating factor is the continuing trend of closures of economically marginal consumer service outlets – shops, post offices, etc. – in rural areas (Clark and Woollett 1990), which means that local residents need to make *more* journeys than before. Poor access to services in the transport sense, as well as in the broader economic sense, contributes to a syndrome of problems known as 'rural deprivation' (e.g. Pacione 1995).

The same situation could be tackled from the 'transport' viewpoint. The ideal mode for conditions of 'dispersed demand' is the privately owned motor car, which rural dwellers in affluent developed countries have adopted in great numbers. Obvious advantages are freedom of choice of route and timing, the ability to carry heavy loads, sheer convenience, and flexibility. In

Plate 35.1 A typical village general store including franchised post office, Suffolk, England. While providing easy access, on foot, for village dwellers, such facilities, where they survive, are inadequate for most people's aspirations and transport to town is normally thought essential. The community bus parked alongside is a recent alternative to declining public bus services, and is run by local volunteers for specific travel needs.

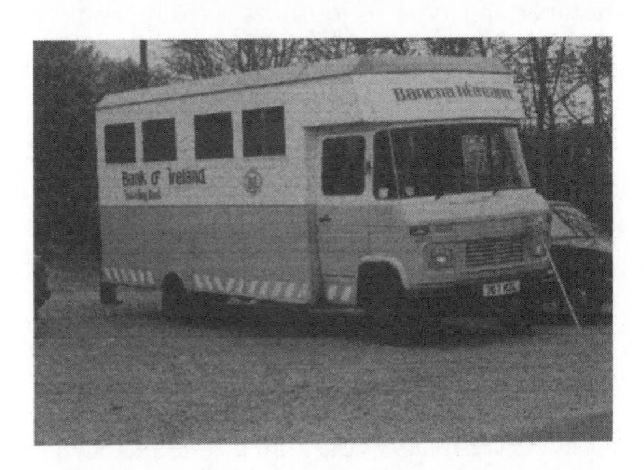

Plate 35.2 A mobile bank, County Clare, Ireland. While access to services for rural dwellers usually means travel into town, there are alternative means of service delivery. One such is 'bringing the service to the people' by mobile or peripatetic means, more common in remoter areas.

the countryside, there are fewer constraints on car use, such as congestion or parking problems, and pollution is rarely an issue. However, it is *need* rather than wealth that causes the higher car ownership rates always found in rural areas, compared with urban, in the more affluent countries. In the United States, car ownership extends to 95 per cent of rural households, including 67 per cent with two or more vehicles. Rural districts of the United Kingdom have car ownership rates between 68 and 80 per cent by household, which are fairly typical of Western Europe. It is therefore easy to assume that car owners have no real difficulties of accessibility in rural areas, apart from longer distances and associated costs, and following this, also to assume that only relatively small minorities deprived of a motor vehicle are left to suffer 'problems'.

As car ownership increases over time, it is very important to refute any complacency over the numbers of people likely to be affected by the rural accessibility (or transport) problem. With saturation coverage of automobiles in the USA, the widespread perception is that no problem exists, except for a small non-car population of elderly people, the disabled and the very poor (Maggied 1982; Nutley 1996). But even here, some authors draw attention to other problematic groups such as young people below driving age, and individuals within households where the number of vehicles is insufficient for everyone's needs (Kidder 1989). Perhaps peculiar to America is the recognition of the burdens of rural car owners: there are constant complaints about costs (ironically in a country where gasoline is notoriously cheap); elderly people who dislike driving are forced to become 'reluctant drivers' in the absence of any public transport alternative (Kihl 1993).

It is unsurprising that concern about rural accessibility is strongest in countries where larger proportions of the population are deprived of the regular use of a car and are dependent on public transport; also, such concern originated in an earlier era, when car ownership was lower. In the UK, the issue arose around 1960, when rural bus and train services began to decline from their previously generous levels (Thomas 1963). Since then, rising car ownership and declining public transport have exposed a widening social gulf between haves and have-nots. It is common to categorise the non-car population as the elderly, the disabled, those on low incomes, the unemployed, those living alone, children and adolescents. However, it is important that identification of transport needs is based upon *individuals* rather than households, such that in a family owning one car, a journey to work might be catered for, while someone else's shopping trip, school journey or evening entertainment would probably require a transport alternative. It has been recognised, belatedly, by governments that such problems can never be 'solved' by rising car ownership, and there will always be a need for at least a basic level of public transport.

From the point of view of public transport operators, rural areas have always been difficult territory. There is a basic incompatibility between the type and scale of transport provision and the dispersed nature of demand. The fixed routes and timetables of trains (especially) and buses, unsuitably large vehicle capacities, and formal operating procedures have left public modes without the flexibility to adapt to rural conditions. For decades, transport companies have found it impossible to operate commercially, and have been forced to withdraw services and/or negotiate subsidies from local authorities. Attempting to reconcile demand that is highly dispersed in time and space with transport supply that is inevitably concentrated has proved as fascinating for geographers as it is frustrating for operators and planners.

CONCEPTS: ACCESSIBILITY AND MOBILITY

Geographers and other researchers interested in analysing problems of this type have a choice of two fundamentally different approaches. *Accessibility* is essentially a measure of 'spatial opportunity', the ability of people to get to places under prevailing conditions, in terms of *what is possible*, whether or not these journeys are actually made. Relevant data can be obtained from secondary sources – maps, the census, transport timetables – and, depending on the technique used, analysis could be done largely as a 'desk study' with relatively little fieldwork in the study area. The alternative approach is to focus on the *travel patterns* of people in the target area, i.e. *mobility* (see below), and from the number and type of journeys actually made to deduce the extent of any disadvantage experienced. Data on trip-making behaviour can only be obtained from household questionnaire surveys, which obviously demand a lot of fieldwork, hiring teams of interviewers, and considerable expense. Although it could be argued that complete understanding requires both strategies to be applied in the same place at the same time, this is hardly ever done in practice.

The contribution of geographers over the years has been to convince policy makers of the importance of accessibility as the central concept, that the problem resides with *people* at local level, and that it is essential to take a consumer viewpoint. This differs crucially from the traditional interpretation, which was overwhelmingly economic, such as a preoccupation with the financial problems of bus companies or railway branch lines. The purpose of transport is to provide accessibility to distant places, where people can obtain goods and services or take part in activities that are not possible at their home location. Transport itself is not the desired product but a means to an end. Its role is to overcome the distance barrier that separates the point of demand (e.g. an isolated house) from the point of supply (e.g. a market town). Accessibility should be seen as a system that coordinates the settlement pattern, the locations of consumer services and transport linkages.

Such a consumer view requires planning to be driven not by demand, as at present, but to be 'needs-based'. Questions of access to what, where, and how often, should not be circumscribed by existing travel patterns but based on an assessment of people's needs. The concept of transport *need* is a difficult one (Bradshaw 1974; Koutsopoulos 1980) and may be expressed by perceptual, comparative or normative methods (see below). Measured in terms of accessibility, a certain proportion of needs will be satisfiable by the facilities and transport in any area, but the residue of *unfulfilled needs* can be taken to represent the 'problem'. It is vital to distinguish needs from 'demand'. As understood by economists, the demand for transport is the amount actually consumed; if this was equivalent to need then no problem would exist. In rural areas, many travel needs are frustrated because of the lack of suitable transport services (*suppressed* or *latent* demand). The whole point is that total demand exceeds transport supply.

There is a similar impasse with studies of mobility. *Mobility* is simply the ability to move around, regardless of destination or purpose. *Potential mobility* is determined first by health and fitness factors, and second by the availability of transport, either private or public. *Actual mobility* refers to the amount of travel undertaken, usually measured by traffic levels or trip rates (trips/person/week). Results from questionnaire surveys on travel behaviour therefore tell us more about the successfully mobile sectors of society, and much less about disadvantaged groups. Trips that cannot be made are not recorded.

TECHNIQUES AND CASE STUDIES

The rest of this chapter is devoted to a review of the techniques developed by geographers to evaluate rural accessibility and related concepts. Policies, on the other hand, have shown less sign of geographical influence. It should be pointed out at this stage that the extent of geographical interest in the subject is by no means evenly spread around the world. Far more work has been done in the UK than in any other country, mainly between 1970 and 1985. This is probably because of the tradition of public services, and great awareness of the decline of public transport, which had previously been abundant. For a general survey of historic trends, transport policies and accessibility applications see Moseley (1979) and Nutley (1992; 1998). In other countries, priorities or cultural values may be different, data might not be available, or the techniques might not be transferable.

Simple methods

The simplest way of indicating accessibility is to present a map depicting 'all areas beyond *x* km of a major road' (or a railway, or a bus route). This can be improved by calculating 'percentage of population beyond *x* km of a transport route'. This is obviously still of limited use. For example, only about 5 per cent of the population of rural Britain is beyond the reach of a bus or train service, which might sound reasonable, but this does not say whether the transport goes to the right places at the right times and with sufficient frequency. Density measures are sometimes seen, for example '*x* km of road per 100 sq km'. These are useful only where there is a lack of further data, such as in a historical situation or in the case of under-developed regions (Box 35.1).

Network methods

Since the 1960s, network analysis has been used to express the accessibility of places relative to others in the same regional system. This involves abstracting a network of roads or public transport routes, converting it to matrix form and working out the shortest paths between each node and every other. Row totals in the matrix then constitute the relative accessibility of each node; these may be mapped as isarithmic surfaces. An example is Lannoy and Oudheusden (1978), but such techniques are now less popular. A common extension is to weight nodal values by population, to give 'population potential', which is a

Box 35.1 The measurement of accessibility in the developing world

In the less developed countries of the world, rural areas are frequently vast in scale and support very large populations, but have extremely sparse networks of surfaced (all-weather) roads, and hence there are severe restrictions on the availability of motorised transport. In such environments, accessibility has a completely different meaning to that in the developed world, and the priorities may be access to clean water supplies, fuel resources, local markets, schools, clinics and hospitals. Even where buses can be operated, they are very infrequent and unreliable; some form of truck, animal-drawn cart or bicycle may be a more common means of transport. Suitable methodologies to deal with this are poorly developed, and data sources are rarely adequate. Simple distance and walking times may have to be used. Hence Ogunsanya (1987) is forced to use road density to indicate accessibility in a part of Nigeria (Figure 35.1); in the circumstances, this is not an unreasonable surrogate for the difficulties in such a region.

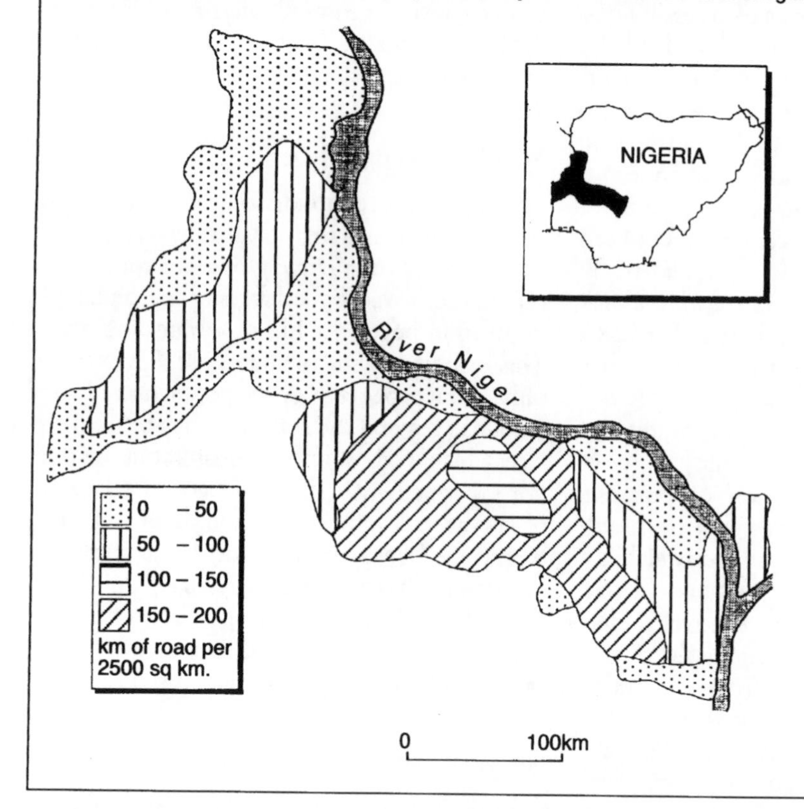

Figure 35.1 Accessibility to motorable roads, Kwara state, Nigeria. (The population density of this region is roughly 40 persons per km^2).

Source: Ogunsanya 1987.

basic simulation of the demand for travel or the 'traffic potential' for transport operators. Another variation is to plot the number of opportunities (destinations such as shopping centres, doctors, etc.) within *x* km, or *x* minutes travel time, via the network, from each origin point. Suitable modifications can deal with a variety of situations, e.g. Joseph and Bantock (1984) on accessibility to doctors in Ontario, Canada. Network-based techniques are aggregative, 'geometric' measures more suited to the regional scale.

Service indices and travel times

There are various simple means of assessing public transport services. Indices of bus or train service frequency (per day or per week) at specific locations represent potential mobility, as they say nothing about access to other places. 'Percentage of villages (or population) with/without a post office, school, etc.' indirectly suggests likely problems of local access (Clark and Woollett 1990; Cloke *et al.* 1994). 'Percentage of population within x minutes travel time of a shopping centre' is a slight improvement. Accessibility by public transport is most easily gauged by scheduled journey times. The relative status of a number of villages, for example, can be discerned by comparing travel times from each to a single regional centre. It is more realistic, however, to calculate timings to the nearest intermediate-sized town (a predictable shopping destination), which also allows 'travel time hinterlands' to be defined around each.

It should be emphasised that at the regional scale, no single measure of accessibility is adequate by itself. Combinations of methods under the last two sub-categories are best. See, for example, Nutley (1979; 1984) on the Scottish Highlands and rural Wales.

Local accessibility and the time–space approach

To be more relevant to the experience of people at grassroots level, traditional techniques need considerable improvement. First, a much higher resolution of study is required, so that localised variations can be detected. Second, it must be possible to distinguish access for specific journey purposes. Third, a time dimension must be included, so that it can be checked that transport arrives at the destination at times suitable for the desired activities. In the first instance, such an approach aims to assess *public* transport for that section of the population dependent on this. If one is dealing with a large area, it must be divided into small units such as parishes or wards. Transport services available to these units are then examined to see whether they connect with suitable towns at times and frequencies appropriate for work, shopping, leisure, etc. Localities with/without the requisite access can then be mapped. Furthermore, by using a points-scoring system, accessibility for different journey purposes can be summed to produce a 'composite access index'. Ideally, the latter should be qualified by some measure of car ownership. For applications, see Nutley (1980) on rural Wales (Box 35.2) and Jordan and Nutley (1993) on Northern Ireland.

However, in order to identify *precisely* where problems of accessibility are occurring (e.g. population sub-group A in village B has inadequate access to facility C in town D), and hence aid policy solutions (e.g. re-timing a bus service), it is necessary to focus on small case study areas. Points of demand are specific settlements such as villages. Destinations (supply points) are also more specific: hence, instead of 'shopping' or 'leisure', access is postulated to different shop types and leisure outlets at different levels in the urban hierarchy, not just the nearest. It is also very important to disaggregate the population into sub-groups likely to have different travel needs, such as working people, housewives, the elderly, schoolchildren, etc. Needs, in terms of destination facilities and desirable frequency of access, are defined by setting normative standards (for other needs measures, see Bird (1981) and Moyes (1989)).

Ensuring adequate accessibility is an exercise in time–space coordination, not merely making connections over physical space but also taking account of available time. Useful techniques are derived from 'time geography', developed by Swedish geographers in the 1970s. All individuals have a 'time budget' such that after domestic duties, work or school, a certain time is left for travel; time in transit is deducted from this, and that remaining must allow sufficient time at the destination and must accord with the facility's opening hours. Standards are pre-defined by the analyst and applied uniformly: times available for travel, maximum walking distance, minimum stay time at destination, and frequency (days per week) that the facility is 'needed'. Car users and public transport users must be studied separately. For each village, there is a matrix of social groups versus needed facilities. In each cell, it is recorded whether or not that group can successfully achieve access to that facility under the defined conditions. Accessibility is the percentage of desirable social group/facility contacts that can be satisfied by the prevailing transport system.

Such methodologies have been used by Moseley *et al.* (1977) in Norfolk, England, and by Nutley (1983) in rural Wales (Box 35.3). They

Box 35.2 Local-scale access demands in rural Wales

This map of rural Wales (Figure 35.2) attempts a representation of local-scale access demands within a regional context. The pattern shown is a generalisation from an original parish basis and was compiled as a summation of a number of specific accessibility indicators, as follows:

1. Availability of public transport in the parish, daily or less.
2. Access from the parish to any centre with employment opportunities at times and frequencies suitable for work.
3. Access from the parish to any low-grade centre for daily shopping.
4. Access from the parish to any high-grade centre for weekly shopping.
5. Access from the parish to any centre at times suitable for evening leisure activities.
6. Availability of Sunday services.
7. Access from the parish to the regional centre (Cardiff) weekly.
8. Dependence on public transport, according to the proportion of population without use of a car, either totally or partially.

A points system was used to score each parish on the above criteria; these were summed to produce the 'composite index' (Nutley 1980). The final map is very complex, showing intense local variations. The bottom two 'deprived' categories contained 8.8 per cent of the population. The map can be used as a rational basis for selecting problematic areas for local-scale case studies.

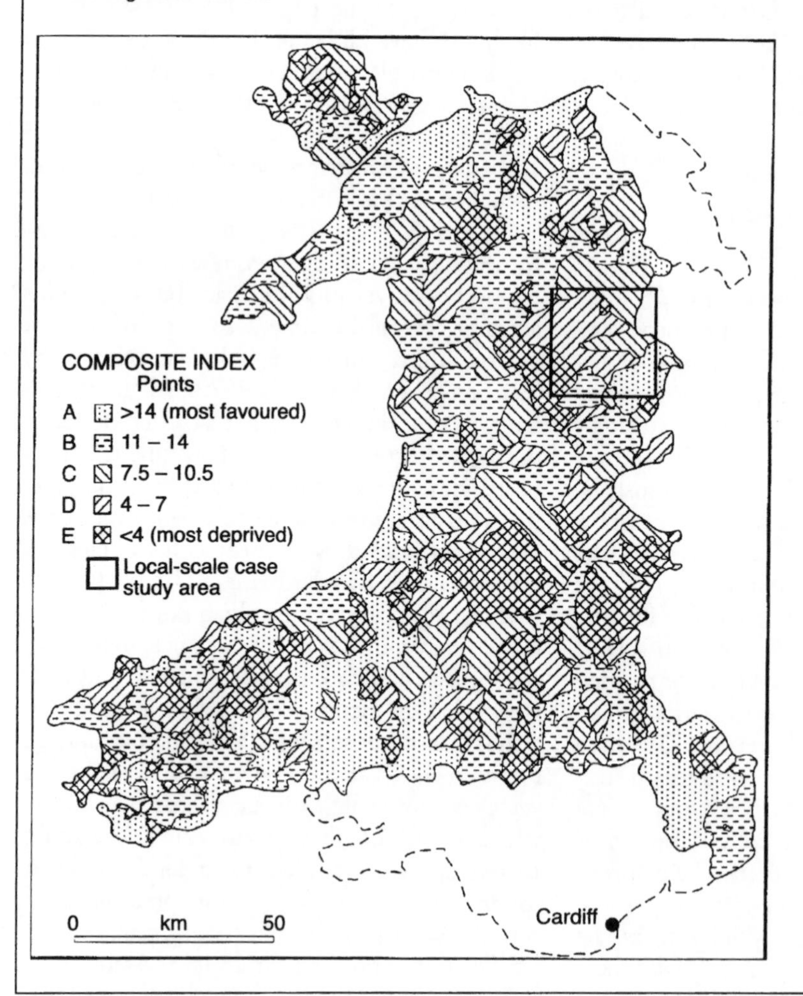

Figure 35.2 Composite index of public transport service and accessibility, Wales, 1979. NB: urban south and northeast excluded.

Source: Nutley 1980.

Box 35.3 The time–space approach to accessibility at the local level

The accompanying map (Figure 35.3) is an example of the type of results that one can get from applying the 'time-space' approach to accessibility at local level. It should be noted that this area, in northern Powys, Wales, was selected largely from the evidence of the regional-scale overview offered in Figure 35.2. Complete description of accessibility in any area is obtained by working at a combination of scales.

For this exercise, access is considered to be 'needed' to a maximum of twenty-eight destination 'functions', although not all are relevant to each of the social groups within the population. The bar charts applicable to each village are divided for convenience into work (three functions), shopping (five), health (five), administration (six), and leisure (nine). The scaling represents the degree of access achievable, under the defined conditions, relative to the optimum (100 per cent), where all people have access to all relevant functions. The advantages of car users over non-car users are immediately apparent. However, shortfalls in the cases of health and administration functions are due to time, not transport, factors, i.e. facility opening hours clash with the times that many people are assumed to be at work. Nowhere is it possible for non-car users to get to work in any of the towns, due to the absence of bus services at the required times and frequencies. Access by non-car users to other function types is extremely variable, dependent upon surviving village-based facilities, proximity to bus routes, and complex timing and frequency allowances. In addition to maps of this type, the methodology can generate many alternative representations of accessibility (Nutley 1983; 1984).

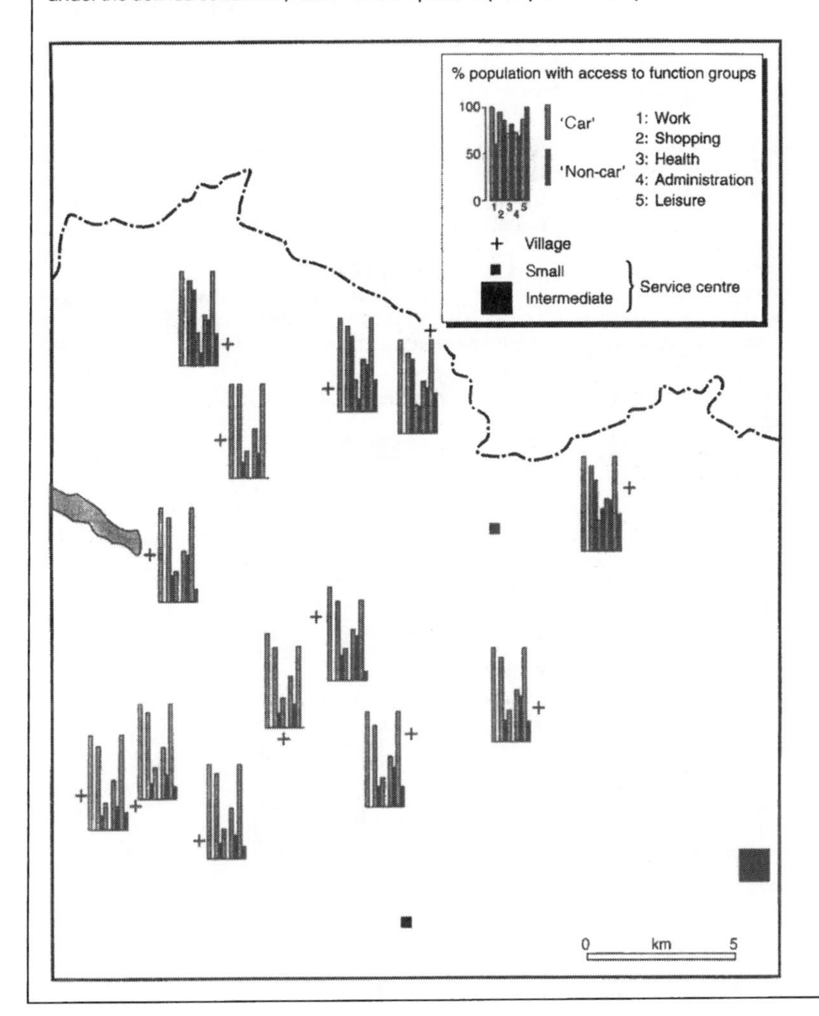

Figure 35.3 Accessibility at the local scale, by village, using time–space methods, North Powys, Wales, 1980.

Source: Nutley (unpublished).

may be extended further by using a 'before and after' approach to evaluate the accessibility benefits of new policy measures and to test the cost-effectiveness of future policy options (Nutley 1985).

Mobility surveys

Questionnaire surveys to discern circulation patterns in rural areas are fairly commonly used by local authorities and consultants as well as by academics, but, as cautioned above, these tend to emphasise mobility. Where the main interest is accessibility, or the detection of transport-related problems, then either a more specific survey is required or conclusions have to be inferred by indirect methods. Thus variations in the popularity of destinations may reflect differential accessibility. Households and persons without cars will be expected to have distinct travel patterns, which should be examined for signs of access difficulties. Variations in travel frequencies among population sub-groups would reveal those with low mobility rates, although whether these are due to choice or hardship is a matter for further enquiry. Case studies employing such approaches include Banister (1980), Smith and Gant (1982) and Nutley and Thomas (1992; 1995) (Box 35.4).

Box 35.4 Rural mobility in Northern Ireland

Questionnaire surveys of rural travel patterns represent a different approach, revealing the extent to which the opportunities to make contacts over space are availed of by people in different circumstances, i.e. how accessibility is actually *used.* In addition, they reveal how people *respond* to conditions of difficulty, such as lack of accessibility. A common expedient is to divide the population into car-owning and non-car-owning groups, as in Table 35.1. Here, grocery stores are used as an example of a low-order destination, and clothing outlets represent a higher-order demand. In this study area, in rural Northern Ireland, non-car owners either walk to local village grocery stores or get lifts into town in neighbours' cars, rather than use the available bus services (Nutley and Thomas 1992). While one might expect car owners to travel with greater frequency, the opposite is the case for 'groceries': non-car owners have to make frequent trips, on foot, to small local outlets, while car owners make less frequent trips to supermarkets in town. Car ownership

Table 35.1 Travel mode and frequency by car ownership: shopping for groceries and clothing, County Londonderry, Northern Ireland, 1988.

	Groceries % of households			*Clothing % of households*		
	No car	*One car*	*Two+ cars*	*No car*	*One car*	*Two+ cars*
Mode						
Walk/cycle	58.1	26.4	20.5	4.8	–	–
Car (own)	–	67.8	77.3	–	88.5	81.8
Car (lift)	30.6	4.6	–	46.8	3.4	2.3
Bus	3.2	1.1	–	17.7	1.1	–
Taxi	8.1	–	2.3	12.9	–	–
Not applicable	–	–	–	14.5	6.9	15.9
Frequency						
1/day	37.1	28.7	29.5	–	–	–
2–4/week	17.7	13.8	20.5	–	–	–
1/week	45.2	54.0	36.4	3.2	4.6	9.1
1/fortnight	–	3.4	9.1	3.2	3.4	13.6
1/month	–	–	4.5	16.1	29.9	22.7
Irreg/yearly	–	–	–	59.7	54.0	40.9
Not applicable	–	–	–	17.7	8.0	13.6
All households	32.1	45.1	22.8	(N = 194)		

Source: Nutley and Thomas 1992.

confers *choice*. In the case of clothes shops, their absence outside towns means that rural non-car owners have to get lifts or use the bus, travelling very rarely, or doing without.

The common assumption that mobility is a desirable attribute leads to a belief that people who make few trips are suffering isolation or hardship. Travel frequencies can be converted to mean trip rates (Table 35.2). Overall rates vary surprisingly little among the population groups extracted here, while excluding walking produces the expected relationships according to car ownership levels and symptoms of disadvantage. The greater number of trips made by car owners is compensated for by more trips on foot made by problematic social groups. The latter have more localised circulation patterns, reflecting lack of choice, and this might reasonably be regarded as a 'problem'.

Table 35.2 Estimated trip rates (work and school excluded), County Londonderry, Northern Ireland, 1988.

	Mean trip rate (trips/household/week)			
	Total	*Excl. walking*	*N*	*% households with car*
All households	8.07	4.21	194	67.5
Non-car households	7.63	2.41	62	0.0
All households with cars	8.27	5.07	131	100.0
One-car households	8.22	4.91	87	100.0
Two+ car households	8.37	5.39	44	100.0
Households with:				
one or more unemployed	8.97	4.00	79	67.1
one or more elderly	6.11	2.88	52	46.1
no employed persons	7.57	3.10	84	45.2
one person alone	7.22	2.93	21	19.0

Source: Nutley and Thomas 1992.

CONCLUSIONS

The geographer's contribution to this issue has been directed not so much towards solving problems through specific policy initiatives but towards elucidating concepts and encouraging a people-centred view. It is realistic to say, at least in the UK, that these efforts have succeeded in communicating to planners the importance of accessibility. Action on the ground, however, such as the maintenance of adequate bus and rail services, cannot escape from economic imperatives and will always be subject to commercial and political decisions at a higher level. Similarly, the rationalisation of consumer services, such as hospitals, continues apace with little thought given to the implications for accessibility of people in low-density areas. Current trends in Western countries towards deregulation and privatisation in the transport sector make appropriate local-scale action considerably more difficult.

As far as techniques are concerned, it will be noticed that the methods illustrated above are by no means new, deriving from the 1970s or even earlier. This reflects their basic simplicity. It is unlikely that any more sophisticated techniques are necessary; simple methods are frequently found to be the most effective. There might be a place for subjective approaches to discern the nature of social and cultural barriers to accessibility, for example in Third World countries. Otherwise, the most promising development would be the application of geographical information systems, which are ideally suited (e.g. Higgs and White 1997).

While public transport in low-density environments will always be problematic, new topics emerging at present stem from contemporary 'car dependency' culture in Western countries. The car-using majority cannot be ignored and, as in North America, many social problems are attributable to the costs of car use by relatively poor people (Farrington *et al.* 1997).

Environmentally motivated transport and energy policies might exacerbate these costs. Also, car dependence is increasingly manifested in cases of localised *rural* traffic congestion, for example in countryside recreation sites in summer (Cullinane *et al.* 1996).

GUIDE TO FURTHER READING

As far as the UK is concerned, the only full-length book on this subject from a geographical viewpoint is still Moseley (1979) and, although the social and policy contexts are outdated, the concepts and techniques described remain valid. Although shorter, a reasonably comprehensive and up-to-date coverage can be found in Nutley (1992; 1998). Note that the first edition (1992) concentrates on the UK and has more detail on accessibility, while the second edition (1998) is more international in scope but has less on methodology. For specific detail on techniques, readers should consult some of the case studies already cited, especially Moseley *et al.* (1977), Nutley (1983; 1984; 1985), and Nutley and Thomas (1992).

Information on other countries is rather scarce and fragmented. For the United States, a review of relevant literature is in Nutley (1996), while the most useful book-length treatment is Due *et al.* (1990). The latter takes a mode-by-mode approach, concentrating on policy and economic issues, principally the effects of deregulation. Rural transport in the developing world is a vitally important problem, only recently given the recognition it deserves. The most useful book on the subject is Barwell *et al.* (1985), which comprises a valuable series of case studies from Africa and Asia. Otherwise, up-to-date and concise synopses can be derived from the 'rural' chapters in Hilling (1996) and Simon (1996).

REFERENCES

Banister, D. (1980) Transport mobility in inter-urban areas: a case study approach in south Oxfordshire. *Regional Studies* 14, 285–96.

Barwell, I., Edmonds, G., Howe, J. and de Veen, J. (1985) *Rural Transport in Developing Countries.* London: Intermediate Technology Publications.

Bird, C. (1981) Analysis of six techniques to identify need for public transport. *Laboratory Report* 1027, Crowthorne: Transport & Road Research Laboratory.

Bradshaw, J. (1974) The concept of social need. *Ekistics* 37, 184–7.

Clark, D. and Woollett, S. (1990) *English Village Services in the Eighties.* London: Rural Development Commission.

Cloke, P., Milbourne, P. and Thomas, C. (1994) *Lifestyles in Rural England.* Salisbury: Rural Development Commission, 110–45.

Cullinane, S., Cullinane, K., Fewings, J. and Southwell, J. (1996) Rural traffic management. The Burrator Reservoir experiment. *Transport Policy* 3, 213–24.

Due, J., Allen, B., Kihl, M. and Crum, M. (1990) *Transportation Service to Small Rural Communities.* Ames: Iowa State University Press.

Farrington, J., Gray, D. and Martin, S. (1997) Rural car dependence and the rising costs of car use. *Town and Country Planning* 66, 214–6.

Higgs, G. and White, S. (1997) Changes in service provision in rural areas. Part 1: the use of GIS in analysing accessibility to services in rural deprivation research. *Journal of Rural Studies* 13, 441–50.

Hilling, D. (1996) *Transport and Developing Countries.* London: Routledge, 157–96.

Hoggart, K., Buller, H. and Black, R. (1995) *Rural Europe. Identity and Change.* London: Edward Arnold, 144–84.

Ilbery, B. (1998) *The Geography of Rural Change.* Harlow: Longman, 13–54.

Jordan, C. and Nutley, S. (1993) Rural accessibility and public transport in Northern Ireland. *Irish Geography* 26, 120–32.

Joseph, A. and Bantock, P. (1984) Rural accessibility of general practitioners: the case of Bruce and Grey Counties, Ontario, 1901–1981. *Canadian Geographer* 28, 226–39.

Kidder, A. (1989) Passenger transportation problems in rural areas. In W. Gillis (ed.) *Profitability and Mobility in Rural America,* University Park: Pennsylvania State University Press, 131–45.

Kihl, M. (1993) The need for transportation alternatives for the rural elderly. In C. Bull (ed.) *Aging in Rural America,* Newbury Park: Sage, 84–98.

Koutsopoulos, K. (1980) Determining transportation needs. *Traffic Quarterly* 34, 397–412.

Lannoy, W. and Oudheusden, D. (1978) The accessibility of nodes in the Belgian road network. *GeoJournal* 2, 65–70.

Maggied, H. (1982) *Transportation for the Poor. Research in Rural Mobility.* Boston: Kluwer Nijhoff.

Marsden, T., Murdoch, J., Lowe, P., Munton, R. and Flynn, A. (1993) *Constructing the Countryside.* London: UCL Press, 129–53.

Moseley, M. (1979) *Accessibility: The Rural Challenge.* London: Methuen.

Moseley, M., Harman, R., Coles, O. and Spencer, M. (1977) *Rural Transport and Accessibility*, 2 vols. Norwich: University of East Anglia.

Moyes, A. (1989) The need for public transport in mid-Wales: normative approaches and their implications. *Rural Surveys Research Unit Monograph* 2, Aberystwyth: University College of Wales.

Nutley, S. (1979) Patterns of regional accessibility in the N.W. Highlands and Islands. *Scottish Geographical Magazine* 95, 142–54.

Nutley, S. (1980) Accessibility, mobility and transport-related welfare: the case of rural Wales. *Geoforum* 11, 335–52.

Nutley, S. (1983) *Transport Policy Appraisal and Personal Accessibility in Rural Wales*. Norwich: Geo Books.

Nutley, S. (1984) Accessibility issues in rural Wales. In P. Cloke (ed.)*Wheels within Wales*. Lampeter: St David's University College, 12–32.

Nutley, S. (1985) Planning options for the improvement of rural accessibility: use of the time–space approach. *Regional Studies* 19, 37–50.

Nutley, S. (1992) Rural areas: the accessibility problem. In B. Hoyle and R. Knowles (eds) *Modern Transport Geography*, London: Belhaven, 125–54.

Nutley, S. (1996) Rural transport problems and non-car populations in the USA. A UK perspective. *Journal of Transport Geography* 4, 93–106.

Nutley, S. (1998) Rural areas: the accessibility problem. In B. Hoyle and R. Knowles (eds) *Modern Transport Geography* 2nd ed. Chichester: Wiley, 185–215.

Nutley, S. and Thomas, C. (1992) Mobility in rural Ulster: travel patterns, car ownership and local services. *Irish Geography* 25, 67–82.

Nutley, S. and Thomas, C. (1995) Spatial mobility and social change: the mobile and the immobile. *Sociologia Ruralis* 35, 24–39.

Ogunsanya, A. (1987) Rural accessibility problems and human resource development: case study from Nigeria. *Journal of Rural Studies* 3, 31–42.

Pacione, M. (1995) The geography of deprivation in rural Scotland. *Transactions of the Institute of British Geographers* NS 20, 173–92.

Simon, D. (1996) *Transport and Development in the Third World*. London: Routledge, 59–92.

Smith, J. and Gant, R. (1982) The elderly's travel in the Cotswolds. In A. Warnes (ed.) *Geographical Perspectives on the Elderly*. Chichester: Wiley, 323–36.

Thomas, D. (1963) *The Rural Transport Problem*. London: Routledge.

36

City marketing as a planning tool

Michael Barke

CITY MARKETING: THE FIELD OF STUDY

Place marketing, of which city marketing is by far the most significant component, has generated a massive literature. Between 1990 and 1994, five major books were published (Ashworth and Voogd 1990; Kearns and Philo 1993; Kotler *et al.* 1993; Gold and Ward 1994; Smyth 1994) containing 242, 560, 248, 633 and 164 references, respectively. Recently, a selective bibliography of city marketing literature consisting of over 280 references was issued (Millington *et al.* 1997). Most of this literature is also very recent. Students of citation will not be surprised to learn that a substantial proportion of this literature is mutually reinforcing but, while city marketing cannot claim to be a discipline in its own right, it is an area of study within which distinctive 'schools' have emerged. In broad terms, the literature divides into three relatively separate groups.

The first category consists of work whose origin lies primarily either in the practice of marketing or in marketing theory. Although place marketing has a host of historical antecedents (Glaab 1967; Jarvis 1994; Ward 1988; 1990; 1994; Zube and Galante 1994) and specific cities have a long history of boosterism, for example Atlanta (Rutheiser 1996) and Syracuse, New York state (Roberts and Schein 1993; Short *et al.* 1993), the application of marketing techniques to cities and other locations stems particularly from two contemporaneous trends in the late 1960s and 1970s. One concerned the development of new marketing approaches, specifically concerned with non-business or non-profit organisations; what came to be known as 'social' marketing (Kotler and Levy 1969; Kotler and Zaltman 1971). The second trend was the onset of an 'urban crisis', which had many manifestations but which was widely perceived at the time as leading to the potential terminal decline of traditional urban economies, with a consequent imperative for economic restructuring (Massey 1984; Lever 1987; Fretter 1993; Holcomb 1993). The latter stimulated the search for new roles for cities and new ways of managing their problems. Initially, this took the form of simple 'promotion' of the city and its attractions but gradually, in some areas, this has evolved into more sophisticated marketing exercises. Whereas 'promotion' is related merely to the idea of trying to sell something, marketing is concerned with finding out what it is that potential consumers wish to buy. Much of the early literature on place marketing was concerned with the USA (Lewis 1978) and the apparently spectacular 'turnarounds' in economic fortunes experienced by cities such as Baltimore, Pittsburgh and Cleveland (Holcomb 1993). Such successes prompted investigations of the 'ways and means' by which decline could be reversed (Guskind 1987; Bailey 1989). Similar ideas were rapidly taken up in European cities (Korn *et al.* 1994), most notably in the Netherlands (Ashworth and Voogd 1988) and, although sporadically at first, also in Britain (Clarke 1986; Wilkinson 1992). Although they are rather different in tone, the epitome of this 'technical' literature on how to achieve best practice in city marketing is found in two of the volumes cited earlier, Kotler *et al.* (1993) and

Ashworth and Voogd (1990), focusing upon the USA and Europe, respectively.

The actual process of marketing is likely to vary from place to place, but various typologies have been suggested (Ashworth and Voogd 1988). The purpose is to bring together 'customers' and 'products' and with this objective, a number of essential elements should be present in the marketing process. First, the functions of the urban area must be commodified and positioned. In other words, they must be examined as if they were products and their relative competitive position studied. Then the actual or potential users of such products or functions need to be considered and their characteristics or market segments determined. Once these two processes have been concluded, the choice of the most appropriate marketing strategy or strategies must be made. The plural is important because, of course, different elements of the urban structure may need quite different marketing strategies. The actual marketing itself could involve a wide range of measures, which could include new developments or activities, some of which may be specific new 'flagship' developments, organisational measures that may improve the functioning of the city and its managers themselves and, finally, specific promotional activities. It is the latter that has attracted most attention and usually involves the creation of new or different images for the place concerned. The latter may be communicated by a variety of means and for a variety of audiences, depending on the market segmentation findings investigated earlier. Outlined in this way, a city marketing strategy is clearly a highly sophisticated operation, and one of the issues to be discussed later concerns the gap between the 'ideal' as described above and as promulgated in the marketing literature and the actual reality of the majority of city marketing.

The second major body of literature, and by far the largest, is concerned with various aspects of place image and place identity and how these may be manipulated for marketing purposes (Burgess 1982; Gold 1994). The tone of much of this literature is critical, with many commentators expressing concern over issues such as the loss of 'authenticity' (Plate 36.1) and the question of in whose interests is such manipulation carried out. Students of cultural studies have found fertile ground in the city marketing phenomenon for their deconstructions and reinterpretations of the meaning of place and the way that place is represented (Burgess and Gold 1985). Therefore, in addition to the study of the nature of 'new' images and what lies behind their content (Watson 1991; Barke and Harrop 1994) an important subset of the literature is concerned with the political context of city marketing and regeneration (Shaw 1993; Boyle 1995; Hall and Hubbard 1996; Strange 1997). A further development in the literature has related to the way that images and other 'marketing' information is communicated. Most obviously this is in various forms of advertising, but it is clear that city marketing also involves the promotion of specific high-profile events and flagship developments (Ley and Olds 1988; Bianchini *et al.* 1992). Cities

Plate 36.1 Catherine Cookson country: South Tyneside's well-known attempt to associate itself with a popular local author.

compete with each other for major international conventions and sporting events such as the Olympic Games (Law 1994; Barke and Towner 1996; Rutheiser 1996) but architecture and other art forms too (Plate 36.2) can also be seen as a form of advertising and promotion (Crilley 1993; Goodey 1994) most obviously in major developments such as Canary Wharf (Brownill 1994), and Paris's spectacular Grands Projects (Kearns 1993). Although often implicit rather than explicit, much of this literature on place imagery and representation challenges many of the fundamental bases of city marketing and, in this sense, stands in contrast to the first group of literature identified above.

The third major area of study and perhaps the least developed is specifically concerned with the empirical assessment of the impact of marketing strategies and promotional activities. Much of this literature tends to be in the form of case studies of particular places. What is striking, however, is how frequently the same places recur. On the international scale, New York, Pittsburgh, Cleveland, Baltimore, Atlanta, London Docklands, Glasgow, Barcelona and Paris are among the most frequently cited, while more specifically in the British urban studies literature Manchester, Tyneside, Birmingham, Sheffield, and London Docklands and Glasgow again appear to be the most popular locations for study. These case studies tend to fall into two groups, one concerned with the 'measurable' impact of marketing exercises, the other focusing rather more on what such an exercise means for local populations.

There is very little evidence that local authorities and other urban managers engage in any systematic evaluation of their city marketing activities (Young and Lever 1997), but even when such evaluation is carried out, its scope is likely to be limited. For example, in assessing flagship events, Bianchini *et al.* (1992) note the positive impacts in terms of visitor numbers, a higher profile for the city and spin-offs into local consumer-related industries. These are the main benefits to which most city marketing agencies would wish to draw attention. However, Bianchini *et al.* recognise that such events have to be judged against wider criteria, including some of the less direct impacts upon local communities. More generally, some commentators note the serial reproduction of images and certain types of urban development (Wilkinson 1992), and the few studies that are specifically concerned with the effectiveness of marketed images tend to indicate very diverse results and to question the efficacy of the promoted image (Burgess and Wood 1988; Young and Lever 1997). Perhaps most damning of all is the findings of some

Plate 36.2 Gateshead's 'Angel of the North'. Gateshead has been famous for many years for using sport and the arts in promotional efforts. The controversial angel seems likely to become a potent symbol for the town.

studies that one of the prime motives for city marketing strategies is that everyone (or every place) else does it (Holcomb 1994), suggesting that the participants are playing a zero-sum game.

Of equal concern are the generally negative conclusions from some case studies of the impact of marketing strategies on local populations, especially in terms of the economic benefits that are claimed for such strategies. A number of studies do, in fact, question the efficacy of the entrepreneurial approach inherent in most city marketing exercises (Levine 1987; Leitner and Garner 1993). The nature of employment generated may have little relevance for pre-existing labour markets, and where marketing strategies are accompanied by large-scale physical redevelopment the impact can be detrimental. For example, 'far from producing a trickle down of economic benefits to local communities, property-led regeneration . . . can make life even more difficult for small local businesses and low-income residents' (Bianchini *et al.* 1992: p. 252). More recently, there have been some direct attempts to analyse the social and political impact of city marketing strategies upon local populations and the main conclusions appear to suggest a diversity of response. For example, in a review of eight case studies of city 'hallmark' events, Boyle (1997) has identified the highly complex range of reactions to such events manifested by local populations.

THE PROBLEMATIC NATURE OF CITY MARKETING

It will be apparent that city marketing is a highly problematic activity that gives rise to a number of contentious issues. Within both the activity itself and its academic study there are a number of paradoxes. Most fundamentally, it may be questioned whether or not a 'place', which is constituted of a multitudinous set of characteristics, many of which are themselves inherently contradictory and in conflict, can be 'marketed' in the same sense as much less diffuse or complex products. Many critics argue that it is only a partial and sanitised version of a multifaceted reality that enters into the consciousness of the marketing experts (Rutheiser 1996).

This relates to a second paradox, which is concerned with the relationship between 'image' and reality and what this means for local populations. Self-evidently, the former has come to dominate city marketing (see Plate 36.1) to the extent that moral questions are raised about the manipulation of places and, more importantly, the people who live in them. The precise meaning and significance of an 'improved' or even altered image soon becomes problematic when one asks in whose interests the 'image' is altered. It has been suggested that the creation of a new image for a place may be a form of social control, designed to create unity or a shared sense of identity within a specific place but also serving to subdue internal discord or polarisation – factors that could be harmful to business confidence (Harvey 1989; Sadler 1993). However, studies of the reactions of local populations that have experienced major civic boosterism campaigns suggest a diversity of outcomes. Although studies of Vancouver (Ley and Olds 1988), Detroit (Neill 1995) and Birmingham, UK. (Hubbard 1996), suggest that local people were broadly positive or, in the case of the latter, neutral in their response to civic boosterism projects. In Belfast (Neill 1993) and Glasgow (Boyle and Hughes 1991), certain groups with strong historical roots actively opposed the 'reading' of history represented by the boosterism agenda. In Milwaukee (Kenney 1995), Indianapolis (Wilson 1993) and Atlanta (Rutheiser 1996), some local groups directly challenged events on the grounds of contemporary economic and political realities.

Continuing with the theme of the projection of the 'image' of places, a further paradox has already become apparent, namely that 'promotional repertoires of many cities . . . bear significant similarities with each other' (Holcomb 1994: p. 125). Even in the absence of such promotional activity, it is the case that a convergence of urban form and structure is leading to a degree of uniformity in the visual environment of many contemporary Western cities. The paradox, therefore, is that while cities perceive it as highly important to engage in marketing strategies – one of the key features of which is to isolate distinctiveness and difference – both the means of communicating that message and the physical entity that is the subject of such communication demonstrate a high degree of convergence. One of the major challenges for the future, therefore, concerns the way in which the activities of promotion and marketing will respond to this convergence. It may be that such agencies will respond by fashioning more and more 'fake heritage' in a vain attempt to create at least the illusion of distinctiveness. Alternatively, the intriguing prospect arises that one way in which standardisation can be prevented is through paying much more attention and respect to 'real' local traditions and identities.

A further paradox, and one that is closely related to the main theme of this book, concerns the management of city promotion strategies, especially through city development and planning departments. The paradox lies in the fact that planning, as an activity practised in local authorities, came into existence in order to prevent the worst excesses of the 'market' from dominating urban development, yet in place promotion, planners are in effect wholeheartedly embracing that 'market'.

Despite this development, a further issue that arises in consideration of the main agencies responsible for city marketing concerns the immense gap between the ideal and the reality in marketing strategies. Although it is not difficult to outline the ideal city marketing strategy, the extent to which it may then be put into practice is another matter. At least one large-scale survey of UK local authorities has demonstrated that, while most authorities claim to be doing some form of marketing, their objectives and target groups are often vague and ill-defined. Furthermore, the frequent fragmentation of responsibility for marketing within a local authority undermines attempts to formulate a strategic approach (Millington 1996). On a larger scale, van der Veen and Voogd (1987) have characterised local authorities as inhabiting two extremes of a spectrum from minimal marketing, simply reflecting a view that the place sells itself, to aggressive marketing; the latter, however, is often a panic response when something has gone wrong. Burgess and Wood (1988) and Wilkinson (1992) document the changes in image presentation for London Docklands and Tyneside, respectively, when it became clear that all was not well with the initial campaigns.

Related to this issue is the fact that there is a relative paucity of genuinely evaluative studies of place marketing. This applies both to local authorities themselves (Millington 1996) and to academic assessments of the activity. There are many studies that are critical of the activity in a general way but few case studies that specifically evaluate promotional campaigns or strategies. For this reason, although most 'places' undertake some form of promotional or marketing activity, many of the same places keep recurring in the academic literature. Where some evaluation has been carried out, it tends to be of a mainly economic nature and, as noted earlier, its conclusions are generally negative, especially when placed in the context of the wider urban economy. For example, a review of one of the prime examples of 1980s urban regeneration (Baltimore) demonstrated that little employment was created for local people (Levine 1987). The jobs created may not be suitable for local populations, local rents and land values may be pushed up with consequent displacement effects, resources may be consumed in promoting the 'flagship scheme' that otherwise may have been used for public services and infrastructure, and there are often problems of accountability in relation to such developments. Several other studies suggest that most of the economic benefits are only marginal, that dual labour markets are often a consequence of flagship developments (Loftman and Nevin 1994), contributing to deepening social polarisation (Boyle and Hughes 1991), and confirm that the opportunity cost of prestige projects is most often borne by the poorer sections of society (Madsen 1992; Sadler 1993). A wide-ranging assessment of one specific aspect of city marketing – cultural policy – in eight major European cities concludes that 'the direct impact of cultural policies on the creation of wealth and employment was relatively small' (Bianchini and Parkinson 1993). Similarly, studies of the impact of the much marketed 'arts' on the economy of Amsterdam have shown that they were far less significant than was generally believed (Griffiths 1993). The paradox here clearly lies with the continuing promotion of flagship developments or events in the interest of city marketing when much of the evidence points in the direction of failure or very limited 'measurable' success. However, with profound irony in the context of 'promotion' and 'image presentation', Bovaird (1994) has noted that the same event may be hailed as a 'symbolic' and perhaps a political success when, in reality, it is an economic or fiscal failure. In some ways, therefore, this may be one of the most fundamental indictments of the 'marketing' approach of the last two decades of urban policy.

CASE STUDIES: EVOLVING MARKETING STRATEGIES

Given the problematic nature of city marketing identified above, it is perhaps hardly surprising that recently the activity has taken on a more fluid and varied identity than formerly. While some local authorities see a reduced significance for the marketing approach – one recent survey suggested that a third of local authorities were doing less marketing than previously (Millington 1996) – others have radically altered their focus and become much more specific, in terms of both their expectations of and targets for city marketing. Conventionally, city marketing has been directed primarily at an external audience. The two main groups of 'customers' have tended to be potential inward investors, especially those with considerable employment generation potential, and the attraction of tourists who will engage in spending within the locality. However, it would be wrong to assume that all local authorities have engaged in a somewhat unthinking quest to attract these particular groups at all costs. Equally, it would be wrong to assume that such city marketing does not produce its own tensions and conflicts (Box 36.1).

Examples of the trend towards less 'all-embracing' and more targeted approaches to city marketing concern a number of urban areas that have a significant image problem in relation to their built environment. In southern England, Slough and Croyden provide two cases, both widely perceived as possessing boring, 1960s or earlier concrete urban fabrics with little cultural vitality. Their marketing initiatives concentrate on urban design, the identification and preservation of existing heritage, general environmental improvement, street events, and public art (Millington 1996; Millar 1998). A very different authority, which, despite its 'left wing' reputation, has wholeheartedly embraced a marketing philosophy is Knowsley on Merseyside. This strategy covers a wide range of issues within the borough. Thus, although the attraction of inward investment is a priority, so too is the requirement to be responsive to the needs and priorities of local residents. This is one of the main identifiable trends in city marketing in the last decade, that is, a shift in the balance of marketing in favour of 'internal' clients, whether these are existing firms or local residents. The second detailed case study, of the Spanish city of Malaga, demonstrates how the latter was a prime objective of a planning strategy adopted in the 1990s (Barke and Newton 1994) (Box 36.2).

CONCLUSION

City marketing, in terms of both its practical application and its academic study, has passed through a series of phases. Initially, the emphasis was on how cities may be promoted (i.e. 'sold'), but subsequently the broader concept of marketing was introduced (i.e. finding out what potential consumers wished to buy). The techniques of such marketing involved the creation of new forms of representation of places and led into a major concern over 'image'. In some places, this

Box 36.1 Carlisle: dealing with peripherality

The city of Carlisle has attempted to turn its peripheral location to its own advantage with its 'Great Border City' campaign which emphasises past border conflicts and has the intention of establishing its local distinctiveness. Such a strategy may appear to be rather narrow and, for example, to possess little that would attract significant inward investors. The reality is that Carlisle city, unlike some neighbouring areas, has little desire or need to attract such investors, as they would be likely to have a major destabilising effect on the local labour market. However, adjacent areas, especially west Cumbria, argue that their surplus labour problems would be assisted by inward investment in the north of the county, including Carlisle city and the city's strategy, based largely on its own history, is a source of some conflict with adjacent areas.

The major thrust of city marketing in Carlisle is to attract more tourism and longer stays by tourists, hence the prominence of history and the 'border' theme. However, this marketing strategy has also created internal tensions. Some local business groups concerned with tourism, for example, local hoteliers, would prefer the city authorities to adopt an image with a broader appeal, the most popular option being the city's proximity to Hadrian's Wall. Their argument is that to highlight the latter in preference, to say, the Border Reivers, would attract a much more international clientele. In response, the city authorities point out that Hadrian's Wall is not unique to Carlisle, while its location as a border city is possibly shared only with the much smaller and distant Berwick-upon-Tweed and that, in any case, the most prominent 'tourist attractions' relating to Hadrian's Wall are in Northumberland, not Cumbria. The city authorities are also much more concerned to 'market' the city and its heritage to local people and to enable them to discover their own history. In their view, this requires an authentic representation of history and one that is based on the specifics of the place, rather than one that is identified (or perhaps even invented) just to attract more visitors and hopefully to make them stay longer.

Therefore, in following a quite specific and apparently narrower marketing strategy, the city authorities in Carlisle have a clear mission, one that is dedicated primarily to local and internal needs and one that appears to be based on an authentic interpretation of their locale. However, in pursuing this mission, conflict has arisen both internally and externally, conflict that, on the one hand, relates to an 'authenticity' versus 'marketability' debate and, on the other, to the political rivalries that arise from the tensions inherent in representing purely local, at the expense of wider, regional, interests. Such tensions and conflicts are not unique to Carlisle over the issue of city marketing.

Box 36.2 Malaga: dealing with the citizenry

The city of Malaga is best known as the location of the airport that provides access to the tourist resorts of the Costa del Sol. Yet the city authorities feel that they have benefited little from the massive development of modern tourism in this part of Andalusia. In order to capture and maintain a larger share of this activity, the city council sought to engage in a process of 'product development' within a sustainable tourism context (Esteve Secall 1991; Barke and Newton 1995). This approach was later incorporated within a wide-ranging strategic plan for the city (Barke and Newton 1994). A central thrust of the promotion of a more sustainable tourism policy and the strategic planning approach was that the citizens of Malaga needed to feel a sense of ownership of both, rediscover a sense of pride in their city and, in a sense, become key agents of change themselves (Granados Cabezas 1992). Thus the city authorities have engaged in an extensive programme of *internal* marketing of tourism and broader planning policies, but a process that also invites the fullest participation with the intention of encouraging confidence and the possibility of indigenous development.

This internal 'marketing' has a number of dimensions, but among the most important have been a carefully phased and publicised programme relating to the development of the city's strategic plan and a series of participatory mechanisms within that plan. The most significant is the constitution of the General Assembly, which is the main channel for citizen participation, responsible for monitoring the guidelines and objectives of the plan. This General Assembly consists of representatives of 250 local organisations, ranging from business groups, to members of various residential neighbourhood organisations, cultural and special interest groups and even gypsy organisations. Various strategic working groups looking at specific issues must contain representatives of the assembly, and 'experts' must present their diagnoses of issues and problems in particular sectors to this assembly for discussion and approval. Thus, in addition to the usual publicity, attempting to raise awareness and inform the public about the plan the 'planners' are, in this case, accountable to wider representatives of the public interest than the normal elected, political ones. The General Assembly is more than a cosmetic acknowledgement of democracy, because it is directly involved in the assessment of alternative strategies.

This internal marketing has involved a substantial additional range of activities, including public meetings, street publicity and exhibitions, the participation of local schools, a special publication (*Ojo a Malaga*), and the full and enthusiastic participation of the local media. The time, energy and cost of this internally directed activity is justified by the chief coordinator of the strategic plan, Vicente Granados, on the grounds that simply to promote the city externally and ignore the local population is not sustainable. The latter must feel that they are a part of what is happening to their city and not peripheralised by it. The concept of city marketing in Malaga has therefore moved a long way from one that views the activity as simply trying to attract more jobs or more tourists.

implied seeking to change an unfortunate image, but in others the activity became more concerned with how to enhance a neutral or already favourable image by, for example, emphasising distinctiveness. The marketing of places then began to transcend mere advertising and started to incorporate hallmark events and specific, high-profile, developments in the built environment. However, as city marketing became an industry in its own right, the content of such images, events and structures have demonstrated convergent tendencies. Furthermore, despite the hopes pinned on city marketing, there appears to be very little evaluation of the benefits of the activity, and what evaluation there is tells a far from clear story of the costs and benefits. The current concerns, therefore, are less with the production of images in the city marketing process but with how such images are consumed and by whom. The current evidence, in the context of the UK and other European cities, is that much more attention is being paid to 'internal' marketing in the sense of attempting to meet the needs of existing local businesses and residents. In conclusion, as a planning contrivance, city marketing has manifested itself in various forms and, in that sense, is not one planning tool but several. Although it is difficult to identify any current consensus, two broad trends may be identified. The first is that the overall expectations of city marketing are almost certainly less than in the mid-1980s. The second is that, in many places and in various forms, city marketing is currently being used as much to serve the specific purposes of local authorities and indigenous populations and businesses as to present an appealing, but possibly superficial, external image.

GUIDE TO FURTHER READING

For an introduction to city marketing from a mainly European perspective and with an emphasis on the external promotion of the city, see Ashworth, G.J. and Voogd, H. (1990) *Selling the City: Marketing Approaches in Public Sector Urban Planning*, London and New York: Belhaven Press. A similar philosophical approach but with examples drawn mainly from US experience is Kotler, P., Haider, D.H. and Rein, I. (1993) *Marketing Places: Attracting Investment, Industry and Tourism to Cities, States and Nations*, New York: The Free Press. For a much more critical and detailed examination of the city marketing experience of one city in the USA see Rutheiser, C. (1996) *Imagineering Atlanta*, London and New York: Verso. Two books that develop more general but equally critical perspectives on city marketing, although with rather different emphases, are Kearns, G. and Philo, C. (1993) *Selling Places: The City as Cultural Capital, Past and Present*, Oxford: Pergamon Press, and Gold, J.R. and Ward, S.V. (1994) *Place Promotion: The Use of Publicity and Marketing to Sell Towns and Regions*, Chichester: John Wiley & Sons. As noted in this chapter, the number of genuinely evaluative case studies remains surprisingly small, although some are included in the texts above. The study of the impact of city marketing upon consumers is a relatively new field, but for a thoughtful account of the variety of such impacts, especially upon local residents, see Boyle, M. (1997) 'Civic boosterism in the politics of local economic development – "institutional positions" and "strategic orientations" in the consumption of hallmark events, *Environment and Planning A* 29, 1975–97.

REFERENCES

Ashworth, G.J. and Voogd, H. (1988) Marketing the city: concepts, processes and Dutch applications. *Town Planning Review* 59(1), 65–79.

Ashworth, G.J. and Voogd, H. (1990) *Selling the City: Marketing Approaches in Public Sector Urban Planning*. London and New York: Belhaven Press.

Bailey, J.T. (1989) *Marketing the Cities in the 1980s and Beyond: New Patterns, New Pressures, New Promises*. Cleveland: American Economic Development Council.

Barke, M. and Harrop, K. (1994) Selling the industrial town: identity, image and illusion. In J.R. Gold and S.V. Ward (eds) *Place Promotion: The Use of Publicity and Marketing to Sell Towns and Regions*, Chichester: John Wiley & Sons, 93–114.

Barke, M. and Newton, M.T. (1994) New approaches to planning in Spain: Malaga's *plan estrategico* (strategic plan). *Planning Practice and Research* 9(4), 415–22.

Barke, M. and Newton, M.T. (1995) Promoting sustainable tourism in an urban context: recent developments in Malaga City, Andalusia. *Journal of Sustainable Tourism* 3(3), 115–34.

Barke, M. and Towner, J. (1996) Urban tourism in Spain. In M. Barke, J. Towner and M.T. Newton (eds) *Tourism in Spain: Critical Issues*, Wallingford: CAB International, 343–74.

Bianchini, F., Dawson, J. and Evans, R. (1992) Flagship projects in urban regeneration. In P. Healey, S. Davoudi, M. O'Toole, S. Tavsanoglu and D. Usher (eds) *Rebuilding the City: Property-Led Urban Regeneration*, London: E. & F.N. Spon, 245–55.

Bianchini, F. and Parkinson, M. (1993) *Cultural Policy and Urban Regeneration*. Manchester and New York: Manchester University Press.

Bovaird, T. (1994) Managing urban economic development: learning to change or the marketing of failure? *Urban Studies* 31, 573–603.

Boyle, M. (1995) The politics of urban entrepreneurialism in Glasgow. *Geoforum* 25(4), 453–70.

Boyle, M. (1997) Civic boosterism in the politics of local economic development – 'institutional positions' and 'strategic orientations' in the consumption of hallmark events. *Environment and Planning A* 29, 1975–97.

Boyle, M. and Hughes, G. (1991) The politics of the representation of 'the real': discourse from the Left on Glasgow's role as European City of Culture, 1990. *Area* 23(3), 217–28.

Brownill, S. (1994) Selling the inner city: regeneration and place marketing in London's Docklands. In J.R. Gold and S.V. Ward (eds) *Place Promotion: The Use of Publicity and Marketing to Sell Towns and Regions*, Chichester: John Wiley & Sons, 133–52.

Burgess, J.A. (1982) Selling places: environmental images for the executive. *Regional Studies* 16, 1–17.

Burgess, J.A. and Gold, J.R. (eds) (1985) *Geography, the Media and Popular Culture*. London: Croom Helm.

Burgess, J.A. and Wood, P.A. (1988) Decoding Docklands: place advertising and the decision making strategies of the small firm. In J. Eyles and D.M. Smith (eds) *Qualitative Methods in Human Geography*, Cambridge: Polity Press, 94–117.

Clarke, A. (1986) Local authority planners or frustrated marketeers? *The Planner* May, 23–6.

Crilley, D. (1993) Architecture as advertising: constructing the image of redevelopment. In G. Kearns and C. Philo (eds) *Selling Places: The City as Cultural Capital, Past and Present*, Oxford: Pergamon Press, 231–52.

Esteve Secall, R. (1991) *Un Nuevo Modelo Turistico para Espana*. Malaga: Universidad de Malaga.

Fretter, A.D. (1993) Place marketing: a local authority perspective. In G. Kearns and C. Philo (eds) *Selling Places: The City as Cultural Capital, Past and Present*, Oxford: Pergamon Press, 163–74.

Glaab, C.N. (1967) Historical perspective on urban development schemes. In L.N. Schnore (ed.) *Social Science and the City*, New York: Praeger, 197–219.

Gold, J.R. (1994) Locating the message: place promotion as image communication. In J.R. Gold and S.V. Ward (eds) *Place Promotion: The Use of Publicity and Marketing to Sell Towns and Regions*, Chichester: John Wiley & Sons, 19–37.

Gold, J.R. and Ward, S.V. (1994) *Place Promotion: The Use of Publicity and Marketing to Sell Towns and Regions*. Chichester: John Wiley & Sons.

Goodey, B. (1994) Art-full places: public art to sell public spaces? In J.R. Gold and S.V. Ward (eds) *Place Promotion: The Use of Publicity and Marketing to Sell Towns and Regions*, Chichester: John Wiley & Sons, 153–80.

Granados Cabezas, V. (1992) El plan estrategico de Malaga. *Technopolis International* 8, 129.

Griffiths, R. (1993) The politics of cultural policy in urban regeneration strategies. *Policy and Politics* 21(1), 39–46.

Guskind, R. (1987) Bringing Madison Avenue to Main Street. *Planning* February, 4–10.

Hall, T. and Hubbard, P. (1996) The entrepreneurial city: new urban politics, new urban geographies? *Progress in Human Geography* 20(2), 153–74.

Harvey, D. (1989) From managerialism to entrepreneurialism: the transformation of urban governance in late capitalism. *Geografiska Annaler* 71B, 13–17.

Holcomb, B. (1993) Revisioning place: de- and re-constructing the image of the industrial city. In G. Kearns and C. Philo (eds) *Selling Places: The City as Cultural Capital, Past and Present*, Oxford: Pergamon Press, 133–44.

Holcomb, B. (1994) City make-overs: marketing the post-industrial city. In J.R. Gold and S.V. Ward (eds.) *Place Promotion: The Use of Publicity and Marketing to Sell Towns and Regions*, Chichester: John Wiley & Sons, 115–31.

Hubbard, P. (1996) Urban design and city regeneration: social representations of entrepreneurial landscapes. *Urban Studies* 33, 1441–61.

Jarvis, B. (1994) Transitory topographies: places, events, promotions and propaganda. In J.R. Gold and S.V. Ward (eds) *Place Promotion: The Use of Publicity and Marketing to Sell Towns and Regions*, Chichester: John Wiley & Sons, 181–94.

Kearns, G. (1993) The city as spectacle: Paris and the bi-centenary of the French Revolution. In G. Kearns and C. Philo (eds) *Selling Places: The City as Cultural Capital, Past and Present*, Oxford: Pergamon Press, 49–102.

Kearns, G. and Philo, C. (eds) (1993) *Selling Places: The City as Cultural Capital, Past and Present*. Oxford: Pergamon Press.

Kenney, J. (1995) Making Milwaukee famous: cultural capital, urban image and the politics of place. *Urban Geography* 16, 440–58.

Korn, J., Lindemann, H., Schultz, A., Woosnam, G. and Woosnam, J. (eds) (1994) *Managing and Marketing of Urban Development and Urban Life*. Berlin: Dietrich Reiner Verlag.

Kotler, P. and Levy, S.J. (1969) Broadening the concept of marketing. *Journal of Marketing* January, 10–15.

Kotler, P. and Zaltman, G. (1971) Social marketing: an approach to planned social change. *Journal of Marketing* July, 3–12.

Kotler, P., Haider, D.H. and Rein, I. (1993) *Marketing Places: Attracting Investment, Industry and Tourism to Cities, States and Nations*. New York: The Free Press.

Law, C.M. (1994) Manchester's bid for the millennium Olympic Games. *Geography* 79, 222–31.

Leitner, H. and Garner, M. (1993) The limits of local initiatives: a reassessment of urban entrepreneurialism for urban redevelopment. *Urban Geography* 14, 57–77.

Lever, W.F. (1987) *Industrial Change in the U.K.* London: Longman.

Levine, M. (1987) Downtown redevelopment as an urban growth strategy: a critical appraisal of the Baltimore renaissance. *Journal of Urban Affairs* 9(2), 103–23.

Lewis, L. (1978) Local government agencies try advertising. In C.H. Lovelock and C.B. Weinberg (eds) *Readings in Public and Non-Profit Marketing*, New York: Scientific Press, 37–54.

Ley, D. and Olds, K. (1988) Landscape as spectacle: world's fairs and the culture of heroic consumption. *Environment and Planning D: Society and Space* (6), 191–212.

Loftman, P. and Nevin, B. (1994) Prestige project developments: economic renaissance or economic myth? A case study of Birmingham. *Local Economy* 8, 307–14.

Madsen, H. (1992) Place marketing in Liverpool: a review. *International Journal of Urban and Regional Research* 16, 633–40.

Massey, D. (1984) *Spatial Divisions of Labour: Social Structures and the Geography of Production*. London: Macmillan.

Millar, S. (1998) Slough launches fightback against image of despond. *The Guardian*, April 18.

Millington, S. (1996) City marketing: issues for local authorities in the 1990s. Paper presented at the European and Regional Studies Conference, Exeter, 11–14 April.

Millington, S., Young, C. and Lever, J. (1997) A bibliography of city marketing. *The Journal of Regional and Local Studies* 17(2), 16–42.

Neill, W.J.V. (1993) Physical planning and image enhancement: recent developments in Belfast. *International Journal of Urban and Regional Research* 17, 595–609.

Neill, W.J.V. (1995) Lipstick on the gorilla: the failure of image led planning in Coleman Young's Detroit. *International Journal of Urban and Regional Research* 19, 639–54.

Roberts, S. and Schein, R.H. (1993) The entrepreneurial city: fabricating urban development in Syracuse, New York. *Professional Geographer* 45, 21–33.

Rutheiser, C. (1996) *Imagineering Atlanta: The Politics of Place in the City of Dreams*. London: Verso.

Sadler, D. (1993) Place-marketing, competitive places and the construction of hegemony in Britain in the 1980s. In G. Kearns and C. Philo (eds) *Selling Places: The City as Cultural Capital, Past and Present*, Oxford: Pergamon Press, 175–92.

Shaw, K. (1993) The development of a new urban corporatism: the politics of urban regeneration in the North East of England. *Regional Studies* 27(3), 251–86.

Short, J.R., Benton, L.M., Luce, W.P. and Walton, J. (1993) Reconstructing the image of an industrial city. *Annals of the Association of American Geographers* 83(2), 207–24.

Smyth, H. (1994) *Marketing the City: The Role of Flagship Developments in Urban Regeneration*. London: E. & F.N. Spon.

Strange, I. (1997) Directing the show? Business leaders, local partnership and economic regeneration in Sheffield. *Environment and Planning C: Government and Policy* 15(1), 1–17.

van der Veen, W. and Voogd, H. (1987) *Gemeentepromotie en Bedrijfsacquisitie*. Groningen: Geopers.

Ward, S.V. (1988) Promoting holiday resorts: a review of early history to 1921. *Planning History* 10(2), 7–11.

Ward, S.V. (1990) Local industrial promotion and development policies 1899–1940. *Local Economy* 5, 100–18.

Ward, S.V. (1994) Time and place: key themes in place promotion in the USA, Canada and Britain since 1870. In J.R. Gold and S.V. Ward (eds) *Place Promotion: The Use of Publicity and Marketing to Sell Towns and Regions*, Chichester: John Wiley & Sons, 53–74.

Watson, S. (1991) Gilding the smokestacks: the new symbolic representations of deindustrialised regions. *Environment and Planning D: Society and Space* 9, 59–70.

Wilkinson, S. (1992) Towards a new city? A case study of image-improvement initiatives in Newcastle upon Tyne. In P. Healey, S. Davoudi, M. O'Toole, S. Tavsanoglu and D. Usher (eds) *Rebuilding the City: Property-Led Urban Regeneration*, London: E. & F. N. Spon, 174–215 .

Wilson, D. (1993) Everyday life, spatiality and inner

city disinvestment in a US city. *International Journal of Urban and Regional Research* 17, 578–94.

Young, C. and Lever, J. (1997) Place promotion, economic location and the consumption of city image. *Tijdschrift voor Economische en Sociale Geografie* 88(4), 332–41.

Zube, E.H. and Galante, J. (1994) Marketing landscapes of the Four Corners States. In J.R. Gold and S.V. Ward (eds) *Place Promotion: The Use of Publicity and Marketing to Sell Towns and Regions*, Chichester: John Wiley & Sons, 213–32.

37

Low-income shelter in the third world city

Rob Potter

INTRODUCTION

A perennial applied development problem is that everybody needs shelter although, viewed globally, not everyone is able to secure what may be regarded as housing of an adequate standard. It is believed that 20 per cent of the world's total population does not have access to decent shelter. Further, it is estimated that in the predominantly poor and middle-income countries that make up what is referred to as the 'third world' or 'South', perhaps as many as one-half live in homes that may be deemed substandard.

In a similar vein, McAuslan (1985) contends that in the majority of major cities to be found in the third world, more than 1 million people live in illegally or informally developed settlements, with little or no piped water, sanitation or services. The occupants are frequently unable to afford even the smallest or cheapest professionally constructed, legal house that possesses basic amenities. The majority of houses have been self-built in so far as their residents have taken responsibility for organising the design and construction of their own homes.

In the early 1960s, Abrams (1964) bemoaned the fact that despite progress in the fields of manufacturing, education and the sciences, the provision of simple shelter affording privacy and protection against the elements was still beyond the reach of the majority of the world's population. At the beginning of the 1990s, it was estimated that 9.47 million people (60 per cent of the population) in Mexico City lived in self-help housing. At about the same time, 1.67 million (61 per cent) of the inhabitants of Caracas, Venezuela were to be found residing in self-help homes.

Such issues are reflected in the simple typology of low-income shelter in third world cities, which is reproduced here as Figure 37.1. First, there are the homeless and the street sleepers. Many urban residents are far too poor to be able to afford any sort of home, whether rented or owned, and are forced to sleep in the streets. In Calcutta in the early 1960s, for example, it was estimated that more than 600,000 dwellers slept on the streets, while in Bombay, one in every sixty-six were homeless and a further 77,000 lived under stairways, on landings and the like (Abrams 1964).

Second, a large group are to be found renting accommodation in slums and tenements. It is

Figure 37.1 A typology of low-income housing in third world cities.

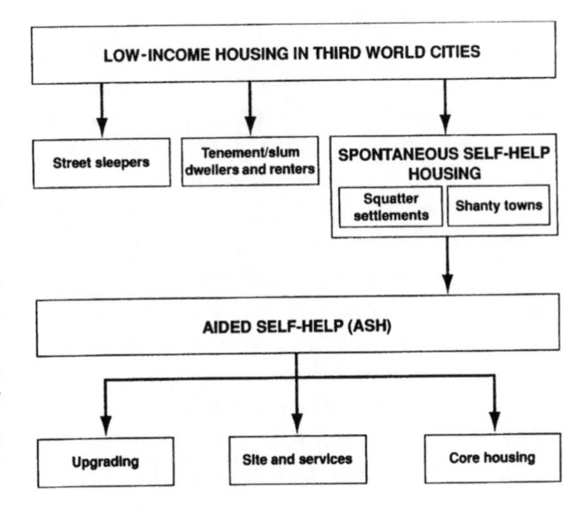

recognised today that there are many renters in self-help settlements (see Gilbert 1983; Gilbert and Varley 1991; Kumar 1996), together with the existence of squatter landlords (see Gilbert 1983; Lee-Smith 1990; Potter 1994). Third, there are the squatters and occupants of shanty towns, who inhabit what may be referred to as spontaneous self-help housing areas.

The inventiveness of low-income residents cannot go unnoticed. In Cairo, for example, the severe housing shortage has led to a number of novel responses. The old city, or *medina*, has become a vast area of tenement slums. Perhaps more surprising are the tomb cities, or cities of the dead, which are to be found located on the eastern edge of the city. Here, the structures built for caretakers or for relatives visiting graves are now occupied by the poor as permanent homes.

Another novel response in Cairo is living on the rooftops of apartments – as long as the structures placed on existing roofs are not constructed of permanent materials, they are legal. It has been estimated that in the region of half a million people live in such rooftop dwellings within the city (Abu-Lughod 1971). The growth of population in low-income settlements is frequently running at between 12 and 15 per cent per annum (Turner 1967; Dwyer 1975). Frequently, shanty towns and squatter settlements account for at least 20–30 per cent of the total urban population, but on occasion the proportion is far higher, as for Bogota (60 per cent), Casablanca (70 per cent) and Addis Ababa (90 per cent).

Clearly, this is an applied problem of vast proportions, and much academic work has been carried out not only by geographers but also by economists, sociologists, anthropologists and planners looking at the issues surrounding such shelter provision. The policy relevance of providing adequate shelter was early recognised by the United Nations conference on Human Settlements (Habitat), which was held in 1976. The follow-up Habitat II or 'City Summit' conference, which was held in Istanbul in June 1996, strongly reiterated this continuing and largely unmet need for adequate low-income shelter (Berghall 1995; Okpala 1996; UNCHS 1996) and the need for housing to be provided on a sustainable environmental basis.

LOW-INCOME HOUSING IN THIRD WORLD CITIES: THE SUBSTANTIVE ISSUES

The housing problem in the developing world has really emerged since the early 1940s. It is the squatter settlement or the shanty town that is the most ubiquitous sign of rapid urban development in this region. Such settlements are also referred to by means of a wide variety of other names, among them spontaneous settlements, informal settlements, uncontrolled settlements, makeshift, irregular, unplanned, illegal, self-help, marginal and peripheral settlements. The wide variety of labels used to describe such settlements points to an important characteristic, namely their extreme diversity with regard to formation, building materials, physical character and the characteristics of their inhabitants. One avenue of applied research, therefore, focuses on the analysis and monitoring of low-income housing conditions (Box 37.1).

The terms 'squatter settlement' and 'illegal settlement' are frequently used but are potentially misleading. Squatter settlements are those where individuals have settled without legal title to land, or alternatively without planning permission. A good example of the latter is provided by the *barrios clandestinos* of Oporto in Portugal. Squatter settlements are frequently located on government or church-owned land. But illegality is not always a characteristic. Many low-income homes are owned, the plots having been subdivided and sold. Similarly, some homes and/or the land on which they are sited are rented. Such rentyards are quite common in Caribbean and Latin American towns and cities (Ward 1976).

Yet another common characteristic is that areas are makeshift settlements or shanties, being constructed from whatever materials are available to hand. Basic shelters made from packing cases and fish barrels, as well as cardboard cartons and even newspapers, have been described in the Moonlight City area of West Kingston, Jamaica,

Box 37.1 The assessment of housing conditions

An applied research task that frequently takes the attention of geographers and other social scientists is the assessment and monitoring of housing conditions. This may be carried out either by means of first-hand field surveys or, alternatively, by means of the analysis of census and/or other data, where these have been collated by the state. These may include planning applications, data on land subdivisions, water connections and the like.

An example of the latter is provided by the work of the present author in relation to the small eastern Caribbean island states (see Potter 1992; 1994; Potter and Conway 1997). Some of the early work on this project involved the mapping of housing data taken from the census for such states at the elemental enumeration district level for the very first time. In recent work, the author prepared a background paper on housing for the government in connection with the preparation of the Revised Physical Development Plan for Barbados, and in this a key input was data on land subdivisions that were vacant – that is they had not been built upon and were being held speculatively, or for future development (Potter 1997).

In the initial work, key housing quality variables such as levels of ownership, houses over twenty years old, houses constructed of wood, the use of pit latrines and drawing water from a public standpipe were mapped for the 121 EDs making up the primate urban area of Bridgetown in Barbados. An example is shown in Figure 37.2, and the proportion of houses drawing water from a public standpipe shows a strong degree of concentricity, being in the upper quartile in respect of the entire inner city area. A small eastern sectoral wedge is also discernible. In the final analysis, the complete data set was factor analysed. Factor I, a measure of housing disamenity (Figure 37.3), accounted for 50.3 per cent and showed clearly the patterning of housing conditions in the city area, picking out the principal inner city problem areas, where despite many declarations of intent, no comprehensive improvement to housing has occurred.

An example of the need for first-hand field observation in the examination of housing conditions is provided by Watson and Potter (1997; 1999) although this example was initially based on rural housing conditions. In 1980, the residents of what are known as the 'plantation tenantries' were given the right to buy the land on which their houses were situated. However, in the meantime, no surveys have been carried out to assess the degree to which housing on the tenantries has been improved or upgraded. In an effort to address this specific issue, over half these housing areas – 150 in all – were sampled. A whole array of characteristics were considered in relation to each house, ranging from house form to upkeep, material of construction, foundations, state of repair, and the visible stockpiling of building materials such as sand and brick. By such means, a clear picture of change was built up and analysed at the level of the entire housing area (Watson and Potter 1999).

Figure 37.2 Percentage of homes obtaining water from a public standpipe, Bridgetown, Barbados.

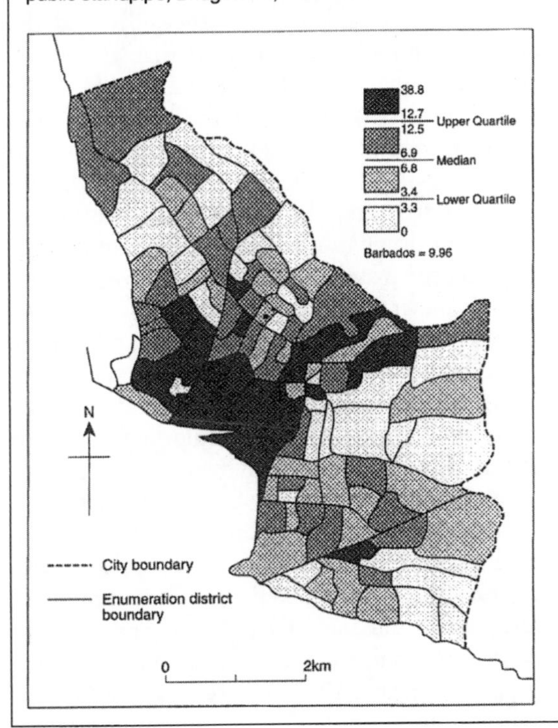

Figure 37.3 Scores on Factor I, a measure of housing disamenity, Bridgetown, Barbados.

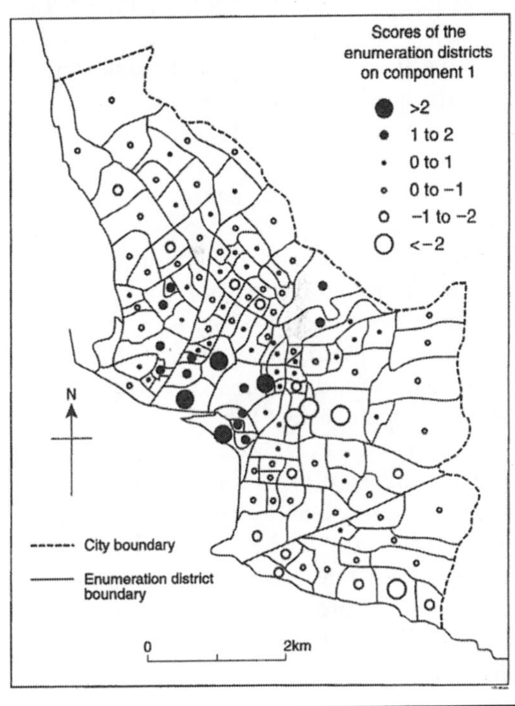

by Clarke (1975). More typically, recycled scraps of wood and corrugated iron may be employed together with printers' drums for both outside walls and roofs. Other materials that are frequently employed in construction include flattened tin cans, straw matting and sacking.

Such settlements may also be makeshift in the sense that they have none of the basic urban services such as water, electricity and sewerage when they are initially developed. But even here caution must be exercised, for with time such services are often acquired, and brick-built houses may come to predominate in a formerly makeshift area. Certainly, the basic but frequently overlooked distinction between squatter settlements characterised by their illegality on the one hand, and shanties identified by virtue of their poor physical fabric on the other, must be fully appreciated and borne in mind.

Yet other settlements are characterised by their unplanned, irregular and informal nature, or by their origin in mass land invasions. Such haphazard or speedy development is epitomised by the description 'spontaneous'. All of these terms are highly appropriate in certain situations but are potentially misleading in others. Thus, while many low-income settlements are unplanned in the professional planning and architectural senses, many are the outcome of much careful forethought on the part of their residents, especially those involving organised land invasions, which may occur at the suggestion of opposition politicians (see Gilbert 1981; Potter 1994). However, in Africa, Asia and the Middle East, the development of low-income housing is typically a much more gradual process, being based on slow infiltration and individual initiative. Such developments are, therefore, based on the very antithesis of spontaneity.

Similar concern may be expressed concerning the lack of universal applicability of descriptions such as peripheral and marginal settlements, whether used in a strictly geographical or an economic sense. Finally, although the terms self-help and autoconstruction are useful in signifying that the building of such dwellings is not normally undertaken by professionals, it would be highly erroneous if the impression were to be given that such houses are built entirely by their present or previous occupants. Indeed, often the assistance of friends and family is enlisted, supplemented by artisans, as in the *coup d'main* system of St Lucia (Potter 1994).

In the 1950s and 1960s, self-help housing was generally viewed with alarm and pessimism, representing a problem that had to be cleared and replaced by regular housing (see Lloyd 1979: 53–7; Conway 1982; 1985). Such negative views were reflected in the writings of the American anthropologist Oscar Lewis. Lewis worked in Mexico, India and Puerto Rico, and he argued that the poor were locked into an inescapable 'culture of poverty' (Lewis 1959; 1966). Lewis maintained that wherever poor groups are found they show traits such as apathy, fatalism, tendencies towards immediate gratification and social disorganisation.

These simple deterministic ideas came to be attacked from a number of quarters (see Lloyd 1979: Drakakis-Smith 1981; Gilbert and Gugler 1992; Potter and Lloyd-Evans 1998). In particular, it was suggested that the culture of poverty is convenient for the wealthy and the powerful in so far as it suggests that 'poverty is the poor's own fault' (Gilbert and Gugler 1982: 84). It is views such as these that are reflected in popular descriptions of squatter settlements and shanty towns as 'urban cancers', 'festering sores', 'urban fungi' and the like. During the period when top-down modernisation was unquestioningly accepted as the route to development, it was almost inevitable that poor housing would be viewed erroneously as the problem, and not the direct outcome and reflection of poverty.

Most authors and governments now agree that third world governments cannot afford high-technology, high-rise monumental responses to their housing problems. In a paper presenting an overall theory of slums, Stokes (1962) drew a clear distinction between what he regarded as successful and unsuccessful poor communities, referring to these as *slums of hope* and *slums of despair*, respectively, a terminology that has stuck (but see Eckstein 1990). Charles Abrams' (1964) book was also influential, for he stressed that urban land costs were soaring, thereby pricing the poor out of the market. But he believed that sufficient land was available if only it could be appropriated by public sector intervention. He noted that in the conditions of a housing shortage, the bulldozing of houses represents a curious policy.

Thus the perspective concerning self-help housing swung from negative to positive in the late 1960s to the early 1970s. Gradually, there was a change in perspective concerning informal sector housing and employment. It was argued that the poor were not indolent, dishonest and disorganised but generally, quite the reverse. The major change in attitudes was to be precipitated by the experiences of two academics-cum-architectural/planning practitioners who were working in Peru in the late 1960s. The first was William Mangin, an American anthropologist, and the second John Turner, a British architect-planner. Both Turner and Mangin advocated self-help housing as a positive force in developing world housing provision. One of the most important papers was written by Mangin (1967), the title of which conveys the essence of the overall argument presented by the two authors: 'Latin American squatter settlements: a problem and a solution'.

In his work, Mangin described most of the then dominant views on low-income residents as myths. They were not disorganised, a drain on the urban economy, populated by criminals and radicals, nor were they made up of a single homogeneous social group. Rather, Mangin stressed that most squatters were in employment, were socially stable and had been residing in the city for a considerable period. Illegal occupancy of land gave them the opportunity to avoid paying high rents and at the same time allowed them to build their own homes at their own pace.

In like manner, Turner worked for over eight years in Peru and for a considerable proportion of that time was involved with self-builders in various *barriadas*. His partly autobiographical account (Turner 1982: 99–103) is very informative in this connection. His overall attitude is clearly summarised in one concise quotation:

> Like the people themselves, we saw their settlements not as slums but as building sites. We shared their hopes and found the pity and despair of the occasional visits from elitist professionals and politicians quite comic and wholly absurd.
>
> (*ibid*.: 101)

Turner argued that all that had to be done to assist self-builders was to approve rough sketch plans and to distribute small amounts of cash in appropriate stages. Turner observed that the economies of self-help were founded upon 'the capacity and freedom of individuals and small groups to make their own decisions, *more* than on their capacity to do manual work' (*ibid*.: 102). As a consequence, Turner articulated the Churchillian cry 'never before did so many do so much with so little' (*ibid*.: 102).

The most positive message promulgated by Turner was that if left to themselves, low-income settlements improve gradually but progressively over time. Thus houses that were originally constructed from straw matting later acquired walls, services and paved streets. In the terminology of Stokes, they were clearly slums of hope, characterised by in situ improvement and the general upward social mobility of their populations (see also Turner 1963; 1967; 1968a; 1968b; 1969; 1972; 1976; 1982; 1983; 1985; 1988; 1990). This was referred to as the process of consolidation. By such means, the use value of the property, reflecting its utility as a basic shelter, is slowly transformed into higher exchange values, reflecting the market valuation of the dwelling.

The major policy implication of Turner's work was that governments are best advised to help the poor to help themselves by facilitating spontaneous self-help, and by fostering and facilitating aided self-help, or what has become known as 'ASH'. There are three principal forms of ASH: (1) the upgrading of existing squatter housing; (2) the provision of site and service schemes; and (3) core housing schemes, where the shell of a house is provided on a site. Examples of these schemes are provided in Martin (1983) and Potter and Lloyd-Evans (1998: Chapter 7).

CASE STUDIES OF URBAN LOW-INCOME HOUSING

The urban poor in the developing world cannot afford houses that are professionally or formally surveyed, built and serviced. Where rental property is available, rents are frequently exorbitantly high. Thus poor citizens often construct peri-urban structures on land that has not previously been used for building purposes. Typical sites include small vacant plots within the old walled section of the city, as in Manila (Dwyer 1975). Another typical site is on steep hillsides, as exemplified in Caracas and Rio de Janeiro. Land that is swampy or subject to flooding offers further opportunities, as shown by the example of Singapore. Similarly, land adjacent to railways is also often occupied, as in the case of Kuala Lumpur. Recently reclaimed areas are also frequently colonised. Thus such settlements are open to a number of environmental and socio-economic risks, and a good deal of recent applied work has focused on this topic (see Main and Williams 1994; Hardoy *et al.* 1992).

The nature of the applied challenges presented by low-income housing in third world cities is well-exemplified in the case of Caracas, Venezuela. This primate capital city grew extremely rapidly following the development of the oil industry in the early twentieth century. In 1950, the city housed a population of just over half a million. By 1981, this had grown to just over 2 million. The city is located in a very narrow west-to-east valley, and sites for new development have become increasingly scarce.

In addition, since the 1950s, a very high proportion of the growth of the city has been accounted for by self-help low-income settlements, referred to locally as *barrios*. The individual dwellings that make up the *barrios* are called *rancho*. Many of these may well start as relatively poor dwellings, but the majority undergo reasonably rapid improvement, upgrading and consolidation (Plate 37.1). By 1985, 61 per cent of all dwellings in Caracas were classified as *barrios*. Inevitably, given the site of the city, an increasing proportion were to be found located on hillsides. Jimenez-Dias (1994) notes that by the mid-1980s, 67 per cent of the total area occupied by *barrios* in Caracas were geomorphologically unstable enough to justify the eviction of the residents.

Before 1950, the record suggests that landslides were rare in Caracas. Although some may have missed being recorded, only twelve were documented between 1800 and 1949. However, the records show that by the 1950s through to the 1960s, such events had increased to an average of one a year. Thereafter, the frequency of landslides

Plate 37.1 Housing in the process of various stages of consolidation, Caracas (*photograph: Rob Potter*).

increased dramatically, from an average of twenty-five per year in the 1970s to thirty-five per year during the early to mid-1980s.

However, the salient fact is that up to the 1960s, the majority of slope failures recorded in Caracas were associated with the incidence of earthquakes as the initiating mechanism; but from 1970, slope failures and mass movements became associated with the occurrence of heavy rainfall rather than seismic activity. Spatially, it is noticeable that they have tended to occur in the *barrio* areas. This is shown clearly in map form in Figure 37.4. The principal areas of landslide activity are almost exactly spatially coincident with the main *barrio* areas. As a recent example of the outcome, when tropical storm Bret hit Caracas in August 1993, over 150 people are thought to have been killed and thousands lost their homes as a result of mudslides on hilly shanty towns.

The example afforded by Hong Kong serves to pinpoint the pressing nature of the more mundane, day-to-day environmental problems that have to be faced by residents and authorities alike in large third world cities. In the case of Hong Kong, 6 million inhabitants are to be found living on a small piece of land that exhibits some of the world's highest population densities (Chan 1994). It is estimated that every day Hong Kong produces 23,300 tonnes of solid waste and 21 tonnes of floating refuse, along with 2 million tonnes of sewage and industrial waste water. In addition, the city has to deal with 100,000 tonnes of chemical waste annually. For the inhabitants of low-income communities, a variety of environmental hazards and safety risks have to be faced on a more or less daily basis. For the denizens of squatter areas, vulnerability to fires, dangerous slopes, landslides, slippery walkways, the effects of excess heat and the lack of public toilets, standpipes and other public facilities, are among the most prominent. For those living in inner city slums, the problems include the illegal construction of dwellings, falling balconies and outer walls, pollution from industry, noise pollution, congestion, and poor ventilation (*ibid.*).

However, despite such day-to-day problems, applied research has shown that Turner and Mangin were right, and that as long as real incomes and a degree of security of tenure are available to the residents of low-income settlements, then social stability, housing improvement and consolidation are frequently the outcome. A very good example of this type is provided by Eyre's examination of the squatter settlements of Montego Bay, Jamaica (Eyre 1972; 1997). The early work demonstrated that the majority of the population of the shanty town were not rootless migrants who had just travelled to the town from the countryside. Rather, they were shown to be urban residents of some standing, On average, household heads had lived in the urban area for eleven years. In addition, it was revealed that over three-quarters of the population living in the ten shanty towns had been born within the city itself. The majority of residents had jobs and were well integrated into the urban economy.

In a recent follow-up study, Eyre (1997) has shown that after ignoring squatters in the urban area for several decades, as part of a changing policy, the Jamaican authorities have started to clear some of these longstanding and well-established low-income housing areas. This has been happening, for example, in connection with the land at the end of the runway at the international airport at Montego Bay. Eyre notes how this reflects the fact that within a context of deregulation and neoliberalism, such land is now wanted for lucrative commercial development.

CONCLUSION

Low-income housing is a topic of continuing applied and practical importance, reflecting the fact that adequate provision must be seen as a basic human right (Chant 1996; Desai 1995). Research has been, and needs to continue to be, pluralist in nature. Empirical-cum-logical positivist approaches are needed in so far as baseline studies of housing conditions and housing change over time will always be needed, in addition to studies examining what it is like to live under such

Figure 37.4 Main areas of landslides in Caracas, 1974–9 (top), and principal *barrio* areas (bottom).

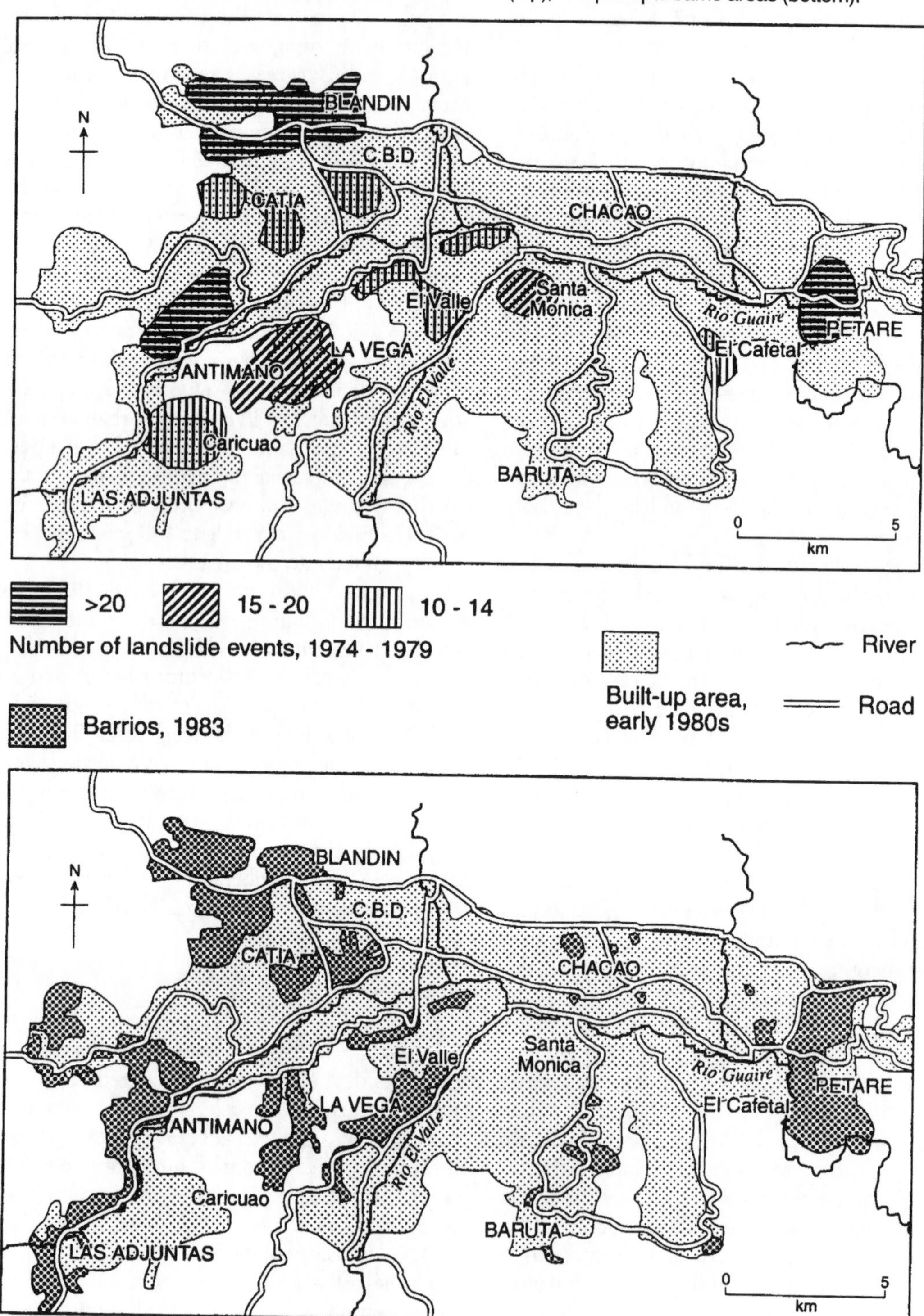

conditions. This latter aspect means that a humanist perspective continues to be required. Third, state and international housing policies need to be critiqued from a structuralist or political economy point of view (Box 37.2).

As argued by Potter *et al.* (1999: 242), in spite of three decades of varied responses to shelter needs, the problems of low-income housing seem to be as widespread as ever. For many years, the practical response to housing problems revolved around promoting self-help in the form of aided self-help, but it is now increasingly being recognised that renting is an important tenurial category (Gilbert and Gugler 1992) that may well

Box 37.2 The critical assessment of housing policies and housing issues

Another area of applied research focuses on housing issues and housing policy. What is being done to improve the supply of adequate and appropriate housing? In this, an issue of academic and practical salience is how the state interacts with citizens in the functioning of the housing market in order to derive appropriate policies for housing. Following the ideas of Turner and others on self-build housing, the nature of the interaction between the state and the citizen can be seen as central to housing provision in poor countries. How much should the state do, and in what domains? Or should the entire process be left to the individual/household?

Given the arguments of Turner, it is now widely accepted that in a housing crisis, the last thing that a government should do is build houses. However, it is equally clear that the responsibility of government does not begin and end there. The government still has responsibilities in relation to controlling the supply of land and providing for jobs and economic well-being. In St Lucia, for example, there is a substantial and growing number of squatters in the main urban areas (Potter 1994). The outcome is areas such as Four à Chaud in the capital, Castries. This area was built by squatters on reclaimed land adjacent to the port area. Although the area is by now well developed, with storm drains, during torrential rains the area is subjected to frequent floods. In many ways, however, this is a thriving and vibrant community, and the area is dotted with workshops such as the one on the main street, which produces excellent furniture (Plate 37.2).

Potter (*ibid.*) shows that in a seemingly paradoxical manner, the Ministry of Housing has little or nothing to do directly with the production of housing. This function is vested in a technically oriented national housing agency known as the Housing and Lands Development Corporation. This was established in 1971 but was largely dormant from the early 1980s to 1989. During its 'active' period, however, it was involved in only one scheme, which resulted in the production of 110 houses in one section of the outer city area. In the final analysis, these houses sold for $EC11,000, and a significant proportion were purchased by middle-income earners.

Potter (*ibid.*) linked this to the articulation of the modes of production. This suggests that the capitalist state interacts with pre-capitalist and capitalist forms in uneven ways. In respect of pre-capitalist or traditional forms, these are preserved where it is not in the interests of capital to intervene and replace them. This gives rise to the idea of no housing policy as a direct housing policy. The implicit housing policy appears to be to let the poor provide for themselves. In such a context, housing is effectively not a political issue, unless social unrest in some shape or form thrusts it onto the political agenda.

Plate 37.2 A thriving furniture workshop in the Four-à-Chaud low-income community, Castries, St Lucia (*photograph: Rob Potter*).

need to be promoted and encouraged.

However, as seen in some of the case studies presented in this chapter, the promotion of self-help can be seen as a practical and very cheap solution on the part of governments and international organisations (see Box 37.2). At the most negative, this view sees self-help housing as exploiting the poor yet further, and that this is happening more and more with the implementation of structural adjustment programmes and neoliberal policies. The multifaceted nature of the housing problems that are faced in most third world cities underlines the need for, and value of, applied research focusing on housing and environmental conditions and the efficacy of housing policies. There remains, therefore, a pressing continuing need for effective applied research on the problems and prospects of low-income housing as an issue that directly affects the quality of life of countless millions.

GUIDE TO FURTHER READING

For a recent general introduction to housing the poor in developing nations, see:
Aldrich, B.C. and Sandhu, R.S. (eds) (1995) *Housing the Urban Poor: Policy and Practice in Developing Countries*. London and New Jersey: Zed Books.

An overview of some of the methodological issues involved in the analysis of housing is provided by:
Jones, G. and Ward, P.M. (eds) (1994) *Methodology for Land and Housing Market Analysis*. London: UCL Press.

If you want to review a selection of different governments' handling of the interface between the individual, self-help imperatives and the responsibilities of the state, set in the context of the Caribbean, and including socialist Cuba, you should find the following useful:
Potter, R. B. and Conway, D. (eds) (1997) *Self-Help Housing, the Poor and the State in the Caribbean*. Knoxville: University of Tennessee Press.

Two excellent collections of general readings on self-help housing are provided by:
Ward, P. (ed.) (1982) *Self Help Housing: A Critique*. London: Mansell.
Mathey, K. (1992) (ed) *Beyond Self-Help Housing*. London and New York: Mansell.

Issues relating to housing and environmental conditions are effectively addressed in:
Main, H. and Williams, S.W. (eds) (1994) *Environment and Housing in Third World Cities*. Chichester: Wiley.
Hardoy, J.E., Mitlin, D. and Satterthwaite, D. (1992) *Environmental Problems in Third World Cities*. London: Earthscan.

A summary of housing conditions, plus chapters on sustainable urban development, the internal structure of cities, basic needs provision and employment and work, can be found in:
Potter, R.B. and Lloyd-Evans, S. (1998) *The City in the Developing World*. Harlow: Longman.

REFERENCES

Abrams, C. (1964) *Housing in the Modern World*. London: Faber & Faber.
Abu-Lughod, J. (1971) *Cairo: 1001 Years of the 'City Victorious'*. Princeton: Princeton University Press.
Berghall, P.E. (1995) *Habitat II and the Urban Economy*. Helsinki: United Nations University and the World Institute for Development Economics Research.
Chan, C. (1994) Responses of low-income communities to environmental challenges in Hong Kong. In H. Main and S.W. Williams (eds) *Environment and Housing in Third World Cities*, Chichester: Wiley, 125–32.
Chant, S. (1996) *Gender, Urban Development and Housing*. New York: United Nations Development Programme.
Clarke, C.G. (1975) *Kingston, Jamaica: Urban Growth and Social Change 1692–1962*. Berkeley: University of California Press.
Conway, D. (1982) Self-help housing, the commodity nature of housing and amelioration of the housing deficit: continuing the Turner–Burgess debate. *Antipode* 14, 40–6.
Conway, D. (1985) Changing perspectives on squatter settlements, intraurban mobility, and constraints on housing choice of the Third World urban poor. *Urban Geography* 6, 170–92.
Desai, V. (1995) *Community Participation and Slum Housing: A Study of Bombay*. New Delhi; Thousand Oaks, Calif., London: Sage.
Drakakis-Smith, D. (1982) *Urbanisation, Housing and the Development Process*. London: Croom Helm.
Dwyer, D.J. (1975) *People and Housing in Third World Cities*. London: Longman.
Eckstein, S. (1990) Urbanisation revisited: inner-city slum of hope and squatter settlement of despair. *World Development* 18, 165–81.
Eyre, L.A. (1972) The shantytowns of Montego Bay,

Jamaica. *Geographical Review* 62, 394–412.
Eyre, A. (1997) Self-help housing in Jamaica. In R.B. Potter and D. Conway (eds) *Self-Help Housing, the Poor and the State in the Caribbean*, Knoxville: University of Tennessee Press, 75–101.
Gilbert, A. (1983) The tenants of self-help housing: choice and constraint in the housing markets of less developed countries. *Development and Change* 14, 449–77.
Gilbert, A.G. (1981) Pirates and invaders: land acquisition in urban Colombia and Venezuela. *World Development* 9, 657–78.
Gilbert, A.G. and Gugler, J. (1982) *Cities, Poverty and Development: Urbanisation in the Third World*. Oxford: Oxford University Press.
Gilbert, A.G. and Varley, A. (1991) *Landlord and Tenant: Housing the Poor in Urban Mexico*. London: Routledge.
Hardoy, J.E., Mitlin, D. and Satterthwaite, D. (1992) *Environmental Problems in Third World Cities*. London: Earthscan.
Jimenez-Diaz, V. (1994) The incidence and causes of slope failures in the barrios of Caracas, Venezuela. In H. Main and S.W. Williams (eds) *Environment and Housing in Third World Cities*, Chichester: Wiley, 33–54.
Kumar, S. (1996) Landlordism in Third World urban low-income settlements: a case for further research. *Urban Studies* 33, 753–82.
Lee-Smith, D. (1990) Squatter landlords in Nairobi: a case study of Korogocho. In P. Amis and P. Lloyd (eds) *Housing Africa's Urban Poor*, Manchester: Manchester University Press, 175–88.
Lewis, O. (1959) *Five Families: Mexican Case Studies in the Culture of Poverty*. New York: Basic Books.
Lewis, O. (1966) The culture of poverty. *Scientific American* 215, 19–25.
Lloyd, P. (1979) *Slums of Hope? Shanty Towns of the Third World*, Harmondsworth: Penguin.
Main, H. and Williams, S.W. (eds) (1994) *Environment and Housing in Third World Cities*. Chichester: Wiley.
Mangin, W. (1967) Latin American squatter settlements: a problem and a solution. *Latin American Research Review* 2, 65–98.
Martin, R.J. (1983) Upgrading. In R.J. Skinner and M. Rodell (eds) *People, Poverty and Shelter: Problems of Self-Help Housing in the Third World*, London and New York: Methuen, 53–79.
McAuslan, P. (1985) *Urban Land and Shelter for the Poor*. London: International Institute for Environment and Development.
Okpala, D. (1996) Viewpoint: the second United Nations conference on human settlements (Habitat II). *Third World Planning Review* 18, iii–xii.
Potter, R.B. (1992) *Housing Conditions in Barbados: A Geographical Analysis*. Mona, Kingston, Jamaica: Institute of Social & Economic Research, University of the West Indies.
Potter, R.B. (1994) *Low-Income Housing and the State in the Eastern Caribbean*. Barbados, Jamaica, Trinidad and Tobago: The Press, University of the West Indies.
Potter, R.B. (1997) *Housing Background Report*. Barbados: Ministry of Health and the Environment, Government of Barbados.
Potter, R.B. and Conway, D. (eds) (1997) *Self-Help Housing, the Poor and the State in the Caribbean*. Knoxville: Tennessee University Press; and Barbados, Jamaica and Trinidad and Tobago: The Press, University of the West Indies.
Potter, R.B. and Lloyd-Evans, S. (1998) *The City in the Developing World*. London: Longman.
Potter, R.B., Binns, T., Elliott, J. and Smith, D. (1999) *Geographies of Development*. London: Longman.
Stokes, C. (1962) A theory of slums. *Land Economics* 38, 187–97.
Turner, J.F.C. (1967) Barriers and channels for housing development in modernizing countries. *Journal of the American Institute of Planners* 33, 167–81.
Turner, J.F.C. (1963) Dwelling resources in South America. *Architectural Design* 37, 360–93.
Turner, J.F.C. (1968a) Housing priorities, settlement patterns and urban development in modernizing countries. *Journal of the American Institute of Planners* 34, 354–63.
Turner, J.F.C. (1968b) The squatter settlement: an architecture that works. *Architectural Design* 38, 355–60.
Turner, J.F.C. (1969) Uncontrolled urban settlement: problems and policies. In G. Breese (ed.) *The City in Newly Developing Countries*, New York: Prentice-Hall, 507–35.
Turner, J.F.C. (1972) Housing as a verb. In J.F.C. Turner and R. Fichter (eds) *Freedom to Build: Dweller Control of the Housing Process*, London: Collier-Macmillan, 148–75.
Turner, J.F.C. (1976) *Housing by People: Towards Autonomy in Building Environments*. London: Marion Boyars.
Turner, J.F.C. (1982) Issues in self-help and self-managed housing. In P.M. Ward (ed.) *Self-Help Housing: A Critique*, London: Mansell, 99–113.
Turner, J.F.C. (1983) From central provision to local enablement: new directions for housing policies. *Habitat International* 7, 207–10.
Turner, J.F.C. (1985) Future directions in housing policies. Paper delivered at the International Symposium on the implementation of a support policy for housing provision, Development Planning Unit, University College London.

Turner, J.F.C. (1988) Issues and conclusions. In B. Turner (ed.) *Building Community: A Third World Case Book*, London: Building Community Books.

Turner, J.F.C. (1990) Barriers, channels and community control. In D. Cadman and G. Payne (eds) *The Living City*, London: Routledge, 181–91.

UNCHS (Habitat) (1996) *An Urbanising World: Global Report on Human Settlements 1996*. Oxford: Oxford University Press.

Ward, P. (1976) The squatter settlement as slum or housing solution: some evidence from Mexico City. *Land Economics* 52, 330–46.

Watson, M. and Potter, R.B. (1993) Housing and housing policy in Barbados: the relevance of the chattel house. *Third World Planning Review* 15, 373–95.

Watson, M. and Potter, R.B. (1997) Housing conditions, vernacular architecture and state housing policy in Barbados. In R.B. Potter and D. Conway (eds) *Self-Help Housing, the Poor and the State in the Caribbean*, Knoxville: University of Tennessee Press, 30–51.

Watson, M. and Potter, R.B. (1999) *Low-Cost Housing in Barbados: Evolution or Social Revolution?* Barbados, Jamaica and Trinidad and Tobago: The Press, University of the West Indies.

38

Informal sector activity in the third world city

Sylvia Chant

INTRODUCTION

Although the term 'informal sector' has been described as an 'unclear . . . but popular shorthand' (Gilbert 1998: p. 65), it is generally used to refer to a wide spectrum of 'precarious' or 'subterranean' employment found in urban areas of developing countries (Portes and Schauffler 1993: p. 33).

The term was first coined back in the early 1970s when, in the wake of massive rural–urban migration and limited labour absorption in the expanding industrial sector of developing economies, academics and policy makers became increasingly interested in how the swelling ranks of the urban poor were managing to 'get by'. On the basis of research in Ghana funded by the International Labour Office (ILO), the anthropologist Keith Hart found that low-income people who were unable to find waged work in import-substitution industries or the public sector devised wide-ranging and resourceful ways to generate income. From this, Hart developed the terms 'formal' and 'informal' employment, to refer to salaried jobs and self-employment respectively. He further distinguished between 'legitimate' and 'illegitimate' activities in the informal economy. The former comprised jobs that made a contribution to economic growth, albeit in small ways, such as petty commerce, personal services and home-based production. 'Illegitimate' informal activities, on the other hand, described occupations that, if not necessarily 'criminal' in nature, were arguably of dubious worth to national development. Activities falling under this heading included prostitution, begging, pickpocketing and scavenging (Hart 1973). Hart's essentially positive conceptualisation of the legitimate wing of the 'informal sector' was taken up by the Kenya Mission of the ILO, with early definitions encompassing ease of entry, small size, individual or family ownership of enterprises, labour-intensive production methods, low levels of skill, capital and technology, and unregulated and competitive markets (ILO 1972; see also Table 38.1).

While it remains true that informal enterprises are often small in scale, use rudimentary technology and are characterised by self-employment or family labour, Roberts (1994: p. 6) asserts that the most generally accepted contemporary definition of the informal sector is 'income-generating activities unregulated by the state in contexts where similar activities are so regulated'. Further noting that the formal sector is also characterised by a growing degree of informal labour arrangements, Roberts argues that 'The persisting interest in the idea of an informal economy lies not in its analytic precision, but because it is a useful tool in analysing the changing basis of economic regulation' (*ibid.*).

Definitions and characterisations of the informal sector have formed a major arena of debate over the years and are discussed in greater detail below. Other issues that have marked discussion

Table 38.1 Stereotypical characteristics of formal and informal employment.

Formal sector	*Informal sector*
Large-scale	Small-scale
Modern	Traditional
Corporate ownership	Family/individual ownership
Capital-intensive	Labour-intensive
Profit-oriented	Subsistence-oriented
Imported technology/inputs	Indigenous technology/inputs
Protected markets (e.g. tariffs, quotas)	Unregulated/competitive markets
Difficult entry	Ease of entry
Formally acquired skills (e.g. school/college education)	Informally acquired skills (e.g. in home or craft apprenticeship)
Majority of workers protected by labour legislation and covered by social security	Minority of workers protected by labour legislation and covered by social security

Sources: Chant 1991b: p. 185, Table 9.4; Drakakis-Smith 1987: p. 65, Table 5.5; Gilbert and Gugler 1992: p. 96.

of the informal sector include the nature of its link with the formal sector of the urban economy, its behaviour during recession and neoliberal economic restructuring, its role as a mechanism of survival versus a 'sector of entrepreneurship', and the question of whether policies should encourage or inhibit its growth. Another key element in research on the informal sector is its association with the increased labour force participation of women in developing countries, especially during the last two decades. The present chapter attempts to summarise the debates that have surrounded these aspects of informal sector enquiry and to identify the likely prospects for informal employment at the beginning of the twenty-first century.

WHAT IS THE INFORMAL SECTOR? CHARACTERISTICS AND CONDITIONS

Although it is primarily an employer of low-income groups, the informal sector is anything but homogeneous. While the bulk of informal workers are found in low-productivity, low-profit commercial and service activities, a substantial number are engaged in manufacturing (see Table 38.2). In some cases, people work alone or in family groups in 'independent' forms of informal activity. Acknowledging that economic activities are rarely completely unconnected with each other (as discussed later), 'independent' enterprises may comprise 'frontroom' eateries, stalls or shops (see Plate 38.1), domestic-based workshops or manufacturing outlets (Plate 38.2), the sale of home-made food on the streets, or the touting of services such as shoe-shining that require little in the way of formally acquired skills or capital assets. In other circumstances, people's work in the informal sector may be linked more directly with the formal economy. This can include the distribution of formal sector goods by informal means (e.g. tricycle transport), or performing tasks subcontracted by the formal sector. The latter is usually referred to as 'outwork' or 'domestic piecework', with workers

Table 38.2 Percentage of production that is informal in selected developing countries.

	Manufacturing	*Transport*	*Services*	*Total*
Africa				
Burundi (1990)	35	8	18	25
Congo (1984)	39	10	36	33
Egypt (1986)	21	29	15	18
Gambia (1983)	48	16	57	51
Mali (1990)	45	45	37	40
Zambia (1986)	41	7	48	39
Latin America and the Caribbean				
Brazil (1990)	12	23	23	18
Costa Rica (1984)	14	9	16	15
Honduras (1990)	26	17	28	26
Jamaica (1988)	19	23	30	25
Mexico (1992)	9	20	20	16
Uruguay (1985)	16	10	16	16
Venezuela (1992)	16	46	22	23
Asia and the Pacific				
Indonesia (1985)	38	44	56	49
Iraq (1987)	15	33	7	12
Republic of Korea (1989)	17	34	44	30
Malaysia (1986)	13	20	23	19
Qatar (1986)	1	3	1	1
Syrian Arab Republic (1991)	21	38	22	24
Thailand (1990)	10	40	18	16
Fiji (1986)	14	21	12	13

Source: United Nations 1995: p. 135, Table 9.

Plate 38.1 Home-based *'sari-sari'* store, Boracay, the Philippines (*photograph: Sylvia Chant*).

paid by the unit of production completed. Outwork is particularly common in industries in which part of a commodity's elaboration lends itself to labour-intensive methods that can be performed with minimal technology in people's own homes. Key industries here include toys, footwear and clothing (see Benería and Roldan 1987; Chant and McIlwaine 1995: Chapter 4; Peña Saint Martin and Gamboa Cetina 1991).

As noted above, a major feature of most informal employment is that it is unregulated by the state. Yet while regulation clearly implies legality, legality itself is a multidimensional concept, with Thomas (1997: p. 6) pointing out that 'being legal usually involves complying with a number of regulations often imposed by a variety of different authorities'. More specifically,

Plate 38.2 Small-scale shoe factory in a residential neighbourhood, León, Mexico (*photograph: Sylvia Chant*).

Tokman (1991: p. 143) identifies three types of legality pertinent to the demarcation between formal and informal sector enterprises.

1 Legal recognition as a business activity (which involves registration, and possible subjection to health and security inspections).
2 Legality in respect of payment of taxes.
3 Legality *vis-à-vis* labour matters such as compliance with official guidelines on working hours, social security contributions and fringe benefits.

In many cases, informal businesses or operators may be legal in some ways but not in others. Social security is often the most costly aspect of legality, so while micro-enterprises are often registered as businesses with the relevant local or national authorities, they may remain illegal on account of their non-observance of legislated labour arrangements. Indeed, with reference to 1989 figures, Roberts (1994: p. 16) notes that up to three-quarters of workers in Mexican micro-enterprises are not covered by social security. In Latin America more generally, it is estimated that only 2 to 5 per cent of self-employed persons (the largest group of the informally employed in the continent) have access to social security, mainly due to high costs, administrative difficulties, lack of incentives due to the eroding value of pensions, and uncertainty in occupational prospects (Tokman 1991: pp. 152–3; see also Lloyd-Sherlock 1997).

Lest it be thought that lack of social security applies only to informal economic operations, it is important to bear in mind that formal sector firms may engage in similar practices. In Mexico in 1989, for example, formal sector employers were not paying social security contributions for up to 17 per cent of their workforce (Roberts 1994: p. 16). This is also found in the Philippines, where many multinational and indigenous factories use loopholes in the law to avoid commitments to social security, fringe benefits and redundancy packages. This often takes the form of hiring workers on a temporary basis (usually up to three months), letting them go when their contracts expire but re-employing them at a later date. This fulfils the twin imperative of capitalising on workers' previous training but denying them the continuous service required for the status of permanent employee and its attendant privileges (Chant and McIlwaine 1995: Chapter 4). In many countries, the pressure exerted by international financial institutions to reduce 'structural rigidities' in the workforce and to encourage greater labour 'flexibility' has led to the passing of labour law revisions to facilitate these processes (Green 1996: pp. 109–10; Tironi and Lagos 1991). In Peru, for example, legislation was introduced in 1991 that gave rights to employers to hire people on 'probationary' contracts with minimal entitlement to fringe benefits and no compensation payable on their release (Thomas 1996: p. 91). In Bolivia, the new economic policy dating from 1985 saw reductions in protection for workers and an elimination of wage indexation, leaving salaries to be bargained within individual firms (Jenkins 1997: p. 113). While such practices may be more frequent now than in the past, they were by no means absent in large firms prior to the onset of crisis and restructuring. As Roberts (1991 p. 118) notes with reference to Mexico, '"implicit deregulation" ... antedates by many years the present policy of explicit deregulation' (see also Standing 1989; 1991).

Aside from a rising incidence of short-term contracts, easier hiring and firing policies and the restriction of trade union activities, formal sector employers have also sought to reduce their operating costs by subcontracting production to enterprises and workers outside the factories (see Eviota 1992; Pineda-Ofreneo 1988). In Bolivia, for example, Jenkins (1997: p. 119) notes that the 1980s gave rise to an increasing concentration of manufacturing in small-scale factories and workshops and a doubling of the percentage of operatives working 49 hours per week or more. In Mexico, too, the crisis of the 1980s forced footwear firms in the city of León to farm out increasing amounts of production to small-scale home-based workshops and individual outworkers. This not only helped to cut labour costs but also facilitated greater flexibility in the face of uncertain demand (Chant 1991a).

PERSPECTIVES ON THE EVOLUTION OF THE INFORMAL SECTOR

Time-series data on informal sector activity need to be treated with extreme caution, not only on account of the irregular and/or clandestine nature of informal work but also because of shifting classificatory schemata by different governments and regional organisations (Salahdine 1991; Thomas 1995). Bearing in mind that this also makes international comparisons difficult, it would appear that in most parts of the world the informal sector has increased in recent decades. In Latin America, for example, one set of estimates suggests that the share of the workforce in informal activities rose from 16.9 to 19.3 per cent between 1970 and 1980 (Tokman 1989: p. 1067). From 1980 onwards, levels of informal employment have grown further, especially in cities. Informal workers in Latin America's urban labour force, for example, rose from an overall average of 25.6 per cent to 30 per cent between 1980 and 1990, vastly outstripping growth in the formal sector (Gilbert 1995b). Over a similar period (1980–1988), disguised (and open) unemployment in sub-Saharan Africa grew by one-sixth, a rate four times higher than in the previous decade (Vandemoortele 1991: p. 84). By 1985, 60 per cent of the urban workforce in the region was informally employed (*ibid.*). In Asian countries such as Thailand, 50 per cent of the urban workforce was in informal occupations in 1987 (Poapongsakorn 1991: p. 113), and for developing countries more generally, micro-enterprises have come to employ between 36 and 60 per cent of the labour force, even if they contribute only 20 to 40 per cent of GDP (Chickering and Salahdine 1991b: p. 3).

While the growth of informal employment could conceivably indicate economic health in the sector (Salahdine 1991: p. 37), increases during the 1960s and 1970s have usually been attributed to labour surpluses in cities arising from rural–urban migration (Portes and Schauffler 1993). In the 1980s, informal sector growth has been more closely linked with the corollaries of recession and restructuring. Important factors here include cutbacks in public employment, the closure of formal sector firms due to increased competition provoked by lowered tariff barriers, and declining labour demand in the formal sector (see, for example, Alba 1989: pp. 18–21; Gilbert 1995a: p. 327). Another significant tendency has been for some smaller firms to join the ranks of the informal sector as a result of declining ability to pay registration, tax and labour overheads (Escobar Latapí 1988; Roberts 1991: p. 129). The latter bears out the argument that 'informality for the self-employed is basically a household survival strategy in the face of unemployment and declining real wages' (Roberts 1995: p. 124; see also below). As Thomas (1996: p. 99) summarises, the 'top-down' informalisation promoted by governments and employers has been matched by a 'bottom-up' informalisation stemming from the need for retrenched formal sector workers and newcomers to the labour market to create their own sources of earnings and/or to avoid the punitive costs attached to legal status. Indeed, it is noteworthy that middle-class as well as poor households are increasingly having to turn to informal sector activities in order to protect incomes and consumption (see Lozano 1997).

THE BEHAVIOUR OF THE INFORMAL SECTOR DURING RECESSION AND RESTRUCTURING

In light of the above, it is hardly surprising that the informal sector has become increasingly competitive during the last two decades. As Miraftab (1994: p. 468) argues with reference to Mexico: 'Poor people have had to concentrate their daily activities with much greater intensity around the issue of survival. This has implied not only longer hours of work, but also the need to be extremely innovative to earn a living at the edges of the urban economy' (see also Escobar Latapí and González de la Rocha 1995). Indeed, although ever more creative strategies to generate income can be witnessed in both the streets and houses of third world cities, competition is such that, according to ILO figures for Latin America and the Caribbean, there was a 42 per cent drop in income in the informal sector between 1980 and 1989 (Moghadam 1995: pp. 122–3). By the same token, it is important to bear in mind that informal incomes may have been better protected than formal sector wages in view of declining demand for imports in favour of cheap food items and locally produced basic consumer goods (see Vandemoortele (1991) on sub-Saharan Africa).

Nonetheless, limits to the continued expansion of informal sector employment arise from lower purchasing power among the population in general and greater numbers of people needing to work (Roberts 1991: p. 135). The latter is in part the legacy of high fertility and declining mortality rates throughout most of the developing world during the 1960s and 1970s, and in part the outcome of increased female labour force participation. Indeed, the saturation of the informal sector is often argued to have hit women the hardest given their disproportionate concentration in the sector and the fact that their limited skills and resources confine them to the lowest tiers of informal activity (see Bromley 1997; Moser 1998; Scott 1994; see also Table 38.3). In Costa Rica, for example, where 41 per cent of the informal workforce is female, low-income women in the northwest province of Guanacaste complain that their limited skills and capital restrict them to petty/part-time ventures such as selling home-made sweets, flavoured ices and pastries outside local schools or on the streets (Chant 1994b). Moreover, since most of their neighbours are forced into the same kind of business, competition is so intense that some abandon the attempt altogether, thereby adding to what is commonly referred to as the 'discouraged worker' effect (Baden 1993: p. 13). Recognising that the trading of basic goods may be one of the few options open to poor women, it is important to note that this has been further threatened by a range of macro-economic policies associated with structural adjustment such as tightened control on credit, the lifting of food subsidies, and greater influxes of imported convenience foods stemming from trade and currency liberalisation (Manuh 1994; Tinker 1997).

GENDER AND THE INFORMAL SECTOR: INFORMALISATION AND THE 'FEMINISATION' OF LABOUR

Although, as indicated by Table 38.3, the informal sector is by no means a 'female sector (Scott 1990), rises in informal employment in recent decades have been linked to a phenomenon commonly referred to as the 'global feminisation of labour' (Standing 1989). Indeed, bearing in mind that women's economic activities often fall outside the net of official data collection because of their informal and/or part-time nature,

Table 38.3 Percentage of male and female labour force in the informal sector in selected developing countries.

	Manufacturing		*Transport*		*Services*		*Total*	
	Men	*Women*	*Men*	*Women*	*Men*	*Women*	*Men*	*Women*
Africa								
Burundi (1990)	31	60	13	0	17	21	21	32
Congo (1984)	39	43	11	0	21	60	25	57
Egypt (1986)	22	5	31	0	18	3	21	3
Gambia (1983)	38	100	13	0	23	60	25	62
Mali (1990)	63	35	50	0	39	33	45	34
Zambia (1986)	31	81	8	0	31	71	29	72
Latin America and the Caribbean								
Brazil (1990)	14	5	24	2	23	24	19	21
Costa Rica (1984)	14	13	11	0	7	22	8	19
Honduras (1990)	15	52	29	0	26	29	21	34
Jamaica (1988)	21	11	29	0	27	32	25	28
Mexico (1992)	8	11	21	2	30	16	22	15
Uruguay (1985)	15	20	12	0	19	14	17	15
Venezuela (1992)	13	30	50	10	25	20	23	21
Asia and the Pacific								
Indonesia (1985)	28	57	44	20	47	68	41	65
Iraq (1987)	15	13	34	0	7	4	5	11
Republic of Korea (1989)	24	21	36	40	78	52	48	41
Malaysia (1986)	9	22	22	5	21	26	17	24
Qatar (1986)	0	0	0	0	1	0	1	0
Syrian Arab Republic (1991)	21	18	39	0	91	4	61	7
Thailand (1990)	8	14	43	14	11	30	12	24
Fiji (1986)	15	20	25	0	13	9	15	10

Source: United Nations 1995: p.135, Table 9.

in all regions except sub-Saharan Africa, women's share of the workforce seems to have risen between the 1970s and the 1990s (Table 38.4).

Although differing within and between regions, prominent reasons for increased female labour force participation in the last twenty to thirty years include a decline in agricultural employment relative to growth in industry and services, rural–urban migration, rising education levels, and declining fertility rates (see, for example, Bullock 1994; Manuh 1994; Safa 1995a, b). More recently, however, and recognising that women's economic activity was already rising before the 1980s debt crisis (Gilbert 1998: p. 74), there is strong evidence to support the notion that neoliberal economic restructuring has played a significant role. At one level, this is revealed by continued, not to mention marked, increases in female labour force participation during the so-called 'lost decade' of the 1980s. In Brazil, for example, Humphrey (1997: p. 171) observes that women's share of employment in the São Paulo metropolitan area grew from 33 to 38 percent between 1980 and 1990. Beyond this, where women in poor urban communities have been interviewed about their reasons for increased involvement in remunerated work, the principal cause cited is almost invariably financial necessity (see, for example, Benería 1991; Chant 1996; García and de Oliveira 1990; González de la Rocha 1988; Moser 1992; 1997). In many cases, this has resulted from the eroding purchasing power of male breadwinners' wages, and in others because men have lost their jobs altogether. In Uruguay, for example, Nash (1995: p. 155) notes that the participation of women in the

Table 38.4 Women's share (percentage) of the labour force in developing regions, 1970–90.

	1970	*1990*
North Africa	9	21
Sub-Saharan Africa	39	38
Latin America	20	34
Caribbean	32	43
East Asia	40	41
Southeast Asia	35	40
South Asia	20	35
West Asia	19	25

Source: UN 1995: p. 109, Chart 5.4A.

workforce increased from 38.7 to 44.2 per cent between 1981 and 1984 due to rising levels of male unemployment. The growing proportion of women responsible for maintaining their households without male partners is also important (Chant 1997: pp. 15–18).

On the demand side of the labour market, the recruitment of female workers in many countries has persisted during economic restructuring due to the expansion of export-oriented industrial production (Chant and McIlwaine 1995; Pearson 1986; Safa 1995a, b). In a climate of increased competition, firms have frequently resorted to increasing their female workforce, either directly by recruitment in industrial plants (especially common in areas where there are multinational export-processing zones such as Mexico, Brazil, Venezuela, Puerto Rico, the Dominican Republic and the Philippines – see Dwyer 1994; Chant and McIlwaine 1995; Moghadam 1995: p. 122), or on a subcontracted piecework basis (see Benería and Roldan 1987; Pineda-Ofreneo 1988). Although on the one hand this may have increased income-generating opportunities for women, there are decidedly more benefits for employers. These include reduced production costs, savings on social security contributions, fragmentation of the workforce, and exploitation of the low 'aspiration wages' of married women (see Miraftab 1994; Peña Saint Martin 1996). At the same time, it is important to recognise that the work offered is often precarious in the short and longer term. In industries such as textiles and electronics, for example, women are not only locked into the least skilled jobs but also face displacement by male workers as automation increases (Acero 1997; Roberts 1991: p. 31; Ward and Pyle 1995: p. 42).

In effect, the worldwide increase in female labour force participation has occurred during a time in which the conditions of employment have deteriorated considerably, and in which future prospects are anything but guaranteed. As Moghadam (1995: pp. 115–16) notes: 'The global spread of flexible labour practices and the supply-side structural adjustment economic package coincide with a decline in labour standards, employment insecurity, increased joblessness, and a rise in atypical or precarious forms of employment'. Women's tenuous position in the labour market is not surprisingly reflected in low earnings, especially in the informal sector. Evidence from Colombia, for example, suggests that women's average incomes in the informal sector are only 74 per cent of men's, compared with 86 per cent in formal occupations (Tokman 1989: p. 1071). In the Dominican Republic, 62 per cent of female informal workers earn below the poverty level compared with only 35 per cent of male informal workers (Lozano 1997: p. 163, Table 6.4). Gender differentials in informal sector earnings are usually due to women's paltry start-up capital, their lower range of formally acquired skills and job experience, the demands placed on their time for income-generating activity by housework and childcare, and their restricted use of space. Lessinger's (1990) study of market traders in Madras, India, for example, shows that women have a much smaller range of spatial operation than men because venturing into communities where no one knows them provokes gossip and social opproprium. Aspersions on the morality of 'mobile' women also prevent them from buying their fruit and vegetables from the male-dominated space of the wholesale market. Women have either to be chaperoned or to buy their goods from intermediaries. Although slightly different reasons for gender divisions of labour and profits apply in other countries, there is ample evidence to suggest that women are at a decided disadvantage in the informal sector in many places (see Box 38.1).

Aside from gender, it is worth noting the importance of age in informal sector divisions. Child workers in developing countries are almost always employed under informal arrangements and are found in the lowest echelons of economic activity. As Green (1998: p. 35) argues with reference to Latin America: 'In this brave new world of "flexible working patterns", children are often perfect employees – the cheapest to hire, the easiest to fire and the least likely to protest.'

LINKAGES BETWEEN THE FORMAL AND INFORMAL SECTORS

Having noted an increase in informal employment in recent years, to what extent is it possible for the informal sector to continue expanding in the wake of changes in the formal economy? This question arises out of recognition that 'the informal sector enjoys a largely symbiotic relationship with the modern manufacturing and service sectors' (Becker and Morrison 1997: p. 93). Detailed empirical studies have revealed that the informal sector is linked to the formal sector in a wide range of subordinate ways, that enterprises with the fewest direct links with the formal sector are likely to be the least dynamic economically, and that, over time, the informal sector is increasingly likely to lose its independent basis for subsistence (see Roberts 1995: pp. 120 *et seq.* for discussion and references).

Taking into consideration the argument that 'the informal sector in Latin America and elsewhere is an adjunct to the large-scale sector of the

Box 38.1 Aspects of the gender division of labour in a Nairobi shanty town

Kenya is one of the most rapidly urbanising countries in the developing world, with one-quarter of its population currently residing in towns and cities. Despite reasonably positive (if variable) rates of economic growth in the post-war period, unemployment and underemployment in the urban economy have risen steadily since the early 1970s. This has forced many of the urban poor to resort to informal modes of shelter provision and to informal income-generating activities. In Mathare Valley, a shanty town that houses around one-fifth of Nairobi's population, around 80 per cent of the population make their living from the informal sector. These occupations fall into five main groups: (1) the entertainment industry; (2) the production of self-built housing for sale or rent; (3) shops; (4) other small businesses, and (5) hawking (petty commerce/ambulant vending).

Although gender divisions are found within and among these sub-sectors, men tend to dominate the upper tiers of the informal economy. Common male activities include running a *duka* (small shop), tailoring and taxi owning. Women, on the other hand, are overwhelmingly concentrated in activities related to their domestic roles, such as domestic service, sex work and/or the sale of small quantities of basic subsistence items such as vegetables, *kimera* (millet flour), and charcoal. The gendered hierarchy of informal employment is partly due to the fact that men's choice of economic activity is wider than women's because they receive more schooling, more instruction in technical subjects and/or have greater opportunity to acquire skills on the fringes of the modern industrialised economy. Differentials also arise from men's greater ability to invest in their businesses. Higher earnings and investment capacity stem not only from the fact that men do not have to divide their time between employment and reproductive labour but also because they can usually count on additional income from their wives. Women workers, on the other hand, are often unpartnered and, as such, solely reponsible for child rearing and household expenditures. Men's greater levels of investment tend to lead to greater returns, thereby giving rise to a widening of male–female income disparities over time. In addition, as pressures on the Nairobi labour market have increased, there has been a tendency for men to colonise niches of the informal sector that have traditionally employed women. A classic example of this is the so-called 'entertainment' industry, which revolves around the sale of alcohol, companionship and sex. As little as twenty-five years ago, the entertainment sector was virtually the exclusive domain of women. Recently, however, some of the more profitable parts of the industry have moved into male hands.

Demand for recreational services and entertainment in Nairobi over the years has been fuelled by the recruitment of migrant male labour in the police, the army and the health service. While many women in the past had a fairly lucrative trade in the illegal brewing of maize beer (*buzaa*), often sold from backroom bars (*shebeens*) in their own houses, from the 1970s onwards, this alternative began to dwindle. This was partly because *buzaa* production was taken on by a large multinational firm (using the brand name 'Chibuku'), and partly because police raids on illegal brewers in the shanty towns grew in number and intensity. Men have tended to take over the running of bars, (a) because they have had more capital to make wholesale purchases of commercially brewed Chibuku, and (2) because they have been able to afford to pay large bribes to the police.

Source: Nelson 1997.

economy, producing those goods for which the market is so reduced and so risky that large-scale enterprises are not interested in entering' (*ibid.*: p. 121), links include the formal sector's use of informal sector enterprises for production, distribution, marketing and retail (Table 38.5). In times of crisis, declining fortunes in the formal sector are likely to cut off valuable sources of contracts and supplies to the informal sector. Thus, although the informal sector has continued to expand during the years of crisis and restructuring (and recovery in some countries in the early 1990s), it has not been able to absorb all the job losses in the formal sector. This undoubtedly accounts for the fact that open unemployment in many places has risen in an unprecedented manner during the last two decades. In Argentina, for instance, urban unemployment in 1991 was 20.2 per cent, compared with 5.6 per cent in 1986 (Bulmer-Thomas 1996: p. 326, Table A9). In the Côte d'Ivoire, where formal sector jobs fell by 12 per cent between 1980 and 1985, unemployment escalated from 2 to 14 per cent (Vandemoortele 1991: p. 94).

Leading on from the above, the behaviour of labour markets in most developing countries over the last twenty years has borne out the importance of eschewing notions of labour market dualism, whereby the formal and informal sectors are conceived as discrete or autonomous entities. Instead, and along the lines of Moser's seminal neo-Marxian exposition on 'petty commodity production', the labour market is more appropriately conceptualised as a continuum of productive activities. This entails recognition of complex gradations of formality and linkages between different enterprises, which may well be more exploitative than benign (Moser 1978; see also Drakakis-Smith 1987: pp. 72–4; Roberts 1994; Thomas 1996). Latterly, these ideas have been worked into the thesis of 'stucturalist articulation', which views urban labour markets as 'unified systems encompassing a dense network of relationships between formal and informal enterprises' (Portés and Schauffler 1993: p. 48). Although links between the large- and small-scale firms are often exploitative, however, it is recognised that some opportunities may be opened up for informal enterprises by globalisation and neoliberal strategies of export promotion (Portés and Itzigsohn 1997: pp. 240–1). In many respects, this has encouraged recommendations for more active and sympathetic policy stances towards the informal sector.

Table 38.5 Economic linkages between the formal and informal sectors of the urban economy.

Backward linkages

- Informal vendors sell products (e.g. soft drinks, cigarettes) obtained from manufacturers, wholesalers and retailers in the UFS.
- Informally produced goods such as cooked foodstuffs, home-made clothing and embroidered items are likely to comprise raw materials supplied by the UFS.

Forward linkages

- The UIS may produce intermediate goods destined for final elaboration and distribution through the UFS. This may occur through subcontracting or purchase on the part of the UFS.

Benefits for UFS of subcontracting to UIS

- Formal employers may avoid paying legal minimum (or above-mininum) wage.
- Formal employers avoid obligations to provide social security contributions and fringe benefits to workers in informal enterprises.
- Formal employers can respond more flexibly (and at lower cost) to fluctuations in product demand.
- The (cheap) goods and services produced in the UIS, when consumed by UFS workers, arguably subsidise the wages of the UFS.

Source: Thomas 1996: pp. 56–9.
Note: UFS = Urban formal sector.
UIS = Urban informal sector.

POLICY APPROACHES TO THE INFORMAL SECTOR: FROM PROBLEM TO PANACEA?

Until relatively recently, there was no explicit policy towards the informal sector in regions such as Latin America. This was partly due to anticipation that labour surpluses would eventually be absorbed by formal industry and services (Tokman 1989: p. 1072). In some respects, this was borne out empirically by the fact that between 1950 and 1980, the informal sector declined in a range of countries, including Argentina, Chile, Colombia, Peru, Mexico and Venezuela (Gilbert 1998: p. 71). Another important reason for policy neglect is that many economists and civil servants construed informal activities as 'parasitic', 'unproductive' and tantamount to 'disguised unemployment'. As such, there was little will to promote growth in the sector (Bromley 1997: p. 124). In effect, the informal sector was (and in many circles still *is*) viewed as an employer of 'last resort' or a fragile means of basic subsistence in situations where social welfare provision for those outside the formal labour force is minimal or non-existent (Cubitt 1995: p. 163; Gilbert 1998: p. 67).

Condemned to a precarious existence on the margins of the urban labour market, it is no surprise that informal sector entrepreneurs have faced serious barriers to survival and growth. The fact that most governments in developing countries have adopted policies to subsidise the large-scale capital-intensive sector (for example via credit transfers, direct and indirect market protection such as tariffs, quotas and so on), has effectively discriminated against the informal sector (Chickering and Salahdine 1991c: p. 186; Grabowski and Shields 1996: p. 172). On top of this, while the process of 'becoming legal' varies across countries and occupations, it is often extremely time-consuming and costly (Tokman 1991: p. 147). With reference to street occupations in Cali, Colombia, for example, Bromley (1997: p. 133), stresses that 'regulations are excessively complex, little known, and ineffectively administered, resulting in widespread evasion, confusion and corruption'. In this light, the productive potential of what some describe as the 'true capitalist entrepreneurs' of developing countries has been 'stunted by excessive and inappropriate interventions on the part of the state via redundant regulation and thwarting red tape' (Tokman 1989: p. 1068). In addition to lack of government support and the prohibitive costs involved in becoming legal, informal entrepreneurs have not uncommonly faced harrassment or victimisation (Thomas 1996: pp. 56–7; see also Box 38.2).

Although the case study presented in Box 38.2 may suggest otherwise, a number of commentators have observed that in the last ten to fifteen years, government authorities, along with planners and social scientists, have come round to the notion that the informal sector is more of a seedbed of economic potential than a 'poverty trap' (Cubitt 1995: p. 175). With reference to Sri Lanka, for example, Sanderatne (1991: p. 96) declares:

> Informal economic enterprises, with the exception of illegal and illicit activities, are accepted as essential components of the Sri Lankan economy. Some of these informal activities have deep social roots and are essential to the life of the community; many are vertically or horizontally linked to, and have a symbiotic relationship with, the formal sector. What is more, informal activity, though continuously changing in character, has grown rather than diminished with the country's increasing development and has become an essential means of meeting the community's economic and social needs.

These more positive constructions hark back to the 'popular entrepreneurship' concept originally emphasised in Hart's work. They are also endorsed by numerous empirical studies showing that some informal workers earn more than salaried workers, that self-employment can be a source of pride or prestige, that informality permits flexibility and ready adaptation to changing demand and family circumstances, and that people often acquire skills in the formal sector that can subsequently be used to advantage in their own businesses (see, for example, Alonzo

Box 38.2 Street traders in Mexico City

In recent years, Mexico City's authorities have attempted to control and clean up street trading in the historic centre of the capital and, more specifically, to relocate street traders to purpose-built markets. This initiative has been motivated, among other things, by the need to resolve traffic congestion, to reduce fire and pollution hazards, and to diminish threats to public hygiene and health, the latter because many street traders operate outside government regulations on health, safety and quality control. These problems have intensified with the huge increase in street trading during the years of debt crisis and structural adjustment.

A major source of opposition to street trading practices has come from formal sector retailers in the centre of Mexico City, represented by the Mexican Chamber of Commerce, CONCANACO. This group claims that street traders block access to their own enterprises and degrade the environment. On top of this, Mexico City authorities are under pressure to protect the '*Centro Histórico*', which they themselves designated as a conservation zone in 1980, and which became a UNESCO world heritage site in 1984.

In 1987, the Department of the Federal District recommended the pedestrianisation of various streets in the Centro Histórico to enhance the environment for residents and tourists, and to relocate a number of street traders to thirty-seven purpose-built markets in the greater downtown area. The new market sites were created to ensure that street trading became more regulated and acceptable, and less visible. Their very creation, however, also represents a tacit acceptance on the part of the authorities that street trading was helping to solve unemployment problems in the Federal District as well as making an important contribution to urban economic life.

In July 1993, the Department of the Federal District passed the enforcement order to clean up and regulate commercial activity and to relocate street traders to the new market areas. As well as representing a positive move to appease the formal retailers, Mexico's imminent accession to NAFTA (North American Free Trade Area) had become an issue. Not only was street trading strongly associated with crime, but the huge number of street traders in a much-visited site in Mexico City was felt to be an embarrassment for the Mexican government, which was concerned about Mexico's image as a secondary member in NAFTA. Understandably, however, street traders were unhappy about the initiative, which meant disrupting links with regular clientele and having to pay higher operating costs for a pitch in one of the new markets. This opposition resulted in demonstrations and marches to the presidential palace.

In light of the above, it is not surprising that a survey conducted in 1995 indicated that there had only been partial take-up of stalls in the new markets and that, aside from an overall increase in ambulant vendors, many of the street traders had just been displaced slightly outwards from the Centro Histórico. Enforcing the order has been limited by the costs of policing and the vigilance and political links of street trader leaders with government officials. A crucial element in the street traders' *de facto* victory has been their vital role in providing low-cost goods during a time of declining incomes.

Source: Harrison and McVey 1997.

1991; Salahdine, 1991). By the same token, the idea that the informal sector may be an arena of 'accumulation', rather than 'subsistence' or 'last resort', differs from Hart's analysis in that the sector is not seen as originating from surplus labour supply but from excessive regulations in the economy (Portes and Schauffler 1993: pp. 39–40).

The work of the Peruvian economist Hernando de Soto has been critical in this shift in perspective, with his key book, *The Other Path*, championing the cause of the informal sector as 'a grass-roots uprising against unjust and excessive regulations' created by governments in the interests of the society's powerful and dominant groups (Bromley 1997: p. 127). In contexts in which legal systems are designed to accord with the interests of the economic elite, de Soto (1989) challenges that illegality is the poor's only, and justifiable, alternative. Thus, although the informal sector is technically 'illegal', it is not criminally so but is more to do with 'non-conformity' with bureaucratic rules and regulations. Emphasising the fundamental ways in which the existence of the informal sector relieves unemployment, provides a gainful alternative to crime, and harnesses the entrepreneurial talent of the disaffected masses, de Soto suggests that the governments would be best advised to stimulate and protect informal entrepreneurs and to grant them greater freedom.

Although the work of de Soto was grounded in the Peruvian experience, his policy recommendations have appealed to both left and right of the political spectrum and have been taken up fairly roundly in a more general development context (Chickering and Salahdine 1991b; Portés and Itzigsohn 1997). Indeed, enthusiasm for deregulation has grown to such an extent that it has virtually dissolved the association between informality and stagnation/marginalisation and shone the spotlight instead on the entrepreneurial potential of informal sector workers (Tokman 1989: p. 1068).

While there is probably much to be said for strengthening support for informal activity, it is nonetheless important to sound a note of caution about de Soto's arguments, one critical one being that if excessive regulations spawn the informal sector, then why has this not proved to be the case in the wealthier economies of the world? (Portés and Schauffler 1993: p. 47). Beyond this, an important implication of advocating decontrol of economic enterprise is that a precedent is set for greater deregulation in the formal sector. This in turn contributes to a broader process propagated by the IMF, the World Bank and others to liberalise production and markets in developing regions. This has often been harmful to low-income groups and, as signalled by Portes and Schauffler (*ibid*.: 55), the repeal of protective regulations on wages, work conditions, health and accident insurance, and unemployment compensation threatens a greater incidence of worker abuse, minimal wages, and disincentives for employers to provide training and/or engage in technological innovation. Greater tolerance of poor working conditions in the informal sector can also be politically expedient in so far as it helps to depress unemployment figures. For these kinds of reasons, a number of commentators have been concerned to stress the importance of remembering that the informal sector basically exists because of poverty and, as such, is not an effective solution to economic and labour market disadvantage. As summed up by Thomas (1995: p. 130), the informal sector 'is a picture of survival rather than a sector full of entrepreneurial talent to be celebrated for its potential to create an economic miracle' (see also Cubitt 1995; Roberts 1991: p. 117).

Although there may clearly be a perverse side to policies geared to assist the informal sector, measures are arguably needed, at least in the short term to help it to operate more efficiently and with better conditions for its workers. On top of this, and with reference to the Caribbean, Portés and Itzigsohn (1997: pp. 241–3), identify the importance of removing constraints to informal sector expansion such as:

1 Lack of working capital due to limited access to mainstream financial institutions.
2 Concentration in highly competitive low-income markets in which there are few possibilities for growth.
3 Oligopsony in some sectors of the informal economy, particularly artisanal craft production, in which barriers to insertion in more dynamic sectors of the formal economy are erected through restrictive agencies or middlemen.
4 Social atomisation of informal entrepreneurs due to the irregular and/or chaotic nature of supplies.
5 The existence of a 'craftsman ethic', which prevents some informal entrepreneurs, particularly in artisanal production, from changing their traditional methods of production.

While specific policy initiatives in different developing countries are discussed in detail elsewhere (see Chickering and Salahdine 1991a; Portés *et al.* 1997; Thomas 1995: Chapter 4), a much-favoured intervention on the institutional/macro-economic side of the labour market is the repeal of regulations and policies that obstruct entrepreneurship without serving any legitimate public regulatory purpose (Chickering and Salahdine 1991b: p. 6). Indeed, even where governments cannot waive regulations to any great extent, they can potentially consider simpler and diminished requirements and/or allow for progressive implementation (Tokman 1991: p. 155).

On the supply side of the labour market, there has been interest in and/or support for policies geared to education and training to promote the diversification of the informal sector, enhanced access to credit, assistance in management, marketing and packaging, and measures to promote greater health and safety. These initiatives are particularly relevant for groups within the informal sector such as ambulant traders and food vendors, where women are often a large percentage of operatives (see Blumberg 1995; Tinker 1997;

Rodgers 1989). There has also been advocacy of decentralised policies to accord with needs and skills in different localities (Portés and Schauffler 1993: p. 56) and the orientation of policies away from individual firms or workers as a means of utilising the social networks and social capital (reciprocity, trust, social obligations between kin, friends, neighbours, and so on) that so frequently fuel the operation of the informal sector (Portés and Itzigsohn 1997: pp. 244–5).

CONCLUDING COMMENTS: THE INFORMAL SECTOR IN THE TWENTY-FIRST CENTURY

Regardless of the degree to which governments and agencies may underpin the viability of the informal sector in developing countries, there are many indications that it will continue to be a significant feature of urban labour markets well into the twenty-first century. One important reason is demographic pressure. The youthful age structure of most developing nations will ensure that population growth remains high, and in the immediate future (probably at least until 2010) will cause a rise in the numbers seeking work. Even if, in the longer run, decreasing fertility rates lead to a fall in new entrants to the labour force, this may be offset by reductions in infant mortality and greater life expectancies. In the context of a paring back of (admittedly exiguous) state welfare, the latter may well lead to greater numbers of older people having to provide for themselves (Thomas 1995: p. 108; also Lloyd-Sherlock 1997). Another important factor is that recent rises in female employment are unlikely to be reversed when, despite the competitive conditions of informal employment and the rigours of their dual labour burden, women are unlikely to want (or be able to afford) to retreat into the home on a full-time basis.

On the demand side of the labour market, the current climate of deregulation is likely to provoke further contraction in public employment and to foster increasingly 'flexible' labour contracts in the formal sector as firms face ever-tougher global competition (Thomas 1995: p. 111). This is perhaps especially so given the the recent explosion of debt crises in the newly industrialised economies of Southeast Asia. On top of this, increased capital intensity in the formal sector is likely to push more people into informal occupations over time (*ibid.*).

Recognising that policies to bolster the informal economy will have to address a wide range of concerns simultaneously, and that it is important to build on existing concerns and capabilities within the sector, care should be taken to avoid perpetuating the social and economic marginalisation that deprives so many of the urban poor from exercising determination over their occupations in the first place. In this regard, it is critical, for example, to protect, extend and enhance systems of public education and training so that young people in particular are in a better position to obtain salaried employment and/or to establish businesses with sturdy prospects of survival. Education that encompasses commercial and managerial training, as well as instruction in cutting-edge developments such as information technology, could bring about greater productivity and employment in many ways. Not only is this likely to enhance indigenous/informal economic activity but it could also play a part in attracting greater foreign investment. While offshore production has not always been in the best interests of developing economies, the youth of populations in the South could place them in a stronger position vis-à-vis multinationals in the future. As Mitter (1997: pp. 26–7) speculates, an increasingly ageing population in the North, set against demands for new technological skills, is likely to spark greater interest in spreading information-intensive aspects of production to Asia, Africa and Latin America in the next few decades.

Policies geared to supporting people's efforts to sustain their livelihoods should also take due steps to consult the groups concerned. As Alonzo (1991: p. 41) has argued: 'Policy reform tends to be more lasting when pressure for it comes from the people themselves, for only then does government see popular will clearly. Institutions imposed from above are likely to die a natural death; the ones that last are born of people's own initiatives.' Testifying to the power of struggles 'from below' is the fact that the informal sector has survived so well through nearly two decades of severe economic crisis in developing regions. In the decades to come, therefore, governments in the South might well be advised to structure their economic and employment policies with respect to two guiding principles. The first is to protect workers from the worst excesses of informality. The second is to foster those elements characterising indigenous micro-enterprises that have contributed to buoyancy and resistance in the face of what might be regarded as misguided state allegiance to an increasingly untenable Western model of economic development.

GUIDE TO FURTHER READING

A.L. Chickering and M. Salahdine (eds) (1991c) *The Silent Revolution: The Informal Sector in Five Asian and Near Eastern Countries*. San Francisco: International Center for Economic Growth. A text primarily concerned with existing and prospective policies for the urban informal sector, including case studies from Morocco, Bangladesh, Sri Lanka, Thailand and the Philippines.

J. Gugler (ed.) (1997) *Cities in the Developing World: Issues, Theory and Policy*. Oxford: Oxford University Press. Contains various chapters on different aspects of the informal sector, including theory and policy in the context of street occupations in Colombia (Bromley), gender divisions of labour in a Nairobi shanty town (Nelson), the changing fortunes of a Jakarta street trader between the 1940s and the 1990s (Jellinek), and child labour (Grootaert and Kanbur).

C. Moser (1978) Informal Sector or Petty Commodity Production? Dualism or Dependence in Urban Development. *World Development* 6, 135–78. A classic paper on the informal sector in developing countries, which, drawing on a neo-Marxist theoretical framework, argues that the informal sector is strongly linked to the formal sector in a range of subordinate ways.

G. Standing and V. Tokman (eds) (1991) *Towards Social Adjustment: Labour Market Issues in Structural Adjustment.* Geneva: International Labour Organisation. A collection of papers that examine the impacts of economic restructuring on the growth of the informal sector and the 'informalisation' of labour arrangements in large firms in a range of European and developing countries.

J.J. Thomas (1995) *Surviving in the City: The Urban Informal Sector in Latin America*. London: Pluto. A thorough account of the nature and behaviour of the informal sector of employment in Latin American cities. It traces the history of conceptualisations of informal sector activity, the growth of informal employment in the wake of urban growth, the debt crisis and economic restructuring, and the underpinnings and implications of different policy interventions.

I. Tinker (1997) *Street Foods: Urban Food and Employment in Developing Countries*. New York/Oxford: Oxford University Press. The selling of foodstuffs in the street occupies up to one-tenth of male and female workers in developing countries. Based on a study organised by the Equity Policy Center, Washington DC, *Street Foods* concentrates on this important segment of informal employment in provincial cities in Egypt, Senegal, Nigeria, Bangladesh, Thailand, the Philippines and Indonesia.

REFERENCES

Acero, L. (1997) Conflicting demands of new technology and household work: women's work in Brazilian and Argentinian textiles. In S. Mitter and S. Rowbotham (eds) *Women Encounter Technology: Changing Patterns of Employment in the Third World*, London: Routledge, 70–92.

Alba, F. (1989) The Mexican demographic situation. In F. Bean, J. Schmandt and S. Weintraub (eds) *Mexican and Central American Population and US Immigration Policy*, Austin: University of Texas Press, 5–32.

Alonzo, R. (1991) The informal sector in the Philippines. In A.L. Chickering and M. Salahdine (eds) *The Silent Revolution: The Informal Sector in Five Asian and Near Eastern Countries*, San Francisco: International Center for Economic Growth, 39–70.

Baden, S. (1993) *The Impact of Recession and Structural Adjustment on Women's Work in Developing Countries*. Sussex: Institute of Development Studies, Bridge Report No. 2

Becker, C. and Morrison, A. (1997) Public policy and rural–urban migration. In J. Gugler (ed.) *Cities in the Developing World: Issues, Theory and Policy*, Oxford: Oxford University Press, 74–87.

Benería, L. (1991) Structural adjustment, the labour market and the household: the case of Mexico. In G. Standing and V. Tokman (eds) *Towards Social*

Adjustment: Labour Market Issues in Structural Adjustment, Geneva: International Labour Office, 161–83.

Benería, L. and Roldan M. (1987) *The Crossroads of Class and Gender: Industrial Homework, Subcontracting and Household Dynamics in Mexico City*. Chicago: University of Chicago Press.

Blumberg, R.L. (1995) Gender, microenterprise, performance and power: case studies from the Dominican Republic, Ecuador, Guatemala and Swaziland. In C. Bose and E. Acosta-Belén (eds) *Women in the Latin American Development Process*, Philadelphia: Temple University Press, 194–226.

Bromley, R. (1997) Working in the streets of Cali, Colombia: survival strategy, necessity or unavoidable evil? In J. Gugler (ed.) *Cities in the Developing World: Issues, Theory and Policy*, Oxford: Oxford University Press, 124–38.

Bullock, S. (1994) *Women and Work*. London: Zed Books.

Bulmer-Thomas, V. (1996) Conclusions. In V. Bulmer-Thomas (ed.) *The New Economic Model in Latin America and its Impact in Income Distribution and Poverty*, Basingstoke: Macmillan, 296–327.

Chant, S. (1991a) *Women and Survival in Mexican Cities: Perspectives on Gender, Labour Markets and Low-income Households*. Manchester: Manchester University Press.

Chant, S. (1991b) National perspectives on third world development. In R. Bennett and R. Estall (eds) *Global Change and Challenge: Geography for the 1990s*, London: Routledge, 175–96.

Chant, S. (1994a) Women, work and household survival strategies in Mexico, 1982–1992. *Bulletin of Latin American Research* 13(2), 203–33.

Chant, S. (1994b) Women and poverty in urban Latin America: Mexican and Costa Rican experiences. In Fatima Meer (ed.) *Poverty in the 1990s: The Responses of Urban Women*, Paris: UNESCO/International Social Science Council: Paris, 87–115.

Chant, S. (1996) Women's roles in recession and economic restructuring in Mexico and the Philippines. *Geoforum* 27(3), 297–327.

Chant, S. (1997) *Women-headed Households: Diversity and Dynamics in the Developing World*. Basingstoke: Macmillan.

Chant, S. and McIlwaine, C. (1995) *Women of a Lesser Cost: Female Labour, Foreign Exchange and Philippine Development*. London: Pluto Press.

Chickering, A.L. and Salahdine, M. (eds) (1991a) *The Silent Revolution: The Informal Sector in Five Asian and Near Eastern Countries*. San Francisco: International Center for Economic Growth.

Chickering, A.L. and Salahdine, M. (1991b) Introduction. In A.L. Chickering and M. Salahdine (eds), *The Silent Revolution: The Informal Sector in Five Asian and Near Eastern Countries*, San Francisco: International Center for Economic Growth, 1–14.

Chickering, A.L. and Salahdine, M. (1991c) The informal sector's search for self-governance. In A.L. Chickering and M. Salahdine (eds) *The Silent Revolution: The Informal Sector in Five Asian and Near Eastern Countries*, San Francisco: International Center for Economic Growth.

Cubitt, T. (1995) *Latin American Society*, 2nd edition. London: Longman.

de Soto H. (1989) *The Other Path: The Invisible Revolution in the Third World*. New York: Harper & Row.

Drakakis-Smith, D. (1987) *The Third World City*. London: Methuen.

Dwyer, A. (1994) *On the Line: Life on the US–Mexico Border*. London: Latin America Bureau.

Escobar Latapi, A. (1988). The rise and fall of an urban labour market: economic crisis and the fate of small workshops in Guadalajara, Mexico. *Bulletin of Latin American Research* 7(2), 183–205,

Escobar Latapi, A. and González de la Rocha, M. (1995) Crisis, restructuring and urban poverty in Mexico. *Environment and Urbanisation* 7(1), 57–76.

Eviota, E. (1992) *The Political Economy of Gender: Women and the Sexual Division of Labour in the Philippines*. London: Zed Books.

García, B. and de Oliveira, O. (1990) El trabajo femenino en México a fines de ochenta. In E. Bautista and H. Dávila (eds) *Trabajo Femenino y Crisis en México*, México DF: Programa Interdisciplinario de Estudios de la Mujer, El Colegio de México, 53–77.

Gilbert, A. (1990) *Latin America*. London: Routledge.

Gilbert, A. (1995a) Debt, poverty and the Latin American city. *Geography* 80(4), 323–33.

Gilbert, A. (1995b) Globalisation, employment and poverty: the case of Bogotá, Colombia. Seminar, Geography and Planning Research Series, London School of Economics, 30 November.

Gilbert, A. (1998) *The Latin American City*, 2nd edition. London: Latin America Bureau.

Gilbert, A. and Gugler, J. (1992) *Cities, Poverty and Development: Urbanisation in the Third World*. Oxford: Oxford University Press.

González de la Rocha, M. (1988) Economic crisis, domestic reorganisation and women's work in Guadalajara, Mexico. *Bulletin of Latin American Research* 7(2), 207–23.

Grabowski, R. and Shields, M. (1996) *Development Economics*. Oxford: Blackwell.

Green, D. (1996) Latin America: neoliberal failure and

the search for alternatives. *Third World Quarterly* 17(1), 109–22.

Green, D. (1998) *Hidden Lives: Voices of Children in Latin America and the Caribbean*. London: Latin America Bureau.

Harrison, M. and McVey, C. (1997) Conflict in the city: street trading in Mexico City. *Third World Planning Review* 19(3), 313–26.

Hart, K. (1973) Informal income opportunities and urban employment in Ghana. In R. Jolly, E. de Kadt, H. Singer and F. Wilson (eds) *Third World Employment*, Harmondsworth: Penguin, 66–70.

Humphrey, J. (1997) Gender divisions in Brazilian industry. In J. Gugler (ed.) *Cities in the Developing World: Issues, Theory and Policy*, Oxford: Oxford University Press, 171–83.

International Labour Office (ILO) (1972) *Employment, Incomes and Inequality: A Strategy for Increasing Productive Employment in Kenya*. Geneva: ILO.

Jellinek, L. (1997) Displaced by modernity: the saga of a Jakarta street trader's family from the 1940s to the 1990s. In J. Gugler (ed.) *Cities in the Developing World: Issues, Theory and Policy*, Oxford: Oxford University Press, 139–55.

Jenkins, R. (1997) Structural adjustment and Bolivian industry. *European Journal of Development Research* 9(2), 107–28.

Lessinger, J. (1990) Work and modesty: the dilemma of women market traders in Madras. In L. Dube and R. Palriwala (eds) *Structures and Strategies*, New Delhi: Sage, 129–150.

Lloyd-Sherlock, P. (1997) *Old Age and Urban Poverty in the Developing World: The Shanty Towns of Buenos Aires*. Basingstoke: Macmillan.

Lozano, W. (1997) Dominican Republic: informal economy, the state and the urban poor. In A. Portés, C. Dore-Cabral, and P. Landoff (eds) *The Urban Caribbean: Transition to a New Global Economy*, Baltimore: Johns Hopkins University Press, 153–89.

Manigat, S. (1997) Haiti: the popular sectors and the crisis in Port-au-Prince. In A. Portés, C. Dore-Cabral and P. Landoff (eds) *The Urban Caribbean: Transition to a New Global Economy*, Baltimore: Johns Hopkins University Press, 81–123.

Manuh, T. (1994) Ghana: women in the public and informal sectors under the economic recovery programme. In P. Sparr (ed.) *Mortgaging Women's Lives: Feminist Critiques of Structural Adjustment*, London: Zed Books, 61–77.

Miraftab, F. (1994) (Re) production at home: reconceptualising home and family. *Journal of Family Issues* 15(3), 467–89.

Miraftab, F. (1996) Space, gender and work: home-based workers in Mexico. In E. Boris and E. Prügl (eds) *Homeworkers in Global Perspective: Invisible No More*, New York: Routledge, 63–80.

Mitter, S. (1997) Information technology and working women's demands. In S. Mitter and S. Rowbotham (eds) *Women Encounter Technology: Changing Patterns of Employment in the Third World*, London: Routledge, 19–43.

Moghadam, V. (1995) Gender aspects of employment and unemployment in global perspective. In M. Simai, V. Moghadam and A. Kuddo (eds) *Global Employment: An International Investigation into the Future of Work*, London: Zed Books, in association with United Nations University, World Institute for Development Economics Research, 111–39.

Monteón, M. (1995) Gender and economic crises in Latin America: reflections on the Great Depression and the debt crisis. In R.L. Blumberg, C. Rakowski, I. Tinker and M. Monteón (eds) *Engendering Wealth and Well-Being: Empowerment for Global Change*, Boulder, Colo.: Westview Press, 39–62.

Moser, C. (1978) Informal sector or petty commodity production? Dualism or dependence in urban development. *World Development* 6, 135–78.

Moser, C. (1992) Adjustment from below: low-income women, time and the triple role in Guayaquil, Ecuador. In H. Afshar and C. Dennis (eds) *Women and Adjustment Policies in the Third World*, Basingstoke: Macmillan, 87–116.

Moser, C. (1997) *Household Responses to Poverty and Vulnerability, Volume 1: Confronting Crisis in Cisne Dos, Guayaquil, Ecuador*, Urban Management and Poverty Reduction Series No. 21, Washington, DC: World Bank.

Moser, C. (1998) The asset vulnerability framework: reassessing urban poverty reduction strategies. *World Development* 26(1), 1–9.

Nash, J. (1995) Latin American women in the world capitalist crisis. In C. Bose and E. Acosta-Belén (eds) *Women in the Latin American Development Processess*, Philadelphia: Temple University Press, 151–66.

Nelson, N. (1997) How women and men got by and still get by (only not so well): the gender division of labour in a Nairobi shanty town. In J. Gugler (ed.) *Cities in the Developing World: Issues, Theory and Policy*, Oxford: Oxford University Press, 156–70.

Pearson, R. (1986) Latin American women and the new international division of labour: a reassessment. *Bulletin of Latin American Research* 5(2), 67–79.

Peña Saint Martin, F. (1996) *Discriminación Laboral Femenina en la Industria del Vestido de Mérida,*

Yucatán, México DF: Instituto Nacional de Antropología e Historia, Serie Antropología Social.

Peña Saint Martin, F. and Gamboa Cetina, M. (1991) Entre telas e hilos de colores: mujeres y confección industrial de ropa en Yucatán. In V. Salles and E. McPhail (eds) *Textos y Pretextos: Once Estudios Sobre la Mujer*, México DF: Programa Interdisciplinario de Estudios de la Mujer, El Colegio de México, 309–80.

Pineda-Ofreneo, R. (1988) Philippine domestic outwork: subcontracting for export-oriented industries. In J.G. Taylor and A. Turton (eds) *Sociology of Developing Societies: Southeast Asia*, Houndmills, Basingtoke: Macmillan, 158–64.

Poapongsakorn, N. (1991) The informal sector in Thailand. In A.L. Chickering and M. Salahdine (eds) *The Silent Revolution: The Informal Sector in Five Asian and Near Eastern Countries*, San Francisco: International Center for Economic Growth, 105–44.

Portés, A., Dore-Cabral, C. and Landoff, P. (eds) (1997) *The Urban Caribbean: Transition to a New Global Economy*. Baltimore: Johns Hopkins University Press.

Portés, A. and Itzigsohn, J. (1997) Coping with change: the politics and economics of urban poverty. In A. Portés, C. Dore-Cabral and P. Landoff (eds) *The Urban Caribbean: Transition to a New Global Economy*, Baltimore: John Hopkins University Press, 227–48.

Portes, A. and Schauffler, R. (1993) Competing perspectives on the Latin American informal sector. *Population and Development Review* 19(3), 33–60.

Roberts, B. (1991) The changing nature of informal employment: the case of Mexico. In G. Standing and V. Tokman (eds) *Towards Social Adjustment: Labour Market Issues in Structural Adjustment*, Geneva: International Labour Office, 115–40.

Roberts, B. (1994) Informal economy and family strategies. *International Journal of Urban and Regional Research* 18(1), 6–23.

Roberts, B. (1995) *The Making of Citizens: Cities of Peasants Revisited*. London: Edward Arnold.

Rodgers, G. (1989) Introduction: trends in urban poverty and labour market access. In G. Rodgers (ed.) *Urban Poverty and the Labour Market*, Geneva: International Labour Office, 1–33.

Safa, H. (1995a) *The Myth of the Male Breadwinner: Women and Industrialisation in the Caribbean*. Boulder, Colo.: Westview Press.

Safa, H. (1995b) Economic restructuring and gender subordination. *Latin American Perspectives* 22 (2), 32–50.

Salahdine, M. (1991) The informal sector in Morocco: the failure of legal systems. In A.L. Chickering and M. Salahdine (eds) *The Silent Revolution: The Informal Sector in Five Asian and Near Eastern Countries*, San Francisco: International Center for Economic Growth, 15-38.

Sanderatne, N. (1991) The informal sector in Sri Lanka: dynamism and resilience. In A.L. Chickering and M. Salahdine (eds) *The Silent Revolution: The Informal Sector in Five Asian and Near Eastern Countries*, San Francisco: International Center for Economic Growth, 71–103.

Scott, A.M. (1990) Informal sector or female sector? Gender bias in labour market models. In D. Elson (ed.) *Male Bias in the Development Process*, Manchester: University of Manchester Press.

Scott, A.M. (1994) *Divisions and Solidarities: Gender, Class and Employment in Latin America*. London: Routledge.

Standing, G. (1989) Global feminisation through flexible labour. *World Development* 17(7), 1077–95.

Standing, G. (1991) Structural adjustment and labour market policies: towards social adjustment? In G. Standing and V. Tokman (eds) *Towards Social Adjustment: Labour Market Issues in Structural Adjustment*, Geneva: International Labour Office, 5–51.

Thomas, J.J. (1995) *Surviving in the City: The Urban Informal Sector in Latin America*. London: Pluto Press.

Thomas, J.J. (1996) The new economic model and labour markets in Latin America. In V. Bulmer-Thomas (ed.) *The New Economic Model in Latin America and its Impact on Income Distribution and Poverty*, Basingtoke: Macmillan, in association with the Institute of Latin American Studies, University of London, 79–102.

Thomas, J.J. (1997) The urban informal sector and social policy: some Latin American contributions to the debate. Paper presented at Workshop for the Social Policy Study Group, Institute of Latin American Studies, University of London, 28 November.

Tinker, I. (1997) *Street Foods: Urban Food and Employment in Developing Countries*. New York/Oxford: Oxford University Press.

Tironi, E. and Lagos, R. (1991) The social actors and structural adjustment. *CEPAL Review* 44, 35–50.

Tokman, V. (1989) Policies for a heterogeneous informal sector in Latin America. *World Development* 17(7), 1067–76.

Tokman, V. (1991) The informal sector in Latin America: from underground to legality. In G. Standing and V. Tokman (eds) *Towards Social Adjustment: Labour Market Issues in Structural Adjustment*, Geneva: International Labour Office, 141–57.

United Nations (1995) *The World's Women 1995: Trends and Statistics*. New York: UN.

Vandemoortele, J. (1991) Labour market informalisation in sub-Saharan Africa. In G. Standing and V. Tokman (eds) *Towards Social Adjustment: Labour Market Issues in Structural Adjustment*, Geneva: International Labour Office. 81–114.

Ward, K. and Pyle, J. (1995) Gender, industrialisation, transnational corporations and development: an overview of trends and patterns. In C. Bose and E. Acosta-Belén (eds) *Women in the Latin American Development Process*, Philadelphia: Temple University Press, 37–64.

39

HIV/AIDS, poverty, exclusion and the third world

Tony Barnett

INTRODUCTION

Recent estimates show that infection with HIV, which causes AIDS, is far more common in the world than was previously thought. UNAIDS (the United Nations agency that coordinates activities to combat the epidemic) and the WHO (the World Health Organisation) estimate that over 30 million people were living with HIV infection at the end of 1997. That is one in every 100 adults in the 'sexually active' age group 15 to 50 years worldwide. Included in this 30 million are 1.1 million children under the age of 15. The overwhelming majority of HIV-infected people – more than 90 per cent – live in the developing world. Most of these do not know that they are infected. Table 39.1 shows the global situation at the end of 1997.

Assuming that currently unbroken global trends continue, UNAIDS estimates that more than 40 million people will be living with HIV in the year 2000. An estimated 2.3 million people died of AIDS in 1997. These deaths represent a fifth of the total 11.7 million AIDS deaths since

Table 39.1 Global summary of the HIV/AIDS epidemic, December 1997.

People newly infected with HIV in 1997	Adults	5.1 million
	Women	2.1 million
	Children <15 years	590,000
	Total	5.8 million
People living with HIV/AIDS	Adults	29.5 million
	Women	12.2 million
	Children <15 years	1.1 million
	Total	30.6 million
AIDS deaths in 1997	Adults	1.8 million
	Women	820,000
	Children <15 years	460,000
	Total	2.3 million
AIDS deaths since the beginning of the epidemic	Adults	9.0 million
	Women	4.0 million
	Children <15 years	2.7 million
	Total	11.7 million

Source: UNAIDS Web Site, January 1998.

the beginning of the epidemic in the late 1970s. Of the adults who died of AIDS in 1997, 46 per cent were women. In addition, 460,000 children died (representing 20 per cent of all deaths).

Unlike many things that we think of as *diseases*, AIDS is not characterised by a set of unique clinical features common to each patient. It is a *syndrome* – a set of symptoms. The particular set of symptoms varies from one part of the world to another. For some regions, there is a *clinical case definition* that provides a checklist of symptoms, allowing clinicians to diagnose the disease.

HIV destroys the ability of the human body to resist infection. The resulting decline in the immune system allows certain (often locally specific) opportunistic bacterial, fungal, protozoal or viral infections to flourish. It may also allow the development of some kinds of cancerous tumour.

The virus is passed between people in blood, semen and vaginal secretions. Thus it may be transmitted through infected blood transfusions, injection syringes, sexual intercourse and between mother and child at the time of birth. Worldwide, the main mode of transmission is through sexual intercourse and to a much lesser, but significant extent, between infected mothers and their newborn babies. Other major channels of transmission are transfusions of unscreened blood and use of blood products derived from unscreened sources and use of infected equipment for injections, including the use of various types of recreational drugs. The rate of sexual transmission is increased if people have active sexually transmitted diseases that result in open wounds on the genitals.

AIDS affects people in all walks of life. Unlike leprosy, smallpox or measles, those who are infected do not show any externally visible signs. In addition, the disease has a very long 'latency period' during which a person shows little or no sign of ill-health, can be normally active, but is also able to pass the infection to other people. This period ranges from a few months up to fifteen or even twenty years. A person may not know that they are infected and thus may infect many others. For this reason, whatever number of people with symptoms appear to the medical services of a country and are diagnosed as having AIDS, there will most probably be a greater number who are HIV-positive. Indeed, the numbers of people who are diagnosed as having AIDS and being HIV-positive will depend on the quality and form of surveillance that exists. This is cost and administration-dependent, and under-recording of both is likely to increase with national poverty, and *vice versa*.

Figure 39.1 shows a projection for an African country of the relationship between 'normal' deaths, AIDS deaths and numbers of people in the population who are HIV-positive and who will therefore become ill and die within a limited period.

AIDS AND HIV DISTINGUISHED

In looking at Figure 39.1, we must be careful not to confuse statements about seroprevalence (the numbers of people in a population who are HIV-positive – usually expressed as a percentage of the population) and statements about the numbers of people who are actually ill with the disease (usually expressed as the number per 100,000 of a national population). It is important to make a clear distinction between someone being 'HIV-

Figure 39.1 Projected normal deaths, AIDS deaths and future HIV infections in an African country, 1985–2005.

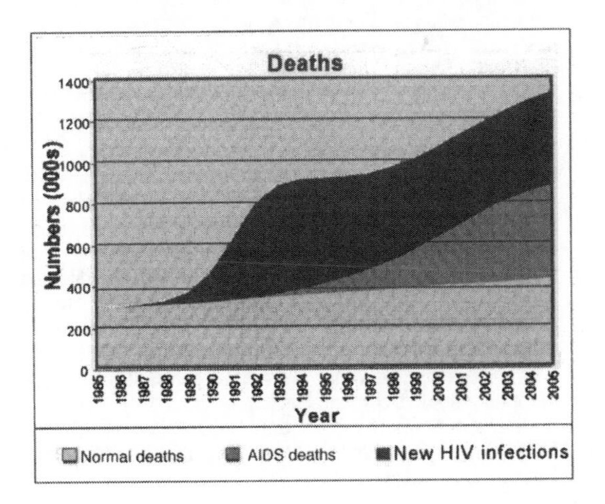

positive' (which means that they are infected and that their blood when tested with two standard tests reacts positively) and someone being ill with AIDS (which means that they are exhibiting a number of clinical symptoms that indicate that the virus has affected the ability of their body to fight infections to which it would normally be resistant). In Africa, these symptoms may include an otherwise rare cancer called *Kaposi's sarcoma*, abrupt and extreme weight loss, severe diarrhoea, *herpes zoster*, and fungal infections of the respiratory tract, as well as some other conditions. An increasingly common associated symptom of AIDS is tuberculosis, and in recent years new forms of drug-resistant tuberculosis have been appearing in close correlation with the epidemic of HIV/AIDS. This is a very serious development indeed, as tuberculosis is spread by droplet infection and could therefore move rapidly into the general population.

In Asia and Africa, the time between onset of illness and death is anything between a few months and two years. This is a shorter survival time than that found typically in Western Europe and North America and reflects the lower health status of the population, the greater number of infectious agents in the environment and the unavailability of some expensive palliative treatments.

In some countries, HIV/AIDS will reverse the declines in infant and child mortality rates observed over recent years, will slow population growth rates, and will mean negative growth rates for some countries (particularly Botswana, Zimbabwe and Guyana); and life expectancies will decline (Stanecki and Way 1997).

In many African countries with major epidemics, the ratio of dependants to supporters, of mouths to hands has been reduced. However, the precise impact of increased death and mortality appears to be quite specific to regions and areas and is mediated through cultural, political, environmental and economic variables, making all but the broadest generalisations difficult to substantiate.

At a local level, there may not be enough hands to grow the crops to feed the family well at satisfactory levels of nutrition. At the district level, it

Box 39.1 Life expectancy and HIV/AIDS

Life expectancy has already been reduced in many countries which have had serious epidemics of HIV/AIDS in the late 1970s and early 1980s. Thus estimated life expectancies in Kenya have fallen from 65 to 55.6 years, in Uganda from 53.2 to 40.3 years, and Zimbabwe from 64.1 to 41.9 years. Such declines are all the more worrying when it is considered that they reflect a decline from what was an *improving* situation. For example, in the case of Zimbabwe, life expectancy had been expected to increase to 70 years by the year 2010.

There will also be reductions in life expectancy – but of a lesser degree – in countries such as Brazil (from 72.5 to 65.1 years) in Haiti (from 58.8 to 52.2 years, and in Guyana (from 67.9 to 49.9 years). In Thailand, it is estimated that the decline will be from 75.1 to 72.9 years.

Box 39.2 The HIV/AIDS epidemic in Africa

In nineteen African countries, HIV prevalence among low-risk urban adult populations (age group 15–49) has now reached 5 per cent or will reach this level within a few years. Such levels of prevalence may mean that a young person has a lifetime risk of around 30 per cent of contracting the disease (Blacker and Zaba 1997). In Zambia and Malawi, some groups of urban ante-natal clinic attenders have been found to have rates of 25 per cent (Stanecki and Way 1997), while in Francistown, Botswana, rates of 40 per cent have been reported (personal communication).

In Uganda, where the epidemic began in the late 1970s / early 1980s, recent reports indicate that the rate of increase of infection may now be levelling off or even declining (Stanecki and Way 1997).

The sub-Saharan African countries currently most affected by the epidemic are Botswana, Burkina Faso, Burundi, Cameroon, Central Africa Republic, Congo, Côte d'Ivoire, Ethiopia, Kenya, Lesotho, Malawi, Nigeria, Rwanda, South Africa, Tanzania, Uganda, Republic of Congo, Zambia and Zimbabwe.

Caldwell and Caldwell (1994) have put forward the hypothesis that the most affected areas of Africa are correlated with areas where men do not practise circumcision.

Box 39.3 The HIV/AIDS epidemic in Asia

Stanecki and Way (1997) suggest that the epidemic in Asia is diverse. Rates of infection are very low in the general population in countries such as Mongolia and South Korea and as high as 2.1 per cent in Thailand. In India, estimates of overall numbers infected are in the range 9 to 10 million, but the distribution is uneven between states and regions. In Mumbai, rates as high as 5l per cent have been recorded among commercial sex workers, but similar rates have not been evident in Calcutta. It is in the poorer countries of Southeast Asia that rates are increasing most rapidly, in Burma, Cambodia and Vietnam. However, in the largest countries of the region, India and China, current knowledge about prevalence is poor and there is probably a tendency to under-report the numbers of cases of people who are sick with AIDS.

Box 39.4 World Bank: Household Impact of adult death

The overall economic impact of adult death on the surviving household members varies according to three sets of characteristics:

- those of the deceased individual such as age, sex, income and cause of death;
- those of the household, such as composition and assets;
- those of the community, such as attitudes towards helping needy households and the availability of resources.

The first set of characteristics determines the basic impact of the death on the surviving household members; the second and third influence how well the afflicted household copes. Although disentangling the three is very difficult, it is nonetheless important when attempting to assess the household impact of an adult death to consider all three factors.

(World Bank 1997: p. 209)

That somebody has died of AIDS may have an effect on each of these three sets of variables for the following reasons: (1) AIDS deaths may take a long time and the search for care may eat into household resources; (2) given the sexual nature of transmission, one death usually suggests the later death of a spouse and thus leads to clustering of deaths; (3) assets will be spent in care and treatment; (4) AIDS is often stigmatising and therefore community assistance may not be forthcoming or community involvement may be rejected; (5) some communities may be affected more than others and therefore be less able to cope with the large numbers of deaths and the extent of illness – a situation that could arise in the case of some communities in Ukraine – Box 39.5.

may mean that there is a decline in the production of cash crops or of cattle products as labour for these tasks is insufficient or is diverted into caring for the sick and other dependants (thus having a major effect on the lives of women); at a national level, the World Bank has estimated that the epidemic is likely to decrease the rate at which GDP increases and the rate of growth of *per capita* income – all against a background in many countries where GDP and *per capita* income growth is either very slow or, as is the case in most of Africa and parts of the Former Soviet Union, where these have been in decline for some years.

Households under stress in Uganda

The following accounts of how households in the Rakai district of Uganda had been affected by HIV/AIDS in the late 1980s dramatically illustrate the effects on rich and poor alike in this most seriously affected region of Africa (Barnett and Blaikie 1994). More recent discussion of how households cope with excess death and illness may be found in World Bank (1997: pp. 206–63).

The first household had been comparatively wealthy and coped by combining members of other related households which were living separately until they themselves began to experience illness and death within their own and related households as a result of the disease.

The man is very old (aged 78) and is married. His wife is also elderly. They had ten children, but a daughter and two sons have died of AIDS. Until the disease struck, this couple was living in an independent household, farming and living off their assets with some help from their son and his wife who lived nearby, as well as hiring farm labour to undertake the heavy cultivation work.

The son died from AIDS in 1986 at the age of 35. His widow brought her three daughters to live with the old couple. One other son, aged 38 has now joined this household. He is ill – probably with AIDS, as his wife died of AIDS within the last year.

The farm work is done in part by the household members, but they also hire Rwandan labourers to plant cassava and cut banana stems. They also grow beans and a range of other annuals, including groundnuts and sorghum. They used to have a coffee plantation but this has been abandoned because of insect infestation.

They now get cash from sale of bananas and beer as well as from licensed distillation of a local alcohol, *waragi*.

This household has been severely affected by the disease. The household was quite wealthy and still has considerable resources, but, predictably, is experiencing stress now and has taken the three grandchildren out of school in order to relieve the pressures of paying for labour and also to help with the domestic work.

The second household is one of the most tragic cases that we encountered. It is hard to describe it as a household; perhaps 'remnant household' would be more accurate.

The man lived alone in a bare hut, sleeping on the floor. His possessions appeared to be little beyond a blanket and a pot over a meagre fire upon which he was cooking some bananas. He was said to be 45 years old but looked considerably older. He was quite clearly very disturbed and could not be interviewed. Information was obtained from others nearby.

Only a few years ago, this was a substantial household with a reasonable farm which was supplemented by fishing. His wife and 8 of his teenage and adult children had died of AIDS within the last few years. He had no relatives living in the village, and supported himself by cultivating and selling some of his bananas. Onlookers said of him 'he is not expected to marry again'. This comment from a bystander can be interpreted as follows. It is widely known that his wife died from the disease, therefore it is quite likely that he is himself infected. Were this not so, in any case he now had no assets or resources which would allow him to re-marry and establish a new household.

This case illustrates how, in an extreme case, the costs of nursing AIDS victims combined with disappearance of the family has led to a state of utter poverty where life is sustained at a bare minimum. The entire family support system having gone, this man was destitute and isolated.

INCIDENCE OF INFECTION, ILLNESS AND DEATH

According to Anderson and May (1991), the rate at which a sexually transmitted disease spreads in a population is largely determined by the following three factors:

- the amount of time a person remains infectious;
- the risk of transmission per sexual contact;
- the rate at which people acquire new sexual partners.

The fact that HIV/AIDS is mainly a sexually transmitted disease means that it affects those who are most likely to (1) be sexually active and (2) have many sexual partners or have sex with somebody who is likely to have sex with many sexual partners. Evidence shows that this includes people in the age range 15–50, the 'sexually active' age group. Another group that is likely to be exposed to infection is children born of mothers who are HIV-positive. These children contract the disease during birth. About 50 per cent of children born to HIV-positive mothers will test HIV-positive and about 50 per cent of these will progress to develop AIDS (Figure 39.2).

In terms of income and wealth, the epidemic is associated with poverty both cross nationally and within nations. Partly this reflects the fact that there are more poor people in the world than there are rich, but it is also the case that the poor

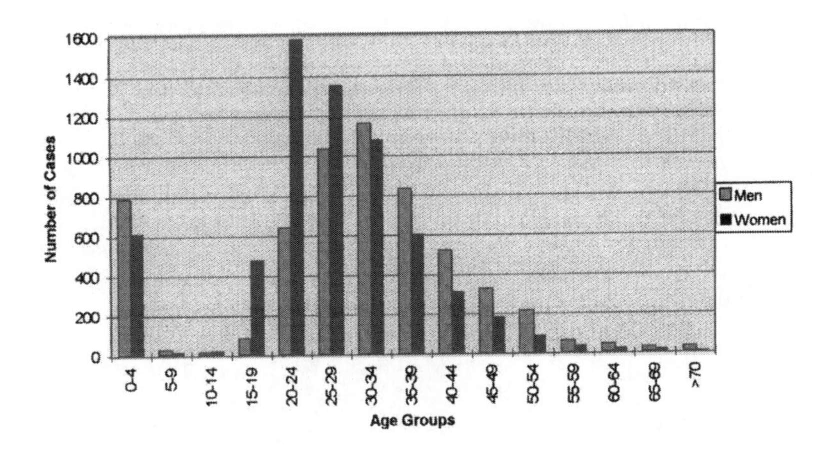

Figure: 39.2 Bar chart showing typical distribution of AIDS cases by age cohort and gender for an east or southern African country.

are likely to have less good health and diet than the rich and lower resistance to infection, and to live more risky lives. Thus, in general, poorer countries have higher rates of infection than do rich countries; poorer people within poor countries have higher rates of infection than do the better off; and finally, in some cases it is true to say that poorer people in rich countries have higher rates than do wealthier people (Table 39.2).

This has three significant implications: (1) the idea that this is a disease of the third world is only true inasmuch as it is a disease associated with poverty; (2) as the livelihoods of poor people often involve spatial mobility, high levels of uncertainty and non-formal contracts and conditions of work, so they are often exposed to high rates of sexual partner change (and may even undertake sex work as part of a survival strategy, sometimes under duress); (3) women, whose rates of infection in generalised heterosexual epidemics such as those in sub-Saharan Africa may sometimes exceed those of men, are at particular risk because of the unequal nature of gender relations in most societies and cultures. It is such factors that have led the World Bank to conclude that:

> For the average developing country a $2000 increase in per capita income is associated with a reduction of about 4 percentage points in the HIV infection rate of urban adults. Reducing the index of inequality from 0.5 to 0.4, the difference in inequality between . . . Honduras and Malawi is associated with a reduction in the infection rate by about 3 percentage points. These findings suggest that rapid and fairly distributed economic growth will do much to slow the AIDS epidemic . . . after controlling for the percentage of the population that is Muslim . . . two measures related to gender inequality are associated with higher HIV infection rates.
>
> (World Bank 1997: pp. 28–9)

PRESENT AND PROSPECTIVE SOCIAL AND ECONOMIC IMPACT OF THE EPIDEMIC

In considering this question, it is important to distinguish between the impact in poor and rich countries and in poor and rich communities in each of these. It is also necessary to distinguish between impact as measured by economists using conventional measures of economic output, growth and well-being, and other measures and other approaches to these questions.

Rich countries have smaller epidemics than do poor countries and are more able to deal with the consequences in terms of both the costs of medical care and the longer term economic and social costs associated with excess illness and death. The first of these statements is supported by Table 39.1, which shows the global and regional distribution of AIDS cases.

It is also important to be aware that economic

Table 39.2 Estimated number of adults with HIV/AIDS by region and characteristics, December 1997 (UNAIDS data).

Region	*Epidemic started*	*Adults and children living with HIV/AIDS*	*Adult prevalence rate (%)*	*Cumulative number of orphans*	*Percentage women*	*Main mode(s) of transmission for people living with HIV*
Sub-Saharan Africa	Late 1970s–Early 1980s	20.8 million	7.4	7.8 million	50	Hetero
North Africa, Middle East	Late 1980s	210,000	0.13	14,200	20	IVDU, Hetero
South and South East Asia	Late 1980s	6.0 million	0.6	220,000	25	Hetero
East Asia, Pacific	Late 1980s	440,000	0.05	1900	11	IVDU, MSM, Hetero
Latin America	Late 1970s–Early 1980s	1.3 million	0.5	91,000	19	MSM, Hetero, IVDU
Caribbean	Late 1970s–Early 1980s	310,000	1.9	48,000	33	Hetero, MSM
Eastern Europe and Central Asia	Early 1990s	150,000	0.07	30	25	IVDU, MSM
Western Europe	Late 1970s–Early 1980s	530,000	0.3	8700	20	IVDU, MSM
North America	Late 1970s–Early 1980s	860,000	0.6	70,000	20	MSM, IVDU, Hetero
Australia and New Zealand	Late 1970s–Early 1980s	12,000	0.1	300	5	MSM, IVDU
Total		30.6 million	1.0	8.2 million	41	

Source: World Bank 1997: Fig 1.1, p. 14.
Note: MSM = men who have sex with men; IVDU = intravenous drug user; Hetero = heterosexual.

measures of the decline attributable to increased illness and death associated with an epidemic such as HIV/AIDS are problematic. These measures take into account only 'economic' transactions and specifically exclude what may be described as 'socially reproductive' activities such as household management, childcare and even subsistence agriculture – what are often but not exclusively women's tasks. Thus there is a likelihood that the true impact of the epidemic is being underestimated as a result of a fundamental and classical measurement error. The social sciences have an important part to play here. Medical epidemiology shows us the broad interaction between disease organisms and human populations, and it enables the epidemic processes to be tracked with a view to prevention of infection. The social sciences, including geography, can build on this knowledge and analysis to indicate the particular and specific nature of those epidemic processes. Such research can tease out the specific social, cultural, locational and economic characteristics of populations that expose them to increased risk of infection – that is the *particular* nature of their risk. It can also provide a second level of analysis, which further explores the *specific* and unique group characteristics that place a population sub-group at risk or that identifies it as having special needs. Two simple illustrations will make the point clear.

Commercial sex workers

All women who take up commercial sex work are not driven by poverty, some have been found to use this as a way to accumulate business capital. The distinction between the groups is important in the development of education and prevention programmes. There are differences between the *particular* population sub-group of 'all women commercial sex workers in a society' and the *specific* characteristics of those who do this work to raise capital (a minority) and those who are driven by debt and poverty (a majority). The former group will not respond to 'income-earning projects' designed to raise them from poverty; the latter group may. The former group may respond to advice and training as to how to negotiate condom use with clients; the latter are unable, in a competitive market, to enter into such negotiations.

Orphans and orphaning

Not all orphans are the same. The number of orphans created by the epidemic runs into many tens of thousands (see Table 39.2). How are they to be helped? In Uganda, there has been considerable debate about institutional versus care in the community and family. It has been in part a matter of national pride to believe that the 'Ugandan family' can cope with parent-less children. To a degree, this has been a success story. However, the experience is not generalisable – perhaps even in one country. In Zimbabwe, there is evidence that some orphans are left to fend for themselves as 'orphan households'. This presents different problems for social policy to the situation in Uganda. Once again, the *particularity* of a sub-group affected by or, as in this case, created by, the epidemic, has to be tempered by its *specific* situation as affected by culture (in this case), economic niche or location. The case study of Ukraine, a very poor country in Eastern Europe, provides more to think about in relation to these questions (Box 39.5).

It is in relation to these problems of particularisation and specification that further work is required in all the social sciences – geography, sociology, anthropology and economics – if we are adequately to understand the diverse impact of the global HIV epidemic.

GUIDE TO FURTHER READING

For those who want information about HIV/AIDS and health and welfare matters, there are few better points of departure than Alcorn, K. (ed.) (1997) *AIDS*

Box 39.5 Orphans and the elderly in the Ukraine

The break-up of the Soviet Union had acute social, political and economic implications for the newly independent Ukraine. The Soviet system was highly centralised and controlled via four main mechanisms: the party, the internal security apparatuses, the official trade unions, and the administration. Entitlements to social, economic and cultural goods were largely administered within this structure.

Evidence and the modelling of the epidemic, which started only in 1994, suggest that AIDS will not have a significant demographic impact. However, there is a likelihood of considerable increase in the number of orphans, with an estimated 111,500 (low scenario) to 317,300 (high scenario) additional orphans by 2016. A second group of concern is the elderly. The population dynamics of Ukraine, which lost about a quarter to a third of its population as a result of Stalin's agricultural collectivisation in the 1920s and the Second World War in the 1940s, and which has resulted in the oldest population in Europe, are such that the elderly comprise a sizeable proportion of the population and numbers are increasing. While AIDS will not directly affect them, it will affect their families – who are increasingly viewed as a source of support. Thus while the demographic consequences may not be great, they will be compounded by the inability of the state to provide the level of social services previously available, and it certainly will not be able to meet the increased demand from either the elderly or the orphaned. The impact of the increased morbidity and mortality is on reproductive labour, which, given the demographic structure of Ukraine, is a shrinking group. Thus the already adverse dependency ratio will be made worse by HIV/AIDS and will affect policy decisions to be made about the care of the young and the old. In such circumstances, institutional care may be the only option – but even so, who are to be the carers? (Barnett and Whiteside 1997)

Reference Manual, 16a Clapham Common Southside, London, SW4 7AB: NAM Publications. This is published annually in an updated form.

An excellent general survey of the global situation in relation to many aspects of HIV/AIDS, from biomedical through clinical to public health, social and economic and legal aspects, is contained in Mann, J. and Tarantola, D. (eds) (1996) *AIDS in the World II: Global Dimensions, Social Roots and Responses*, New York and Oxford: Oxford University Press. This is updated and renumbered every two years or so.

The World Bank has been concerned about the social and economic implications of HIV/AIDS for many years and has now published a wide-ranging review of the health economic, health system and public policy issues in World Bank (1997 and 1998) *Confronting AIDS: Public Priorities in a Global Epidemic*, New York and Oxford: Oxford University Press for the World Bank.

A useful source of up-to-date information about the state of the epidemic globally and in particular regions and countries is the Internet site of *UNAIDS*. The URL is http://www.unaids.org/

In addition, the *United States Bureau of the Census* is also a source of information on rates of infection and illness in countries other than the USA. The address of its Internet site is http://www.census.gov/

People seeking advanced information about the epidemiology of human infectious diseases, and the modelling of that epidemiology, are referred to Anderson, R. and May, R. (1991) *Infectious Diseases of Humans: Dynamics and Control*, Oxford: Oxford University Press. Particular discussion relevant to HIV and AIDS may be found between pages 236 and 303, as well as elsewhere in the book.

Early but still useful discussion of the macro-economic impact of HIV/AIDS is to be found in Over, M. (1992) The Macro-economic Impact of AIDS in Sub-Saharan Africa, *Technical Working Paper No. 3*, Washington: World Bank, Population, Health and Nutrition Division, Africa Technical Department.

An early case study of the impact of HIV/AIDS on communities in Africa is to be found in Barnett, T. and Blaikie, P. (1992 and 1994) *AIDS in Africa: Its Present and Future Impact*, Chichester: Wiley.

REFERENCES

Barnett, T. and Whiteside, A. (1997) *The Social and Economic Impact of HIV/AIDS in Ukraine*, Kyiv: British Council and UNAIDS.

Blacker, J. and Zaba, B. (1997) HIV prevalence and lifetime risk of dying of AIDS. *Health Transition Review* 7 (Supp. 2), 45–62.

Caldwell, J.C. and Caldwell, P. (1994) The neglect of an epidemiological explanation for the distribution of HIV/AIDS in sub-Saharan Africa: exploring the circumcision hypothesis. *Health Transition Review* (4 supplement), 23–46.

Khodakevich, L. (1997) Development of HIV epidemics in Belarus, Moldova and Ukraine and response to the epidemics. Summary of a presentation at the 8th International Conference on the Reduction of Drug Related Harm, Paris, 23–27 March.

McNeill, W.H. (1977) *Plagues and People*, Oxford: Basil Blackwell.

Stanecki, K.A. and Way, P.O. (1997)The demographic impact of HIV/AIDS: perspectives from the world population profile: 1996. *IPC Staff Paper No. 88*, International Programs Center, Population Division, US Bureau of the Census, Washington, DC.

Part IV

Techniques of spatial analysis

40

GIS, remote sensing and the problem of environmental change

Roy Haines-Young

INTRODUCTION

People cannot help but change their environment. It is a characteristic we share will all other species. Change comes about because we need to acquire energy and materials to live, and these activities have an inevitable consequence for the environment. What sets us apart from all other organisms, however, is the scale and speed of the changes that we have initiated. There are few places on the Earth today that do not bear some trace of human activity. The impacts of modern lifestyles can now be seen by environmental changes going on at local, national and global scales.

Concern about the impacts of people on the environment has stimulated a good deal of scientific work directed at understanding the nature of environmental change, its consequences and what needs to be done to avoid its worst effects. In this chapter, we will look at some important computer-based technologies that have been used in such work. Our goal will be to consider how the technology has been used to monitor environmental change and help us to develop appropriate management strategies for the future.

The technologies that we will consider are those of geographical information systems (GIS) and remote sensing (RS). We need to understand these technologies because they have become important tools that can be used by geographers and other scientists to apply their knowledge to many of the environmental problems that confront modern societies. Figure 40.1, for example, shows the recent deforestation history of Madagascar. It is important to understand these types of change, because the world's tropical forests are resources in their own right and because of the wide-ranging effect that such changes might have on global climates through the release into the atmosphere of the carbon stored in such ecosystems. Figure 40.2 shows data from a very different type of environment, namely that of central Asia. This map comes from a study that attempted to monitor land cover at a time when there had been a change in land holding and grazing patterns in some areas. These case studies are but two examples of scientists using map data to understand the nature of environmental change. In this chapter, we will consider some of the technical and scientific issues that lie behind such work.

It would clearly be a difficult and time-consuming task if data such as those shown in Figures 40.1 and 40.2 had to be collected by ground survey and drawn up by hand. Fortunately, we now have access to computer-based tools that can help to speed up the process. On the one hand, we have access to satellite and airborne remote-sensing systems that can automate the collection of data about the Earth's surface. On the other, we have available computer-based systems, in the form of GIS, that can store, analyse and display such data, and link them with other sources of information. Figure 40.3 sets out the relationships between the processes of *data*

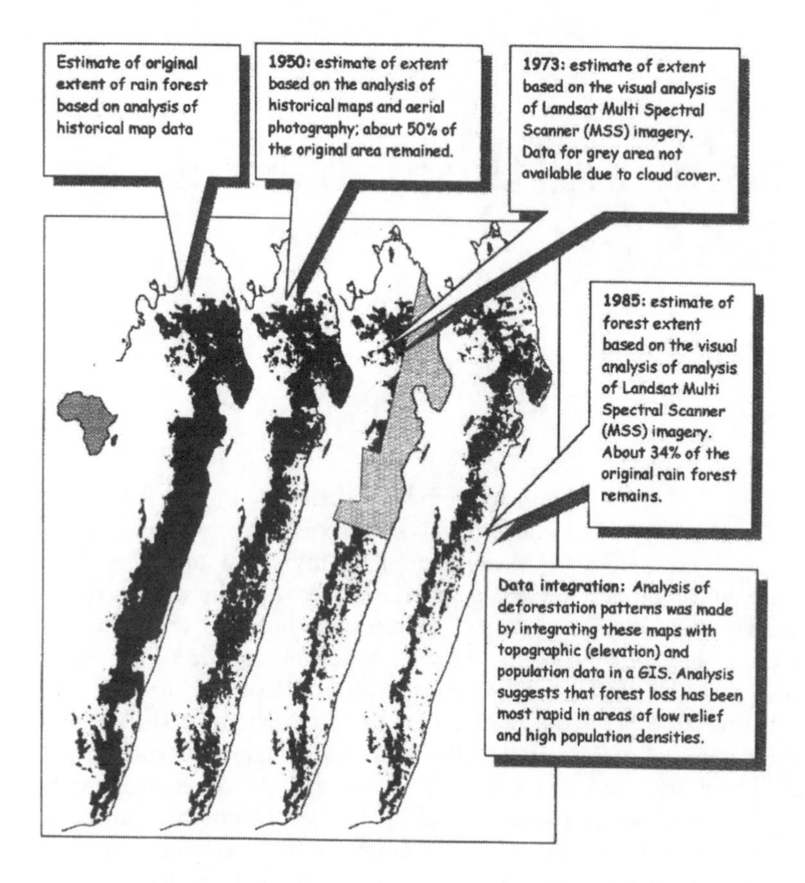

Figure 40.1 Mapping loss of tropical rain forest in Madagascar.

Source: Green and Soussman 1990.

capture and *data analysis* in a little more detail. This diagram is the framework around which the rest of this chapter is built. We will show how remote-sensing systems can help us to acquire information about the Earth's surface and the way it is changing. We will also see how these data can be linked with other information in a GIS to understand what is happening and to develop sound management strategies. Hinton (1998) has recently highlighted the importance of a closer integration of remote-sensing and GIS technologies.

At the same time as describing how remote sensing and GIS can be used to monitor and understand environmental change, this chapter will also look at some of the basic scientific concepts that underlie the study of environmental change. We will need to consider, for example, the idea of setting up a *base line* against which we can make judgements about the changes taking place in an environmental system (Figure 40.3). We also need to consider whether the change is measured in terms of a simple *change of state* in the system or whether it represents a more complex transformation in the *behaviour or dynamics* of the system over time.

PUTTING THE WORLD INTO A COMPUTER

If we think of a GIS as a computer-based system that can be used to store, analyse and display spatial or geographical data (Burrough and McDonnell 1998; Chrisman 1996), then clearly the first step in setting up such a system is to acquire or capture the information that we need. The process of data capture involves the collection or conversion of data in a digital, machine-readable format.

We cannot describe all the sources of digital

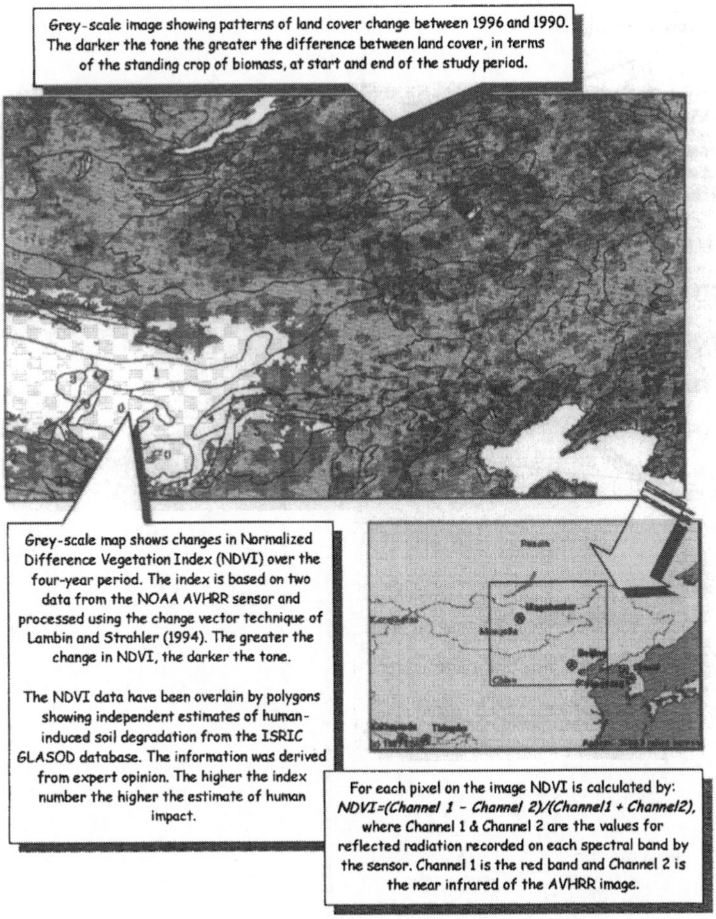

Figure 40.2 Land cover change in Central Asia.

Source: NOAA AVHRR data for calculation of NDVI by Mark Chopping; GLASOD data from ISREC, via UNEP; base map from ESRI, constructed using online world base map facility.

data that might be used in a GIS to study environmental change. However, it is useful to highlight three general types, namely:

- remotely sensed imagery;
- digital maps; and
- digital data collected in the field.

We will look at each of them in turn.

Remote sensing

The value of remotely sensed imagery can be illustrated by looking at the way in which satellite data can be used to map land cover. Remote sensing is defined as the process of acquiring information about an object using the electromagnetic energy that is reflected or transmitted by the target. The process is 'remote' in the sense that we do not have to physically make contact with the target to find out about it. Indeed, with satellite systems the sensor can be many hundreds of kilometres away from the target.

Many different types of sensor are available. They range from ground-based instruments held on towers or elevated platforms through to systems carried on aircraft or satellites. An introduction to the different types of sensor can be found in Lillesand and Kiefer (1994). In this chapter, the potential of remote sensing will be illustrated by reference to systems carried on some of the current generation of Earth observation satellites, namely Landsat Thematic

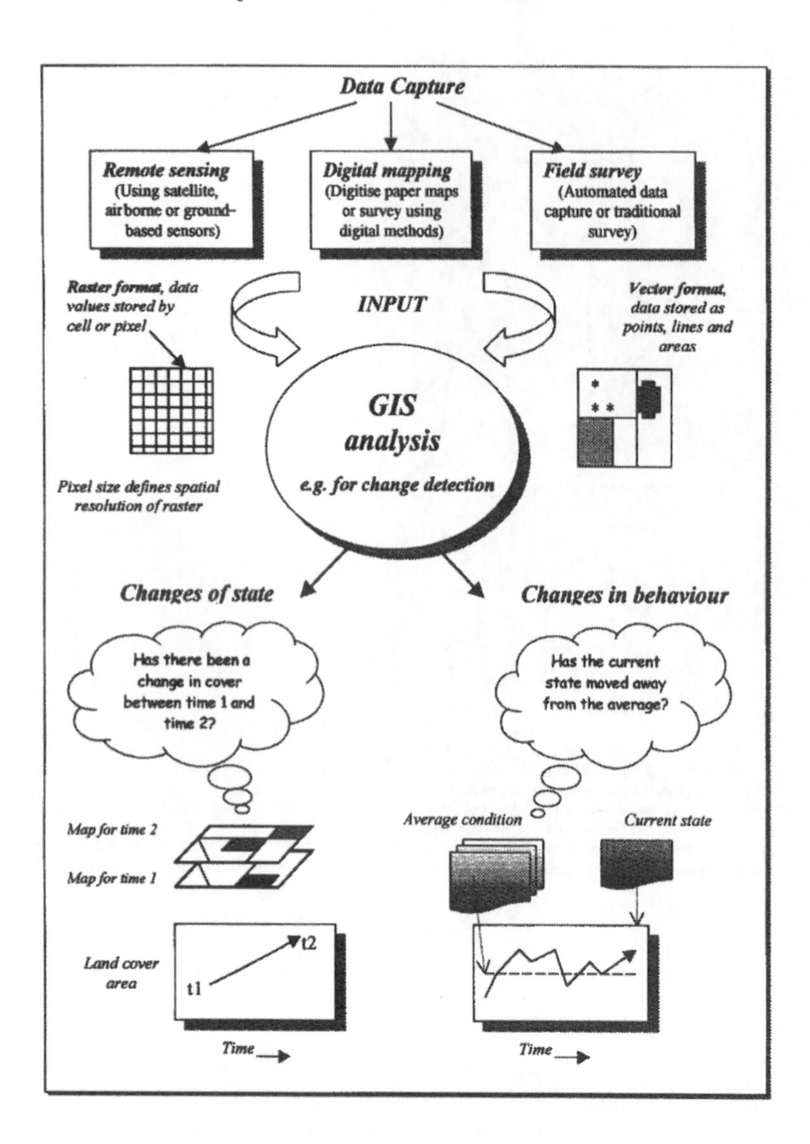

Figure 40.3 Data capture and data analysis: using GIS and remote sensing to understand environmental change.

Mapper (TM), SPOT, and AVHRR.

When the Sun's energy reaches the Earth's surface, it is partly reflected back into space. This reflected radiation may be detected using an optical sensor and used to characterise the nature of the reflecting surface. The TM sensor, for example, can detect this reflected radiation in seven different parts of the electromagnetic spectrum. These are called the *spectral bands* or *channels* on which the sensor produces data; three are in the visible, two are in the near infrared and one is in the thermal part of the spectrum (Figure 40.4). SPOT-3, by contrast, records information for only three parts of the spectrum, while the most recent satellite in the SPOT series records four. As the sensors pass across an area, they record the reflected radiation in each band for a small patch, or *pixel*, on the ground. The size of the pixel depends on both the design of the sensor and its altitude. For TM, the pixel size is equivalent to an area of about 30 × 30 m on the ground. SPOT, by contrast, has a spatial resolution of between 10 and 20 m, depending on the channel. A full image is thus made up of a large

Table 40.1 A summary of some of the types of data that will be available as part of NASA's Earth Observing System (EOS).

TRMM	The Tropical Rainfall Measuring Mission (TRMM) is a joint mission between NSASA and the National Space Development Agency (NASDA) of Japan designed to monitor and study tropical rainfall and the associated release of energy that helps to power the global atmospheric circulation shaping both weather and climate around the globe.
EOS	The EOS AM-1 satellite is the flagship of EOS. It will provide global data on the state of the atmosphere, land, and oceans, as well as their interactions with solar radiation and with one another.
Landsat-7AM	The Landsat Program is the longest-running enterprise for acquisition of images of the Earth from space. The first Landsat satellite was launched in 1972. The next in the series, Landsat-7, is scheduled for launch in late 1998.
SAGE III	The SAGE-III mission is to enhance our understanding of natural and human-derived atmospheric processes by providing accurate long-term measurements of the vertical structure of aerosols, ozone, water vapour, and other important trace gases in the upper troposphere and stratosphere.
EOS PM	The focus for the EOS PM satellite is the multidisciplinary study of the Earth's interrelated processes (atmosphere, oceans and land surface) and their relationship to Earth system changes.
ICESat	ICESat is a small satellite mission to fly the Geoscience Laser Altimeter System GLAS) in a near-polar orbit in 2001. GLAS will accurately measure the elevation of he Earth's ice sheets, clouds and land.
EOS Chemistry	The EOS CHEM-1 satellite will focus on measurements of atmospheric trace gases and their transformations. The CHEM-1 instruments and other facts and mission goals can be found by accessing this link.
ACRIM	The SAGE III mission is to enhance our understanding of natural and human-derived atmospheric processes by providing accurate long-term measurements of the vertical structure of aerosols, ozone, water vapour and other important trace gases in the upper troposphere and stratosphere.

Source: http://eospso.gsfc.nasa.gov/eos_homepage/missions.html
Note: See also the home pages for EIOS and ENVISAT.

correspond to real objects on the ground.

Most modern mapping agencies now produce maps in digital format. One can see the kinds of product that are available by visiting, for example, the web site of the Ordnance Survey of Great Britain (see section on additional sources, below). In the past, such maps have been produced by converting paper maps into machine-readable formats. Today, however, modern survey techniques are such that information can be collected directly in digital format and used to create a computer database from which various map products are derived.

In addition to the digital mapping produced by national mapping agencies, a range of more specialist products is also available. In Britain, for example, we can obtain information on soils from the Soil Survey and Land Research Centre, geological information from the British Geological Survey, and species records data from the Biological Records Centre. A review of UK rural data resources is given by Haines-Young and Watkins (1996). At the international level, spatially referenced environmental data can be obtained under the aegis of the United Nations Environment Programme (UNEP-GRID). The web addresses of these organisations are also given at the end of this chapter.

Automating data capture in the field

Automated techniques for data capture in the field are now being widely used. Thus, for example, hydrological or climatological measurements

can be made on a continuous basis and the data transmitted back to the laboratory or base station. Such data can be input directly into a GIS and the information used to monitor the state of the environment to give early warning of extreme weather events and other hazards. Figure 40.5 shows part of the system set up to monitor the El Niño oscillation. It consists of an array of buoys that record changes in sea temperatures across the Pacific. This information is relayed via satellite link back to the monitoring centre for processing. As we will see later, the information is used to monitor the dynamics of the atmosphere–ocean system in the area and to give warning of the occurrence of an El Niño event.

A particularly important recent development in the area of automated data capture has been the use of the Navstar GPS to record the spatial location of objects (see additional information). The GPS system consists of a constellation of satellites that orbit the Earth. These satellites broadcast radio signals that can be picked up by GPS receivers on the ground and the information used to locate the position of the receiver with accuracies down to a few millimetres, depending on the equipment and methods used. The location is fixed by using the information from three or more satellites to triangulate the position of the receiver. The positional information is usually stored by the field worker using a small

Figure 40.5 Array of buoys used to monitor sea temperatures across the Pacific Ocean.

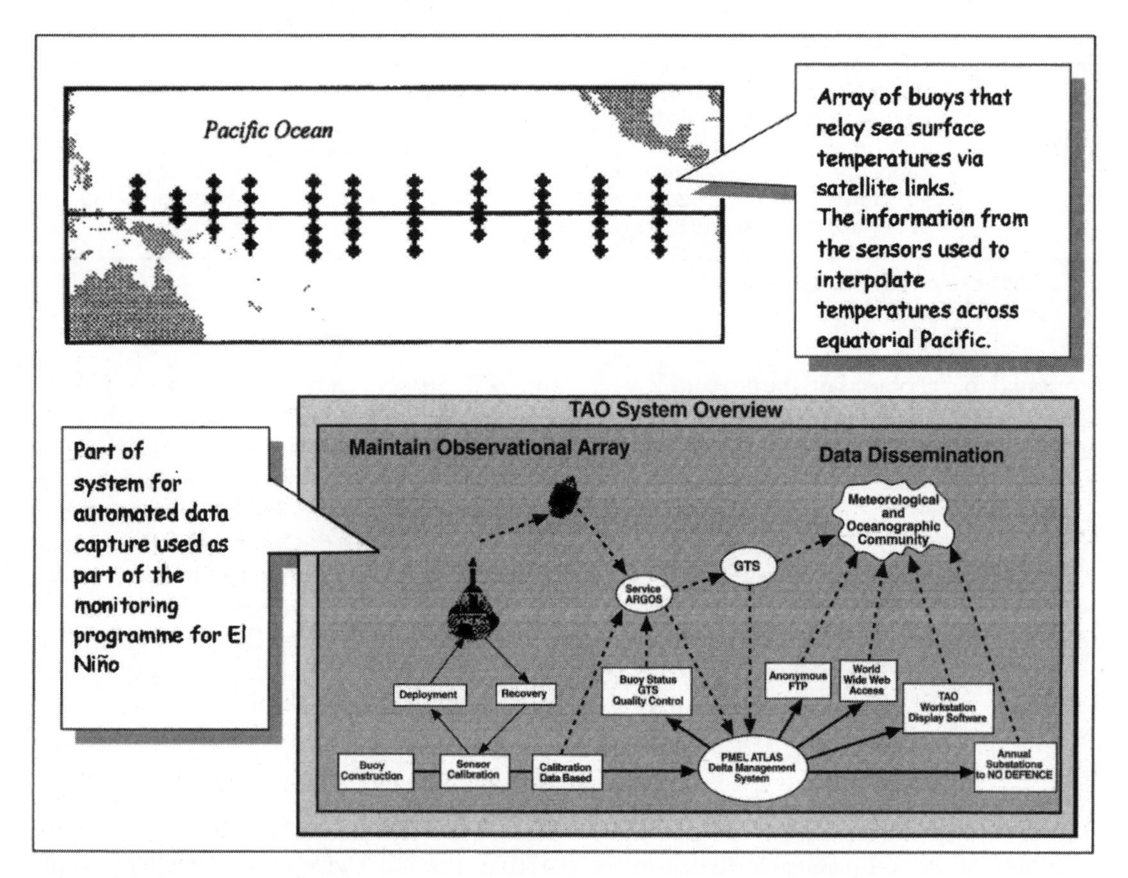

Source: TAO Project. Acknowledgements to TAO Project Office, Director Dr Michael J. McPhaden.

data logger, and the information is later downloaded into a database back in the laboratory.

The availability of GPS not only enables the more accurate recording of survey information in the field but it can also speed up the collection of data, because surveyors do not have to determine their location in relation to a map, which can be difficult in areas without clear landmarks. Moreover, GPS can help to relocate sample points quickly and accurately so that resurvey can be undertaken. The ability to record positional information rapidly also enables more dynamic systems and processes to be studied. GPS tracking techniques have been used for a wide variety of wildlife studies, for example to follow the movement of organisms (e.g Moen *et al.* 1996).

UNDERSTANDING OUR COMPUTER WORLDS

In the first part of this chapter, we have looked at some of the issues that surround the collection of the information that we need in order to set up our GIS. We must now consider how GIS can help us to analyse these data to understand how the environment might be changing, and what importance we might attach to such changes. We will use the framework set out in Figure 40.3 as the basis for our discussion. On the one hand, we will consider studies in which GIS has been used to set up some baseline against which changes in the overall condition of the environment can be measured. On the other, we will consider the more complex situation, when we need to consider whether it is the dynamic behaviour of the system itself that has changed for the better or worse. In both cases, we are looking at the problem of *change detection*.

Detecting changes in state

One of the easiest ways to detect change is in relation to some baseline or initial set of conditions. The baseline is used as a reference (or control) against which differences in the condition or *state* of the environment can be judged. Although we may recognise that there is no 'time-zero' or starting point, it is often possible to agree that the state of the environment at some given instant or period will stand as a reference against which further change can be measured. In this way, appropriate monitoring systems can be established.

An example of work that illustrates how both GIS and remote sensing can be used to establish a baseline and to detect change is provided by the various surveys of the British countryside undertaken by the Institute of Terrestrial Ecology (ITE) in the UK. Countryside Survey 1990 (CS1990), for example, was an expanded and refined programme of work that built on two earlier surveys in 1978 and 1984 (Barr *et al.* 1993). Collectively, these studies have provided information on the stock and change of rural land cover in Great Britain, together with data on the changes in abundance of the more common plant species that characterise the wider countryside. The 1990 survey also set down a baseline for the monitoring of freshwater biota in rural areas.

GIS was used as a basic tool for the storage, integration and analysis of countryside survey data (Figure 40.6). The field survey component of CS1990 mapped land cover, the character and condition of linear landscape features and the abundance of terrestrial plant species in a sample of the 1 × 1 km grid squares from the Ordnance Survey National Grid. The sample was stratified by major landscape type, so that national estimates could be made for each parameter recorded. Of the 508 squares surveyed in 1990, 256 had been surveyed in 1984 and 1978, so change could be determined. GIS was used as the basic tool for this analysis. The field maps for 1984 and 1990, for example, were compared using *overlay* procedures within the GIS. Thus differences in the area of each land cover type could be determined, or differences in the length of linear features such as hedge-rows or walls could be calculated. For the purposes of dissemination, the summary data derived from CS1990 has been set up on a simple GIS, known as the Countryside Information System (Haines-Young *et al.* 1994).

Since the 1990 survey was more extensive than

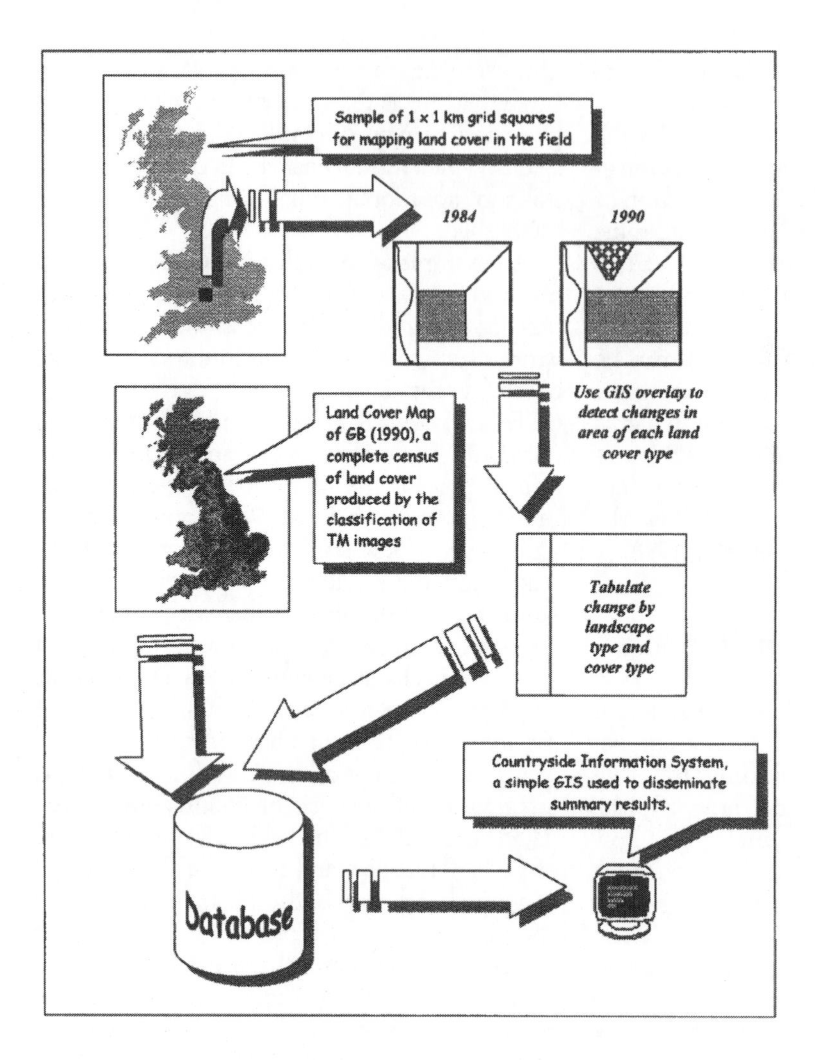

Figure 40.6 Integration and analysis of countryside survey data.

the earlier ones, CS1990 was regarded as the baseline against which past change was calculated. It was also designed as the baseline for monitoring future change. The results of the next countryside survey will be reported in around 2000; work is already in progress and you can find updates by visiting the ITE web site shown in the section on additional sources of information at the end of this chapter.

In addition to the field survey component of CS1990, remotely sensed satellite data were used to make a complete census of land cover in Great Britain. The output from this component of CS1990 was the LCMGB, which we discussed above. Howard *et al.* (1996) described some of the ideas behind linking these data with those of the field survey. Since no earlier images were available, this map was regarded as part of the 1990 baseline against which future change might be determined. A complete re-census of land cover using remotely sensed images is, however, also planned as part of Countryside Survey 2000, and so in the future it may well be used to help us to understand where changes in land cover within the British countryside are taking place.

Map overlay is a simple and effective change detection technique that can be used within a GIS. It is fairly straightforward, for example, using the type of vector data produced by the field survey techniques used in CS1990. The comparison of classified remotely sensed, raster data can also be made using overlay-type techniques. This is essentially the approach used for the analysis of deforestation in Madagascar described above. However, the analysis of satellite data to determine change can be more complex. Johnson and Kasischke (1998) have recently given an overview of the problem. They suggest that change detection using remotely sensed images can be approached in two general ways.

The first approach identified by Johnson and Kasischke depends on making a spectral classification of the images either separately for different time periods or by combining the multi-data imagery to identify areas of change. In the future, it is likely that much more sophisticated classificatory approaches will be developed with the use of classification routines that are based on the idea of spatial objects rather than pixels (Hinton 1998). It is likely, for example, that such methods will be used for the census of land cover using remotely sensed data in Countryside Survey 2000.

The second general approach to change detection identified by Johnson and Kasischke (1998) depends on the analysis of radiometric differences between images. Differences between bands for different dates can, for example, be compared once images have been corrected radiometrically to ensure that direct comparison between them can be made. Alternatively, band ratioing, principal component analysis or methods based on the calculation of indices based on particular band combinations can be used (see discussion of NDVI below). With these methods, some form of radiometric correction is required to take account of different atmospheric or illumination conditions at the times the images were acquired, or any differences in sensor characteristics. Examples of the application of these methods have recently been provided by Prakash and Gupta (1998), who looked at the problem of mapping land-use change in a coal-mining area of Jharia, India; Sunar (1998), who has considered the problem of mapping land cover change in Istanbul, Turkey; and, Mino *et al.* (1998), who have used multi-date image data to monitor grassland improvement in Japan.

Detecting changes in dynamics

While some problems can be reduced to the analysis of change in relation to some baseline state, other environmental problems are often more complex. Particular difficulties arise, for example, in the context of environmental systems that are *dynamic* in character. In these situations, it is not a simple change of state that we need to consider but a change in the *behaviour* of the system over time (Figure 40.3).

The importance of understanding environmental change in terms of change in system dynamics can be illustrated by recent work on El Niño, which is a disruption in the ocean–atmosphere system in the tropical Pacific. The El Niño effect can be seen, for example, by the analysis of sea-surface temperatures recorded from an array of buoys distributed across the Pacific on each side of the equator (Figure 40.5). These point measurements can be used to interpolate surface temperature patterns using a GIS. Under normal conditions, the western side of the Pacific is warmer than the east. However, with the El Niño oscillation, a warm tongue of water extends eastwards to the coast of Equador and Peru, disrupting the upwelling of the cold waters that enrich the Pacific coast of South America. There are many web sites that discuss El Niño, with excellent graphics and animations that illustrate how the dynamics of the atmosphere–ocean system change (see additional sources below). The importance of understanding and detecting such changes lies in the widespread economic and social consequences that arise as a result of the fluctuation in this environmental system.

On land, the problems posed by the need to monitor the behaviour of an environmental system can be illustrated by work that uses the Normalised Difference Vegetation Index (NDVI). In Figure 40.2, we have already seen how NDVI can be used to analyse the dynamics of land cover using the 'change vector analysis' approach described by Lambin and Strahler (1994). Analyses based on NDVI are now routinely used throughout the world to help to resolve problems related to fluctuations in terrestrial productivity. Such work can be illustrated by recent work undertaken by Sannier *et al.* (1998). These workers have used NOAA-AVHRR data for real-time vegetation monitoring as a tool for wildlife management and food security assessment in southern Africa.

NDVI is an indicator of the level of photosynthetic activity in crops and other vegetation. Using data from the FAO ARTMIS NDVI archive, Sannier *et al.* (1998) processed the information to construct the statistical distribution of NDVI values based on each ten-day period of the year; data for ten years were used to construct the distributions. Using these data, it was possible to develop a 'template' that indicated the thresholds between years with very low, low, average, high and very high productivity. Such templates describe the seasonal dynamics of the ecosystem. Sannier *et al.* suggest that the templates can be used to plot the trajectory of productivity for an area for any one year to determine whether it represented an extreme event that may be significant for people or wildlife. There is now widespread interest in the application of AVHRR data to give early warning of poor harvests, so that steps can be taken to secure food supplies in areas at risk.

Modelling change

When seeking to understand the nature of environmental change, we are fortunate if we can measure directly the particular parameter of interest, as in the case of land cover. In some situations, particular environmental characteristics cannot be assessed or measured directly. Instead, we have to use a surrogate measure and infer what might be happening to some underlying variable that really interests us. In other words, we may have to *model* changes using our GIS.

Indices like NDVI are examples of such models. The index is not a direct measurement of photosynthetic activity, but it is hypothesised that it is highly correlated with such activity. Thus NDVI can be used to predict productivity. There are many other examples in the literature that illustrate how GIS can be used to model and make predictions about the types of change that occur within environmental systems.

In the late 1980s and early 1990s, for example, the impact of commercial forestry on the ecology of wading bird populations of the Flow Country of Caithness and Sutherland in Scotland was a matter of concern. Unfortunately, large areas of forests had been planted before the scale of the potential impacts were recognised, so GIS and remote-sensing techniques were used to try to gain an insight into what had happened. The approach employed a habitat suitability model to evaluate the quality of the land that had been lost to forestry and to make some judgement about the importance of the loss.

The Flow Country is a large area of upland peat. The flat, open landscapes are characterised by the presence of small pool systems, which make these areas particularly attractive to wading birds. In fact, the Flow Country is one of the most important breeding areas in Europe for wading birds such as the dunlin, golden plover and green shanks. The fear over commercial forestry in the late 1980s was that the ecological value of these areas would be undermined by habitat loss and disruption of the peat hydrology.

A simple ecological model was used to estimate the quality of dunlin habitat that had been lost to forestry (Avery and Haines-Young 1989; Lavers *et al.* 1996; Lavers and Haines-Young 1997). The model was refined in several stages and was eventually based on the integration of remotely sensed satellite images with ground-based measurements of the distribution of pools across the peat surface. The first stage in the analysis was to calibrate the model with sites at which breeding densities of dunlin had been estimated by field survey (Figure 40.7). The second stage was to test the model against another set of sites to see if the model correctly predicted the numbers of birds observed there. Once the model had been corroborated, GIS could be used to apply the model to the landscapes of the Flow Country as a whole to map the variations in habitat quality across the area. The study used archive satellite data from 1978 to make the

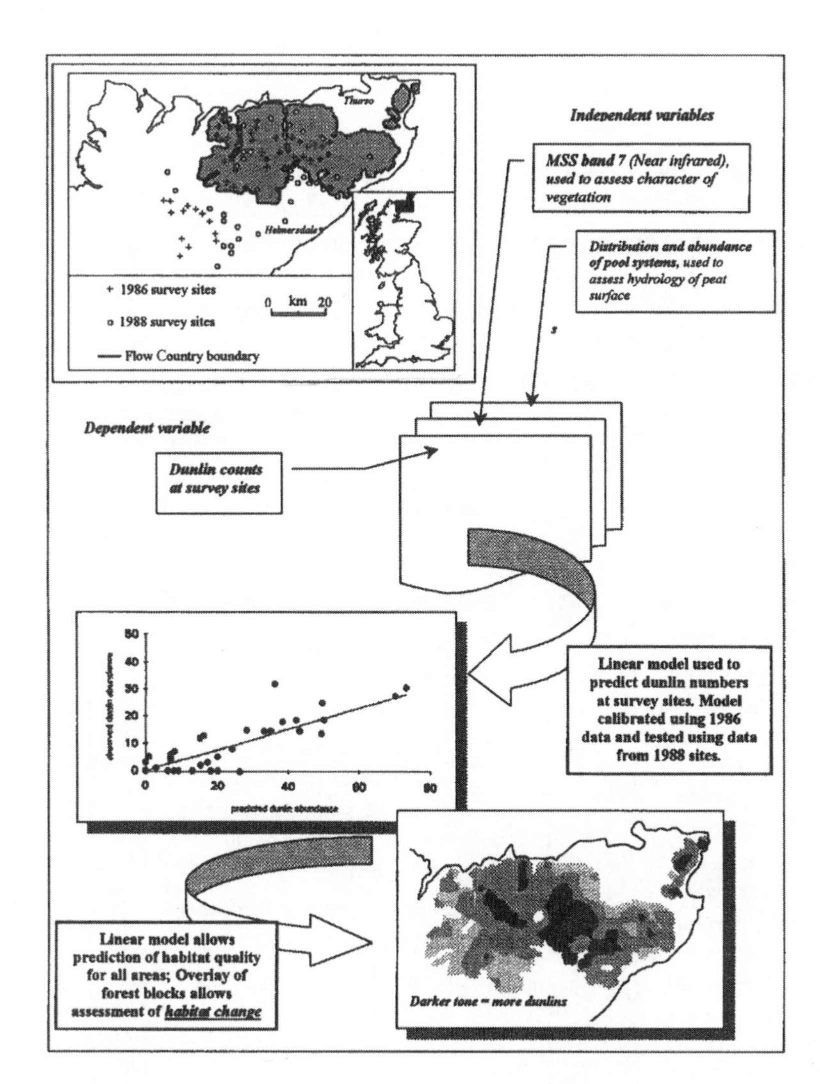

Figure 40.7 Modelling the impact of environmental change in the Flow Country.

assessment of the whole area, because this predated most of the commercial forestry that had taken place during the 1980s. GIS overlay techniques were then used to remove the areas that were previously available to the dunlin, to make an assessment of the quality and significance of the habitat that had been lost.

Habitat suitability models are now widely used to help to analyse the relationship between organisms and their environment, and they can lead to important applications because they enable the implications of different management strategies to be evaluated. A detailed account of the issues that arise in their construction has been provided by Morrison *et al.* (1992). Recent work that elegantly illustrates how remotely sensed satellite data can be used to model habitat suitability for tsetse flies in southern Africa is that of Robinson *et al.* (1997a, b).

MAKING A DIFFERENCE

There is little point investing in technologies such as GIS and remote sensing unless they can be used to make a difference to the way in which people

Box 40.1 Issues of data quality

In making a judgement about ***data quality*** always consider the ***fitness of those data relation to the purpose*** of the application. Some applications will require higher quality data than others.

Issues of data quality can arise in terms of both their ***content*** and the ***context*** in which they are used.

CONTENT

Micro level aspects of quality concern individual data items

- **Positional accuracy**: does the recorded position in database differ from the true position?
- **Attribute accuracy**: is the information recorded about the data item correct?
- **Logical consistency**: are the relationships between data elements consistent throughout the data set?
- **Resolution**: is the smallest size of the spatial unit for which data are held appropriate?

Macro level aspects of data quality concern the data set as a whole

- **Completeness**: does the data set cover all areas and items?
- **Time**: were all elements recorded or updated at the same time so that the data set is temporally consistent?
- **Lineage**: what errors were introduced in creating the data set?

CONTEXT

While the data set itself may be of good quality, one must ask whether the data items are being used appropriately. This is particularly the case when using one variable as a ***surrogate*** measure of another.

Source: Adapted from the discussion by Aronff (1994).

use and manage resources. The case studies described in this chapter certainly illustrate that the technologies can be useful, but their application also poses a number of wider scientific and institutional issues that have to be overcome before the value of the technologies can be fully realised.

An important area of scientific concern, for example, must be the quality of the data used to construct our GIS and the resulting quality of the output. Data quality issues represent some of the most challenging problems that face the GIS community at present and should not be overlooked by anyone who seeks to apply GIS or remote-sensing techniques to solve a geographical or ecological problem. The framework provided by Aronff (1989) is still a useful one to begin to think about some of the issues that relate to data quality. Duggin and Robinove (1990) discuss issues that specifically concern the analysis of remotely sensed data. When assessing data quality, it is useful to begin by asking questions about the *fitness* of a given data set for the particular application and then moving on to consider issues in relation to *content* and *context*. An explanation of these terms and some examples that illustrate them are given in Box 40.1. The important message here is that only if data of appropriate quality are used in appropriate ways can sound decisions be made when we seek to apply the technology.

In addition to the scientific issues posed by the use of GIS and remote sensing, we also need to understand something of the institutional context in which the technologies are applied. Experience suggests that the way in which information flows into and around organisations also controls how effective these technologies are in helping us to make a difference to the way we manage the environment. Knowledge cannot be used, for example, unless one can gain access to it. Thus despite the potential value of GIS and remote sensing, organisations or groups may not have the resources to buy into the technology or to train their staff to use it. Moreover, societies may not have the relevant administrative procedures in place to implement suitable management strategies even if they know what should be done.

Davis and Medyckyj-Scott (1996) provide some recent insights into GIS users in the developed world, while Sahay and Walsham (1996) document a useful case study that focuses on the implementation of GIS in India. Perhaps the most challenging commentary, however, is still that provided by Pickels (1995), who with others has looked at the wider social and political implications of the technologies. A key issue posed in this work is whether the technology excludes people from decision making or whether it forces us to look at the work in particular sorts of ways.

There is little doubt that access to the technologies of GIS and remote sensing will make a difference to the way in which we understand and manage our environment. Such technologies have the potential to speed up and improve the flow of information and can clearly benefit our decision-making strategies. However, more sustainable forms of environmental management will not evolve automatically. Like any other technologies, they can be used for good or ill. This chapter has shown that GIS and remote sensing have come of age and can be used in an operational context. The urgent tasks that face us are to understand their wider capabilities and to find ways of using them to inform and involve people in the decisions we have to make, given the pressing problems of environmental change.

GUIDE TO FURTHER READING

The following texts provide a sound introduction to the expanding fields of GIS and remote sensing and guidance in the principles and practice of each.

Burrough, P.A. and McDonnell, R.A. (1998) *Principles of Geographical Information Systems.* Oxford: Oxford University Press.

Foody, G. (1996) *Environmental Remote Sensing from Regional to Global Scales.* Chichester: John Wiley & Sons.

Goodchild, M. (1996) *GIS and Environmental Modelling: Progress and Research Issues.* New York: John Wiley & Sons.

Mather, P.M. (1999) *Computer processing of Remotely-Sensed Data,* second edition. Chichester: John Wiley & Sons.

Power, C.H. (ed.) (1996) *Remote Sensing for Natural Resource Management.* Natural Resources Institute.

Star, J.L. (1996) *Integration of Geographic Information Systems and Remote Sensing.* Cambridge: Cambridge University Press.

Other sources of useful information:

Information on AVHRR: http://edcwww.cr.usgs.gov/glis/hyper/guide/avhrr

Information on British Geological Survey http://www.bgs.ac.uk/

Information on CORINE: http://www.ecnc.nl/doc/servers/land.html

Information on Countryside Information System: http://www.nmw.ac.uk/ite/software/cisflier.html

Information on CS2000: http://www.cs2000.org.uk

Information on Environmental Change Network: http://www.nmw.ac.uk/ite/edn2.html

Information on ENVISAT: http://envisat.estec.esa.nl/

Information on Landsat: http://edcwww.cr.usgs.gov/glis/hyper/guide/landsat_tm

Information about Land Cover Map of Great Britain: http://www.nmw.ac.uk/ite/lcm.html

Information on SPOT: http://edcwww.cr.usgs.gov/glis/hyper/guide/spot#spot18

Information on the Biological Records Centre: http://www.nmw.ac.uk/ite/edn2.html#brc

Information on the Navstar system and GPS: http://www.navstar-systems.com/

Information on the Ordnance Survey of Great Britain: http://www.ordsvy.gov.uk/

Information on Soil Survey and Land Research Centre: http://www.cranfield.ac.uk/sslrc/publics/digital/

Information on United Nations Environment Programme (UNEP): http://www.unep.org/home.htm

and UNEP GRID (Global Resource Information Database):

http://grid2.cr.usgs.gov/
http://www.unep.org/unep/eia/ein/grid/web/document/grid.htm
http://www.unep.org/unep/eia/ein/grid/web/document/grid.htm

Information on El Niño: http://www.pmel.noaa.gov/toga-tao/el-nino/

NASA's Earth Observing System Home Page: http://eospso.gsfc.nasa.gov/

NASDA Earth Observation Data and Information System of Japan: http://www.eoc.nasda.go.jp/homepage.html

REFERENCES

Aronff, S. (1989) *Geographic Information Systems: A Management Perspective*. Ottawa: WDL.

Avery, M. and Haines-Young, R.H. (1989) Population estimates derived from remotely sensed imagery for *Calidris alpina* in the Flow Country of Caithness and Sutherland. *Nature* 344, 860–2.

Barr, C.J., Bunce, R.G.H., Clarke, R.T., Fuller, R.M., Fruse, M.T., Gillespie, M.K., Groom, G.B., Hallam, C.J., Hornung, M., Howard, D.C. and Ness, M.J. (1993) *Countryside Survey 1990, Main Report.* Department of the Environment, Countryside Survey 1990 Series, Vol. 2.

Burrough, P.A. and McDonnell, R.A. (1998) *Principles of Geographical Information Systems.* Oxford: Oxford University Press.

Chrisman, N. (1996) *Exploring Geographic Information.* New York: Wiley.

Cruickshank, M.M. and Tomlinson, R.W. (1996) Application of CORINE land cover methodology to the UK – some issues raised from Northern Ireland. *Global Ecology and Biogeography Letters* 5(4–5), 235–48.

Davis, C. and Medyckyj-Scott, D. (1996) GIS users observed. *International Journal of Geographical Information Systems* 10(4), 363–84.

Duggin, M.J. and Robinove, C.J. (1990) Assumptions implicit in remote sensing data acquisition and analysis. *International Journal of Remote Sensing* 11, 1669–94.

EEA (1998) *Proceedings from Workshop on Land Cover Applications – Needs and Use*. European Topic Centre on Land Cover, Copenhagen, 12–13 May 1997. Copenhagen: European Environment Agency.

Fuller, R.M., Groom, G.B. and Jones, A.R. (1994a) The Land Cover Map of Great Britain: an automated classification of Landsat Thematic Mapper data. *Photogrammetric Engineering and Remote Sensing* 60, 553–62.

Fuller, R.M., Groom, G.B. and Willis, S.M. (1994b) The availability of Landsat TM images of Great Britain. *International Journal of Remote Sensing* 15, 1357–62.

Green, G.M. and Soussman, R.W. (1990) Deforestation history of the eastern rain forests of Madagascar from satellite imagery. *Science* 248, 212–15.

Haines-Young, R.H., Bunce, R.G.H. and Parr, T. (1994) Countryside Information System: an information system for environmental policy development and appraisal. *Geographical Systems* 2, 329–45.

Haines-Young, R.H. and Watkins, C. (1996) The rural data infrastructure. *International Journal of*

Geographical Information Systems 10(1), 21–46.
Hinton, J.C. (1998) GIS and remote sensing integration for environmental applications. *International Journal of Geographical Information Systems* 10, 877–90.
Howard, D.C., Fuller, R.M. and Barr, C.J. (1996) Linking ecological information recorded from ground, air and space: examples from Countryside Survey 1990. *Global Ecology and Biogeography Letters* 5(4–5), 227–34.
Johnson, R.D. and Kasischke, E.S. (1998) Change vector analysis: a technique for the multispectral monitoring of land cover and condition. *International Journal of Remote Sensing* 19(3), 411–26.
Lambin, E.F. and Strahler, A.H. (1994) Change vector analysis in multispectral space: a tool to detect and categorise land-cover change processes using high temporal-resolution satellite data. *Remote Sensing of Environment* 48, 231–44.
Lavers, C.P. and Haines-Young, R.H. (1997) Displacement of dunlin (*Calidris alpina schinzii*) by forestry in the Flow Country and estimation of the value of moorland adjacent to plantations. *Biological Conservation* 79, 87–90.
Lavers, C.P., Haines-Young, R.H. and Avery, M.I. (1996) The habitat associations of dunlin *Calidris alpina)* in the Flow Country of northern Scotland and an improved model for predicting habitat quality. *Journal of Applied Ecology* 33, 279–90.
Lillesand, T.M. and Kiefer, R.W. (1994) *Remote Sensing and Image Interpretation*. New York: John Wiley & Sons.
Mather, P.M. (1999) *Computer Processing of Remotely-Sensed Data*, second edition. Chichester: John Wiley & Sons.
Mino, N., Saito, G. and Ogawa, S, (1998) Satellite monitoring of changes in improved grassland management. *International Journal of Remote Sensing* 19(3), 439–52.
Moen, R., Pastor, J., Cohen, Y. and Schawartz, C.C. (1996) Effects of moose movement and habitat use on GPS collar performance. *Journal of Wildlife Management* 63, 659–68.
Morrison, M.L., Marcot, B.G. and Mannan, R.W. (1992) *Wildlife–Habitat Relationships*. University of Wisconsin Press.
Prakash, A. and Gupta, R.P. (1998) Land-use mapping and change detection in a coal mining area – a case study in the Jharia coalfield, India. *International Journal of Remote Sensing* 19(3), 391–410.
Pickels, J. (ed.) (1995) *Ground Truth: The Social Implications of Geographic Information Systems*. New York: Guilford Press.
Robinson, T., Rogers, D. and Williams, B. (1997a) Mapping tsetse habitat suitability in the common fly belt of southern Africa using multivariate analysis of climate and remotely sensed vegetation data. *Medical and Veterinary Entomology* 11(3), 235–45.
Robinson, T., Rogers, D. and Williams, B. (1997b) Univariate analysis of tsetse habitat in the common fly belt of southern Africa using climate and remotely sensed vegetation data. *Medical and Veterinary Entomology* 11(3), 223–34.
Sannier, C.A.D., Taylor, J.C., du Pleiss, W. and Campbell, K. (1998) Real time vegetation monitoring with NOAA AVHRR in southern Africa for wildlife management and food security. *International Journal of Remote Sensing* 19(4), 631–40.
Sahay, S. and Walsham, G. (1996) Implementation of GIS in India: organisational issues and implications. *International Journal of Geographical Information Systems* 10(4), 385–404.
Sunar, F. (1998) An analysis of change in a multi-date data set: a case study in the Ikitelli area, Istanbul, Turkey. *International Journal of Remote Sensing* 19(2), 225–36.

41

Cartography: from traditional to electronic and beyond

David Green and Stephen King

INTRODUCTION

Mapping is a fundamental geographical activity, and as both an academic and a practical subject cartography has a very long and very well-documented history (see any textbook on cartography, e.g. Keates 1973; Cuff and Mattson 1982; Dent 1985; Robinson *et al.* 1995; Jones 1997; Dorling and Fairbairn 1997; http://www.geosys.com/cgi-bin/genobject/cartography/tig.3eed/). A great deal of theoretical and practical research has been undertaken over the years in all areas of the subject, including work on map projections and coordinate systems, map design (e.g. use of colour, symbols, legibility and composition), visualisation and communication, cartograms, journalistic and propaganda maps, 3D terrain models, and generalisation, to mention but a few.

With the development of mainframe computers in the late 1950s and early 1960s, computer-assisted cartography (CAC) or mapping soon began to develop quite rapidly (see, for example, publications such as Peucker 1972; International Cartographic Association 1980; Taylor 1980; Monmonier 1982; and Carter 1984). However, until the late 1980s and early 1990s, CAC remained largely the province of the mainframe, and subsequently minicomputers, and the realm of the large institution and the computer specialist. This is despite the emergence of the early microcomputers e.g. the Apple II/IIE and Macintosh, the IBM PC, and others, e.g. BBC, Atari.

However, the continuing and very rapid evolution of both microcomputer hardware and software technology over the years has since permitted cartography to become another software application on the desktop computer system. In many ways, cartography has now become akin to everyday applications such as word processing, spreadsheets and databases, and one that no longer requires the expertise of the specialist cartographer. With the advent of increasingly user-friendly systems, it has been possible for more people to design, create and produce maps without the aid of a cartographer or indeed the requirement to possess any cartographic knowledge. Cartography has become a communication tool available to all, whether it is for drawing maps to illustrate a journal paper or book chapter, a screen map display on a public information system, or presentational purposes in a seminar.

As we move towards the millennium, it is becoming evident that the current and future computer technology available will provide even more new and exciting opportunities for both cartography and cartographers (Green 1994). The technology will enable a new form of cartography – one that is less of an electronic equivalent of the more traditional form of cartography (a problem that has tended to overshadow the visual and innovative potential of cartography in an electronic medium) – but more flexible in the way that it makes use of information technology to take the subject – both theory and practice – forward into a new era of map design, creation, visualisation and use.

This chapter focuses on some of the developments in computer or digital mapping, from the early days of mainframe software packages, through Windows-based graphics and cartographic software, to the rapidly emerging and powerful visualisation tools to be found on the desktop computer, CD-ROM and the Internet.

EARLY COMPUTER CARTOGRAPHY

Early examples of electronic mapping or cartography, often referred to as computer-assisted cartography (CAC) or computer cartography (CC), were largely attempts to duplicate the practices of traditional cartography whereby the design, creation and production of a map, and even the final hard copy or printed map, was derived with the aid of computer technology, e.g. computer input (digitiser/scanner), storage (computer disk), and output (screen/printer/plotter). Initially, the 'new' cartography was undertaken with the aid of large computer systems and subsequently the minicomputer and latterly the microcomputer (PC and Apple Macintosh).

While it was possible with the aid of the early software packages to duplicate some traditionally produced map products, much of the early CC output was relatively simple and not entirely satisfactory from the traditional cartographer's point of view, or indeed the computer cartographer's. The sophistication of the hardware and software technology at the time left a great deal to be desired. Although some of the basic products (both in terms of the potential to design the map and to provide soft copy (screen display) and hard copy (monochrome or colour printed output) were acceptable and novel, the potential was limited by the graphics display and printer hardware available. It was soon evident that map design and production on a computer was far from easy and that it was not possible for the computer system to take on and replace the role of the cartographer, as some traditionalists had initially feared.

As time went by, with the development of both computer hardware and software, map design, creation and production became much easier with new mainframe software packages such as GIMMS (Waugh and McCalden 1982) and SAS/GRAPH (Baker *et al.* 1983; 1985); Carter 1984; Green *et al.* 1985; Green and Schwartz 1988). These new software packages provided greater flexibility to both design and create maps using smaller computer systems, accessed via remote terminals, and were used as the basis for producing a number of atlases (Thom 1990; Buckland *et al.* 1990; Maguire *et al.* 1987; Bourne *et al.* 1985; 1986).

MAPPING SOFTWARE FOR MICROCOMPUTERS

The first microcomputers had relatively limited processing power and capabilities for storing and handling large volumes of data, and most had low-resolution displays. Very few had colour monitors, and output devices provided relatively low-quality output. As the power of the microcomputer grew, so products such as GIMMS (Apple Macintosh), and SAS/GRAPH (PC) were able to migrate to the smaller platforms. Another similar package was SPSS Mapmaker (Green *et al.* 1990). Topographer (CLARES) and Navigator (Topologika) were also available for the Acorn Archimedes series of microcomputers (Anon 1993; Bennett 1994).

DOS-BASED GRAPHICS DESIGN AND CAD SOFTWARE

To a large extent, the growth in computer cartography was aided by the growing number of graphics design software packages that soon became available, at first running under DOS and later the Windows operating systems. At the most basic level, graphics software packages, e.g. Dr. Halo, EGA-Paint, were often used as the basis for drawing simple maps (see Green *et al.* 1990; Green and Calvert 1998; Peterson 1995). Map outlines could be input directly by hand, traced from a transparent overlay on the screen, digitised or scanned, and the graphics toolboxes used to add text, symbols, shade areas and other annotation to create a map (see Green *et al.* 1990; Green and Calvert 1998). For the most part, it was the simplicity and availability of this basic software, either at a low cost or bundled with a computer, that encouraged people to make use of them for mapping tasks. The power of the early CAD packages running under DOS, e.g. AutoCAD, also provided a useful way for cartographers to produce maps (see, e.g., Bedell 1989; Green *et al.* 1990; Green and Calvert 1998; and Peterson 1995); albeit a quite expensive and slightly complicated 'toolbox' at the time.

THE WINDOWS INTERFACE AND GRAPHICS SOFTWARE

The emergence of the user-friendly Windows graphical user interface (GUI) or the Windows Icons Mouse Pointer (WIMP) work environment, however, had a huge impact upon the so-called 'usability' of computer hardware and software. It is the Apple Macintosh (and subsequently PCs and workstations running Windows 3.1/95/98/NT and Open Windows, respectively) that perhaps can be credited most with opening up the way for the cartographer and non-cartographer of the 1980s. What was once difficult to undertake, requiring programming experience or computer literacy, was replaced by a simple-to-operate, intuitive and quick-learning-curve environment, the complexity of operation being concealed behind multifunctional menus in the Windows interface.

Graphics design software provided a far wider range of possible mapping applications and flexibility than the existing map packages and offered cartographers a major new tool for map design, creation, and production. Although many of the early software packages were not connected to cartography, the combination of both vectors and raster-based graphics software products provided excellent map-making toolboxes for both the practising cartographer and the novice. The software was simple to use, extremely powerful, and in fact still has a vital role in allowing the cartographer and non-cartographer alike to produce both traditional and illustrative cartography quickly and easily. Good examples of the software that was and still is in use are Superpaint, Canvas, Aldus Freehand, Adobe Illustrator and subsequently, Corel Draw (Figure 41.1). These packages are still widely used today in many cartographic units and offices in educational institutions, albeit in more recent versions, on either Macs or PCs (http://www.mun.ca/geog/muncl/manual.htm). The Ordnance Survey(OS)/UK Hydrographic Office collaboration, resulting in the experimental Coastal Zone map series of the Southampton area, used an Apple Macintosh running Adobe Illustrator (Eden 1993). Monmonier (1989) also mentions the widespread use of Apple-based software packages in newspaper offices for journalistic cartography (Figure 41.2) (see also Whitehead and Hershey 1990).

With the growing number of GIS applications, and demands for more sophisticated map generation tools, packages such as Adobe Illustrator (version 7.0) now come with additional plug-ins such as Map-Publisher from Avenza Software. Many maps produced in GIS software are in fact exported to graphics software for 'finishing' (e.g. DiBiase 1991).

CARTOGRAPHIC DEVELOPMENTS IN REMOTE SENSING AND GIS SOFTWARE

In today's world, many different types of map are produced in different ways (Green 1996). Many are freehand-drawn maps, e.g. journalistic products, which simply require a good graphics package that is capable of adding a recognisable 'cartographic' element e.g. scale bar, north arrow, border, to the graphic. Others are maps produced through the analysis of GIS data sets and overlays, and digital image processing and classification. While it was once the case that map production capabilities within GIS and RS software were relatively limited, allowing little freedom or functionality to produce a good end

Figure 41.1 Perth city services route map.

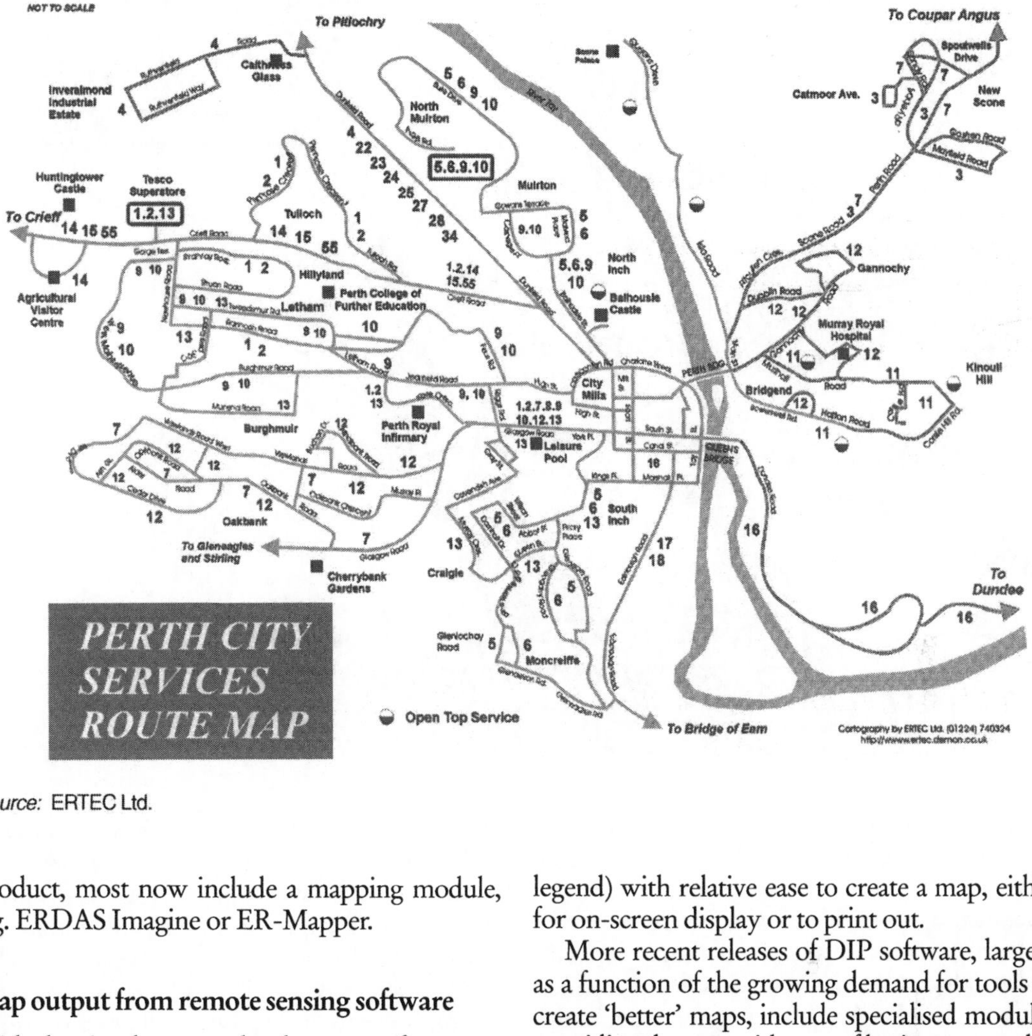

Source: ERTEC Ltd.

product, most now include a mapping module, e.g. ERDAS Imagine or ER-Mapper.

Map output from remote sensing software

With the simultaneous development of remote-sensing digital image processing (DIP) software, much of the output being a map, e.g. an image map, or classified scene, graphics software was an ideal environment in which to create a finished map product. At first, this took the form of either an 'in-house' module, e.g. ERDAS 7.4/5, or the adoption of a graphics package, e.g. the CHIPS DIP system made use of a low-cost graphics package known as Dr. Halo. Classified images could be annotated (e.g. through the addition of a north arrow, border, scale, title, subtitle, and legend) with relative ease to create a map, either for on-screen display or to print out.

More recent releases of DIP software, largely as a function of the growing demand for tools to create 'better' maps, include specialised modules providing the user with a set of basic cartographic tools. For example, ERDAS Imagine includes a product called Map Composer, and more recently for the Windows 95/NT platform has enhanced the cartographic tools available still further with MapSheets. ER-Mapper also includes a map output product called Map Composition.

Map output from GIS software

Well-known GIS products such as Idrisi, GRASS, Arc/Info and MapInfo among many others offer

Figure 41.2 Journalistic cartography.

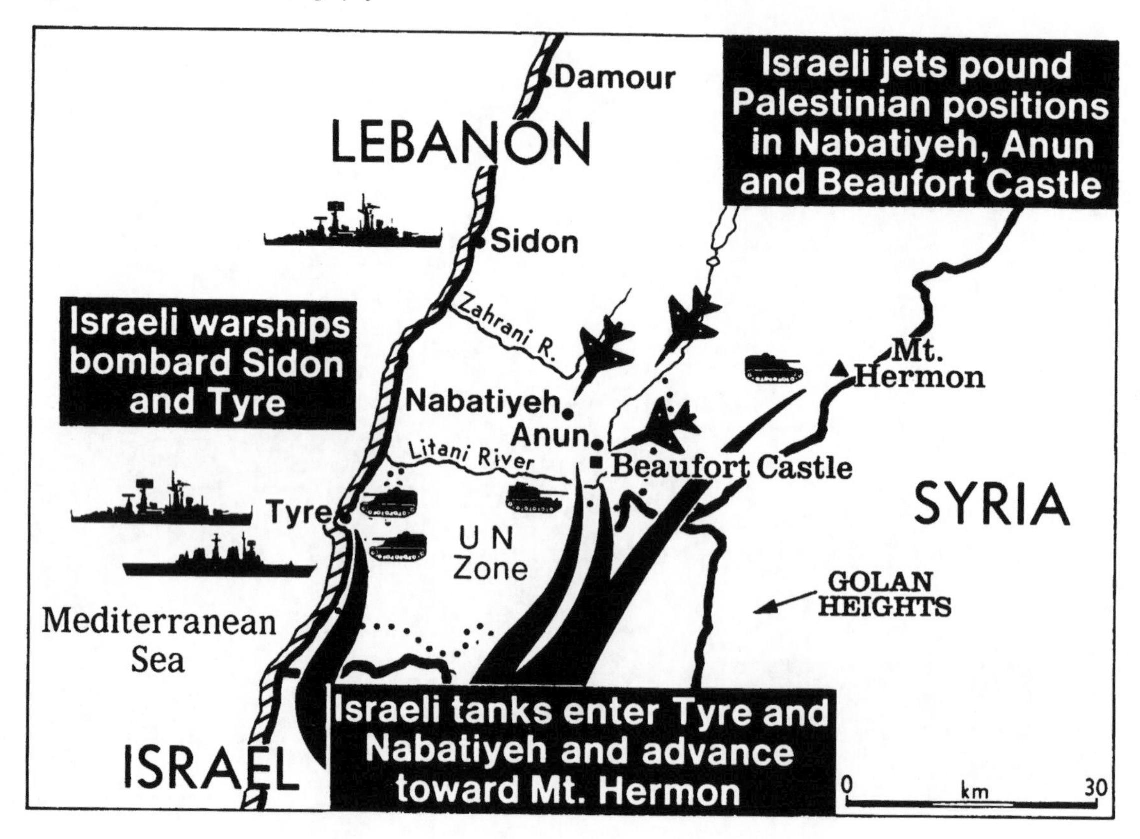

a user the means of retrieving and displaying spatial data in the form of maps. Despite the obvious historical links to cartography and digital mapping, many early GIS software products offered little scope for the production of the cartographic end product. Only subsequently have products such as ESRI's Arc/Info provided additional cartographic tools in the form of modules to help to improve the map output. Another development has been the more user-friendly interface provided by ArcView (http://www.esri.com) and similar software, which is now widely used as a visual and exploratory mapping tool. Recently Maplex, a cartographic name placement system, has been added to the user toolbox (http://www.esri.com/maplex).

Despite this, many GIS-produced maps are still ultimately 'finished off' using graphics software that allows for customisation of the map. One reason for this is that many GIS software packages still do not really allow the user freedom to do anything more than perhaps create an electronic version of a paper map, usually within the constraints of the software's perception of what cartography entails. The additional flexibility offered by graphics software allows for greater customisation of the map. More recent map-creation and -production products such as DRY 2.0 (a product of Lorik), Map Maker Pro 2.1 (Price 1998), and Map Publisher offer an even closer link between cartography, graphics and GIS.

CARTOGRAPHIC SOFTWARE

With time, there have been more and more examples of cartographic software appearing on the market for desktop computer systems. A well-known example was the Atlas software. For the Windows-based work environment, Golden Software's MapViewer and Surfer from the United States, and Maps 35/40 (European Standard Software 1995) from Denmark are good examples. These are custom cartographic packages providing the user with a toolbox that lies somewhere between a pure digital mapping package and a graphics design package, as well as having the benefit of a spreadsheet window to help to plot map data input by the user. For the non-cartographer, the toolbox in, for example, MapViewer, offers a simple way to plot spatial data, and the tools a means to add to and customise the map. The package provides some cartographic guidance for the novice in the sense that it constrains the user to a fixed set of map types. However, each item on the map, whether it is an area, a symbol, or text, is an 'object' and its properties can therefore be altered. This provides a great deal of flexibility for the experimental cartographer, allowing them to cut, paste, move, change, add and delete objects that make up the map. At the same time, too much design freedom without appropriate cartographic guidance means that the user can create a relatively poor cartographic end product.

PRESENTATION SOFTWARE

Other software products such as Microsoft's Powerpoint presentation software (part of the MS Office suite) provides users with a selection of very basic map outlines and clip art as the basis for creating simple maps to add variety to overhead slide presentations (both manual and electronic) (Figure 41.3). Annotation can be added to complete the map, and since each item drawn is an object, it is possible to alter the appearance of the maps to suit.

Figure 41.3 Powerpoint-prepared presentation graphics.

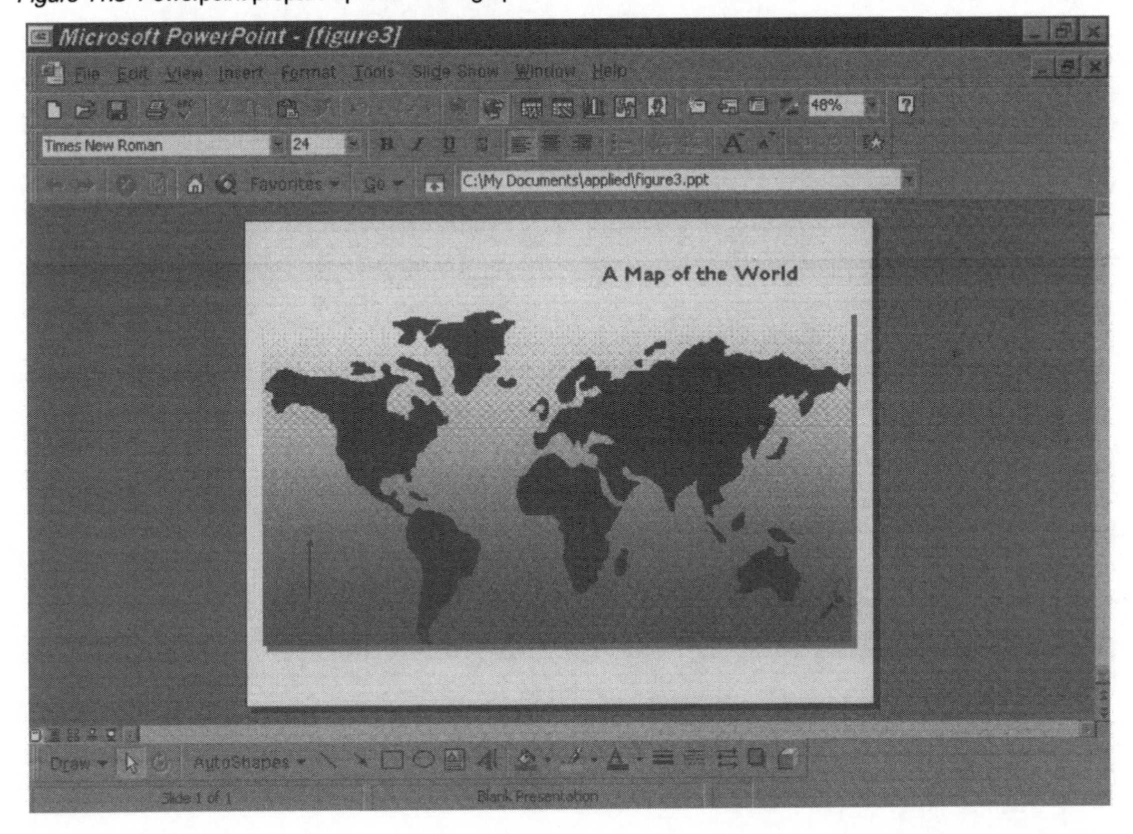

SPREADSHEETS

Growing realisation of the importance of mapping software (largely through the development of geographic information systems (GIS) and digital image processing (remote sensing)) has led software companies, e.g. Microsoft, to offer desktop mapping capabilities as part of products such as the spreadsheet MS Excel (see Whitener and Creath 1997; Figure 41.4). Not only have these developments been a direct response to new desktop GIS products but they also recognise the growing demand for what were previously separate software products to be combined into one for practical desktop applications: the sort of applications that need more functionality than a presentation package but less than a full-blown GIS.

POTENTIAL PROBLEM AREAS

Freedom and lack of training

One area in which many, if not all, of these above-mentioned opportunities for computer mapping are currently lacking is that of guidance in map design. Freedom to design one's own map and the growth in applications of spatial data (geographic information (GI) and GIS) have made the map the domain of the user, who is, more often than not, not a cartographer or cartographically trained. What this means is that tools to facilitate map design, creation and production are now widely available in a variety of different guises, for anyone to use, but without the provision of any guidelines for the user.

While some software packages assist users

Figure 41.4 Desktop mapping via Excel.

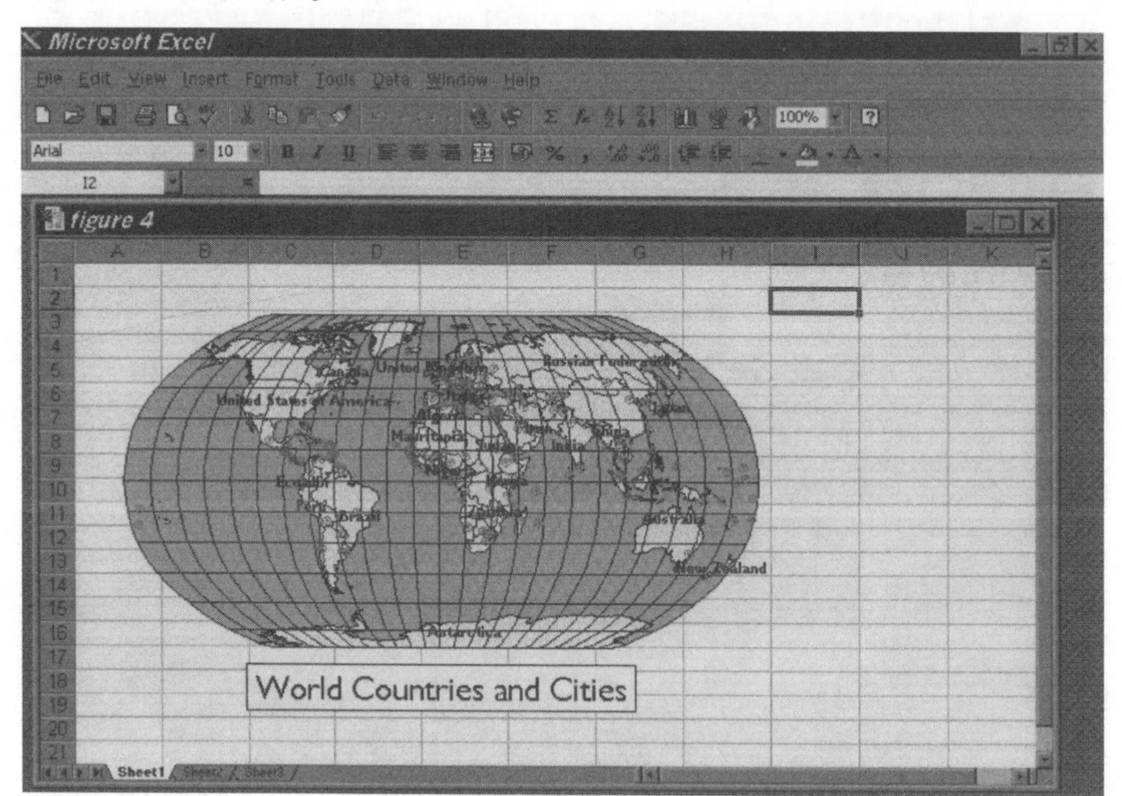

through the inclusion of manual-based tutorials (e.g. MapViewer), templates (e.g. MapSheets), or specialised books (e.g. Whitener and Creath 1997 (Excel)), more often than not the help offered is merely practical guidance in using the software to help to create a map but with little emphasis being placed on the visual quality of the end product. The MapViewer tutorial, for example, leads to the production of a rather garish and colourful choropleth map. The software also allows the user to create a wide variety of different types of map using the same data, some of which may not be appropriate. Other mapping software simply does not offer any guidance at all, which can lead to what is ultimately a very powerful toolbox being used to create a poor map product that does not effectively communicate the information contained within it.

Although attempts have been made to provide specific guidance in the form of reference colour charts to aid in the use of colour, map composition templates to aid in map layout, and even knowledge-based systems that run alongside the mapping package designed to offer appropriate guidance, very few commercially available systems have been developed, and as a result the software user is still left with a powerful toolbox to create whatever they want. The situation is made worse by the fact that many of the software packages encourage the user to simply duplicate the traditional paper map product in an electronic medium, while others that at least offer greater flexibility in the electronic medium do not offer sufficient control over the end product. The electronic medium offers a great deal of potential for the modern-day map maker, which goes far beyond simply recreating a paper map 'on screen'. Nevertheless, there are elements of traditional cartography that are very relevant to the electronic medium, and at the very least the end-user should be aware of these.

Education

Cartography still offers many graduating students a job at the end of their programme, either in more traditional roles as a practising cartographer, in the national mapping agencies, or in GIS. In many geography undergraduate degree programmes, cartography (either traditional or computer-assisted) was and still is taught as a 'skills course' e.g. University of Aberdeen – second year component of mapping science (topographic science, surveying, remote sensing, aerial photointerpretation, GIS, GPS (global positioning systems) and fourth year honours module (http://www.abdn.ac.uk/geography). Cartography and digital mapping are also taught as postgraduate courses in a number of institutions, e.g. Department of Geography and Topographic Science, University of Glasgow (http://www.geog.gla.ac.uk/courses/cartogit.htm).

Today numbers in cartography options are often small, with more and more students choosing the more popular GIS modules or degree courses. Relatively few of the GIS courses, however, seem to offer students (especially those with a limited background or exposure to cartography from non-geographical backgrounds) an opportunity to explore cartography in the context of map design, communication, visualisation and use in an electronic medium. Furthermore, there are still relatively few texts on maps in an electronic medium.

Map design – colour

Colour is a key element in visual communication and one that is very important in map design. Colour choice and selection, however, is difficult at the best of times. Although the use of colour is well documented in both the art and cartographic literature, much of the latter concerns colour use in a paper medium. To date, less has been written about colour choice and use in an electronic medium (see, e.g., Brewer 1994; Green 1991; 1993; Brown and van Elzakker 1993). While knowledge and understanding about colour use in a paper medium is important, it is very different in a computer environment. The problem has been exacerbated by the significant improvements in colour display and output devices (monitors, graphics cards, printers and plotters), providing the user (once again unlikely to have a great deal of knowledge about the use of colour) with almost unlimited colour choice, sometimes with the capability to swap between different colour models at the click of a button. Unfortunately, more options often lead to more temptation and the potentially arbitrary use of colour. While it may be aesthetically appealing, the end result is often little more than a very colourful graphic. A colour map designed and created on-screen may also look very different when it is printed out on paper.

SOLUTIONS

Map-production software

Some attempts have been made to try to provide both the computer cartographer and the GIS map producer with suites of tools to help to generate maps that are suitable for widespread use. MA Publisher from Avenza is described as 'The recognised standard for mapmakers' and comprises a suite of GIS filters for the graphics design software Adobe Illustrator and Macromedia Freehand. This software is specifically geared to map production and as such goes a step further by emphasising the cartographic element but also recognising the value of graphics software such as Adobe Illustrator.

CAL

Help for the non-cartographer has been addressed in part by the Computers in Teaching Initiative Centre for Geography (CTICG) Map Design CAL module from the University of Leicester (Computers in Teaching Initiative Centre for Geography 1994). The objective has been to raise awareness about maps and cartography, and to emphasise the importance of map design in an electronic medium, tackling such issues as text placement, the use of colour and map composition, among other things. It provides a way of increasing awareness among a generation where freedom to create maps has increased (i.e. desktop mapping software, GIS, remote sensing and visualisation), at a time when cartography as a subject is less popular, by placing an emphasis on the key elements of map design that facilitate the creation of a 'good' map (Figure 41.5). Other solutions have been:

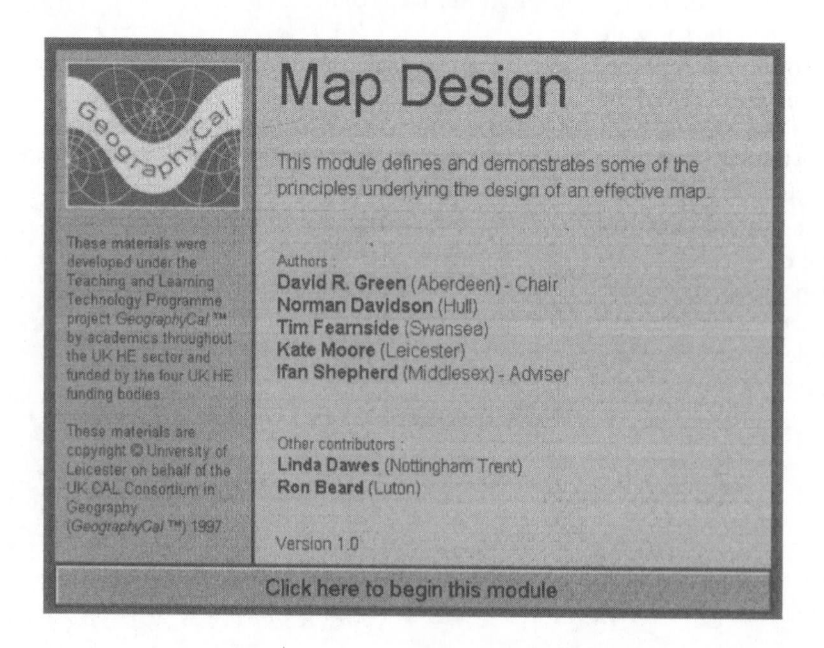

Figure 41.5 Teaching map design.

- On-line tutorials, templates and 'wizards' within map composition modules in ER-Mapper and Erdas Imagine have also helped to provide an element of guidance for the user to (1) use the product and (2) produce a reasonable map end product.
- A map design tutorial on the Geographer's Craft pages at the University of Texas at Austin also provides guidance (http://www.utexas.edu/depts/grg/gcraft/notes/cartocom/cartocom_ftoc.html).
- Journal papers (see Brown 1993).

Colour

Some attempts have been made to provide the means to help users with colour choice in an electronic medium. These have taken the form of various academic research studies resulting in paper-based colour reference charts, computer-based colour tutors, and the use of artificial intelligence (AI) in the form of knowledge-based and expert systems (e.g. Gill and Trigg 1988; Brown and Schokker 1989; Wang and Brown 1991; Schettini *et al.* 1992; Brown and van Elzakker 1993).

DEVELOPMENTS IN COMPUTER-BASED CARTOGRAPHY

Visualisation

The development of computer technology and also GIS, together with growing use of spatial data as a whole, has seen much greater emphasis placed upon the use of the computer as a means of visualising digital spatial data in a wide variety of different ways, of which maps are but one form of representing and examining spatial distributions and patterns, trends and processes (Kraak and Ormeling 1996; Kraak 1998b). While traditional and computer-based visualisation techniques are not new, the computer tools now available with PCs offers even greater potential than before to examine, explore, analyse and to display spatial data (see http://www.itc.nl/~carto/showcase/index.html).

Dynamic maps, animation and virtual reality

Computer-based visualisation has recently provided a number of new possibilities for cartography that were previously somewhat difficult to make use of despite widespread recognition of their value. A number of authors have discussed the value of animation in the display and communication of spatial information and changing patterns over time (e.g. Monmonier 1990; DiBiase *et al.* 1992; Dorling 1992; Peterson 1995, see also http://maps.unomaha.edu/books/iacart/book.html; http://www.utexas.edu/depts/grg/gcraft/notes/cartocom/section8.html; MacEachren and Kraak 1997; Kraak 1998a). The concept and practice of animation dates back to the 1960s (Kraak 1998b; see also Campbell and Egbert 1990), although obviously not in the sophisticated form that it is today. While many maps in use are typically of the static kind, there is a clearly defined role for the animated map or the use of animation techniques (both temporal and non-temporal) to display, e.g., spatial changes over time and sequences of maps (Kraak 1998a). Kraak (*ibid.*) provides a simple but effective example showing the spatial growth over time of Enschede in the Netherlands. DiBiase *et al.* (1992) outline a project involving the design of animated maps for a multimedia encyclopedia on CD-ROM (http://www.deasy.psu.edu/deasy/mepaper.html).

Simple change detection over time can be achieved using the 'swipe tools' within ERDAS Imagine to alternate between a base map layer and a current image layer (aerial photograph/satellite image). Similarly, a simple form of map animation can be achieved using the slide-show tools in MS Powerpoint.

Other more sophisticated examples are 'fly-bys' through terrain generated by draping a map, aerial photograph or satellite images over a digital elevation model (DEM). Good examples of the potential are provided in most digital image-processing software, e.g. ERDAS Imagine, ER-Mapper and PCI EasiPace, as demonstrations. More than one layer can be draped over a terrain model, e.g. the combination of a classified satellite image, map vector data and place names. While many early examples generated by computers were quite crude visualisations, developments in computer technology, e.g. Pentium II processor power, have provided opportunities to provide visually 'smooth' fly-bys over and through the terrain with the aid of navigation controls.

Interactive cartography also allows users to study various spatially related issues. A good example is NAISMap2 (http://atlas.gc.ca/schoolnet/issuemap) (see also Cartwright 1994).

Virtual reality offers yet another interesting and challenging dimension to cartography, and as noted on the Geographer's Craft homepage at the University of Texas poses a question as to how cartographers will be employed in cyberspace (see also Jiang and Ormeling 1997).

Electronic maps and atlases

With the development of the desktop computer and peripheral technology such as optical and compact discs (CDs), it has become increasingly easy to take advantage of the potential of multimedia innovations to create and distribute digital maps and atlases. While electronic atlases are not new (the first appeared in 1979), and some of the early ones such as the United Kingdom Digital Marine Atlas (UKDMAP – British Oceanographic Data Centre (BODC)) were initially distributed on multiple sets of floppy disks, CDs, CD writers and CD-ROMS have come down in price to the extent that most new computer hardware specifications include a 24- or 32-speed CD-ROM drive as standard. This has meant that many more opportunities are available to market both map- and image-based atlases on CDs at relatively low cost.

There are many examples of varying quality and type, e.g. De Lorme Global Explorer and Microsoft Encarta 97 (Box 41.1) (see Kraak 1997). Not only do these atlases provide static maps but also a range of tools to allow the user to interact with the maps, to search for places and to query the maps, with results being displayed in the form of text, images, graphics, video clips and so on. Other products include Microsoft Autoroute, which provides a means of planning a route from the point of departure to the termination of the journey. It is based upon a digital database, and the user is able to establish the optimum and optional routes, the distance, and the time from start to finish. The user specification is produced in the form of a map and a text-based description of the route numbers. A similar product is the DELORME AAA Map'n'Go 4.0 travel planner (http://www.delorme.com/mapngo/), which allows the user to generate a detailed travel plan with stops, driving time, mileage, road names and exit information, together with tourist information (e.g. place descriptions, hotels, museums, etc.). It also provides a slide-show facility to include user-acquired images. Other examples include:

- American Digital Cartography Inc. (ADC) produces a family of digital atlas products, including ADC WorldMap for use with MapInfo software (http://www.dci.com/adcworld.htm).
- MSAT has recently produced an interactive satellite image atlas of Europe, which allows a user to zoom in to the image and to add map annotation (MSAT 1998).
- The A-Z of London, AA StreetMaster London and Bartholomew London Maps are three examples of digital maps of London on CD-ROM (see Shepherd 1998).
- See also http://www.geod.ethz.ch/karto/projects/mmatlas/mmatlas.html.

In-car and other navigation systems

With the availability of more and more spatial data from, e.g., the Ordnance Survey (OS), Bartholomews and the Automobile Association (AA), in-car navigation systems are now becoming a far more frequent sight in cars, either as a standard fitting or as an optional extra. Although many are still relatively simple systems, with crude displays, and costs are still relatively high, there is growing awareness and use of digital maps in cars as a potential replacement for the more familiar paper map or road atlas. Ocean-

Box 41.1 Electronic atlas: the Microsoft Encarta 97

The user interface to MS Encarta is a large world map in the centre of the screen with a much smaller globe at the top left of the screen to help to locate the user. There are options relating to both the map and operation of the software, allowing the user to change the map style, return to the original map, switch any of the tools on or off, select the main program options, view the map only, or to obtain help.

Every map in the atlas has an option to show the key and other options to provide information about the land and climate, facts and figures, culture, family portraits, sights and sounds, animals, panoramic views, geography and Web links for further up-to-date information. Search and distance-measuring tools are also provided.

Figure 41.6 Detail from Encarta 97.

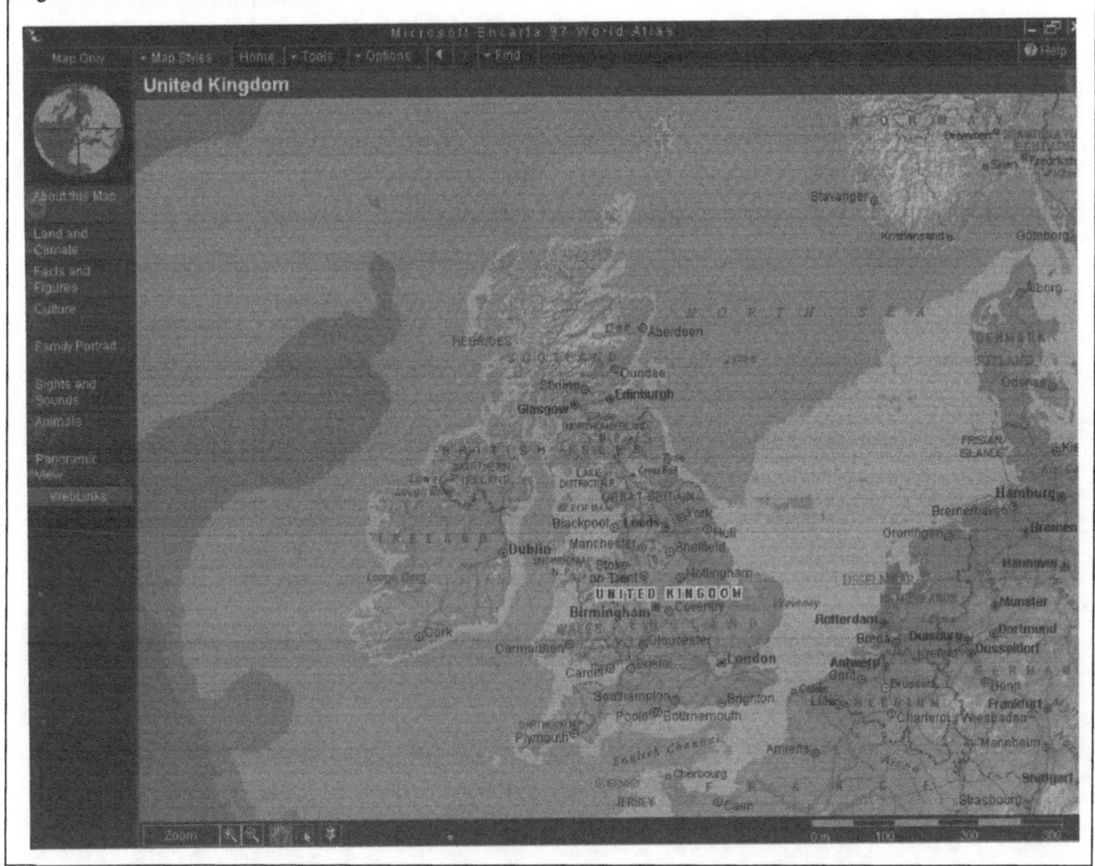

Routes has developed an on-board ship-routing and weather-tracking system, ORION, which integrates satellite images, weather data and world charts.

Computer mapping and GPS

Hand-held GPS (global positioning systems), palmtop computers and a variety of other similar microprocessor-based products also make use of digital spatial data (see Chapter 43). The DeLorme PalmPilot can make use of the Street Atlas product for address-to-address directions. This can also be used in conjunction with the Tripmate GPS receiver to track latitude and longitude, altitude and speed. Garmin recently launched a hand-held receiver with electronic mapping capabilities.

Artistic maps

With the growing number of low-cost multimedia computer systems now available, together with powerful graphics software, there is even more scope to create artistic and realistic representations of the world. Collinson (1997) has examined the potential to utilise graphics and landscape visualisation software to create more visually realistic examples of the world (Figure 41.7).

Cartography on the Internet: multimedia mapping

In recent years, the Internet has evolved very rapidly. This has since been joined by the Intranet and the Extranet. The Internet and its derivatives are essentially a multimedia information resource, delivering all sorts of different data and information to a wide end-user base, in a wide variety of different formats, ranging from text to images and graphics, video, and sound. Access to this vast information resource is via the use of a browser, e.g. Netscape or Internet Explorer among others, a relatively simple Windows-based interface, which includes tools to search, download, cut, copy and paste, display and save retrieved information, as well as launch other software applications, e.g. MPEG player, spreadsheet, graphics software, etc.

Among the different types of data and information available on the Internet are maps. The Internet currently offers a considerable

Figure 41.7 Three-dimensional computer mapping.

amount of scope for different types of map access and delivery. What is delivered to the end-user is very much dependent upon the objective of the map and the provider. In the simplest cases, this is simply an electronic equivalent of the paper map, while in the most extreme example it is a completely new form of interactive mapping that offers the end-user a whole new environment in which to work (see Plewes 1997).

Maps or spatial information have appeared in a number of different forms and for different purposes on the Internet in the last five years. While there are still many constraints to presenting and delivering map information on the Internet at present, e.g. standards, quality, confidentiality, copyright and legality, among others, this has not deterred the individual or organisation from providing many different types of map on web pages and commercial suppliers from selling map data and information (http://www.mapsunlimited.com/; htmlsite1/samples.html; http://www.elstead.co.uk/attwn5.htm; http://www.on-linebusinesscom/shops/graphic_maps.html). In some cases, sample maps (usually raster maps) or catalogues of maps available, e.g. the Ordnance Survey (OS) (http://www.ordsvy.gov.uk), are provided.

Many commercial web sites simply display static location maps, e.g. for customers (Figure 41.8). These usually take the form of either a computer-drafted or scanned map saved as a .GIF (normal or interlaced) or .JPG file. The Multi-Media Maps (http://www.multimap.com) site provides both colour and black and white maps at different scales for the UK, which can be viewed on screen. Using a place name (e.g. city) search tool the user is able to retrieve the appropriate map sheet for the area. Other sites provide maps as an additional home page user interface. In their simplest form, these maps are created as Mapedit-generated GIFs, allowing the creator to place so-called hotspots on a map outline

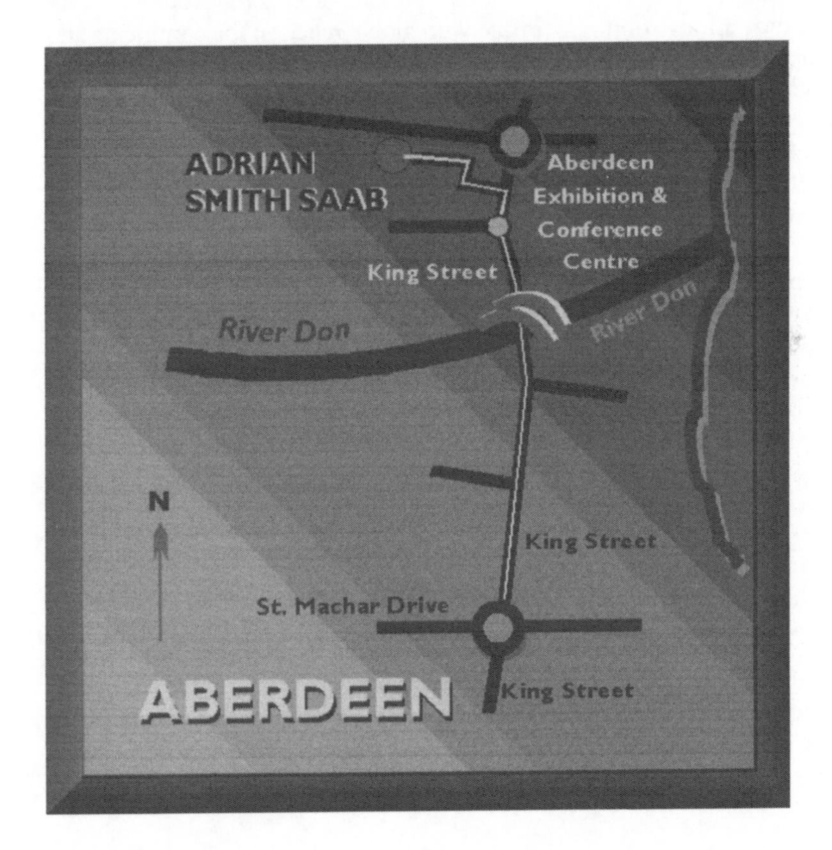

Figure 41.8 Cartography on the Internet: a static location map.

providing links to other web pages, which contain either further maps or text (see, for example, the Washington DC Sightseeing Map (http://sc94.ameslab.gov/TOUR/tour.html). More elaborate examples of these interactive maps are the search interfaces to, e.g., image databases (http://cs6400.mcc.ac.uk/maps/spot/spotuk.html) or, for example, the Weather Visualizer (http://covis.atmos.uiuc.edu/covis/visualizer).

Vector-based maps can also be delivered in a .PDF format using the Adobe Acrobat viewer, allowing users to roam and zoom without loss of map detail – something not possible with raster maps. However, this does not allow the user to access the data itself but merely to display and use the map information. A good example of a site that delivers map information in this form is the London transport web site (http:/www.londontransport.co.uk).

With the development of Internet technology, especially in the context of (GIS), more sophisticated forms of map data and information delivery have rapidly become possible (Toon 1997; Ireland 1998; Wagner 1998). With the evolution of Internet-based GIS, more and more software suppliers have developed Internet-based map delivery software, e.g. ESRI's Map Objects, Autodesk's Mapguide. Simple examples allow for the delivery of raster-backdrop maps, while other more complex examples offer more tools allowing, e.g., raster and vector maps (Figure 41.9). Some systems allow for the creation of 'on-the-fly' maps. An Internet-based system, developed as a joint venture by Laser-Scan and Adhoc (http://www.lsl.co.uk), has resulted in a tourist-based information system using vector maps. The advantage of such a system is the capability to query the map and to link other information to a particular location. The objective is to provide a publicly accessible map information system.

Some further examples are:

- The Florida Online GIS Mapping facility (http://www.fmri.usf.edu/sori/pages/views.html) provides users with a menu to access different marine and coastal maps, e.g. shorelines. A direct link to the ESRI Map Cafe server is then made. The interface is similar to that provided by ArcView and allows the user to switch different map layers on and off, to zoom in, and so on (Box 41.2).
- The Interactive Map of Kansas (http://www.ukans.edu/heritage/towns/kanmap.html).
- A similar product is the Interactive California Environmental Management, Assessment, and Planning System (http://ice.ucdavis.edu:8080/ice_maps/).
- At the extreme of cartography is the Atlas of Cyberspace (http://www.geog.ucl.ac.uk/casa/martin/atlas/atlas.html) (see also Jiang and Ormeling 1997).
- A novel site is the SkiMaps site with a Trail Map Archive and a link to the Holmes Linette Online Digital Skimaps (http://www.skimaps.com/).
- Examples of animation on the Internet include a site that provides access to a rotating Earth image with global relief on the continents and the sea floor (http://agcwww.bio.ns.ca/earth/earth.html) and that uses either a .GIF animation format or an MPEG viewer.
- The Color Landform Atlas of the United States is an example of an electronic atlas on the Web (http://fermi.jhuapl.edu/states/states.html).
- The Digimap Project (http://digimap.ed.ac.uk:8081) developed by Edinburgh University has provided limited access to Ordnance Survey maps and digital data (Digital Elevation Models (DEMS)), either as printed hard copy output or on disk for the purposes of teaching and research (Box 41.3).

Beyond simply accessing and printing out maps displayed on a home page, the newer visualisation tools allow users to search a spatial database and to create a map selectively related to a specific query. This allows the user effectively to design their own map, and as Kraak (1998a) points out, to use cartography in an exploratory mode, with a range of tools to allow a user to analyse and to experiment.

Figure 41.9 Cartography on the Internet: (A) raster and (B) vector maps.

A

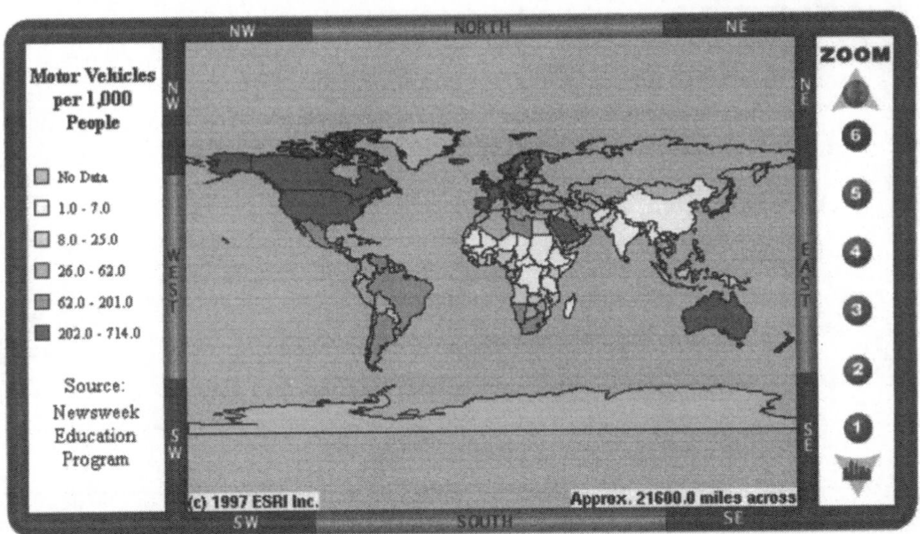

B

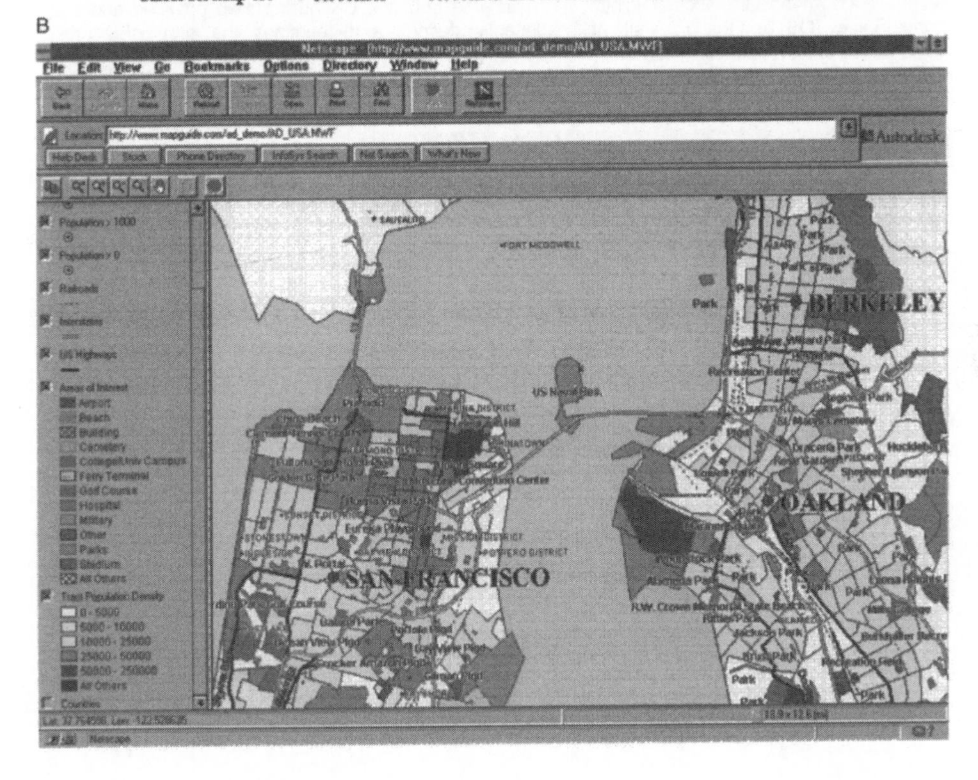

Box 41.2 Internet (global): Florida on-line

The Florida On-Line GIS Mapping tool provides interactive Internet map access and delivery for spatial data and information on the Florida coastline. Users select a map topic from a menu. Using ESRI's Map Cafe, the selected map is delivered to the screen 'on the fly' by the server, and a range of tools allow the user to interact with the map, e.g. switching layers on and off, zooming in and out of the map.

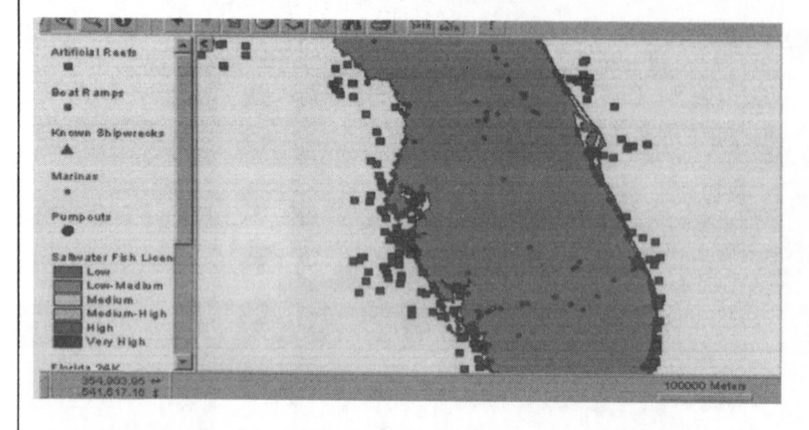

Figure 41.10 Map of Florida selected from ESRI's Map Cafe.

Box 41.3 Digimap

Digimap is an Internet-based map data access jointly hosted by the Ordnance Survey (OS) and the University of Edinburgh and provides students and researchers with direct access to Ordnance Survey digital map and elevation data. Access requires a userid and password. The user selects a map area from the UK outline map provided and subsequently defines and redefines the area of interest, eventually choosing to download the NTF (OS National Transfer Format) map data in a zipped file for conversion to a GIS format.

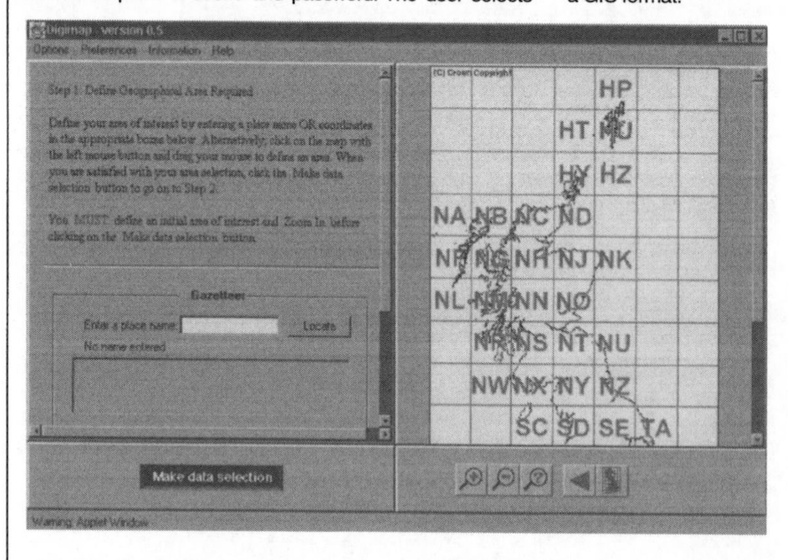

Figure 41.11 Digimap search screen.

WHERE IS COMPUTER MAPPING GOING?

The Internet is now probably the best place to obtain a clear picture of the direction in which computer and digital mapping technology is currently going, simply because it is a medium for the non-cartographer and cartographer alike to communicate, display and apply the technology, which raises awareness of digital maps (see http://www.maps.unomaha.edu/NACIS/paper.html). Digital maps and map data (in both raster and vector format) are now widely available across the Internet to browse, interact with, download (http://www.deasy.psu.edu/deasy/download.html) or purchase. There are a wide variety of different examples to examine, from the static map, to the animated and the interactive map, and the image map. The Internet has provided a multimedia information resource and delivery medium for many different types of data and information, the map (as data or information) being just one example. More than anything else, the Internet reveals modern cartography in practice.

The rapid developments in computer, networking and communications technology have successfully brought cartography back to the forefront of applied geography in recent years. The growth in GIS applications and now visualisation tools has helped to both raise and to reawaken current awareness in cartography and to re-emphasise the importance of the map as a tool for exploration, spatial analysis and visual communication. But with the advent of the new generation of powerful multimedia computer technology, maps have (fortunately or perhaps unfortunately) become something that everyone can design and create for themselves. People can select the data and information they want to display. They can use maps in an electronic medium, on a computer. Maps can be static, animated, interactive, or linked to other maps and information. They can be simple, complex or multi-dimensional. It is clear that the map, albeit still widely used in a paper medium, has increasingly become a very powerful and attractive, even persuasive, medium for examining data, information and patterns and ultimately communicating information to a wide range of people. The technology – the graphics display cards, the printers, the innovative tools and the WWW have all made the cartography of today 'come alive'. While it is now possible to do a great deal more with the computer than was at first possible, static maps still feature in the map vocabulary. However, the electronic map has been extended to new dimensions, providing many more options to examine spatial data and information in new and exciting ways.

As noted by Todd (http://maps.unomaha.edu/peterson/methods/Research/Todd/Todd.html):

> As we near the year 2000, we continue to see greater and greater advances in technology every day. It seems that no matter what discipline one is in, there is no escape for the furthering of use, technology and knowledge. Cartography is not immune to this and in fact, cartography is thriving because of this. In the not too distant past, we often would wonder at the marvel of what these 'new' computers could do – now it seems that we are often disappointed at what a computer can not do. Now that we have been given a glimpse of what is possible, we want more. Again, that is the case in cartography.
>
> Products such as Quick Time VR can serve as a medium to deliver cartographic work into another level. . . . And can be used effectively to tell the story better. . . . I believe that interactive maps are the future of cartography and that mapping is entering a new interactive age. QuickTime VR and cartography have an unlimited potential together and soon we will be treated to a whole new world of interactive maps which are both easy to use . . . and also easy to create. Virtual cartography has begun.

The advances in computer cartography, together with the related and integrated technologies of GIS, GPS, remote sensing, visualisation and the Internet, provide the applied geographer with a powerful set of tools to aid in the analysis and representation of the contemporary world. Taken together, these technologies should be seen as a core and compulsory component of the applied geography curriculum, and one that is essential for the geographer to acquire greater knowledge and understanding of the environment, and vital to practise the geographer's craft.

GUIDE TO FURTHER READING

Journals

The Cartographic Journal (http://www.cartography.org.uk/)
Cartographica (http://www.utpress.utoronto.ca/journal/jour5/car_lev5.htm)
Cartography and Geographic Information (http://www.landsurveyor.com/acsm/commun42/cagis00.htm)

Magazines

GIS Europe (http://www.geoplace.com/print/ge/Default.asp)
Mapping Awareness (http://www.geoplace.com/print/ma/)

Newsletters

Maplines (see British Cartographic Society entry under Societies)

Internet addresses

General

http://www.utexas.edu/depts/grg/gcraft/notes/cartocom/
http://www.mapping.com/index.shtml
http://math.rice.edu/~lanius/pres/map/
http://loki.ur.utk.edu/ut2kids/maps/map.html
http://tdc-www.harvard.edu/maps.html
http://icg.harvard.edu/~maps/Incont.htm
http://www.deasy.psu.edu/
http://www.geosys.com/cgi-bin/genobject/maproom/tig.3eed/
http://www.athena.ivv.nasa.gov/curric/land/geograph/carto/index.html
http://www.mapquest.com
http://www2.lib.udel.edu/subj/maps/internet.htm
http://www.geog.psu.edu/maceachren/maceachrenhtml/maceachrentop.html
http://sparky.sscl.uwo.ca/carto.html
http://kartoserver.frw.ruu.nl/html/staff/oddens/mapsatl4.htm#online
http://imagiware.com/via/Maps
http://geog.gmu.edu/projects/maps/cartogrefs.html
http://maps.unomaha.edu/books/IACart/book.html
http://nvkserver.frw.ruu.nl/html/nvk/ica/madridproc.html
http://www.cgrer.uiowa.edu/servers/servers_references.html#interact.anim
http://www.colckwk.com/

Societies

http://www.cartography.org.uk/ (British Cartographic Society)
http://www.shef.ac.uk/uni/projects/sc/ (Society of Cartographers)

Contacts

The British Cartographic Society (BCS), c/o Royal Geographical Society, 1 Kensington Gore, London, England. Internet: http://www.cartography.org.uk/
The Society of Cartographers (http://www.shef.ac.uk/uni/projects/sc/)
The International Cartographic Association (ICA) (http://www.abdn.ac.uk/ica/)
International Institute for Aerospace Survey and Earth Sciences (ITC), Hengelosestraat 99, PO Box 6, 7500 AA Enschede, The Netherlands. Tel. +31 53 4874444; Fax. +31 53 4874400; Internet: http://www.itc.nl/~carto/

Map data suppliers

Ordnance Survey (http://www.ordsvy.gov.uk)
Bartholomew (http://www.harpercollins.co.uk/maps/)
AA (http://www.theaa.co.uk/travel/bookshop/maps/maps.asp)
Philip's (http://www.philips-maps.co.uk/)
Hydrographic Office (http://www.hydro.gov.uk/)
Streetmap (http://www.streetmap.co.uk/)

REFERENCES

Anon. (1993) Topographer. Mapping has never been easier. *Archimedes World* April, 15–16.
Baker, A.M., Faludi, R. and Green, D.R. (1983) An evaluation of SAS/GRAPH for computer

cartography. Paper presented to the Association of American Geographers Annual Convention, Denver, Colorado.

Baker, A.M., Faludi, R. and Green, D.R. (1985) An evaluation of SAS/GRAPHTM for computer cartography. *Professional Geographer* 37(2), 204–14.

Bedell, R. (1989) The use of autocad for digital mapping. Paper presented at a workshop on GIS. 24 May 1989. The Remote Sensing Society Remote Sensing Group, Department of Geography, University of Nottingham.

Bennett, A. (1994) Course work. Education: navigator. *Archimedes World* 2.

Bourne, L.S., Baker, A.M., Kalbach, W., Banon, W., Richard, M., Cressman, R., Matthews, B. and Schulte, S. (1985) *Ethnocultural Data Base Materials Series III: Special Report No. 3: Ontario's Ethnocultural Population, 1981: Socio-economic Characteristics and Geographical Distributions.* Ontario Ministry of Citizenship and Culture/ Center for Urban and Community Studies, University of Toronto.

Bourne, L.S., Baker, A.M., Kalbach, W., Cressman, R. and Green, D. (1986) Canada's ethnic mosaic: characteristics and patterns of origin groups in urban areas. *Major Report No. 24.* Center for Urban and Community Studies, University of Toronto, Ontario, Canada.

Brewer, C.A. (1994) Color Use Guidelines for Mapping and Visualization. In A.M. MacEachren and D.R. Fraser-Taylor (eds) *Visualization in Modern Cartography*, Volume 2. Oxford: Pergamon, 123–47.

Brown, A. (1993) Map design for screen displays. *The Cartographic Journal* 30, 129–35.

Brown, A. and Schokker, P.W.M. (1989) Offset colour charts for use with a Macintosh II microcomputer. *ITC Journal* 3/4, 225–8.

Brown, A. and van Elzakker, C.P.J.M. (1993) The use of colour in the cartographic representation of information quality generated by GIS. *Proceedings of the 16th International Cartographic Conference, Vol. 2*, Cologne, Germany, 707–20.

Buckland, S.T., Bell, M.V. and Picozzi, N. (1990) (eds) *The Birds of North-East Scotland*, Aberdeen: North-East Scotland Bird Club.

Campbell, C.S. and Egbert, S.L. (1990) Animated cartography: thirty years of scratching the surface. *Cartographica* 27(2), 24–46.

Carter, J.R. (1984) *Computer Mapping: Progress in the '80s.* AAG Resource Publications in Geography, Washington DC: AAG.

Cartwright, W. (1994) Interactive multimedia for mapping. In A.M. MacEachren and D.R. Fraser-Taylor (eds) *Visualization in Modern Cartography Volume 2.* Oxford: Pergamon, 63–89.

Collinson, A. (1997) Virtual Worlds. *The Cartographic Journal* 34(2), 117–24.

Computers in Teaching Initiative Centre for Geography (CTICG) (1994) CAL PROJECT. *Map Design T20*, University of Leicester.

Cuff, D.J. and Mattson, M.T. (1982) *Thematic Maps: Their Design and Production.* New York: Methuen.

Dent, B.D. (1985) *Principles of Thematic Map Design.* Reading, Mass.: Addison Wesley.

DiBiase, D. (1991) Linking illustration and mapping software through Postscript. *Cartography and Geographic Information Systems* 18(4), 268–74.

DiBiase, D., MacEachren, A.M., Krygier, J.B. and Reeves, C. (1992) Animation and the role of map design in scientific visualization. *Cartography and Geographic Information Systems* 19(4), 201–14.

Dorling, D. (1992) Stretching space and splicing time: from cartographic animation to interactive visualization. *Cartography and Geographic Information Systems* 19(4), 215–27.

Dorling, D. and Fairbairn, D. (1997) *Mapping: Ways of Representing the World.* London: Addison Wesley Longman.

Eden, B. (1993) Personal communication.

European Standard Software (ESS) (1995) *MAPS – User's Manual.* Denmark.

Gill, G.A. and Trigg, A.D. (1988) Canvas: an intelligent colour selection tool for VDU images. *Proceedings of IGARS'88 Symposium*, Edinburgh, Scotland, 1785–877.

Green, D.R. (1991) Colour ad infinitum! – creating expensive wallpaper or functional maps from GIS, remote sensing and cartographic systems? In K. Rybaczuk and M. Blakemore (eds) *Proceedings 15th Conference Mapping The Nations, Volume 2*, ICA, Bournemouth, September–October 1991, 871–6.

Green, D.R. (1993) Map output from geographic information and digital image processing systems: a cartographic problem. *The Cartographic Journal* 30, 91–6.

Green, D.R. (1994) An exciting new role in GIS for the cartographer of the future. In D.R. Green and D. Rix (eds) *The AGI SourceBook for Geographic Information Systems 1995*, J. Wiley/AGI, 63–72.

Green, D.R. (1996) The GIS map: new opportunities for artistic and scientific visualization. Visualisatie in het GIS-onderwijs. *NVK publikatiereeks* 20, 5–16.

Green., D.R. and Calvert, L. (1998) The cartographic potential of graphics design software for education. *The Cartographic Journal* 35(1) (forthcoming).

Green, D.R., Deeth, A.L., Faludi, R. and Nuzzo, A. (1985) SAS/GRAPH for cartography – map projections and labelled choropleth maps.

Cartographica 22(2), 63–78.

Green, D.R., McEwen, L.J. and Taylor, S. (1990) The cartographic potential of art, CAD and graphics software for teaching and research in geography. *Society University Cartographers Bulletin* 23(2), 21–30.

Green, D.R. and Schwartz, D. G. (1988) Creating cartographic symbols with SAS/GRAPHTM for specialised mapping tasks. *Society University Cartographers Bulletin* 22(1), 9–15. December 1988.

International Cartographic Association (ICA) (1980) *Glossary of Terms in Computer Cartography*. Falls Church, Va.: ACSM.

Ireland, P. (1998) How to serve data on the web. *Mapping Awareness* 12(5), 32–5.

Jiang, F.J. and Ormeling, F.J. (1997) Cybermap: the map for cyberspace. *The Cartographic Journal* 34(2), 111–16.

Jones, C.B. (1997) *Geographical Information Systems and Computer Cartography*. London: Longman.

Keates, J.S. (1973) *Cartographic Design and Production*. London: Longman.

Kraak, M.J. (1997) Microsoft Encarta97 World Atlas, 2nd edition, software review. *GIS Europe* 6(7), 42–3.

Kraak, M.J. (1998a) *Exploratory cartography: maps as tools for discovery*. Inaugural Address. ITC, The Netherlands. 26pp. (http://www.itc.nl/~kraak/address)

Kraak, M.J. (1998b) Cartographic animation – a challenge. *Geoinformatic* April/May, 18–21.

Kraak, M.J. and Ormeling, F.J. (1996) *Cartography, the Visualization of Spatial Data*. London: Addison Wesley.

MacEachren, A.M. and Kraak, M.J. (1997) Exploratory cartographic visualization: advancing the agenda. *Computers and Geosciences* 23(4), 335–44.

Maguire, D.J., Brayshay, W. and Chalkley, B.S. (1987) *Plymouth in Maps: A Social and Economic Atlas*. Plymouth Polytechnic.

Monmonier, M. (1982) *Computer-Assisted Cartography: Principles and Prospects*. Englewood Cliffs, NJ: Prentice-Hall.

Monmonier, M. (1989) *Maps with the News: The Development of American Journalistic Cartography*. London: University of Chicago Press.

Monmonier, M. (1990) Strategies for the Visualization of Geographic Time-Series Data. *Cartographica* 27(1), 30–45.

MSAT (1998) *Photospace – The Atlas of Europe from Space*. PC and Mac Version. UK/France.

Peterson, M.P. (1995) *Interactive and Animated Cartography*. Englewood Cliffs, NJ: Prentice-Hall.

Peucker, T.K. (1972) *Computer Cartography*. AAG Resource Paper 17, Washington, DC.

Plewes, B. (1997) *GIS Online: Information Retrieval, Mapping and the Internet*. Santa Fe, NM: Onword Press.

Price, C. (1998) Map Maker Pro 2.1 – a review. *Mapping Awareness* 12(1), 34–5.

Robinson, A.H., Morrison, J.L., Muehrcke, P.C., Kimerling, A.J. and Guptill, S.C. (1995) *Elements of Cartography*, 6th edition. New York: John Wiley & Sons.

Schettini, R., Della Ventura, A. and Arteset, M.T. (1992) Color specification by visual interaction. *The Visual Computer* 9, 143–50.

Shepherd, I. (1998) The Streets of London – a review. *Mapping Awareness* 12(1), 30–3.

Taylor, D.R.F. (1980) (ed.) *The Computer in Contemporary Cartography*. Chichester and New York: John Wiley & Sons.

Thom, J. (1990) Personal communication. *Grampian Region Census Atlas* (1986) (monochrome): produced as a joint effort between Grampian Regional Council and University of Aberdeen.

Toon, M. (1997) The world by your window. *GIS Europe* 11(6), 38–41.

Wagner, M.J. (1998) Data joins the way of the web. *Mapping Awareness* 12(4), 25–7.

Wang, Z. and Brown, A. (1991) A knowledge-based system for selection of map-area colours from a colour chart. *ITC Journal* 3, 122–6.

Whitehead, D.C. and Hershey, R. (1990) Desktop mapping on the Apple Macintosh. *The Cartographic Journal* 27(2), 113–18.

Whitener, A. and Creath, B. (1997) *Mapping with Microsoft Office: Using Maps in Everyday Business Operations*. Santa Fe, NM: OnWord Press.

Waugh, T.C. and McCalden, J. (1982) *GIMMS Reference Manual*. Edinburgh: GIMMS Ltd.

42

Geodemographics, marketing and retail location

Graham Clarke

INTRODUCTION: QUANTITATIVE GEOGRAPHY AND BUSINESS LOCATION

If you browse the geography student textbooks of the 1960s and 1970s, you may be surprised now to see how many were focused on the broad subject of 'location analysis'. The classic works of Haggett (1965), Harvey (1969), Haggett *et al.* (1977) and Wilson (1970; 1974) are perhaps the most cited of this group and were the core readings in many geography departments at this time. They built on classic foundations that now seem to be rarely taught at undergraduate level – Christaller, Lösch, von Thunen, Weber, Palander, Reilly, etc. Although often perceived as theoretically and mathematically complex (Peter Gould (1972) termed Wilson's 1970 modelling text as 'the most difficult I have ever read'), they were arguing for an essentially applied geography – the use of these models to help industrialists, farmers, land-use planners, retailers, etc. to find optimal sites to locate and reach their markets given spatial variations in the cost of various factors of production (remember von Thunen himself was a farm estate manager interested in improving the efficiency of production and distribution). Unfortunately, this type of economic geography rapidly fell from grace. As Pacione notes in the introduction to this volume, in relation to geography as a whole, the moral high ground was captured by 'new' geographies that emphasised the wider theoretical frameworks concerned with underlying capitalist social relations and shifts in national and international political economies at the expense of location models *per se* (see Massey and Meegan 1985 for a good illustration). All the sub-fields of economic geography were subsequently influenced by this theoretical shift. Retail geography, for example, provides a good illustration of this transformation. Despite notable exception (see Fenwick 1978, Davies and Rogers 1984, and Ghosh and MacLafferty 1987 on site location methods), the number of new texts on quantitative methods for site location rapidly diminished, while newer retail geographies based on culture, consumption and capital came to the fore (Wrigley and Lowe (1996) provide the best illustration). While these new retail geographies are in themselves important additions to the literature, they deliberately omit location models, GIS, etc. as if they are no longer important. Smith (1989) summarises this 'progress': 'instead of trying to identify the optimum site for a new supermarket many geographers turned to identify the broader processes in which whole landscapes were made and remade' (p. 142). Research on site location was now seen as politically incorrect. Clarke (1996) provides some explanation for this. He suggests that GIS and models produce knowledge that proves most useful to those already powerful groups in society within a descriptive neo-classical economic geography still largely underpinned by central place theory (CPT).

It could be argued that much of this damning critique is simply wrong. First, models can be used by, or for, all groups in society (for example, they would still be valuable in a social justice framework for assessing the impacts of planning applications made by the powerful multiple retailers on accessibility and welfare for different consumer groups). Second, the use of (descriptive) models does not and should not preclude explanation – having suggested an optimal location, there is an obvious need to justify that decision and to analyse the expected impacts. Lastly, although models such as spatial interaction models can be linked to CPT (cf. Wilson 1978), they are deliberately preferred to CPT because they are not bound by the limiting assumptions made by CPT concerning uniform population distribution and nearest centre choice hypotheses.

However, there is good news! The modelling community is fighting back. It is fair to say that important lessons were learned during the years in the wilderness, which has enabled the new generation of GIS and spatial modellers to be better equipped to handle the second era, an era characterised by better theoretical models and more widely available data sets and, consequently, in tune with a more applied operational environment. The early success of this new era of applied modelling at the University of Leeds encouraged the university to form a consultancy company (GMAP) in 1987, which by 1996 was turning over £5 million per annum and employing over 100 geography graduates (mainly from Leeds of course!). [Its rapid growth inevitably attracted predators, and two-thirds of the company was sold to the American marketing company Polk in 1997.] The speed and extent of this development is a result of the ever-growing demand for geographical analysis. Despite the critiques of the 'radical' geographers, location problems never went away in the business world. The discipline of operational research, and to some extent computing, filled the gaps that geographers had left in the search for radicalism (for example the growth of GIS was driven entirely by computer scientists in its early days). The success of companies like GMAP (and a number of 'regional research laboratories' that have also had success; see Clarke *et al.* 1995) are testimony to the fightback undertaken by quantitative geographers. But why do businesses require such techniques to help them to run their operations? The aim of the rest of this chapter is to answer that question by focusing on the need for geographical research and the principal methods available to retailers to analyse their markets geographically.

FINDING CUSTOMERS AND LOCATING STORES

For any organisation that sells directly to the general public, geography is important. First is the need to find customers. For mail order companies, this exercise may be sufficient in its own right. For most other retail organisations, a second requirement is the need to locate stores in such a way that these customers can be reached and sales subsequently captured. To add to the complexity, these search procedures are taking place in an increasingly competitive retail environment (see Clarke and Clarke 1998). For most products, customer profiling and segmentation are important. The grocery market provides a good illustration. All households in the UK require groceries. However, the market can be segmented geographically. The lowest-income groups (along with those with low mobility such as householders without a car, mothers with young children and old age pensioners) are most likely to use (or need) discount stores, where price may be a more important store attribute than quality or choice. Therefore, if you were a market analyst for Kwik Save, Netto or Aldi you would be interested in finding where such customers lived and shopped. Given the social geography of most UK industrial cities, it is not surprising to find these stores concentrated in great numbers in the inner areas of Birmingham, Manchester, Liverpool, Leeds and Newcastle (see Wrigley and Clarke 1999). On the other hand, companies like Sainsbury and Tesco have more commonly targeted the higher-income groups, who are more prepared to travel longer distances to enjoy quality and choice, which is the hallmark of these two organisations. Their market domination of the south and southeast of the UK (along with stores in the affluent northern suburbs) largely reflects the location of this high-income market and the deliberate strategy of targeting these types of customer (see the market share maps in Langston *et al.* 1998).

Many high street retail organisations are even greater masters of market segmentation practices. This is especially true in fashion and footwear. Dawson and Broadbridge (1988) summarise the target market of Sears as it was in the late 1980s (Table 42.1). A number of these markets are clearly very specific. The key question is where are the best locations in the UK to maximise sales from these target groups? Answering such a question requires methods for customer spotting and then for predicting the revenue and impacts of new store development. We shall explore each of these themes in turn below.

First, the ability to find the location of key customer types has been aided by the development of new geographical tools such as geodemographics and geographical information systems or GIS. Geodemographic systems attempt to profile geographical areas by identifying key customer segment types within them. Usually, this profile is a single classification based on the majority population type present in the area. As we noted above, this is especially useful for those retailers whose customers are concentrated in certain geodemographic segments (i.e. by age, sex, social class) and are keen to find localities of the 'right' type for their products. Although geodemographic systems in the UK have been available since the late 1970s (CACI's ACORN system being the earliest commercial application in the UK), they proliferated after the publication of the 1981 census and following their availability within GIS packages (which encouraged the routine mapping of such classifications). These small area profiles are based on a multivariate analysis of a large number of census variables associated with that area. Other areas that fall within the same profile classification can be considered alike and to contain similar types of household (at least in theory!). Table 42.2 shows the ACORN classifications, which are now available for both census geographies (the smallest areal unit being the enumeration district) and postcode geographies. For a detailed history and review of geodemographics, see Beaumont (1991a, b), Sleight (1993) and Batey and Brown (1995).

Some retailers are able to identify their customer base even more narrowly (usually referred to as niche marketing) and thus require more disaggregated information on customer types or profiles. From the late 1980s, geodemographic systems have thus become increasingly more sophisticated by linking population

Table 42.1 Sears' target markets in the late 1980s (*cf.* Dawson and Broadbridge 1988).

Operation	*Sector*	*Target/Position*
Fosters	Meanswear	Middle market, 15–30
Your price	Menswear	Keen-priced fashion
Bradleys	Menswear	30–50, traditionalist
Hornes	Menswear	25–40, upmarket
Zy-Jargon	Menswear	Younger men
Wallis	Womenswear	25–35 professional
Miss Selfridge	Womenswear	Young fashion
Curtess	Footwear	High volume, low price
Trueform	Footwear	Family
Dolcis/Tiptoe/Bertie	Footwear	Fashion
Saxone/Manfield	Footwear	Quality/high price

Table 42.2 ACORN geodemographic classification.

Category		
Category A	'thriving'	1 Wealthy achievers, suburban areas
		2 Affluent greys, rural communities
		3 Prosperous pensioners, retirement areas
Category B	'expanding'	4 Affluent executives, family areas
		5 Well-off workers, family areas
Category C	'rising'	6 Affluent urbanites, town and city areas
		7 Prosperous professionals, metropolitan areas
		8 Better-off executives, inner-city areas
Category D	'settling'	9 Comfortable middle agers, home-owning areas
		10 Skilled workers, home-owning areas
Category F	'aspiring'	11 New home owners, mature communities
		12 White collar workers, multi-ethnic areas
Category G	'striving'	13 Older people, less prosperous areas
		14 Council house residents, better-off homes
		15 Council house residents, high unemployment
		16 Council house residents, greatest hardship
		17 People in multi-ethnic, low-income areas

demographics with lifestyle information obtained from companies such as NDLI and CMT (Openshaw 1995; Birkin 1995). Lifestyle data is information collected about individual households through the use of self-completed questionnaires. Since the data are collected at the household level, this avoids the problem encountered by using census data, namely the fact that data are available only at an aggregate level. Using lifestyle data, it is possible to identify the number of households that have specific combinations of characteristics such as 'Volvo-owning golfers with an interest in fine wine'. A good example of the use of lifestyle data comes from the UK wine merchants known as 'Bottoms-up'. They have identified their main target group not as a single age or single social-class group but as persons with a special type of lifestyle. They call these persons 'serious piss artists' – a crude terminology for persons 25–40 with higher incomes who spend most of their drinking and socialising time now at home, so are usually married persons with young families and hence may have less time or opportunity to visit the pub (see Belchamber 1997). A further example is provided in Box 42.1.

It is probably true that lifestyle data offers a more precise way of targeting particular customer groups, as well as being a more powerful tool for direct marketing. Its main drawback, however, is that it is not a complete census of the UK population and has under- and over-representation of some consumer groups. However, the largest lifestyle database (collated by NDLI) contains over 10 million households, nearly 50 per cent of the UK total. These systems are set to have a large impact on geodemographic marketing tools in the next century.

GIS packages have increasingly been used in retail site assessment research to supplement geodemographic analysis. They first allow information relating to stores or shopping centres, and the populations within their catchment areas, to be *geocoded* (that is, placed on the computer with a spatial referencing point) and visually *displayed* through maps and graphs. Once the information is stored in the computer, the user can then attempt to estimate store revenues. Take, for example, the problem of predicting the revenue of a new grocery store. The GIS enables the user to *buffer* (or demarcate) travel times around the new store and then calculate the population within each time band using the standard *overlay* procedure available in most GIS packages (see Beaumont 1991a, b; Howe 1991; Elliot 1991; Ireland 1994). An example of this will be given in the next section. Once an estimate has been made concerning the demand within the likely buffered areas, then a variety of methods may be used to

Box 42.1 Finding customers for a large cinema chain

Imagine you are a marketing analyst for a large cinema chain. From various surveys of your customers, you are able to build up profiles of their individual and/or household characteristics. Most new cinemas are now built out of town, with large car-parking facilities, multi-screen complexes and in association with many other leisure facilities such as ten-pin bowling and fast food restaurants. Who do you think is most likely to visit such cinemas? Using ACORN (see Table 42.2), Thompson (1997) identified the following target groups:

- wealthy achievers
- affluent urbanites
- prosperous professionals
- better-off executives
- comfortable middle ages
- skilled workers.

The next question is where are these groups located? The purchase of ACORN would allow the analyst to search through the database and highlight all postal districts in the UK that have one of the above customer groups associated with it. In Leeds, this would produce the following list: LS6, LS7, LS8, LS15, LS16, LS17, LS18, LS19, LS20, LS21, LS22, LS23, LS29.

Thus the analyst could now either (1) target these areas of Leeds through localised advertising (say local newspapers) or through direct mail shots to addresses in these areas; (2) if the total numbers were large enough, maybe encourage the company to locate somewhere within such neighbourhoods.

translate these population totals into estimates of individual outlet sales. The most common method is the so-called 'fair-share' approach (Beaumont 1991b). Hence, if there are three other competing stores in the buffered catchment area of the new store, then the new store may be expected to obtain 25 per cent of the revenue generated in that catchment area. This simple fair-share allocation could be weighted by store size or by retail brand to increase realism. The alternative is to assume that the consumer will simply travel to the nearest store within the catchment area (*dominant store analysis*: see Ireland 1994).

Many retailers now use geodemographics and GIS for marketing and site appraisal. In the UK, these include most of the major grocery retailers as well as large international groups such as Marks & Spencer and Kingfisher. A good example of a linked GIS/geodemographic system in the UK is CACI's 'Insite System', which has been targeted specifically at retail businesses wishing to match catchment area profiles (based on census and geodemographics) with those obtained from their customer databases. CACI is working with a number of UK high street retailers, including Norweb, Britdoc, Budgens, Woolworth and Yorkshire Building Society (CACI 1993). Such bespoke geodemographic systems are also increasingly available within general GIS packages for other regions of Europe (Hinton and Wheeler 1992; Reynolds 1993). Indeed, 'Experian' (formerly CCN) has recently launched a pan-European version of its popular MOSAIC system (Webber 1993; Birkin 1995).

Although the GIS/geodemographics approach is popular with retailers, there are two principal drawbacks in relation to estimating store turnovers. First, there is the problem of how to define the catchment area and, second, how to adequately treat the competition. As we saw above, the former is normally represented by distance or drive time bands, and it is often assumed that the store will capture trade uniformly in all directions. Even when drive time bands are drawn in relation to transport networks (Reynolds 1991), there is still the assumption of equal drawing power in all directions. Second, the treatment of the competition is wholly inadequate. As suggested above, the method most often used is 'fair share', with the potential revenue of the catchment area simply being divided between all retailers on some *ad hoc* basis (type of retailer, amount of floorspace, etc.). This does not allow for the complex set of real interactions between residential areas and retail locations, which are distorted in the real world by intervening opportunities. As Elliot (1991) acknowledges, the presence of competing centres will restrict the catchment boundary of a new store in some directions. Her response is to 'override the drive time where it seems appropriate' (p. 171). Such subjectivity is the precise reason why such methods are not as accurate as alternative modelling methods discussed below.

Hence, this methodology (while offering a useful overview of potential catchment area revenue) is unreliable due to the inadequate treatment of spatial interactions and the inadequate treatment of competitor impacts (for more details, see Benoit and Clarke 1997). The alternative is to build models of interactions or flows.

Let us label any residential zone such as a postal sector or enumeration district (i) and any facility location such as a centre or supermaket (j). Then the number of people travelling between i and j can be labelled S_{ij}, and modelled using a spatial interaction approach:

$$S_{ij} = A_i \times O_i \times W_j \times f(c_{ij}) \qquad (42.1)$$

where

S_{ij} is the flow of people or money from residential area i to supermarket j;
O_i is a measure of demand in area i;
W_j is a measure of the attractiveness of supermarket j;
c_{ij} is a measure of the cost of travel or distance between i and j;
A_i is a balancing factor that takes account of the competition and ensures that all demand is allocated to centres in the region.

Formally, it is written as:

$$A_i = \frac{1}{\sum_j W_j \times f(c_{ij})} \qquad (42.2)$$

The model allocates flows of expenditure between origin and destination zones on the basis of two main hypotheses:

1 Flows between an origin and destination will be proportional to the relative attractiveness of that destination *vis-à-vis* all other competing destinations.
2 Flows between an origin and destination will be proportional to the relative accessibility of that destination *vis-à-vis* all other competing destinations.

The model works on the assumption that, in general, when choosing between centres that are equally accessible, shoppers will show a preference for the more attractive centre (which can be measured by size or other attributes such as car-parking availability, price, etc.). When centres are equally attractive, shoppers will show a preference for the more accessible centre. Note, however, that these preferences are not deterministic. Thus, when choosing between equally accessible centres, shoppers will not always choose the most attractive. The models are therefore able to represent the stochastic nature of consumer behaviour. Neighbouring households would not be expected to behave in exactly the same way, even though their characteristics are similar. Equally, particular individuals and households will not always use the same retail centres.

These models can be *disaggregated* in a number of ways. First, recognition of different types of consumer such as car owners and non-car owners is important in most real-world applications. Second, as mentioned above, the destination attractiveness term can be disaggregated to include all sorts of centre or store attributes (Pacione 1974; Spencer 1978; Timmermans 1981; Wilson 1983). Third, various forms of the distance deterrence terms may be used and different transport modes introduced. Wilson (1983) provides a useful summary of the degree to which retail models can be disaggregated, while other authors have looked at new formulations of spatial interaction models that incorporate additional behavioural variables. Fotheringham (1986) has argued that the models need to be modified to allow stores in close proximity to other stores to have greater attractiveness to consumers. These competing-destination models measure relative accessibility of stores to one another to measure the degree to which stores located close to each other have a locational advantage over isolated outlets. This may be particularly important in comparison shopping.

THE USE OF GEODEMOGRAPHICS AND MODELS IN GROCERY RETAILING

Having discussed the basic principles of geodemographics and spatial models, the aim in this section is now to show how they can be applied. The grocery sector will be used to illustrate these techniques (see other illustrations in Birkin *et al.* 1996). First, we will examine the use of geodemographics and GIS in more detail.

In the previous section, we introduced the idea that retailers such as Kwik Save, Aldi and Netto would be most likely to target less affluent, less mobile customers. Sainsbury and Tesco, on the other hand, would be most likely to target the more affluent consumer groups. In terms of the ACORN classification shown in Table 42.2, therefore, Kwik Save might target the 'striving', while Sainsbury would look for areas containing 'thriving' residents. In the following example, we draw upon a different geodemographic package, a system called 'GB Profiles' developed at the University of Leeds. Figure 42.1 maps the distribution of postal districts in Leeds that have been assigned the lowest affluence category, 'struggling'. Thus all the shaded postal districts are characterised as having the greatest percentage of their populations within this category. The map shows the cluster of such postal areas surrounding the city centre (the inner city) and a skewed pattern to the south and east, which reflects the location of outer council estates. Figure 42.2 shows the location of Kwik Save stores and its market share within these postal districts. Although there is not a direct match between the patterns shown in Figure 42.1 and 42.2, Kwik Save seems to be well represented in many areas of 'struggling' Leeds (it is also represented in some of the more distant suburban areas, where it has identified geographical market gaps). However, within the core cluster of 'struggling' Leeds, there does seem to be a gap in Kwik Save's distribution network to the south of the city centre. We will examine the use of GIS to predict the impact of a new store in this locality below.

Figure 42.3, by contrast, maps the distribution of the most affluent 'GB Profiles' geodemographic group in Leeds labelled

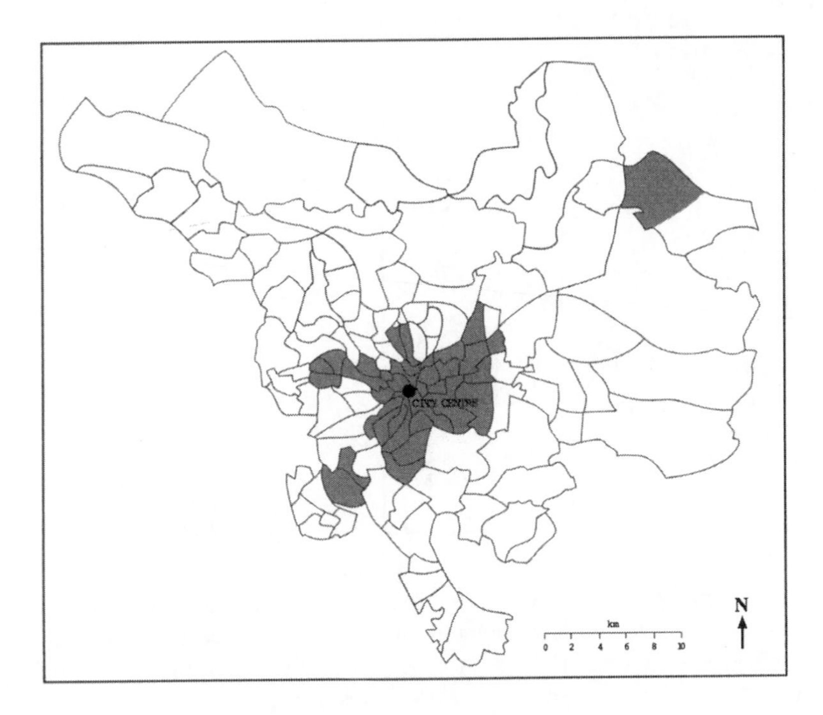

Figure 42.1 Location of 'struggling' residents of Leeds.

Source: GB Profiles, School of Geography, University of Leeds.

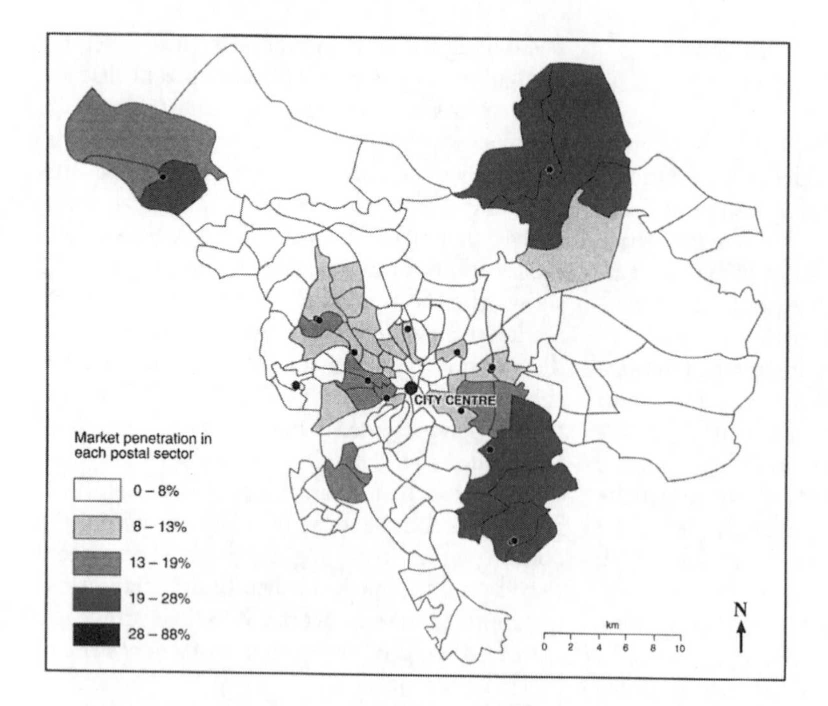

Figure 42.2 Market penetration of Kwik Save.

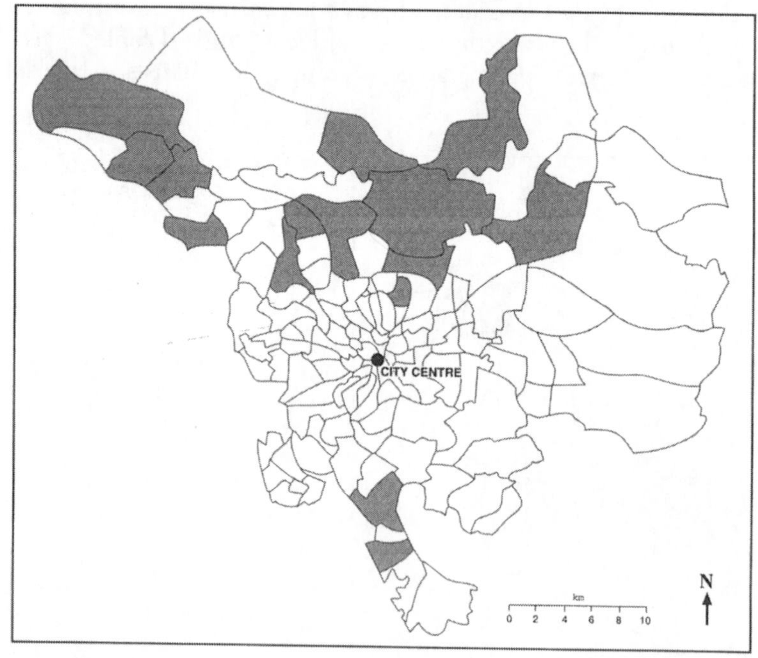

Figure 42.3 Location of 'prosperous' residents of Leeds.

Source: GB Profiles, School of Geography, University of Leeds.

'prosperous'. Here we can see a cluster of postal districts in the northern and northwestern suburbs of Leeds. Figure 42.4 shows the location of the one longstanding Sainsbury store in Leeds. There is a clear correlation here between its location and the location of 'prosperous' customers. A second Sainsbury store may well be suitable for northwest Leeds.

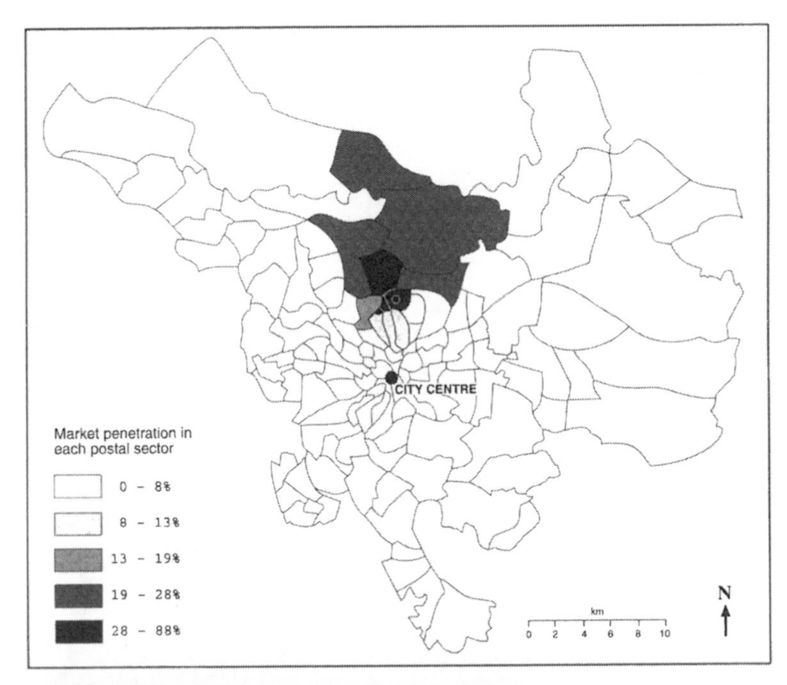

Figure 42.4 Market penetration of Sainsbury.

Plate 42.1 Dorothy Perkins and Burton's stores: who are their targeted customers and where do these persons live?

Having used geodemographics to locate customer types and identify possible locations for new stores, GIS is often used to predict the revenues of these new stores. Figure 42.5 shows a classic buffer-and-overlay analysis for a potential new Kwik Save store in south Leeds. As shown in Figure 42.5, the GIS buffers a primary, secondary and tertiary circular catchment area based on assumptions about how far customers are likely to travel to the new store. Then the population that lives within these buffers is estimated by overlaying census data on to the spatial backcloth of the newly created buffers. The problems with this methodology were discussed in the previous section. To recap, the difficulties appear when we try to estimate how much revenue from within these buffers will actually go to the new store (i.e. the problem of dealing with in- and outflows from this buffered region).

The alternative is to build a retail model (as in equations 42.1–42.2) for Leeds. Demand is estimated by calculating the spending power available in each postal district (a function of population size and affluence). The attractiveness of the retail outlets is usually given as a simple function of size. The distance deterrence term is a function of drive times between each postal district and each shopping centre or supermarket. (Birkin and Clarke (1991) offer a simple introduction to retail models for those new to the subject.) Having built the model itself, the next stage is to calibrate it (the procedure to estimate the model parameters) to reproduce existing interaction patterns between populations (either at home or at work) and the grocery stores of

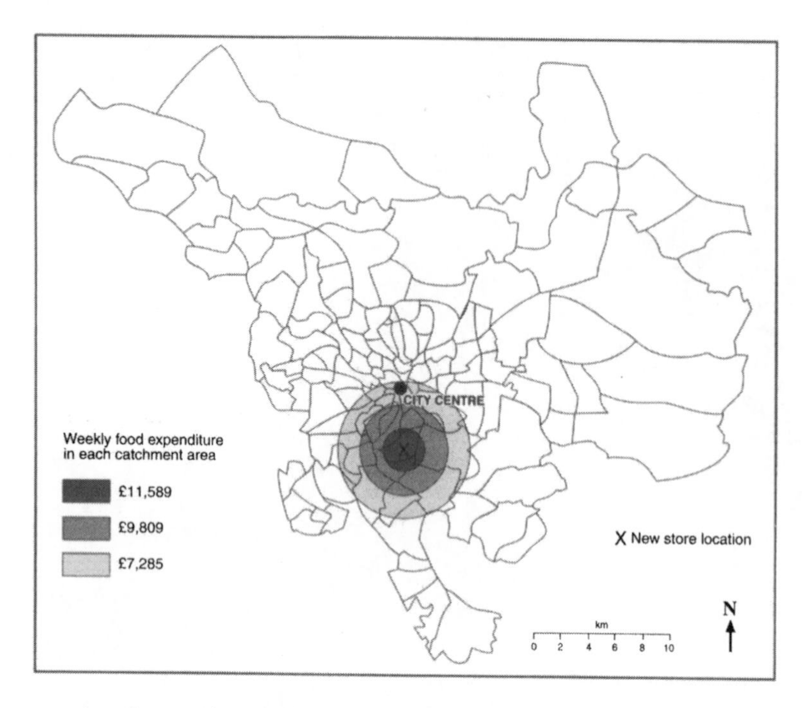

Figure 42.5 Catchment area of a potential new supermarket for Kwik Save using GIS.

Leeds. The results of this process are shown for one store in Figure 42.6. Note the success of the model in estimating the shape and size of the catchment area for that store. The successful calibration of the model in turn facilitates the accurate estimation of store turnovers (which may not be available from published sources: few retailers have turnover estimates for their competitors, for example). Having allocated expenditures between all postal districts and all grocery

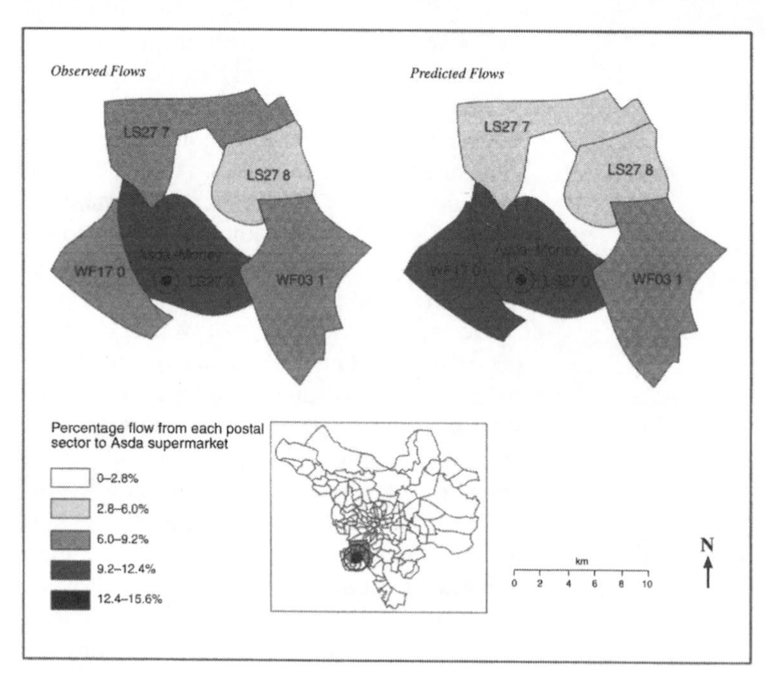

Figure 42.6 Observed versus predicted flows for Morley supermarket.

stores in this way, then the most obvious new geographical indicator is that of local market penetration. Figure 42.7 shows a map of market penetration for Asda in Leeds. As with Figure 42.2 and 42.4, we can clearly see the relationship between store location and market share. This is an interesting relationship in terms of advertising and marketing. Asda runs a very high-profile advertising campaign based on the 'Asda price' slogan. Yet for persons living in the centre of Leeds (see Figure 42.7), this could be deemed a waste of time. However enticing the advertising campaign is, few customers are likely to travel long distances (and pass many intervening opportunities) to buy their groceries. Perhaps such advertising is successful locally only when the consumer is faced with many choices.

In addition to new information such as local market shares, the models can also be used to compare actual turnovers to model predictions – that is, given a certain population size and type and the nature of the distribution of all competitor outlets, what would the model expect a certain outlet to be achieving in sales terms? This helps to provide a more objective picture of store potential. Is a store that turns over £3 million per annum doing well or badly in relative rather than absolute terms?

The models are most often used in 'what-if?' fashion. One example would be the use of the model to evaluate the impacts of a new store opening. In this sense, they compete directly with GIS as a methodology for new store revenue predictions. Figure 42.8 shows the model's predictions for the new Kwik Save store in south Leeds. Since this model is based on real interactions, it can be seen that the estimated catchment area for the new store is not circular but is skewed to the east and south, where competition is more scarce (compare these patterns with those shown in Figure 42.5). This we argue is far more realistic and therefore accurate than GIS techniques (see Birkin *et al.* 1996 and Benoit and Clarke 1997 for more discussion). Box 42.2 gives a simple illustration of model results based on all competitor locations being considered.

The use of spatial interaction modelling (still often referred to as gravity modelling in the literature) has increased since the late 1980s. Some organisations now develop these models in-

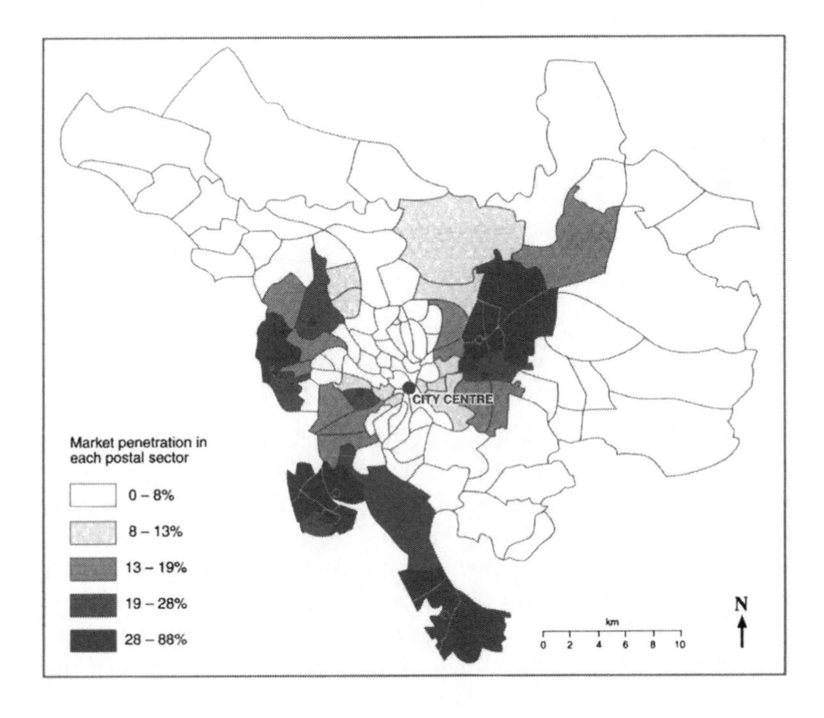

Figure 42.7 Market penetration of Asda superstores.

Plate 42.2 A large out-of-town Asda store: how does Asda locate optimally?

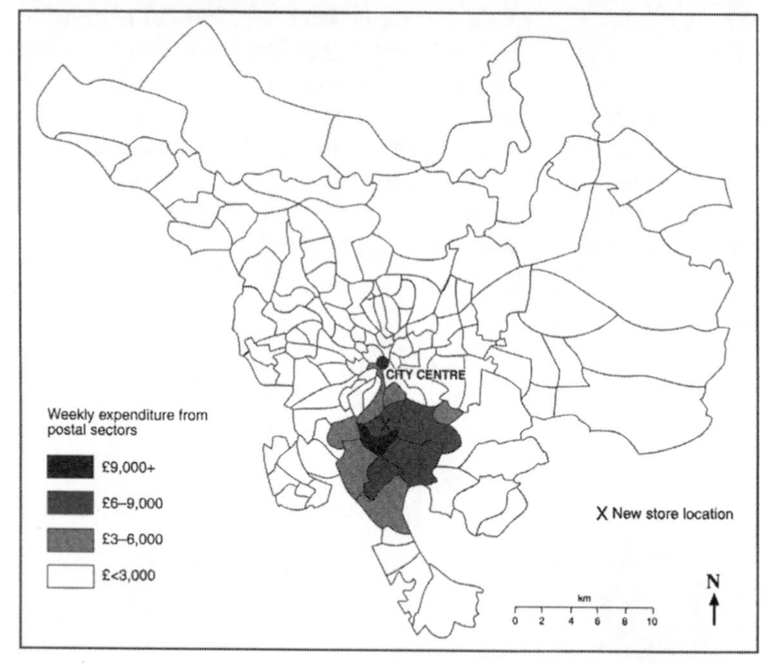

Figure 42.8 Catchment area of the Kwik Save supermarket based on model predictions.

house. Marks & Spencer is a good example of a recent convert to this methodology, especially as it searches for sites outside the UK (see Bond 1997). Elsewhere, such modelling is carried out by consultancy groups such as GMAP at the University of Leeds. By using such consultancy groups, retailers can tap into years of university research into the development and calibration of these models. GMAP's client list in 1996 included a number of blue chip clients working in many countries of Europe, America and Australasia (including Ford, Toyota, Mazda, Halifax, Barclays, Asda and Smith Klein Beecham).

The power of spatial intereaction modelling is being increased through optimisation procedures. In the long term, a company may be

Box 42.2 Estimating new store turnovers

Imagine you are a marketing analyst for a major car company, say Fiat. You have identified Blackpool in Lancashire, UK, as a large urban area without a major Fiat dealer. Having built and run a spatial interaction model (as given in equations 42.1–42.2), you now need to analyse the results. The results below give the predicted sales of your new Fiat dealer in Blackpool, along with the sales lost from other Fiat dealers nearby:

	Existing sales	*New sales*	*Impact*
Blackpool Fiat	0	510	+510
Lytham Fiat	412	291	−121
Preston Fiat	805	696	−109
Overall	1217	1497	+280

With profit levels in the UK car industry at approximately 17 per cent for the manufacturers, and the average price around £12,000, then the net gain in profit alone to Fiat is 280 × 12,000 × 0.17 = £571,200 (assuming land costs, etc. are not included in operating profits at this stage).

This example emphasises the importance of modelling techniques that deal with all competitor outlets. The prediction for the new site must be offset by the predicted loss of sales from surrounding outlets. Clearly, the company must examine these costs and benefits (and also examine sales gained from the competition: in this case, other Blackpool dealers of other major manufacturers).

interested to know what the optimal locations for its local network should be, given the objectives of maximising either total sales or market share, and how this compares with the existing distribution network. Formally, the spatial interaction model can be rewritten as a mathemtical programming formulation. The following may be a typical objective and set of constraints (given here in words rather than equations!):

Maximise: Market share in region X for organisation Y
Subject to: Maximum number of stores
Minimum number of stores
Mininum store sales of £Y
Minimum inter-store drive time of *T* minutes
No consumer to be more than *M* minutes drive time from a store

The problem can be solved either with existing outlets *in situ* or with all outlets free to relocate (at least theoretically). A heuristic algorithm has been developed that solves this complex problem on a PC (for a full description of the detail, see Birkin *et al.* 1995). Clarke and Clarke (1995) describe an application of such optimisation techniques by GMAP for Toyota in the US car market. These models were asked to find two new locations for Toyota in Seattle/Tacoma such that the impact on existing Toyota dealers was minimised. The two new locations produced a net gain in sales to Toyota of 1000 units. Given an average profitability of $1500 per unit, the overall increase in profits in Seattle/Tacoma alone is of the order of $1.5 million.

CONCLUSIONS

It is interesting to speculate on the future of store location research. Some authors are convinced that the increasing saturation of many retail markets (in theory meaning no more new constructions are possible or feasible) means that store location research will become redundant. Clarkson *et al.* (1996: p. 31) suggest:

> As the UK grocery market becomes increasingly saturated, the development of new stores on new sites would diminish in importance. The need for more sophisticated location assessment procedures would then become significantly less important to retailers in their pursuit of growth strategies.

However, there are two major responses to such an argument. First, in many markets the notion of saturation can be challenged, even in the sophisticated markets of the grocery sector in the late 1990s (see Langston *et al.* 1997; 1998; Guy 1996). Second, it could be argued that the increasing sophistication of retailing may result in a greater need for store location research rather than less. Clarkson *et al.* claim that the response of retailers to home saturation will either be internationalisation or store refurbishments. However, they miss the point concerning alternative distribution systems. In the UK grocery market, for example, we have witnessed the search for sites for a new wave of medium and smaller-sized supermarkets, thus making their store location teams even busier (Wrigley 1998). In addition, retailers are looking for ways of analysing the impacts of new trends that threaten the traditional nature of distribution. These include a new wave of large corporate mergers and acquisitions, advances in information technology, and the process of disintermediation (this refers to the elimination of layers of added cost from the distribution network by companies that have previously relied on other organisations to sell their products: for example British Airways). These trends cast increasing doubts over branch viability, and the threat of rationalisation is very real in many retail sectors. The simple removal of selected branch outlets, however, may not be the optimal strategy. What branch closures do create is a network that may be at odds with the existing spatial demands of consumers. The lack of consideration as to the link between supply and demand remains a major drawback in business development strategies. What will be required is more flexible, local responses to these trends. That will require greater and more subtle store location research, not less. That has to be good news for future geography graduates, provided they are skilled in spatial analysis rather than radical Marxism!

GUIDE TO FURTHER READING

For more information on retail marketing, try:

McGoldrick, P.J. (1990) *Retail Marketing*. McGraw-Hill, London.

For geodemographics and lifestyles, the best introductions are:

Birkin, M. (1995) Customer targeting, geodemographic and lifestyle approaches. In P.A. Longley and G.P. Clarke (eds) *GIS for Business and Service Planning*. Cambridge: Geoinformation, 104–49.

Batey, P.J. and Brown, P. (1995) From human ecology to customer targeting: the evolution of geodemographics. In P.A. Longley and G.P. Clarke (eds) *GIS for Business and Service Planning*. Cambridge: Geoinformation, 77–103.

For retail models, try:

Birkin, M. and Clarke, G.P. (1991) Spatial interaction in geography. *Geography Review* 4(5), 16–24.

Birkin, M., Clarke, G.P., Clarke, M. and Wilson, A.G. (1996) *Intelligent GIS: Location Decisions and Strategic Planning*. Cambridge: Geoinformation.

REFERENCES

Batey, P.J. and Brown, P. (1995) From human ecology to customer targeting: the evolution of geodemographics. In P.A. Langley and G.P. Clarke (eds) *GIS for Business and Service Planning*. Cambridge: Geoinformation, 77–103.

Beaumont, J.R. (1991a) *An Introduction to Market Analysis*. CATMOG 53, Norwich: Geo-Abstracts.

Beaumont, J.R. (1991b) GIS and market analysis. In D. Maguire, M. Goodchild and D. Rhind (eds) *Geographical Information Systems: Principles and Applications*. London: Longman, 139–51.

Belchamber, J. (1997) Store planning strategy – a case study from Thresher. Paper presented to 'The Art of Store Location' conference, Henry Stewart Conference Studies, 28/30 Little Russell Street, London, WC1A 2HN.

Benoit, D. and Clarke, G.P. (1997) Assessing GIS for retail location planning. *Journal of Retail and Consumer Services* 4(4), 239–58.

Birkin, M. (1995) Customer targeting, geodemographic and lifestyle approaches. In P.A. Longley and G.P. Clarke (eds) *GIS for Business and Service Planning*. Cambridge: Geoinformation, 104–49.

Birkin, M., Clarke, G.P., Clarke, M. and Wilson, A.G. (1996) *Intelligent GIS: Location Decisions and Strategic Planning*. Cambridge: Geoinformation.

Birkin, M., Clarke, M. and George, F. (1995) The use of parallel computers to solve non-linear spatial optimisation problems: an application to network planning. *Environment and Planning A* 27, 1049–68.

Bond, S. (1997) Gravity modelling and its applicability to the internationalisation of business. Paper presented to 'The Art of Store Location' conference, Henry Stewart Conference Studies, 28/30 Little Russell Street, London, WC1A 2HN.

CACI (1993) CACI's Insite system in action.

Marketing Systems Today 8(1), 10–13.
Clarke, D.B. (1996) The limits to retail capital. In N. Wrigley and M. Lowe (eds) *Retailing, Consumption and Capital: Towards the New Retail Geography*. Essex: Longman, 284–301.
Clarke, G.P. and Clarke, M. (1995) The developments and benefits of customised spatial decision support systems. In P.A. Longley and G.P. Clarke (eds) *GIS for Business and Service Planning*. Cambridge: Geoinformation, 227–45.
Clarke, G.P. and Clarke, M. (1998) Trends in UK retailing and the implications for network planning. Working paper, School of Geography, University of Leeds.
Clarke, G.P., Longley, P. and Masser, I. (1995) Business, geography and academia in the UK. In P.A. Longley and G.P. Clarke (eds) *GIS for Business and Service Planning*. Cambridge: Geoinformation, 272–81.
Clarkson, R.M., Clarke-Hill, C.M. and Robinson, T. (1996) UK supermarket location assessment. *International Journal of Retail and Distribution Management* 24(6), 22–33.
Davies, R.L. and Rogers, D.S. (1984) *Store Location and Store Assessment Research*. Chichester: Wiley.
Dawson, J. and Broadbridge, A.M. (1988) *Retailing in Scotland 2005*. Stirling: Institute for Retail Studies.
Elliott, C. (1991) Store planning with GIS. In J. Cadeau-Hudson and D.I. Heywood (eds) *Geographic Information 1991*. London: Taylor & Francis, 169–72.
Fenwick, I. (1978) *Techniques in Store Location Research: A Review and Applications*. Corbridge: Retail and Planning Associates.
Fotheringham, A.S. (1986) Modelling hierarchical destination choice. *Environment and Planning A*. 18, 401–18.
Ghosh, A. and MacLafferty, S. (1987) *Locational Strategies for Retail and Service Firms*. Lexington, Mass.: Lexington Books.
Gould, P.R. (1972) Pedagogic review: entropy in urban and regional modelling. *Annals of the Association of American Geographers* 62, 689–700.
Guy, C.M. (1996) Grocery store saturation in the UK – the continuing debate. *International Journal of Retail and Distribution Management* 24(6), 3–10.
Haggett, P. (1965) *Locational Analysis in Human Geography*. London: Edward Arnold.
Haggett, P., Cliff, A.D. and Frey, A. (1977) *Locational Analysis in Human Geography*, 2nd edition. London: Edward Arnold.
Harvey, D. (1969) *Explanation on Geography*. London: Edward Arnold.
Hinton, M.A. and Wheeler, K. (1992) GIS in 1992 – towards a single financial European market. *Mapping Awareness* 6(1), 19–22.
Howe, A. (1991) Assessing potential of branch outlets using GIS. In J. Cadeau-Hudson and D. Heywood (eds) *Geographic Information 1991*. London: Taylor & Francis, 173–5.
Ireland, P. (1994) GIS: another sword for St. Michael. *Mapping Awareness* April, 26–9.
Langston, P., Clarke, G.P. and Clarke, D.B. (1997) Retail saturation, retail location and retail competition: an analysis of British food retailing. *Environment and Planning A* 29, 77–104.
Langston, P., Clarke, G.P. and Clarke, D.B. (1998) Retail saturation: the debate in the mid 1990s. *Environment and Planning A* 30, 49–66.
Massey, D. and Meegan, R. (1985) *Politics and Method: Contrasting Studies in Industrial Geography*. London: Methuen.
Openshaw, S. (1995) Marketing spatial analysis: a review of prospects and technologies. In P.A. Longley and G.P. Clarke (eds) *GIS for Business and Service Planning*. Cambridge: Geoinformation.
Pacione, M. (1974) Measures of the attraction factor. *Area* 6, 279–82.
Reynolds, J. (1991) GIS for competitive advantage: the UK retail sector. *Mapping Awareness* 5(1), 33–6.
Reynolds, J. (1993) The role of GIS in European cross-border retailing. *Mapping Awareness* 7(2), 20–5.
Sleight, P. (1993) *Targeting Customer: How to Use Geodemographic and Lifestyle Data in your Business*. Henley-on-Thames: NTC Publications.
Spencer, A.H. (1978) Deriving measures of attractiveness for shopping centres. *Regional Studies* 12, 713–26.
Smith, N. (1989) Uneven development and location theory: towards a synthesis. In R. Peet and N. Thrift (eds) *New Models in Geography*. London: Unwin Hyman, 142–63.
Thompson, A. (1997) Location strategies for the leisure industry: case studies from Top Rank and Odeon Cinemas. Paper presented to 'The Art of Store Location' conference. Henry Stewart Conference Studies, 28/30 Little Russell Street, London, WC1A 2HN.
Timmermans, H. (1981) Multi-attribute shopping models and ridge regression models. *Environment & Planning A* 13, 43–56.
Webber, R. (1993) Building geodemographic classifications. Paper presented to the Market Research Society Census Interest Group, London, 5 November.
Wilson, A.G. (1970) *Entropy in Urban and Regional Modelling*. London: Pion.
Wilson, A.G. (1974) *Urban and Regional Models in Geography and Planning*. Chichester: Wiley.
Wilson, A.G. (1978) Spatial interaction and settlement structure: towards an explicit central place theory. In A. Karlquist, L. Lundquist, F. Snickars and J.W.

Weibull (eds) *Spatial Interaction Theory and Planning Models*. Amsterdam: North-Holland, 137–56.

Wilson, A.G. (1983) A generalised and uniform approach to the modelling of service supply structures. Working paper 352, School of Geography, University of Leeds.

Wrigley, N. (1998) Understanding store development programmes in post-property-crisis UK food retailing. *Environment and Planning A* 30, 15–35.

Wrigley, N. and Clarke, G.P. (1999) The market expansion of the limited-line deep discounters and their impacts on UK food retailing, *International Review of Retail Distribution and Consumer Research* (forthcoming).

Wrigley, N. and Lowe, M. (eds) (1996) *Retailing, Consumption and Capital: Towards the New Retail Geography*. Harlow: Longman.

43

Global positioning systems as a practical fieldwork tool: applications in mountain environments

Ian Heywood, Graham Smith, Bruce Carlisle and Gavin Jordan

INTRODUCTION

For centuries, geographers have found fascination in *where* things are in relation to each other. This obsession has manifested itself in the geographer's desire to map everything and more recently (within the last century) to seek out and understand spatial relationships both in and between human and physical systems. Knowing where something is located in both space and time is therefore a prerequisite for almost all geographical research. Traditionally, geographers have used numerous techniques to determine both the relative and actual location of spatial phenomena. These have ranged from the very accurate and precise techniques of the surveyor to the more general methods of the social scientist. Recently, however, a new technique for locating spatial phenomena has found particular favour with geographers. This is the use of a satellite navigation system or global positioning system (GPS). These portable locational devices can be mounted on a vehicle, carried in a backpack or held in the hand and used to record location at almost any point on the Earth's surface (Plate 43.1). Location information is obtained literally at the push of a button, with accuracy ranging from 150 m to within a few millimetres, depending upon the quality and number of receivers used and whether a military or civilian version of the system is accessed. Originally designed for real-time navigation purposes, all GPS receivers will store collected coordinates and associated information in their internal memory

Plate 43.1 GPS in the field.

so they can be downloaded directly into a computer. The ability to walk or drive around collecting coordinate information whenever it is required has obvious appeal for the geographer involved in applied field-based research.

There is nothing complex in developing an appreciation of how a GPS works. The technique is based on the principles of trilateration, which have been used by surveyors for centuries. However, like all new technologies GPS has developed its own mystique and jargon. This chapter explores the basics of GPS, how they work, their practical value and their limitations. GPS receivers are now available at costs similar to those of good-quality compasses. In the same way that the geographer needs to know how the accuracy of a compass is affected by local environmental magnetism, the GPS user needs to have an appreciation of the factors that can influence the quality of the locational information provided by a GPS.

To illustrate the value and limitations of a GPS for geographical research, three case studies, reviewing their use as a field tool for supporting research in mountain environments, are introduced in this chapter. These are:

- geomorphological mapping with GPS (Box 43.1);
- the use of GPS to assess the data quality of digital elevation models (Box 43.2); and
- applying GPS in community forest resource mapping (Box 43.3).

To conclude the chapter a closer look is taken at the advantages and disadvantages associated with using GPS technology for geographical field research.

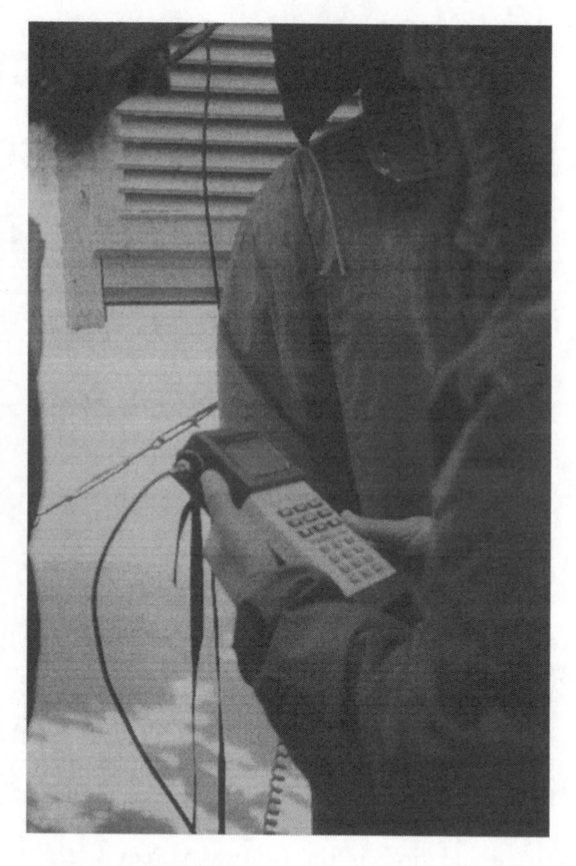

Plate 43.2 GPS receiver.

GPS BASICS

GPS is a set of satellites and control systems that allow a specially designed receiver (Plate 43.2) to determine its location anywhere on Earth 24 hours a day (Barnard 1992). Two main systems exist, the American NAVSTAR (*NAV*igation *S*ystem with *T*ime *A*nd *R*anging), and the Russian GLONASS systems. In this chapter, we focus on the workings of the American NAVSTAR system, as this is the most widely used.

The GPS framework

To understand how a GPS works, it is important to appreciate that the system has five components. These are:

- the NAVSTAR satellite constellation;
- the satellite control and monitoring stations;
- the national and regional GPS base station networks;
- the GPS receivers; and
- the users.

The NAVSTAR GPS satellite constellation consists of twenty-four satellites orbiting the Earth at a distance of more than 20,000 km. NAVSTAR was developed and implemented by the US Department of Defense (DoD), which is responsible for the maintenance of the system. The first of the GPS satellites was launched in 1978, and by mid-1994 all twenty-four satellites were in operation. Each satellite weighs approximately 860 kg and is 8.7 m in height (Kennedy 1996). The satellites are solar-powered, and only twenty-one of the twenty-four satellites transmit positional information. The three remaining satellites are spare units that can be brought into immediate use should the need arise. The twenty-four satellites are arranged on six orbital planes, with four satellites on each plane. Each plane is inclined to the equator by 55°. This design ensures complete coverage over the entire surface of the Earth and that a GPS receiver can see multiple satellites from the majority of locations (van Sickle 1996). For many locations, it is often possible for a receiver to see ten or more satellites at particular times of the day. There is, however, a recognised problem in the polar regions, where the poor geometrical alignment of the satellites at certain periods reduces positional accuracy.

The NAVSTAR satellites have a twelve hour orbit time and pass over control stations during each orbit. This allows close monitoring of the orbit and position. This locational information can be transmitted from the monitoring stations to the satellite so that each satellite has up-to-date information concerning its own location. The control stations also monitor the accuracy of atomic time clocks on board the GPS satellites and transmit information regarding the accuracy of these clocks to the satellite. It is essential that each NAVSTAR satellite provides complete and accurate information regarding position and that the accuracy of its atomic clock is known.

A large number of national and regional GPS base station networks exist that monitor satellite positions over time. These provide continuous information about the positional accuracy of the GPS signals at specific geographic locations. Many of these networks transmit this information to users of stand-alone GPS receivers. The information from base station networks allows users to improve the positional accuracy of the data from their own GPS receivers.

GPS receivers are capable of detecting the signals transmitted from the NAVSTAR satellites and converting this information into location information. A wide range of receivers are now available on the market at prices from a few hundred dollars to well over $10,000. The cheapest receivers will usually have an accuracy, when used individually, of 100 m for 95 per cent of the time. Receivers costing over $10,000 should be capable of at least centimetre and in some cases millimetre accuracy. No matter what the cost or positional accuracy a receiver can achieve all receivers have the following components in common (adapted from Kennedy 1996; and van Sickle 1996):

- a microcomputer, which processes the information from the GPS satellites and uses this information to determine position;
- an aerial and associated electronics, which receives the signal from which position is calculated;
- a user interface (usually an LCD screen and key pad), which allows the user to program the receiver and view the output); and
- computer memory, which allows the user to record positional information. More expensive receivers have greater memory and allow the user to record additional information, for example, what is present at a given location.

The last component of the GPS framework is the user. Historically, the main users of GPS were the military. Today, however, civilian applications of the technology far exceed military use. The majority of civilian users fall into one of two categories, although the boundary between is fuzzy. The first group uses GPS for navigation. The second group uses the technology to record, map and survey the position of features such as the location of an oil well, the boundary of a forest or the footprint of a new building. From a geographer's perspective, it is this latter application that is perhaps the most interesting. Therefore, the remainder of this chapter focuses on the use of GPS technology for recording survey locations and mapping.

HOW GLOBAL POSITIONING WORKS

GPS works by using a variation on trilateration, a technique used by surveyors for many years. Kennedy (1996) and Barnard (1992) provide a comprehensive explanation of the trilateration method. In summary, the location of an *unknown* position is calculated by measuring the distance between the unknown position and that of several *fixed*, known positions. In the context of GPS, the ground receiver represents the unknown position and the constellation of GPS satellites the known fixed positions.

In trilateration, the distances between a minimum of three locations, or *ranges*, are measured by monitoring a signal between the satellites and ground-based receivers. Both the GPS receiver and the satellites transmit similarly coded radio signals, and the time delay between receipt of the signals is used to determine distance between the satellite and receiver. To measure time delay, the GPS signal from the satellite must contain information about the exact time that the signal left the satellite. In addition, the signal from the satellite must contain information about the exact location of the satellite at the moment that the signal was sent. Without this, the GPS receiver would not know the location of the satellite, since GPS satellites are continually orbiting the Earth. Therefore, all GPS satellites continuously transmit two signals: L1 at 1575.42 MHz and L2 at 1227.60 MHz. Both signals carry coded information in binary format (0 and 1 strings). Three basic codes are transmitted: the *navigation* code, the *coarse acquisition* (C/A) code and the *precise* (P) code (van Sickle 1996). The C/A code is available to all civilian users and is used for the standard positioning service (SPS), whereas the use of the P code is restricted, and encrypted, by the US DoD. There are several important elements to the three codes, including satellite position or *ephemeris*, *atmospheric correction* (required to correct for delays to the signal as it travels through the ionosphere), *antispoofing* (intentional degrading of the signal), satellite *almanac* (the location of all other satellites) and satellite *health* (the quality of the signal). Van Sickle (*ibid.*) provides a comprehensive review of the different code elements. It is these codes that provide the information necessary for a receiver to determine position.

A GPS receiver can calculate its position using either *pseudorange* or *carrier phase* observations. Pseudoranges are used primarily for instantaneous positioning in navigation (Ashkenazi and Dodson 1992). In pseudorange observations, distance between satellite and receiver is calculated by measuring the time delay between the transmission and receipt of a unique segment of the C/A code. However, because the receivers have less accurate clocks than those on board the satellites, there will be a degree of unknown error in the distance calculated. This uncorrected observed distance is known as pseudorange. In pseudoranging, three satellites are used to determine an uncorrected location for the receiver, and a fourth satellite is used to remove the errors associated with the receiver clock. This ranging method is typically capable of positioning to within 15–30 m. However, this accuracy is degraded because of the problem of *selective availability* (SA).

SA was introduced in March 1990 by the US DoD to deliberately degrade the positional accuracy possible with the civilian system (until 1990, 15-metre accuracy was possible). SA can take two forms, dither, which affects the satellite clock, and epsilon, which refers to the errors introduced into GPS satellite *orbital parameters* (Gilbert 1995). The effect of both errors is to reduce positional accuracies to approximately 100 m horizontally and 150 m vertically (Ashkenazi and Dodson 1992) (Figure 43.1). Technological developments, including new differential solutions enabled improvements in the positional accuracy available to civilian users. These solutions are discussed below. SA can also be countered by the use of carrier phase observations.

Carrier phase observations improve upon the accuracy of the pseudoranging method by working with the carrier signals L1 and L2 rather than the C/A or P codes. In carrier phase observations, the nature of the signal carrier is compared rather than the coded message. Therefore, as SA affects the C/A code and not the carriers L1 and L2,

Figure 43.1 A plot of a position measurement over time with selective availability (SA) in action.

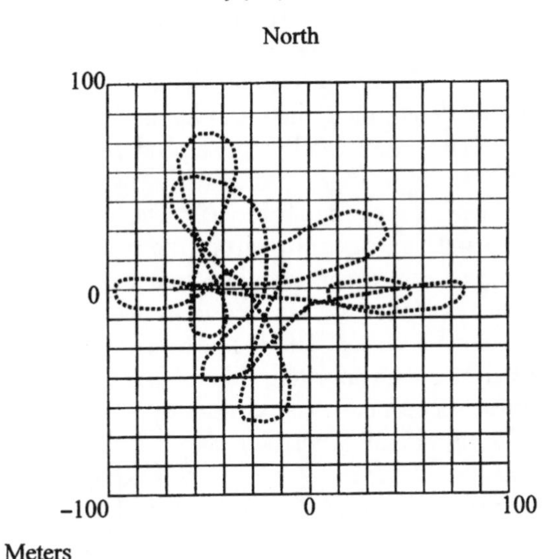

carrier phase observations are immune from SA.

Finally, the GPS receiver also needs to receive information about any atmospheric effects and satellite geometry that may result in a degraded signal. This gives an indication of the quality and reliability of location information received.

OPERATIONAL ISSUES IN GPS: SIGNAL QUALITY AND RELIABILITY

Both pseudorange and carrier phase observations are subject to a number of potential errors. These include *multipath* errors, *ionospheric delay* and *poor satellite geometry* (Gilbert 1995). These errors are significant and often determine the accuracy of the data collected.

Multipath error is caused by signal propagation where the signal transmitted to the ground receiver is bounced or reflected during its journey from the satellite. The error usually occurs when the GPS unit receives signals from very low-elevation satellites (<15°) (*ibid.*). This error may be significant when operating GPS technology in very steep mountainous terrain or urban environments, where many surfaces are present which can reflect the satellite signal. Often the only way to deal with multipath signal error is to implement minimisation methods such as mask angles (which reduce the angle of the field of view in which a receiver will search for satellites) or to avoid multipath-prone sites (*ibid.*).

Ionospheric delay results from atmospheric disruption of the GPS signal. The ionosphere disperses the GPS signal, thus delaying the arrival of the signal. This can result in miscalculation of distance. One of the primary causes of delay is the amount of water vapour in the atmosphere.

The quality of positional information is also influenced by the geometry of the satellites in the user's field of view at the time of observation. GPS receivers provide the user with several quality indicators relating to the geometry of the satellites. The most important is the geometric dilution of precision (GDOP). Higher GDOP values indicate poor satellite geometry, which may give rise to poor positional information. Figure 43.2 shows examples of good

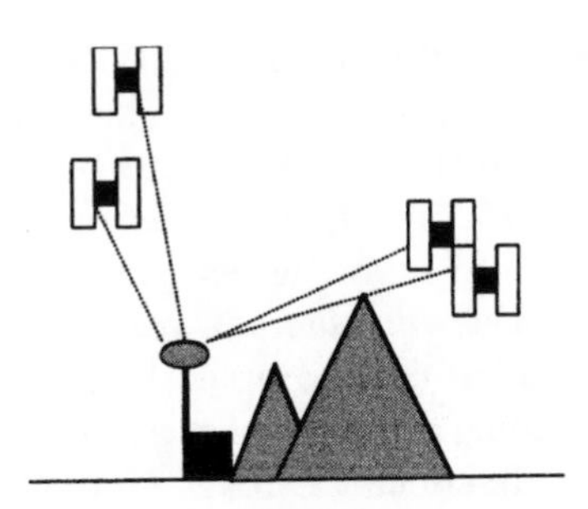

Poor satellite geometry (high PDOP)

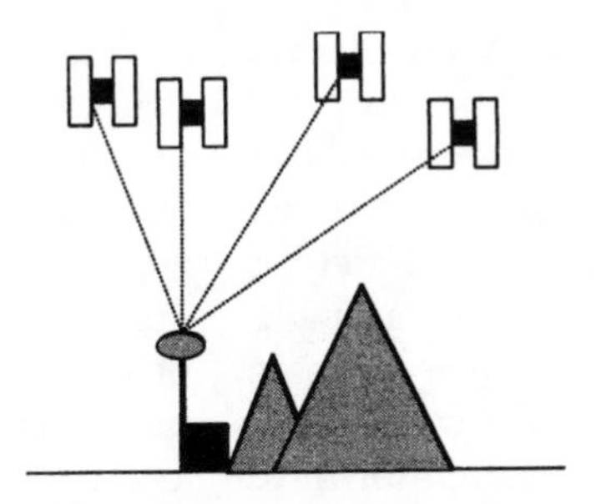

Good satellite geometry (low PDOP)

Figure 43.2 Good and bad satellite geometry.

and bad satellite geometry in relation to receiver position.

HOW TO USE GPS TO RECORD POSITION

There are two main techniques for determining the position of a GPS receiver: kinematic (mobile) GPS and static (stationary) GPS. With kinematic GPS, one or more GPS receivers is in motion during the recording of positional information (van Sickle, 1996). Kinematic techniques are used for the collection of positional information while on the move, for example in the survey of a footpath or in onboard car, boat or aeroplane navigation systems. With static GPS, the receiver (or receivers) remains at a fixed point and the position recorded averaged from a large number of readings taken over a specified time period. Static GPS is the technique used by the surveying community to survey sample points or the precise location of buildings. Both kinematic and static techniques have found practical application in geographical research, as the case studies described below show. Kinematic and static techniques need subdividing into two further approaches: absolute positioning and differential positioning.

Absolute techniques require only one receiver and are frequently used for navigation, where accuracy of tens of metres is acceptable. In kinematic GPS, absolute techniques could be used to record the position of a forest or landscape feature by walking around its perimeter. In static GPS, absolute techniques might be used to survey the location of a sampling site. The inherent inaccuracies in this method (± 100 m), caused by SA, can be reduced by averaging the positions recorded over a given time period. Despite the lack of positional accuracy, absolute kinematic and static techniques have found widespread use. The advantage of absolute positioning is the ease with which a position can be recorded and the speed of measurement. In addition, a single, low-cost receiver is required. When only an approximate position is required, absolute techniques may be appropriate. However, for most users a more accurate method is necessary. This is offered by the use of differential GPS (DGPS).

DGPS works on the principles of differential positioning, which assumes that two receivers tracking the same satellites will be subject to similar errors. Therefore, if the errors for one receiver can be determined, these can be used to correct the errors recorded by the second receiver. In DGPS, one receiver is placed at a known location. The position of the location is determined over time and any deviations from the true position monitored. A second receiver, tracking the same satellites, is then used to record the position of an unknown location. The position recorded by this receiver is then adjusted, either in real time or by post-processing, to cancel out any errors in the position obtained by the receiver at the known location. Differential correction of GPS data can be used to counter SA and some of the other errors that may affect the position. Using DGPS techniques, accuracy is usually between 2 and 15 metres; however, it is possible to obtain millimetre accuracy with high quality equipment and an appropriate technique.

It is not necessary to own two receivers to perform differential correction. A number of 'base station' networks have been established around the globe. These transmit correction information, often in real time. This permits those with an appropriate receiver and decoder to correct their GPS positions in the field. This real-time correction of positional information is of particular use to the navigator.

GPS IN PRACTICE: RESEARCH IN MOUNTAIN ENVIRONMENTS

In order to appreciate the potential of GPS as a field tool, it is useful to examine how it has been used in a variety of research situations. In this chapter, to provide context, we focus on the application of GPS in mountain environments. However, there is a growing literature on the use of GPS in other areas of geographical research (see the guided reading at the end of this chapter). The applications described in Boxes 43.1, 43.2 and 43.3 cover many of the standard tasks typically undertaken as part of a geographical research project. These include field mapping (Boxes 43.1 and 43.3), verification of spatial models (Box 43.2) and georeferencing of spatial information (Box 43.3). The case studies described in the boxes also show how a range of

Box 43.1 Field mapping: geomorphological interpretation with GPS

Glacial terrain is usually mapped from aerial photography and large-scale topographic maps. However, there are many formerly glaciated mountain environments where such data are unavailable or of poor quality. Also, the difficult terrain often prevents the use of traditional instrumental techniques. GPS, on the other hand, is unaffected by these problems and possesses several advantages to the geomorphologist. For this study, field mapping was carried out in the Llanberis area of Snowdonia, North Wales. Snowdonia was chosen as the study site because of easy field access to its well-known glacial landforms. This has permitted controlled testing of GPS. The study site is located in steep terrain, with several peaks in excess of 1000 metres. The area was previously glaciated during the Pleistocene and contains a series of erosional and depositional landform systems (see Addison *et al.* 1990; Gray 1982; Gray and Lowe 1982). The work described here concerns the mapping of the smaller-scale glacial landforms associated with glaciation during the Loch Lomond Stadial, such as moraines. Two GPS strategies were employed to map the glacial landforms: kinematic DGPS (using pseudoranging) for delineating landform boundaries and static DGPS (using carrier phase) to determine the location of sites for further investigation (Smith *et al.* 1997).

Figure 43.3 shows the kinematic DGPS data overlain on an orthophotograph of the same area. The lines on the map represent the boundaries of different relict glacial features. Walking around the perimeter of the landforms and recording their position with a GPS receiver produced these boundaries. The DGPS method provided positional accuracies of a few metres (2 m at best and typically 5–10 metres). However, as the boundaries between glacial landforms are 'fuzzy' and the surveyor had to use field cues to determine the boundary of a particular feature, this level of accuracy was considered acceptable (Smith *et al.* 1998). A major advantage of the kinematic DGPS method was that it allowed fieldworkers to map features continuously. In addition, as the data had been collected digitally, they were easily transferred into a geographical information system (GIS), where they were checked for accuracy and used in further quantitative analysis.

Carrier phase DGPS was used to locate field investigation sites where moraine soil samples could be taken or former ice flow direction recorded from striae (*ibid.*). Carrier phase observations were accurate to within a metre (Magellan 1994). These data were also used to record the location of good exposures for use in later sedimentological analyses.

Figure 43.3 GPS landscape feature map, Snowdonia, North Wales.

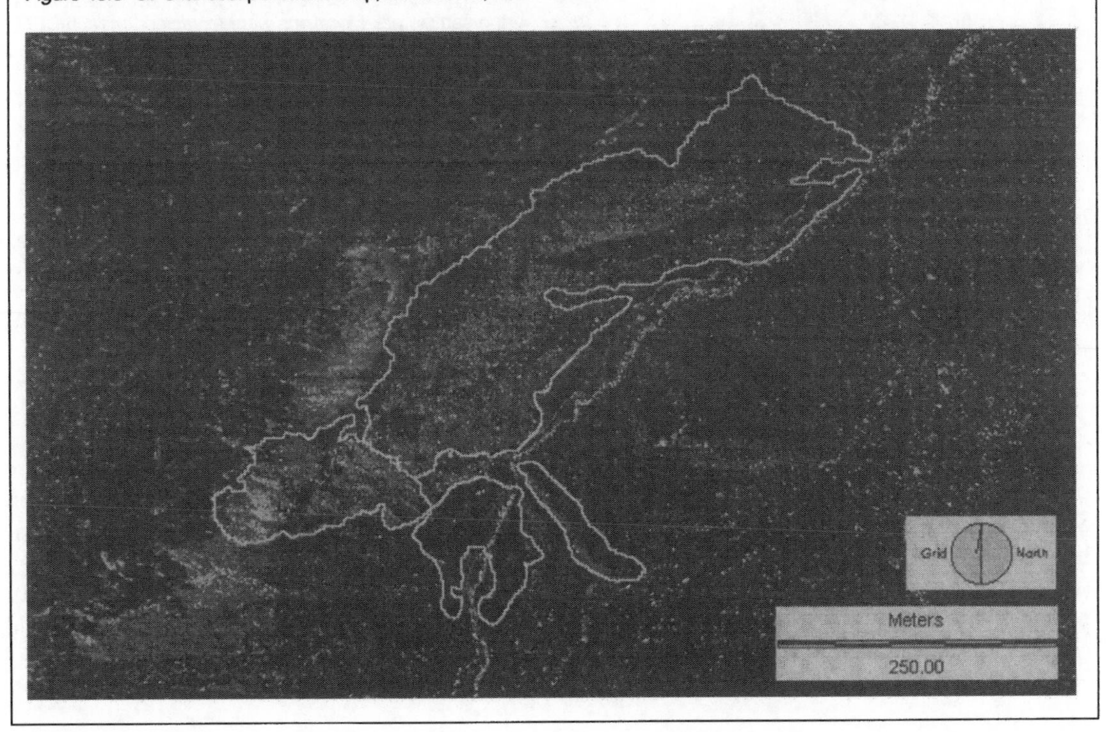

Box 43.2 Model verification: using GPS to assess data quality of digital elevation models

Given the influence of altitude and relief on the characteristics of mountain environments, a high-quality model of the topographical surface, a digital elevation model (DEM) (Figure 43.4), forms the basis of the successful development of computer-based modelling tools for mountain environments (Stocks and Heywood 1994; Heller and Weibel 1991). However, the potential use of readily available DEMs, often derived from national mapping agency data, is limited by inadequate or insufficient knowledge about the quality of these models.

The factors determining DEM data quality can be classed under the two headings: digital elevation data production and interpolation from these data to create the DEM. Measures of digital elevation data accuracy may be available. For example, the Ordnance Survey (1996), the UK's national mapping agency, states that its 1:10,000-scale digital contour data has a root mean squared error (RMSE) of 1.5 m. However, little is known about the influence of the DEM generation process, through data interpolation on the quality of the resulting DEM. To provide this information, accurate altitude measurements can be collected using GPS and compared with altitude values derived from a DEM. To test this approach, fieldwork was conducted in an area of approximately 2 km^2 within the Snowdonia massif, North Wales, UK. This area was chosen for its accessibility, geological and geomorphological variety and extremes of slope, aspect and relative relief.

Figure 43.4 Digital elevation model of Snowdonia, North Wales.

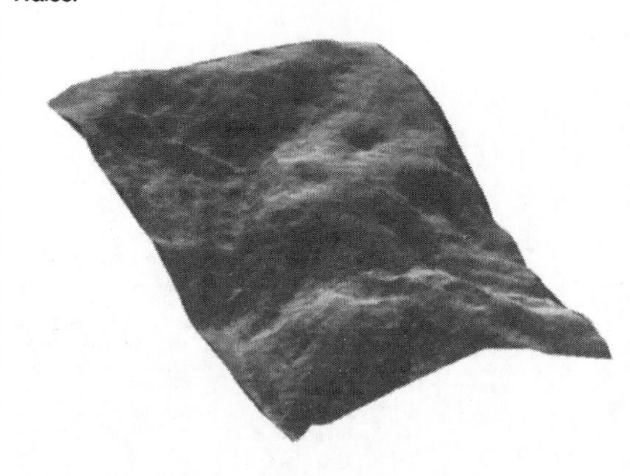

The basis for the estimation of the quality of a DEM was the collection of a sample of control points at which elevation had been determined to a high order of accuracy. GPS was used to collect these control data because of its suitability for use in the more inaccessible and data-poor areas of mountain environments. Static DGPS-based measurements of altitude and location were taken using carrier phase observations. Magellan (1994) states that of sub-metre accuracy data (± 0.9 m) can be acquired with these receivers. Field testing showed that such accuracy positions were possible approximately 70 per cent of the time, i.e. similar accuracies to those stated by Magellan.

In the field, one receiver was used to collect data at a known base station location. The second receiver was taken to the sample points. At each sample point, 10 minutes of GPS data were collected. For each sample point, the data collected by the mobile receiver were differentially corrected to provide estimates at sub-metre accuracy of the sample point's location and elevation. Position and elevation data for 106 points within the Snowdon study area were recorded. Points were chosen to provide a representative sample of the area's varying terrain.

Digital contour data at 1:10 000 scale (RMSE = ±1.5 m) were used to generate twenty-six DEMs of the study area. Various combinations of different data patterns (contour lines, random points, regular grids of points), data densities, interpolation algorithms and interpolation parameters were sed to produce this range of DEMs. The elevations at each the sample point locations were extracted for each DEM nd compared with the elevations from the GPS survey. The ature of the resulting elevation error distributions was used describe the quality of each DEM. The mean elevation ror and the standard deviation of the elevation error for ach of the twenty-six DEMs was calculated. The absolute ean error ranged from ± 3.12 to 7.11 m and the standard eviation of the error from ± 3.77 to 10.32 m. Therefore, it is ear that the elevation accuracy of a DEM varies conderably and depends on the pattern and density of source ata and the interpolation method used. It is also apparent at the quality of the original digital elevation data with a cumented RMSE of ± 1.5 m gives little indication of the uality of the resultant DEM (Carlisle and Heywood 1997).

different GPS methods have been used to provide positional information. The examples are from Snowdonia, Wales, and Nepal and look at geomorphological mapping using kinematic DGPS (Box 43.1); the verification of the quality of data provided by digital elevation models (DEMs) (Box 43.2); and the use of GPS to assist mapping of forest resources (Box 43.3). In the Nepal case study, GPS was used to help the local population to map the forest resource for use in local management strategies. This example illustrates how GPS can be used not only for geographical research but also for more pragmatic and practical applications.

Box 43.3 Georeferencing for resource assessment: the role of GPS in community forest resource mapping

Community forestry involves returning state-owned and managed forests to villagers and local forest users, who manage them for their subsistence needs of timber, fuelwood, fodder, grazing and medicinal plants. There is a growing interest in Nepal in obtaining information regarding the quality of these resources: does the forest improve or degenerate under village management? In order to be able to gauge this reliable, baseline information on the forest resource is required. Two critical elements of this baseline information are an accurate estimate of the position of the forest boundaries and georeferenced air photographs that show the status of the resource. In both cases, GPS was identified as a cost-effective means of providing information (Jordan and Shrestha 1998).

Forest boundary mapping used GPS as a survey instrument. The receiver was taken around the forest boundary and a geographical location recorded every second. This information was stored in the GPS receiver, and later downloaded into a computer, post-processed and put into a GIS. In many cases, GPS data could be used to map forest boundaries accurately and rapidly. The feasibility (in terms of speed and accuracy) of this depends largely on terrain: forests that have very steep slopes and no clear delineation between community forest and other land uses are difficult to map. This is due partly to the physical difficulties and dangers of using GPS in this terrain (in this region of the mid-hills of Nepal, 70 per cent of the land has a slope angle greater than 30 degrees, and slope angles in excess of 50 degrees are common), and partly due to the obstructed view of satellites caused by the hillsides.

To obtain useful results, kinematic DGPS was necessary and ± 5–20 m accuracy was obtained. For resource mapping in Nepal this is more than adequate (and far more accurate than other spatial information available). Absolute GPS did not provide a sufficient level of accuracy, with positional fixes at one point being highly variable. This was partially overcome by recording fixes for 15–20 minutes at each point, and averaging the fixes, but this is prohibitively slow.

To georeference air photographs, GPS techniques were used to obtain an accurate geographical location for an identifiable feature (such as a bend in the road, house, temple or hilltop). This was used as a control point. With georeferenced control points the spatial information from the air photographs could be corrected geometrically and entered into a GIS. This technique provided a means for mapping approximate forest boundaries rapidly (they are only approximate due to the spatial distortion inherent in standard aerial photographs). Georeferenced control points were established using DGPS. In this project, five minutes of readings were found to be adequate for an accuracy of ± 2–5 m. The key was to identify control points that could be identified easily on aerial photographs, and in positions where the satellite signal is not obstructed.

GPS users in developing countries face a lack of georeferenced base stations in convenient and secure positions. This problem was overcome in Nepal by locating the GPS base station at a project building in the same watershed as the mapping and resource assessment work. The location was determined by averaging over 30,000 individual readings. Statistical analysis of the data collected determined that the location was accurate to within approximately ± 5 m, more than adequate for this type of work.

THE ADVANTAGES AND DISADVANTAGES OF USING GPS TECHNOLOGY

GPS technology is proving to be a valuable tool in a wide range of geographical studies. DGPS static and kinematic methods are the most widely used. The specific benefits of DGPS for the geographer involved in field research are summarised below.

Kinematic DGPS data collected using pseudo-range observations facilitate the creation of digital maps delineating, for example forest and glacial landform boundaries. Positional accuracies of ± 3 m can be achieved. Carrier phase DGPS allows point locations to be defined to sub-metre accuracy. This range of accuracy allows collection of field data at a variety of spatial scales for a variety of applications. The digital format of GPS data allows rapid transfer of information to GIS or other computer mapping software. This in turn permits data to be checked for accuracy while still in the field (Carver *et al.* 1995; Smith *et al.* 1998). In addition, work in Snowdonia has shown that DGPS methods can supplement information derived from orthophotography or other remotely sensed data sets. Better-quality delineation of features is possible by avoiding the limitations of data resolution and allowing field knowledge to be incorporated into the mapping process. DGPS makes field mapping and higher-accuracy surveying possible in difficult sites such as remote and rugged mountain terrain and forested areas, where traditional surveying techniques would be impractical due to the weight of the equipment and problems with line of sight.

Clearly DGPS can be a highly valuable mapping tool. However, the user must be aware of several operational issues and potential problems if this technology is to be used in appropriate ways with maximum effectiveness. Kinematic GPS allows rapid data collection while on the move. However, collection of more accurate carrier phase observations is time-consuming, taking up to ten minutes at each individual point. Additionally, high volumes of data stored on the receiver may be problematic. One day's fieldwork can produce tens of megabytes of data, which must then be downloaded from both the mobile GPS and the base station receivers to a PC for differential correction. The time and operational problems that this may cause can be avoided by the use of data logging PCs connected to the receivers or real-time correction of the GPS data using a transmitter to send the base station data to the rover. However, both these options can be prohibitively expensive and add to the weight of field equipment. Also, real-time correction may not be possible due to obstruction of the broadcast correction signal caused by the nature of the terrain, for example mountains, high buildings or extensive tree cover.

Mountainous terrain, buildings and tree cover can also impede post-processing correction of GPS data. Both mobile and base station receivers must be tracking the same set of satellites for differential correction to succeed. This will not always happen when large areas of the sky are obstructed. However, the chances of success increase the closer the base station receiver is to the mobile receiver. Using GPS to determine the position of an unknown base station location right in the area of study is useful (Carlisle and Heywood 1997). However, in Snowdonia more than 25 per cent of data could not be corrected at first attempt, despite the mobile GPS being no more than 2 km from the base station.

In forested and built-up areas, satellites rapidly disappear from and then return to view as they pass behind tree branches, trunks and buildings. To minimise the time spent by the receivers searching for lost satellites or locating new ones, rather than recording position data, receivers with ten or more channels are beneficial. Ten-channel receivers allow the tracking of more satellites at one time. Therefore, when the signal from one satellite is lost the receiver uses the next best satellite, which is already being tracked on another channel.

There can be significant variation in performance between different makes of receiver, in different types of terrain and at different points in time. This uncertainty over the exact accuracy of the collected GPS data can be off-putting to the novice. It is vital to verify the performance of the equipment and the data collection method, particularly when using GPS for the collection of data with sub-metre accuracy. Testing equipment by surveying known locations, such as the Ordnance Survey's triangulation points in the UK, to assess the effectiveness of the methodology and define constant inaccuracies such as datum transformation errors can be an invaluable precursor to fieldwork and also allow the user to develop a better understanding of GPS accuracy issues. It is equally important to look at signal strength, GDOP and the number of points processed during differential correction as indicators of the accuracy of the data.

Overall, GPS technology is feasible and effective in a variety of research areas where other surveying and mapping techniques would be impractical. The summary of potential problems given above is not meant to deter use of GPS but to encourage sensible application of the technology by users with an awareness of the limitations of the technology. With such an understanding, users will find themselves able to map new types of feature efficiently and use the data in new ways to discover new areas of geographical research.

GUIDE TO FURTHER READING

There is no GPS text specifically aimed at geographers. The best introductory text is by Kennedy (1996). This text also provides an excellent practical overview of how GPS data can be used with a GIS. The book is of particular value if the reader has access to a Trimble Pathfinder GPS receiver and the ArcView GIS. The book comes with sample data sets on CD-ROM as well as self-study exercises. Leick (1995) and van Sickle (1996) provide more technical overviews and look in detail at many of the geodetic issues and recent advances in GPS. Geographical examples of GPS technology, particularly its physical applications, can be found in journals such as *Photogrammetric Engineering and Remote Sensing*, and in the Technical Bulletin series of *Earth Surface Processes and Landforms*. For example; August *et al.* (1994) consider the accuracy and precision of locational data derived using GPS, and Cornelius *et al.* (1994) and Fix and Burt (1995) GPS applications in geomorphology. *GPS World* provides information on the types of GPS products available and has review and research examples of GPS technology applications. Peter Dana of the University of Texas (a geographer) maintains one of the best and most comprehensive web sites on GPS technology. It is often the first call for new users of GPS technology and can be found at:

http://www.utexas.edu/depts/grg/gcraft/notes/gps/gps.html

There are also several e-mail discussion lists dedicated to GPS. The best of these are GPS-L and CANSPACE. GPS-L is a heavily subscribed general discussion list and is a good place to seek help with any problems encountered with your GPS. CANSPACE provides technical information, data concerning NAVSTAR and GLONASS system operation and updates on the contents of GPS-related journals.

REFERENCES

Addison, K., Edge, M.J. and Watkins R. (1990). *The Quaternary of North Wales: Field Guide.* Coventry: Quaternary Research Association.

Ashkenazi, V. and Dodson, A. (1992) Positioning with GPS. *Mapping Awareness and GIS in Europe* 6, 7: 7–9.

August, P., Michaud, J., Labash, C. and Smith, C. (1994) GPS for environmental applications: accuracy and precision of locational data. *Photogrammetric Engineering & Remote Sensing* 60, 1: 41–5.

Barnard, M.E. (1992) The global positioning system. Institution of Electrical Engineers, *IEE Review* 99–102.

Carlisle, B.H. and Heywood, D.I. (1997) The accuracy of a mountain DEM: research in Snowdonia, North Wales, UK. In D. Bax (ed.) *Proceedings of the 4th International Symposium on High Mountain Remote Sensing Cartography*, Sweden: University of Karlstad, 71–86.

Carver, S.J., Cornelius, S.C., Heywood, D.I and Sear D.A. (1995) Using computers and geographical information systems for expedition fieldwork. *The Geographical Journal* 161, 2: 167–76.

Cornelius, S.C., Sear, D.A., Carver, S.J. and Heywood, D.I. (1994) GPS, GIS and geomorphological field work. *Earth Surface Processes and Landforms* 19: 777–87.

Fix, R.E. and Burt, T.P. (1995) Global positioning system: an effective way to map a small area or catchment. *Earth Surface Processes and Landforms* 20: 817–27.

Gilbert, C. (1995) What's wrong with this picture? GPS error – what you can fix and what you can't. *Mapping Awareness* 9, 4: 26–8.

Goodchild, M.F. and D.W. Rhind (eds) (1991) *Geographical Information Systems: Principles and Applications*. London: Longman, 269–97.

Gray, J.M. (1982) The last glaciers (Loch Lomond Advance) in Snowdonia, North Wales. *Geological Journal* 17, 2: 111–33.

Gray, J.M. and J.J. Lowe (1982) Problems in the interpretation of small-scale erosional forms on glaciated bedrock surfaces: examples from Snowdonia, North Wales. *Proceedings of the Geologists Association* 93: 403–14.

Heller, M. and Weibel, R. (1991). Digital terrain modelling. In D.J. Maguire, M.F. Goodchild and D.W. Rhind (eds) *Geographical Information Systems Principles and Applications.* London: Longman, 269–97.

Jordan, G. and Shrestha, B. (1998) *Integrating Geomatics and Participatory Techniques for Community Forest Management: Case Studies from the Yarsha Khola Watershed, Dolakha District, Nepal.* Nepal: ICIMOD.

Kennedy, M. (1996) *The Global Positioning System and GIS: An Introduction.* Michigan: Ann Arbor Press.

Leick, A. (1995) *GPS Satellite Surveying.* New York: John Wiley & Sons.

Magellan (1994) *ProMARK X User Guide.* California: Magellan Systems Corporation.

Ordnance Survey (1996) *Land-Form PROFILE User Guide.* Southampton: Ordnance Survey.

Smith, G.R., D.I. Heywood and J.C. Woodward (1997) Integrating remote sensing, global positioning systems and geographic information systems for geomorphological mapping in mountain environments. In D. Bax (ed.) *Proceedings of the 4th International Symposium on High Mountain Remote Sensing Cartography*, Sweden: University of Karlstad, 221–36.

Smith, G.R., J.C. Woodward and D.I. Heywood (1998) Mapping glaciated terrain using remote sensing and DGPS: examples from the mountains of North Wales and Northwest Greece/*Working Paper School of Geography Series* 98/2, Leeds: University of Leeds.

Stocks, A.M. and Heywood, D.I. (1994) Terrain modelling for mountains. In M.F. Price and D.I. Heywood (eds) *Mountain Environments and Geographical Information Systems*, London: Taylor & Francis, 25–40.

Van Sickle, J. (1996) *GPS for Land Surveyors.* Michigan: Ann Arbor Press.

44

Computer simulation and modelling of urban structure and development

Paul Longley

INTRODUCTION

Today's urban settlement structures are developing and changing at a faster rate than ever before. Applied geographers and planners are frequently posed questions such as: What will be the effect upon the existing retail structure of a town of the development of a new out-of-centre grocery store? What might be the effects of renewal policies on house values in an urban area? How will the density of population across a city change if more central 'brown-field' sites are developed in preference to peripheral ones? And what would be the implications of either of these developments for non-car-based transport policy? Computer-based urban models and simulations provide answers to these and other important, explicitly geographical, questions.

These questions are fundamentally about changes in the *form* of urban areas and also the ways in which they *function* (Batty and Longley 1994). Applied and other geographers have long sought to understand these problems by thinking of towns and cities as *systems*, in which interrelated geographical units or objects form a 'set' that together function as a whole (Haggett *et al.* 1977: 6). In reality, individual towns and cities (like most all other geographical systems) do not develop in isolation from the rest of reality, but it is often helpful to screen out these more general considerations in order to 'bound' the specific problems that we are interested in. Limiting the extent of what we consider for practical purposes to be an 'urban system' is an important first step in *urban modelling*.

Many of the other chapters in this book have used models as practical tools to understand aspects of real-world systems. In the most general terms, a 'model' can be defined as a 'simplification of reality', nothing more, nothing less. In order to answer the sorts of specific and focused questions that were posed at the start of this chapter, it is necessary to disregard information about aspects of the urban system that are irrelevant (or only partially or indirectly relevant) and focus on those remaining system characteristics that have the greatest impact upon outcomes. Thus a good model *selectively* retains all of the aspects of a system that are important from a particular point of view and discards those that are not. This process of selection is central to the art of model building.

In practice, the selected aspects of the system are represented in an urban model principally using quantitative data. Although such data are used to represent system characteristics, they rarely if ever do this in a way that is either perfectly accurate or precise. This arises for a variety of reasons which are common to all quantitative analyses of socio economic systems. There is not enough space here to explore these in much detail (but see Martin 1997 for an introductory overview of some important aspects). Briefly, socio-economic data (such as those from censuses) are usually averaged across administrative zones prior to being made available to researchers for secondary analysis, and this creates problems of scaling and aggregation with respect to the precision with which characteristics can be ascribed to points and places on the Earth's surface (Openshaw 1984). Even if data were geographically precise, this does not necessarily mean that they accurately portray relevant aspects of socio-economic reality in a model, since in practice many facets of the urban system are measured and represented using surrogate information – an often-used example is the absence of income data from most censuses, which makes it necessary to represent spending power using surrogate *indicators*, such as car ownership rates (Bracken 1981: 281–5).

Nevertheless, urban models can be used in a *static* sense, to depict the state of the system, or in a *dynamic* sense to illustrate how system characteristics change over time. A static application of an urban model might be to represent the size and shape of an urban system. Dynamic urban models use representations of static states along with information about the nature and rate of change in order to develop *simulations* (or scenarios) of what might happen in a future stage of the system's development. These scenarios can be used to answer the types of question posed at the start of this chapter. Dynamic urban modelling presumes some understanding of the ways in which an urban system *functions*, and such understanding is often predicated upon what we know about the spatial *forms* of urban settlements. If a model abstracts all that is important and discards what is unimportant, then there is a good chance that we can use it to summarise, recreate and forecast the salient characteristics of an urban area.

Each of the questions posed at the start of this section requires us to investigate some aspect of the urban system by modelling selective aspects of urban development. There is a vast range of ways in which urban models and simulations may be developed, and in a chapter of this length it is not possible to classify, let alone summarise, all of applied urban modelling. Instead, therefore, we will focus upon the foundations to urban modelling – that is, we will describe the computer environment in which urban modelling and simulation takes place, and describe how urban models may be used to abstract the essential characteristics of urban systems and develop simulations in applied analysis.

THE CHANGING NATURE AND REMIT OF URBAN MODELLING AND SIMULATION

Most students are introduced to urban geography through the classic models of urban structure that were developed by Burgess, Hoyt and the other 'Chicago human ecologists' during the 1920s and 1930s. There can be few of today's geography students who are not familiar with the classic concentric ring diagram of 1920s Chicago, and one of its central characteristics, residential zoning, has been borrowed in generations of school projects to describe the mosaic of urban land uses of hundreds of urban areas. Other models often encountered by students at the beginnings of their undergraduate careers include those of Lösch and Weber in the realm of industrial location, Christaller's central place theory, the agricultural bid rent theory of von Thünen, and the 'bid rent' micro-economic theories of urban land use (developed in the work of Alonso 1964; Muth 1969). Of these, the micro-economic formulations of bid rents for different land uses, and 'density gradients' of different land-use categories, will be most relevant to our discussion of urban structure here. Such models are enduring because of the powerful yet simple way that they abstract (*model*) essential characteristics of real-world urban systems – the spirit of the very best applied geography. (Just because a model is simple does not mean that it cannot be misused, however. For example, the Burgess model is too often portrayed as a static representation of urban *form* rather than a dynamic representation of the changing *function* of residential areas – in which the inflow of successive waves of immigrants fuels waves of outward movement of established communities, in the same way that a pebble dropped into a pool of water generates ripples.)

These, then, are the widely known entry points to a generalised understanding of human geography. Yet as students progress through the discipline, so such models appear increasingly redundant and irrelevant to structuring and representing the real world. Over time, the regularised geometry of urban land use becomes too strained and implausible to sustain further investigation and interest, and more 'advanced' urban morphology research tends to refocus on more ideographic ('one-off') studies of the historiographies of particular urban forms. The consequence is that some students find themselves discouraged from seeking to draw generalisations about urban morphologies across space and through time.

The quest to generalise inductively has not always been so readily ignored. In the heyday of applied geography in the 1970s, the concepts of *social physics* and *spatial interaction* were widely used by geographers and planners to model the functioning of entire urban systems (Batty 1976; Birkin 1996; Wilson 1974). This has been part of an adherence to the basic tradition of *rational planning* in which urban problems and their solutions are conceived in largely physical terms (Batty and Longley 1997). The urban models of the 1970s were avowedly ambitious, in that they attempted to model entire urban systems. They were developed around the best technologies and data sources that were available at the time, yet in practice this inevitably meant 'making do' with crude and inappropriate data at coarse scales of spatial resolution, crude specifications of flows, and analysis on mainframe computers less powerful than the lap-top word processor on which this chapter was written. In short, the quality of the data and the computer technology of the time were not commensurate with the power of the underlying geographical concepts and ideas that were developed. As the 1970s progressed, increasing numbers of geographers came to the view that such models were gross *over*-simplifications of all that was interesting about urban systems, and as such that the early urban models were largely irrelevant to any but the most trite understanding of them (e.g. Sayer 1979). Disillusionment with the approach set in, and the quest to develop quantitative depictions of the state and dynamics of urban areas was all but abandoned.

The tide has since turned. In this chapter, we will illustrate some of the ways in which the modelling and simulation of city systems is being reinvigorated. We will focus upon two themes: first, the implications of the revolution that is taking place in the creation and availability of digital data about urban areas; and second, the ways in which these data are being reworked into sophisticated data models of urban systems, which in turn can inform further quantitative analysis. These themes are each specific outgrowths of the 'geographical information systems (GIS) revolution' in geography, specific aspects of which are described by Heywood and by Green elsewhere in this book.

GIS provide a framework for the orderly input, storage, manipulation and output of geographical data, and the field has developed rapidly over the last two decades. There are now many excellent textbooks (e.g. Burrough and McDonnell 1998; Martin 1996) that provide guides to their use in geography. The history of GIS is in some aspects the history of an important and extraordinarily successful branch of applied geography (Foresman 1998; Longley *et al.* 1999). GIS today is a huge software industry probably worth $1 billion world-wide, with about 1.6 million users. At the current rate of expansion, there could be 8 million 'applied geographers' worldwide using GIS by the year 2000! Over time, precipitous falls in the real costs of computing have made it possible to process larger data sets, much faster than ever before, while developments in computer graphics and networking have encouraged the use of information systems in an ever-wider range of new and novel ways. GIS are thus central to any discussion of applied geography at the end of the twentieth century, and they provide an ideal environment for a reinvigorated approach to urban modelling and simulation. This is a software environment in which niche applications of urban modelling are once again flourishing – as in the fine-scale analysis of retail location problems using geodemographics and other detailed sources as described by Clarke in this book.

REINVENTING URBAN MODELLING AND SIMULATION

There is not space here to do justice to the development of urban modelling within GIS, although a wide-ranging overview is made by the various contributions in Longley and Batty (1996). These provide convincing evidence that GIS are reinvigorating our thinking about urban modelling in an applied way, by virtue of the increasingly data-rich depictions of reality that they facilitate. An inherent characteristic of urban modelling is that as a model becomes more complex, so it more closely resembles the real-world system that it seeks to approximate. We can imagine 'complexity' here to be a composite index of the amount and detail of data about urban systems, and the nature of the way in which relationships between the components of the urban system are simplified. It has long been recognised (Haggett 1978) that the rewards to increasing complexity tail off, in the kind of way illustrated in Figure 44.1, and the innovation of GIS has undoubtedly made it possible to move further up this 'learning curve' than hitherto. This is because much of the complexity of urban modelling arises out of the increased burden of integrating, managing and processing data, and the environment of GIS has begun to trivialise these problems. In this way, new, firmer and more substantial foundations to urban modelling are being established. We will develop this idea by developing two extended illustrative case studies – the first in the context of predicting the values of housing across an urban area, and the second revisiting some of the classic concepts underpinning micro-economic models of housing demand. It is not possible in the space available here to do more than review the principles and methods used in urban modelling, although the guide to further reading contains further details of the ways in which the case study models can and have been applied. A shared theme between our two examples is that modelling (and by extension, simulation) within GIS can reinvigorate the ways in which we think about generalising the characteristics of urban systems across space and time.

Figure 44.1 The trade-off between model complexity and fidelity to real-world representation.

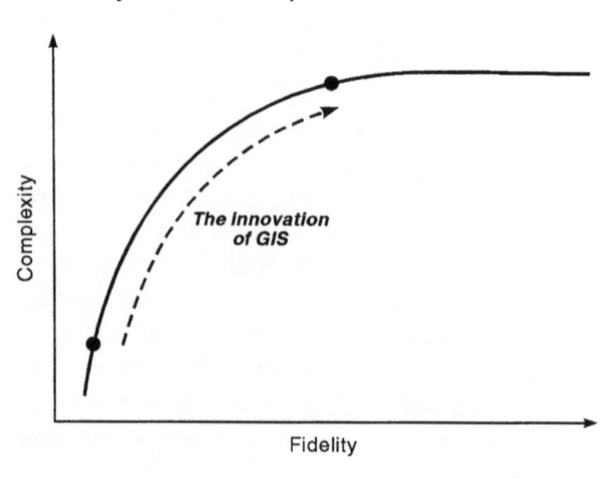

These developments are timely, because urban theory itself is being forced to change. The demise of the 'first-generation' urban models was also in no small part because the assumptions of single centred cities (focused upon a dominant central business district area, often dominated by manufacturing industry, etc.) were increasingly strained by the rapid changes in the structuring of contemporary urban areas. Yet the consequence, the relative demise of urban modelling as an area of applied geography, has also led to a slackening in the pursuit of the rational planning goals that have dominated the post-war era. It remains a largely uncontested tenet of urban geography that the notions of economic bid rents, urban hierarchies and such-like remain of enduring relevance to the subject, and they also provide a means of comparing different urban systems. As cities have developed and become more complex, varied and specialised, so the old certainties about urban structure are being swept away and new approaches to the measurement of urban structure and form must be developed. Batty and Longley (1994) have written on this broader theme at length, but here we will restrict our discussion to the data foundations to reinvigorated urban modelling.

Modelling property values using GIS

Our first example concerns the variation in property values across an urban area in a single time period. Although much has been written on so-called 'bid rent' models (Alonso 1964; Muth 1969), we actually seem to know remarkably little about the detailed ways in which value is constituted in the built environment (Orford 1997). The classic bid rent model postulates that households make trade-offs between the distance that they live from the centre of the city and the size (in terms of, say, floorspace) of the properties that they occupy. Yet in reality, today's cities are not monocentric: tenure subdivides housing markets into owner-occupied, private rented and public rented sectors; residents perceive distance in a wide variety of metrics (e.g. travel time, public transport cost); and there is much more to urban living than the journey to work (e.g. availability of good schools, proximity to parks). The bid rent formulation of property values seems intuitively plausible yet insufferably simplistic in such circumstances. Moreover, in many countries (including the UK) we know rather little about the distribution of property values across towns and cities, in the absence of publicly available land-use information systems.

In this context, Longley *et al.* (1994) developed an urban model of the capital values of all 45, 889 properties in the inner area of Cardiff, UK – the extensive and heterogeneous core area of the city. The UK has no national comprehensive and publicly accessible land-use/valuation information system, and UK readers should note that at the time of their study the council tax valuation list was only in the process of being compiled for the first time. (This valuation list classifies properties into eight broad categories only for local taxation purposes and is not usually made available to researchers for geographically extensive areas). The following disaggregate data sources were available to them:

1 A digital copy of the 1989 domestic rates register. This provided historical information on the 'rateable values' of every property – that is, the annual rent that a property would historically have commanded in a rental market. Until the late 1980s, such values provided the basis for local taxation in the UK. Such valuations can provide only a rather crude estimate of capital values in a housing market, which today is dominated by owner-occupied housing, supplemented by local authority rented housing (which historically have been built by the state and let at subsidised rents, and today have often been bought under 'right to buy' legislation: see Balchin *et al.* 1998). The domestic rates register also contains limited information on property type.
2 A digitised network of all streets in the study area.
3 Digitised boundaries of eighty-three 'house condition survey areas', devised for a local authority-funded house condition survey and designed to identify small areas within which there was general homogeneity of house structure and type (but which in practice were often far from homogeneous).

The study area, together with the digital GIS layers, is shown in Figure 44.2. Together, these represented a rich range of data sources and provided the data foundations for development of a simple spatial model of house price variation. The basis for the model is essentially that of Tobler's (1970) famous 'First Law of Geography': 'all things are related but nearby things are more related than distant things'. Thus, for example, values of properties around an attractive park are likely to be uniformly high, and values will tail off with distance from this 'positive externality'; and, by contrast, property values around a waste tip are likely to be low, and values will recover only with distance from this 'negative externality'. But this assumes that we hold other variables such as construction type (house, flat, etc.), size and so forth constant, and in reality we will have to model the effects of more than just space alone.

These principles were put into practice in the

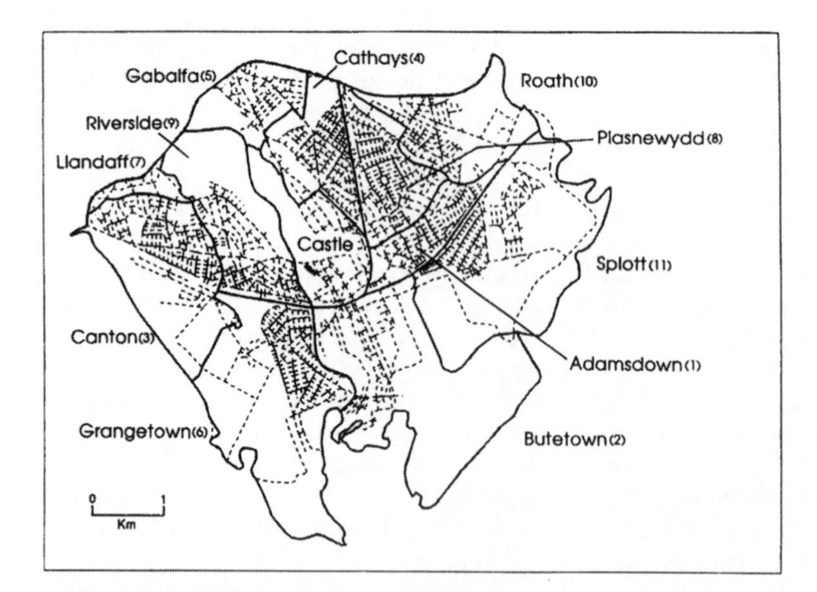

Figure 44.2 Cardiff's inner area.

following way. A survey of estate agents was undertaken to identify the asking prices of properties for sale at the time of the study. This house price survey identified the asking prices of 796 properties (2.1 per cent of all properties in the study area – 296 of them identified as individual properties, the remainder identified to the street level of resolution), which were spatially distributed even across the study area. On the assumption that asking prices bore a correspondence with actual values, a geographical model was developed in order to allow computation of the capital values of all properties that were not actually on sale at the same time.

The stages in this model are summarised in Figure 44.3. Essentially, the rates register was used as the address frame, although quite a large number of its entries required aggregation from constituent rooms and bed-sits into whole properties. The first stage entailed identifying the precise geographical locations of the 269 'price survey properties'. Next, those properties that were in the same street and were of the same type as a price survey property were allocated the same capital value – or an average value if there were two or more price survey properties in a given street. This allowed capital values to be computed for 20,982 properties. In the third stage, the digitised house condition survey boundaries were used to assign values to properties located in streets in which there was no price survey property. However, at this scale it was considered that 'street effects', even of physically identical properties, could conspire to invalidate use of values from a neighbouring street. Consequently, the rateable value of the nearest price survey property was compared with the distribution of rateable values in the adjoining street in order to ascertain whether the value of a price survey property could legitimately be used to compute the values of some or all of its more distant neighbours – with the acceptance criterion based on a standard deviation measure. This allowed values to be computed for a further 22,523 properties. This left just 1884 of the 45,658 without assigned capital values. These were overwhelmingly idiosyncratic properties, so values were assigned using a simple regression relationship between the asking prices (denoted CVAL) and the rateable values (denoted RVAL) of the price survey properties. This empirical relationship was statistically estimated as:

$$\mathrm{CVAL} = 4003 + 358.64\,(\mathrm{RVAL})$$

Housing exhibits a very wide range of characteristics in terms of structure, age and condition. Additionally, each individual property is different

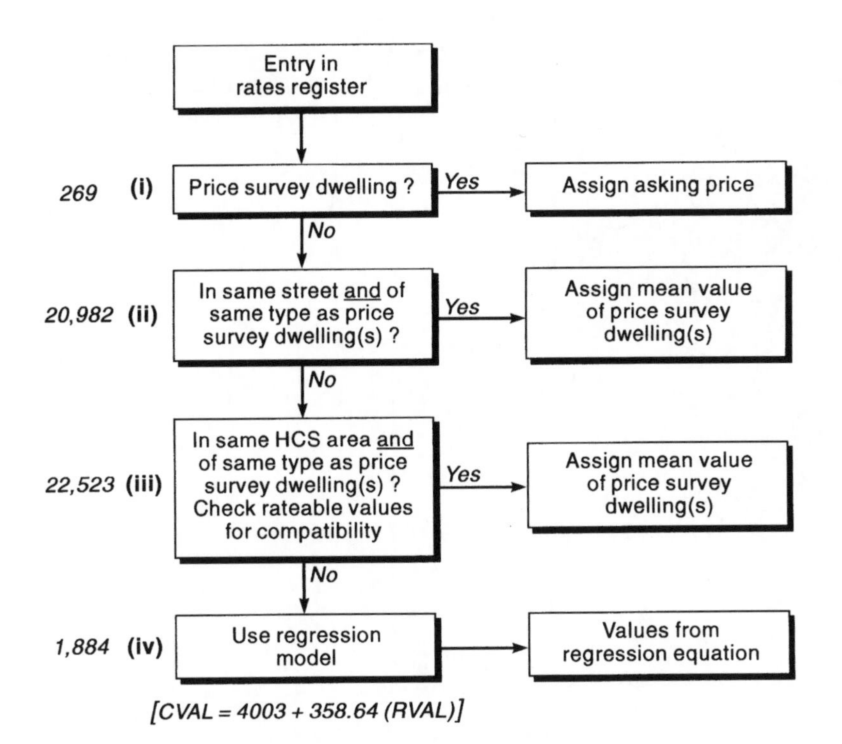

Figure 44.3 The sequence of GIS-based operations in the modelling of house prices.

for the simple reason that it is located on a unique part of the Earth's surface – and even at fine scales of resolution this can lead to dramatic differences between the values of two properties if one happens to be located 'on the wrong side of the tracks'. There is not space here to discuss the findings in any great detail, although the reader is referred to Longley *et al.* (1994) in the Guide to Further Reading. Briefly, the modelled results were compared with the official council tax valuation list (used to determine how much council tax every household should pay) and found to be similar, both in absolute values and distribution of values. There were, however, streets in which either positive or negative differences were observed, and the geographical distribution of such streets is shown in Figure 44.4. Such patterns beg interesting questions in both methodological and substantive terms. Was the model too simple to replicate 'official' values in these areas, and might it have been improved by adding in other factors – for example, the localised effects of gentrification upon house prices in specific neighbourhoods? Or might the official valuations themselves have been systematically wrong? – property valuation is at best an inexact science, and there is no real way of ascertaining the 'true' value of housing in a city, except perhaps if every house were put up for sale in order to let the market decide! And, third, if house prices have indeed been modelled successfully in one time period, could the model be updated with new house price data in order to reflect the geographical dynamics of house price change? At some point, government will have to produce comprehensive revaluations of all properties in order to retain fairness in the sharing of the local tax burden, and this kind of GIS-based model would seem to offer a way forward that is transparent, cost-effective and fair. These issues are summarised in Box 44.1. The results of this simple geographical model suggest that GIS provide an ideal medium for assessing the relative importance of structural and locational attributes in creating value. Generalising from this, Orford (1997) has also described how a wide range of neighbourhood and structural attributes may be incorporated into sophisticated GIS-based

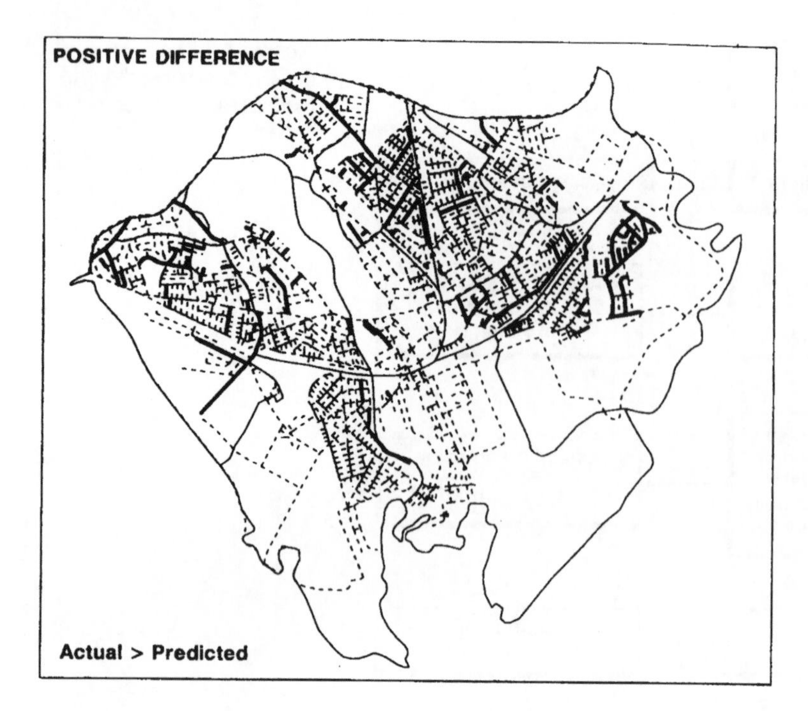

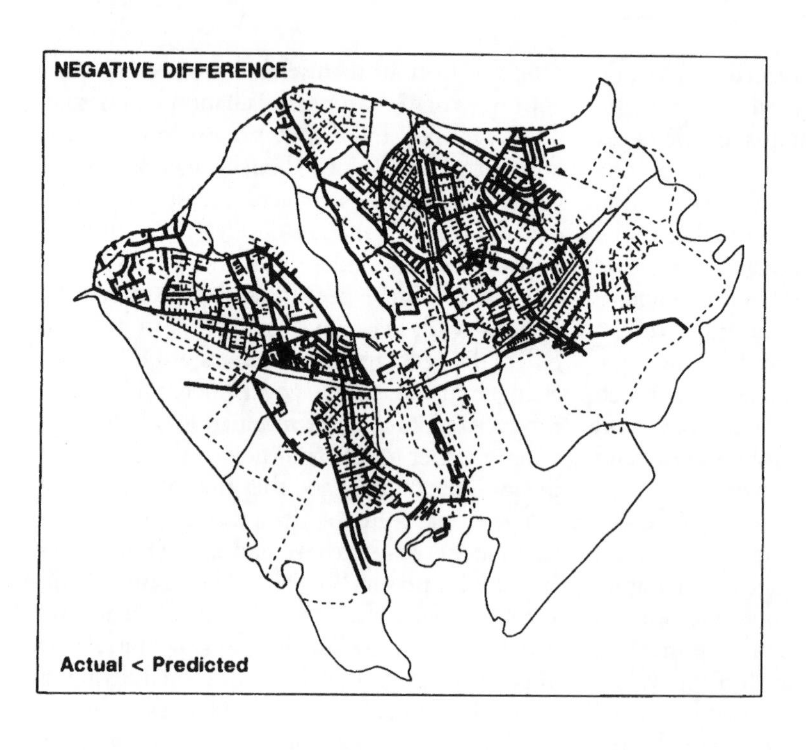

Figure 44.4 Council tax valuations: positive and negative differences.

Box 44. 1 Using GIS to model property valuations

The problem: what is the value of every property in an urban area?
Modelling tools and data: GIS; digitised streets and small areas; historic rating information; own survey
Principles behind model: 'Tobler's First Law of Geography' (= neighbourhood effects), plus the effects of individual property attributes
Model validation: banded comparison with official council tax valuation list
Results: (1) high prediction success when compared with official council tax valuation list (as detailed in Longley *et al.* 1994); (2) systematic spatial patterning to under- and over-predictions; (3) GIS provides a potential means for updating the council tax valuation list.

statistical explanations of housing prices using so-called 'hedonic' house price models.

This example does not model property prices from first principles in so far as a central ingredient to the predictive model was pre-existing rateable value information. Yet it does provide evidence of the ways in which GIS provides detailed and accurate predictions across geographically extensive areas. The predictions largely replicate the painstaking survey measurements of a large number of field surveyors. Perhaps more important, it also lays the basis for future updating. House price dynamics are geographically very variable, reflecting, for example, processes of obsolescence and decay on the one hand, and gentrification on the other. Once an explanatory model has been successfully created for an urban area, it should prove possible to update the model predictions to take into account general house price inflation and within-area variation attributable to known urban processes. In applied geography terms, this is thus an instance of a detailed model of the current form of the city that might be adapted to incorporate a range of urban dynamics.

Measuring and simulating urban structure

Our first example provided as an output a spatially extensive valuation of the urban environment that, by extension, might be used as an input to dynamic modelling of relative property price changes, or even simulation of the spatially variable effects of changes to local or national planning policy. A related and established theme in urban analysis has been the changes in the density of residential, industrial, commercial and public open space land uses in towns and cities. This problem lies at the heart of applied geography and rational planning policy: for example, to what extent is it reasonable to expect 'brownfield' sites within cities to accommodate the demands for new house building? Or if demographic information and projections suggest that most such households will be small, is it realistic to expect the demand to be met only by subdivision of existing properties (e.g. into flats)? There is much verbiage in the literature about the quest for more sustainable cities, yet little comparative work has been carried out on the ways in which space is filled and quantification of the effects of population, retail and employment decentralisation. Indeed, we seem to know very little about the density of urban living, and the way that residential areas juxtapose with employment, leisure and retail land uses. Previous empirical research clings to vague notions of apparent declines in the density gradients of cities over time – which may be no more than an artefact of the crude ways in which density has been measured.

In essence, the problem is that there has been too little detailed work on reconciling the carcass of the city – that is its built *form* in terms of housing, industrial, retail, other commercial, and public open space land uses – with the human activities that take place in it in terms of journeys to work, leisure pursuits and the full panoply of human activities. It follows that the best way of proceeding is to integrate a data model of the physical form of an urban area with a data model that gives some indication of the way in which human settlement is configured around it. The most obvious source of information about the physical form of the city is satellite remote-sensing information, and the most widely used socio-economic source in the UK is the Census of Population. Table 44.1 identifies some of the important characteristics of each of these sources.

Remote sensing is an established means of monitoring a wide range of environmental phenomena, but the recent innovation of high-resolution satellite images is making this source increasingly appropriate to monitoring the morphology of urban areas (Mesev *et al.* 1995). Raw spectral signals are converted into a land cover image using a classification technique that results in a statistical model of the most likely mosaic of land covers. GIS has developed to the point where such classified images can be readily analysed within most mainstream packages. The morphology of a typical urban area (Norwich, UK), as revealed by a Landsat image, is shown in Figure 44.5D.

The UK Census of Population is conventionally made available at enumeration district scale and in choropleth map form (Figure 44.5B). Such representations present the misleading impression that within-area densities are uniform and hence that the only changes in density occur across boundaries. The innovation of GIS has made a much wider range of transformations and projections possible, and one such manipulation is to think of the distribution of populations as a continually varying density surface (Bracken and Martin 1989: Figure 44.5C). Figure 44.5A shows one further choropleth representation of the same settlement, but this time broken up into the rather different geography of postcode sectors.

It is important to remember that all of the city maps shown in Figure 44.5 are *models* of reality. The satellite image is derived from a statistical classifier of reflected solar radiation, measured and classified at the scale of the pixel (picture element), and different classifiers would yield different results; the choropleth map models the distribution of a census variable as a mosaic of uniform areas; and the surface model allocates the data (in this case population) across a continuous space by presuming that density decreases with distance from each enumeration district centroid. Applied urban geography has never had much to do with remote sensing, because the resolution of data generated from early satellites was too coarse to discriminate between elements of the built

Table 44.1 Some characteristics of satellite imagery and UK census data.

Satellite imagery	*Census*
Used to derive land *cover* information	Measures socio-economic characteristics, activities and land *use*
Comprehensive coverage	Comprehensive coverage subject to (illegal) non-response
Frequently updateable	Usually carried out at ten-year intervals
Longitudinal information available since 1984 (first Landsat satellite)	Long historical time series but only since 1971 in digital form
Pixel the basic element of spatial resolution (size 30 × 30 m in the case of SPOT images)	Census enumeration district (ED: typically 160 households) the basic unit of aggregation and data availability
Consistent over time, subject to minor variation in sensor characteristics	ED boundaries variable over time
Similar physical forms have different socio-economic functions ecological	Clearer information on socio-economic function, but available at ED scale only and vulnerable to fallacy

Source: Adapted from Martin 1996: 144–5.

Figure 44.5 Different data models of Norwich, UK: (A) postcode sector geography; (B) a choropleth census map; (C) a population surface map; and (D) a Landsat image.

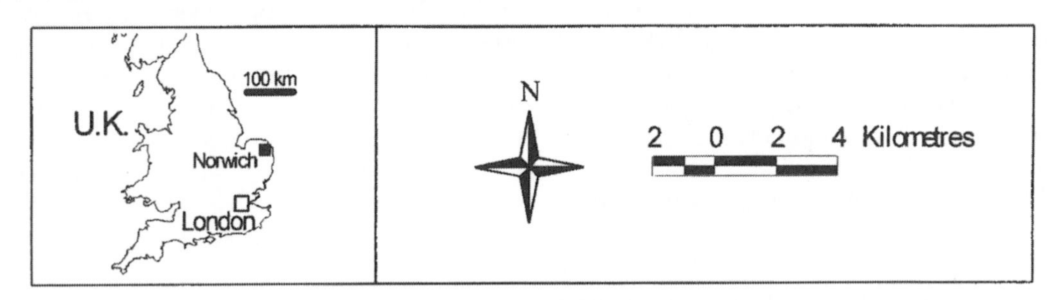

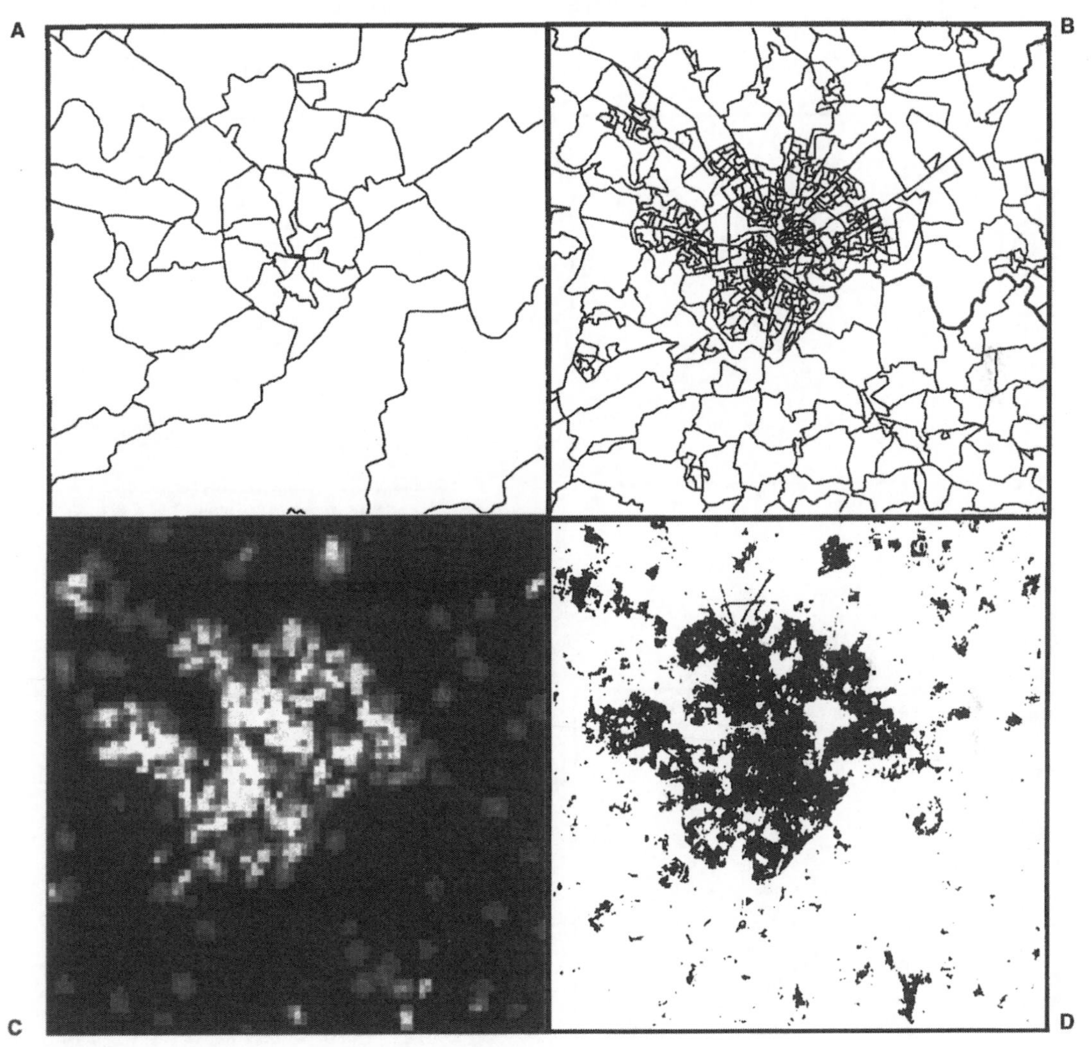

environment – although this situation is changing rapidly with the launch of a new generation of high-resolution satellites. Moreover, there is no reason why census data from digital models of spatial distributions, such as that shown in Figure 44.5C, should not be incorporated as ancillary information in the classification of urban satellite images.

The most usual method of classifying thematic coverages from satellite images is a statistical procedure known as maximum likelihood estimation. In conventional remote-sensing applications, this technique is used to classify raw spectral images, yet the technique can be extended to incorporate population surface model information of the type shown in Figure 44.5C. Indeed, there is no reason why a wide range of socio-economic sources should not be incorporated

Figure 44.6 The density gradients of four different facets to urban form, Norwich, UK.

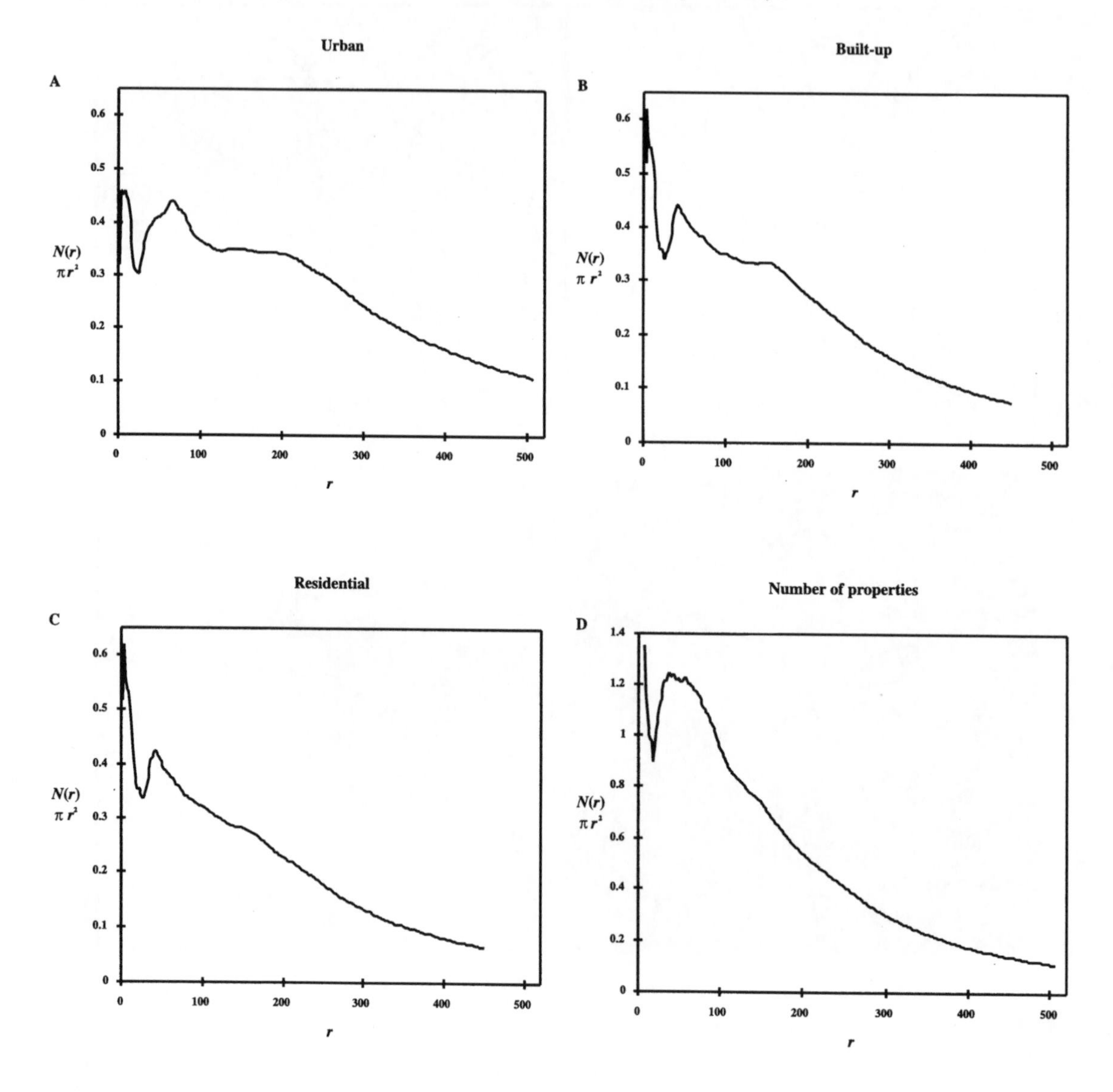

in order to create tailor-made classifications of urban land use for particular purposes. For example, we might begin by dividing land use into 'rural' versus 'urban' on the basis of the results of a standard maximum likelihood classification of a satellite image; a narrower definition of urban land use might be as urban land that is 'built-up', based on supplementary information from the census or other socio-economic sources. Such sources might also be used to discriminate between residential and other (commercial, industrial) built-up land uses, and/or information might be incorporated about the numbers of properties in the area. The resultant classified digital images might then be used to create a range of graphical and numerical indices of urban morphology.

One such graphical tool that has been widely used in applied urban analysis is the 'density gradient', which measures the rate of change in the proportion of land that is filled with 'urban' land uses with increased distance from the historic centre of the settlement. This is measured as $(N(r)/(\pi r^2))$, that is, the number of occupied pixels in each distance band r divided by the annular area of that distance band. Longley and Mesev (1997) use this kind of approach to identify the density gradients of four different facets to urban form of Norwich, UK, illustrated in Figure 44.6 – 'urban', 'built-up', 'residential' and 'numbers of properties'.

As this figure shows, the differences between these apparently similar categories are more than semantic and can heavily condition whether and to what extent we might consider that the profile is characterised by discontinuities and 'density craters'. However, the optimistic message contained in this figure is that, once the differences between different conceptions of 'urban-ness' have been clearly grasped, it is possible to develop a range of customised indicators of urban morphology. These GIS-based data models are informing our thinking about the ways in which urban settlements fill space, as well as providing detailed information as to the morphology of particular settlement structures.

CONCLUSION

The shape, form and dimension of cities are today changing faster than ever before, as processes of retail decentralisation, demographic change and ongoing industrial restructuring exact changes in urban morphology of unprecedented scale. Yet at the same time our ability to measure, model and simulate change is developing too, thanks to the GIS and digital data revolutions, which now allow us to invoke a full range of modelling assumptions and simplifications at different scales. This is important because in all the best classical theories, scale plays an important role and empirical modelling requires us to generalise between scales. The resulting 'real-world' indicators of urban morphology will rarely exhibit the clear geometry and orderliness of high school geography models: however, just as GIS makes it possible to depict real-world spatial distributions with greater precision than ever before, so developments in spatial analysis are allowing traditional models of urban structure to be recast in new and exciting ways (see Box 44.2).

The early models of urban structure suggested that cities were clearly organised, simply ordered, and thus predictable. It followed that they could be planned in such a way that the quality of life of their residents could be directly improved by manipulating their physical form. During the last twenty years, many geographers have been overwhelmed by the complexity that characterises

Box 44.2 Modelling in a GIS environment

SUMMARY

- Models are simplifications of reality that can be used to create scenarios.
- The ways in which we decide to simplify reality (e.g. what *is* urban land?) conditions the way in which we measure geographical phenomena. This in turn conditions the way we analyse them and (depending on the particular application) the likely outcome of analysis
- Problems of the spatial scale of data and the level of geographical aggregation pose particular problems for geographical modelling (see Openshaw 1984)
- GIS is a useful applied technology for creating digital models that can be used to explore the world around us (see Longley *et al.* 1998).

real-world city systems, and disillusionment with this rational planning view has taken hold. Yet the thrust of the ideas and examples set out here suggests that this has been only a temporary setback, and that the applied modelling and simulation environment of GIS is today providing a renewed stimulus to the applied analysis of urban systems.

GUIDE TO FURTHER READING

Batty, M. and Longley, P.A. (1994) *Fractal Cities: A Geometry of Form and Function*. London and San Diego: Academic Press. A fairly recent book on the state of the art of one approach to urban modelling, with colour computer graphics.

Birkin, M. (1996) Retail location modelling and GIS. In P.A. Longley and M. Batty (eds) *Spatial Analysis: Modelling in a GIS Environment*, Cambridge: Geo-Information International, 207–25. A very lucid account of the recent history of urban modelling, with applications in retail analysis.

Longley, P.A., Goodchild, M.F., Maguire, D.J. and Rhind, D.W. (1999) Epilogue. In P.A. Longley, M.F. Goodchild, D.J. Maguire and D.W. Rhind (eds) *Geographical Information Systems: Principles, Techniques, Management and Applications*. New York: John Wiley & Sons. A prospective view of the general impacts of geographical information systems. It is the last chapter of the second edition of this major GIS reference work, which is probably the most detailed current statement of the remit of GIS.

Longley, P.A., Higgs, G. and Martin, D. (1994) The predictive use of GIS to model property valuations. *International Journal of Geographical Information Systems* 8, 217–35. Contains full details of the first case study described in this chapter.

Openshaw, S. (1984) *The Modifiable Areal Unit Problem*. Concepts and Techniques in Modern Geography 38, Norwich, UK: GeoBooks. Still probably the clearest statement of the effects of scale and aggregation effects upon geographical modelling.

Martin, D. (1997) Geographical information systems and spatial analysis. In R. Flowerdew and D. Martin (eds) *Methods in Human Geography*. Harlow: Longman, 213–29. A clear and succinct overview of the ways in which GIS and spatial analysis might be used in undergraduate dissertations and project work.

REFERENCES

Alonso, W. (1964) *Location and Land Use*. Cambridge, Mass: Harvard University Press.

Balchin, P., Isaac, D. and Rhoden, M. (1998) Housing policy and finance. In P. Balchin and M. Rhoden (eds) *Housing: The Essential Foundations*, London: Routledge, 50–106.

Batty, M. (1976) *Urban Modelling: Algorithms, Calibrations, Predictions*. Cambridge: Cambridge University Press.

Batty, M. and Longley, P.A. (1997) The fractal city. *Architectural Design Profile* 129, 74–83.

Bracken, I. (1981) *Urban Planning Methods: Research and Policy Analysis*. London: Methuen.

Bracken, I. and Martin, D. (1989) The generation of spatial population distributions from census centroid data. *Environment and Planning A* 21, 537–43.

Burrough, P.A. and McDonnell, R.A. (1998) *Principles of Geographical Information Systems*. Oxford, Oxford University Press.

Foresman, T.W. (1998) *The History of Geographic Information Systems: Perspectives from the Pioneers*. New Jersey: Prentice-Hall.

Haggett, P. (1978) Spatial forecasting: a view from the touchline. In R.L. Martin, N.J. Thrift and R.J. Bennett (eds) *Towards the Dynamic Analysis of Spatial Systems*. London: Pion, 236–47.

Haggett, P., Cliff, A.D. and Frey, A.E. (1977) *Locational Analysis in Human Geography*, 2nd edition. *Volume 1: Locational Models*. London: Edward Arnold.

Longley, P.A. and Batty, M. (1996) *Spatial Analysis: Modelling in a GIS Environment*. Cambridge: GeoInformation International.

Longley, P.A. and Mesev, V. (1997) Beyond analogue models: space filling and density measurement of an urban settlement. *Papers in Regional Science* 76, 409–27.

Longley, P.A., Goodchild, M.F., Maguire, D.J. and Rhind, D.W. (1999) Introduction. In P.A. Longley, M.F. Goodchild, D.J. Maguire, D.W. Rhind (eds) *Geographical Information Systems: Principles, Techniques, Management and Applications*, New York: Wiley, 1–20.

Martin, D. (1996) *Geographic Information Systems: Socioeconomic Applications*. London: Routledge.

Mesev, V., Longley, P.A., Batty, M. and Xie, Y. (1995) Morphology from imagery: detecting and measuring the density of urban land use. *Environment and Planning A* 27, 759–80.

Muth, R.F. (1969) *Cities and Housing: The Spatial Pattern of Urban Residential Landuse*. Chicago: University of Chicago Press.

Orford, S. (1997) *Valuing the Built Environment: A*

GIS Approach to the Hedonic Modelling of Housing Markets. Unpublished Ph.D. dissertation: University of Bristol.
Sayer, A. (1979) Understanding urban models versus understanding cities. *Environment and Planning A* 11, 853–62.
Tobler, W.R. (1970) A computer movie simulating growth in the Detroit region. *Economic Geography* 46, 234–40.
Wilson, A.G. (1974) *Urban and Regional Models in Geography and Planning*. London: John Wiley.

Place index

Subject index